Yield Gains in Major U.S. Field Crops

Yield Gains in Major U.S. Field Crops

Stephen Smith, Brian Diers, James Specht, and Brett Carver, editors

CSSA Special Publication 33

American Society of Agronomy, Inc.
Crop Science Society of America, Inc.
Soil Science Society of America, Inc.
5585 Guilford Road, Madison, WI 53711-5801 USA
agronomy.org | crops.org | soils.org

dl.sciencesocieties.org
societystore.org

ISBN: 978-0-89118-619-9 (print)
ISBN: 978-0-89118-620-5 (electronic)
doi:10.2135/cssaspecpub33
Library of Congress Control Number: 2014937884

CSSA Special Publications
ISSN: 2165-9745 (online)
ISSN: 0895-9978 (print)

Cover design: Patricia Scullion
Cover photos: USDA-ARS, USDA-NRCS, ThinkStock, Istock

Printed in the United States of America.

Contents

Foreword | vii
Preface | ix
Contributors | xi

Chapter 1 | 1
Barley
Tom Blake, Victoria Blake, and Jochim Wiersma

Chapter 2 | 13
Cotton
B. Todd Campbell, Debbie Boykin, Zaid Abdo, and William R. Meredith, Jr.

Chapter 3 | 33
Cool-Season Forages
E. Charles Brummer and Michael D. Casler

Chapter 4 | 53
Lettuce and Spinach
Ivan Simko, Ryan J. Hayes, Beiquan Mou, and James D. McCreight

Chapter 5 | 87
Edible Grain Legumes
George J. Vandemark, Mark A. Brick, Juan M. Osorno, James D. Kelly, and Carlos A. Urrea

Chapter 6 | 125
Maize
Stephen Smith, Mark Cooper, Joseph Gogerty, Carlos Löffler, Delos Borcherding, and Kevin Wright

Chapter 7 | 173
Peanut
C. Corley Holbrook, Tim B. Brenneman, H. Thomas Stalker, W. Carroll Johnson III, Peggy Ozias-Akins, Ye Chu, George Vellidis, and Duncan McClusky

Chapter 8 | 195
Potato
Shelley H. Jansky, David M. Spooner, and Paul C. Bethke

Chapter 9 | 219
Rangeland and Warm-Season Forage Grasses
Kevin B. Jensen and William F. Anderson

Chapter 10 | 267
Rice
K.S. McKenzie, X. Sha, K.A.K. Moldenhauer, S.D. Linscombe, N.B. Lyman, and L.L. Nalley

Chapter 11 | 293
Sorghum
Roger Monk, Cleve Franks, and Jeff Dahlberg

Chapter 12 | 311
Soybean
J.E. Specht, B.W. Diers, R.L. Nelson, J.F.F. de Toledo, J.A. Torrion, and P. Grassini

Chapter 13 | 357
Sugarbeet
Lee Panella, Stephen K. Kaffka, Robert T. Lewellen, J. Mitchell McGrath, Mike S. Metzger, and Carl A. Strausbaugh

Chapter 14 |397
Sugarcane
Serge J. Edmé, Andru Suman, and Collins Kimbeng

Chapter 15 | 433
Sunflower
Brent S. Hulke and Larry W. Kleingartner

Chapter 16 | 459
Wheat
Robert Graybosch, Harold E. Bockelman, Kimberly A. Garland-Campbell, David F. Garvin, and Teshome Regassa

Foreword

Few books in history have been planned with the focus and detail included in *Yield Gains in Major U.S. Field Crops*. From the beginning the book was based on a critical assessment of the yield gains across the crops essential for feeding a future population of 9 billion. Crafted as an intellectual challenge impossible for an individual scientist or team to answer, it relied on bringing together 65 scientists and their teams to develop the background data for the book.

There is no doubt that this book will be fundamental to predicting future gains from plant breeding and will be an instant classic for its powerful assessment of breeding efforts to this point in time. While each crop is a story in itself, the unique coverage of so many crops in one book provides a vision of the power of plant breeding never before assembled.

We hope that this CSSA Special Publication is a difference-maker for our scientific societies and our society at large.

David D. Baltensperger, 2014 CASA President
David B. Mengel, 2014 ASA President
Jan W. Hopmans, 2014 SSSA President

Preface

Some 8,000 to 10,000 years ago humankind crossed a crucial threshold as groups began to save seed to plant for harvest during the next season. A bond of codependence between domesticated cultivated varieties and not only farmers, but also the whole of society, was created. Today, the dependence of society on this bond of codependence has become more critical due to increasing human populations and the need to farm more sustainably by making the most effective use of available resources while conserving biodiversity and protecting the environment from pollution. However, this bond becomes increasingly tenuous as an increasing proportion of society loses touch with even an elementary understanding of how or where food is created. Consequently, policy makers, the public at large, and even prospective plant breeders, agronomists, soil scientists, statisticians, information management experts, conservationists, physiologists, pathologists, and experts in sequencing and reading genomic information—to name but a few—fail to engage positively in the field of plant breeding and agriculture. Yet the quality and sustainability of life on Earth are in very large measure connected with how food, fiber, and fuel are produced. It is therefore imperative to encourage a greater understanding and informed debate regarding how food and fiber are produced to ensure continued productivity through long-term stewardship and the most effective use of critical resources such as water, soil, genetic resources, and human intellect.

In 1984 the Crop Science Society of America (CSSA) published the proceedings of a 1981 symposium as Genetic Contributions to Yield Gains of Five Major Crop Plants. These crops were corn, cotton, potato, sorghum, and wheat. This was a seminal publication stimulating the publication of further research in numerous other countries and on additional crops to determine the elements contributing to advances in crop productivity, their interactions, and to monitor the rate of progress. In 2012 a group of us decided that it was time to revisit the status of productivity including genetic gain in each of these five major crop species. But why stop there? We developed a proposal for a book that would include all major U.S. field crops. We were fortunate to gain the enthusiastic support of numerous authors who agreed to contribute the chapters, many other experts who agreed to be reviewers, and CSSA, who agreed to sponsor and to publish the book.

This book provides a fascinating snapshot-in-time account of the productivity status of all major U.S. field crops. Each crop has a different story to tell. Much has changed in the field of plant breeding, biotechnology, and agronomy during the 30 years that have passed since the previous CSSA book was published. We earnestly hope that these chapters will provide insights for many from various fields into how each crop is being bred and produced today, with some historical perspective and also a look at future challenges and opportunities.

In closing, we are well aware that we have not covered productivity of crops using organic farming methods. This subject is a gap that should be filled. However, we would make a plea that methods of farming be examined more scientifically and less from the perspective of marketing. The more that is known about how food is produced and the resources necessary to sustain a productive agriculture, the better the quality of debate. Quality discourse is imperative to rebuild and strengthen the bond of codependence we have with the cultivated plant species upon which each of us depends for health, economic, and political security.

Stephen Smith, Brian Diers, James Specht, and Brett Carver, Editors

Contributors

Z. Abdo	USDA-ARS, Richard B. Russell Agricultural Research Ctr., 950 College Station Rd., Athens, GA 30605 (zaid.abdo@ars.usda.gov)
W.F. Anderson	USDA-ARS, Crop Genetics and Breeding Research Unit, 115 Coastal Way, Tifton, GA 31793 (Bill.anderson@ars.usda.gov)
P.C. Bethke	USDA-ARS, Vegetable Crops Research Unit, 1575 Linden Dr., Madison, WI 53706-1590 (paul.bethke@ars.usda.gov)
T. Blake	Montana State Univ., Dep. of Plant Science & Pathology, Room 109, Agric. BioSci Building, Bozeman, MT 59717 (blake@montana.edu)
V. Blake	USDA-ARS, Western Regional Research Ctr., 800 Buchanan Street, Albany, CA 94710 (victoria.blake@ars.usda.gov)
H.E. Bockelman	Small Grains and Potato Germplasm Research, 1691 S 2700 W, Aberdeen, ID, 83210 (Harold.Bockelman@ars.usda.gov)
D. Borcherding	DuPont Pioneer, 205 Fair View Dr., Dallas Ctr., IA 50063 (Delos.Borcherding@Pioneer.com)
D. Boykin	USDA-ARS, Jamie Whitten Delta States Research Ctr., 141 Experiment Station Rd., Stoneville, MS 38776 (debbie.boykin@ars.usda.gov)
T.B. Brenneman	Univ. of Georgia, Plant Sci. Bldg., 2360 Rainwater Rd., Tifton, GA 31793-5766 (arachis@uga.edu)
M.A. Brick	C113 Plant Sciences Building, Colorado State Univ., Fort Collins, CO 80523 (Mark.Brick@ColoState.edu)
E.C. Brummer	Univ. of California, Dep. of Plant Sciences, One Shields Ave., Davis, CA 95616; formerly, Forage Improvement Division, The Samuel Roberts Noble Foundation, 2510 Sam Noble Pkwy., Ardmore, OK 73401 (ecbrummer@ucdavis.edu)
B.T. Campbell	USDA-ARS, Coastal Plains Soil, Water, and Plant Research Ctr., 2611 W. Lucas St., Florence, SC 29501 (todd.campbell@ars.usda.gov).
B. Carver	Oklahoma State Univ., 368 North Ag Hall, Plant & Soil Science, Stillwater, OK 74078-6028 (brett.carver@okstate.edu)
M.D. Casler	USDA-ARS, U.S. Dairy Forage Research Ctr., 1925 Linden Dr., Madison, WI 53706-1108 (mdcasler@wisc.edu)
Y. Chu	Univ. of Georgia, NESPAL, 2360 Rainwater Rd., Tifton, GA 31793-5766 (ye.chu.test@gmail.com)
M. Cooper	DuPont Pioneer, 7250 NW 62nd Ave., P.O. Box 552, Johnston, IA 50131-0552 (Mark.Cooper@Pioneer.com)
J. Dahlberg	Kearney Agric. Research and Ext. Ctr., 9240 S. Riverbend Ave., Parlier, CA 93648 (jadahlberg@ucanr.edu)
J.F.F. de Toledo	Avenida dos Imigrantes, 2715, Apt. 104, Bairro Centro, 78890-000, Sorriso, MT, Brazil (JToledo@nidera.com.br)
B.W. Diers	Dep. of Crop Sciences, 268 National Soybean Research Ctr., 1101 W. Peabody Dr., Univ. of Illinois, Urbana, IL 61801 (bdiers@illinois.edu)
S.J. Edmé	USDA-ARS Sugarcane Field Station, 12990 U.S. Hwy. 441 N., Canal Point, FL 33438 (serge.edme@ars.usda.gov)
C. Franks	Dupont Pioneer, 2260 CR 60, Plainview, TX 79072 (cleve.franks@pioneer.com)

K.A. Garland-Campbell	Wheat Genetics, Quality Physiology and Disease Research, Johnson Hall, 291D, Washington State Univ., Pullman, WA 99164-6420 (kim.garland-campbell@ars.usda.gov)
D.F. Garvin	USDA-ARS, 411 Borlaug Hall, 1991 Upper Buford Circle, Saint Paul, MN 55108-6026 (David.Garvin@ars.usda.gov)
J. Gogerty	DuPont Pioneer, 1920 East McGregor Street, P.O. Box 557, Algona, IA 50511-0557 (Joseph.Gogerty@Pioneer.com)
P. Grassini	Dep. of Agronomy and Horticulture, 387 Plant Science Hall, East Campus, 1875 N. 38th Street, Univ. Nebraska, Lincoln, NE 68583-0963 (pgrassini2@unl.edu)
R. Graybosch	USDA-ARS, 137 Keim Hall, East Campus, UNL, Lincoln, NE 68583-0937 (Bob.Graybosch@ars.usda.gov)
R.J. Hayes	USDA-ARS, Crop Improvement and Protection Research Unit, 1636 E. Alisal St., Salinas, CA 93905 (ryan.hayes@ars.usda.gov)
C.C. Holbrook	USDA-ARS, 115 Coastal Way, Tifton, GA 31793 (Corley.holbrook@ars.usda.gov)
B.S. Hulke	USDA-ARS, Northern Crop Science Lab., 1307 18th St. N., Fargo, ND 58102-2765. *Corresponding author (brent.hulke@ars.usda.gov).
S.H. Jansky	USDA-ARS, Vegetable Crops Research Unit, 1575 Linden Dr., Madison, WI 53706-1590 (shelley.jansky@ars.usda.gov; shjansky@wisc.edu)
K.B. Jensen	USDA-ARS, Forage and Range Research Lab, 695 North 1100 East, Utah State Univ., Logan, UT 84332-6300. (Kevin.Jensen@ars.usda.gov)
W.C. Johnson III	USDA-ARS, 2747 Davis Rd., Tifton, GA 31793 (Carroll.johnson@ars.usda.gov)
S.R. Kaffka	Dep. of Plant Sciences, Univ. of California, Davis, CA 95616-8780 (srkaffka@ucdavis.edu).
J.D. Kelly	Dep. of Plant, Soil, and Microbial Sciences, Plant and Soil Science Bldg., Rm. 370, 1066 Bogue St., East Lansing, MI 48824 (kellyj@msu.edu)
C. Kimbeng	Louisiana State Univ. Agricultural Ctr., 5755 LSU Ag Rd., St. Gabriel, LA 70776 (CKimbeng@agcenter.lsu.edu).
L.W. Kleingartner	National Sunflower Association, 2401 46th Ave. SE., Suite 206, Mandan, ND 58554-4829 (larry.kleingartner@icloud.com)
R.T. Lewellen	USDA-ARS, Crop Improvement and Protection Research Unit, 1636 E. Alisal St., Salinas, CA 93905 (Robert.Lewellen@ars.usda.gov)
S.D. Linscombe	Rice Research Station, Louisiana State Univ. Agricultural Ctr., 1373 Caffey Rd., Crowley, LA 70578 (SLinscombe@agcenter.lsu.edu).
C. Löffler	DuPont Pioneer, 7250 NW 62nd Ave., P.O. Box 552, Johnston, IA 50131-0552 (Carlos.Loeffler@Pioneer.com)
N.B. Lyman	Dep. of Agricultural Economics and Agribusiness, Univ. of Arkansas, Fayetteville, AR 72701 (nblyman@uark.edu).
D. McClusky	Univ. of Georgia, Library, 2360 Rainwater Rd., Tifton, GA 31793-5766 (Mcclusky@uga.edu)
J.D. McCreight	USDA-ARS, Crop Improvement and Protection Research Unit, 1636 E. Alisal St., Salinas, CA 93905 (jim.mccreight@ars.usda.gov)
J.M. McGrath	USDA-ARS, Sugarbeet and Bean Research Unit, Dep. of Plant, Soil, and Microbial Sciences, 494 Plant & Soil Sci. Bldg., 1066 Bogue St., Michigan State Univ., East Lansing, MI 48824-1325 (mitchmcg@msu.edu)

K.S. McKenzie	Rice Experiment Station, California Cooperative Rice Research Foundation, Inc., P.O. Box 306, 955 Butte City Highway, Biggs, CA 95917 (KMcKenzie@crrf.org)
W.R. Meredith, Jr.	USDA-ARS, Jamie Whitten Delta States Research Ctr., 141 Experiment Station Rd., Stoneville, MS 38776 (bill.meredith@ars.usda.gov)
M. Metzger	Minn-Dak Farmer's Cooperative, 7525 Red River Rd., Wahpeton, ND 58075 (mmetzger@mdf.coop)
K.A.K. Moldenhauer	Rice Research and Exten. Ctr., Univ. of Arkansas, Div. of Agriculture, 2900 Hwy 130 E, Stuttgart, AR 72160 (kmolden@uark.edu)
R. Monk	Dupont Pioneer, 115 Meyer St., P.O. Box 97, Taft, TX 78390 (roger.monk@pioneer.com)
B. Mou	USDA-ARS, Crop Improvement and Protection Research Unit, 1636 E. Alisal St., Salinas, CA 93905 (beiquan.mou@ars.usda.gov)
L.L. Nalley	Dep. of Agricultural Economics and Agribusiness, Univ. of Arkansas, Fayetteville, AR 72701 (llnalley@uark.edu)
R.L. Nelson	Soybean/Maize Germplasm, Pathology & Genetics Research Unit, USDA-ARS, 1101 W. Peabody Dr., Urbana, IL 61801 (randall.nelson@ars.usda.gov)
J.M. Osorno	Dep. of Plant Science, Loftsgard Hall 166, P.O. Box 6050, North Dakota State Univ., Fargo, ND 58108 (Juan.Osorno@ndsu.edu)
P. Ozias-Akins	Univ. of Georgia, NESPAL, 2360 Rainwater Rd., Tifton, GA 31793-5766 (pozias@uga.edu)
L. Panella	USDA-ARS, NPA, Crop Research Lab., 1701 Centre Dr., Fort Collins, CO 80526 (Lee.Panella@ars.usda.gov)
T. Regassa	Dep. of Agronomy & Horticulture, 175 Keim Hall, Univ. of Nebraska, East Campus, Lincoln, NE 68583 (Tregassa2@unl.edu)
X. Sha	Rice Research and Exten. Ctr., Univ. of Arkansas, Div. of Agriculture, 2900 Hwy 130 E, Stuttgart, AR 72160 (xsha@uark.edu)
I. Simko	USDA-ARS, Crop Improvement and Protection Research Unit, 1636 E. Alisal St., Salinas, CA 93905 (ivan.simko@ars.usda.gov)
S. Smith	DuPont Pioneer, 7300 NW 62nd Ave., P.O. Box 1004, Johnston, IA 50131 (Stephen.Smith@Pioneer.com)
J.E. Specht	Dep. of Agronomy and Horticulture, 322 Keim Hall, East Campus, 1875 N. 38th St., Univ. of Nebraska, Lincoln, NE 68583-0915 (jspecht1@unl.edu)
D.M. Spooner	USDA-ARS, Vegetable Crops Research Unit, 1575 Linden Dr., Madison, WI 53706-1590 (david.spooner@ars.usda.gov)
H.T. Stalker	North Carolina State Univ., Dep. of Crop Sci., Greenhouse Unit 3, 214 Method Rd., Raleigh, NC 27695-7629 (tom_stalker@ncsu.edu)
C.A. Strausbaugh	USDA-ARS, Northwest Irrigation and Soils Research Lab., 3793 North 3600 East, Kimberly, ID 83341-5076 (Carl.Strausbaugh@ars.usda.gov)
A. Suman	DuPont-Pioneer, 8305 NW 62nd Ave., Johnston, IA 50131 (sumangene@yahoo.com)
J.A. Torrion	Crop Physiology, Northwestern Agricultural Research Ctr. (NWARC), Dep. of Research Centers, Montana State Univ., 4570 MT Hwy 35, Kalispell, MT 59901 (jessica.torrion@montana.edu)
C.A. Urrea	Univ. of Nebraska, Panhandle Research and Exten. Ctr., 4502 Avenue I, Scottsbluff, NE 69361 (currea2@unl.edu)

G.J. Vandemark — USDA-ARS, Grain Legume Genetics and Physiology Research Unit, 303 Johnson Hall, Washington State Univ., Pullman, WA 99164 (george.vandemark@ars.usda.gov)

G. Vellidis — Univ. of Georgia, Engineering Bldg., 2329 Rainwater Rd., Tifton, GA 31793 (yiorgos@uga.edu)

J. Wiersma — Univ. of Minnesota, NW Research and Outreach Ctr., 2900 University Ave., Crookston, MN 56716 (wiers002@umn.edu)

K. Wright — DuPont Pioneer, 7300 NW 62nd Ave., P.O. Box 1004, Johnston, IA 50131-1004 (Kevin.D.Wright@Pioneer.com)

1

Barley

Tom Blake,* Victoria Blake, and Jochim Wiersma

ABSTRACT

From 1912 through 1927, U.S. harvested barley (*Hordeum vulgare* L. ssp. *vulgare*) area and yield were roughly constant, averaging 3.1 million hectares (about 7.7 million acres) per year and approximately 1.2 t ha^{-1}, or 22 bu ac^{-1}. From 1928 through 1942, yield remained at about 1.2 t ha^{-1}, and barley area grew to a peak of 6.9 million ha harvested, about 17 million acres. From 1943 through 1986, barley area across the United States dropped to about 4.3 million ha, but with considerable improvement of yield, from 1.2 t ha^{-1} in 1942 to 2.7 t ha^{-1} in 1986 (about 51 bu ac^{-1}). 1988 and 1989 were intense drought years across much of the U.S. small grains production region, and in those 2 yr national yields significantly declined. In 1990, barley yield recovered, but barley's area decline continued, falling to 3.0 million ha. In the next decade the barley area decline accelerated to a low of 1.05 million ha nationally (2.6 million acres) in 2011. In 2012, barley area recovered slightly, with 1.3 million ha harvested. Between 1990 and 2012, yields grew from 3.0 to 3.7 t ha^{-1}. Overall, from 1942 to 2012 average gain in yield per year was a reasonable 43 kg ha^{-1}. While this indicates good progress in yield, it may not indicate good progress in yield gain through higher-yielding cultivars. The 1985 funding of the Conservation Reserve Program (CRP) permitted millions of marginal cropping hectares in the Midwest and upper plains states to be rested. This contributed to yield increases by decreasing the contributions of marginally productive barley hectares. Loss of barley area in the Red River Valley due to scab (*Fusarium oxysporum* Schlecht.) and due to competition from Roundup Ready short-season soybean [*Glycine max* (L.) Merr.] and maize (*Zea mays* L.) varieties during this period also contributed to this dramatic acreage loss. The U.S. malting and brewing industries are America's largest consumers of barley, purchasing more than one-half of the U.S. barley grain crop. More than 70% of the hectares seeded to barley are seeded to cultivars recommended by the American Malting Barley Association (AMBA). The malting and brewing industries are more concerned with crop quality and consistency than yield and are slow to adopt new cultivars. Since barley selected for malting commands a substantially higher farmgate price than does feed barley (often a two-fold difference), most barley growers prefer the recommended malting barley cultivars. The growth of low- and no-till cropping systems coupled with effective management of soil nutrition contributed to barley yield gains over the past two decades. These gains, together with the CRP-mediated reduction of production in truly marginal hectares, are likely responsible for more of the yield gains observed over the past 50 yr than is cultivar improvement.

Abbreviations: AMBA, American Malting Barley Association; CRP, Conservation Reserve Program; FHB, Fusarium head blight.

Tom Blake, Montana State Univ., Dep. of Plant Science & Pathology, Rm. 109, Agric. BioSci Bldg., Bozeman, MT 59717. *Corresponding author (blake@montana.edu).
Victoria Blake, USDA-ARS, Western Regional Research Center, 800 Buchanan Street, Albany, CA 94710 (victoria.blake@ars.usda.gov).
Jochim Wiersma, Univ. of Minnesota, NW Research and Outreach Center, 2900 University Ave., Crookston, MN 56716 (wiers002@umn.edu).

doi:10.2135/cssaspecpub33.c1

Yield Gains in Major U.S. Field Crops. CSSA Special Publication 33.
Stephen Smith, Brian Diers, James Specht, and Brett Carver, editors.

Globally, barley remains the fourth most important cereal crop and is primarily used for animal feed (about 80% for feed, 18% for malt, and 2% for food, http://faostat.fao.org). In the United States, an average of about 56% of the barley harvested in the years 2006 through 2012 was malted. About 71% of the U.S. barley crop is produced in the three main barley producing states—North Dakota, Montana, and Idaho. In these three states, more than 70% of the area seeded to barley is seeded to cultivars recommended by the AMBA (http://ambainc.org/, http://www.nass.usda.gov/). While agronomic performance is one criterion leading to varietal recommendation, malting quality is of primary importance to the U.S. malting and brewing industry. One principal concern of the U.S. malting and brewing industry is to ensure that sufficient high quality malting barley is available to meet industry demand. This has led to slow adoption of higher-yielding cultivars in favor of cultivars with malting quality similar to those already accepted by the U.S. malting and brewing industries. Since 1990, malt barley consumption by the U.S. brewing industry has declined from 132.2 million bushels to 110.7 million bushels. Much of this decline can be attributed to improvements in brewing efficiency by the major breweries. During the same timeframe, U.S. barley production dropped from 422.2 million bushels to 220.3 million bushels. Each year about 80,000 ha of barley are harvested for hay, approximately 150,000 t are used for food, and about 1.02 million tonnes are used as livestock feed. That more than one-half of the U.S. barley crop is acceptable to maltsters and brewers suggests that we are now growing barley on land that more reliably produces high quality grain and that breeders' efforts to develop barley cultivars that more reliably produce high quality malting barley in these environments have been successful.

Although several small seed companies bred barley cultivars over the past 50 yr, only two (the Coors breeding program and the Busch Agricultural Resources program) focused on the development of malting barley cultivars. Their impact on barley production in the main barley producing states has been significant. The Coors Moravian cultivars are in large part responsible for the remarkably high barley yields in Colorado and Idaho. These semidwarf cultivars have exceptional yield potential in irrigated production systems. The Busch Ag Resources six-rowed cultivar Tradition dominates North Dakota's barley production, and its two-rowed cultivar Conrad leads production in Idaho. Most of the remaining cultivars in current production derive from U.S. or Canadian public sector breeding programs (Table 1–1). The lifespan of dominant malting barley cultivars tends to be long, largely due to the needs and demands of the malting and brewing industries (e.g., 'Robust,' recommended in 1984 and still in production). Since the price difference between recommended malting barley cultivars and feed barley has historically been about twofold, more than 70% of the barley hectares in the United States are seeded to malting barley cultivars. These tend to have lower yield potential and weaker agronomic characteristics than do the best adapted feed-barley cultivars.

Table 1–1. Dominant malting barley varieties in production as of 2102 and year of American Malting Barley Association recommendation.

State	Cultivar (and source)	State area in 2012	Date recommended
		%	
Idaho	Conrad (Busch Agric. Res.)	24.4	2007
	AC Metcalfe (Agric. Canada)	13.6	2005
	Moravian 69 (Coors)	11.8	2010
	Merit 57 (Busch Agric. Res.)	8.4	2012
	Merit (Busch Agric. Res.)	5.0	2000
Montana	AC Metcalfe (Agric. Canada)	26.7	2005
	Hockett (Montana State Univ.)	14	2010
	Moravian 115 (Coors)	9.4	2012
	Conlon (North Dakota State Univ.)	4.2	2000
North Dakota	Tradition (Busch Ag. Res.)	42	2004
	Lacey (Univ. Minnesota)	20.2	2000
	Robust (Univ. Minnesota)	5.8	1984

National Trends

Barley production in the United States first exceeded 1 million ha in 1883. Until 2010, production remained well above that value. Barley production peaked in 1942 and has since declined (Fig. 1–1).

During this period from 1942 through 2012, yields grew from about 1 t ha^{-1} to about 3.6 t ha^{-1}, an average gain of about 37 kg ha^{-1} per year (Fig. 1–2). In Canada, over a similar 78-yr timespan, genetic yield gain equaled 30 kg ha^{-1} yr^{-1} (Bulman et al., 1993). Ortiz et al. (2002) found genetic gain equaled 13 kg ha^{-1} yr^{-1} in Nordic

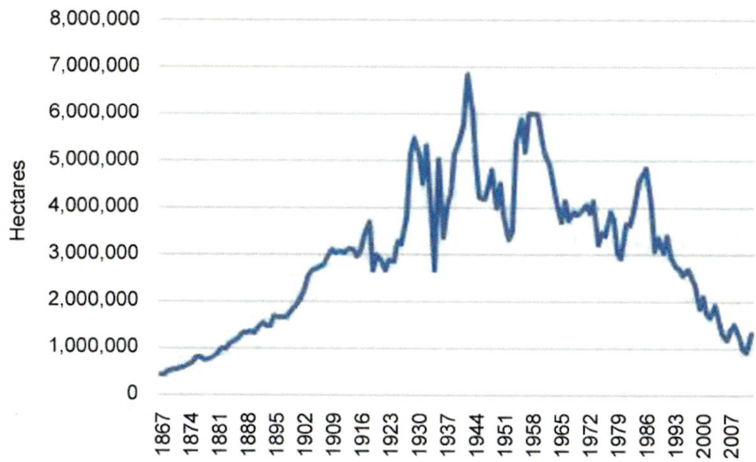

Fig. 1–1. Area in hectares devoted to barley production in the United States from 1867 through 2012.

barley germplasm. Similarly, a genetic gain of 21 kg ha^{-1} yr^{-1} was documented for six-rowed barley cultivars in Minnesota from 1920 through 1982 (Wych and Rasmussen, 1983). As with all crops, barley yield gain in the United States is not solely the result of improved cultivars.

Regional Trends

The Red River Valley of North Dakota and Minnesota is the lake bed of glacial Lake Agassiz. Nicknamed the "Breadbasket of the World" it has historically been an important production region of hard red spring wheat (*Triticum aestivum* L.) and six-row barley (Hanson, 1987). This region accounted for almost a quarter of the total U.S. barley area in 1985, or about 1.2 million ha (http://www.nass.usda.gov/Statistics_by_State/North_Dakota/Publications/County_Estimates/). Since that time, barley area in the Red River Valley region has declined sharply (Fig. 1–3). While U.S. acreage in recent years is only one-fifth of the acreage 25 yr ago, the Red River Valley region now accounts for less than 10% of the remaining U.S. acreage. The six-row barley areas in central and western North Dakota have seen less precipitous declines.

The reasons why six-row barley acreage declined so sharply in Minnesota and North Dakota are varied yet intertwined. McMullen et al. (1997) described how the re-emergence of Fusarium head blight (FHB) had a devastating impact on wheat and barley production in western Minnesota and eastern North Dakota in 1993 and 1994; FHB is primarily caused by *Fusarium graminearum* Schwabe, Group II [anamorph]. The authors estimated that in those 2 yr alone yield losses in barley equaled 2.16 million metric tons (99 million bushels). Nganje et al. (2004) estimated that between 1993 and 2001 barley yield losses due to FHB totaled 2.7 million tonnes (124 million bushels). These losses were aggravated by the presence of deoxynivalenol (DON), a mycotoxin commonly referred to as vomitoxin that is mainly produced by *F. graminearum*, further reducing barley marketability and price. The total economic losses for barley over the same period were estimated at $485 million (Nganje et al., 2004).

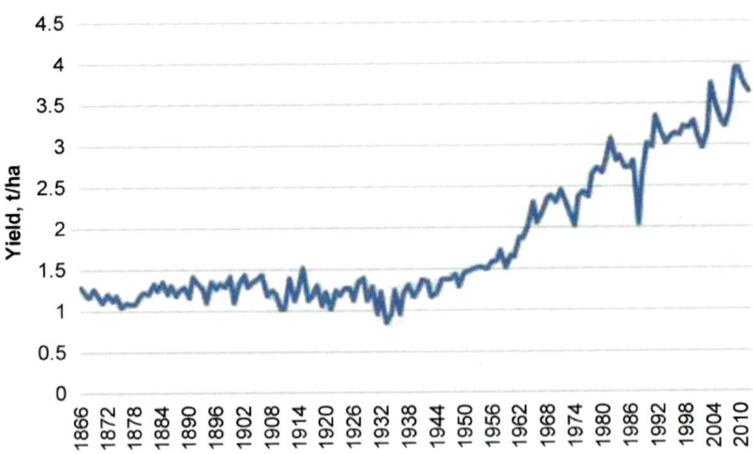

Fig. 1–2. U.S. barley yield from 1866 through 2012.

The decline in barley acreage, however, predates the FHB epidemics of the 1990s and accelerated after 1999. These observations point to other reasons why acreage shifted away from barley in favor of other crops. The decline in acreage is nearly mirrored by the increase in soybean acreage (Fig. 1–3). The Federal Agricultural Improvement and Reform Act of 1996 (P.C. 104-127), informally known as the Freedom to Farm Act, revised and simplified the direct payment programs for a number of commodities, including barley. Although base acreages for commodities were frozen, producers were allowed the flexibility to plant up to 100% of their contract acres to any crop. This flexibility allowed producers to transition to soybeans and away from less profitable, higher-risk choices, including barley. The advent of Roundup Ready soybean cultivars that were launched onto the market in 1996 further amplified the shift away from barley. In just 2 yr, nearly 50% of the total U.S. acreage had adopted this technology (Carpenter and Gianessi, 1999). The adoption of Roundup Ready soybean started a few years later in the Red River Valley as early maturing Roundup Ready soybean cultivars were initially not available. The adoption rate, however, was high once cultivars became available, and by 2001, 49 and 63% of all soybean hectares in North Dakota and Minnesota, respectively, had converted to Roundup Ready soybean (Fernandez-Cornejo, 2012).

From 1970 through 2010, yield improvements in North Dakota averaged 25 kg ha^{-1} yr^{-1} (Fig. 1–4a). In addition to genetic improvements, cultural practices have contributed to the yield gains achieved. Whereas less than 5% of the cultivated land area was in no-till production in 1989 in the Northern Great Plains, by 2004 this had increased fivefold to nearly 25% of the area (Hansen et al., 2012). Although no data are available for barley, Cochran et al. (2000), for example, found that continuous wheat using no-till practices yielded about 25% more than a conventional spring wheat–summer fallow rotation in northeast Montana. Lanning et al. (2010) found

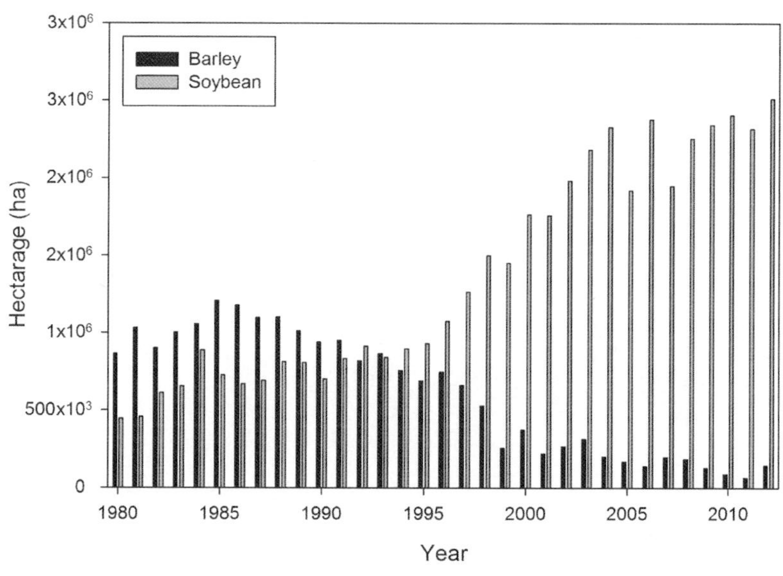

Fig. 1–3. Barley and soybean acreage in the Red River Valley from 1980 through 2012.

that warming temperatures in Sidney, MT, just west of the North Dakota–Montana border, in early spring had allowed for significantly earlier planting dates since 1950, resulting in higher grain yields of 'Thatcher' spring wheat. These changes have likely impacted barley productivity in western North Dakota similarly.

Two scenarios have developed in the intermountain states of Montana, Idaho, Colorado, and Utah. In both Utah and Colorado, barley acreage was never large, but declined to near insignificance after 1990. In Montana and Idaho, barley

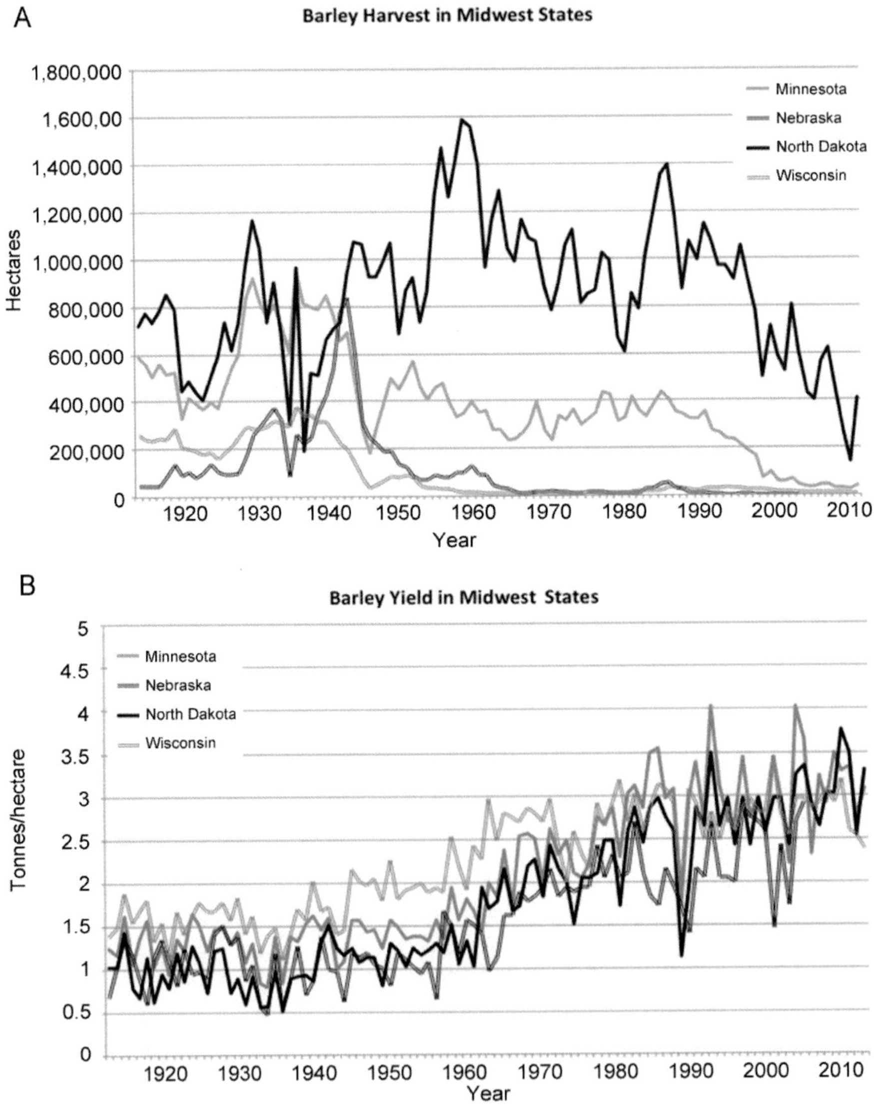

Fig. 1–4. (A) Barley production in harvested hectares for Minnesota, Nebraska, and North Dakota and (B) on-farm barley yield trends for Minnesota, Nebraska, North Dakota, and Wisconsin.

remains a significant, although reduced, component of the grain production system. Although barley production peaked in terms of area harvested in the 1980s, afterward the area devoted to barley rapidly declined (Fig. 1–5a). Harvested area peaked in Idaho in 1985 at 502,000 ha but declined to 240,000 ha by 2012. In Montana, the area harvested peaked in 1986 at 802,000 ha and now stands at about 320,000 ha. Nearly all the barley crop is irrigated in Idaho, while in Montana approximately one-third of the crop is irrigated. In Colorado and Utah, barley peaked at 141,000 and 64,000 ha in 1986 and 1985, respectively, and now stands at 22,000 and 10,000 ha. Commercial barley production is effectively restricted now to western North Dakota, Montana, and Idaho, accounting for 71% of the U.S. crop. In these three states, malting barley cultivars recommended by the

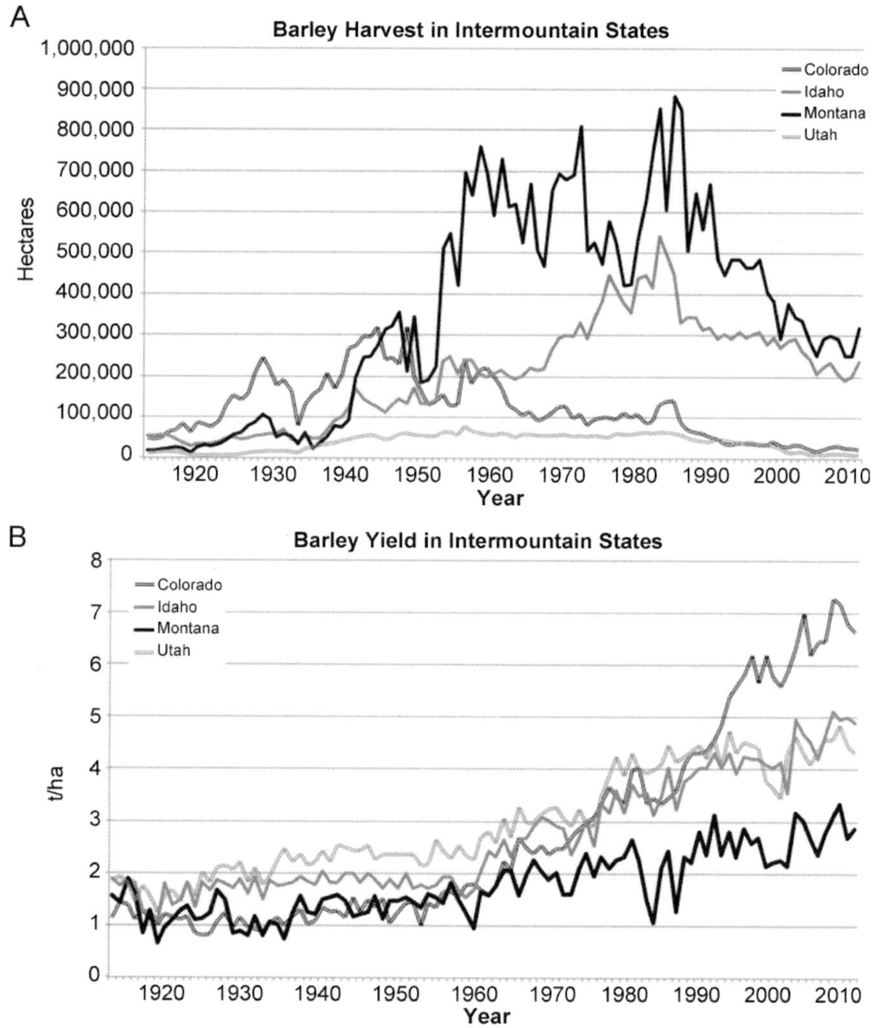

Fig. 1–5. Barley production in (A) harvested hectares and (B) on-farm barley yield trends for Montana, Idaho, Colorado, and Utah.

AMBA are dominant. Ninety percent of North Dakota's barley land was seeded to AMBA-recommended cultivars in 2012, while 77% of Idaho's barley land and 69% of Montana's barley land was seeded to AMBA-recommended cultivars (see Table 1–1 for partial list). Yield gains paralleled those observed nationwide (Fig. 1–5b).

On the West Coast, barley production has begun to rebound in Oregon and Washington after dropping to near insignificance in 2010 (Fig. 1–6). Washington's barley crop is seeded largely to the WestBred cultivar Champion, a high-yielding feed barley. A nascent local malting industry is developing in Oregon and

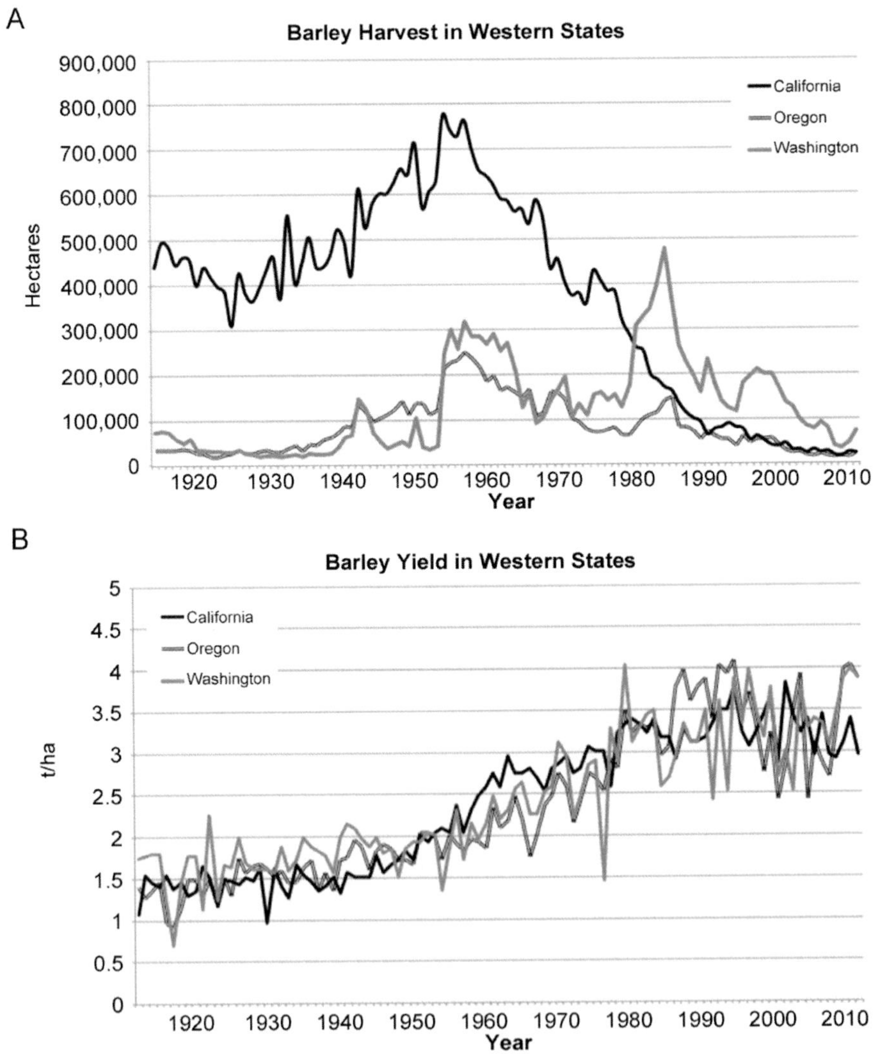

Fig. 1–6. Barley production in (A) harvested hectares and (B) on-farm barley yield trends for California, Oregon, and Washington.

Washington (Thomas, 2013), and barley cultivars not yet recommended by AMBA are being malted and brewed by local craft maltsters and brewers.

In the East, barley production appeared to show potential for growth when a barley-focused fuel ethanol plant was built in Virginia. Unfortunately, equipment malfunction during testing forced the plant to close, and barley production continued its decline (Fig. 1–7). Limited production remains in New York, but barley production along the Eastern Seaboard has severely declined.

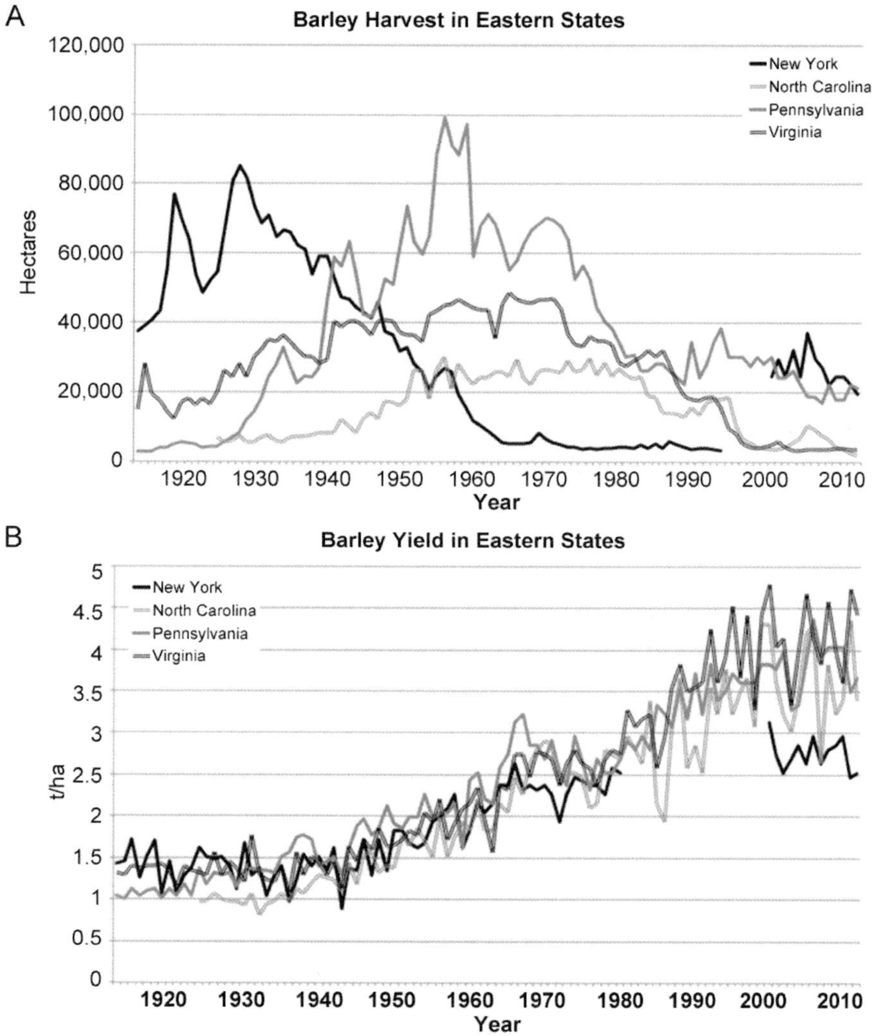

Fig. 1–7. Barley production in (A) harvested hectares and (B) on-farm barley yield trends for New York, North Carolina, Pennsylvania, and Virginia.

Moving Forward

The past three decades were challenging for barley as a crop. High grain prices have made enrollment in the Conservation Reserve Program less attractive, contributing to the recent decline in CRP enrollment (Fig. 1–8). Over the next 3 yr, approximately 40,000 ha in Idaho will see their CRP contract expire as will 290,000 ha in Montana and 180,000 ha in North Dakota. We expect a portion of this area to return to production, and for much of that increased production to be feed and hay barley.

The two major brewers and the large malting companies that provide their malt have focused malt barley production in Idaho, Montana, and North Dakota. This is due in part to the concentration of malt production facilities in the upper Midwest, Montana, and Idaho, but also to the relatively favorable weather during the growing seasons from 1990 through 2012. Barley's yield improvement in the United States during the same period reflects its concentration in the more amicable production regions of North Dakota, Montana, and Idaho. Barley has long been considered a crop of last resort for farmers with poor soil, limited rainfall and irrigation, and a short growing season. It's long been known that breeding barley for favorable environments results in selection of lines with limited stress adaptation (Ceccarelli et al., 1992, 1998). The current concentration of U.S. barley breeding resources in these favorable environments may result in a germplasm base with limited value in more stressful environments.

The more than 2200 craft brewers in the United States produce about 6% of the beer consumed by American beer drinkers, but utilize about 20% of the malt used to make beer in the United States. Currently, 26 small malting companies in 14 states have been created to help serve the needs of their local craft brewers (Thomas, 2013). While craft brewers have spurred innovation in the brewing industry, we hope that craft maltsters will help spur innovation in the evaluation and adoption of new barley cultivars suited to their local environments.

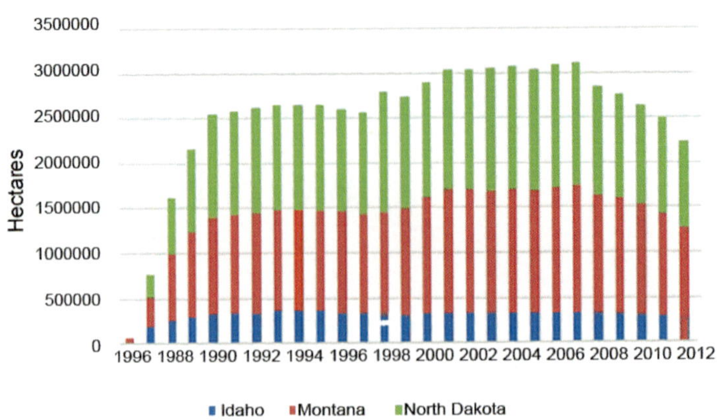

Fig. 1–8. Conservation Reserve Program enrollment of cropland in the three major barley producing states.

References

Bulman, P., D.E. Mather, and D.L. Smigh. 1993. Genetic improvement of spring barley cultivars grown in eastern Canada from 1910 to 1988. Euphytica 71:35–48. doi:10.1007/BF00023465

Carpenter, J., and L. Gianessi. 1999. Herbicide tolerant soybeans: Why growers are adopting Roundup Ready varieties. AgBioForum 2(2):65–72.

Ceccarelli, S., S. Grando, and A. Impiglia. 1998. Choice of selection strategy in breeding barley for stress environments. Euphytica 103:307–318. doi:10.1023/A:1018647001429

Ceccarelli, S., S. Trando, and J. Hamblin. 1992. Relationship between barley grain yield measured in low- and high-yielding environments. Euphytica 64:49–58.

Cochran, V.L., T.C. Ceasar, and R. Kolberg. 2000. Effects of tillage and cropping intensity on soil quality in the semi-arid northern Great Plains. In: 2000 Agronomy abstracts. ASA, CSSA, and SSSA, Madison, WI.

Fernandez-Cornejo, J. 2012. Adoption of genetically engineered crops in the U.S. (database). USDA-ERS. http://www.ers.usda.gov/Data/BiotechCrops/ (accessed 18 Nov. 2013).

Hansen, N.C., B.L. Allen, R.L. Baumhardt, and D.J. Lyon. 2012. Research achievements and adoption of no-till, dryland cropping in the semi-arid U.S. Great Plains. Field Crops Res. 132:196–203. doi:10.1016/j.fcr.2012.02.021

Hanson, N.E. 1987. Bread basket of the world: Fargo, Grand Forks, and the Red River Valley. Dakota Graphic Society, Fargo, ND.

Lanning, S.P., K. Kephart, G.R. Carlson, J.E. Eckhoff, R.N. Stougaard, D.M. Wichman, J.M. Martin, and L.E. Talbert. 2010. Climatic change and agronomic performance of hard red spring wheat from 1950 to 2007. Crop Sci. 50:835–841. doi:10.2135/cropsci2009.06.0314

McMullen, M., R. Jones, and D. Gallenberg. 1997. Scab of wheat and barley: A re-emerging disease of devastating impact. Plant Dis. 81:1340–1348. doi:10.1094/PDIS.1997.81.12.1340

Nganje, W.E., D.A. Bangsund, F.L. Leistritz, W.W. Wilson, and N.M. Tiapo. 2004. Regional economic impacts of Fusarium head blight in wheat and barley. Rev. Agric. Econ. 26:332–347. doi:10.1111/j.1467-9353.2004.00183.x

Ortiz, R., M. Nurminiemi, S. Madsen, O.A. Rognli, and A. Bjornstad. 2002. Genetic gains in Nordic spring barley breeding over sixty years. Euphytica 126:283–289. doi:10.1023/A:1016302626527

Thomas, D. 2013. Malting: The latest craft. Brewer and Distiller International July:47–50.

Wych, R.D., and D.C. Rasmussen. 1983. Genetic improvement in malting barley cultivars since 1920. Crop Sci. 23:1037–1041. doi:10.2135/cropsci1983.0011183X002300060004x

2

Cotton

B. Todd Campbell,* Debbie Boykin, Zaid Abdo, and William R. Meredith, Jr.

ABSTRACT

Cotton (*Gossypium hirsutum* L.) is produced in more than 30 countries and provides a major fiber source for textile manufacturers. In the United States, upland cotton is produced along the southern-most portion of the country in 17 states from California to Virginia. In 2012, the direct market value of 17.0 million bales of U.S. cotton equated US$8.1 billion. Our objective was to document trends in U.S. upland cotton productivity since 1980. A second objective was to document the impact of genetic gain on productivity. Analyses of on-farm and replicated variety trial data suggest that productivity and genetic gain have occurred since 1980. Production systems during this time saw major advances; most notably among these was the successful eradication of the boll weevil (*Anthonomous grandis* Boheman). Beginning with the adoption of transgenic cultivars in 1996, productivity increased significantly. The rate of genetic gain for yield between 1981 and 1995 was significantly lower than the rate of genetic gain between 1996 and 2011. This indicates the shift to transgenic cotton production systems paralleled substantial yield productivity and genetic gains realized through long-term breeding efforts. Significant genetic gains occurred through efforts to transfer stably inherited transgenes into improved cultivars. Traditional germplasm enhancement programs, new transgenic technology, and molecular and genomic breeding technologies offer exciting opportunities for future productivity increases.

Abbreviations: AMS, Agricultural Marketing Service; NCVT, National Cotton Variety Trial.

B.T. Campbell, USDA-ARS, Coastal Plains Soil, Water, and Plant Research Center, 2611 W. Lucas St., Florence, SC 29501. *Corresponding author (todd.campbell@ars.usda.gov).
D. Boykin (debbie.boykin@ars.usda.gov) and **W.R. Meredith, Jr.** (bill.meredith@ars.usda.gov), USDA-ARS, Jamie Whitten Delta States Research Center, 141 Experiment Station Rd., Stoneville, MS 38776.
Z. Abdo, USDA-ARS, Richard B. Russell Agricultural Research Center, 950 College Station Rd., Athens, GA 30605 (zaid.abdo@ars.usda.gov).

doi:10.2135/cssaspecpub33.c2

Yield Gains in Major U.S. Field Crops. CSSA Special Publication 33.
Stephen Smith, Brian Diers, James Specht, and Brett Carver, editors.

Cotton is produced in more than 30 countries and provides a major fiber source for textile manufacturers. In the United States, cotton is produced along the southern-most portion of the country in 17 states from California to Virginia. In this chapter, we focus on documenting productivity changes occurring in U.S. cotton production systems since 1980. During this time, the U.S. cotton industry experienced several notable changes, including: (i) the globalization of textile manufacturing, (ii) enhanced production methods, and (iii) new genetic technologies. In 1981, 15.6 million bales (218 kg equivalents) of cotton fiber were produced, equating to US$4.1 billion direct market value. In 2012, 17.0 million bales were produced equating to US$8.1 billion direct market value (Table 2–1). Although additional economic value is captured from cottonseed and its associated products, cotton fiber represents about 90% of cotton's total economic value (Campbell and Hinze, 2010).

Since 1980, consumption of U.S. cotton in the domestic textile industry has decreased, while exports have increased. In 1981, 44.5% of U.S. cotton was used by the domestic textile manufacturing industry, and 55.5% was exported (Table 2–1). For the 2012 crop, estimates suggest 17.8% will be used by the domestic textile manufacturing industry, and 78.2% will be exported (Table 2–1). The globalization of the textile manufacturing industry has not directly affected yield gains over this period of time. As reported by Meredith (1991) and Bowman (2000), yield has and continues to be a primary breeding objective in cultivar development programs. However, the globalization of the textile industry has resulted in increased emphasis on fiber quality in an effort to meet global market demands. Due to the well-known negative relationship between yield and fiber quality, simultaneously improving yield and fiber quality has been difficult (Campbell et al., 2013).

As is the case with many other agricultural commodities, U.S. cotton production systems have seen significant enhancements since 1980. One of the greatest enhancements resulted from dedicated programs to eradicate the boll weevil. For many areas in the U.S. Cotton Belt, cotton production was only feasible following this successful eradication program. Carlson et al. (1989) noted that boll weevil eradication in North and South Carolina increased yield amounting to US$84 per hectare, insecticide savings of US$74 per hectare and an increased land value of US$35 per hectare. Advances in weed, insect, and disease control measures resulted in increased cotton productivity. Other agronomic practices such as water and nutrient management also have advanced during this period of time. Another significant advance was realized with the adoption of short season production systems in many parts of the U.S. Cotton Belt. Short season production systems were readily adopted in an effort to decrease production risks, such as environmental and pest pressures, commonly associated with full season production systems (Hake and Meredith, 1990). Short season production systems were especially helpful in overcoming insect damage in regions yet to eradicate the boll weevil (Bourland, 2013).

Advances in genetic technologies through traditional breeding efforts have continued since 1980. The development of high yielding, early maturing cultivars such as DES 119 (Bridge, 1986) provided excellent genetic potential for increasing productivity. Such genetic advances provided the foundation for short season production systems. In addition to the development of early maturing cultivars,

Table 2–1. United States upland cotton production statistics from 1981–2012. Source: USDA-NASS.

Year	Planted Area	Production	Mill use	Exports	Value	Export bales	Productivity
	thousands of ha	— thousands of 214 kg bales —			billion US$	%	bales ha^{-1}
1981	7,130	15,646	5,264	6,567	4.08	55.5	2.2
1982	5,711	11,963	5,513	5,207	3.42	48.6	2.1
1983	3,960	7,771	5,921	6,786	2.48	53.4	2.0
1984	5,619	12,982	5,538	6,215	3.67	52.9	2.3
1985	5,338	13,432	6,413	1,960	3.63	23.4	2.5
1986	4,801	9,731	7,452	6,684	2.45	47.3	2.0
1987	5,048	14,760	7,617	6,582	4.55	46.4	2.9
1988	6,046	15,411	7,782	6,148	4.19	44.1	2.5
1989	4,936	12,196	8,759	7,694	3.88	46.8	2.5
1990	5,808	15,505	8,657	7,793	5.08	47.4	2.7
1991	6,424	17,614	9,613	6,646	4.91	40.9	2.7
1992	6,073	16,218	10,250	5,201	4.27	33.7	2.7
1993	6,206	16,134	10,418	6,862	4.52	39.7	2.6
1994	6,358	19,662	11,198	9,402	6.80	45.6	3.1
1995	7,715	17,900	10,647	7,675	6.57	41.9	2.3
1996	6,643	18,942	11,126	6,865	6.41	38.2	2.9
1997	6,282	18,793	11,349	7,500	5.98	39.8	3.0
1998	5,862	13,918	10,401	4,298	4.12	29.2	2.4
1999	6,476	16,968	10,194	6,750	3.81	39.8	2.6
2000	6,882	17,188	8862	6,740	4.26	43.2	2.5
2001	6,863	20,303	7696	11,000	3.12	58.8	3.0
2002	5,996	17,209	7273	11,900	3.78	62.1	2.9
2003	5,870	18,255	6266	13,758	5.52	68.7	3.1
2004	5,943	23,251	6691	14,436	4.85	68.3	3.9
2005	6,082	23,890	5871	17,673	5.69	75.1	3.9
2006	6,362	21,588	4935	12,959	5.01	72.4	3.4
2007	4,506	19,207	4584	13,634	5.65	74.8	4.3
2008	3,937	12,815	3541	13,261	3.54	78.9	3.3
2009	3,793	12,188	3550	12,037	3.45	77.2	3.2
2010	4,577	18,104	3900	14,376	7.13	78.7	4.0
2011	6,138	15,573	3300	11,714	7.18	78.0	2.5
2012	5,118	17,010	3400	12,200	8.16	78.2	3.3

beginning in the mid to late 1990s, transgenic herbicide and insect tolerance traits were available in commercial cultivars. These transgenic technologies allowed for the implementation of environmentally sound production practices in the form of reduced tillage and fewer pesticide applications. These practices resulted in better weed and insect control and improved soil nutrient and water conservation. According to the USDA-ERS, since 1996 transgenic cotton cultivars have seen rapid adoption (http://www.ers.usda.gov/data-products/adoption-of-genetically-engineered-crops-in-the-us, accessed 28 Mar. 2013). In 1997, 10.5 and 15% of the U.S. cotton crop was planted to transgenic herbicide and insect tolerant cultivars, respectively. By 2000, 61% of the U.S. crop was planted to transgenic cotton cultivars. Most recently in 2012, transgenic cultivars accounted for 94% of the U.S. crop.

In this chapter, our goal is to document changes in productivity occurring in U.S. cotton production systems since 1980. Grower supplied yield data from 1981 to 2012 will be assessed to provide a measure of changes in on-farm productivity. Historical U.S. cotton cultivar trial data from 1981 to 2011 will be assessed to provide a measure of the genetic contribution to increased productivity. Prospects for future productivity increases will de discussed.

Materials and Methods

On-Farm Productivity Trends

To assess on-farm productivity trends, upland cotton production data from 1981 through 2012 were obtained from the USDA-NASS (http://www.nass.usda.gov/Data_and_Statistics/index.asp, accessed 28 Mar. 2013). Planted area and lint yield data were obtained on an individual state basis. Although *Gossypium barbadense* L. production also exists in the United States (primarily in the West), only upland cotton production data were considered in these analyses. Individual state data were combined into four major U.S. production regions: (i) Eastern (Virginia, North Carolina, South Carolina, Georgia, Florida, and Alabama), (ii) Mid-South (Arkansas, Louisiana, Mississippi, Missouri, and Tennessee), (iii) Plains (Kansas, Oklahoma, and Texas), and (iv) Western (Arizona, California, and New Mexico).

For each year from 1981 through 2012, a mean lint yield was calculated for each of the four major U.S. cotton production regions. For each of the four regions, the regional lint yield mean (y axis) was plotted against production year (x axis). To represent a U.S. wide estimate as well, the mean lint yield across regions was plotted against production year. Simple linear regressions of lint yield over time were fitted. The first linear regression examined the yield trend for the entire range of 1981 through 2012. The second linear regression examined the yield trend for the time period before (1981–1995) and the third following (1996–2012) the introduction of herbicide and insect resistant cotton cultivars into commercial production. In each case, slopes were used as indicators of yearly productivity increases.

Genetic Contributions to Productivity Increases

To assess genetic contributions to productivity increases, lint yield data were obtained from the USDA-ARS National Cotton Variety Trial (NCVT) administered from Stoneville, MS. The NCVT was first organized in 1960 to establish a uniform system of reporting data from cotton yield trials across the U.S. Cotton Belt (Meredith et al., 2012). To facilitate data continuity for historical reference and comparisons across years, the program established 3-yr testing cycles consisting of several (3–5) national standard cultivars in each cycle. The program also established criteria to carry over at least one national standard cultivar between two consecutive testing cycles. Using these criteria, regional trials were organized such that commercial cultivars could be tested in targeted regions of the U.S. Cotton Belt. Hence, variety trials were established for four major U.S. production regions: Eastern, Mid-South, Plains, and Western. Since 1960, individual testing locations within each regional trial, with few exceptions, have been maintained from year to year.

Table 2–2 provides a general description of the regional NCVTs used in the current analysis. Three locations within each region were considered representative of that particular region in a given year. Within a region, each trial contained the same entries. Each year, some entries differed between regions. However, each year all trials (Eastern, Mid-South, Plains, and Western) contained the same national standard cultivars. At each location, trials were conducted using a randomized complete block design with four replications. Agronomic practices were standard for the particular trial location. At each trial location, weed and insect pests were controlled using conventional herbicide and insecticide methods as needed. This was the case for trials conducted before and after the adoption of transgenic cultivars.

Procedures to analyze lint yield data obtained from the NCVT archive were as follows. In an effort to standardize data collected in each year, overlapping national standard cultivars across trial cycles were used to create a standardized national standard cultivar (Table 2–3). This standardization was initiated and based on McNair 235, which was an overlapping national standard present in Cycles 8 and 9 of the NCVT. The mean yield difference between McNair 235 and the second national standard cultivar Paymaster 145 in Cycle 9 was used to adjust the yield of Paymaster 145 during Cycle 10. The mean yield difference between Paymaster 145 and Deltapine 50 was added to the mean difference obtained between McNair 235 and Paymaster 145. This was used to adjust the yield of Deltapine 50 during Cycles 11 and 12. The mean difference between the yield of Deltapine 50 and Paymaster HS 26 over Cycles 11 and 12 was added to the combined mean difference from the previous two steps and used to adjust the yield of Paymaster HS 26 during Cycle 13. This iterative process progressed toward Cycle 18 resulting in standardization of each carryover national standard cultivar

Table 2–2. Description of National Cotton Variety Trials used to estimate the genetic contribution to increased productivity.

Trial	Test years	Locations	Avg. number of experimental cultivars/year	Avg. total number of entries/year
Eastern	1981–2011	Florence, SC	10.5	14.7
		Belle Mina, AL		
		Auburn, AL		
Delta	1981–2011	Clarkedale, AR	5.8	10.0
		Saint Joseph, LA		
		Stoneville, MS		
Plains	1981–2011	Altus, OK	8.3	12.5
		Chickasha, OK		
		Lubbock, TX		
Western	1981–2011	Artesia, NM	6.3	10.5
		Pecos, TX		
		University Park, NM		

Table 2–3. Cultivars forming the base for the Standardized National Standard.

Cultivar	Cycle	Years
McNair 235	8, 9	1981–1986
Paymaster 145	9, 10	1984–1989
Deltapine 50	10, 11, 12	1987–1995
Paymaster HS 26	11, 12, 13	1990–1998
Acala Maxxa	13, 14	1996–2001
All Tex Atlas	14, 15	1999–2004
Stoneville 4892 BR	15, 16	2002–2007
Phytogen 72	16, 17	2005–2010
FiberMax 9058F	17, 18	2008–2011

to that of McNair 235. This standardization controlled for possible genetic differences between the national standard cultivars over time and thus reflected changes due to environmental factors.

By controlling for environmental effects in regional trials, per year and within each location, the standardized national standard cultivar provided a reference point that increased precision of genetic gain measurements. This was done by subtracting the yield of the standardized national standard from the mean of each of the three highest yielding experimental cultivars per year. The deviation for each trial was used as the *y* variable in a mixed effects model analysis of variance to evaluate trends in yield over time. Region, year, year group, and their interactions were considered fixed effects. Region represented the four regional NCVTs. Year group was a dummy variable that was used to separate 1981 through 2011 into one period, before transgenic technology (1981–1995), and a second period following transgenic technology (1996–2011). Year was modeled as a linear trend or covariate in the analysis. Locations within a region and their interactions with year and year group were considered random effects. Three locations were used and considered replications within a given region. The PROC MIXED module of SAS version 9.2 (SAS Institute, 2008) was used for this analysis. A scatter plot of studentized residuals was plotted against predicted values according to the regression and was used to assess the lack of fit.

Contrast statements were used to estimate regression coefficients (slopes) for each region before and after the adoption of transgenic technology. *t* tests were used to test if each slope was different than zero. In addition, a pairwise *t* test was used to test if slopes were different before and after transgenic technology. The slopes represented estimates of genetic gain per year. Slopes were also estimated from mean data across regions and compared before and after the adoption of transgenic technology. For each regression equation, *F* values were calculated to provide a measure of fit.

Results

On-Farm Productivity Trends

From 1981 through 2012, total U.S. upland cotton production averaged 16.3 million bales and ranged from 7.8 (1983) to 23.9 (2005) million bales (Table 2–1).

During this time, shifts occurred in total production among the four predominant U.S. production regions (Eastern, Mid-South, Plains, and Western). As noted previously by Meredith and Bridge (1984), regional shifts in cotton hectarage confound estimates of U.S. wide yield changes over time. Hence, care must be taken in developing inferences from U.S. wide on-farm yield estimates. From 1981 through 2012, total production decreased in the Mid-South and Western regions, while increasing in the Eastern and Plains regions (Fig. 2–1). By far, the greatest shift in production occurred in the Western region, where upland production recently fell below 500,000 ha. Taking this into account, yield trends from 1981 through 2012 will be discussed in both a United States and regionalized context. These trends will also be discussed as they relate to the time period before and after the adoption of transgenic cotton technology occurring in 1996.

In 1981, the mean U.S. yield approached 700 kg ha^{-1}. In 2012, the mean U.S. yield was nearly 1100 kg ha^{-1}. During this 30-yr period, on-farm yields increased by 400 kg ha^{-1}, or 57%. On-farm U.S. upland cotton yields increased at a linear rate of 13 kg ha^{-1} yr^{-1} (Fig. 2–2). For comparison purposes, on-farm U.S. cotton yields from 1870 to 1980 ranged from 150 to nearly 600 kg ha^{-1} (Meredith and Bridge, 1984). From 1936 to 1960, on-farm yields increased at a linear rate of 10.4 kg ha^{-1} yr^{-1} and decreased at a rate of 0.92 kg ha^{-1} yr^{-1} from 1960 to 1980. Current on-farm data (1981–2012) suggest that since 1981, yield increases have been achieved at a higher rate than reported previously by Meredith and Bridge (1984). Remarkably, from 1870 to 2012 average on-farm yields increased seven fold from 150 to 1100 kg ha^{-1}.

Figure 2–3 provides a regionalized view of on-farm productivity trends from 1981 through 2012. Slopes for the regression of yield over time indicated productivity increases ranged from 9.4 kg ha^{-1} yr^{-1} for the Eastern region to 16.3 kg ha^{-1} yr^{-1} for the Western region. Not surprisingly, R^2 values for these linear regressions were higher in the primarily irrigated Plains and Western regions. More year-to-year variation (environmental) was present in the Eastern region where rainfed production systems were more prevalent. Nonetheless, for each region, on-farm productivity trends were consistent with the U.S. wide trend.

Interestingly, after the adoption of transgenic technology (1996), there was a notable increase in yearly estimates of on-farm productivity. Considering the linear trends fitted for U.S. wide on-farm data, the period before transgenics showed increases of 4.9 kg ha^{-1} yr^{-1} whereas the period after transgenics showed increases of 22.4 kg ha^{-1} yr^{-1} (Fig. 2–2). Upon examination of regionalized on-farm productivity, a similar trend was apparent (Fig. 2–3). Before the adoption of transgenic technology, slopes estimated from linear regressions ranged from 3.0 kg ha^{-1} yr^{-1} for the Western region to 7.8 kg ha^{-1} yr^{-1} for the Mid-South region. Following adoption of transgenic technology, slopes ranged from 17.2 kg ha^{-1} yr^{-1} for the Plains region to 31.6 kg ha^{-1} yr^{-1} for the Western region. These data suggest remarkable increases in yearly on-farm productivity following the adoption of transgenic technology. On the basis of U.S. wide on-farm data, this translates to a 4.6-fold increase in productivity (kg ha^{-1} yr^{-1}) following the adoption of transgenic technology. Most noticeable among the regionalized data, adoption of transgenic technology in the Western region followed with a 10.5-fold increase in productivity (kg ha^{-1} yr^{-1}).

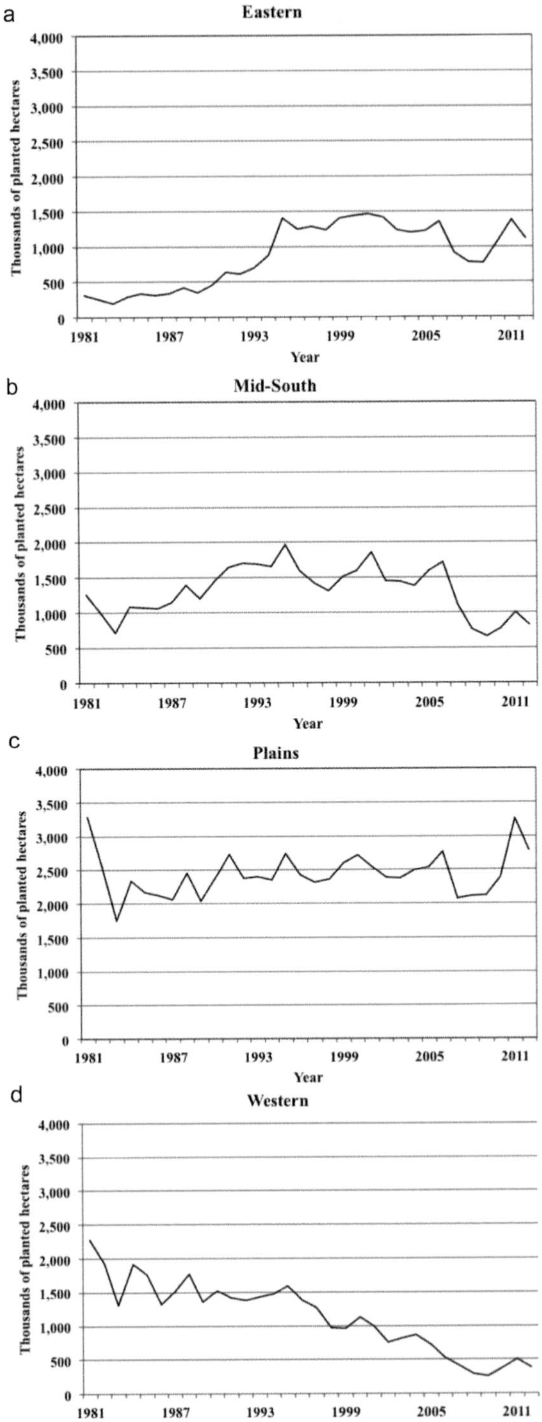

Fig. 2–1. Total area planted within regionalized U.S. production areas (a) Eastern, (b) Mid-South, (c) Southwest, and (d) Western) from 1981–2012. Source: USDA-NASS.

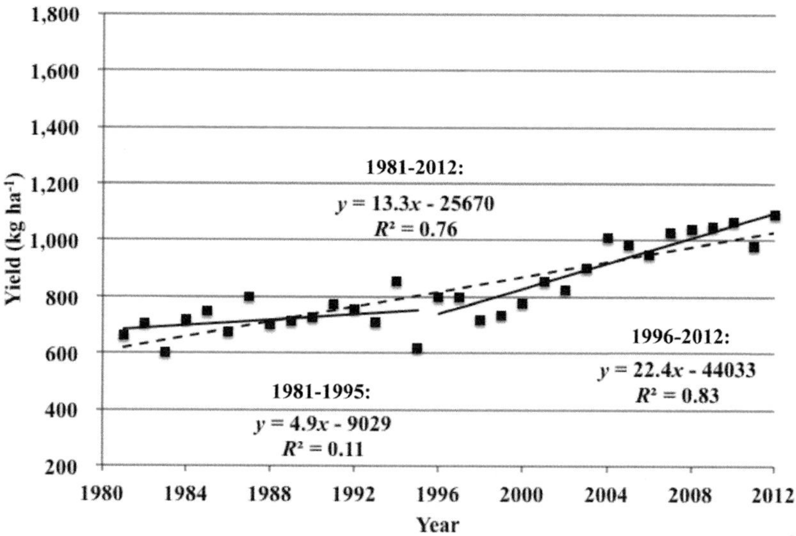

Fig. 2–2. Regression of yield over time for U.S. production areas for 1981 through 2012. Source: USDA-NASS.

Genetic Contributions to Productivity Increases

On-farm productivity increases have occurred from 1981 through 2012. However, as noted above, there is great difficulty in deciphering the relative contributions of environment, production practices, genetics, and their interactions to changes in productivity. To alleviate the confounding nature of on-farm productivity data, replicated yield data collected in the USDA-ARS NCVT were used to estimate genetic contributions to productivity increases.

The confounding effect of environment was removed in this analysis by standardizing the experimental data according to the mean of an overall standardized national standard cultivar. Using standardized data significantly reduced variability by explaining the portion of residual error due to the environmental trend (data not shown). Any confounding effects due to changes in production practices were negligible in this analysis since all trials were conducted using conventional cotton production methods. For example, all trials including transgenic insect, herbicide, or stacked resistant cultivars were managed conventionally using insecticide and/or herbicide applications routine for nontransgenic production systems.

Table 2–4 shows the results of the analysis using the linear mixed effects model of standardized lint yield obtained from regional NCVTs conducted from 1981 through 2011. The significant year × year group effect indicates the regression of the yield deviation over time averaged across regions was different before and after the adoption of transgenic technology. The nonsignificant year × region × year group effect suggests that the regression of the yield deviation over time before and after the adoption of transgenic technology was consistent across regions. The plot of the studentized residuals against the predicted values (Fig. 2–4) did not show any possible trends that were not captured by the model and did not provide signs of a major lack of fit problem. These analyses supported

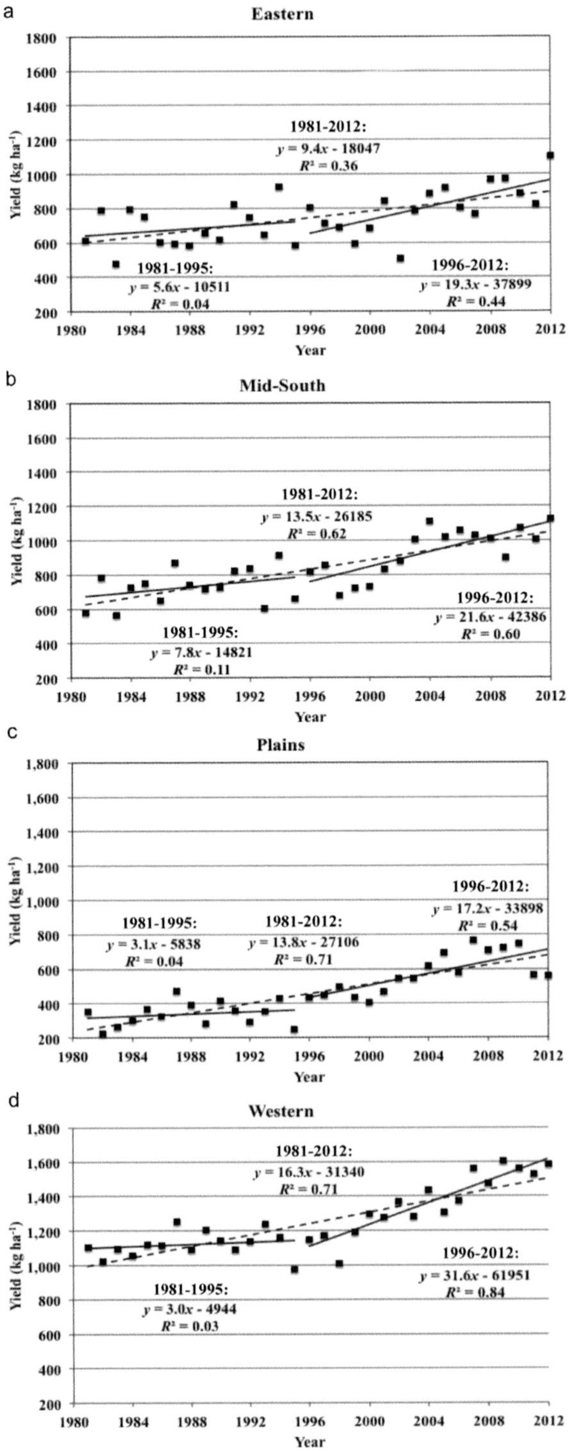

Fig. 2–3. Regression of yield over time for regionalized U.S. production areas (a) Eastern, (b) Mid-South, (c) Plains, and (d) Western for 1981 through 2012. Source: USDA-NASS.

Table 2–4. Tests of fixed effects and random effects from linear mixed model analysis of lint yield obtained from Eastern, Mid-South, Plains, and Western region national cotton variety tests conducted from 1981–2011.

Fixed effects			
Effect	df	*F* value	*P* > *F*
Region	3	6.1	0.0008
Year group†	1	20.6	<0.0001
Region × year group	3	3.7	0.0119
Year	1	28.5	<0.0001
Year × year group	1	11.1	0.0015
Year × region	3	2.9	0.0411
Year × region × year group	3	1.9	0.1488
Random effects			
Effect	Variance component		
Location (region)	957		
Location × year group (region)	0		
Year × location (region year group)	44		
Residual	20060		

† Year groups are 1981–1995 and 1996–2011.

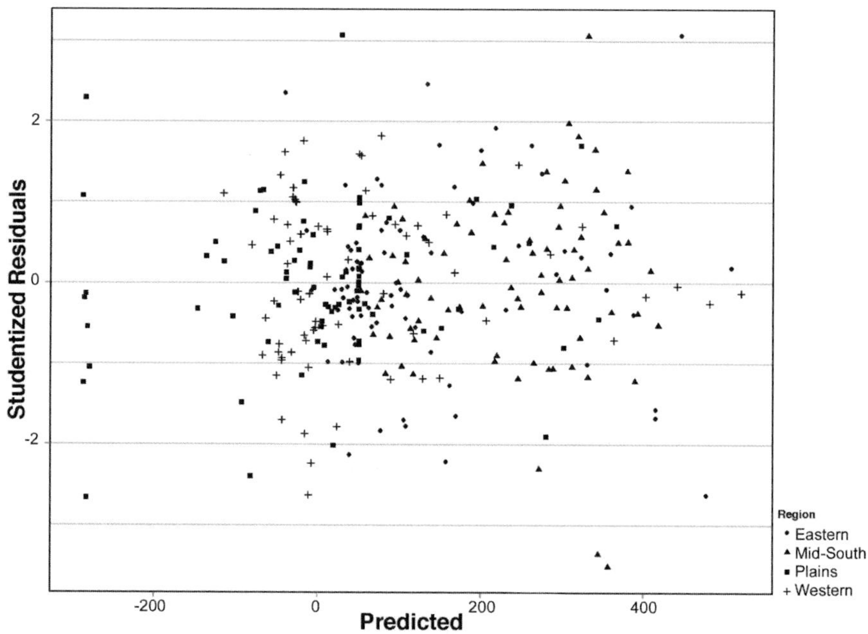

Fig. 2–4. Plot of studentized residuals versus predicted values to assess goodness of fit of the linear mixed effects model.

Table 2–5. Regression coefficients (*b*) and standard errors (SE_b) obtained from linear regression of cotton cultivar means on year of national variety test.

Region	Year group	$b \pm SE_b$		*t* value†	$P > t$
		—— kg ha^{-1} yr^{-1} ——			
East	Before 1996	5.8	6.3	0.91	0.3667
	After 1996	28.4	7.1	3.97	0.0002
Mid-South	Before 1996	11.7	6.5	1.81	0.0762
	After 1996	10.1	7.0	1.45	0.1535
Plains	Before 1996	−5.0	6.4	−0.78	0.4366
	After 1996	10.0	7.0	1.44	0.1560
West	Before 1996	7.5	6.3	1.18	0.2440
	After 1996	38.0	9.3	4.08	<0.0001
Avg.	Before 1996	5.0	3.2	1.56	0.1258
	After 1996	21.6	3.8	5.64	<0.0001

† *t* value corresponds to test $b = 0$.

efforts to estimate and compare individual regression coefficients (slopes) within and across regions.

Table 2–5 shows the regression coefficients and their standard errors for each region and averaged across regions estimated before and after the adoption of transgenic technology. Averaged across regions, there was a significant trend following adoption of transgenic technology indicating genetic progress of 21.6 kg ha^{-1} yr^{-1} (Fig. 2–5). On a regional basis, trends for the Eastern, Mid-South, and Western regions before the adoption of transgenic technology were positive (Fig. 2–6). The trend for the Plains region was negative. Each of these trends was not statistically different than zero (Table 2–5). Following the adoption of transgenic technology, each regional trend was positive (Fig. 2–5). Trends for the Eastern and Western regions were highly significant and indicated genetic progress of 11.7 and 38.0 kg ha^{-1} yr^{-1}, respectively (Table 2–5). Collectively, these data suggest very little genetic gain for yield from 1981 up to the adoption of transgenic technology. In contrast, for the time period following the adoption of transgenic technology to the present, genetic gains have been achieved at a remarkably high rate per year. It is important to note that the current analysis of USDA-ARS NCVT used transgenic cultivar yield data beginning in 1998 (data not shown). With one exception in each year following 1998, the number of transgenic cultivars included in the analyses increased. The exception was the Western dataset in which no transgenic cultivars were included in the high three cultivars selected for yield analysis from 2000 to 2004.

Table 2–6 provides a summary of *t* tests conducted to test for differences between rates of genetic gain (regression coefficients) before and after the adoption of transgenic technology. Averaged across regions, rates of genetic gain before and after the adoption of transgenic technology were significantly different. The difference in genetic gain between the two time periods was 16.6 kg ha^{-1} yr^{-1}. Each of the regions, except for the Mid-South, showed a similar trend. For the Mid-South region, the two estimates of genetic gain were not significantly different. Estimates of genetic gain for the Plains region did not differ at the $P <$

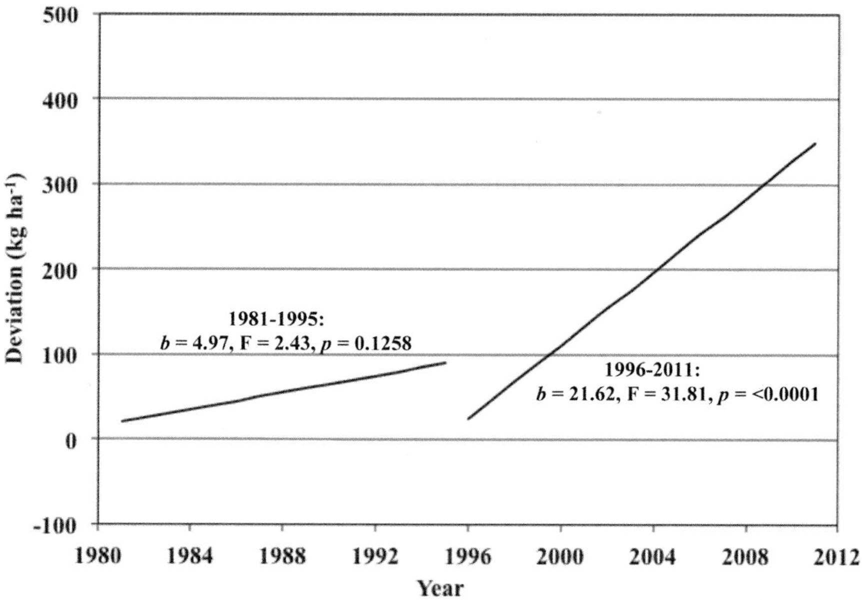

Fig. 2–5. Regression of the deviation from the standardized national standard yield averaged across four regional national cotton variety trials conducted from 1981 to 2012.

0.05 level of significance; however, the difference in genetic gain was in the positive direction. For the Eastern and Western regions, the difference in genetic gain between the two time periods was highly significant and indicated a remarkable increase in genetic gain (22.6 and 30.5 kg ha^{-1} yr^{-1}, respectively) after transgenic technology was adopted. Collectively, these data suggest the impact of transgenic technology on genetic gain has been significant across the United States. It should be noted, however, the impact on genetic gain in the Mid-South region has not been as great.

Discussion

Even though 1.9 million fewer hectares were planted in 2012, 1.4 million more bales were produced translating to 3.3 bales ha^{-1} (2.2 bales ha^{-1} in 1981). Since 1981, U.S. wide on-farm cotton productivity increased at a linear rate of 13.3 kg ha^{-1} yr^{-1}. This increase in on-farm productivity resulted from a combination of factors, including genetics and enhanced production practices. In particular, during this time the boll weevil eradication program played a major role in increasing production. Also, corresponding with the adoption of transgenic technology packaged in cultivars beginning in 1996, the increase in on-farm productivity dramatically increased to 22.4 kg ha^{-1} yr^{-1}. This is in contrast to the estimate of 4.9 kg ha^{-1} yr^{-1} from 1981 to 1995.

Regardless of whether increased on-farm productivity has resulted more or less from enhanced production practices and/or genetics, transgenic technologies have contributed to increased yield and decreased the amount of pesticides

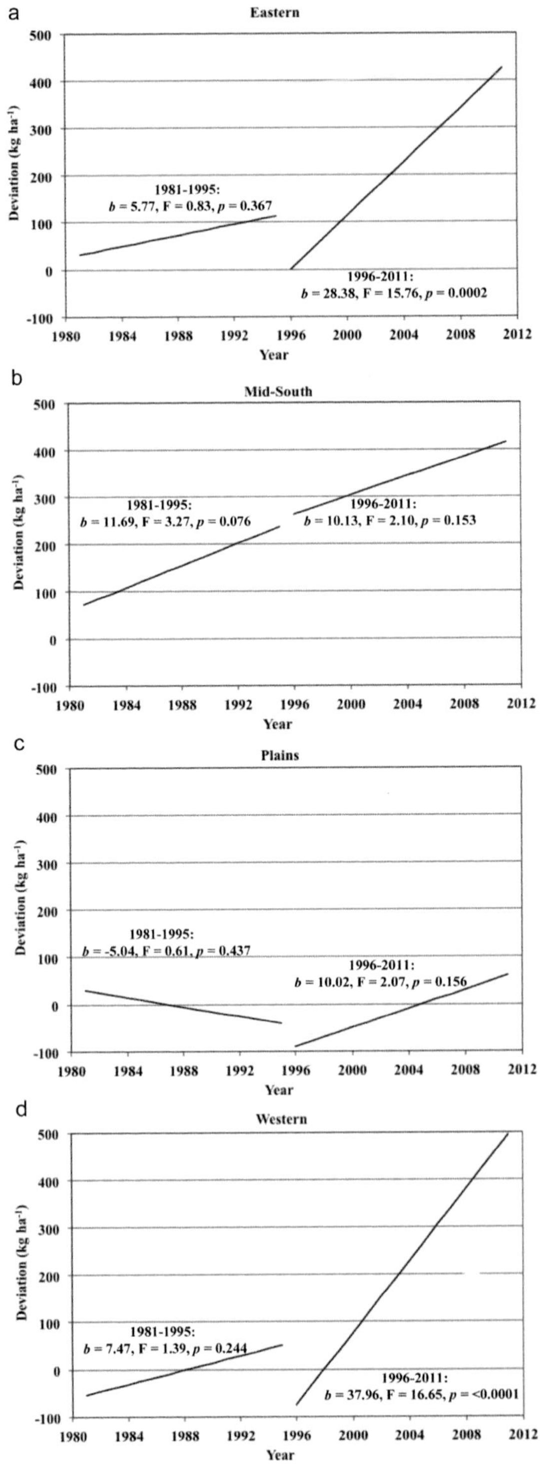

Fig. 2–6. Regression of the deviation from the standardized national standard yield for regionalized U.S. national cotton variety trials (a) Eastern, (b) Mid-South, (c) Plains, and (d) Western for 1981 through 2012.

Table 2–6. Comparison of regression coefficients (b_{diff}) and their standard errors (SE_b) obtained before and after 1996.

Region	b_{diff} ± SE_b		*t* value	*P* > *t*
	—— kg ha⁻¹ yr⁻¹ ——			
Eastern	22.6	9.5	2.36	0.0222
Mid-South	−1.6	9.5	−0.16	0.8699
Plains	15.1	9.5	1.59	0.1180
Western	30.5	11.3	2.70	0.0086
Avg.	16.6	5.0	3.34	0.0015

applied (Cattaneo et al., 2006; Fernandez-Cornejo and Caswell, 2006; Sankula, 2006). Bryant et al. (2008) demonstrated the economic benefit associated with the use of transgenic cotton technologies; these were especially beneficial in environments with high pest pressure. Kathage and Qaim (2012) demonstrated economic benefits associated with the adoption of transgenic insect resistant cotton in India, in which yield increased 24% and profit increased 50% among smallholders. In the United States, production practices including tillage, weed control, and insect control have changed with the adoption of transgenic technology. Conservation tillage practices have been adopted which conserve water and reduce soil erosion (Bauer and Busscher, 1996). Weed and insect control has dramatically improved with the availability of herbicide and insect resistance technologies. According to the ERS agency of the USDA, the adoption of transgenic cotton cultivars since 1996 has been rapid (http://www.ers.usda.gov/data-products/adoption-of-genetically-engineered-crops-in-the-us, accessed 28 Mar. 2013). In 1997, transgenic herbicide and insect tolerant cultivars were planted on 10.5 and 15% of the total U.S. cotton area, respectively. By 2000, 61% of the U.S. crop was planted to transgenic cotton cultivars and increased to 94% in 2012.

Previously, in their analysis of cotton productivity from 1866 to 1980, Meredith and Bridge (1984) estimated on-farm productivity changes for 1866 through 1935, 1936 through 1960, and 1961 through 1980. From 1866 to 1935, productivity was stagnant. From 1936 to 1960, yields increased at an annual rate of 10.4 kg ha^{-1} yr^{-1}. Obsolete and modern cultivar comparison studies conducted by Bridge et al. (1971) and Bridge and Meredith (1983) estimated genetic gains of 10.2 kg ha^{-1} yr^{-1} between 1922 and 1962 and 9.5 kg ha^{-1} yr^{-1} between 1910 and 1978. However, from 1961 to 1980 they estimated yields declined at a rate of 0.9 kg ha^{-1} yr^{-1} (Meredith and Bridge, 1984). Paterson et al. (2004) attributed decreased yield improvement and genetic gain from 1985 to 1998 to increased genetic vulnerability. Genetic vulnerability resulted from an erosion of the U.S. gene pool due to both genetic bottlenecks imposed by polyploid formation, domestication, and migration, and recent overexploitation of few genetic backgrounds in breeding programs (Paterson et al., 2004).

In the current assessment, analysis of 1981 to 2011 USDA-ARS NCVT data suggested increased productivity due to genetic gain following the adoption of transgenic technology averaged 21.6 kg ha^{-1} yr^{-1}. The estimate of genetic gain from 1981 to 1995 averaged 5.0 kg ha^{-1} yr^{-1}. Although difficult to separate, genetic improvements have resulted directly from both the transgenic trait(s) per se and improvements in the remaining plant genetic architecture. A closer look at Fig. 2–2 shows the average U.S. yield fell on or below the regression line from 1998 to 2003, while being on or above from 2004 to 2012 (excluding 2011). This observation supports the idea that little or no yield increases resulted from the first generation of transgenic cotton cultivars. This is consistent with the hypothesis that yield stagnation occurred in the first generation of transgenic cotton cultivars as a consequence of backcross breeding. Backcross breeding methods were employed to quickly commercialize the first transgenic cultivars and focused primarily on adding transgenic traits to existing recurrent parents. The increased yields observed beginning in 2004 likely reflect the commercialization of transgenic cultivars developed with simultaneous recurrent parent improvements. Hence, recent transgenic cultivars, in addition to the transgenic trait(s) alone, represent genetic improvements realized through forward breeding efforts.

Compared to previously reported estimates of genetic gain for cotton yield, genetic gain estimates over a 40-yr time frame ranged from 3.7 to 10.2 kg ha^{-1} yr^{-1} (Schwartz and Smith, 2008). Campbell et al. (2011) conducted a study designed to document genetic change in the Pee Dee cotton germplasm enhancement program over eight cycles of organized breeding. Yield gains were estimated at 28 kg ha^{-1}, or 3% per breeding cycle. Liu et al. (2013) conducted a study in Australia designed to estimate the contribution of genetics, management, and cultivar × management interactions to yield gains occurring from 1980 to 2009. Genetics accounted for 48%, management 28%, and cultivar × management interactions 24%. From 1980 to 1994, genetic gain was 7.0 kg ha^{-1} yr^{-1}, while it was 18.3 kg ha^{-1} yr^{-1} from 1995 to 2009. Interestingly, these genetic gain estimates (over similar time periods) were similar to those estimated in the current study. Liu et al. (2013) attributed increased genetic gain from 1995 to 2009 to increased yield stability.

As noted by Schwartz and Smith (2008), varying planting densities and levels of interplant competition also affect productivity and genetic gain estimates. Similar to what was found in maize (Duvick and Cassman, 1999), Schwartz and Smith (2008) concluded that genetic gains have been achieved not only for yield potential per se but also for tolerance to interplant competition. Since tolerance to interplant competition has not been a major objective in contemporary cotton breeding programs, it is possible that unintentional selection for tolerance to interplant competition has contributed to genetic gains for yield over time. The current study did not address the impact of interplant competition tolerance on yield genetic gain.

Another important point worthy of consideration is the effect of increased yield on cotton fiber quality. The long-standing negative relationship between yield and fiber quality has made efforts to simultaneously maximize both traits difficult (Campbell et al., 2012). Figure 2–7 provides the on-farm fiber length and fiber strength data reported by the Florence, SC (Eastern) and Visalia, CA (Western) Cotton Classing Offices of the Agricultural Marketing Service agency of the USDA (http://www.ams.usda.gov, accessed 28 Mar. 2013). Linear trends for the time period following the adoption of transgenic technologies (1996–present) show very little increase in fiber length and strength. In an analysis of Australian and U.S. variety trial data, Clement et al. (2012) demonstrated that the negative association between yield and fiber quality continues to persist.

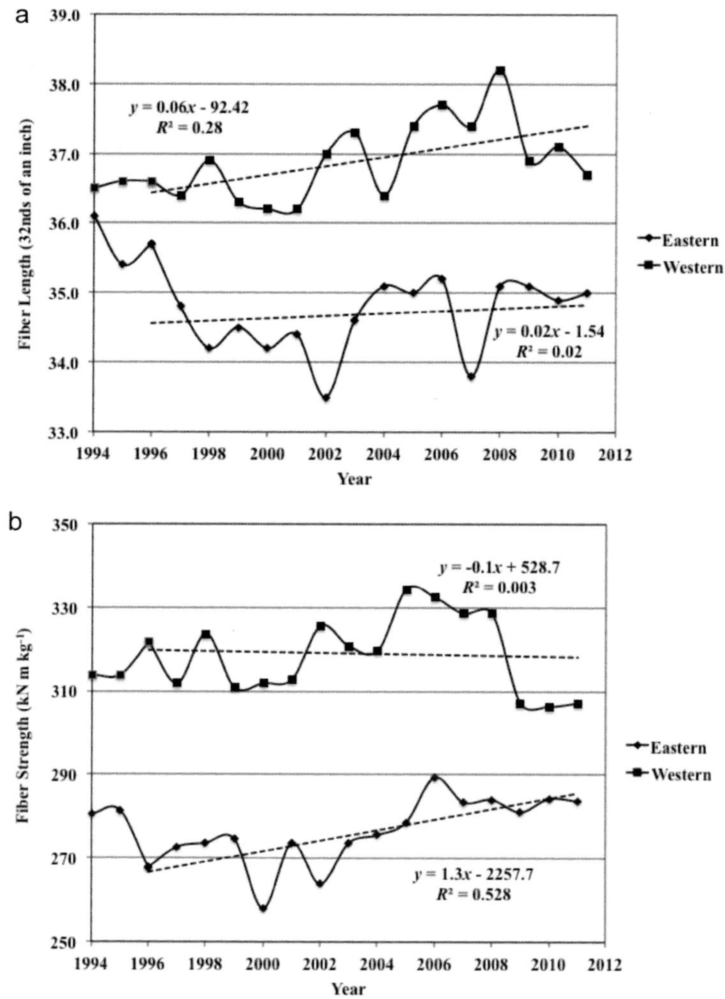

Fig. 2–7. Regression of High Volume Instrument fiber quality parameters for (a) fiber length and (b) fiber strength over time for Eastern and Western U.S. production areas between 1994 and 2011. Source: USDA-AMS.

Future Prospects

Results of the current study and the other studies reviewed here suggest reduced productivity and genetic gains occurred from the 1970s to the mid 1990s. Clearly, various reports articulating concerns over cotton yield stagnation were warranted during this time period (Meredith and Bridge, 1984; Paterson et al., 2004). However, the current analysis, focused on 1981 to the present, provides evidence to support the conclusion that cotton productivity has dramatically increased following the adoption of transgenic technology. Genetic gain increases have been realized through efforts to transfer stably inherited transgenes and improve genetic backgrounds. Clearly, since the advent of the first generation of transgenic cultivars, forward breeding efforts have increased productivity. In addition, increased investment in cultivar development programs, largely by private corporations, has dramatically increased the scale of evaluation data collected across locations spanning the entire U.S. cotton production area. Similar to the conclusion of Liu et al. (2013) for Australia, it is plausible that this increase in data collection has resulted in increased yield stability. The genetic gain estimate of 21.6 kg ha^{-1} yr^{-1} provides evidence that a genetic yield plateau has not yet been reached in U.S. cotton production. Evidence suggests that significant genetic gains have occurred and will continue in the future.

Public sector cotton breeding and germplasm enhancement programs have and will continue to play a significant role in improving cotton productivity by developing new genetic resources that broaden the genetic base. Due to the proprietary nature of germplasm and traits today, quantitative information on the extent public sector germplasm that has been used in recent cultivar development programs is unknown. However, it is known that from 2000 to 2009 public sector programs released and registered over 200 germplasm lines and/or cultivars in *Crop Science* or *Journal of Plant Registrations* (Campbell and Hinze, 2010). Since 2010, nearly 40 additional germplasm lines and/or cultivars have been released and registered in the *Journal of Plant Registrations*. Public sector programs play an especially important role in developing germplasm that overcome the negative relationship between yield and fiber quality (Campbell et al., 2013). Due to the continued globalization of cotton textile manufacturing, improving fiber quality is essential to meet future global market demands.

In addition to traditional germplasm enhancement programs and new transgenic technology, molecular breeding technologies offer new opportunities for future increases in productivity (Eathington et al., 2007; Heffner et al., 2009). In 2012, the first *Gossypium* genome sequencing projects were completed (Paterson et al., 2012; Wang et al., 2012). These projects and others provide the foundation for increasing our knowledge of the genetic architecture of the cotton genome. In future breeding programs, this knowledge will be harnessed and exploited to increase cotton productivity.

Acknowledgments

The NCVT Program is supported by funding from CRIS No. 6402-21000-032-00D of the U.S. Department of Agriculture. The authors would like to acknowledge all past and current leaders, cooperators, and participants of the USDA-ARS NCVT Program. Special thanks to Ellen Keene for her work in the administration of the program. Mention of trade names or commercial products in this publication is solely for the purpose of providing specific information and does not imply recommendation or endorsement by the USDA.

References

Bauer, P.J., and W.J. Busscher. 1996. Winter cover and tillage influences on coastal plain cotton production. J. Prod. Agric. 9:50–54. doi:10.2134/jpa1996.0050

Bourland, F. 2013. Another challenge for cotton. Cotton Farming, April, p. 22. http://www.cottonfarming.com/home/issues/2013-04/My-Turn.html (accessed 31 Mar. 2014).

Bowman, D.T. 2000. Attributes of public and private cotton breeding programs. J. Cotton Sci. 4:130–136.

Bridge, R.R. 1986. Registration of 'DES 119' cotton. Crop Sci. 26:646–647.

Bridge, R.R., and W.R. Meredith. 1983. Comparative performance of obsolete and current cotton cultivars. Crop Sci. 23:949–952. doi:10.2135/cropsci1983.0011183X002300050032x

Bridge, R.R., W.R. Meredith, and J.F. Chism. 1971. Comparative performance of obsolete varieties and current varieties of upland cotton. Crop Sci. 11:29–32. doi:10.2135/cropsci1971.0011183X001100010010x

Bryant, K.J., J.M. Reeves, R.L. Nichols, J.K. Greene, C.H. Tingle, G.E. Studebaker, F.M. Bourland, C.D. Capps, Jr., and F.E. Groves. 2008. Valuing transgenic cotton technologies using a risk/return framework. J. Agric. Appl. Econ. 40:767–775.

Campbell, B.T., P.W. Chee, E.L. Lubbers, D.T. Bowman, W.R. Meredith, J. Johnson, and D.E. Fraser. 2011. Genetic improvement of the Pee Dee cotton germplasm collection following seventy years of plant breeding. Crop Sci. 51:955–968. doi:10.2135/cropsci2010.09.0545

Campbell, B.T., P.W. Chee, E. Lubbers, D.T. Bowman, W.R. Meredith, J. Johnson, D. Fraser, W. Bridges, and D.C. Jones. 2012. Dissecting genotype ´ environment interactions and trait correlations present in the Pee Dee cotton germplasm collection following seventy years of plant breeding. Crop Sci. 52:690–699. doi:10.2135/cropsci2011.07.0380

Campbell, B.T., and L.L. Hinze. 2010. Cotton production, processing, and uses of cotton. In: B. Singh, editor, Industrial crops and uses. CABI, Egham, UK. p. 259–276.

Campbell, B.T., D.B. Weaver, R. Sharpe, J. Wu, and D.C. Jones. 2013. Breeding potential of elite Pee Dee germplasm in upland cotton breeding programs. Crop Sci. 53:894–905.

Carlson, G.A., G. Sappie, and M. Hamming. 1989. Economic returns to boll weevil eradication In: Agriculture Economics Report. Agric. Econ. Rep. 621. USDA-ERS, Washington, DC.

Cattaneo, M.G., C. Yafuso, C. Schmidt, C. Huang, M. Rahman, C. Olson, C. Ellers-Kirk, B.J. Orr, S.E. Marsh, L. Antilla, P. Dutilleul, and Y. Carriere. 2006. Farm-scale evaluation of the impacts of transgenic cotton on biodiversity, pesticide use, and yield. Proc. Natl. Acad. Sci. USA 103:7571–7576. doi:10.1073/pnas.0508312103

Clement, J.D., G.A. Constable, W.N. Stiller, and S.M. Liu. 2012. Negative associations still exist between yield and fibre quality in cotton breeding programs in Australia and USA. Field Crops Res. 128:1–7.

Duvick, D.N., and K.G. Cassman. 1999. Post-green revolution trends in yield potential of temperate maize in the north-central United States. Crop Sci. 39:1622–1630. doi:10.2135/cropsci1999.3961622x

Eathington, S.R., T.M. Crosbie, M.D. Edwards, R.S. Reiter, and J.K. Bull. 2007. Molecular markers in a commercial breeding program. Crop Sci. 47:S-154–S-163. doi:10.2135/cropsci2007.04.0015IPBS

Fernandez-Cornejo, J., and M. Caswell. 2006. The first decade of genetically engineered crops in the United States In: Electronic Information Bull. 11. USDA-ERS. p. 1–30.

Hake, K., and W.R. Meredith. 1990. Full season yield from short season weather. Cotton Physiology Today 1(9):1–4.

Heffner, E.L., M.E. Sorrells, and J. Jannink. 2009. Genomic selection for crop improvment. Crop Sci. 49:1–12. doi:10.2135/cropsci2008.08.0512

Kathage, J., and M. Qaim. 2012. Economic impacts and impact dynamics of Bt (*Bacillus thuringiensis*) cotton in India. Proc. Natl. Acad. Sci. USA 109:11652–11656. doi:10.1073/pnas.1203647109

Liu, S.M., G.A. Constable, P.E. Reid, W.N. Stiller, and B.R. Cullis. 2013. The interaction between breeding and crop management in improved cotton yield. Field Crops Res. 148:49–60.

Meredith, W.R. 1991. Contributions of introductions to cotton improvement. In: H.L. Shands and L.E. Wiesner, editors, Use of plant introductions in cultivar development. Part I. CSSA, Madison, WI. p. 127–146.

Meredith, W.R., D.L. Boykin, F.M. Bourland, W.D. Caldwell, B.T. Campbell, J.R. Gannaway, K. Glass, A.P. Jones, L. May, C.W. Smith, and J. Zhang. 2012. Genotype ´ environment interactions over seven years for yield, yield components, fiber quality, and gossypol traits in the regional high quality tests. J. Cotton Sci. 16:160–169.

Meredith, W.R., and R. Bridge. 1984. Genetic contributions to yield changes in upland cotton. In: W. Fehr, editor, Genetic contributions to yield gains of five major crop plants. CSSA, Madison, WI. p. 75–86.

Paterson, A.H., R.K. Boman, S.M. Brown, P.W. Chee, J.R. Gannaway, A.R. Gingle, O.L. May, and C.W. Smith. 2004. Reducing the genetic vulnerability of cotton. Crop Sci. 44:1900–1901. doi:10.2135/cropsci2004.1900

Paterson, A.H., J.F. Wendel, H. Gundlach, H. Guo, J. Jenkins, D. Jin, D. Llewellyn, K.C. Showmaker, S. Shu, J. Udall, et al. 2012. Repeated polyploidization of Gossypium genomes and the evolution of spinnable cotton fibres. Nature 492:423–427. doi:10.1038/nature11798

Sankula, S. 2006. Quantification of the impacts on US agriculture of biotechnology-derived crops planted in 2005. National Center for Food and Agriculture Policy, Washington, DC.

SAS Institute. 2008. SAS 9.2 for Windows. SAS Inst., Cary, NC.

Schwartz, B.M., and C.W. Smith. 2008. Genetic gain in yield potential of upland cotton under varying plant densities. Crop Sci. 48:601–605. doi:10.2135/cropsci2007.01.0049

Wang, K., Z. Wang, F. Li, W. Ye, J. Wang, G. Song, Z. Yue, L. Cong, H. Shang, S. Zhu, C. Zou, et al. 2012. The draft genome of a diploid cotton *Gossypium raimondii*. Nat. Genet. 44:1098–1103. doi:10.1038/ng.2371

3

Cool-Season Forages

E. Charles Brummer* and Michael D. Casler

ABSTRACT

Yield improvements of most temperate forage crops have been limited. Data from on-farm USDA statistics, cultivar evaluations, and experiments designed to estimate yield improvement show that alfalfa yield increases have likely been due to improvements in persistence and disease resistances. Little advance in yield per se is evident, particularly in recent decades. Estimates of yield improvement with clover and grass breeding programs that used yield as a selection criterion range up to 6% per decade, although in many cases limited improvement has been noted. Numerous reasons for the lack of forage yield gain can be suggested, including the need for long cycle times, the necessity of high seed yield, and the need to select multiple traits, including forage nutritive value, which is negatively correlated with biomass production. We discuss several ways to improve yield in the future, including testing in sward plots, focusing on smaller population sizes with more highly selected individuals, capturing heterosis, and incorporating genetic markers, particularly using genomic selection. Although breeding gains for yield have been modest in most forage crops, we are optimistic about greater advances in the future.

Abbreviations: BFDP, U.S. Department of Energy Bioenergy Feedstock Development Program; G × E, genotype × environment [interactions]; GS, genomic selection; QTL, quantitative trait loci.

E. Charles Brummer, Univ. of California, Dep. of Plant Sciences, One Shields Ave., Davis, CA 95616; formerly Forage Improvement Division, The Samuel Roberts Noble Foundation, 2510 Sam Noble Pkwy., Ardmore, OK 73401. *Corresponding author (ecbrummer@ucdavis.edu). **Michael D. Casler,** USDA-ARS, U.S. Dairy Forage Research Center, 1925 Linden Dr., Madison, WI 53706-1108 (mdcasler@wisc.edu).

doi:10.2135/cssaspecpub33.c3

Yield Gains in Major U.S. Field Crops. CSSA Special Publication 33.
Stephen Smith, Brian Diers, James Specht, and Brett Carver, editors.

The cool-season forage crops represent a host of legume and grass species. The main cool-season forage legumes grown in the United States are alfalfa (*Medicago sativa* L.), red clover (*Trifolium pratense* L.), white clover (*T. repens* L.), and birdsfoot trefoil (*Lotus corniculatus* L.), as well as a number of annual clovers and medics. Cool-season grasses of note include tall fescue [*Lolium arundinaceum* (Schreb.) Darbysh. = *Festuca arundinacea* Schreb.], orchardgrass (*Dactylis glomerata* L.), smooth bromegrass (*Bromus inermis* Leyss), timothy (*Phleum pretense* L.), perennial ryegrass (*Lolium perenne* L.), Italian ryegrass (*Lolium multiflorum* Lam.), festulolium [*Festulolium loliaceum* (Hudson) P.V. Fournier], reed canarygrass (*Phalaris arundinacea* L.), and numerous others in niche regions. That this list is rather extensive hints at the main conclusion of this chapter—that yield gains have been limited in most forage crops, partly because many breeders have active programs on several species. We include switchgrass (*Panicum virgatum* L.), a warm-season grass, largely due to its long-term status as a hay, pasture, and biomass crop in temperate regions of the United States and because of the close similarity of its breeding methods and procedures to those developed for cool-season forage crops.

The yield trends for all hay as well as for alfalfa hay, as compiled by the USDA-NASS (http://www.nass.usda.gov/Data_and_Statistics/index.asp, accessed 13 Jan. 2014), show three distinct phases in terms of yield (Fig. 3–1). First, little yield improvement was obvious from the early 20th century through 1950. From 1950 until 1987, yield increased from about 3.0 to 5.5 t ha^{-1}, for an average increase during this time frame of 0.069 t ha^{-1} yr^{-1}, or 2.3% yr^{-1}. However, following the severe drought in 1988, yields appeared to stagnate, if not slightly decline, through 2012, which represented another severe drought year in many hay producing regions.

What do we make of these data? Do they reflect the extent of genetic progress? We believe that drawing concrete conclusions from these data is fraught with difficulty for several reasons. First, hay is produced in all states, and the data therefore represent an agglomeration of multiple species growing in vastly different environments, with production that can range over nearly an order of magnitude. On-farm forage yield data are inherently somewhat suspect, given that yields are not easily monitored, unlike the case of cereal or oilseed crops, and that yields of highly managed forages are mixed with hay crops that receive limited or no management. Finally, hay yields only tell part of the yield story—silage yields are generally not included, nor are estimates of pasture productivity. The best data on hay production probably are those for alfalfa, which tends to be more actively managed than many forages, but even for alfalfa, we see a similar trend. For this reason, we caution that the agricultural statistics may not tell the full story. Nevertheless, these data show that while an overall yield increase occurred between 1950 and 1987, no widespread hay yield increase across species and locations has occurred since the late 1980s.

One of the questions that confront forage breeders is that of what "yield" figure should be used in assessing gain? Yield can be measured as the total production in the first full year of production (i.e., after the always low yielding seeding year), the total production across two or more years of production, the total production in the final year of a trial (which then conflates yield with persistence), or the yield of any given harvest. Because of these variations in what constitutes yield, the entire topic is hard to assess. In this chapter, we attempt to address the improvement of yield per se, independent of improvements in persistence or other traits. We recognize that better persistence—possibly through increased winter hardiness or improved disease resistance—will result (and has resulted) in higher yields of new cultivars, particularly in later years of stand life. We are not attempting to minimize the importance of yield improvement derived from better persistence, but rather we are trying to assess the yield trait itself, and whether improvement in yield per se has been, or can be, realized in forage crops.

Cool-Season Legumes

The most important legume grown for hay is alfalfa. On-farm yield of alfalfa, showing a stagnation overall (Fig. 3–1), shows a different trend depending on the growing region (http://www.nass.usda.gov/Data_and_Statistics/index.asp,

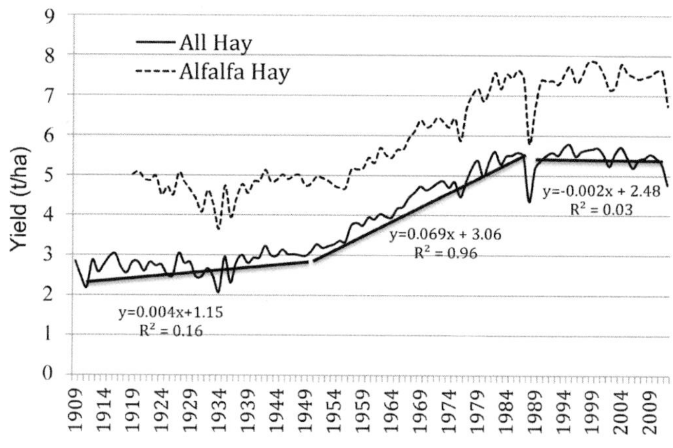

Fig. 3–1. U.S. average on-farm yield trends for all hay and for alfalfa hay from 1909 through 2012 based on USDA historical yield data.

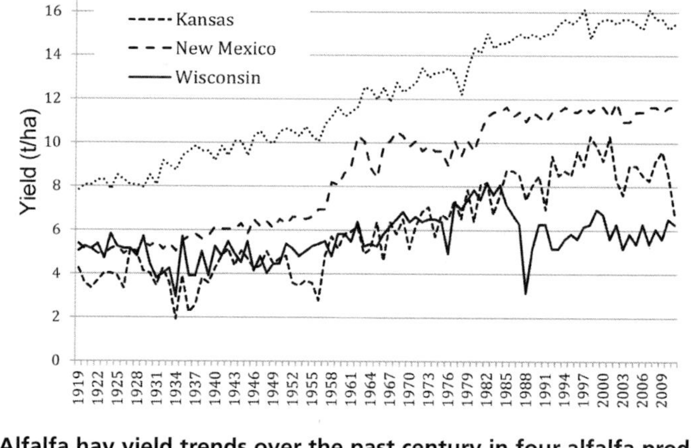

Fig. 3–2. Alfalfa hay yield trends over the past century in four alfalfa producing states based on USDA historical yield data.

accessed 13 Jan. 2014) (Fig. 3–2). First, California production leads the country by a large amount, thanks to long growing seasons and intensive management, and yields appear to have continued to increase evenly over the past century, unlike the overall trend and the trend in other growing regions such as Wisconsin and Kansas. Second, in Wisconsin, more than any other state, a significant yield depression has occurred since the late 1980s, according to on-farm data.

Several lines of evidence suggest that yield gains shown in the on-farm results between 1950 and the late 1980s were real. An experiment assessing yield progress across year of cultivar release found a yield increase of about 2% per decade, based on the total production across 3 yr at one Wisconsin location (Holland and Bingham, 1994). They had divided cultivars into three eras, with the larger increase coming between Era 2 (1940–1953) and Era 3 (1979–1985) than between Era 1 (1898–1910) and Era 2. Summarizing disparate evaluations or inferences, yield gains in the range of 2 to 3% per decade appear to have been obtained throughout the 1950 to 1980 period (Hill et al., 1988; Elliott et al., 1972; and references within them). As a percentage of the cultivar Vernal, yield increases of nearly 9% were observed by the early 1980s, or about 3% per decade (Hill et al., 1988). Benchmarking against an older cultivar may not represent an increase in yield potential per se. New cultivars may perform better than older ones because increased disease or pest pressure is reducing older cultivar performance over time (e.g., as seen in rice [*Oryza sativa* L.]; Peng et al., 1999).

Nevertheless, evidence that alfalfa yield gains were obtained through breeding into the 1980s seems quite strong, even if that yield gain has not been on par with gains in the major grain and oilseed crops. If we accept that an increase due to genetic improvement is between 2 and 4% per decade, we then can infer that only 10 to 20% of the yield increase observed on-farm can be attributed to genetic gains.

For the period when on-farm trends indicated that yields have not improved substantially (since the late 1980s), we compared data from small plot, replicated yield trials grown in the upper midwestern United States and Ontario, Canada from 1989 to 2011 (data compiled by D. Undersander, University of Wisconsin, 2013, personal communication). We assessed yield as the total production in the first full production year (i.e., the first year after establishment) and computed the average of the top five cultivars in each trial. To be included, a location needed to have at least nine trials during that period as well as one trial before 1992 and one trial after 2008. This resulted in 14 locations in Iowa, Minnesota, Wisconsin, Michigan, and Ontario, and a total of 242 trials. Across all trials, no relationship was observed between yield and year of trial ($R^2 = 0.005$), but some differences were observed at individual locations. All three Wisconsin locations showed positive yield increases across years. No increase was observed in Iowa, and decreases were seen in three of four Ontario test sites. The largest increase was at Lancaster, WI, which had the fewest trials, of 0.44 t ha^{-1} yr^{-1}. A generally similar analysis of data from 1978 through 1996 also showed a lack of progress (Wiersma et al., 1997).

The cultivar trial data obviously do not reflect typical hay production environments, with yields being considerably higher than average on-farm figures. Cultivar trials are typically grown on better soil and under higher management than many hay fields, even with alfalfa, and the yield from small plots probably does not reflect field-scale yields quite as well as the relationship between grain and oilseed trials and farm production. Nevertheless, the yield trends from cultivar trials show that yield increases are occurring in some locations, but whether this is due to genetics or management is unclear. These results suggest that on-farm data may reflect shifts in production environments; for example, as more corn (*Zea mays* L.) silage is grown on the best ground, alfalfa hay production occurs on less productive ground. Additionally, as alfalfa haylage is not included in the hay figures, hay yields may reflect the fact that haylage production is done on more highly productive land.

In the midwestern United States, one major experiment has assessed the yield relationship with year of cultivar release (Lamb et al., 2006). In this experiment, more recently released cultivars had higher yields than older cultivars in Ohio and Wisconsin, but no trend was observed in Minnesota and Iowa. In concluding their discussion, Lamb et al. (2006) summarized:

> We concluded that evidence for changes in forage yield for cultivars released between 1940 and 1995 was environmentally dependent. In environments where conditions lead to plant stand losses, recently released cultivars with multiple disease resistance had a yield advantage over older cultivars, but in environments where no differences in plant density occurred, older cultivars yielded the same as the improved cultivars.

In the first full production year, when disease-related stand loss was likely not a factor, however, yield differences were small or absent at all four locations. Depending on location and year of stand life, the yield improvement across decades of cultivar release has been between zero and 10% per decade, with a skew toward the lower end. However, the evidence for improved yield per se—that is, yield improvements not due to improved persistence—is not evident from this experiment.

Collectively, the on-farm data, independent yield trial results, and an evaluation of historical cultivars all show that improvement in yield per se probably has not occurred in alfalfa over the past several decades. This does not mean that new cultivars are not offering value to hay growers, of course; the improvement in pest resistances, for example, clearly resulted in yield improvements of new cultivars relative to older ones (Lamb et al., 2006), particularly in later years of a stand, in locations with disease pressure. These results mirror those of Hill et al. (1988), discussed above, showing improvement relative to the older cultivar Vernal, released in 1953. Improvements in yield due to better persistence, unrelated to disease resistance, have likely also been made, although these may not be fully obvious under relatively short-term variety trials grown under ideal managements.

Finally, most of the experiments testing alfalfa yield gains have not assessed single breeding programs, where gains may be more clearly defined, but rather simply compared cultivars released by any program in different eras. Thus, while some programs may have made progress for yield, others may have focused on other traits—e.g., nutritive value, grazing tolerance, potato leafhopper tolerance, or others—that may have had detrimental effects on yield. Hence, the yield increases possibly realized by the former would be diluted by the latter.

Yield improvements in red clover have been documented within breeding programs in Wisconsin and in New Zealand. Yield comparisons of three cultivars released by the USDA-ARS–University of Wisconsin joint breeding program (Lakeland, released in 1953; Arlington, released in 1973; and Marathon, released in 1987) showed an average gain of 0.4% yr^{-1} for first post-establishment year,

0.5% yr^{-1} for the second, and 1.0% yr^{-1} for the third (Riday, 2010; Smith, 2000). These data show that while some improvement has been made in yield per se, yield increases across the life of a stand have improved due to better persistence. In New Zealand, gains of 0.2 to 1.4% yr^{-1} have been documented for improvements between older and newer cultivars (Woodfield, 1999). These two examples represent gains within breeding programs, rather than generalized improvements in commercially available cultivars.

White clover is typically grown in a mixture with grasses for use as a grazed forage. A major experiment evaluating 110 cultivars released between 1920 and 1990 found that the average genetic gain for biomass yield was 6% per decade, considerably higher than other forage species (Woodfield and Caradus, 1994). This yield gain was measured as grams per square meter on mixed clover–grass swards, and a concomitant increase of 6% per decade was also realized for clover percentage in the sward during this period. Two other assessments of yield improvement of white clover under grazing by sheep or cattle showed improvements of 1.2 to 1.5% yr^{-1} (Woodfield, 1999). Whether this can be partially attributed to improved stand percentage is not clear.

Many other legumes are used in forage systems, including birdsfoot trefoil and a range of annual clover and medic species. To our knowledge, genetic gains in forage yield not been estimated for these species. Very few cultivars have been released for most of these minor species.

Cool-Season Grasses

Long-term gains in forage yield of cool-season grasses have been measured in field evaluations of cultivars developed over many years of breeding. These field evaluations demonstrate a wide range of estimates, depending on species and experiment (Table 3–1). In general, gains in forage yield are directly proportional to the amount of effort and resources devoted to genetic improvement. Worldwide resources devoted to genetic improvement are directly proportional to the order of species as listed in Table 3–1. The most successful and sustained breeding programs have resulted in increases in forage yield of up to 6% per decade. In some cases, these gains have been sustained for 40 to 50 yr. Conversely, some species have demonstrated no evidence of increased forage yield, despite 50 yr of breeding efforts (e.g., Casler et al., 2000b). Typically, forage yield has not been a significant selection criterion for these species.

The ability to measure these positive gains over many years derives from pooling resources across several breeding programs, increasing the probability that the best cultivars in a region will be identified and compared to older check cultivars. Thus, field-evaluation networks that are fed by numerous breeding programs will have the highest probability of detecting long-term genetic improvements in forage crops. Field-evaluation networks exist for all of the important species throughout the European Union, largely in support of the European DUS system that requires cultivars to be registered on national lists and that cultivars must be "distinct, uniform, and stable" to become registered. Both commercial companies and public DUS-evaluation agencies must have field-trial systems that are sufficiently extensive and powerful to detect small and incremental genetic improvements that occur over relatively short periods of time. Such a system favors the ability to quantify genetic gains through the repeated use of older check cultivars and combined analyses across trials, sites,

Table 3–1. Long-term genetic gains in forage yield of cool-season forage grass species during 1950 to 2000.

Grass species	Forage yield gain	References
	% decade^{-1}	
Perennial ryegrass (*Lolium perenne* L.)	2 to 6	Aldrich (1987); Allerit (1986); van Wijk and Reheul (1991); Veronesi (1991)
Italian ryegrass (*Lolium multiflorum* Lam.)	2 to 5	Allerit (1986); van Wijk and Reheul (1991)
Tall fescue [*Lolium arundinaceum* (Schreb.) Darbysh. = *Festuca arundinacea* Schreb.]	1 to 6	Veronesi (1991)
Orchardgrass (*Dactylis glomerata* L.)	0 to 6†	Casler et al. (2000a, 2001); Veronesi (1991)
Timothy (*Phleum pretense* L.)	2	Casler (2001, 2005)
Smooth bromegrass (*Bromus inermis* Leyss)	0	Casler et al. (2000b)
Reed canarygrass (*Phalaris arundinacea* L.)	0	Casler et al. (2009)

† Varies among maturity groups and breeding programs.

and years (Vargas et al., 2013). Conversely, no such system exists in the United States, so that quantification of genetic gains for forage crops requires focused and dedicated efforts specifically designed to meet that goal (e.g., Casler et al., 2000a, 2000b, 2001).

Wilkins and Humphreys (2003) also pointed out that estimates of genetic gains summarized from these networks of field evaluations (Table 3–1) are likely conservative, underestimating true gains for some species. Because these field evaluations are rarely conducted for more than two harvest years, they seldom represent detailed or accurate evaluations of genetic differences in stress tolerances, unless those stresses were present during the first 2 yr of the experiment. Persistency is a significant issue for many forage crops, often manifested by gradual declines in stands or ground cover, and often associated with specific management or environmental factors. Use of short-term field-evaluation networks with optimal management and environmental conditions will seldom allow genetic differences in persistency to be expressed or measured as a component of forage yield.

Many of the most significant breeding advances for forage crops derive from breeding efforts conducted in fringe or extreme environments, expanding the species range by improving abiotic stress tolerance and/or increasing persistency. Both perennial and Italian ryegrass have ranges that are limited largely by heat stress in one extreme and cold stress in the other extreme. Selection for survivorship following harsh winter conditions resulted in a significant increase in forage yield of festulolium [*Festulolium loliaceum* (Hudson) P.V. Fournier] when evaluated in USDA Hardiness Zones 3 and 4 (mean annual minimum temperature = −33°C), directly associated with a 32% increase in survivorship (Casler et al., 2002). Conversely, there were no differences in either forage yield or survivorship when the same populations were evaluated in less stressful environments represented by USDA Hardiness Zones 5 through 7 (mean annual minimum temperature = –19°C).

Switchgrass

Efforts to improve switchgrass as a forage crop, largely for grazing livestock, have been ongoing since the 1950s. Breeding objectives have focused largely on quality-related traits, such as digestibility and disease resistance, and establishment-related traits, such as seedling vigor. Breeding for increased forage yield has generally been a secondary objective in these efforts.

Breeding switchgrass for biomass yield became a primary breeding objective after the species was chosen as an herbaceous model for the U.S. Department of Energy Bioenergy Feedstock Development Program (BFDP). Realized and projected gains from several improvement activities and strategies are illustrated in Fig. 3–3. Early efforts were focused on development of regional trials to determine the importance of genotype × environment (G × E) interactions and to identify the adaptation limits of forage cultivars and local land races (e.g., Casler et al., 2007). Accurate prediction of cultivar adaptation regions based on G × E interactions was one of the principal factors contributing to a 25% decrease in predicted production costs for switchgrass biomass during the 10-yr lifetime of the BFDP (Sanderson et al., 2007).

Three population-improvement strategies are currently being employed for increasing biomass yield of switchgrass. First, intrapopulation improvement of switchgrass has been accomplished by several cycles of recurrent selection, based on either phenotypic or family-based selection strategies. The rate of genetic gain for biomass yield has averaged 1 to 4% yr^{-1} in several breeding programs that represent a wide array of germplasm across diverse environments (Casler, 2010;

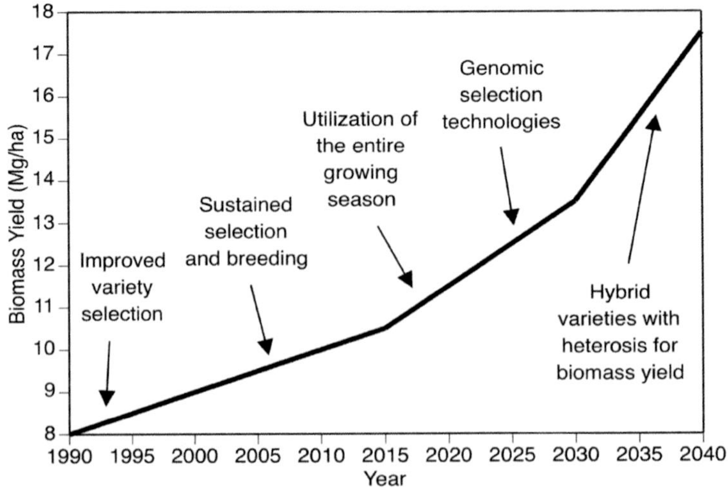

Fig. 3–3. Actual and projected timelines for switchgrass improvement using classical and modern breeding strategies.

Missaoui et al., 2005; Rose et al., 2007). In some breeding programs, these gains have been sustained for periods of 15 to 20 yr, resulting in gains of up to 30% in biomass yield (Casler and Vogel, 2013).

Second, breeders in the northern range of the species are taking advantage of the strong photoperiod sensitivity of the species to extend the growing season. Southern populations are considerably later in flowering than northern populations, but possess a low frequency of plants that are capable of surviving harsh winters of northern climates. A single cycle of phenotypic selection for survivorship may be sufficient to broaden the adaptation of a southern switchgrass population by two hardiness zones (Casler and Vogel, 2013). Thus, we expect that only two or three cycles of selection will be required to adapt germplasm from the Gulf Coast region to biomass production regions of the northern United States (Casler, 2012). Delaying flowering by up to 6 wk allows these populations to continue accumulating dry matter well past the peak-biomass time for local populations from prairie and savanna habitats (Casler, 2012).

Third, the feasibility of interpopulation improvement methods has also been demonstrated. First-generation hybrids between contrasting upland and lowland ecotypes had 30 to 35% heterosis for biomass yield over the high parent (Vogel and Mitchell, 2008). Much of these heterosis effects were preserved through three cycles of selection for biomass yield within the hybrid progeny, resulting in advanced-generation hybrid progeny that were superior to both parent cultivars (Casler and Vogel, 2013). More importantly, the advanced-generation hybrids were much more broadly adapted than either the northern-adapted upland parent or the southern-adapted lowland parent. Advanced-generation hybrid progeny demonstrated significant heterosis effects and >90% survivorship in USDA Hardiness Zones 3 through 5, a range of adaptation not observed in either parent alone (Casler and Vogel, 2013).

Forces Driving Objectives and Decisions in Forage Breeding

Issues Associated with Breeding for Forage Yield

The seed industry and the livestock industry are the principal drivers of breeding objectives in forage crops. The seed industry requires a minimum level of seed production for economic viability of a cultivar. Breeders ignore this requirement at their peril. Cultivars that have low or unreliable seed production will be nonviable economically, regardless of their potential value to the livestock industry. While there are numerous examples of superior forage grass and legume cultivars that do not meet industry standards for seed production, very few are capable of overcoming this obstacle. 'Satin' orchardgrass is one exception to this rule, a cultivar with superior quality and palatability and sufficient demand for seed to support a small amount of annual seed production. Seed yield, however, is a distinct trait from forage biomass yield, and because seed production is typically conducted in an environment different from the targeted forage production environments, selection for seed production could very well negatively impact forage yield.

A second concern of the seed industry, after seed production per se, is the ability to sell a differentiated product in the market, one that has a distinguishing trait or suite of traits. In alfalfa, for example, the incorporation of multiple pest resistances, multifoliolate leaflets, lodging tolerance characteristics, and others have likely diverted attention from biomass yield to develop a product that would be more easily marketed. Thus, while yield is an important trait for growers, incremental improvements are not necessarily highly marketable, so concentrating on other, more visible traits is usually a pragmatic breeding decision.

A second reason for lack of biomass yield gain in some forages is the presence of serious deficiencies that must be fixed before they are acceptable to the livestock industry. The presence of toxic alkaloids in reed canarygrass (*Phalaris arundinacea* L.) herbage is a prominent example of a trait that prevented full and efficient utilization of this species in grazing systems. Once these alkaloids were removed or reduced in concentration, demand in the livestock industry rapidly increased, and the seed industry responded with significant efforts to improve both genetic and management aspects of seed production (Casler, 2010). Some of these deficiencies require many years of breeding to solve, with no guarantee of a satisfactory resolution. For example, genetic improvements in seedling vigor of switchgrass have been difficult to achieve, prompting agronomists to focus on management and establishment methods to enhance seedling vigor (Casler et al., 2012). Low seedling vigor in kura clover (*T. ambiguum* M. Bieb.) and birdsfoot trefoil limit their establishment, and breeding to improve vigor has only partially been successful.

Third, for many forage species, simultaneous improvement of multiple traits is required, complicating the breeding program on several levels. First, adding traits significantly reduces the average selection intensity that can be applied to each trait. For example, with selection of the "best" 10 plants in a 1000-plant nursery, as the number of traits increases from one to four, the average proportion of plants selected for each trait changes from 0.01 to 0.32 (Fig. 3–4). The net result

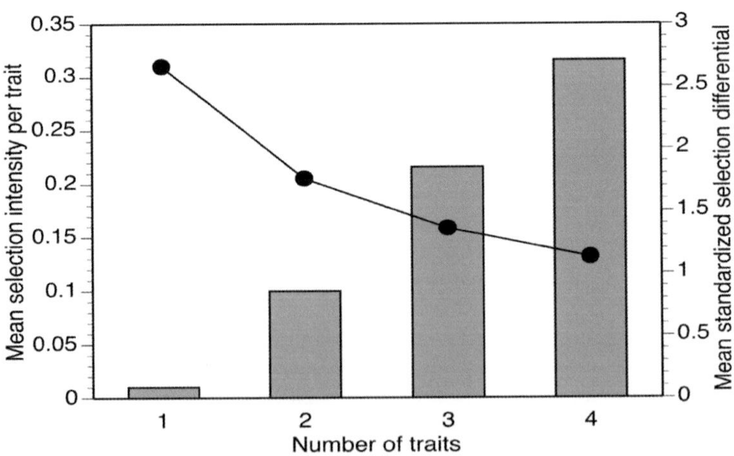

Fig. 3–4. Increase in proportion of plants selected (decreased selection intensity) for each individual trait (vertical bars) and subsequent decrease in average standardized selection differential per trait (line graph) as a function of number of traits in an independent culling selection program. Values in the graph are average selection differentials on a per-trait basis, assuming 10 plants out of 1000 are selected as parents of the next generation, e.g., 0.01 selection intensity for one trait, but an average per-trait selection intensity of 0.32 when four traits are used as joint selection criteria.

of this is a significant decrease in the average selection differential for each trait, decreasing the potential gain for each trait. Second, multi-trait selection is complicated by the issue of exactly how to incorporate multiple traits into the program, for example, tandem selection, independent culling, or some form of selection index, and by the requirement that some of these traits will require multiple year evaluations. Third, multi-trait selection may require the use of multiple selection environments. For example, seed of many grass and legume cultivars are produced in arid environments in the western United States where water and pollinators can be controlled and delivered as needed. In these situations, seed production traits cannot be reliably enhanced in humid forage environments, and, likewise, forage traits cannot be reliably improved by selection in arid seed production environments. The most successful forage breeding programs are those that have developed links between forage and seed-production environments, utilizing some form of tandem selection for each trait under the proper environmental conditions.

Further complicating matters, some forage traits cannot be reliably measured on spaced-plant nurseries, where the majority of forage breeding is conducted (Casler et al., 1996; Wilkins and Humphreys, 2003). Spaced-plant nurseries are highly valued because many forage species are perennial, and individual genotypes must be observed numerous times over several years of lateral tillering and growth before reliable selection pressure can be applied to certain traits, especially those related to survivorship (Vogel et al., 2002). Many attempts have been made to identify traits measured on spaced plantings for the purpose of predicting sward-plot or drill-plot forage yield. With very few exceptions, these attempts at indirect selection for forage yield have been unsuccessful (Casler et al., 1996; Wilkins and Humphreys, 2003; Waldron et al., 2008), although they have been useful to improve winterhardiness and persistency, thereby indirectly affecting yield. In contrast, however, Riday (2009) has found that indirectly selecting for vigor was highly correlated with forage yield in red clover.

Management is an additional factor that contributes to G × E interactions in forage breeding programs. Genotypes and families generally do not rank similarly under widely different managements, such as rotational grazing vs. a two- or three-harvest hay management (Casler et al., 2000a, 2001; Wilkins and Humphreys, 2003). For example, short-leaf genotypes tend to perform better under frequent defoliation, while the converse is true for infrequent defoliation managements (Hazard and Ghesquére, 1997). In alfalfa, cultivar rankings can vary widely between grazing and hay evaluation, and in many cases, the best grazing cultivars are not the highest yielding under hay managements, as was observed with 'Alfagraze' alfalfa.

Genetic variation in flowering time is another factor that complicates the breeding process, because forage yield increases progressively with later flowering. As such, populations are frequently developed to limit their genetic variability for flowering time, simplifying the process of identifying genotypes and families with superior breeding value for forage yield that is independent of flowering time. This concept is routinely used in cultivar evaluation trials for many species, simplifying the process of identifying the best cultivars with similar flowering times, but complicating both the breeding and evaluation processes by creating more potential cultivars to evaluate and more field trials to conduct.

The Role of Forage Quality

While the seed industry is a principal driver and an important stakeholder of forage breeding programs, the vast majority of clients and end users are livestock producers. In most livestock production systems, forage quality is at least as important as forage yield, often more important. The major legumes are generally high in forage nutritive value, with relatively small differences among species and cultivars within some species. However, among the grasses, variation among cultivars and species in forage quality are very obvious to livestock producers, often observable on a daily basis when dairy cattle are moved from high- to low-quality pasture or vice versa. As such, much of the demand from the livestock industry is for persistent and stress-tolerant cultivars that possess documented superiority in forage quality. While breeders are accustomed to using high-throughput laboratory methods to predict forage-quality traits, the most valuable documentation of genetic improvements in forage quality derive from grazing or feeding trials (Chapter 9, Jensen and Anderson et al., 2014; Casler and Vogel, 1999). The related trait of palatability, which can affect dry matter intake, cannot be estimated from typical forage quality analyses, adding further impetus to conduct direct grazing trials to measure performance.

In general, forage yield and quality are negatively correlated with each other. Forage yield increases and forage quality decreases as both grasses and legumes mature. Removing reproductive maturity from this equation moderates this negative relationship, so that breeders can improve both forage yield and quality simultaneously, but only when placing direct selection pressure on both types of traits (Chapter 9, Jensen and Anderson et al., 2014; Casler and Vogel, 1999). Breeding objectives that weight forage-quality traits too heavily at the expense of forage yield and fitness traits result in fitness crashes, such as loss of persistency, pest resistance, or yield potential (Casler, 2008, 2014b; Vogel et al., 2002).

Digestibility and related traits represent the most commonly utilized selection criteria to improve forage quality (Casler and Vogel, 1999). The development of a high-throughput in vitro digestibility assay with sufficient repeatability and heritability to be of value in breeding programs is one of the most significant achievements in the history of grassland research (Hopkins and Wilkins, 2006). The use of near-infrared reflectance spectroscopy to reduce costs and increase sample throughput represents another significant achievement, allowing breeders to process tens of thousands of forage-quality samples per year. These technological advances have resulted in numerous genetic improvements in forage quality, many of which have been documented at the livestock level (Casler and Vogel, 1999). Breeding for increased digestibility is highly efficient because G × E interactions are relatively small compared to other traits. Genetic progress can be achieved from a single observation made at a single point in time from unreplicated spaced plants, with realized heritability estimates in the range of 0.3 to 0.4 (Casler and Vogel, 1999). As such, and partly due to the multi-trait selection phenomenon (Fig. 3–4), forage yield is very typically a secondary objective of most breeding programs.

Future Prospects to Increase Yield

Improving yield is of importance to all crops, including forage crops. As demonstrated above, yield improvements have been irregular or nonexistent. While we can point to a number of reasons for this lack of gain, we may instead consider what we can do about it (Brummer, 2005). Further advances in machinery able to measure yield accurately on small plots, the adoption of experimental designs to control field-based spatial variation, and application of advanced statistical methods based on best linear unbiased predictors allow routine testing of large numbers of families that may have small or highly variable amounts of seed available (Casler, 2014a; Resende et al., 2013; Wilkins and Humphreys, 2003) would all contribute to better genetic gains in yield. Beyond those advances that can be incorporated into programs immediately, we also suggest some fundamental changes in breeding programs.

Measure yield, ideally under sward conditions. Undoubtedly, many breeding programs fail to improve yield because yield is not a target of selection. However, in many programs, "yield" is selected but only indirectly, based on visual scores, plant height, or another (hopefully) correlated trait. Further, measurements of yield, or of proxies for yield, are often measured on spaced plants or on seeded rows, neither of which reflect "real-world" field conditions. When yield is measured under these conditions, the breeder is, in effect, measuring a correlated trait to the ultimately desired trait of sward yield. Therefore, measuring yield per se in sward plots should be the goal of all breeding programs. Direct selection for forage yield on some type of sward plot or drill plot, or some form of spaced-plant-in-swards nursery (Casler et al., 1996; Wilkins and Humphreys, 2003), remains the most reliable and well documented mechanism to improve forage or biomass yield. Elliott et al. (1972) indicated that alfalfa selection based on progeny evaluations in seeded swards produced genetic gain for yield, but that progeny evaluation based on seeded rows did not result in increased yield.

Historically, family-based selection methods have been dismissed as inefficient due to our inability to utilize within-family genetic variability effectively in the selection process (Breese and Hayward, 1972). More recently, efforts have focused on improving breeding methods to utilize both among- and within-family sources of genetic variation effectively as selection pressure on traits that are closely related to forage yield (Casler and Brummer, 2008). Traits such as survivorship and competitive ability have the potential to increase the rate of gain by allowing meaningful selection pressure on individual genotypes within families (Casler, 2008; Resende et al., 2013; Wilkins and Humphreys, 2003). The use of DNA markers to evaluate paternity of half-sib progeny (Riday, 2011) can boost gain by selecting on both parents, rather than simply the maternal lineage.

Focus on a few carefully analyzed, highly selected parents. One of the most successful forage breeding programs is the perennial ryegrass program at IBERS, Aberystwyth (Wilkins and Humphreys, 2003). This program has realized substantial gains over time for biomass yield, and yet the initial population derived from only seven genotypes, with two additional parental plants added to the program over time. In addition, selection for yield has been made on the basis of progeny test yield data obtained from sward plots. This breeding strategy has not been widely adopted in most other breeding programs of any forage crop.

In contrast, many breeding programs maintain large populations and recurrently introduce new genetic diversity. The reasons for this are ostensibly sensible to avoid inbreeding depression and to increase desirable allele frequencies for multiple disease and insect resistances. However, in doing so, breeders limit their ability to concentrate desirable alleles for complex traits like yield and, perhaps more importantly, prevent the development of desirable multi-locus allele complexes. (One could also argue that maintaining large population sizes also conserves variation for other traits that could be selected in the future; whether this variation should remain in active, elite breeding programs or in germplasm repositories can be discussed elsewhere.)

Explore heterosis and mechanisms to capture heterosis in outbreeding forage crops. The development of true hybrid cultivars has remained elusive for the outbreeding forage crops. In fact, for alfalfa or other autopolypoid crops, single-cross hybrids will not result in maximum heterosis expression because these species express progressive heterosis when using inbred parental genotypes (Bingham et al., 1994). Regardless, capturing some amount of heterosis is possible through the use of semihybrids (Brummer, 1999) or by capitalizing on self-incompatibility, especially in grasses (Aguirre et al., 2012). A self-incompatibility system, in which clonally multiplied genotypes would interpollinate but not self-pollinate, would result in complete hybridity of the resulting seed population. Male sterility systems are not well characterized in any forage species, but these could be incorporated into future programs. In alfalfa, a cytoplasmic male-sterility system is being used to develop so-called "hybrid" cultivars (Viands et al., 1988; Davis, 1971; Sun et al., 2004), which are really more like narrow-based synthetics. For hybrid-oriented breeding programs, the identification of heterotic groups, or more usefully, the development of heterotic groups from existing germplasm resources, would be beneficial. Reciprocal recurrent selection is an ideal method to develop and improve heterotic populations (Scotti and Brummer, 2010; Tanaka et al., 2011), but it has rarely been used or even evaluated in forage crops, the one exception being in alfalfa (Katepa-Mupondwa et al., 2002), which demonstrated it could be effective.

Integrate molecular markers into breeding programs. The development of molecular markers in forage crops began in the late 1980s, but only recently has their true value to assist yield improvement become obvious. Until now, linkage maps have been used to identify the locations of quantitative trait loci (QTL) for yield in alfalfa (Robins et al., 2007; Li et al., 2011), but not in most other forage species. Manipulating individual QTL tends to be difficult, due to the Beavis effect (Beavis, 1994) and other issues (Bernardo, 2001) such as large QTL × environment interactions. The result has been that, while markers may be useful to characterize germplasm (and possibly to identify heterotic groups, although this is not well established) and to identify and select for single gene traits, their ability to make major improvements in yield has been limited.

The major issue confronting forage breeders is the length of time per cycle; hence, a marker strategy that cannot substantively decrease cycle time will not be useful. Therefore, marker-only selection is needed to decrease the cycle time as short as possible, ideally converging on the time necessary for a reproductive cycle. The development of methods to conduct genomic selection (GS) (Jannink et al., 2010; Heffner et al., 2010; Lorenz et al., 2011; Hayes et al., 2013), offer the possibility to realize this goal. In most forages, at least two cycles per year could be conducted, possibly more (a schematic program is described in Li and Brummer, 2012). If a genomic selection model could predict yield reasonably well, then genetic gain could be considerably accelerated compared to conventional selection programs that require 3 to 6 yr per cycle.

In order for GS models to accelerate gain, they need to be based on high quality phenotypic data. Thus, we return to the earlier comment about measuring yield per se on sward plots in multilocation trials. Not only will selecting for yield likely lead to yield improvements, adding genomic selection has the potential to further advance yield. However, accurate measurement of forage or biomass yield is of critical importance, and this is something that has not been well done in the past in forage breeding programs. Genomic selection can also be used for other traits, of course, and selection indices to facilitate multi-trait selection could be developed and used in a straightforward manner. However, as noted earlier (Fig. 3–4), adding traits to a selection index will limit gain in any one of them; markers do not enable breeders to avoid this dilemma.

With GS as a possible tool for selection, forage breeders may consider redesigning breeding programs to maximize the value of marker-only selection. First, to acquire high quality phenotypes, recombination should be conducted in a way that generates enough seed for sward plot evaluations. This will often require an additional year to generate the necessary seed. Second, multilocation trials of breeding material should be conducted to develop robust GS models, and perhaps regional evaluations among public and private sector breeders could be developed to facilitate these trials. In fact, if seed is limiting, it is more critical to employ multiple locations than to replicate within locations (Casler, 2014a). Moving from the traditional four replicate randomized complete block experimental designs to advanced designs, such as unreplicated, augmented designs, would enable breeders to generate useful information using fewer plots of any given entry.

Third, rather than exploring large, diverse breeding populations, breeders may improve gain by concentrating on a few, highly selected parents, generating robust data on combining ability, disease, and pest resistances, etc. on those genotypes, which may then form the training population for development of a GS model. Concentrating on fewer parental genotypes, keeping populations closed, and admitting new germplasm only after extensive testing would help concentrate desirable alleles and provide strong estimated breeding values for the alleles in the population. Fourth, selection could be conducted in the framework of a reciprocal recurrent selection program, facilitating the development of hybrids as feasible. Extending GS models to include dominance components (e.g., Denis and Bouvet, 2013) would further aid in identifying potential hybrid parents.

Finally, on a more practical, but highly critical level, improved plot equipment capable of high throughput and highly sensitive yield measurements (especially in the presence of wind) is urgently needed. Current harvesting equipment works well for larger, variety trial plots, but getting accurate and precise values for small breeding nursery sward plots is essential, and further equipment improvements are needed. Similarly, precision planting and harvesting equipment tied to global positioning units could make data collection both faster and more accurate. Remote sensing methods to predict forage yield (and quality!) may enable the testing of more genotypes; initial research on incorporating these technologies has been done in maize (e.g., Weber et al., 2012). However, prediction equations would need to be quite strong for these data to be usefully incorporated into a GS model.

Ultimately, the cost of breeding for increased yield must be captured through seed sales. While alfalfa has a well-developed seed market, most of the other forage species do not have a seed market in the United States that could recover the significant added expense of breeding for yield. With GS, we are not eliminating the need to measure phenotype, which must continue on a year-to-year basis to allow frequent updates and improvements to the GS model. Hence, genotyping costs are almost certainly an added expense to the field program, and this expense will need to be recouped. A culture among growers that values incremental improvements for various traits needs to be developed through extension and/or marketing programs before the full potential of genomic tools can be realized for most forage species.

Conclusions

We conclude that across the cool-season perennial forage species, little to no yield gain has been observed in general. However, programs that have focused directly on yield as a selection criterion have improved yield. A greater commitment from all breeders to improving yield is needed to keep forage-based production systems competitive with annual grain and oilseed systems. Perennial forage crops provide numerous positive services to ecosystems—improving water infiltration, limiting soil erosion, for instance—and contribute to better performance of other crops in a rotation through minimizing pest and weed pressures, among other benefits. Therefore, these systems are essential for a resilient agricultural system, and improving yield will continue to be a major target trait. Although yield increases have been modest to absent in forages, we think there is reason to be optimistic.

References

Aguirre, A.A., B. Studer, U. Frei, and T. Lübberstedt. 2012. Prospects for hybrid breeding in bioenergy grasses. BioEnergy Res. 5:10–19. doi:10.1007/s12155-011-9166-y

Aldrich, D.T.A. 1987. Developments and procedures in the assessment of grass varieties at NIAB 1950–1987. J. Natl. Inst. Agric. Bot. 17:313–327.

Allerit, R. 1986. Espéces fouragères perennes: Progrès réalisès depuis 25 ans, apprécies à travers 11 expérimentation officielle. Fourages 107:17–33.

Beavis, W.D. 1994. The power and deceit of QTL experiments: Lessons from comparative QTL studies. In: 49th Annual Corn and Sorghum Research Conference American Seed Trade Association, Washington, DC. p. 252–268.

Bernardo, R. 2001. What if we knew all the genes for a quantitative trait in hybrid crops. Crop Sci. 41:1–4. doi:10.2135/cropsci2001.4111

Bingham, E.T., R.W. Groose, D.R. Woodfield, and K.K. Kidwell. 1994. Complementary gene interactions in alfalfa are greater in autotetraploids than diploids. Crop Sci. 34:823–829. doi:10.2135/cropsci1994.0011183X003400040001x

Breese, E.L., and M.D. Hayward. 1972. The genetic basis of present breeding methods in forage crops. Euphytica 21:324–336. doi:10.1007/BF00036773

Brummer, E.C. 1999. Capturing heterosis in forage crop cultivar development. Crop Sci. 39:943–954. doi:10.2135/cropsci1999.0011183X003900040001x

Brummer, E.C. 2005. Thoughts on breeding for increased forage yield. In: F.P. O'Mara et al., editors, 20th International Grassland Congress. Wageningen Academic Publishers, Dublin, Ireland. p. 63.

Casler, M.D. 2001. Patterns of variation in a collection of timothy accessions. Crop Sci. 41:1616–1624. doi:10.2135/cropsci2001.4151616x

Casler, M.D. 2005. Agronomic performance of timothy germplasm from forage and turf sods under two harvest managements. Crop Sci. 45:1990–1996. doi:10.2135/cropsci2004.0659

Casler, M.D. 2008. Among-and-within-family selection in eight forage grass populations. Crop Sci. 48:434–442. doi:10.2135/cropsci2007.05.0267

Casler, M.D. 2010. Changes in mean and genetic variance during two cycles of within-family selection in switchgrass. BioEnergy Res. 3:47–54. doi:10.1007/s12155-009-9071-9

Casler, M.D. 2012. Switchgrass breeding, genetics, and genomics. In: A. Monti, editor, Switchgrass. Springer, New York. p. 29–54.

Casler, M.D. 2014a. Fundamentals of experimental design: Guidelines for designing successful experiments. Agron. J. (in press). doi:10.2134/agronj2013.0114

Casler, M.D. 2013b. Partial decomposition of the genetic correlation between forage yield and fiber using semihybrids. Crop Sci. 53:1403–1411. doi:10.2135/cropsci2013.01.0056

Casler, M.D., and E.C. Brummer. 2008. Theoretical expected genetic gains for among-and-within-family selection methods in perennial forage crops. Crop Sci. 48:890–902. doi:10.2135/cropsci2007.09.0499

Casler, M.D., J.H. Cherney, and E.C. Brummer. 2009. Biomass yield of naturalized populations and cultivars of reed canarygrass. BioEnergy Res. 2:165–173. doi:10.1007/s12155-009-9043-0

Casler, M.D., S.L. Fales, A.R. McElroy, M.H. Hall, L.D. Hoffman, and K.T. Leath. 2000a. Genetic prograss from 40 years of orchardgrass breeding in North America measured under hay management. Crop Sci. 40:1019–1025. doi:10.2135/cropsci2000.4041019x

Casler, M.D., S.L. Fales, D.J. Undersander, and A.R. McElroy. 2001. Genetic Progress from 40 Years of orchardgrass breeding in North America measured under management intensive rotational grazing. Can. J. Plant Sci. 81:713–721. doi:10.4141/P01-032

Casler, M.D., R.B. Mitchell, and K.P. Vogel. 2012. Switchgrass. In: C. Kole et al., editors, Handbook of bioenergy crop plants. Vol. 2. Taylor & Francis, New York. p. 563–590.

Casler, M.D., J.F. Pedersen, G.C. Eizenga, and S.D. Stratton. 1996. Germplasm and cultivar development. In: L.E. Moser et al., editors, Cool-season forage grasses. Agron. Monogr. 34. ASA, CSSA, and SSSA, Madison, WI. p. 413–469.

Casler, M.D., P.R. Peterson, L.D. Hoffman, N.J. Ehlke, E.C. Brummer, J.L. Hansen, M.J. Mlynarek, M.R. Sulc, J.C. Henning, D.J. Undersander, P.G. Pitts, P.C. Bilkey, and C.A. Rose-Fricker. 2002. Natural selection for survival improves freezing tolerance, forage yield, and persistence of festulolium. Crop Sci. 42:1421–1426. doi:10.2135/cropsci2002.1421

Casler, M.D., and K.P. Vogel. 1999. Accomplishments and impact from breeding for increased forage nutritional value. Crop Sci. 39:12–20. doi:10.2135/cropsci1999.0011183X003900010003x

Casler, M.D., and K.P. Vogel. 2014b. Selection for biomass yield in upland, lowland, and hybrid switchgrass. Crop Sci. 54 (in press). doi: 10.2135/cropsci2013.04.0239.

Casler, M.D., K.P. Vogel, J.A. Balasko, J.D. Berdahl, D.A. Miller, J.L. Hansen, and J.O. Fritz. 2000b. Genetic progress from 50 years of smooth bromegrass breeding. Crop Sci. 40:13–22. doi:10.2135/cropsci2000.40113x

Casler, M.D., K.P. Vogel, C.M. Taliaferro, N.J. Ehlke, J.D. Berdahl, E.C. Brummer, R.L. Kallenbach, C.P. West, and R.B. Mitchell. 2007. Latitudinal and longitudinal adaptation of switchgrass populations. Crop Sci. 47:2249–2260. doi:10.2135/cropsci2006.12.0780

Davis, W.H. 1971. Hybrid alfalfa production. U.S. Patent No. 3570,181. Issued March 16, 1971.

Denis, M., and J.-M. Bouvet. 2013. Efficiency of genomic selection with models including dominance effect in the context of Eucalyptus breeding. Tree Genet. Genomes 9:37–51. doi:10.1007/s11295-012-0528-1

Elliott, F.C., I.J. Johnson, and M.H. Schonhorst. 1972. Breeding for forage yield and quality. In: C.H. Hanson, editor, Alfalfa science and technology. Agron. Monogr. 15. ASA, CSSA, and SSSA, Madison, WI. p. 319–333.

Hayes, B.J., N.O.I. Cogan, L.W. Pembleton, M.E. Goddard, J. Wang, G.C. Spangenberg, and J.W. Forster. 2013. Prospects for genomic selection in forage plant species. Plant Breed. 132:133–143. doi:10.1111/pbr.12037

Hazard, L., and M. Ghesquére. 1997. Productivity under contrasting cutting regimes of perennial ryegrass selected for short and long leaves. Euphytica 95:295–299. doi:10.1023/A:1003048316012

Heffner, E.L., A.J. Lorenz, J.L. Jannink, and M.E. Sorrells. 2010. Plant breeding with genomic selection: Gain per unit time and cost. Crop Sci. 50:1681–1690. doi:10.2135/cropsci2009.11.0662

Hill, R.R., Jr., J.S. Shenk, and R.F. Barnes. 1988. Breeding for yield and quality. In: A.A. Hanson, et al., editors, Alfalfa and alfalfa improvement. Agron. Monogr. 29. ASA, CSSA, and SSSA, Madison, WI. p. 809–825.

Holland, J.B., and E.T. Bingham. 1994. Genetic improvement for yield and fertility of alfalfa cultivars representing different eras of breeding. Crop Sci. 34:953–957. doi:10.2135/cropsci1994.0011183X003400040022x

Hopkins, A., and R.J. Wilkins. 2006. Temperate grassland: Key developments in the last century and future perspectives. J. Agric. Sci. 144:503–523. doi:10.1017/S0021859606006496

Jannink, J.L., A.J. Lorenz, and H. Iwata. 2010. Genomic selection in plant breeding: From theory to practice. Brief. Funct. Genomics 9:166–177. doi:10.1093/bfgp/elq001

Jensen, K.B., and W.F. Anderson. 2014. Rangeland and warm-season forage grasses. In: S. Smith, B. Diers, J. Specht, and B. Carver, editors, Yield gains in major U.S. field crops. CSSA Spec. Publ. 33. ASA, CSSA, and SSSA, Madison, WI. p. 219–266. doi:0.2135/cssaspecpub33.c9

Katepa-Mupondwa, F.M., B.R. Christie, and T.E. Michaels. 2002. An improved breeding strategy for autotetraploids alfalfa (*Medicago sativa* L.). Euphytica 123:139–146. doi:10.1023/A:1014488307000

Lamb, J.F.S., C.C. Sheaffer, L.H. Rhodes, M. Sulc, D.J. Undersander, and E.C. Brummer. 2006. Five decades of alfalfa cultivar improvement. Crop Sci. 46:902–909. doi:10.2135/cropsci2005.08-0236

Li, X., and E.C. Brummer. 2012. Applied genetics and genomics in alfalfa breeding. Agronomy 2:40–61. doi:10.3390/agronomy2010040

Li, X., Y. Wei, K.J. Moore, R. Michaud, D.R. Viands, J.L. Hansen, A. Acharya, and E.C. Brummer. 2011. Association mapping of biomass yield and stem composition in a tetraploid alfalfa breeding population. Plant Genome 4:24–35. doi:10.3835/plantgenome2010.09.0022.

Lorenz, A.J., S. Chao, F.G. Asoro, E.L. Heffner, T. Hayashi, H. Iwata, K.P. Smith, M.E. Sorrells, and J.L. Jannink. 2011. Genomic selection in plant breeding: Knowledge and prospects. Adv. Agron. 110:77–123. doi:10.1016/B978-0-12-385531-2.00002-5

Missaoui, A.M., V.A. Fasoula, and J.H. Bouton. 2005. The effect of low plant density on response to selection for biomass production in switchgrass. Euphytica 142:1–12. doi:10.1007/s10681-005-0149-y

Peng, S., K.G. Cassman, S.S. Virmani, J. Sheehy, and G.S. Khush. 1999. Yield potential trends of tropical rice since the release of IR8 and the challenge of increasing rice yield potential. Crop Sci. 39:1552–1559. doi:10.2135/cropsci1999.3961552x

Resende, R.M.S., M.D. Casler, and M.D. Resende. 2013. Selection methods in forage breeding: A quantitative appraisal. Crop Sci. 53:1925–1936. doi:10.2135/cropsci2013.03.0143

Riday, H. 2009. Correlations between visual biomass scores and forage yield in space planted red clover (*Trifolium pratense* L.) breeding nurseries. Euphytica 170:339–345. doi:10.1007/s10681-009-9991-7

Riday, H. 2010. Progress made in improving red clover (*Trifolium pretense* L.) through breeding. Int. J. Plant Breed. 4:22–29.

Riday, H. 2011. Paternity testing: A non-linkage based marker-assisted selection scheme for outbred forage species. Crop Sci. 51:631–641. doi:10.2135/cropsci2010.07.0390

Robins, J.G., D. Luth, T.A. Campbell, G.R. Bauchan, C. He, D.R. Viands, J.L. Hansen, and E.C. Brummer. 2007. Mapping biomass production in tetraploid alfalfa (*Medicago sativa* L.). Crop Sci. 47:1–10. doi:10.2135/cropsci2005.11.0401

Rose, L.W., IV, M.K. Das, R.G. Fuentes, and C.M. Taliaferro. 2007. Effects of high- vs. low-yield environments on selection for increased biomass yield in switchgrass. Euphytica 156:407–415. doi:10.1007/s10681-007-9390-x

Sanderson, M.A., P.R. Adler, A.A. Boateng, M.D. Casler, and G. Sarath. 2007. Switchgrass as a biofuels feedstock in the USA. Can. J. Plant Sci. 86:1315–1325. doi:10.4141/P06-136

Scotti, C., and E.C. Brummer. 2010. Creation of heterotic groups and hybrid varieties. In: C. Huyghe, editor, Sustainable use of genetic diversity in forage and turf breeding. Springer, Berlin, Germany. p. 509–518.

Smith, R.R. 2000. Red clover in the "twenty-first" century. In: Proc. Wisconsin Forage Council. Wisconsin Dells, Wisconsin, 25 January 2000. Available at http://www.uwex.edu/ces/forage/wfc/proceedings2000/smith.htm (accessed 17 Jan. 2014).

Sun, P., M. Velde, and D. Gardner. 2004. Alfalfa hybrids having at least 75% hybridity. U.S. Patent 6774,280. Date filed 31 Jan. 2001. Issued 10 Aug. 2004.

Tanaka, T., H. Tamaki, K. Ashikaga, H. Fujii, and T. Yamada. 2011. Proposal to shift to reciprocal recurrent selections in "clone and strain synthesis" timothy breeding using molecular marker diversity. Crop Sci. 51:2589–2596. doi:10.2135/cropsci2011.02.0105

van Wijk, A.J.P., and D. Reheul. 1991. Achievements in fodder crop breeding in maritime Europe. In: A.P.M. den Nijs and A. Elgersma, editors, Fodder crops breeding: Achievements, novel strategies. and biotechnology. Proc. Eucarpia Fodder Crops Section, 18–22 Nov. 1990, Wageningen, The Netherlands. Pudoc, Wageningen. p. 13–18.

Vargas, M., E. Combs, G. Alvarado, G. Atlin, K. Mathews, and J. Crossa. 2013. META: A suite of SAS programs to analyze multienvironment breeding trials. Agron. J. 105:11–19. doi:10.2134/agronj2012.0016

Veronesi, F. 1991. Achievements in fodder crop breeding in Mediterranean Europe. In: A.P.M. den Nijs and A. Elgersma, editors, Fodder crops breeding: Achievements, novel strategies, and biotechnology. Proc. Eucarpia Fodder Crops Section, 18–22 Nov. 1990, Wageningen, The Netherlands. Pudoc, Wageningen. p. 25–30.

Viands, D.R., P. Sun, and D.K. Barnes. 1988. Pollination control: Mechanical and sterility. In: A.A. Hanson et al., editors, Alfalfa and alfalfa improvement. Agron. Monogr. 29. ASA, CSSA, and SSSA, Madison, WI. p. 931–960.

Vogel, K.P., A.A. Hopkins, K.J. Moore, K.D. Johnson, and I.T. Carlson. 2002. Winter survival in switchgrass populations bred for high IVDMD. Crop Sci. 42:1857–1862. doi:10.2135/cropsci2002.1857

Vogel, K.P., and R.B. Mitchell. 2008. Heterosis in switchgrass: Biomass yield in swards. Crop Sci. 48:2159–2164. doi:10.2135/cropsci2008.02.0117

Waldron, B.L., J.G. Robins, M.D. Peel, and K.B. Jensen. 2008. Predicted efficiency of spaced-plant selection to indirectly improve tall fescue sward yield. Crop Sci. 48:443–449. doi:10.2135/cropsci2007.06.0354

Weber, V.S., J.L. Araus, J.E. Cairns, C. Sanchez, A.E. Melchinger, and E. Orsini. 2012. Prediction of grain yield using reflectance spectra of canopy and leaves in maize plants grown under different water regimes. Field Crops Res. 128:82–90. doi:10.1016/j.fcr.2011.12.016

Wiersma, D.W., D.J. Undersander, and J.G. Lauer. 1997. Lack of alfalfa yield progress in the Midwest. In: Central Alfalfa Improvement Conf. Abstracts, LaCrosse, WI, 16–18 July 1997. p. 36.

Wilkins, P.W., and M.O. Humphreys. 2003. Progress in breeding perennial forage grasses for temperate agriculture. J. Agric. Sci. 140:129–150. doi:10.1017/S0021859603003058

Woodfield, D.R. 1999. Genetic improvements in New Zealand forage cultivars. Proc. New Zealand Grassl. Assoc. 61:3–7.

Woodfield, D.R., and J.R. Caradus. 1994. Genetic improvement in white clover representing six decades of plant breeding. Crop Sci. 34:1205–1213. doi:10.2135/cropsci1994.0011183X003400050011x

4

Lettuce and Spinach

Ivan Simko, Ryan J. Hayes, Beiquan Mou, and James D. McCreight*

ABSTRACT

Lettuce (*Lactuca sativa* L.) and spinach (*Spinacia oleracea* L.) are cultivated in many countries around the world. Their production in the United States is concentrated mostly in California and Arizona, where they are grown year-round. U.S. lettuce products are categorized into three market uses: whole heads, bulk harvest (for salad processing, food service, or value-added products), and "baby leaf" or "spring mix." The three main lettuce types produced in the United States are iceberg (*L. sativa* L. var. *capitata* L.), romaine (*L. sativa* L. var. *longifolia* L.), and leaf (*L. sativa* var. *crispa* L.). Iceberg yield increased 183% from 1950 to 2012, romaine yield increased 12% from 1992 to 2012, and leaf yield increased 14% from 1992 to 2012. Comparative yield trials of old and modern lettuce cultivars have not been done. Yield components were identified for each market use. Spinach market uses are fresh, which may be bunched, bagged [with small (baby) or intermediate sized leaves], or blended into spring mix, and processed (frozen or canned). Spinach yield varies with locations and seasons, but gains in yield have largely been attributed to the development of hybrid cultivars that exhibit heterosis. Fresh market spinach yield in California increased 50% from 1961 to 2011. Increased lettuce and spinach yield due to changes in agronomic practices (e.g., wide beds and high plant density) and more uniform and disease- and pest-resistant cultivars are offset to some extent by shifts in market uses (e.g., reduced proportion of the head harvested for romaine lettuce hearts) and early harvest for baby leaf lettuce and spinach. Sustained and increased lettuce and spinach yield depend on further optimized agronomic practices and productive cultivars with desirable quality, uniformity, and disease and pest resistance that are adaptable to variable environmental conditions and have increased water and nitrogen use efficiency.

Abbreviations: DPP, days post-planting; LIYV, Lettuce infectious yellows virus; QTL, quantitative trait locus; SPWF; Sweet potato whitefly.

I. Simko (ivan.simko@ars.usda.gov), **R.J. Hayes** (ryan.hayes@ars.usda.gov), **B. Mou** (beiquan.mou@ars.usda.gov), and **J.D. McCreight**, USDA-ARS, Crop Improvement and Protection Research Unit, 1636 E. Alisal St., Salinas, CA 93905. *Corresponding author (jim.mccreight@ars.usda.gov). The authors contributed equally to this review.

doi:10.2135/cssaspecpub33.c4

Yield Gains in Major U.S. Field Crops. CSSA Special Publication 33.
Stephen Smith, Brian Diers, James Specht, and Brett Carver, editors.

Cultivated lettuce is a self-fertilizing diploid species ($2n = 2x = 18$) from the *Compositae* family. Lettuce is the most popular, commercially produced leafy vegetable and is cultivated mainly in moderate climates in many countries around the world. Over 79% of the world production that totaled 24.3 million tonnes in 2011 (FAOSTAT, 2013) originated from four countries: China, with 55.2% of the total production (by weight), the United States (16.0%), India (4.4%), and Spain (3.6%).

Lettuce cultivars are divided into eight horticultural types on the basis of the shape and size of the head; the shape, size, and texture of leaves; stem length; and seed size (Fig. 4–1). The leaves of crisphead (which includes two subtypes, iceberg and Batavia), romaine, butterhead, leaf, and Latin types are typically eaten

Fig. 4–1. Lettuce types frequently grown in the United States: (top row, left) iceberg; (top row, right) Batavia; (second row, left) romaine; (second row, right) speckled romaine; (third row, left) green leaf; (third row, right) red leaf; (bottom row, left) red leaf; and (bottom row, right) butterhead.

fresh. The stem type is cultivated mainly for edible stems, fresh or cooked. The oil type is used for cooking oil from its relatively large seeds. There are no modern cultivars of the stem or oilseed types commercially grown in the United States. Detailed descriptions of each lettuce type were provided by Rodenburg (1960) and Ryder (1999) and are only summarized here.

Crisphead cultivars are known for their thick, crisp leaves that cup and form a spherical head weighing approximately 0.8 to 1 kg (1.8–2.2 lb). Butterhead cultivars produce spherical heads that weigh approximately 0.2 to 0.4 kg (0.44–0.88 lb); the leaves of this type are known for their pliability and oily texture. Romaine cultivars exhibit leaves that are much longer than wide and are typically upright to form an elongated head. Heads of commercially produced romaine weigh approximately 0.75 kg (1.65 lb). The top of the romaine head may or may not close over the inner leaves. Latin types produced quite small heads, less than 0.3 kg (<0.66 lb). The Latin heads have upright leaves that are longer than wide, similar to a romaine, but their leaf texture is similar to those of butterhead lettuce. Leaf types are the most variable, with a large number of leaf color, shape, size, texture, leaf margin, and leaf blistering combinations. The features in common for leaf lettuces are leaves that are shorter than romaine and heads that do not close to cover younger leaves. The most popular types of lettuces cultivated in the U.S. are iceberg, romaine, and leaf. Production of butterhead, Latin, and a range of specialty or artisan types occur, but the limited amounts are too small to quantify.

Lettuce production in the United States is concentrated in California (~75–80%) and Arizona (~20–25%) (USDA-ERS, 2013), where lettuce is produced year-round (Fig. 4–2). The remainder of lettuce is grown in Colorado (July–September), Florida (December through April), New Mexico (April and May), and Washington (June–September) (Vance Publishing, 2013). Seasonal commercial

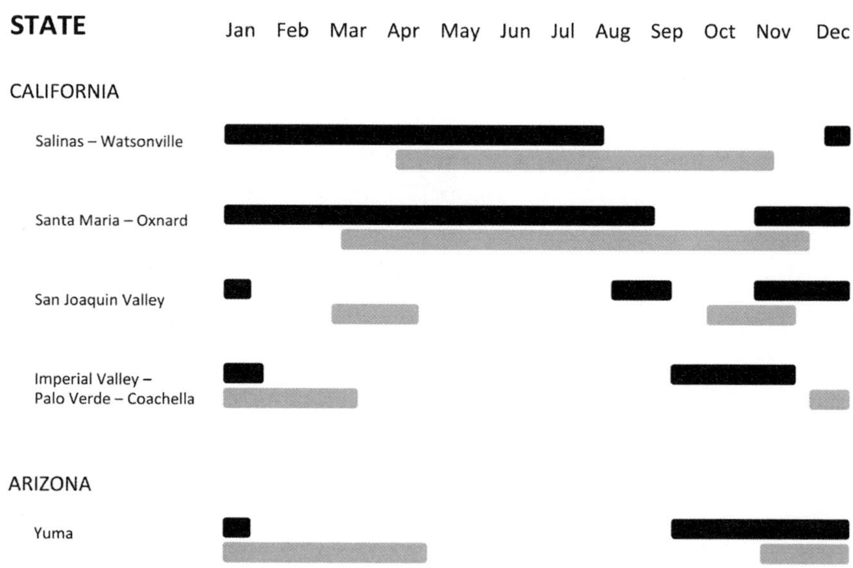

Fig. 4–2. Planting and harvesting schedules of major lettuce production areas for lettuce; bars represent periods of time, black for planting and gray for harvesting. (Courtesy of the California Leafy Greens Research Program.)

lettuce production for local and regional markets occurs in many states that are not presently included in the Economic Research Service Reports: Connecticut, Hawaii, Maine, Massachusetts, Michigan, New Jersey, New York, Ohio, North Carolina, Oregon, Pennsylvania, Texas, and Wisconsin (see USDA-ERS, 2013 for a list of reports).

California summer lettuce production (April–October) occurs in three coastal districts of California: Salinas-Watsonville, Santa Maria, and Oxnard, which produce approximately 66% of the total production (Fig. 4–3). Huron to Bakersfield in the west side of the San Joaquin Valley of California serves as a fall (October–November) and spring (March–April) transition period. Winter production occurs in the low deserts of California (Coachella, Imperial, and Palo Verde Valleys) and Arizona (Yuma Valley area). Rotation of production among these distinct areas (coastal districts, San Joaquin Valley, low desert) ensures that successive crops are grown in the range of temperatures to which lettuce is adapted. These areas and production dates encompass extensive variation for photoperiod, soil characteristics, and prevailing diseases and pests.

Lettuce products in the United States can be categorized into three broad areas: whole heads, bulk harvest (for salad processing, food service, or value-added products), and baby leaf or spring mix. The initial plant population, which determines a theoretical or potential maximum yield for lettuce grown for head production, is a major determinant of yield for lettuce grown for bulk harvest and spring mix. Lettuce in California and Arizona is produced on raised beds that are typically approximately 10 to 25 cm (4–10 in) high and are narrow or wide. Narrow beds are 102 to 107 cm (40–42 in) wide, center to center, with two parallel seed lines that are 35 cm (14 in) apart (Fig. 4–4). Growers in Arizona may, for leaf and romaine types, plant a third seed line down the middle of narrow beds. Wide beds are 203 to 213 cm (80–84 in) wide with five or six seed lines. Growers

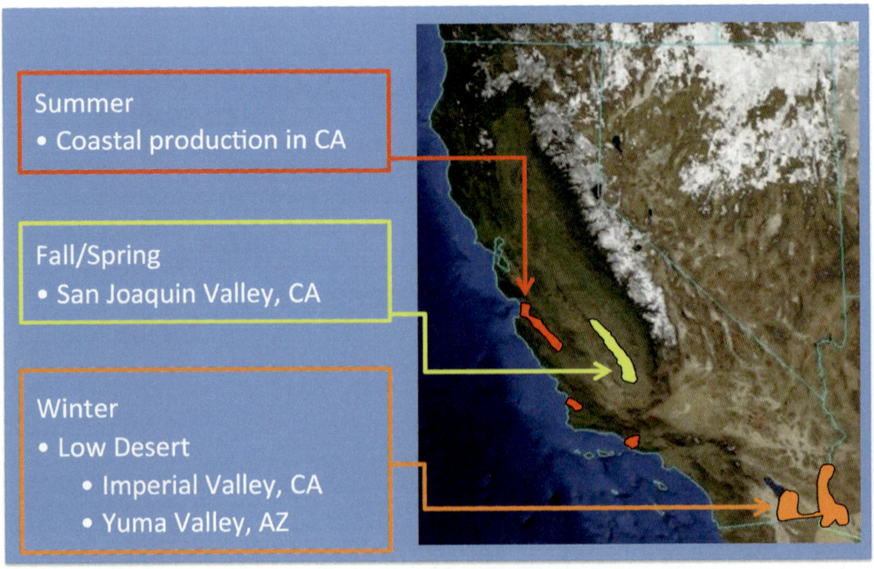

Fig. 4–3. Main lettuce production areas of the United States; courtesy of K.V. Subbarao.

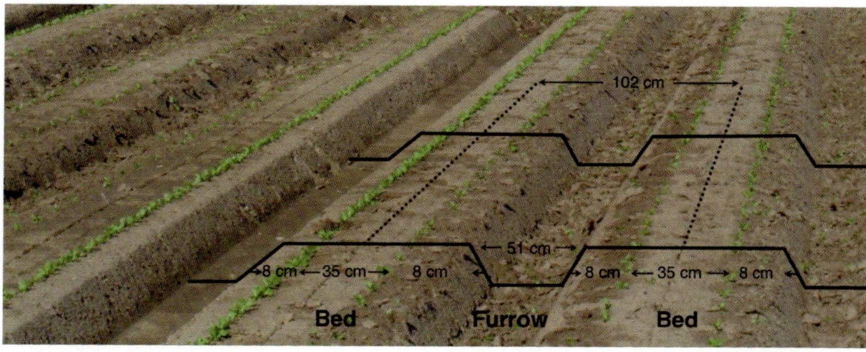

Fig. 4–4. Traditional 102-cm (40-in) bed. Modified from Whitaker et al., (1974); photograph is courtesy of K. Nolte.

can increase the number of plants per hectare by 25 or 50% with minimal cost by changing the bed width and number of seed lines per bed (Table 4–1) (Smith, 2012; Whitaker et al., 1974). Coated lettuce seeds are typically planted just below the surface, approximately 2.5 to 7.5 cm (1–3 in) apart, and are later thinned to the final distance of 25 to 31 cm (10–12 in) apart within the seed lines (Fig. 4–5). Romaine for hearts, some leaf types, and specialty types may be as close as 13 cm (5 in) within seed lines.

Mature plants of marketable quality are individually harvested. Iceberg lettuce for whole head production has traditionally been harvested at market maturity,

Table 4–1. Potential numbers of plants and yield of lettuce per hectare under ideal conditions on 1.0-m (40-in) and 2.0-m (80-in) wide beds, with 2 to 32 seed lines per bed and three plant spacings within the seed lines.

Bed width	No. seed lines	Plant spacing		Potential†	
		Between lines	In lines	Plants ha^{-1}	Yield‡
m		cm			t ha^{-1}
1	2	35.6	25.4	77,500	70.32
			30.48	64,583	58.60
			35.56	55,357	50.23
	3	20.32	25.4	116,250	105.48
			30.48	96,875	87.90
			35.56	83,036	75.34
2	5	30.48	25.4	96,875	87.90
			30.48	80,729	73.25
			35.56	69,196	62.79
	6	25.4	25.4	116,250	105.48
			30.48	96,875	87.90
			35.56	83,036	62.79
	24	2.54	2.54	4,650,002	16.48
	32	1.88	2.54	6,200,003	21.97

† Perfect stand and 100% marketable plants.

‡ 2, 3, 5, and 6 seed lines: whole heads (907 g per head); 24 and 32 seed lines, 2.8 g per plant.

60 to 130 d post-planting (DPP), depending on the planting date. Romaine reaches market maturity as fast or faster than iceberg, and leaf reaches market maturity faster than romaine. Plants that are below market quality are left in the field and contribute nothing to the final yield. Crops harvested for heads are packed in the field into cartons or some other container that holds a specified number of heads. A typical carton holds 24 heads, but other configurations are available (Vance Publishing, 2013). Iceberg lettuce may be packed "naked" or as wrapped lettuce with the outer leaves removed. Growers are paid per carton. Lettuce for processing is harvested into palletized bins; growers are paid by harvested weight or area. Numerous trimming strategies are used to eliminate lettuce tissue that cannot be used in processing. Removal of the core (stem) along with diseased or blemished leaves at harvest is common for lettuce that will be processed or chopped for salad

Fig. 4–5. Lettuce production: (top row, left) thinning of lettuce to final distance; (top row, right) furrow irrigation in Arizona; (middle row, left) tape irrigation of romaine lettuce in California, (middle row, right) labor intensive harvest of iceberg lettuce; (bottom row, left) automated harvest of red baby leaf lettuce; and (bottom row, right) automated harvest of spinach (all but romaine lettuce photograph are courtesy of K. Nolte).

mixes. The top few inches of romaine plants are trimmed off when they exhibit tip-burn, a physiological defect (Jenni and Hayes, 2010), or marginal leaf burn.

Lettuces for baby leaf and spring mix are grown on 203- to 213-cm wide (80–84 in) raised beds with 24 to 32 seed lines per bed (Smith et al., 2009). Approximately 15 to 20 different cultivars with unique combinations of leaf shape and color are currently used for production of baby leaf. These crops are planted at extremely high densities, up to 7.4 million seeds per hectare (3 million seeds per acre; Fig. 4–5) (Nolte, 2010), and mechanically harvested when a desirable size is reached, that is, when the first four true leaves are about 5 to 13 cm (2–5 in) long, approximately 20 to 30 DPP.

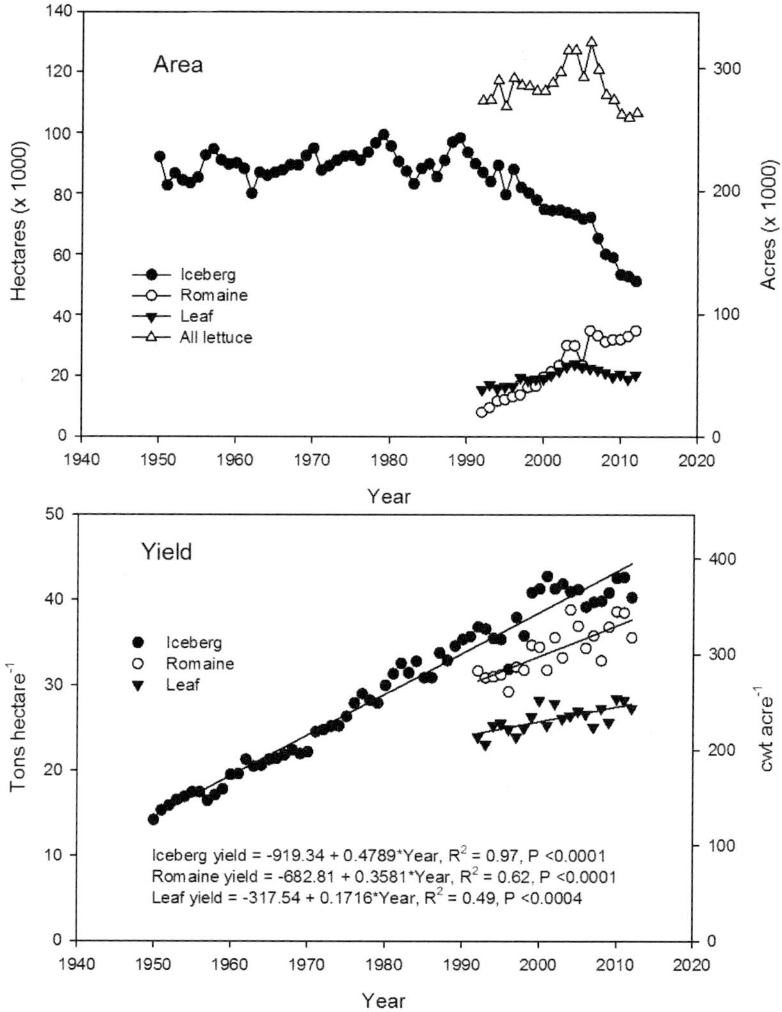

Fig. 4–6. Lettuce harvested area and yield in the United States, based on data compiled from USDA-ERS (2013). Yield equations in this and the following figures are calculated from t ha^{-1} (1 t ha^{-1} equals approximately 8.92 cwt ac^{-1}).

Lettuce Yield Gain

Iceberg was the major type of lettuce produced in the United States through the 1980s (Fig. 4–6). The increased popularity of salad bars and Caesar salads led to higher demand for leaf and romaine type lettuces as well as other leafy vegetables such as spinach (*S. oleracea* L.) and arugula (*Eruca sativa* L. Mill). Technological improvements in packaging material in the 1990s enabled the increased availability and shelf life of ready-to-eat salads, which are various mixtures of iceberg, romaine, and leaf lettuce types, as well as endive and escarole (*Cichorium endiva* L.) and radicchio (*Cichorium intybus* L.) (Vance Publishing, 2013). This led to increased romaine and leaf production and further eroded the dominance of iceberg lettuce. In 2012, romaine and leaf lettuce accounted for approximately 52% of the total U.S. lettuce crop.

Yield of iceberg lettuce in the United States increased 183% from 1950 through 2012 from 14.2 to 40.3 t ha^{-1} (127–360 cwt ac^{-1}). The maximum average yield of 41.2 t ha^{-1} (368 cwt ac^{-1}) was reached in 2001. The mean yield gain, calculated from linear regression analysis, was 0.48 t ha^{-1} yr^{-1} (4.3 cwt ac^{-1} yr^{-1}). The yield in the last 7 yr (2006–2012) was, however, below the fitted regression line, which indicated that the yield gain may have reached a plateau, though a pattern of periodic "plateaus" is also evident (Fig. 4–6). Monterey County, CA iceberg yield surpassed the maximum U.S. average every year with one exception (1995) since 1987 (Fig. 4–7). Iceberg yields in Imperial Valley, CA and Arizona tend to be near or below the U.S. means (Fig. 4–8 and 4–9).

Yield of romaine lettuce in the United States increased from 31.7 t ha^{-1} (283 cwt ac^{-1}) in 1992 to 35.6 t ha^{-1} (318 cwt ac^{-1}) in 2012 (Fig. 4–6). This 12% increase in yield through a 20-yr period corresponds to an increase of 0.36 t ha^{-1} yr^{-1} (3.2 cwt ac^{-1} yr^{-1}) calculated from linear regression analysis. The maximum average yield of 38.9 t ha^{-1} (347 cwt ac^{-1}) was recorded in 2004. Monterey County romaine yield surpassed the maximum U.S. average every year from 2004 to the present, with a high of 55.3 t ha^{-1} (493 cwt ac^{-1}) in 2005 (Fig. 4–7). Romaine yield in Arizona is generally below the U.S. mean (Fig. 4–9).

Yield of leaf lettuces in the United States increased 14% from 24.0 t ha^{-1} (214 cwt ac^{-1}) in 1992 to 27.3 t ha^{-1} (244 cwt ac^{-1}) in 2012 (Fig. 4–6). The maximum average yield of 28.4 t ha^{-1} (254 cwt ac^{-1}) was recorded in 2010. Linear regression analysis indicated an increase in yield of 0.17 t ha^{-1} yr^{-1} (1.5 cwt ac^{-1} yr^{-1}). Monterey County leaf lettuce yield surpassed the maximum U.S. average every year from 2007 to the present, with a high of 30.2 t ha^{-1} (269 cwt ac^{-1}) in 2011 (Fig. 4–7). Leaf lettuce yield in Arizona is generally below the U.S. mean (Fig. 4–9).

Statistical analysis showed that absolute yield increase in leaf lettuces during 1992 through 2012 period was significantly slower ($P \leq 0.016$) than yield gains in romaine cultivars. During the same period, yield of iceberg cultivars increased 0.33 t ha^{-1} yr^{-1} (2.9 cwt ac^{-1} yr^{-1}), which is not significantly different from the romaine ($P \leq 0.792$) or leaf type cultivars ($P \leq 0.068$).

Lettuce crop reports for Monterey County, which encompasses the Salinas Valley, are prepared by the County Agricultural Commissioner and are available online for the years 1929 to 2011 (ag.co.monterey.ca.us). Yield data were calculated using the total acres harvested and total cartons harvested for each year. In some years, production data were reported as tons harvested, whereas in most years production was reported in number of cartons harvested. Data were converted

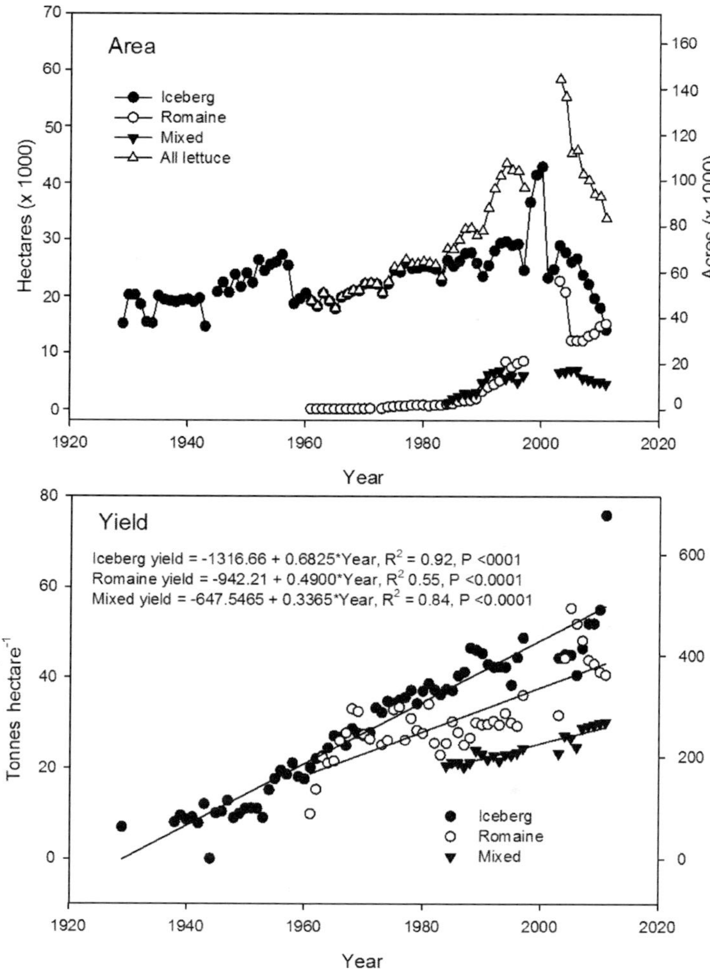

Fig. 4–7. Lettuce harvested area and yield in the Monterey County (Salinas Valley), California, based on data compiled from Monterey County Agricultural Commissioner Crop Reports (ag.co.monterey.ca.us).

from tons to cartons, assuming cartons of iceberg were 22.6 kg (50 lb) and cartons of romaine, leaf, and butterhead were 15.9 kg (35 lb). Harvesting operations have changed over time and are not uniform across producers. In older data sets, there is no description of what constituted a carton or how much of the plant was collected during harvest. This is particularly problematic for bulk harvests, where current practices leave more of the plant (core, outer leaves, and leaf tips) in the field than was done 10 yr ago. Alternative products such as romaine hearts have also increased in popularity over time and may use unique harvesting procedures. In the case of romaine hearts, most of the plant is left in the field, which will suppress romaine yield estimates. Single harvests are currently a standard production practice, although this was not always the case. Before the 1980s,

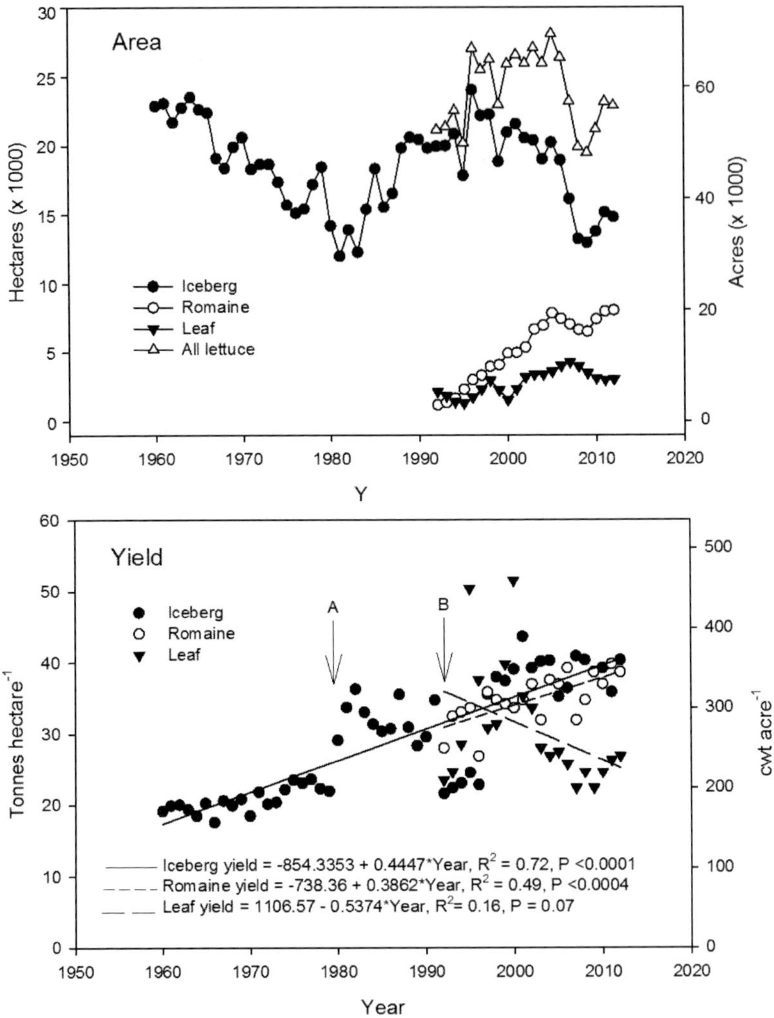

Fig. 4–8. Lettuce harvested area and yield in Imperial Valley, California, based on data compiled from Imperial County Agricultural Commissioner Crop Reports (http://www.co.imperial.ca.us/ag/). Vertical arrows indicate appearance of sweet potato whitefly biotype A and B, respectively.

fields were harvested up to three times. We are assuming that the data in the Monterey County reports represent the sum of all lettuce harvested from each hectare and that a second or third harvest was not counted as a second or third hectare of harvested lettuce.

Statistics for bulk harvest and spring mix are also available from Monterey County. Approximately 33% of the crop in Monterey County is for bulk harvest; spring mix represents <10% of lettuce acreage (County of Monterey Agricultural Commissioner, 2011) (Fig. 4–7). Monterey County produces approximately 50% of U.S. lettuce, thus the lettuce crop data for Monterey County generally mirrors

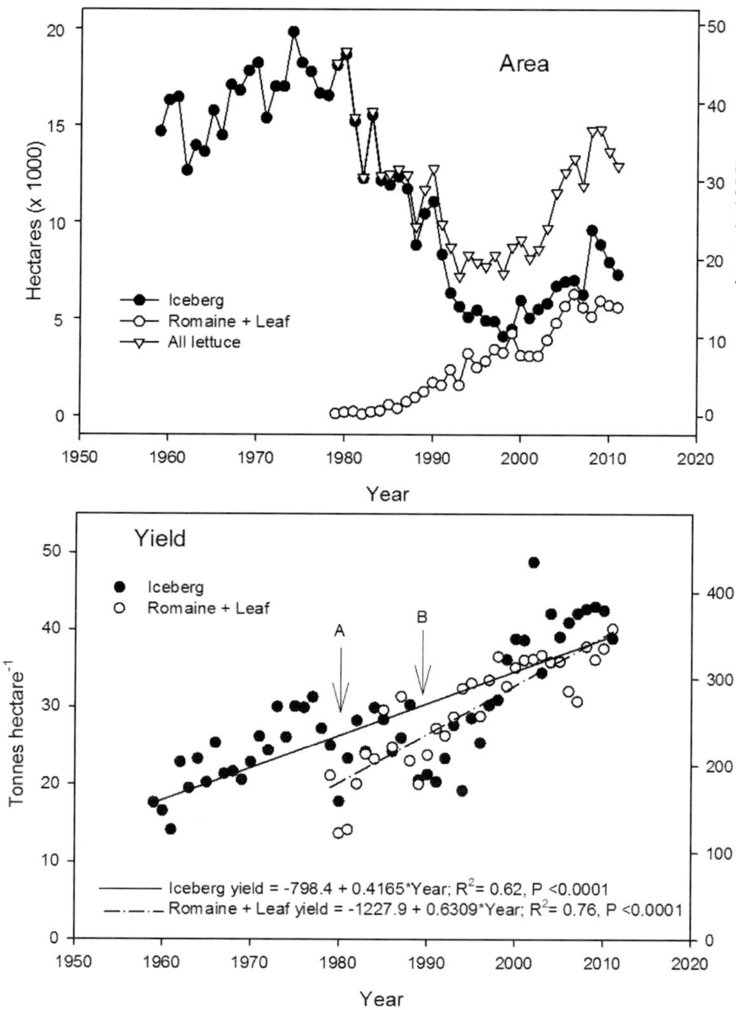

Fig. 4–9. Lettuce harvested area and yield in Yuma, AZ, based on data compiled from the U.S. Economic Research Service (USDA-ERS, 2013). Vertical arrows indicate appearance of sweet potato whitefly biotype A and B, respectively.

national trends. Production (area harvested) was dominated by iceberg type lettuce for decades, with increases in romaine and mixed types occurring in the 1990s (Fig. 4–7). In 2011, the most recent year with reported data, production of romaine surpassed iceberg. Comparing the 1929 to 2011 production years, iceberg lettuce yield increased from 6.9 t ha^{-1} (61.6 cwt ac^{-1}) to 75.9 t ha^{-1} (677.2 cwt ac^{-1}), a 1000% increase. However, the 1929 and 2011 data seem to be outliers. Using the average yield from 1938 to 1942 [8.6 t ha^{-1} (76.7 cwt ac^{-1})] and from 2006 to 2010 [49.2 t ha^{-1} (439.0 cwt ac^{-1})], the percent increase in yield is 472%, which is still substantial. Iceberg yields increased 102% from 1970 to 2010. A linear regression of yield on the production year was significant and indicates that yield increased at a rate of 0.68 t ha^{-1} yr^{-1} (6.1 cwt ac^{-1} yr^{-1}). Inspection of the data points around the regression line reveals two important time frames where yields are less than predicted by the regression. As with the U.S. data, yields in recent years (2003–2010) appear to be leveling off. Yield increases before 1954 appear to be relatively flat, and a regression using data only for the years before 1954 results in a slope that is not significantly different from zero (a separate regression for years prior 1954 is not shown).

Yield of mixed types (red leaf, green leaf, and butterhead) show consistent increases, averaging 0.34 t ha^{-1} yr^{-1} (3.0 cwt ac^{-1} yr^{-1}) (Fig. 4–7). This represents a 47% increase from 20.5 t ha^{-1} (183 cwt ac^{-1}) in 1984 to 30.2 t ha^{-1} (270 cwt ac^{-1}) in 2011. Yield of romaine lettuce increased from 9.9 t ha^{-1} (88 cwt ac^{-1}) in 1961 to 40.7 t ha^{-1} (363 cwt ac^{-1}) in 2011, a 311% increase. Linear regression analysis indicated a yield increase of 0.49 t ha^{-1} yr^{-1} (4.4 cwt ac^{-1} yr^{-1}). The coefficient of determination (R^2) for the regression with the romaine yield data was 0.55, noticeably lower than those for mixed and iceberg. The relatively poor fit to a linear regression is apparent in Fig. 4–7. Yield of romaine lettuce did not increase from 1966 to 2003; sharp increases in yield are apparent before and after these time points.

The low desert produces lettuce for approximately 4.5 mo, from mid November through March. The total area of iceberg lettuce harvested in Imperial Valley and Arizona in 1960 was 39,155 ha (96,754 ac; Fig. 4–8 and 4–9). The area of iceberg lettuce harvested in the low desert decreased from the mid-1970s through 2010, which reflected in part the gain in market share of the mixed lettuce types (Fig. 4–8 and 4–9). Romaine and leaf area harvested and yield data were combined in the Imperial County crop reports (http://www.co.imperial.ca.us/ag, accessed 30 Jan. 2014). Arizona romaine and leaf lettuce data were reported separately. The total area of all lettuce types harvested in the low desert was less in 2010 than in 1960.

The harvested area of iceberg lettuce in the low desert appeared more variable on an annual basis than in Monterey County through 1974. The area of iceberg lettuce harvested in Imperial Valley trended upward from 1964 through 1974, when it peaked at 19,829 ha (49,000 ac), and then turned downward through 2011, when it was 7300 ha (18,039 ac) (Fig. 4–8). Harvested area of iceberg lettuce in Arizona was essentially stable from 1960 to 1966 with a mean of 22,697 ha (56,086 ac) per year, steadily decreased to 11,978 ha (29,598 ac) in 1981, and began an unsteady climb to 23,038 ha (56,929 ac) in 1996 (Fig. 4–9). As romaine and leaf lettuce production increased from 1991 through 2012, the harvested area of iceberg decreased to 14,811 ha (36,599) (Fig. 4–9).

The decreased area of iceberg harvested in Imperial Valley and Arizona was accompanied by volatile yields. Yields in Imperial Valley tended to be at or below the regression line (Fig. 4–8), while Arizona ranged from well above to well below the regression line (Fig. 4–9). Sweet potato whitefly (SPWF; *Bemisia tabaci* Gennadius) biotype A became an important pest of lettuce during the 1980s as a vector of *Lettuce infectious yellows virus* (LIYV) (Bakker et al., 1998). As growers gained control of this pest, lettuce production increased (area harvested and yield) until 1991, when the B biotype (SPWF-B) appeared and LIYV disappeared (Bakker et al., 1998). Yield decreased in both low desert areas in response to SPWF-B feeding (Fig. 4–8 and 4–9) but then increased through the 1990s as growers gained more control over SPWF-B. The area of iceberg lettuce harvested in Imperial Valley continued to decrease, whereas it remained steady in Arizona.

Yield of iceberg lettuce in the low desert, like in Monterey County (Fig. 4–7), increased steadily from 1960 to the present (Fig. 4–8 and 4–9). The rate of yield gain for iceberg was virtually the same in Imperial Valley [0.44 t ha^{-1} yr^{-1} (3.9 cwt ac^{-1} yr^{-1})] and Yuma [0.41 t ha^{-1} yr^{-1} (3.7 cwt ac^{-1} yr^{-1}); Fig. 4–8 and 4–9] but was approximately 66% of the rate in Monterey County [0.68 t ha^{-1} yr^{-1} (6.1 cwt ac^{-1} yr^{-1}) Fig. 4–7].

Gain for the combined romaine and leaf lettuce in Imperial Valley was 0.63 t ha^{-1} yr^{-1} (5.6 cwt ac^{-1} yr^{-1}), approximately 114% of the gain for romaine in Monterey County [0.55 t ha^{-1} yr^{-1} (4.9 cwt ac^{-1} yr^{-1})], but the rate of gain for romaine in Yuma was 0.49 t ha^{-1} yr^{-1} (4.4 cwt ac^{-1} yr^{-1}), approximately 90% of Monterey County. In contrast, the reported yield of leaf lettuce in Yuma appeared to decrease from 1990 to the present at a rate of −0.54 t ha^{-1} yr^{-1} (−4.8 cwt ac^{-1} yr^{-1}; Fig. 4–9). This may be an artifact of higher yields from 1995 through 2002. Remarkably high leaf lettuce yields of 50.4 t ha^{-1} (450 cwt ac^{-1}) and 51.5 t ha^{-1} (460 cwt ac^{-1}) were reported in 1995 and 2000, respectively. High leaf lettuce yields occurred as well through the intervening years (1996–1999) and in 2001 and 2002 when compared with the preceding (1992–1994) and succeeding (2003–2012) years. Thus, yield of leaf lettuce in Arizona was flat from 1992 through 2012 (yield in t ha^{-1} = 75.25 − 0.0248 × Year, $R^2 = -0.007$, $P = 0.78$) when corrected for the period of high yields, that is, the data from 1995 through 2002 were ignored.

Components of lettuce yield depend on market use and can be characterized as follows.

- Whole head lettuce yield = Initial number of plants × Proportion of surviving plants × Proportion of plants of marketable weight × Proportion of plants of desired type.
- Bulk harvested lettuce yield = Initial number of plants × Proportion of surviving plants × Average weight of harvested plant × Average harvested proportion of plant.
- Baby leaf lettuce yield = Initial number of plants × Proportion of surviving plants × Number of leaves per plant × Average weight of harvested leaves.

Specific traits to target for breeding efforts aimed at improving yield can be deduced from examination of these formulas. To increase yield, lettuce cultivars are bred for early vigor, size, earliness (head maturation), uniformity, shape, texture, late bolting, small core (stem), and resistance to pests, diseases, and physiological disorders. Other important factors are adaptability to different conditions and quality of product. Breeding-improved forms of these traits may affect at least one yield component to a certain extent, with the exception of initial number of plants, although breeding for disease resistance may facilitate closer spacing. Uniformity and close adherence to market standards are crucial for high yield of lettuce harvested as whole heads and packed into cartons. However, the culling levels for head weight and horticultural quality are not consistently applied throughout the season. Quality standards are often lowered when lettuce is in short supply, such as during extreme disease epidemics, or when adverse environmental conditions reduce yield across large areas. Conversely, the quality of lettuce is highly scrutinized when a surplus of product is available. For example, if tipburn is observed on more than 5% of plants at harvest, the whole field may be rejected. The main disorders caused by adverse environmental conditions are tipburn, rib discoloration, premature bolting, and cold injury on mature plants (Davis et al., 1997).

The increased popularity of packaged salads that are processed from bulk harvested lettuce opens up new strategies to increase yield for this market segment. Two complementary approaches can be considered. The lettuce core and the darker green outer leaves of the plant are not used in salad processing. Reducing the core size and the number of darker green or nonblanched leaves of the plant decreases the proportion of the plant biomass that is unusable for salad processing. Blanched leaves, which are yellow to yellow-green in color, are normally produced when older leaves envelope the younger inner leaves such as in iceberg or hearting romaine type cultivars. Head closure is to a variable degree temperature dependent. Periods of unexpectedly high temperatures increase the rate of growth and may result in a more open head with darker green leaves. The proportion of blanched leaves may be increased by the incorporation of genes for chlorophyll deficiency, light green leaf color, or yellow-green leaf color. These genes are expected to produce blanched leaves across a range of environments. Increasing leaf length and the number of leaves per plant are a potential approach to increase the useable portion of each plant. This is important, as the number of plants per acre cannot be significantly increased any further. The combination of these two approaches will result in taller plants because of longer leaves with shorter cores, more leaves, and lighter green coloration. This should increase the weight of the harvested portion of each plant compared with present cultivars. Information on the genetics of these traits and the genetic relationships of these traits is incomplete. Robinson et al. (1983) reviewed lettuce genes affecting leaf color and other morphological characters. Hayes et al. (2011a) identified *short leaf 1*, a major gene effecting leaf length in romaine × Latin crosses. More research is needed to determine the feasibility of breeding these types of cultivars.

Economically important diseases and pests in the California and Arizona lettuce production regions reduce yield by decreasing plant survival, reducing plant vigor and mature plant weight, increasing the number of leaves that must be trimmed off at harvest, or causing the architecture or appearance of the plant to be unmarketable (Table 4–2). Information on most economically significant lettuce diseases can be found in Davis et al. (1997). Disease resistance does not need to be complete or absolute to prevent economic damage, although single resistance genes of major effect are known in lettuce for many pathogens (McHale et al., 2009) and have been widely used in lettuce breeding. Host resistance to foliar pathogens that limits damage to the lower frame leaves of the plant, which are trimmed off the head and left in the field as part of normal harvesting, is typically sufficient to prevent economic damage. Examples of this form of resistance occur with downy mildew (*Bremia lactucae* Regel) and bacterial leaf spot (*Xanthomonas campestris* pv. *vitians*) (Hayes et al., 2008a; Simko et al., 2012). Lettuce is harvested in the vegetative stage before reproductive development begins, and genes that delay disease onset past harvest maturity can also be used to reduce disease loss. Resistance to the soilborne viral disease big-vein in the cultivar Pacific is attributed to this mechanism, as is resistance in some accessions to *Sclerotinia* spp., which causes lettuce drop (Hayes et al., 2011a; Smith et al., 2009).

Table 4–2. Lettuce yield components affected by economically important diseases and insect pests in California and Arizona.

Disease, pathogen, or insect pest	Plant yield component: Survival	Weight†	Architecture/ appearance	Reference
Downy mildew (*Bremia lactucae* Regel)		X		McHale et al. (2009); Simko et al. (2012)
Verticillium wilt (*Verticillium dahlia* Kleb)	X	X		Hayes et al. (2007, 2011c)
Fusarium wilt *(Fusarium oxysporum* f. sp. *lactucae*)	X			Matheron et al. (2005); Scott et al. (2010)
Lettuce drop (*Sclerotinia minor*, and *S. sclerotiorum*)	X			Hayes et al. (2010)
Powdery mildew (*Golovinomyces cichoracearum sensu stricto*)		X		Whitaker and Pryor (1941); Lebeda et al. (2012)
Phoma basal rot (*Phoma exigua*)	X			Grube et al. (2003)
Bottom rot (*Rhizoctonia solani*)	X			Davis et al. (1997)
Bacterial leaf spot (*Xanthomonas campestris* pv. *vitians*)		X		Bull et al. (2007)
Corky root (Rhizomonas suberifaciens Mosley)		X		Brown and Michelmore (1988)
Big-vein (*Mirafiori lettuce big-vein virus*)		X	X	Ryder and Robinson (1995); Hayes et al. (2008b)
Dieback (*Tomato bushy stunt virus,* and *Lettuce necrotic stunt virus*)	X	X	X	Simko et al. (2009)
Lettuce mosaic virus		X	X	German-Retana et al. (2008)
Tomato spotted wilt virus and *Impatient necrotic spot virus*	X	X	X	O'Malley and Hartmann (1989)
Lettuce aphid (*Nasonavia ribisnigri* Mosely)			X	McCreight (2008)
Leafminers (*Liriomyza langei*, *L. trifolii*, and *L. sativae*)			X	Mou and Liu (2004)

† Plant weight is reduced by removal of disease leaves, or reduction in plant vigor. This reduces the proportion of plants of marketable weight and the average plant weight.

Lettuce Genetic Yield Gain

The Imperial group of iceberg cultivars represented a significant proportion of lettuce production before the mid-1950s (Ryder, 1999). The cultivar Great Lakes was released in 1941 (Barrons and Whitaker, 1943) and according to Ryder (1999) was widely grown through the mid-1970s. During this period at least 60 cultivars with adaptation to a broad range of environments were developed from Great Lakes (Ryder, 1999), a process that seems to coincide with a dramatic increase in lettuce yields in Monterey County starting in the mid 1950s (Fig. 4–7). The release of other landmark iceberg cultivars Vanguard in 1959 (Thompson and Ryder, 1961), Calmar in 1960 (Welch et al., 1965), and Salinas in 1975 (Ryder, 1979b) do not appear to have significantly affected the rate of yield increase. 'Salinas' was reported to exhibit more uniform maturity compared to previous cultivars. This should increase yield by increasing the proportion of plants having a marketable weight and a desirable type within a single harvest. It is likely that 'Salinas' and its descendants contributed to the reduction in the number of harvests per crop. However, data from multi-environment trials that compare modern and historical lettuce cultivars for uniformity are not available, so this hypothesis is speculative.

Lettuce cultivars for commercial production are, in general, narrowly adapted to specific planting seasons. The range of recommended planting dates for a cultivar is referred to as a "planting slot." Seed companies conduct extensive field trials to determine the correct "slot" for a cultivar. Planting slots are typically location specific, since lettuce production rotates to different locations in California and Arizona during the year. The plantings slots for iceberg cultivars are narrower than those for romaine and leaf. Planting slots for winter production are generally narrower than those for summer production. For example, recommended seeding dates for summer-grown iceberg lettuce in coastal California can begin in March and continue until July. Iceberg crops harvested in January and March in the low desert production region use cultivars with planting slots spanning about 20 d during October and November. Iceberg crops harvested in December use cultivars with planting slots of about 10 d during September. Breeders may be exploiting favorable genotype × environment interactions to achieve improved yields. For iceberg cultivars, adaption is closely related to a cultivar's ability to uniformly produce well-shaped, firm heads of marketable size. Head formation and higher head weight are promoted by high light intensity and lower temperatures (Wurr and Fellow, 1991), although the genes governing the interaction with these environmental factors are unknown. Higher temperatures can accelerate maturity in iceberg lettuce but result in lower head weight (Wurr et al., 1996). Consequently, lettuce production in high temperature conditions using slower maturing cultivars may be advantageous. Resistance in iceberg, romaine, and leaf to physiological defects such as tipburn (Jenni and Hayes, 2010), rib discoloration (Jenni, 2005; Ryder, 1999), and premature bolting (Ryder, 1999) enables planting where these problems are common.

Genetic diversity contributes significantly to development of cultivars with high levels of host plant resistance to diseases, insects, and abiotic stresses, adaptability, and increased yield and quality potential of the harvested product. Lettuce cultivars developed by hybridization usually originate from mating accessions within horticultural type, to a smaller extent among horticultural types, and infrequently by mating cultivated lettuce with compatible wild

Lactuca species (*L. serriola* L., *L. saligna* L., and *L. virosa* L.). The types of breeding populations used for new cultivar development during the 1970 to 2004 era were two parent (52%), within cultivar selection (31%), three parent (7%), two or more backcrosses (5%), single backcross (2%), four or more parents (2%), and interspecific cross (1%) (Mikel, 2007). Mating of closely related accessions may have adverse effects on the genetic diversity of lettuce. Analysis of lettuce cultivars with molecular markers (Simko, 2009; Simko and Hu, 2008) indicates very limited genetic variability in modern iceberg cultivars and to a lesser extent in romaine cultivars. These results were confirmed by pedigree analyses of lettuce cultivars registered in the United States from 2000 through 2010 (Mikel, 2013). Cultivars Vanguard, Salinas, and Calmar provided 23.8, 12.1, and 10.9%, respectively, of genetic contribution to iceberg cultivars. The role of 'Vanguard' in the development of iceberg lettuce is the most extensive, contributing nearly 25% of the genes, and its descendants represent >66% of the registered cultivars (Mikel, 2013). 'Vanguard' was released as an F_4 line (Thompson and Ryder, 1961). The combination of residual heterozygosity found in the F_4 generation of a diverse pedigree enabled selection of numerous new cultivars from 'Vanguard.' The diverse pedigree of 'Vanguard' is a result of interspecific crosses between *L. virosa* (PI 125130) and *L. serriola* (PI 125819, PI 114535, and PI 114512) that led to development of breeding line 5192. This breeding line is one of the grandparents of 'Vanguard' and 'Salinas' (Thompson and Ryder, 1961). Calmar, a Great Lakes type cultivar bred for resistance to downy mildew, is one of the parents of Salinas. This shows the close genetic relationship among iceberg cultivars that have the highest genetic contribution to modern iceberg lettuces.

Romaine cultivars Parris Island Cos and Tall Guzmaine contributed 25.9 and 23.4% of the genetic background for the romaine type lettuce (Table 4–3) (Mikel,

Table 4–3. Principal iceberg, romaine, and leaf lettuce cultivars grown in the California and Arizona by season and district, 2004–2006, listed in alphabetic order.†

Area	Season	Iceberg	Romaine	Leaf
Coastal	Spring	Coach Supreme Honcho II Westland Yuma	Capri Darkland Cos Parris Island Cos P.I.C. Bos	Genecorp Green Glossy Green Royal Green Waldmann's Green
	Summer	Bronco El Dorado Salinas Target	Capri Darkland Cos Parris Island Cos P.I.C. Bos	Genecorp Green Glossy Green Royal Green Waldmann's Green
Huron	Fall	Acacia Desert Queen Empire Marvel	Capri Darkland Cos Parris Island Cos P.I.C. Bos	Genecorp Green Glossy Green Royal Green Waldmann's Green
Imperial Valley	Winter	Empire Fall Green Tres Equis Niner	Capri Darkland Cos Parris Island Cos P.I.C. Bos	Genecorp Green Glossy Green Royal Green Waldmann's Green
Yuma	Winter	Diamondback Enterprise Heat Master Sun Devil Winter Gold	Fresh Heart Green Towers King Henry Rubicon Sunbelt	Big Star Martin North Star Two Star

† Data were obtained from the USDA-ERS (2013).

2013). Pedigree analysis indicates that these two cultivars are not directly related to each other (Mikel, 2007). Leaf lettuce cultivars show more genetic diversity than iceberg or romaine cultivars. Generally, new iceberg and romaine lettuce cultivars originate from mating within each type, whereas leaf cultivars are developed by either mating within the type or with other types of lettuces (Mikel, 2007, 2013). Pedigree analysis shows that largest genetic contribution to leaf cultivars in the 2000 to 2010 era came from romaine cultivar Parris Island Cos (13.5%), followed by leaf cultivars Malibu (6.4%), Waldmann's Green (6.1%), and Salad Bowl (3.5%) (Mikel, 2013). In the 1970 to 2004 era, the most frequently used leaf cultivars for development of new cultivars were Prize Head, Waldmann's Green, and Slobolt (Mikel, 2007). Compared with iceberg and romaine lettuces, leaf cultivars show larger phenotypic diversity, pedigree diversity (Mikel, 2013), and genetic diversity (Simko, 2009; Simko and Hu, 2008).

The breeding process frequently changes allele frequency in populations, particularly in the genomic regions that are related to the traits of interest. Phenotypic and molecular data provide evidence of the selection in specific genomic regions of lettuce. In the cases described here, the genomic regions under selection include a gene or genes that increase yield by conferring resistance to diseases. Modern iceberg cultivars were developed through several cycles of recombination (Whitaker, 1974). Cycle 0 includes New York type cultivars, Cycle 1 includes Imperial type cultivars, Cycle 2 includes Great Lakes type cultivars, and Cycle 3 includes iceberg cultivars of today (Mikel, 2007).

In the early 1920s, brown blight disease was threatening to seriously curtail iceberg lettuce production in the Imperial Valley of California (Jagger, 1940). Brown blight was an unknown disease until lettuce culture became important in the Imperial Valley. New York type cultivars, which were almost exclusively grown in the area, were highly susceptible to this soilborne disease (Jagger, 1940). In Cycle 1 of recombination, the Imperial type cultivars were developed primarily through selection within the cultivar New York (Jagger, 1940) or through recombination of New York with the romaine type cultivar Blonde Lente A Morter. The Imperial type cultivars exhibited resistance to the brown blight disease (Whitaker, 1974). Because only iceberg cultivars resistant to the brown blight disease were subsequently grown in California and Arizona, the disease virtually disappeared and was no longer of economic concern to the lettuce industry. Since the early 1990s, when romaine and leaf lettuce became widely grown, a previously uncharacterized dieback disease of lettuce has been observed throughout the main producing areas of California and Arizona. The disease affects romaine and leaf type lettuce, often leading to almost complete crop loss. The disease is caused by two closely related soilborne viruses from the family *Tombusviridae—Tomato bushy stunt virus* and *Lettuce necrotic stunt virus* (Obermeier et al., 2001). Available evidence suggests that dieback disease and the previously described brown blight disease are caused by the same or a similar *Tombusvirus* (Wisler and Duffus, 2000). Similar to brown blight disease, dieback was never observed in any modern type iceberg cultivar that has 'Imperial' in the pedigree (Grube et al., 2005). Genetic analysis and molecular mapping revealed that resistance to the disease is conferred by a single dominant gene (*Tvr1*) located on Linkage Group 2 (Simko et al., 2009). Testing with molecular markers proved that all modern iceberg cultivars possess resistance alleles against dieback, whereas most of the pre-Imperial type iceberg cultivars harbor alleles associated with susceptibility to the disease (Simko et al., 2009). Similarly, a majority of romaine and leaf cultivars do not possess resistance alleles. A combination of classical breeding and marker-assisted selection led to the development of romaine breeding lines resistant to dieback and with an improved shelf life (Simko et al., 2010). Development of other romaine and leaf lettuces with resistance to the disease is in progress. Because the *Tvr1* gene appears to provide durable resistance against dieback, only lettuce cultivars possessing the resistance gene will be grown in California and Arizona.

Similar selection in a specific genomic region could happen in the area linked to the Verticillium wilt resistance gene. The dominant *Vr1* gene for resistance to Verticillium wilt (*Verticillium dahlia* Kleb.) race 1 was mapped on Linkage Group 9 (Hayes et al., 2011d). Iceberg breeding lines with resistance to this race of the pathogen were subsequently released (Hayes et al., 2011b). It is expected, however, that growing lettuce cultivars with the *Vr1* gene will increase the frequency of the race 2 strains of the pathogen and render the *Vr1* gene ineffective at controlling the disease.

Two cellular biology techniques were utilized to introgress disease resistance genes identified in wild *Lactuca* species into modern lettuce cultivars (Maisonneuve et al., 1995). In vitro rescue of immature embryos was successfully applied to produce hybrid plants between *L. sativa* and *L. virosa*. Protoplast fusion allowed production of somatic hybrids between *L. sativa* and *L. tatarica* L., which cannot be produced via conventional plant pollination.

No genetically engineered lettuce cultivars are grown in the United States. Molecular marker techniques are used in cultivar identification (fingerprinting), development of molecular linkage maps, and marker-assisted breeding. Selection of desirable genotypes with the aid of molecular markers is being used by both private and public breeding sectors for numerous traits, including resistance to downy mildew, corky root (*Rhizomonas suberifaciens*), *Lettuce mosaic virus*, lettuce dieback, lettuce aphid (*Nasonovia ribisnigri* Mosley), selection against late bolting, and male sterility (Simko, 2013). Genomic locations of genes or quantitative trait loci (QTLs) for resistance against Verticillium wilt, *Turnip mosaic virus*, downy mildew, powdery mildew (*Golovinomyces cichoracearum sensu stricto*), big-vein (*Mirafiori lettuce big-vein virus*), Fusarium wilt (*Fusarium oxysporum* f. sp. *lactucae*), and anthracnose [*Microdichium panattonianum* (Berl.) B. Sutton, Galea, & T.V. Price] have been mapped. Genes for resistance to lettuce drop [*Sclerotinia minor* Jagger, and *S. sclerotiorum* (Lib.) de Bary] and bacterial leaf spot are currently being mapped. Assays for marker-assisted selection are under development for several of these resistance genes, as well as for genes involved in postharvest decay of fresh-cut lettuce (Simko, 2013).

An ultrahigh density map of lettuce was recently developed that consists of approximately 14,000 molecular markers (Truco et al., 2013). Segregation of marker polymorphisms was analyzed using recombinant inbred lines generated by mating the iceberg cultivar Salinas with a *L. serriola* accession UC96US23. In addition, the whole genome of 'Salinas' has been sequenced and assembled into scaffolds (Reynes-Chin-Wo et al., 2012). The most comprehensive and recent information regarding the lettuce genome project is available at the Compositae Genome Project website (compgenomics.ucdavis.edu).

Spinach Production

Spinach is a cross-pollinated cool season crop ($2n = 2x = 12$) that is one of the most desirable dark green leafy vegetables because of its high content of β-carotene (provitamin A), lutein, folate, vitamin C, calcium, iron, phosphorous, and potassium. World spinach production exceeded 20 million tonnes in 2010, of which about 90% was produced in China (FAOSTAT, 2010). The United States was the second largest spinach producer, with about 2% of the production in the world. The development of baby leaf and spring mix in bagged salad makes it more convenient to consume this vegetable, and it has contributed to the rapid increase in demand and production of spinach in recent years (Fig. 4–10). About 426,000 t

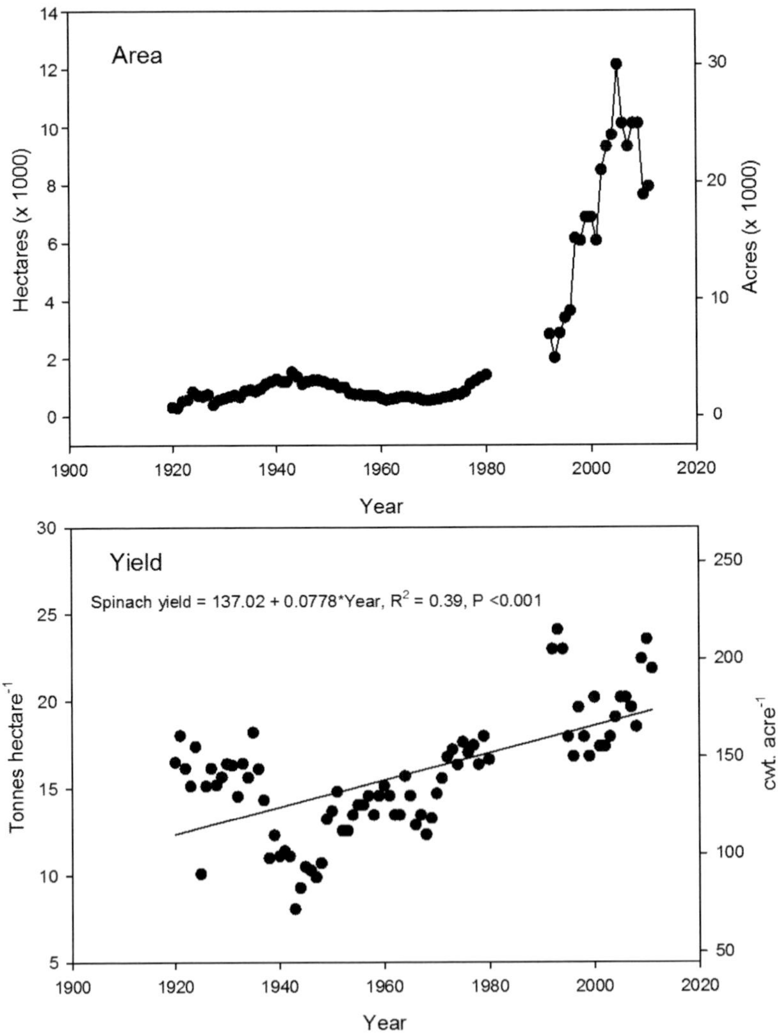

Fig. 4–10. Spinach harvested area and yield in the United States, based on data compiled from U.S. National Agricultural Statistics Service (USDA-NASS, 2012a).

(9,391,534 cwt) of spinach are produced on approximately 17,400 ha (43,000 ac) in the United States annually for fresh and processed markets, with a crop value of about US$270 million (USDA-NASS, 2011). Fresh market spinach accounted for 66 and 93% of the total production and value, respectively, from 77% of the total area of harvested spinach. Approximately 90% of the processing crop is for frozen products, whereas the remaining 10% is canned. Most U.S. spinach is harvested in California, which produces about 70% of the fresh market and 65% of the processing products in the country (USDA-NASS, 2011).

Spinach is normally harvested before the bolting stage to avoid the development of the seed stalk and to reduce yellowing, breakage, and other leaf deterioration. Most spinach is harvested mechanically. Ready-to-eat, bagged spinach is cut at the crown just above the root, elevated into bins or totes on a trailer, and transported to a packing facility where it is cooled, washed, sorted, and bagged. Some spinach crops are harvested manually, especially if the whole plant with leaves attached to the root is desirable for fresh markets. In such cases, leaves are cut below the crown with a sickle, and the harvested plants are tied into bunches of 8 to 12 plants. About 20% of fresh market spinach is packed in the field as bunched produce, whereas the other 80% is bulk harvested for bagged products. Processing spinach is cut above the growing point so that only leaves are harvested, allowing the plant to continue growing for a second and perhaps a third crop. Sometimes the first harvest of a field is used for fresh market, and subsequent harvests are for processing since the crops harvested later are less uniform. The area for organic spinach is steadily increasing and is routinely overplanted in anticipation of losing 20 to 30% of production to downy mildew caused by *Peronospora farinosa* f. sp. *spinaciae*.

Spinach yield has increased steadily in the United States over the years. According to historic data from the USDA-NASS (2012b), yield of spinach for fresh market and processing was 8.7, 8.4, and 12.1 t ha^{-1} and for processing only 12.3 t ha^{-1} (78, 75, 108, and 110 cwt ac^{-1}) in Washington State in 1926, 1936, 1946, and 1956, respectively. The yield of California fresh market spinach in the past 90 yr is shown in Fig. 4–10.

Spinach yield varies with locations and seasons. For example, fresh market spinach yields varied in several states from 1972 to 1974: Texas (fall and winter), 3.9 t ha^{-1} (35 cwt ac^{-1}); California (all year), 16.8 t ha^{-1} (150 cwt ac^{-1}); New Jersey (spring and fall), 7.4 t ha^{-1} (66 cwt ac^{-1}); Maryland and Virginia (spring), 3.8 t ha^{-1} (34 cwt ac^{-1}); and Colorado (summer) 7.1 t ha^{-1} (63 cwt ac^{-1}) (Sackett, 1975).

In the past decade, clipped, bagged, and ready-to-eat spinach products (including spring mix with spinach as a component) have become popular in the United States, which constitutes at least 80% of the fresh spinach industry. The bagged spinach is the dominant product produced in California and contains either very small young leaves ("baby spinach") or slightly older, medium-sized leaves ("teenage spinach"). Baby spinach can be harvested 21 to 40 DPP, while teenage spinach can be harvested in 26 to 50 DPP, depending on the season (Koike et al., 2011). The tender baby leaves have lower yield but receive a higher price than teenage crops. In contrast to fresh market spinach, processing spinach is allowed to grow for longer periods of time (48 to 90 DPP) so that leaf size and thickness as well as yield are significantly greater. In 1998, fresh market spinach yield was 14.7 t ha^{-1} (131 cwt ac^{-1}), and processing spinach yield was 21.4 t ha^{-1} (191 cwt ac^{-1}) in the United States. In 2011, the U.S. fresh market spinach yield reached 20.8 t ha^{-1} (186 cwt ac^{-1}), whereas processing spinach yielded 36.2 t ha^{-1} (323 cwt ac^{-1}) (USDA-NASS, 2011).

In 2010, the spinach yield in the United States averaged 24.5 t ha^{-1} (218 cwt ac^{-1}), as compared with the world average of 26.3 t ha^{-1} (235 cwt ac^{-1}), 17.4 t ha^{-1} (155 cwt ac^{-1}) in Europe, and 30.4 t ha^{-1} (271 cwt ac^{-1}) in China (FAOSTAT, 2010). In China, spinach is mostly consumed as stir-fried dishes or soup, and plants normally grow for longer periods of time, thus achieving larger leaf sizes and resulting in higher yield than is achieved by the prevailing salad crops in the United States. Since China accounts for about 90% of the world spinach production, the high yield of Chinese crops lifts the world average.

Spinach Genetic Gains

Spinach yield may be affected by the cultivars planted. Bradley et al. (1975) found that smooth-leaf 'Norgreen' out-yielded savoy-leaf 'Bounty' in spring trials. But the savoy or semi-savoy types, 'Hybrid 612,' 'Dixie Market,' and 'Chesapeake,' had higher yields than the popular smooth-leaf 'Resistoflay' and 'Viroflay'. This suggests that spinach yield may not be associated with leaf types. Thicker leaves contribute directly to spinach yield. Savoy leaves are harder to wash clean, so processors generally prefer semi-flat spinach for salad bags. Completely flat leaves are undesired as they tend to stack together and are thus not easily washed.

Gains in spinach yield have largely been attributed to the development of hybrid cultivars. Field observations indicated that heterosis occurs in spinach crosses. Thompson (1956) measured heterosis by comparing 12 unselected hybrids, 3 standard cultivars, and 8 inbred lines. The hybrids had 20.1 and 16.3% greater yields than the cultivars on average in two experiments. The hybrids averaged 22.6 and 23.5% higher yields over the inbreds. Mukhanova et al. (1979) found 30 to 40% higher yield in F_1 hybrids than in the parent plants.

USDA-ARS initiated breeding for hybrid spinach in 1947. The first spinach hybrid, 'Early Hybrid 7' was released in 1955 (Jones et al., 1956). Most spinach cultivars are dioecious, although some are monoecious. Spinach hybrids were initially produced using six to eight rows of female (seed) parents alternated with one to two rows of the male (pollinator) parent. Male plants in the dioecious seed rows were manually removed before pollen was shed. After pollination, the pollinator rows were rototilled or disced to facilitate the mechanical harvest of hybrid seeds and to prevent the mixing of hybrid and pollinator seeds. The pollinator was in some cases planted multiple times to ensure synchrony of flowering of the seed and pollinator plants. This method is still used by some seed companies today.

Although hybrid spinach seeds can be produced in good quantity and quality this way, it does have some drawbacks. It is a labor-intensive process because large numbers of male plants have to be removed by hand from the seed parent rows, and multiple trips through the fields are needed during the pollination season. Incomplete removal of the males could lead to self-pollination in the female rows and reduced percentage of hybrid seeds in the final product. All-female (gynoecious) lines have, therefore, been used as seed parents. This procedure significantly reduced the labor costs and self-pollination rate and improved the quality and profitability of hybrid seed production. Gynoecious lines are weakly monoecious (staminate and pistillate flowers on same plant) and are maintained by a process called sex reversion. When the female plants flower in isolation without male pollen, they often eventually produce a small amount of pollen to self- or sib-pollinate themselves. This modified hybrid seed production method has likely contributed to the increased use of hybrid spinach cultivars in the United States, Europe, Japan, and China. Hybrid spinach currently makes up 85 to 90% of production (area harvested) in the West and contributes to the yield increase worldwide (Morelock and Correll, 2008).

A potential improvement in spinach yield, vitamin A and C content, and delayed bolting may eventually come from the use of polyploids. Bragdø (1962) compared diploid, triploid, and tetraploid versions of cultivars Forste, Snitt, and de Gaudry. When harvested at the same age, diploids showed greater total yield, leaf blades, and dry matter total than polyploids, as the polyploids are slower in growth. When compared at the same stage of development, tetraploids produced greater yield than the diploids, with the triploids intermediate. These differences occurred two out of 3 yr, with the third year showing no significant differences.

Spinach Environmental and Management Factors

Zink (1956) studied growth rates of spring spinach in the Salinas Valley of California. Mean number of leaves increased linearly through the entire growing period in five trials. Leaf area and fresh weight increased slowly initially, then very rapidly. In a 56-d growth period in March and April, the plants acquired 68% of their fresh weight in the 21 d before harvest.

Spinach yield decreases with very early spring plantings, increases in mid-spring plantings, then is reduced again with still later planting dates. Both the cold weather at early plantings and the high temperatures and long daylengths at later plantings promote development (bolting or seedstalk formation) at the expense of growth (leaf formation) (Parlevliet, 1968).

Bradley et al. (1975) investigated the effects of various factors on yield in spring, fall, and overwinter trials. Yields of canned product increased in all seasons with a decrease in between-row spacing from 25 to 12.5 cm (10–5 in). Within-row spacing closer than 5 cm (2 in) showed no advantage. Nitrogen fertilizer generally increased yield, but the effect varied according to season, spacing, and method of application.

In California, spinach is grown on the standard raised beds described for lettuce. Elsewhere, seeds are planted in rows on flat ground or on slightly raised, very wide beds or are broadcast. Growers have gradually increased the plant density over the years to increase yield. In the 1980s, California growers planted fresh market spinach on 102-cm (40-in) wide beds in two rows near the bed shoulders. This method allowed easy cutting of the whole plant for bunching. Processing spinach was seeded with a multiple shoe planter, which sowed four rows on each side of the bed (Ryder, 1979a). Bunching and processing spinach are now sown on 102-cm (40-in) beds with 6 to 9 seed rows or on 203-cm (80-in) beds with 12 to 21 seed lines, with 8.6 to 9.9 million seeds per hectare (3.5–4.0 million seeds per acre) for baby leaf production (Koike et al., 2011). In general, clipped baby and teenage spinach is planted only on 203-cm (80-in) beds with 21 to 48 seed lines (Koike et al., 2011).

Spinach Pests

High seeding rates result in dense crop canopies that prolong leaf wetness and high humidity and favor development of various foliar diseases. Downy mildew has been a problem in spinach for decades. Currently available downy mildew resistance is not durable; 10 new downy mildew races appeared in the last 15 yr that overcome all known resistance genes. Other major diseases of spinach include white rust [*Albugo occidentalis* (G.W. Wilson) Sacc. & Trotter], Fusarium wilt [*Fusarium oxysporum* f. sp. *spinaciae* (Sherb.) W.C. Snyder & H.N. Hansen], and a host of virus diseases (Correll et al., 1994). There are several emerging diseases that pose new threats to spinach production, such as Stemphylium leafspot (*Stemphylium botryosum* f. sp. *spinacia*) (Mou et al., 2008), *Impatiens necrotic spot virus* (Liu et al., 2009), and *Beet necrotic yellow vein virus* (Mou et al., 2012). Important insect pests of spinach include leaf miners (*Liriomyza* spp.), green peach aphid (*Myzus persicae* Sulzer), loopers worms (*Trichoplusia ni* Hübner), beet armyworm (*Spodoptera exigua* Hübner), whiteflies (*Bemisia argentifolii* Bellows & Perring), thrips (*Frankliniella occidentalis* Pergande), and mites (*Rhizoglyphus* spp.) (Koike et al., 2011). Disease and insect problems can develop quickly and reduce yield and quality and even lead to total loss of a field. Besides management measures, greater efforts are needed to find effective and durable genetic resistances against these diseases and insects.

Sustainable Lettuce and Spinach Production: Adaptability to Climate Change

Lettuce and spinach production are vulnerable to abiotic stresses brought on by climate change. Adapting leafy vegetable production to future conditions is essential to meet the need of a growing population and increasing demand for these products. Current forecasts predict that global climate change over the next several decades will take place irrespective of any remedial action taken today (Karl and Trenberth, 2003). Summer temperatures in California are projected to be 2 to 8°C (3.6–14.4°F) higher by 2049 and will also be more variable within a season, which may cause heat-induced physiological defects to become more frequent with respect to currently available lettuce cultivars (Cavagnaro et al., 2006). Breeding cultivars that can cope with heat and other climate extremes may well be the single most important first step towards adaptation to the changing environment. In addition, leafy vegetable production is losing acreage to housing and industrial development and is facing competition from other crops such as strawberry and, of course, increased land costs in the major production areas of coastal California. Heat-tolerant varieties would be adapted to hotter conditions of inland regions of California and could expand the growing season in hot, but lower land cost areas such as the Central Valley of California or Imperial Valley, assuming that there is adequate water supply in these areas. Much of the baby leaf spinach is currently grown in the warmer parts of the Salinas Valley to lessen leafminer and downy mildew pressure. Higher temperatures in the future may cause the production to move farther north in this major production district. Climate changes lead to less predictable weather, which will likely negatively impact production schedules for these crops that are planted nearly every day of the year.

Production of leafy vegetables at high temperature ranges results in yield and quality losses because of thermo-inhibition of germination, premature bolting, and internal quality defects such as tipburn. Most of these defects are either caused, or exacerbated, by high temperatures, plus other environmental factors that influence their severity and incidence. The USDA breeding program in Salinas, CA is screening lettuce and spinach germplasm collections to find cultivars that are resistant to these constraints to mitigate the adverse effects of these high temperature induced problems. Breeding for higher levels of resistance to these disorders or combining these resistances into individual cultivars may be needed to develop cultivars adapted to warmer and more variable temperatures. Understanding the environmental conditions needed to reproduce these disorders and understanding the genetics of resistance are crucial for achieving this goal.

Poor seed germination due to thermo-inhibition is a major problem associated with both lettuce and spinach production. Depending on genotype, lettuce and spinach seed germination is inhibited at temperatures higher than 28 to 32°C (82–90°F) (Argyris et al., 2008; Sackett, 1975). This inhibition of seed germination at high temperatures causes a reduction in field emergence, stand establishment, and yield. Seed priming (controlled hydration followed by drying) in osmotic solutions may improve germination rates and seedling emergence in high temperature production scenarios. This is a viable option for lettuce but has not proved logistically feasible for spinach given the bulkiness of the seed, the high seeding densities, and the corkiness of the outer portions of the seed. Some accessions of *L. serriola*, a wild relative of lettuce, have thermo-inhibitory temperature limits up to 10°C (18°F) higher than cultivated lettuce (Argyris et al., 2005). Quantitative trait loci associated with seed germination, early seedling growth, and seed priming were detected on the molecular linkage map of lettuce (Argyris et al., 2005; Schwember and Bradford, 2010). Linkage mapping revealed that the genetic variation for thermo-inhibition is associated with expression of abscisic acid (ABA) (Argyris et al., 2008). A major QTL (*Htg6.1*) accounted for ~25% of the total phenotypic variance in high-temperature germination and was mapped to ~1 centimorgan that contained a gene encoding a key regulated enzyme in ABA biosynthesis, 9-*cis*-epoxycarotenoid dioxygenase4 (NCED4). It was demonstrated that NCED4 expression is required for thermo-inhibition of lettuce seed germination (Huo et al., 2013). The research results indicate the possibility of developing lettuce cultivars with higher tolerance to thermo-inhibition. Cultivars resistant to thermo-inhibition may reduce the need and cost of seed priming.

Cultivated lettuce is an annual plant that produces a vegetative rosette. Stem elongation (bolting) begins at a time dependent on environment and plant genotype, and culminates with flowering and seed set. The time from planting to induction of bolting in lettuce is considered a quantitative, long-day character of lettuce, and extensive genetic variation and genotype × environment interactions are known to influence bolting (Ryder, 1999). Photoperiod is a critical factor influencing bolting, and temperature appears to mediate or accelerate the interaction (Waycott, 1995), possibly through the accumulation of gibberellins (Fukuda et al., 2009). Bolting before harvest renders plants unmarketable, thus resulting in fewer harvested plants and reduced yield. Premature bolting may occur any time of the year but is most common in summer plantings in the Salinas Valley and in August and September plantings in the Central Valley and low desert, respectively. A specific group of interrelated, slow-bolting cultivars are grown in the low desert September plantings, despite the fact that their overall quality may be inferior to other cultivars adapted to summer production in Salinas Valley.

The genetics of early bolting has been discussed by several authors. The *T* allele confers early bolting under long days but does not act by simple dominance alone (Bremer, 1931; Bremer and Grana, 1935; Waycott, 1995). Six early flowering (*Ef*) genes shorten the time to flowering by 7 to 80 d (Ryder, 1983, 1988; Ryder and

Milligan, 2005). The effects of each gene may be dependent on other *Ef* genes. The genes *Ef-1*, *Ef-3*, *Ef-5*, and *Ef-6* express fast bolting that is dominant to slow bolting, whereas the *Ef-2* and *Ef-4* genes confer dominant slow bolting (Ryder and Milligan, 2005). Quantitative trait loci for early bolting or flowering were reported on Linkage Groups 2 and 7 (Hartman et al., 2012). Fukuda et al. (2011) identified a putative lettuce homolog of *Flowering Locus T* that has greater numbers of transcripts in lettuce leaves when grown at high temperatures.

Tipburn is a physiological defect manifested as necrotic or "burned" areas on leaf margins (Davis et al., 1997) in the spring (Imperial Valley and Yuma) and summer (coastal districts of California). Consumers consider the tipburn objectionable, and the damage may predispose lettuce to head rot. Tipburn occurs on all lettuce types, typically on inner leaves within enclosed heads where the damage is not readily observed. Lettuce growers and packing companies experience severe economic damage when tipburn occurs. Some companies reject entire fields when tipburn incidence is >5%. Lettuce processed into salad requires even higher stringencies. Tipburn occurs during a narrow period of time immediately preceding harvest.

The occurrence of tipburn is correlated with environmental conditions that either decrease transpiration from the leaf surface or promote rapid growth, both of which impair the plants ability to supply sufficient calcium to growing tissues, which then results in tipburn (Barta and Tibbitts, 2000; Thibodeau and Minotti, 1969). These conditions include low air movement (Goto and Takakura, 1992), artificial or natural enclosure of the leaves (Barta and Tibbitts, 1986), high relative humidity during the daytime (Collier and Tibbitts, 1982), high temperatures (Cox et al., 1976; Misaghi and Grogan, 1978; Yanagi et al., 1983), high radiation levels (Tibbitts and Rao, 1968), long photoperiods (Koontz and Prince, 1986), and high fertilization levels. Growers cannot control all of the environmental factors involved in tipburn, and breeding for tipburn resistance is the best long-term solution. Most breeding for tipburn resistance is conducted by selective culling of plants, families, and lines with symptoms. Using this approach, iceberg lettuce cultivars have been bred with levels of resistance that are adequate for most planting slots (Ryder and Waycott, 1998), although no cultivars are completely resistant (Nagata and Stratton, 1994). Genetic improvements for tipburn resistance are needed most in the romaine gene pool (Jenni and Hayes, 2010). Screening cultivars, breeding lines, and segregating families for tipburn resistance can be performed in the field or in controlled environments, but the results are not always consistent (Cox et al., 1976; Nagata and Stratton, 1994). Tipburn incidence is subject to genotype × environment interaction (Jenni and Hayes, 2010), and extensive field testing is needed to fully characterize resistance in cultivars or breeding lines.

Little is known regarding the genetics of tipburn resistance. While cultivars exhibited extensive genetic variation for the incidence of the disorder, plant architecture is a major factor influencing the prevalence of the disorder, with open top and slower maturing cultivars expressing lower tipburn incidence (Jenni and Hayes, 2010). The dependence of tipburn incidence on plant architecture makes comparison of resistance across market types virtually impossible. Resistance genes that are independent of morphology are most valuable, as they could improve resistance in all market types. The high level of resistance in iceberg cultivars such as Salinas (Ryder, 1979b) and Tiber (Ryder and Waycott, 1998) have been proposed as sources of resistance to breed resistance in other types (Jenni and Hayes, 2010). Six QTLs for tipburn resistance were discovered in two intraspecific recombinant inbred line populations tested in Arizona, and four of the loci cosegregated with head closure, core height, or head maturity (Hayes and Simko, 2010).

Global climate change may affect agriculture more through its impact on water availability than on temperature. All lettuce and spinach production in California and Arizona are surface irrigated, but future water availability for growers is expected to decrease. Water is a precious resource and has become increasingly scarce because of population growth, environmental needs, and frequent drought. Climate change has resulted in less predictable precipitation, less snow pack, and earlier snowmelt, leading to periods of water shortages in California (Weare, 2009). Court orders limit the pumping of Northern California water to farms in San Joaquin Valley, severely restricting the leafy vegetable production. Coastal Monterey County faces water shortage and seawater intrusion and has planned the construction of several desalination plants for residential uses. Conversely, ever tighter regulations on farm runoffs call for crop varieties with reduced irrigation requirement and better water use efficiency (WUE). Cultivars with improved WUE will tolerate water deficit stress to the degree that product quality and yield are maintained, significantly reducing growers' irrigation costs. As the costs of land, labor, fuel, fertilizer, pesticides, seeds, packing material, cooling, transportation, and overhead including food safety continue to rise, it is essential to reduce production costs of leafy vegetables to benefit producers as well as consumers. With the limited water supply, decreases in crop water requirements could result in proportional increases in lettuce and spinach production area. In addition to the conservation of water resources, the decreased irrigation needs of drought-tolerant lettuce and spinach may reduce farm runoff and pesticides leaching into groundwater, helping growers with regulatory compliance and alleviating the pressure on the environment.

Research to increase yield in production systems that use limited irrigation is a priority for lettuce breeding but is still in its infancy. Cultivated lettuce is reliant on frequent irrigation and high input of nutrients because of its short taproot and prolific lateral branches in the upper layers of the soil (Jackson, 1995). Contrary to cultivated lettuce, *L. serriola,* its wild progenitor, develops a long taproot and relies on water from deep soil zones during periods of surface soil drought (Jackson, 1995). The wild *L. serriola* is, therefore, a potential source of agriculturally important alleles to optimize water and fertilizer use efficiency (Johnson et al., 2000) when integrated with improved irrigation and fertilizer practices. A single QTL was identified in an *L. sativa* × *L. serriola* population controlling increased taproot length per gram of plant biomass. This QTL was independent of QTL for shoot biomass and the percentage of biomass allocated to roots (Johnson et al., 2000). These results suggest that breeding for a reallocation of root biomass may develop cultivars that can access nutrients from deeper soil depths. Selection for increased root biomass may not be advantageous, as gains are likely to come at the expense of shoot growth and therefore yield (Johnson et al., 2000).

Weed problems will become more prominent with global warming, as weeds are expected to adapt to the changing environment better than crops (Clements and DiTommaso, 2012). Weeds in leafy vegetables increase production costs and reduce yields and quality. Weed cover of 25% reduces lettuce yield by 20 to 40%, and >25% weed cover may cause complete yield loss. Baby lettuce, a primary component of spring mix, and spinach have zero tolerance for weed contamination, yet high plant densities prevent mechanical cultivation. Herbicides are essential for profitability, but few herbicides are registered for lettuce and spinach in California, and they are subject to cancellation by manufacturers because of their relatively small acreage and low sales volumes, marginal profitability, and high regulatory costs. The lettuce industry has depended on pronamide (Kerb) as the major herbicide. The U.S. Environmental Protection Agency stopped the use of Kerb on leaf lettuce in 2009 because of crop group reclassifications. There is no certainty that Kerb will be reregistered on leaf lettuce, as the herbicide has been classified as a B2 carcinogen and found to leach into groundwater in Virginia (Bruggeman et al., 1995). Herbicide producer Helm Agro suspended in 2008 to 2009 the production of RoNeet, the primary herbicide for spinach. Bayer Co. recently stopped production of Spin-Aid, an herbicide for processing spinach. The prospect of losing effective herbicides is a constant threat for California leafy greens industry. A decade of searching for new lettuce and spinach herbicides has failed, as the crops are very sensitive to herbicide damage. Kerb does not effectively control some weeds, for example, sowthistle (*Sonchus* spp.) and groundsel (*Senecio* spp.), and there is no postemergence broadleaf herbicide available for these crops. The loss of Kerb or RoNeet with no adequate herbicide replacement may have a considerable adverse effect on the industry. The loss of Kerb for leaf lettuce has significantly increased production costs by >US$371 per hectare (>US$150 per acre) because of hand weeding (Haar and Fennimore, 2003). Increased hand weeding costs often mark the difference between profit and loss for leafy greens producers.

A simple, reliable, and sustainable weed management system might be developed through the use of herbicide-tolerant lettuce and spinach cultivars. A herbicide-tolerant crop is an attractive option, since a broad-spectrum herbicide can remove weeds without injuring the tolerant crop. Herbicide tolerance is a proven approach to weed control as 73% of corn (*Zea mays* L.), 93% of soybean [*Glycine max* (L.) Merr.], and 80% of cotton (*Gossypium hirsutum* L.) crops in the United States were planted with herbicide-tolerant varieties in 2012 (USDA-ERS, 2012). No transgenes are currently used in lettuce and spinach production. On the basis of feedback from the industry, lettuce and spinach cultivars developed by conventional breeding, the same method used successfully in the creation of herbicide-tolerant sunflower (*Helianthus annuus* L.), rice (*Oryza sativa* L.), and wheat (*Triticum aestivum* L.), will be more easily and widely accepted by growers, the industry, and consumers than transgenic crops. Such a breeding program for herbicide-tolerant lettuce and spinach is currently underway in our project by screening mutants in these crops (Fennimore et al., 2012). This project may provide efficient and much more effective weed control tools than growers have now and improve the profitability and sustainability of lettuce and spinach production.

Acknowledgments

The authors thank K. Nolte for providing photographs of lettuce and spinach production in Arizona, K.V. Subbarao for the figure of the lettuce production areas, and A. Atallah for help with editing figures and the revised version of the manuscript.

References

Argyris, J., P. Dahal, E. Hayashi, D.W. Still, and K.J. Bradford. 2008. Genetic variation for lettuce seed thermoinhibition is associated with temperature-sensitive expression of abscisic acid, gibberellin, and ethylene biosynthesis, metabolism, and response genes. Plant Physiol. 148:926–947. doi:10.1104/pp.108.125807

Argyris, J., M.J. Truco, O. Ochoa, S.J. Knapp, D.W. Still, G.M. Lenssen, J.W. Schut, R.W. Michelmore, and K.J. Bradford. 2005. Quantitative trait loci associated with seed and seedling traits in *Lactuca*. Theor. Appl. Genet. 111:1365–1376. doi:10.1007/s00122-005-0066-4

Bakker, M.I., W.J. Baas, D.T.H.M. Sijm, and C. Kollöffel. 1998. Leaf wax of *Lactuca Sativa* and *Plantago major.* Phytochemistry 47:1489–1493. doi:10.1016/S0031-9422(97)01084-4

Barrons, K.C., and T.W. Whitaker. 1943. Great Lakes, a new summer head lettuce adapted to summer conditions. Mich. Agric. Exp. Stn. 25:1–3.

Barta, D.J., and T.W. Tibbitts. 1986. Effects of artificial enclosure of young lettuce leaves on tipburn incidence and leaf calcium concentration. J. Am. Soc. Hortic. Sci. 111:413–416.

Barta, D.J., and T.W. Tibbitts. 2000. Calcium localization and tipburn development in lettuce leaves during early enlargement. J. Am. Soc. Hortic. Sci. 125:294–298.

Bradley, G.A., W.A. Sistrunk, E.C. Baker, and J.N. Cash. 1975. Effect of plant spacing, nitrogen, and cultivar on spinach (*Spinacia oleracea* L.) yield and quality. J. Am. Soc. Hortic. Sci. 100:45–48.

Bragdø, M. 1962. Breeding of polyploidy spinach. Euphytica 11:143–148. doi:10.1007/BF00033786

Bremer, A.H. 1931. Einfluss der tageslänge auf die wachstumsphasen des salats. Genetische Untersuchungen I. Gartenbauwissenschaft 4:469–483.

Bremer, A.H., and J. Grana. 1935. Genetische Untersuchungen mit Salat II. Gartenbauwissenschaft 9:231–245.

Brown, P.R., and R.W. Michelmore. 1988. The genetics of corky root resistance in lettuce. Phytopathology 78:1145–1150. doi:10.1094/Phyto-78-1145

Bruggeman, A.C., S. Mostaghimi, G.I. Holtzman, V.O. Shanholtz, S. Shukla, and B.B. Ross. 1995. Monitoring pesticides and nitrate in Virginia's groundwater–a pilot study. Trans. ASAE 38:797–807. doi:10.13031/2013.27894

Bull, C.T., P.H. Goldman, R.J. Hayes, L.V. Madden, S.T. Koike, and E.J. Ryder. 2007. Genetic diversity of lettuce for resistance to bacterial leaf spot caused by *Xanthomonas campestris* pv. *vitians*. Plant Health Prog. 10.1094/PHP-2007–0917–02-RS.

Cavagnaro, T., L. Jackson, and K. Scow. 2006. Climate change: Challenges and solutions for California agricultural landscapes. California Energy Commission, Sacramento, CA.

Clements, D.R., and A. DiTommaso. 2012. Predicting weed invasion in Canada under climate change: Evaluating evolutionary potential. Can. J. Plant Sci. 92:1013–1020. doi:10.4141/cjps2011-280

Collier, G.F., and T.W. Tibbitts. 1982. Tipburn of lettuce. J. Am. Soc. Hortic. Sci. 109(2):128–131.

Correll, J.C., T.E. Morelock, M.C. Black, S.T. Koike, L.P. Brandenberger, and F.J. Dainello. 1994. Economically important diseases of spinach. Plant Dis. 78:653–660. doi:10.1094/PD-78-0653

County of Monterey Agricultural Commissioner. 2011. Monterey County Crop Report, County of Monterey Agricultural Commissioner. http://ag.co.monterey.ca.us/assets/resources/assets/252/cropreport_2011.pdf (accessed 24 July 2012).

Cox, E.F., J.M.T. McKee, and A.S. Dearman. 1976. The effect of growth rate on tipburn occurrence in lettuce. J. Hortic. Sci. 51:297–309.

Davis, R.M., K.V. Subbarao, R.N. Raid, and E.A. Kurtz. 1997. Compendium on lettuce diseases. American Phytopathological Society, St. Paul, MN.

FAOSTAT. 2010. Production. Food and Agricultural Organization of the United Nations. http://faostat.fao.org/site/339/default.aspx (accessed 19 Dec. 2012).

FAOSTAT. 2013. Crops data for 2011. Food and Agricultural Organization of the United Nations. http://faostat3.fao.org (accessed 21 Feb. 2013).

Fennimore, S.A., R. Smith, and B. Mou. 2012. Weed management systems for leafy greens. Annual report. California Leafy Greens Research Program, Salinas, CA.

Fukuda, M., S. Matsuo, K. Kikuchi, Y. Kawazu, R. Fujiyam, and I. Honda. 2011. Isolation and functional characterization of the *FLOWERING LOCUS T* homolog, the *LsFT* gene, in lettuce. J. Plant Physiol. 168:1602–1607. doi:10.1016/j.jplph.2011.02.004

Fukuda, M., S. Matsuo, K. Kikuchi, W. Mitsuhashi, T. Toyomasu, and I. Honda. 2009. The endogenous level of GA_1 is upregulated by high temperature during stem elongation in lettuce through *LsGA3ox1* expression. J. Plant Physiol. 166:2077–2084. doi:10.1016/j.jplph.2009.06.003

German-Retana, S., J. Walter, and O. Le Gall. 2008. *Lettuce mosaic virus*: From pathogen diversity to host interactors. Mol. Plant Pathol. 9:127–136. doi:10.1111/j.1364-3703.2007.00451.x

Goto, E., and T. Takakura. 1992. Prevention of lettuce tipburn by supplying air to inner leaves. Trans. Am. Soc. Agric. Eng. 35:641–645.

Grube, R.C., E.J. Ryder, S.T. Koike, J.D. McCreight, and W.M. Wintermantel. 2003. Breeding for resistance to new and emerging lettuce diseases in California. In: Th.J.L. van Hintum et al., editors, Proceedings of the EUCARPIA Meeting on Leafy Vegetables Genetics and Breeding, Noordwijkerhout, The Netherlands. 19–21 March. Centre for Genetic Resources, Wageningen, The Netherlands. p. 25–30.

Grube, R.C., W.M. Wintermantel, P. Hand, R. Aburomia, D.A.C. Pink, and E.J. Ryder. 2005. Genetic analysis and mapping of resistance to lettuce dieback: A soilborne disease caused by tombusviruses. Theor. Appl. Genet. 110:259–268. doi:10.1007/s00122-004-1825-3

Haar, M.J., and S.A. Fennimore. 2003. Evaluation of integrated practices for common purslane management in lettuce. Weed Technol. 17:229–233. doi:10.1614/0890-037X(2003)017[0229:EOIPFC]2.0.CO;2

Hartman, Y., D.A.P. Hooftman, M.E. Schranz, and P.H. van Tienderen. 2012. QTL analysis reveals the genetics architecture of domestication traits in crisphead lettuce. Genet. Resour. Crop Evol. 10.1007/s10722–012–9937–0.

Hayes, R.J., W. BoMing, and K.V. Subbarao. 2011a. A single recessive gene conferring short leaves in romaine × Latin type lettuce (*Lactuca sativa* L.) crosses, and its effect on plant morphology and resistance to lettuce drop caused by *Sclerotinia minor* Jagger. Plant Breed. 130:388–393. doi:10.1111/j.1439-0523.2010.01822.x

Hayes, R.J., K. Maruthachalam, G.E. Vallad, S.J. Klosterman, I. Simko, Y. Luo, and K.V. Subbarao. 2011b. Iceberg lettuce breeding lines with resistance to Verticillium wilt caused by race 1 isolates of *Verticillium dahliae*. HortScience 46:501–504.

Hayes, R.J., K. Maruthachalam, G.E. Vallad, S.J. Klosterman, and K.V. Subbarao. 2011c. Selection for resistance to Verticillium wilt caused by race 2 isolates of *Verticillium dahliae* in accessions of lettuce (*Lactuca sativa* L.). HortScience 46:201–206.

Hayes, R.J., L.K. McHale, G.E. Vallad, M.J. Truco, R.W. Michelmore, S.J. Klosterman, K. Maruthachalam, and K.V. Subbarao. 2011d. The inheritance of resistance to Verticillium wilt caused by race 1 isolates of *Verticillium dahliae* in the lettuce cultivar La Brillante. Theor. Appl. Genet. 123:509–517. doi:10.1007/s00122-011-1603-y

Hayes, R.J., E.J. Ryder, and C.T. Bull. 2008a. Notice of release of six F6:8 and one F4:5 iceberg breeding lines of lettuce germplasm with resistance to bacterial leaf spot caused by *Xanthomonas campestris* pv. *vitians*. USDA Agricultural Research Service, Washington, DC.

Hayes, R.J., E.J. Ryder, and W.M. Wintermantel. 2008b. Genetic variation for big-vein symptom expression and resistance to *Mirafori lettuce big-vein virus* in *Lactuca virosa* L., a wild relative of cultivated lettuce. Euphytica 164:493–500. doi:10.1007/s10681-008-9738-x

Hayes, R.J., and I. Simko. 2010. Spring harvested lettuce tipburn resistance research. Arizona Department of Agriculture. http://www.azda.gov/act/Spring_Harvested_Lettuce.pdf (accessed 13 Mar. 2013).

Hayes, R.J., G.E. Vallad, Q.-M. Qin, R.C. Grube, and K.V. Subbarao. 2007. Variation for resistance to Verticillium wilt in lettuce (*Lactuca sativa* L.). Plant Dis. 91:439–445. doi:10.1094/PDIS-91-4-0439

Hayes, R.J., B.M. Wu, B.M. Pryor, P. Chitrampalam, and K.V. Subbarao. 2010. Assessment of resistance in lettuce (*Lactuca sativa* L.) to mycelial and ascospore infection by *Sclerotinia minor* Jagger and *S. sclerotiorum* (Lib.) de Bary. HortScience 45:333–341.

Huo, H., P. Dahal, K. Kunusoth, C.M. McCallum, and K.J. Bradford. 2013. Expression of 9-cis-EPOXY-CAROTENOID DIOXYGENASE4 is essential for thermoinhibition of lettuce seed germination but not for seed development or stress tolerance. Plant Cell 28:884–900.

Jackson, L.E. 1995. Root architecture in cultivated and wild lettuce (*Lactuca* spp.). Plant Cell Environ. 18:885–894. doi:10.1111/j.1365-3040.1995.tb00597.x

Jagger, I.C. 1940. Brown blight of lettuce. Phytopathology 30:53–64.

Jenni, S. 2005. Rib discoloration: A physiological disorder induced by heat stress in crisphead lettuce. HortScience 40:2031–2035.

Jenni, S., and R.J. Hayes. 2010. Genetic variation, genotype × environment interaction, and selection for tipburn resistance in lettuce in multi-environments. Euphytica 171:427–439. doi:10.1007/s10681-009-0075-5

Johnson, W.C., L.E. Jackson, O. Ochoa, R. van Wijk, J. Peleman, D.A. St. Clair, and R.W. Michelmore. 2000. Lettuce, a shallow-rooted crop, and *Lactuca serriola*, its wild progenitor, differ at QTL determining root architecture and deep soil water exploitation. Theor. Appl. Genet. 101:1066–1073. doi:10.1007/s001220051581

Jones, H.A., D.M. McLean, and B.A. Perry. 1956. Breeding hybrid spinach resistant to mosaic and downy mildew. Proc. Am. Soc. Hortic. 68:304–308.

Karl, T.R., and K.E. Trenberth. 2003. Modern global climate change. Science 302:1719–1723. doi:10.1126/science.1090228

Koike, S.T., M. Cahn, M. Cantwell, S. Fennimore, M. Lestrange, E. Natwick, R.F. Smith, and E. Takele. 2011. Spinach production in California. Publication 7212. University of California, Agriculture and Natural Resources. http://anrcatalog.ucdavis.edu/VegetableCropProductioninCalifornia/7212.aspx (accessed 3 Jan. 2013).

Koontz, H.V., and R.P. Prince. 1986. Effect of 16 and 24 hours daily radiation (light) on lettuce growth. HortScience 21:123–124.

Lebeda, A., B. Mieslerová, I. Petrželová, P. Korbelová, and E. Česneková. 2012. Patterns of virulence variation in the interaction between *Lactuca* spp. and lettuce powdery mildew (*Golovinomyces cichoracearum*). Fungal Ecol. 5:670–682. doi:10.1016/j.funeco.2012.03.005

Liu, H.Y., J.L. Sears, and B. Mou. 2009. Spinach (*Spinacia oleracea*) is a new natural host of *Impatiens necrotic spot virus* in California. Plant Dis. 93:673. doi:10.1094/PDIS-93-6-0673C

Maisonneuve, B., M.C. Chupeau, Y. Bellec, and Y. Chupeau. 1995. Sexual and somatic hybridization in the genus *Lactuca*. Euphytica 85:281–285. doi:10.1007/BF00023957

Matheron, M.E., J.D. McCreight, B.R. Tickes, and M. Porchas. 2005. Effect of planting date, cultivar, and stage of plant development on incidence of Fusarium wilt of lettuce in desert production fields. Plant Dis. 89:565–570. doi:10.1094/PD-89-0565

McCreight, J.D. 2008. Potential sources of genetic resistance in *Lactuca* spp. to the lettuce aphid, *Nasanovia ribisnigri* (Mosely) (Homoptera: *Aphididae*). HortScience 43:1355–1358.

McHale, L.K., M.J. Truco, A. Kozik, T. Wroblewski, O.E. Ochoa, K.A. Lahre, S.J. Knapp, and R.W. Michelmore. 2009. The genomic architecture of disease resistance in lettuce. Theor. Appl. Genet. 118:565–580. doi:10.1007/s00122-008-0921-1

Mikel, M.A. 2007. Genealogy of contemporary North American lettuce. HortScience 42:489–493.

Mikel, M.A. 2013. Genetic composition of contemporary proprietary U.S. lettuce (*Lactuca sativa* L.) cultivars. Genet. Resour. Crop Evol. 60:89–96. doi:10.1007/s10722-012-9818-6

Misaghi, I.J., and R.G. Grogan. 1978. Effect of temperature on tipburn development in head lettuce. Phytopathology 68:1738–1743. doi:10.1094/Phyto-68-1738

Morelock, T.E., and J.C. Correll. 2008. Spinach. In: J. Prohens and F. Nuez, editors, Handbook of plant breeding, Vegetables I, Asteraceae, Brassicaceae, Chenopodicaceae, and Cucurbitaceae. Springer, New York. p. 189–218.

Mou, B., S. Koike, and L. du Toit. 2008. Screening for resistance to leaf spot diseases of spinach. HortScience 43:1706–1710.

Mou, B., and Y.B. Liu. 2004. Host plant resistance to leafminers in lettuce. J. Am. Soc. Hortic. Sci. 129:383–388.

Mou, B., K. Richardson, S. Benzen, and H.Y. Liu. 2012. Effects of *Beet necrotic yellow vein virus* in spinach cultivars. Plant Dis. 96:618–622. doi:10.1094/PDIS-09-11-0748

Mukhanova, Y.I., K.A. Trebukhina, and O.I. Kozlova. 1979. The production of heterotic hybrids of spinach. Ref. Zh. 8:310.

Nagata, R.T., and M.L. Stratton. 1994. Development of an objective test for tipburn evaluation. Proc. Fla. State Hortic. Soc. 107:99–101.

Nolte, K. 2010. Baby leaf or spring mix. University of Arizona. http://cals.arizona.edu/fps/sites/cals.arizona.edu.fps/files/cotw/Baby Leaf or Spring Mix.pdf (accessed 21 Feb. 2013).

Obermeier, C., J.L. Sears, H.Y. Liu, K.O. Schlueter, E.J. Ryder, J.E. Duffus, S.T. Koike, and G.C. Wisler. 2001. Characterization of distinct tombusviruses that cause diseases of lettuce and tomato in the western United States. Phytopathology 91:797–806. doi:10.1094/PHYTO.2001.91.8.797

O'Malley, P.J., and R.W. Hartmann. 1989. Resistance to *Tomato spotted wilt virus* in lettuce. HortScience 24:360–362.

Parlevliet, J.E. 1968. The influence of sowing date and germination temperature on yield of spinach and chervil. Neth. J. Agric. Sci. 16:53–57.

Reynes-Chin-Wo, S., Z. Wang, C. Beitel, A. Kozik, S. Chi, W. Chen, M.J. Truco, X. Xu, L. Froenicke, D. Lavelle, B. Yang, I. Korf, J. Wang, and R. Michelmore. 2012. Lettuce genome update. Paper presented at: Plant & Animal Genomes XX Conference. Town & Country Convention Center, San Diego, CA 12–16 Jan. Abstract W188.

Robinson, R.W., J.D. McCreight, and E.J. Ryder. 1983. The genes of lettuce and closely related species. In: J. Janick, editor, Plant breeding reviews. Vol. 1. John Wiley & Sons, Hoboken, NJ. p. 267–293.

Rodenburg. 1960. Varieties of lettuce: An international monograph. Varietal Descr. 3, Inst. voor de Verdeling van Tuinbouwgewassen, Wageningen, the Netherlands.

Ryder, E.J. 1979a. Leafy salad vegetables. AVI Publ. Co., Westpoint, CT.

Ryder, E.J. 1979b. 'Salinas' lettuce. HortScience 14:283–284.

Ryder, E.J. 1983. Inheritance, linkage, and gene interaction studies in lettuce. J. Am. Soc. Hortic. Sci. 108:985–991.

Ryder, E.J. 1988. Early flowering time in lettuce as influenced by a second flowering time gene and seasonal variation. J. Am. Soc. Hortic. Sci. 113:456–460.

Ryder, E.J. 1999. Lettuce, endive and chicory. CAB Int., New York.

Ryder, E.J., and D.C. Milligan. 2005. Additional genes controlling flowering time in *Lactuca sativa* and *L. serriola*. J. Am. Soc. Hortic. Sci. 130:448–453.

Ryder, E.J., and B.J. Robinson. 1995. Big-vein resistance in lettuce: Identifying, selecting, and testing resistance cultivars and breeding lines. J. Am. Soc. Hortic. Sci. 120:741–746.

Ryder, E.J., and W. Waycott. 1998. Crisphead lettuce resistant to tipburn: Cultivar Tiber and eight breeding lines. HortScience 33:903–904.

Sackett, C. 1975. Spinach. Fruit and vegetable facts and pointers. United Fresh Fruit and Vegetable Association, Washington, DC.

Schwember, A.R., and K.J. Bradford. 2010. A genetic locus and gene expression patterns associated with the priming effect on lettuce seed germination at elevated temperatures. Plant Mol. Biol. 73:105–118. doi:10.1007/s11103-009-9591-x

Scott, J.C., S.C. Kirkpatrick, and T.R. Gordon. 2010. Variation in susceptibility of lettuce cultivars to fusarium wilt caused by *Fusarium oxysporum* f.sp. *lactucae*. Plant Pathol. 59:139–146. doi:10.1111/j.1365-3059.2009.02179.x

Simko, I. 2009. Development of EST-SSR markers for the study of population structure in lettuce (*Lactuca sativa* L.). J. Hered. 100:256–262. doi:10.1093/jhered/esn072

Simko, I. 2013. Marker-assisted selection for disease resistance in lettuce. In: R.K. Varshney and R. Tuberosa, editors, Translational genomics for crop breeding, Vol. I: Biotic stresses. Wiley-Blackwell Publ., Hoboken, NJ.

Simko, I., R.J. Hayes, and M. Kramer. 2012. Computing integrated ratings from heterogeneous phenotypic assessments: A case study of lettuce post-harvest quality and downy mildew resistance. Crop Sci. 52:2131–2142. doi:10.2135/cropsci2012.02.0111

Simko, I., R.J. Hayes, K.V. Subbarao, and R.G. Sideman. 2010. SM09A and SM09B: Romaine lettuce breeding lines resistant to dieback and with improved shelf life. HortScience 45:670–672.

Simko, I., and J. Hu. 2008. Population structure in cultivated lettuce and its impact on association mapping. J. Am. Soc. Hortic. Sci. 133:61–68.

Simko, I., D.A. Pechenick, L.K. McHale, M.J. Truco, O.E. Ochoa, R.W. Michelmore, and B.E. Scheffler. 2009. Association mapping and marker-assisted selection of the lettuce dieback resistance gene *Tvr1*. BMC Plant Biol. 9:135.

Smith, R. 2012. Changing vegetable crop production practices. Growing Produce. University of California Agriculture and Natural Resources. http://www.growingproduce.com/article/32166/changing-vegetable-crop-production-practices (accessed 26 Feb. 2013).

Smith, R.F., S.A. Fennimore, and M. LeStrange. 2009. UC IPM Pest Management Guidelines: Lettuce. Weed management for organic lettuce production. University of California Agriculture and Natural Resources. http://www.ipm.ucdavis.edu/PMG/r441700511.html (accessed 26 Feb. 2013).

Thibodeau, P.O., and P.L. Minotti. 1969. The influence of calcium on the development of lettuce tipburn. J. Am. Soc. Hortic. Sci. 94:372–376.

Thompson, A.E. 1956. The extent of hybrid vigor in spinach. Proc. Am. Soc. Hortic. Sci. 67:440–444.

Thompson, R.C., and E.J. Ryder. 1961. Descriptions and pedigrees of nine varieties of lettuce. Tech. Bull. 1244, USDA, Washington, DC.

Tibbitts, T.W., and R.R. Rao. 1968. Light intensity and duration in the development of lettuce tipburn. Proc. Am. Soc. Hortic. Sci. 93:454–461.

Truco, M.J., H. Ashrafi, A. Kozik, H. van Leeuwen, J. Bowers, S. Reyes Chin Wo, K. Stoffel, H. Xu, T. Hill, A. Van Deynze, and R.W. Michelmore. 2013. An ultra high-density, transcript-based, genetic map of lettuce. G3 (Bethesda) 3:617–631.

USDA-ERS. 2012. Recent trends in GE adoption 2012. http://www.ers.usda.gov/data-products/adoption-of-genetically-engineered-crops-in-the-us/recent-trends-in-ge-adoption.aspx (accessed 14 Mar. 2013).

USDA-ERS. 2013. U.S. lettuce statistics 2011. http://usda.mannlib.cornell.edu/MannUsda/viewDocumentInfo.do?documentID = 1576 (accessed 21 Feb. 2013).

USDA-NASS. 2011. National statistics for spinach. http://www.nass.usda.gov/Statistics_by_Subject/index.php (accessed 19 Dec. 2012).

USDA-NASS. 2012a. California fresh market spinach, 1920–2011. http://www.nass.usda.gov/Statistics_by_State/California/Historical_Data/index.asp (accessed 10 Feb. 2013).

USDA-NASS. 2012b. Vegetable historic data. Spinach 1923–1956. http://www.nass.usda.gov/Statistics_by_State/Washington/Historic_Data/vegetables/spinach.pdf (accessed 19 Dec. 2012).

Vance Publishing. 2013. The Packer produce market guide. Vol. CXX. No. 54. Vance Publ., Lenexa, KS.

Waycott, W.W. 1995. Photoperiodic response of genetically diverse lettuce accessions. J. Am. Soc. Hortic. Sci. 120:460–467.

Weare, B.C. 2009. How will changes in global climate influence California? Calif. Agric. 63:59–66. doi:10.3733/ca.v063n02p59

Welch, J.E., R.G. Grogan, F.W. Zink, G.M. Kihara, and K.A. Kimble. 1965. Calmar: ...A new lettuce variety resistant to downy mildew. Calif. Agric. 19:3–4.

Whitaker, T.W. 1974. Lettuce: Evolution of a weedy Cinderalla. HortScience 9:512–514.

Whitaker, T.W., and D.E. Pryor. 1941. The inheritance of resistance to powdery mildew (*Erysiphe cichoracearum*) in lettuce. Phytopathology 31:534–540.

Whitaker, T.W., E.J. Ryder, V.E. Rubatzky, and P.V. Vail. 1974. Lettuce production in the United States. Agric. Handb. 221. U.S. Gov. Print. Office, Washington, DC.

Wisler, G.C., and J.E. Duffus. 2000. A century of plant virus management in the Salinas Valley of California, 'East of Eden'. Virus Res. 71:161–169. doi:10.1016/S0168-1702(00)00196-9

Wurr, D.C.E., and J.R. Fellow. 1991. The influence of solar radiation and temperature on the head weight of crisp lettuce. J. Hortic. Sci. 66:183–190.

Wurr, D.C.E., J.R. Fellow, and K. Phelps. 1996. Investigating trends in vegetable crop response to increasing temperature associated with climate change. Sci. Hortic. (Amsterdam) 66:255–263. doi:10.1016/S0304-4238(96)00925-9

Yanagi, A.A., R.M. Bullock, and J.J. Cho. 1983. Factors involved in the development of tipburn in crisphead lettuce in Hawaii. J. Am. Soc. Hortic. Sci. 108:234–237.

Zink, F.W. 1956. Growth and nutrient absorption in spring spinach. Proc. Am. Soc. Hortic. Sci. 83:675–679.

5

Edible Grain Legumes

George J. Vandemark,* Mark A. Brick, Juan M. Osorno, James D. Kelly, and Carlos A. Urrea

ABSTRACT

Edible grain legumes, including dry bean (*Phaseolus vulgaris* L.), dry pea (*Pisum sativum* L.), chickpea (*Cicer arientinum* L.), and lentil (*Lens culinaris* Medikus), have served as important sources of protein in the human diet for thousands of years. In the United States, these crops are consumed nationally and produced for export markets. The objectives of this study were to examine yield gains in edible grain legume crops over the past 25 yr. Genetic gain in dry bean during the past 30 yr based on common trials was 13.9 kg ha^{-1} yr^{-1} (0.77% yr^{-1}) and 17.4 kg ha^{-1} yr^{-1} (0.85% yr^{-1}) for navy and pinto bean cultivars, respectively. Data from national yield trials on research sites indicates that yield gains were 0.4, 0.7, 0.9, and 1.7% for pinto, navy, black, and kidney beans, respectively. The results also suggest that dry bean cultivars have not reached a yield plateau for most market classes. Continued introgression of germplasm from other races of common bean should provide new sources of genetic diversity to enhance yield in the future. Over the past 25 yr, the production of cool season food legumes (pea, lentil, and chickpea) in the United States has increased dramatically; however, yields of dry pea in the United States have decreased by 0.3% per year, lentil yields have increased by only 0.1% per year, and chickpea yields have increased by 2.8% per year. Pea and lentil production has increased dramatically in Montana and North Dakota, but the cultivars grown in this region were originally developed in the U.S. Pacific Northwest (PNW) and Canada and are likely not well adapted for Montana and North Dakota. Several currently grown cultivars are at least 20 yr old, but new cultivars have been released that are superior to these older cultivars.

Abbreviations: CDBN, Cooperative Dry Bean Nursery; NGP, Northern Great Plains; PNW, U.S. Pacific Northwest.

George J. Vandemark, USDA-ARS, Grain Legume Genetics and Physiology Research Unit, 303 Johnson Hall, Washington State Univ., Pullman, WA 99164. *Corresponding author (george.vandemark@ars.usda.gov).
Mark A. Brick, C113 Plant Sciences Bldg., Colorado State Univ., Fort Collins, CO 80523 (Mark.Brick@ColoState.edu).
Juan M. Osorno, Dep. of Plant Science, Loftsgard Hall 166, P.O. Box 6050, North Dakota State Univ., Fargo, ND 58108 (Juan.Osorno@ndsu.edu).
James D. Kelly, Dep. of Plant, Soil, and Microbial Sciences, Plant and Soil Science Bldg., Rm. 370, 1066 Bogue St., East Lansing, MI 48824 (kellyj@msu.edu).
Carlos A. Urrea, Univ. of Nebraska, Panhandle Research and Extension Center, 4502 Avenue I, Scottsbluff, NE 69361 (currea2@unl.edu).

doi:10.2135/cssaspecpub33.c5

Yield Gains in Major U.S. Field Crops. CSSA Special Publication 33.
Stephen Smith, Brian Diers, James Specht, and Brett Carver, editors.

Dry edible bean is arguably the most important grain legume in the human diet and has been characterized as the almost perfect food (Broughton et al., 2003; Câmara et al., 2013). Dry beans were widely cultivated in Mexico and the United States during pre-Columbian times. Kaplan (1965) reported that dry beans found at an archeological site in the southwestern United States were cultivated 2300 yr ago and that they likely originated from Middle America. New World settlers cultivated dry beans from European introductions in the eastern United States and from landraces of small red, pink, pinto, and great northern beans in the western United States that were cultivated by Native Americans. The early landraces were grown on small acreages in the United States until state and federal governments initiated improvement programs during the late 1800s and early 1900s. The first large-scale production of dry edible beans in the United States occurred in Orleans County, NY in 1839 (Bowen, 1898). New York State became an important producer of dry beans and maintained its dominance until the early 1900s when Michigan became the leading producer (http://www.agmrc.org/commodities__products/grains__oilseeds/dry-edible-bean-profile, accessed 15 Oct. 2013). Michigan was the largest producer of commercial dry bean in the United States until 1991 when North Dakota became the top-ranking producer. The leading market classes of dry beans produced in the United States are pinto (~42%), navy or pea (~17%), black (~11%), and great northern (~5%) (USDA-ERS, 2010). Other important market classes produced include light red kidney, dark red kidney, pink, small red, cranberry, and small white, with lesser amounts of specialized market types such as yellow eye, Appaloosa, Anasazi, and others.

Total world production of dry bean increased 65% from the 10-yr period of 1961 to 1970 to 169 million tonnes during 2001 to 2010 (USDA-ERS, 2013). During this time span, U.S. dry bean production increased 71.0% to 11.5 million tonnes and accounted for 5.8% of world production.

U.S. production is limited by many environmental conditions, including heat and drought stress, early fall frost and late spring frost, soil compaction, hail, etc. (Brick and Grafton, 1999). In addition, pathogens that cause rust (*Uromyces appendiculatus* Pers:Unger), common bacterial blight [*Xanthomonas axonopodis* pv. *phaseoli* (Smith) Dye], bacterial brown spot [*Pseudomonas syringae* pv. *syringae* (van Hall)], halo blight [*Pseudomonas syringae* pv. *phaseolicola* (Burkholder)], white mold [*Sclerotinia sclerotiorum* (Lib.) de Bary], and soilborne fungi that causes root rots can further reduce yield of beans (Brick and Grafton, 1999).

Basis for Yield Improvement in Dry Beans

In recent years, scientists in developing countries and the United States have made major advances in dry bean disease resistance, stress tolerance, and increased yield (Kelly, 2004). Agronomic and biotechnological tools have contributed to these achievements. Dry bean breeding programs in the United States were initiated to improve specific market classes and production systems in their region. Because environmental conditions such as nighttime temperatures, rainfall pattern, relative humidity, and soil types are so different among production areas, genetic improvements in one region have little more than academic interest to colleagues working with different market classes in other regions. For example, gains made in breeding pinto beans for the irrigated semiarid production zones of southern Idaho have little application to breeding small seeded determinate navy or Andean kidney beans for the high humidity, rainfed conditions of the midwestern United States. As a result, breeders in each region developed individual strategies and focused on specific market classes for local production. Breeders in the 1960s through the 1990s became known for the specific market class of dry bean on which they focused: M.W. Adams in Michigan focused on navy bean; D.H. Wallace in New York focused on kidney bean; D.R. Wood in Colorado focused on pinto bean; D.P. Coyne in Nebraska focused on great northern bean; M.J. Lebaron and J.J. Kolar in the PNW focused on pinto bean; D.W. Burke in Washington focused on small red and pink beans; and F.L. Smith in California focused on pink bean. A summary of the achievements and limitations of contemporary bean breeding programs was recently reviewed by Beaver and Osorno (2009).

In the early 1900s, breeding programs at Cornell University focused largely on disease resistance, including the earliest work on the genetic race structure of anthracnose [*Colletotrichum lindemuthianum* (Sacc. & Magnus) Lams.- Scrib.] and differential responses of this pathogen to host genotypes (Burkholder, 1918). Breeding efforts also focused on physiological traits and improving yield through efficient partitioning of plant metabolites (Wallace et al., 1993).

Breeding programs in Michigan extend back to the early 1900s (Michigan State University, 2009), and those at the University of Idaho were initiated in 1925 (Singh et al., 2007). Research in Michigan focused largely on improving plant architecture and yield (Adams, 1982; Kelly, 2000). In the intermountain west, breeders focused largely on developing resistance to a range of pathogens, including rust, white mold, bacterial blights, plant viruses, and root pathogens (Singh and Schwartz, 2010). Exchange of germplasm among programs was limited to major disease resistance traits because more complex traits such as yield, plant architecture, and energy partitioning were not directly transferable because of genotype × environment interactions among commercial seed classes. Fundamental concepts such as yield component compensation, first described by Adams (1967), prevented breeders from combining the high pod load potential of small seeded navy beans with very large seed size of a kidney bean (White and Gonzales, 1990). In addition, breeders were constrained by differences in growth habit, maturity, and seed traits demanded by the market place. Not only do these traits limit access to broader genetic variability, but they also lock breeders into genetically narrow breeding populations that limit yield gains (Kelly et al., 1998).

Singh (1982) classified plant architecture of dry bean germplasm into four classes, including Type I upright bush with determinate plant growth, Type II upright with indeterminate growth, Type III semi-vine with indeterminate growth, and Type IV strong climbing vine with indeterminate growth. Breeding for upright Type II architecture has been an important component of breeding pinto bean cultivars since 1980 because upright growth habit provides a level of disease avoidance for some fungal pathogens such as white mold (Park, 1993) and enables the crop to be direct harvested, thus reducing seed loss during cutting and field curing. Early breeding efforts to develop upright architecture in pinto beans utilized navy and black bean germplasm from race Mesoamerica. The pinto cultivar Sierra was the first pinto bean cultivar released with improved architecture (Kelly et al., 1990), followed by additional cultivars with upright architecture (Osorno et al., 2010; Brick et al., 2011). Improvements in yield have also been achieved in some cases by extending maturity by 10 to 14 d. Upright plant architecture usually requires more time to produce a larger plant that sets pods on later nodes in contrast to early season highly efficient bush navy cultivars and semi-vine pinto landraces that lack upright plant structure.

Michigan released the first U.S. navy bean cultivar Robust in 1915 as a selection from locally grown landraces, followed by Michelite, the first cultivar bred and released by Down and Thayer (1938). Later, mutation breeding was used to introduce determinate bush growth habit into the traditional prostrate semi-vine indeterminate growth habit of Michelite (Down and Andersen, 1956). The navy bean cultivar Sanilac released in 1957 was the first bush navy bean cultivar that had the advantage of an erect bush habit combined with early maturity, white mold avoidance, and harvest ease, but it initiated an era of underperforming navy bean cultivars through the 1980s that lacked adequate genetic variability to generate future yield gains (Andersen et al., 1960). The narrow base of genetic variability in navy beans was corrected by introducing genetic variability from black beans from Central America (Adams, 1982), and yields have increased 50% over the last three decades (Kelly and Cichy, 2012). Early cultivars from these crosses lacked canning quality, but this has been corrected, along with refinements in maturity, lodging resistance, and partitioning, such that over 75% of the bean crop in Michigan is currently direct harvested. A valuable spin-off of this work has been the development of new high-yielding locally adapted black bean cultivars, many of which were derived from the same crossing program as navy beans (Kelly and Cichy, 2012).

In Nebraska, early yield gains were achieved in the great northern market class by selection for a more vigorous vine that produced higher biomass than traditional landraces. These gains came at a cost of increasing the risk of white mold due to denser plant canopies that retained humidity (Steadman, 1983). Major breeding efforts were focused on improving levels of resistance to white mold and seedborne bacterial diseases, including common bacterial blight, halo blight, and bacterial wilt [*Curtobacterium flaccumfaciens* pv. *flaccumfaciens* (Hedges) Collins & Jones]. Resistance introgressed from tepary bean (*P. acutifolius* A. Gray) in the tertiary gene pool has led to a number of new cultivars with high levels of resistance to common bacterial blight. Current gains in yield are being achieved through interracial crosses to improve architecture (Coyne, 1980), disease avoidance from white mold, and water use efficiency as water restrictions expand in the western United States (Urrea et al., 2009).

In New York, breeding efforts focused on improving physiological traits and efficient partitioning of nutrients and energy in light red kidney beans. The early cultivars were highly photoperiod sensitive and as a result produced plants with large vegetative biomass that were inefficient in partitioning energy to the seed. Wallace (1985) and colleagues developed the Redkloud cultivar that was photoperiod neutral, matured 2 wk earlier, and outyielded the older Redkote cultivar. 'Redkloud' was widely used as a parent, but the processing industry did not prefer it because of its dark seed coat color. Progress in breeding light red kidney bean was made at Sacramento Valley Milling Co. and the University of California, Davis with the releases 'Sacramento' in 1976 and 'California Early Light Red Kidney' ('CLERK') in 1989. These cultivars replaced the low-yield and long-season cultivar California Light Red Kidney. The cultivars Sacramento and CELRK are still predominant cultivars because of their excellent seed quality and early maturity.

Root pathogens caused by *Fusarium, Rhizoctonia,* and *Pythium* species were the major constraint to bean production in the PNW where the USDA program in Washington led efforts to introgress resistance to these pathogens into a wide array of pink, small red, and pinto seed types. The main source of resistance came from a small black seeded bean introduction 'N203' (PI 203958) collected by Oliver Norvell in Mexico (Wallace and Wilkinson, 1965). Introgression of resistance was challenging given the diversity and unadapted nature of the source material. Screening for resistance was conducted in infertile soils with limited irrigation that were infested with several different root pathogens (Burke and Miller, 1983). As a result, many of the lines and cultivars derived from these beans carry high levels of root rot resistance and have a yield advantage when grown in regions where root rots are a problem. In addition, many of these materials are recognized for higher levels of drought resistance that are, in part, due to improved root systems. In the early 1980s, a number of the pinto lines from this program were released as cultivars, and the germplasm was used to establish the bean breeding program at North Dakota State University. This germplasm and material mostly from the Michigan State University program have allowed the release of several pinto and navy cultivars that have excellent adaptation to the harsh conditions in North Dakota such as disease pressure, short growth cycle, frost, and flooding. Viral diseases, principally *Bean common mosaic virus* and *Beet curly top virus,* are also problematic in the PNW because of high insect populations that vector the pathogen. Resistance breeding was actively pursued and improved germplasm from this region continues to provide valuable resistance sources. Despite these disease problems, bean yields in the central basin of Washington are among the highest in the country and set the standard as some of the highest yields found.

Pinto bean production has long been associated with Idaho, largely because of the high-quality seed produced, the general overall adaptation of race Durango bean types, and favorable weather combined with adequate irrigation water. The landrace Common Pinto is well adapted to the region because of its overall robustness and favorable growing conditions. The breeding program at the University of Idaho has a long history of releasing improved pinto, great northern, pink, and small red cultivars (Singh et al., 2007). The preferred growth habit has been the early-season, semi-vine, Type III growth habit cultivars that respond well to cool nighttime temperatures, furrow irrigation, and low humidity. During harvest, plants are cut at maturity and allowed to dry in the field for 3 to 5 d before threshing to ensure even drying and excellent seed quality. The focus of the breeding program in Idaho has been largely on multiple disease resistance, yield improvement, and adaptation to western growing conditions.

Pinto beans have been produced in Colorado since the early 20th century. Production systems are similar to Idaho, but foliar diseases such as rust, bacterial diseases, and Fusarium wilt (*F. oxysporum* Schlechtend:Fr. f. sp. *phaseoli* J. B. Kendrick & W. C. Snyder) are more problematic (Brick et al., 2006). The program at Colorado State University (CSU) was initiated in 1948 and has focused on breeding for rust resistance, improved seed quality, upright architecture, and staying ahead of the emerging races of the rust pathogen (http://beans.agsci.colostate.edu/, accessed 13 Nov. 2013). Wood and Keenan (1982) developed the first rust resistant pinto cultivar Olathe and the first bush pinto bean cultivar Ouray (Wood and Keenan, 1982). Ouray was never widely grown because its determinate growth habit lacked the yield potential of indeterminate cultivars. In recent years, the program at CSU has incorporated upright Type II architecture in the cultivars Croissant and Long's Peak to improve plant structure and further enhance yield potential (Brick et al., 2011). In addition, CSU has a breeding program for rainfed pinto production in the Four Corners region of Colorado, New Mexico, Arizona, and Utah. The cultivars Fischer (Fisher et al., 1995), Cahone, and San Juan Select were released for this region. Although 'Fisher' is recognized for its drought tolerance in the Four Corner's region, it has shown little adaptation elsewhere.

Several breeding methodologies have been reported to improve dry beans. Interracial hybridization between beans from races Durango and Mesoamerica have been used to improve pinto, great northern, small red, and pink beans (Singh et al., 1993). Urrea and Singh (1994) compared breeding methods for beans and found that the pedigree breeding produced lines with the highest mean yield, followed by the mass selection. They suggested that early generation testing and selection should be used to more efficiently manage populations from interracial crosses. Urrea and Singh (1995) reported on the use of recurrent and congruity backcrossing for wide interracial hybridization in common bean.

The development of cultivars with multiple disease and pest resistance has been achieved in many bean breeding programs. Pinto rust resistant breeding lines BelDakMik-RMR 14 to BelDakMik-RMR 23 carry pyramided resistance to all known races of rust in the United States and resistance to all strains of *Bean common mosaic virus* and *Bean common mosaic necrotic virus* (Terán et al., 2009). Resistance to anthracnose in common bean is conditioned by at least nine major independent genes, *Co-1* to *Co-10* (Kelly and Vallejo, 2004). Molecular markers linked to the majority of *Co* genes have been reported that provide the opportunity to enhance disease resistance through marker-assisted selection (MAS) and gene pyramiding (Kelly and Vallejo, 2004). Miklas et al. (2006) highlighted examples of breeding using molecular markers that revealed the role and success of MAS in gene pyramiding. They discussed the ability to rapidly deploy resistance genes via marker-assisted backcrossing, enabling simpler detection and selection of resistance genes in the absence of pathogens, which lead to simplified breeding of complex traits. Marker-assisted selection has been well documented to speed up the selection process or reduce the selection cost during early generations of breeding program (Yu et al., 2000).

Yield Gains in Dry Bean: On-Farm Yields vs. Potential Yield

Several authors have discussed the difference between on-farm yields and potential yield (also known as realized yield) (Bingham, 1967; MacKey, 1979; Evans and Fisher, 1999). On-farm yield estimates are taken from production statistics and takes into consideration all factors affecting crop production, including cultivar potential, the environment, management practices, biotic factors, and their interactions. Potential yield trials primarily focus on the expression of the genetic potential under optimum or near optimum environmental and management conditions as achieved in trials conducted on research stations or where management practices are carefully applied in a selected environment (site) to allow plants to express their yield potential. The difference between on-farm yield and potential yield is known as the "yield gap" (Fisher et al., 2009; Godfray et al., 2010; Licker et al., 2010; Neumann et al., 2010). In most cases, the yield gap develops because of the impact of biotic and abiotic factors such as water regime, soil physical and chemical conditions, disease and pest pressure, weed competition, suboptimum plant stands, and harvest loss, as well as other factors that reduce on-farm yield. Crop management and other agronomic practices also play a crucial role (Duvick and Cassman, 1999). Economic factors such as fluctuations in commodity prices, marketing, and availability of seed stocks may also influence the pace in which yield gains are achieved. Plant breeders primarily focus on increasing potential yield while also attempting to optimize the interaction between genotype, environment, and agronomic practices to increase on-farm yield and reduce the yield gap.

During the last 20 yr, much of the pinto bean production moved from the western United States into the Midwest and Northern Great Plains (NGP). These ecogeographic regions are very different; consequently, cultivars originally developed for the western region did not perform well when grown in the latter areas. Such differences have been shown to impact estimates of yield gains as shown in the three major crops rice (*Oriza sativa* L.), wheat (*Triticum aestivum* L.), and maize (*Zea mays* L.) by using high-resolution geospatial databases and statistical tools (Ray et al., 2012).

Reliable estimates of on-farm yields can be obtained from surveys and statistical sources. Estimating potential yield is more challenging because of the difficulty of finding an optimum environment that would ensure little or no limitations for production resources. Potential yield is often estimated using data obtained from field experiments at research stations where environmental and management practices are maximized but are still not optimum for genetic expression of potential yield. Weather alone as a factor guarantees variation in yield expression below genetic optimum for every crop, and management practices cannot be applied to fully compensate for environmental variability.

For this section, production data from the USDA-NASS (2013) were used to estimate on-farm yields, and data obtained from the Cooperative Dry Bean Nursery (CDBN) was used to estimate potential yield. The CDBN has been grown every year since 1950 with an average of 10 locations each year across the most important dry bean production regions in North America (http://www.ars.usda.gov/pandp/docs.htm?Docid=21679, accessed 15 Nov. 2013). Different market classes are included in this trial with pinto, navy, black, and kidney beans the most frequent. For our comparisons, only data between 1981 and 2012 have been used to provide estimates of potential yields during the last 30 yr.

The earliest year in which on-farm yield values are available from the USDA-NASS is 1909 (USDA-NASS, 2013). Overall, on-farm yield across all market classes of dry bean grown in the United States show a seed yield increase of 12.9 kg ha^{-1} yr^{-1} (2.4% yr^{-1}) between 1909 and 2012 (Fig. 5–1). Potential yield data from the CDBN also show a positive trend with an average seed yield increase of 7.3 kg ha^{-1} yr^{-1} (0.5% yr^{-1}) during the last 31 yr (data not shown). Interpreting these results is difficult because of the large genetic diversity among market classes, yield potential, disease resistance, seed characteristics, and other traits. Therefore, we evaluated individual yield gains among four market classes, namely pinto, navy, black, and kidney beans during the past ~30-yr period. These four market classes account for approximately 95% of the total U.S. production.

Pinto bean yield gains between 1981 and 2012 showed a positive trend for both on-farm yield and potential yield, with yield gains ranging between 12.2 and 15.2 kg ha^{-1} yr^{-1} (0.4% yr^{-1} for potential yield) (Fig. 5–2). There was a yield gap between on-farm and potential yield, which suggests that the genetic poten-

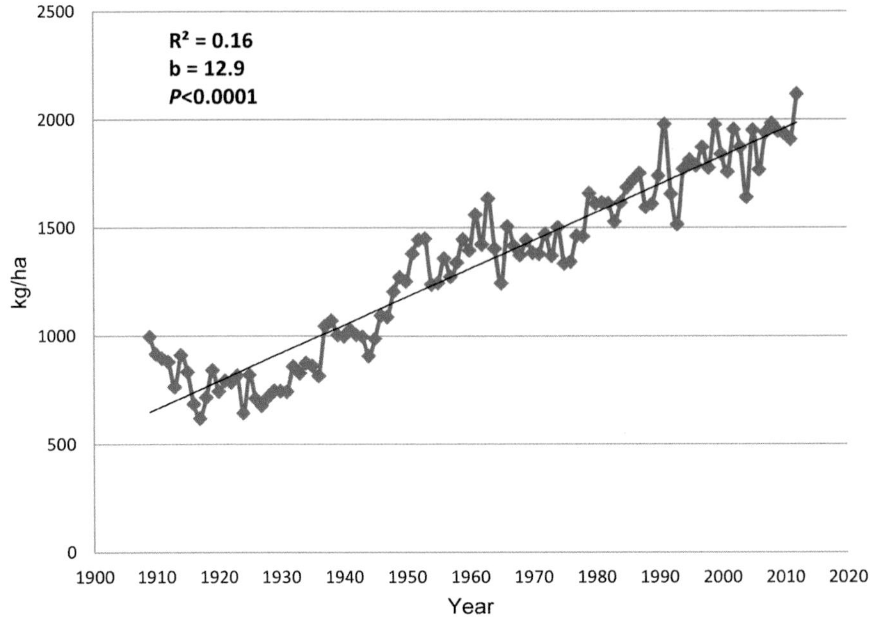

Fig. 5–1. On-farm seed yield averaged across dry bean market classes in the United States between 1909 and 2012 using production data from USDA-NASS.

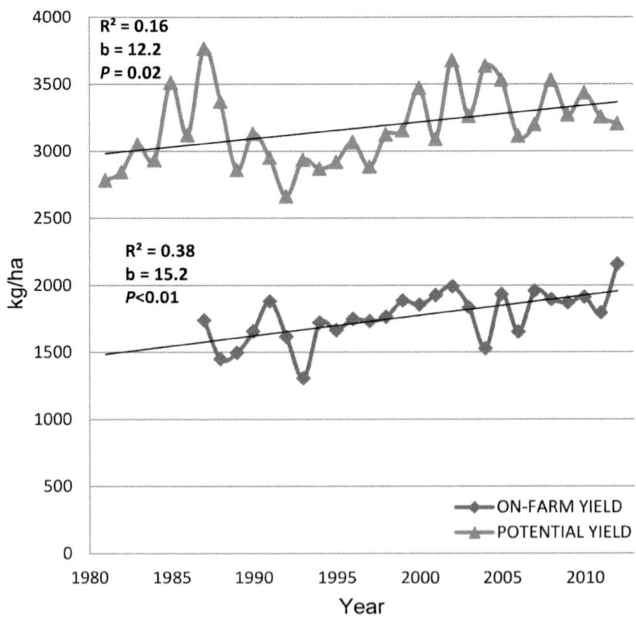

Fig. 5–2. On-farm yield and potential yield of pinto beans from 1981 to 2012 in the United States using production data from both USDA-NASS and the Cooperative Dry Bean Nursery.

tial of pinto bean cultivars was not fully expressed in on-farm trials. Pinto bean can be greatly affected by foliar bacterial and fungal diseases, root pathogen, and weed competition, which may explain a portion of the differences observed between on-farm and potential yield. Because the CDBN sites are often selected to represent high-yielding environments with adequate fertilization and optimum cultural practices, it is not surprising that potential yield was higher than on-farm yields. The gap between potential yield and on-farm yields clearly demonstrates that there are opportunities to improve production and management practices to increase on-farm yields and more fully exploit the genetic yield potential of new cultivars.

Navy and black beans show similar yield gains, which is not unexpected because they have similar genetic background (race Mesoamerica) and plant growth characteristics. Navy bean had yield gains between 18.5 and 25.9 kg ha^{-1} yr^{-1} (0.7% yr^{-1} for potential yield) (Fig. 5–3), which is greater than pinto bean. Black bean show a more variable range in yield gains between the on-farm and potential yield, with 5.7 and 24.3 kg ha^{-1} yr^{-1}, respectively (0.9% yr^{-1} for potential yield) (Fig. 5–4). These results may reflect more variation among cultivars and production regions in the black bean market class. Because fewer black bean cultivars are tested each year in the CDBN in comparison to other market classes, the results may have higher variability across years and locations. In addition, black bean cultivars had a larger yield gap than navy bean. On-farm results indicate that black bean may be reaching a yield plateau, which suggests that opportunities to improve production and management practices may be needed to increase on-farm yields for black bean.

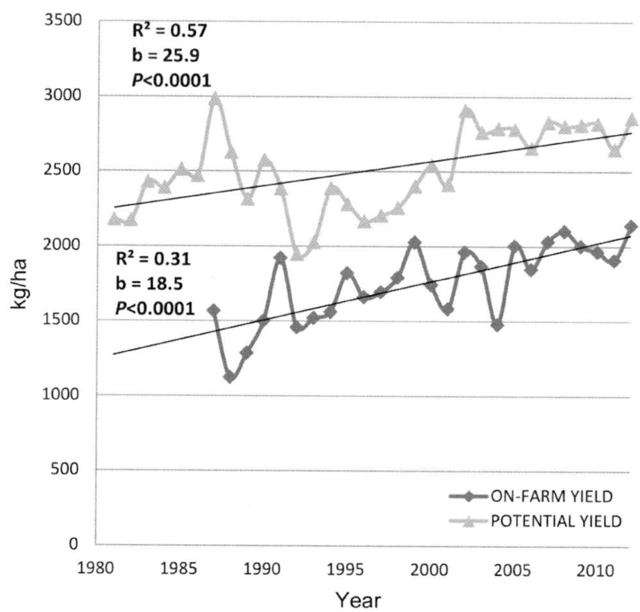

Fig. 5–3. On-farm yield and potential yield of navy beans from 1981 to 2012 in the United States using production data from both USDA-NASS and the Cooperative Dry Bean Nursery.

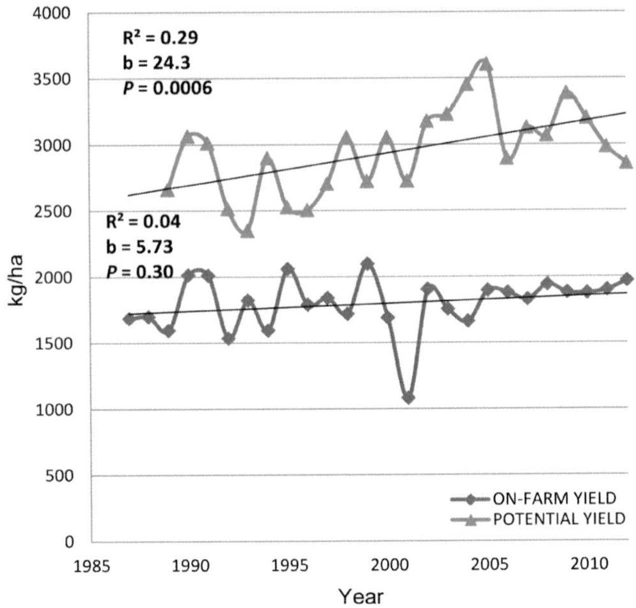

Fig. 5–4. On-farm yield and potential yield of black beans from 1987 to 2012 in the United States using production data from both USDA-NASS and the Cooperative Dry Bean Nursery.

Combined data from both light and dark red kidney bean show a yield increase between 19.1 and 39.9 kg ha^{-1} yr^{-1} (1.7% yr^{-1} potential yield) (Fig. 5–5). This is the highest gain among all market classes, which is surprising given the challenges associated with production and genetic improvement of kidney bean. Yield gains in kidney bean have been a challenge because they have a narrow genetic base and high susceptibility to biotic and abiotic factors (Beaver and Osorno, 2009; Kelly and Cichy, 2012). In addition, kidney bean has the smallest yield gap among those compared. The small difference between on-farm and potential yield in kidney bean may result from the fact that producers of kidney bean tend to manage the crop better than for other market classes because market prices are typically higher than for other market classes.

Yield Gains as Measured in Common Dry Bean Trials

Nuland and Taylor (1996) reported that yield levels of dry bean have increased 4.0 kg^{-1} ha^{-1} yr^{-1} from 1940 to 1994. They analyzed yield data from four major bean-producing counties in the United States to compare yield changes due to the introduction of new cultivars. They reported that changes in yield were largely driven by the introduction of new cultivars that possessed novel genes for resistance to pathogens. Singh et al. (2007) conducted studies on yield gain by using common trials in Idaho in which yield ranged between 2904 kg ha^{-1} for pinto cultivar UI 111 released in 1945 and 3921 kg ha^{-1} for cultivar Bill Z released in 1986 (Wood et al., 1989). This yield difference represents a 35% gain in 52 yr for these cultivars. Common trials that test cultivars released over time in the same environment have

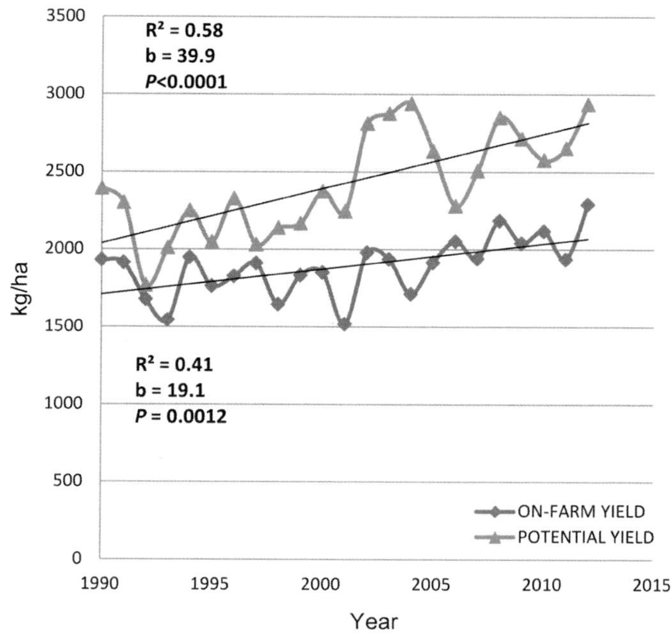

Fig. 5–5. On-farm yield and potential yield of kidney beans from 1990 to 2012 in the United States using production data from both USDA-NASS and the Cooperative Dry Bean Nursery.

been suggested by Cox et al. (1988) and others to specifically measure the genetic contribution of yield gain over time without confounding environmental or cultural effects. The results of trials reported in this report were conducted by four participants in the Bean Coordinated Agriculture Project in Colorado, Michigan, North Dakota, and Nebraska during 2011 and 2012. The trials included cultivars released since 1956 for navy bean and since 1965 for pinto bean. Field trials at all sites were managed following standard recommendations for tillage, herbicides, and insecticides. In Colorado, the trials were grown on the Agricultural Research Development and Extension Center in Fort Collins. The trials were furrow irrigated and received 506 mm of rainfall plus irrigation during the growing seasons in 2011 and 464 mm in 2012. Yield levels were normal in both years, with growing degree days 6% higher in 2012 than 2011. In Michigan, trials were grown near Frankenmuth in 2011 and 2012. In 2011, plots received 206-mm rainfall during the growing season and 263 mm in 2012. Growing degree days were 8 and 10% higher than normal for 2011 and 2012, respectively. In North Dakota, plots were grown at the Carrington Research Station under limited irrigation in 2011 (395-mm rainfall + irrigation during the growing season), and in 2012 plots were grown in a growers field near Hatton under rainfed conditions (257-mm rainfall). In Nebraska, plots were planted in a grower's fields near Mitchell and irrigated with an overhead sprinkler both years. Total rainfall + irrigation was 390 and 411 mm in 2011 and 2012, respectively. The 2012 season was warmer than 2011, with the maximum temperature 2°C higher during the growing season. Seed yield was measured using standard field plot practices at each location. Depending on the protocols used by each breeding program, seed yield was measured in two or four row plots, 5 to 7 m long. Two replicates per location were used in 2011, and three replicates per location were used in 2012. Plant maturity was the number of days from planting to visually estimated 30% pod discoloration (yellow to brown), seed weight (g 100 seed^{-1}) was measured by weighing 100 random seeds from each replicate, and plant canopy height was measured from the soil surface to the uppermost plant part by using a meter stick. Analysis of variance was used to determine interactions between location and cultivar, and linear regression was used to estimate genetic progress over time by regressing the four response variables on the year of cultivar release. Years and replicates were considered random model factors and location and cultivars considered fixed factors. A combined analysis of variance over locations was used independently for navy and pinto bean cultivars with replicates nested in locations and years.

Seed yield increased for both navy and pinto bean cultivars over the time periods of cultivar release (Fig. 5–6a, 5–6b, and 5–6c). For both navy and pinto beans, the analysis of variance indicated that the main effects of cultivar and the location × cultivar interaction were significant ($P < 0.001$); however, the main effects of location were not. The analysis indicated that navy cultivars released since 1956 had a similar linear response across the Midwest region (Michigan and North Dakota) and western region (Colorado and Nebraska); therefore, the response was summarized over both regions (Fig. 5–6a). The overall regression of yield on year of release indicated that mean seed yield in navy cultivars increased from 1832 kg ha^{-1} in 1956 to 2558 kg ha^{-1} in 2008. The regression equation indicated that yield increased 13.9 kg ha^{-1} yr^{-1} ($P = 0.003$) and the regression equation accounted for 54% of the variation in response. Yield response for pinto bean cultivars differed between the Midwest and West regions. In the Midwest (Fig. 5–6b), mean seed yield increase over time was not significant ($b = 2.6$; $P = 0.68$; $R^2 = 0.01$) and likely occurred because most of the pinto cultivars released between 1965 and 2008 were developed by breeding programs located in the western United States. In the West region (Fig. 5–6c), mean seed yield of pinto cultivars increased from 2041 kg ha^{-1}in 1965 to 2880 kg ha^{-1} in 2008. The regression equation indicated that mean seed yield increased 17.4 kg ha^{-1} yr^{-1} ($P = 0.009$) and explained 36% of the variation in yield (Fig. 5–6c). Only the test cultivars Sierra, Stampede, and Lariat were developed by breeding programs in the Midwest, hence the set of cultivars in this study are not representative of progress in yield gains in the Midwest. It is noteworthy that the three cultivars developed in the Midwest fell above the Midwest region regression line (Fig. 5–6b). The average yield increase for navy beans was 0.77% yr^{-1} [13.9 kg ha^{-1} yr^{-1}; and for pinto beans developed and tested in the West, 0.85% yr^{-1} (17.4 kg ha^{-1} yr^{-1})]. These results are similar to the yield increases reported for soybean [*Glycine max* (L.) Merr.] (Specht and Williams, 1984; Voldeng et al., 1997), wheat (Lopes et al., 2012), and dry beans (Singh et al., 2007) and to findings based on on-farm and potential yield data from the USDA-NASS and the CDBN, respectively.

Seed weight, harvest maturity, and plant height were also measured in the common trials. Seed weight is an important criterion that defines market classes for both navy and pinto bean cultivars. In general, navy bean cultivars range in seed weight from 18 to 22 g 100 seed^{-1} and pinto cultivars from 35 to 40 g 100 seed^{-1}. Therefore, breeders do not select for seed weight outside of the range that is commercially acceptable. Consequently, mean seed weight did not change

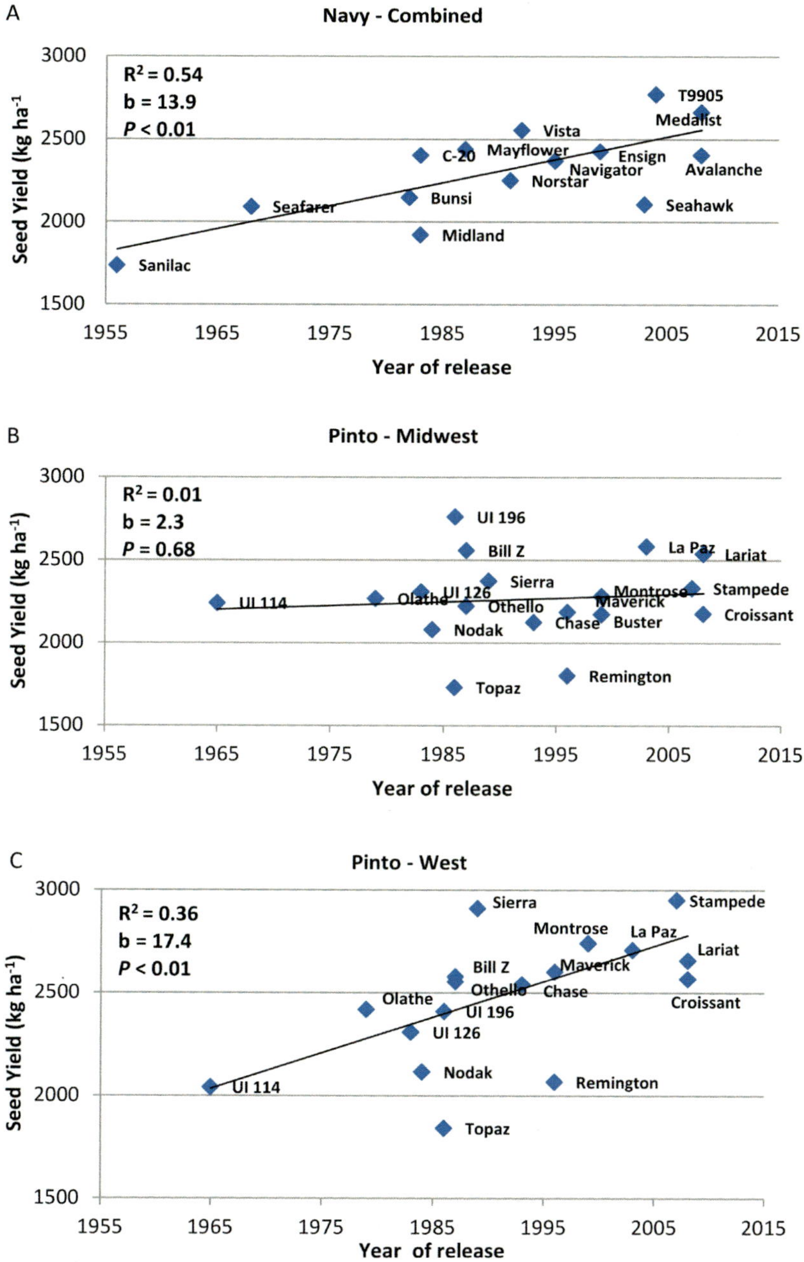

Fig. 5–6. Seed yield (kg ha^{-1}) among (a) navy bean cultivars averaged over Midwest and Western region trials, (b) pinto bean cultivars in the Midwest region, and (c) pinto cultivars in the western region regressed on year of release.

($P > 0.05$) for navy cultivars and only increased 0.09 g yr^{-1} ($P < 0.05$) for pinto bean cultivars over time (Fig. 5–7a and 5–7b). This small increase in seed weight likely reflects some pressure on pinto breeders to increase seed size because a small premium price is placed on large seed in pinto markets. Mean harvest maturity (days to pod maturity) did not change for either navy ($P = 0.61$) or pinto bean cultivars over the time of release ($P = 0.40$) (data not shown). This lack of relationship is not surprising, because dry bean harvest in the United States and Canada must occur in September when the air temperature is high and humidity is low

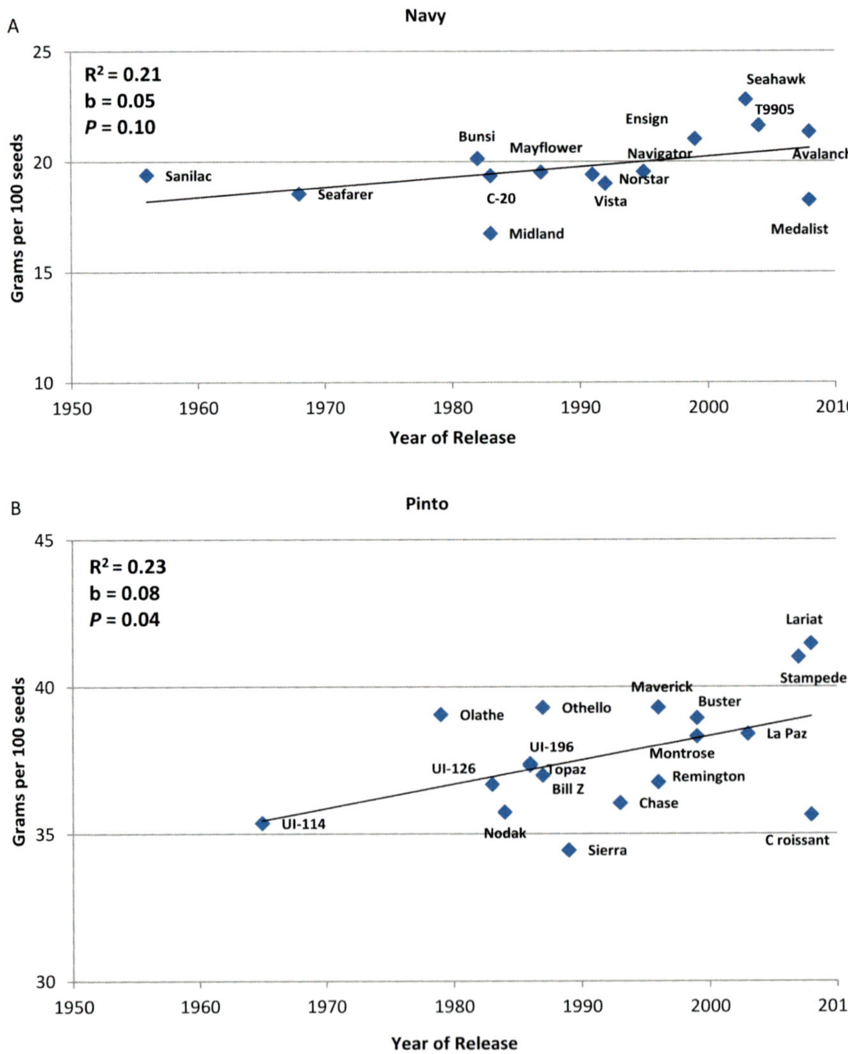

Fig. 5–7. Seed weight (g 100 seed⁻¹) among (a) navy bean cultivars and (b) pinto bean cultivars averaged over Midwest and western region trials regressed on year of release.

to allow the pods to dry for threshing. This drying requirement is important to produce bright colored seed color because high humidity or rainfall on the crop can cause dark or stained seed color and reduce the market value of the crop. In fact, in many regions of the West that produce high-quality dry beans, September is the driest month of the year. In some regions, early harvest is also important to allow producers to timely harvest other full-season crops such as corn, soybean, sugarbeet (*Beta vulgaris* L.), and sunflower (*Helianthus annuus* L.) and to prepare the field for fall planting of winter cereal or oil seed crops. These concerns have prevented breeders from extending plant maturity to increase seed yield as in other crops.

Plant height increased significantly in pinto bean cultivars over time; however, plant height did not significantly increase in navy bean cultivars (Fig. 5–8a) even though the plant height of the earliest released navy bean cultivars Sanilac and Seafarer is well below that of more recent cultivars Medalist and T9905. A few of the more recent navy bean cultivars such as Seahawk and Avalanche have a tendency to lodge, which is reflected in lower plant height. Mean plant height for pinto beans in the Midwest trial (Fig. 5–8b) increased from 38 cm for cultivars released in 1965 to 50 cm for cultivars released in 2008 ($P = 0.02$); for pinto bean in the West trial (Fig. 5–8c), plant height increased from 32 cm for cultivars released in 1965 to 64 cm for cultivars released in 2008 ($P = 0.02$). For navy bean cultivars, mean plant height across regions was not significantly different over the time of release (from 52 cm in 1956 to 57 cm in 2008) ($P = 0.39$). This large difference in response is due to the fact that all navy bean cultivars tested in this research had upright architecture; however, pinto cultivars had prostrate vine architecture before 1980.

Summary and Conclusions Regarding Dry Bean Yields

Genetic gain in dry bean based on common trials was 13.9 kg ha^{-1} yr^{-1} (0.77% yr^{-1}) and 17.4 kg ha^{-1} yr^{-1} (0.85% yr^{-1}) for navy and pinto bean cultivars, respectively. These results are similar to the yield increases reported for soybean (Specht and Williams, 1984; Voldeng et al., 1997), wheat (Lopes et al., 2012), and dry bean (Singh et al., 2007) in previous studies. They are also similar to findings in this report that evaluated potential yield data within market classes from the CDBN trials over the past ~30 yr (0.9% yr^{-1} across market classes). Data from potential yield gains within each market class showed a gain of 0.4, 0.7, 0.9, and 1.7% for pinto, navy, black, and kidney beans, respectively. The fact that similar gains are found by using different sources of data suggest that the results provided here are robust and accurate. The results also suggest that yield increases have been linear since 1956 and suggest that dry bean cultivars have not reached a yield plateau for most market classes. Continued introgression of germplasm from other races of common bean should provide new sources of genetic diversity to enhance yield into the foreseeable future. For navy bean, only seed size has not changed since 1956. Plant height for pinto bean cultivars increased from 34 to 57 cm over time because of the introgression of Mesoamerican germplasm with upright Type II architecture. Seed size increased slightly in pinto bean but did not change in navy bean cultivars over time. In addition to breeding efforts to reduce the yield gap found in all market classes, there is a need to keep improving management practices that allow for higher production efficiency.

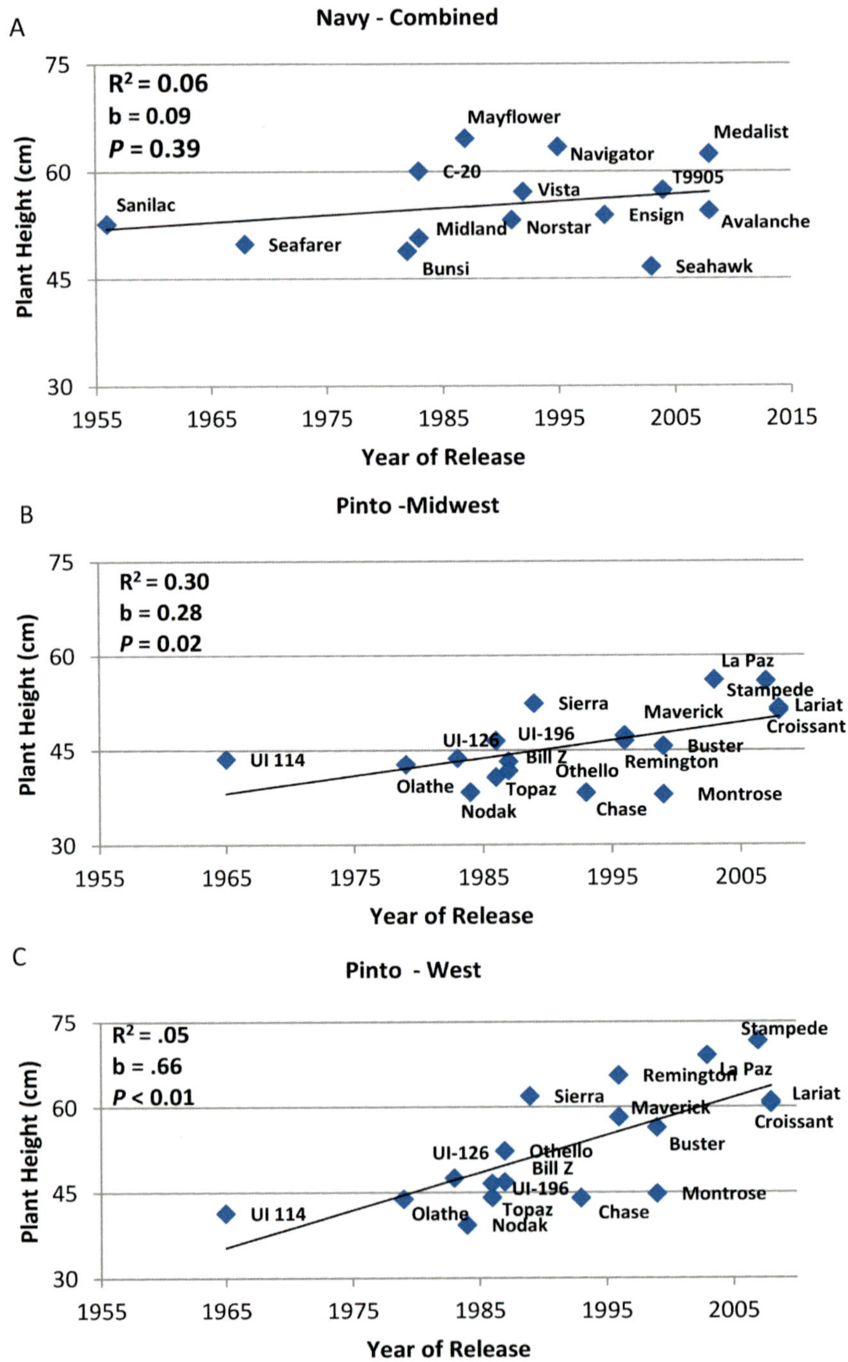

Fig. 5–8. Plant height (cm) among (a) navy bean cultivars averaged over the Midwest and western region trials, (b) pinto bean cultivars in the Midwest trial, and (c) pinto bean cultivars in the West region trial regressed on year of cultivar release.

Yield Gains in Cool Season Food Legumes

Cool season food legumes, including pea, lentil, and chickpea, were among the first crops to be domesticated during the advancement of Neolithic culture that occurred 8,000 to 12,000 yr ago in riparian societies along the Euphrates and Tigris rivers (Ladinsky and Alder, 1976). These crops have been historically grown in rotations with small cereal grains and confer several benefits to small grain production, including disruption of cereal disease cycles and increasing availability of residual nitrogen produced by fixation of atmospheric N by rhizobacteria in root nodules. Cool season food legumes also complement small cereal grains in nutritional qualities, with food legumes having relatively high seed concentrations of lysine and low concentrations of the sulfur-containing amino acids methionine and cysteine, while cereal grains tend to have low concentrations of lysine but high concentrations of sulfur-containing amino acids (Bressani and Elias, 1988).

Dry Peas

Pea was domesticated at least 8000 yr ago in areas of present-day Turkey through Iran (Zohary and Hopf, 1973). Currently, dry pea ranks second only to common bean as the most widely grown grain legume in the world. Global pea production during 2006 to 2010 averaged nearly 9.9 million tonnes, with Canada being the largest producer followed by the Russian Federation, China, India, and France (FAOSTAT, 2013). The National Center for Biotechnology Information (NCBI) currently recognizes *P. sativum* with five subspecies (*abyssinicum, asiaticum, elatius, sativum* and *transcaucasicum*) and *P. fulvum*.

Commercial dry pea production in the United States began in the 1910s in eastern Washington and northwestern Idaho, a region that is known as the "Palouse," as growers tried to identify an alternative rotational crop to oats (*Avena sativa* L.) and barley (*Hordeum vulgare* L.) for dryland wheat production systems. Initial cultivars recommended for the Palouse were the green pea Alaska and the yellow pea Canadian White (Shinn, 1916). By 1926, yield benefits to wheat production using pea rotations were recognized (Schafer, 1926). The PNW continued to be the primary area for producing dry pea until the mid 1990s, when pea production expanded rapidly in the dryland regions of the NGP, especially in eastern Montana and North Dakota.

In the United States, dry pea production consists primarily of yellow and green market classes (FAOSTAT, 2013). Green and yellow peas are typically dried and eaten split or whole and are also used to make flour or for animal feed (Wright, 1985). The largest pea-producing states in the United States now are Montana, followed by Washington, which is closely followed by North Dakota, and finally, Idaho (USDA-NASS, 2013). Although the USDA-NASS does not separate yellow and green peas when reporting dry pea statistics, the U.S. Dry Pea and Lentil Council estimates that 2012 U.S. dry pea acreage consisted of approximately 45% green peas and 55% yellow peas (U.S. Dry Pea and Lentil Council, 2012). The U.S. Dry Pea and Lentil Council estimated that in 2013 approximately 36% of the total U.S. green pea crop was produced in the PNW and 64% in the NGP. However, in 2013 approximately 95% of the total U.S. yellow pea crop was produced in the NGP (U.S. Dry Pea and Lentil Council, 2012).

Over the past 25 yr (1987–2011), the area of peas harvested annually in the United States has increased considerably. During the most recent 4 yr (2008–

2011) of the past 25-yr (1987–2011) period, the average area of peas harvested in the United States was 277,364 ha, which is a 300% increase over the average of 69,356 ha harvested in the United States during the first 4 (1987–1990) of the past 25 yr (USDA-NASS, 2013). The rapid increases over time in the area of peas harvested in the NGP compared with the PNW can be seen in Fig. 5–9. During the recent 4-yr period (2008–2011) the average area of dry pea harvested in the PNW was 42,120 ha, a 38% decrease from the average of 68,141 ha harvested from this region during the first 4 (1987–1990) of the past 25 yr (USDA-NASS, 2013). In contrast, during the most recent 4 yr (2008–2011), the average area of dry pea harvested in the NGP was 232,976 ha, a 472% increase over the average of 40,753 ha for the first 4 yr (1998–2001) for which data are available from NASS for pea production in the NGP.

On-Farm Productivity Trends for Dry Peas

Although both the area harvested and total production of peas in the United States have greatly increased over the recent 25-yr period from 1987 through 2011, yields per hectare have actually decreased by 0.3% per year (Fig. 5–10). This change from an average of 2274 kg ha^{-1} (2031 lb ac^{-1}) during 1987 to 1991 to an average of 2084 kg ha^{-1} (1861 lb ac^{-1}) during 2007 to 2011 represents an average annual yield decrease of 8 kg ha^{-1} (7.1 lb ac^{-1}). Considering the PNW separately, over the past 25 yr pea yields have decreased by approximately 0.4% per year. Dry pea yields in the U.S. Northern Plains have also decreased over time, by approximately 1.5% per year over the past 14 yr for which data are available.

Basis for Yield Improvement in Dry Peas

The biological basis of crop yield is the final result of genetic, environment, and genetic × environment interaction effects that take place in a dynamic system that includes the crop, beneficial and detrimental microorganisms, insect pests, herbivores, and the growing environment. Pea improvement programs must consider a range of traits that impact yield and consumer acceptance. Quality traits of

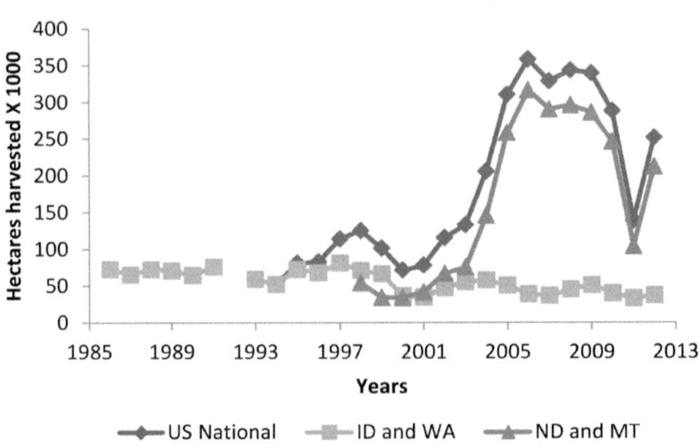

Fig. 5–9. Historical changes in the area (ha) of peas harvested in the total United States, the U.S. Pacific Northwest (Idaho and Washington), and the U.S. Northern Plains (Montana and North Dakota) during 1987 to 2011.

importance to consumer acceptance include cotyledon color, bleaching resistance (green cotyledon only), seed coat and hilum color, seed size and shape, splitting quality, cooking time, and organoleptic characteristics (Hawtin et al., 1988). New cultivars that have improved seed quality traits must also have acceptable yield levels to be grown commercially. Several reproductive traits, including seed size, number of pods per plant, and number of seeds per pod, contribute to yield in pea. A significant positive correlation was observed between grain yield and seed size, number of pods per plant, and plant harvest index, which is the ratio of seed yield per plant divided by total biomass per plant (Sarawat et al., 1994). Significant general combining ability and specific combining ability effects have been detected for yield and several yield components of yield in pea, including pods per plant, plant harvest index, seeds per pod, and seed weight, suggesting these traits are conditioned by both additive and nonadditive genetic variance (Sarawat et al., 1994). A narrow sense heritability (h^2_n) estimate of 0.55 has been calculated for seed yield of dry pea (Timmerman-Vaughan et al., 2005). Several developmental and morphological traits also impact dry pea yield. Early flowering is desirable because high temperatures can occur later in the growing season and reduce pod set. A negative correlation has been observed between days to maturity and grain yield in dry pea (Tar'an et al., 2004). The leaf type of a pea cultivar can also impact havestability. Cultivars with conventional leaves, such as Columbian, tend to have long vines and are susceptible to lodging. Newer cultivars are predominately *afila* leaf types, which are also known as "semi-leafless." *Afila* type plants only produce stipules while producing many more tendrils than normal leaf type plants, which produce stipules, leaflets, and tendrils. The profuse

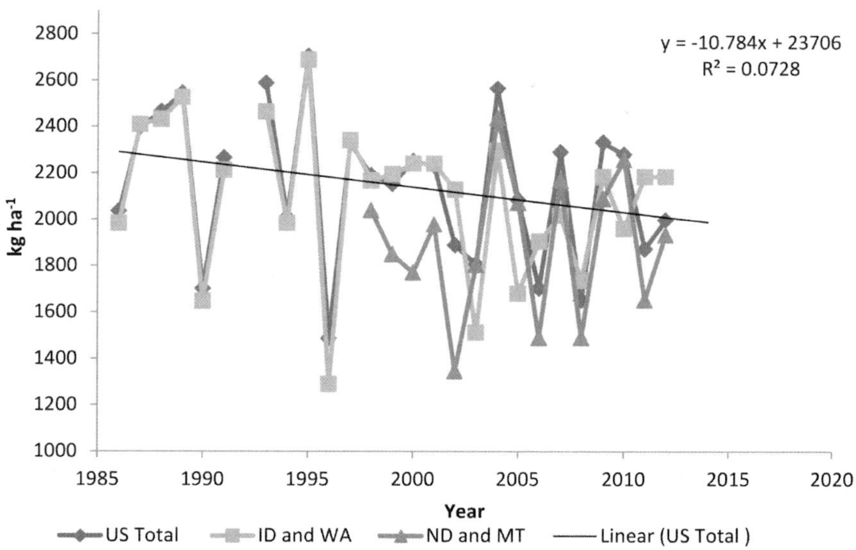

Fig. 5–10. Historical changes in on-farm pea yields (kg ha^{-1}) across the total United States, the U.S. Pacific Northwest (Idaho and Washington), and the U.S. Northern Plains (Montana and North Dakota) during 1987 to 2011.

tendrils that develop from *afila* lines intertwine in the field and serve to support the crop, which reduces lodging and allows for more efficient harvesting.

Pea production in the United States is impacted by several diseases that remain a focus for breeding efforts. Aphanomyces root rot, caused by *Aphanomyces euteiches* Dresch., is the most destructive disease of pea in the United States and can cause complete crop loss when cool and wet spring conditions promote disease development (Pfender, 2000). Fungicides are not effective at controlling this disease and resistance is lacking in commercial dry pea cultivars (Pfender, 2000). Progress in improving levels of resistance in pea to Aphanomyces root rot has been impeded by linkage between resistance genes and loci responsible for undesirable production traits, such as tall plants (*Le*), colored flowers (*A*), and colored seed (*Pl*) (Marx et al., 1972). Heritability estimates for resistance in peas to *A. euteiches* have been less than 0.50 (Pilet-Nayel et al., 2002; Shehata et al., 1983), which suggests environmental variance contributes considerably to disease reaction. Other important diseases of pea in the United States include wilt caused by *Fusarium oxysporum* Schlech. (Haglund, 2000), blight caused by *Mycosphaerella pinodes* Berk. & Bloxam (Bretag and Ramsey 2000), and white mold disease caused by *Sclerotinia sclerotiorum* (Lib.) de Bary (Biddle, 2000).

Pea is a self-pollinated diploid ($2N = 2X = 14$) crop that produces cleistogamous flowers. Commercial dry pea cultivars are pure lines developed either through pedigree, backcross, or modified bulk-pedigree descent breeding approaches (Gritton, 1986). The initial sources of resistance to several soilborne diseases, including Aphanomyces root rot and Fusarium wilt, were PI accessions or pea cultivars developed for canning. PH-14-119 (Kraft et al., 1972), a canning pea with resistance to both Fusarium wilt and Aphanomyces root rot, was crossed with an *afila* type line to produce the pea germplasms 792022 and 792024, the first high-yielding dry pea lines with field tolerance to soilborne diseases of the PNW (Kraft, 1981). The germplasm 792022, PI 180693, and PI 189171 were then used as sources of resistance to develop pea germplasm lines 90-2079 and 90-2131, and 90-2322, which were released based on enhanced resistance to Aphanomyces root rot, Fusarium wilt, and Fusarium root rot (Kraft, 1992). Pea germplasm 90-2031 was subsequently crossed to the cultivar Dark Skin Perfection, and several lines were recently released that have improved resistance to multiple soilborne diseases coupled with desirable seed and plant architecture characteristics (McGee et al., 2012).

Major efforts in the United States to develop improved dry pea cultivars for U.S. production began in the early 1960s by the USDA-ARS in cooperation with Washington State University. These two research programs, one located in Prosser, WA and the other in Pullman, WA, remained the primary public breeding programs in the United States for dry pea until 2008, when a dry pea improvement program began at North Dakota State University. Currently, the majority of dry pea cultivars grown in the United States are derived from private breeding programs. These include green pea cultivars Columbian, Aragorn, and Banner and yellow pea cultivars Delta and Universal.

Yield Gains Measured in Recent Dry Green Pea Cultivar Trials

Only a few new green pea cultivars have been released over the past decade in the United States, and the majority of these cultivars are not intended for human consumption. Specter (McPhee and Muehlbauer, 2007) and Windham (McPhee et al., 2007) are two cultivars of peas that can be sown in the fall and produce grains that are used for animal feed. The recently released cultivar Lynx (McGee et al., 2012) is intended for use as a component of wildlife food plots. The most successful spring green pea cultivars grown in the United States have been developed by private breeding programs and include Ariel (Progene Plant Research 2013a), which was released in 2002 on the basis of its high yield and bleaching resistance, Aragorn (Progene Plant Research 2013b), released in 2007 on the basis of its high yield, large seed size, and early maturity, and Banner (Progene Plant Research 2013c), released in 2008 on the basis of its high seedling vigor, yield, and early maturity. 'Aragorn,' 'Ariel,' and 'Banner' were compared with 'Columbian' for yield and other metric traits to determine what traits have been markedly changed between older and newer materials (Table 5–1). Comparisons are based on the results of field trials conducted at four locations in the PNW (Genesee, ID; Kendrick, ID; Pullman, WA; and Fairfield, WA) during 2011 and 2012. The Pullman, WA trials were at the WSU–Spillman Research Farm, and all other trials were conducted on fields provided by cooperating growers. The average yield of 'Columbian' was significantly less than the average yield of all three newer cultivars. The three new lines had an average yield of approximately 2242 kg ha^{-1} (1996 lb ac^{-1}), which is 35% higher than the observed average yield of 'Columbian.' These results suggest that considerable gains in yield of green peas are being realized though breeding efforts.

Considering selected traits of agronomic importance to pea, 'Ariel' produced significantly smaller seed than 'Columbian,' while the seed sizes of 'Banner,' 'Aragorn,' and 'Columbian' were not significantly different (Table 5–1). 'Columbian' reached 50% flowering in a significantly shorter time than any of the newer cultivars. The early flowering of 'Columbian' is a consequence of its indeterminate pod development compared to the newer entries. 'Columbian' had a significantly lower canopy height and plant height index than any of the newer cultivars, which suggests that gains are also being made through breeding in improving traits associated with harvesting ease.

Table 5–1. Yield and agronomic characteristics of green pea cultivars based on field trials conducted in Washington and Idaho in 2011 and 2012.

Entry	Release	Yield†	Seed size‡	Days to 50% flower‡	Canopy height‡	Plant height index‡¶
		kg ha^{-1}	g 100 $seeds^{-1}$		cm	
Banner	2008	2325 A	19.4 A	54.2 A	55.5 A	0.86 A
Aragorn	2007	2273A	19.8 A	55.0 A	54.6 A	0.89 A
Ariel	2002	2129 A	17.7 B	55.6 A	56.7 A	0.89 A
Columbian	~1980	1666 B	19.7 A	47.8 B	40.4 B	0.49 B
LSD (α = 0.05)		332	0.9	2.9	6.7	0.08

† Based on trials conducted at Pullman, WA; Fairfield, WA; Genesee, ID; and Kendrick, ID in 2011 and 2012 (eight location years).

‡ Based on trials conducted at Pullman, WA in 2011 and 2012.

¶ Plant height index = plant height mature/green vine length.

Lentils

Lentil was domesticated approximately 10,000 yr ago in areas of present-day Syria, Turkey, Iran, and Iraq (Harlan, 1992). Recent taxonomic studies (Galasso, 2003; Sonnante et al., 2003) have identified six species in the genus *Lens*, which includes *L. culinaris* (spp. *culinaris* and spp. *orientalis*), *L. odemensis, L. tomentosus, L. nigricans, L. ervoides,* and *L. lamottei*. Cultivated lentil (*L. culinaris* spp. *culinaris*) is proposed to have been developed from *L. culinaris* spp. *orientalis,* with domestication associated with increases in plant height, number of leaves and total leaf area, along with larger pods and seeds (Cubero et al., 2009).

Global lentil production during 2006 to 2010 averaged nearly 3.6 million tonnes, with Canada being the largest producer, followed by India, Turkey, the United States, and Nepal (FAOSTAT, 2013). Cultivated lentils may be divided into market classes that are based predominately on seed size, the color of the seed coat, and cotyledon color. The majority of lentils produced in the United States are exported, and specific market classes are targeted to distinct global markets (Muehlbauer et al., 2009). Red lentils are typically exported to nations in southwest Asia, including India, Sri Lanka, and Bangladesh. Spanish Brown lentils, which have small seed ($\approx$4 g 100 seed^{-1}), a dark seed coat, and yellow cotyledons, are primarily exported to Spain. A broader market exists for medium and large yellow lentils, which includes the cultivars Brewer (Muehlbauer, 1987) and Riveland (McPhee and Muehlbauer, 2009), and they are exported to Latin America, Europe, and North Africa. Another market class grown in the United States is small yellow lentils, such as the cultivar Eston, which produces small seed ($\approx$4.5 g 100 seed^{-1}) with a green seed coat and yellow cotyledons.

Similar to dry pea, commercial lentil production in the United States began in the1930s in the Palouse region of eastern Washington and northern Idaho (Youngman, 1967). This region remained the primary production area in the United States until the 1990s, when lentil production began to expand to the dryland regions of central and eastern Montana and North Dakota. The largest lentil producing state in the United States is Montana, followed closely by North Dakota, then Washington and Idaho (USDA-NASS, 2013). Currently, approximately 78% of the total lentil production in the United States occurs in Montana and North Dakota, and the remaining 22% is produced in Washington and Idaho (USDA-NASS, 2013). The area of lentils harvested annually in the United States has increased considerably over the past 25 yr (1987–2011) (Fig. 5–11). During the most recent 4 yr (2008–2011) of the 25-yr (1987–2011) period, the average area harvested in the United States was 173,340 ha, a 326% increase over the average of 40,661 ha harvested during the first 4 (1987–1990) of the past 25 yr (USDA-NASS, 2013). The rapid increase over time in the area harvested in the NGP compared with the PNW is evident in Fig. 5–11. During the most recent 4 yr (2008–2011) the average area harvested in the PNW was 44,348 ha, only a 7% increase over the average of 41,411 ha harvested from this region during the first 4 (1987–1990) of the past 25 yr (USDA-NASS, 2013). In contrast, during the most recent 4 yr, the average area of lentils harvested in the NGP was 128,993 ha, a 510% increase over the average of 21,162 ha for the first 4 yr (1997–2000) of data available from USDA-NASS for lentil production in the NGP.

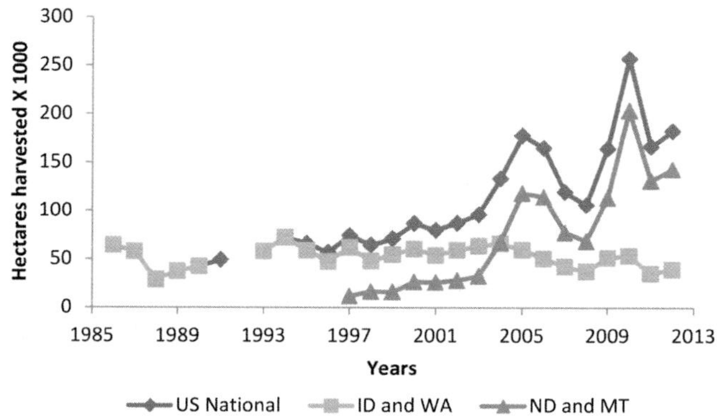

Fig. 5–11. Historical changes in the area (ha) of lentils harvested in the total United States, the U.S. Pacific Northwest (Idaho and Washington), and the U.S. Northern Plains (Montana and North Dakota) during 1987 to 2011.

On-Farm Productivity Trends for Lentils

Although both the area harvested and total production of lentils in the United States has increased considerably over the 25-yr period from 1987 through 2011, lentil yields have been essentially unchanged, increasing only 0.1% per year (Fig. 5–12). This change from an average of 1345 kg ha^{-1} (1201 lb ac^{-1}) during 1987 to 1991 to an average of 1368 kg ha^{-1} (1222 lb ac^{-1}) during 2007 to 2011 represents an average yield gain per year of only 1 kg ha^{-1} (0.9 lb ac^{-1}). In the PNW over the past 25 yr, lentil yields have remained unchanged, decreasing approximately 0.1% per year. Yields in the NGP have decreased by approximately 0.1% per year over the past 14 yr of available data.

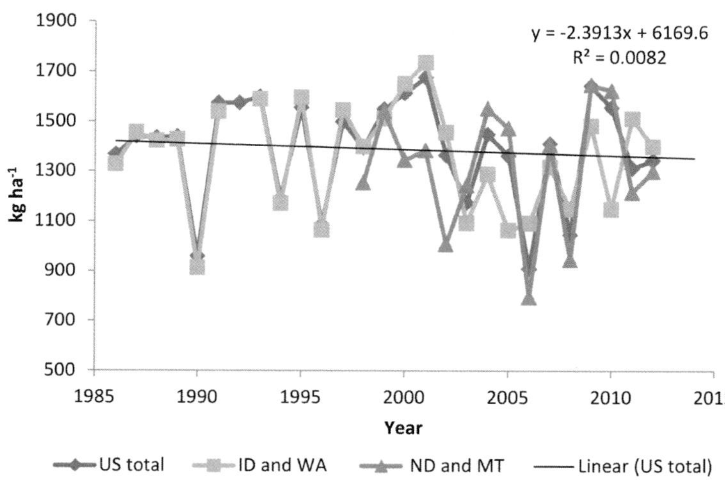

Fig. 5–12. Historical changes in on-farm lentil yields (kg ha^{-1}) across the total United States, the U.S. Pacific Northwest (Idaho and Washington), and the U.S. Northern Plains (Montana and North Dakota) during 1987 to 2011.

Basis for Yield Improvement in Lentil

Successful lentil cultivars must have traits necessary for field production and market acceptance. Critical production traits include yield, plant height, lodging tolerance, pod height, flowering time, number of seeds per pod, and reduced pod dehiscence (Sharma, 2009). Although pod dehiscence serves as a trait to promote survival in wild *Lens* spp., it is a very undesirable trait in cultivated lentil because pod drop reduces yields. Increased plant height, tolerance to lodging, and high pod height facilitates mechanical harvesting and reduces pod damage. Low pod height at harvest causes damage to harvesting combines through contact with rocks and soil along with contamination of harvested grains with soil, rocks, and plant stubble. Inheritance of plant height in lentil appears to be conditioned by both genetic and environmental sources of variance, with narrow sense heritability (h^2_n) estimates for plant height ranging from 0.35 (Khodambashi et al., 2012) to 0.65 (Tullu et al., 2008).

Considerable yield losses in lentil occur annually because of diseases, and cultivar development programs place great emphasis on improving resistance to important diseases. Lentils in the PNW are affected by the soilborne pathogens *Aphanomyces euteiches* Drechsler (Vandemark and Porter, 2010) and *Rhizoctonia solani* Kühn (Porter et al., 2011). Foliar diseases caused by fungi also reduce lentil yields in the United States, particularly in the NGP, where humidity during the growing season averages higher than in the PNW. These include anthracnose caused by *Colletotrichum truncatum* Schwein. (Buchwaldt, 2011) and Stemphylium blight caused by *Stemphylium botryosum* Wallr. (Banniza, 2011). Lentils in the PNW are also impacted by a number of viruses, including *Pea streak virus* and *Bean yellow mosaic virus* (Schwinghamer and Larsen, 2011).

Lentil is a self-pollinated diploid ($2N = 2X = 14$) crop that produces cleistogamous flowers. All commercial cultivars are pure lines that have been developed through a modified bulk-pedigree descent breeding approach (Muehlbauer et al., 2009). The initial seed stocks of lentil introduced into the United States in the 1930s were medium green (Persian) and large green (Chilean) landraces brought from Europe (Youngman, 1967). Harlan (1950) collected lentil accessions in Turkey that were maintained by the USDA Plant Introduction System. Beginning in 1961, these accessions were evaluated in lentil yield trials conducted by Washington State University. During the 1930s through the 1970s, growers in the United States produced their own common seed or purchased common seed from processors.

Commercial lentil cultivars grown in the United States have been developed from PI accessions, landraces, and breeding materials from other countries. The small yellow cultivar Eston is a pure line derived from PI 179307 (Slinkard, 1981). The popular medium yellow cultivar Brewer was developed using a black seeded landrace (no name available) that conferred earlier flowering as compared to the landrace Chilean (Muehlbauer, 1987). The large yellow cultivar Palouse was developed from a population derived from a cross between the Canadian lentil cultivar Laird and the Argentine cultivar Precoz (Muehlbauer, 1992), with large seed size coming from Laird and early flowering coming from Precoz. Laird also served as a parent in the cross used to develop Riveland, an exceptionally large yellow lentil cultivar (McPhee and Muehlbauer, 2009). Richlea, a high-yielding medium yellow lentil developed in Canada, was crossed to PI 297754, which has tolerance to lodging and tall height, to produce the recently released medium yellow cultivar Essex (Vandemark et al., 2011).

Major efforts to develop improved lentil cultivars for U.S. production began in the 1970s by the USDA-ARS in cooperation with Washington State University, located in Pullman, WA. Selections from this breeding program served as sources for the majority of lentil cultivars grown in the United States, including Brewer (Muehlbauer, 1987) and Merrit (Muehlbauer and McPhee, 2004). Currently, the majority of lentil cultivars grown in the United States are derived from either the USDA-ARS breeding program, the lentil breeding program from the University of Saskatchewan, or landraces introduced from Europe.

Yield Gains Measured in Recent Lentil Cultivar Trials

Older lentil cultivars and newer cultivars and advanced breeding lines were compared for yield and other metric traits. The comparisons are based on the results of field trials conducted at three locations in the PNW (Genesee, ID; Pullman, WA; and Fairfield, WA) during 2011 and 2012. All three field sites are located in areas of commercial lentil production. The results are presented in Table 5–2 for three different lentil market classes. For each market class, the breeding lines selected for comparisons were the highest yielding lines across all six location years. The newer cultivars and breeding lines in the small yellow market class were compared to the cultivar Eston (Slinkard, 1981). The results suggest that

Table 5–2. Yield and agronomic characteristics of recent lentil breeding lines and older cultivars based on field trials conducted in Washington and Idaho in 2011 and 2012.

Entry	Release	Yield†	Seed size‡	Days to mature‡	Canopy height‡	Plant height index‡¶
		kg ha^{-1}	g 100 seeds^{-1}	d	cm	
			Small yellow lentils			
Essex	2009	1593 A	4.4 B	94.7	34.4 A	0.93
LC08600005E	Line	1438 AB	4.8 A	94.7	31.0 B	0.90
LC01602273E	Line	1375 AB	3.3 D	92.5	29.1 C	0.90
LC09600054E	Line	1299 AB	3.8 C	93.3	28.4 C	0.96
Eston	1981	1191 B	3.2 D	94.7	28.5 C	0.92
LSD (α = 0.05)		383	0.2	ns	1.8	ns
			Spanish brown lentils			
LC08600116P	Line	1643	5.0 A	94.3	25.5	0.77
LC08600113P	Line	1579	4.5 B	93.3	26.6	0.74
LC0860115P	Line	1556	4.6 B	94.7	26.5	0.77
Pardina	< 1980	1469	4.0 C	92.0	24.8	0.75
Morena	2012	1353	3.9 C	94.7	28.6	0.77
LSD (α = 0.05)		Ns	0.3	ns	ns	ns
			Large yellow lentils			
LC07600541L	Line	1369	7.1 B	96.5	36.5 A	0.91
LC06601734L	Line	1356	7.3 AB	96.5	35.8 A	0.89
LC0860B130L	Line	1352	7.4 AB	96.5	35.4 AB	0.91
Riveland	2009	1239	7.5 A	94.7	35.5 AB	0.89
Brewer	1987	1192	6.2 C	92.0	32.3 B	0.91
LSD (α = 0.05)		Ns	0.4	ns	3.3	

† Based on trials conducted at Pullman, WA; Genesee, ID; and Kendrick, ID in 2011 and 2012 (six location years).

‡ Based on trials conducted at Pullman, WA in 2011 and 2012.

¶ Plant height index = plant height mature/green vine length.

considerable gains in yield have been made over time in breeding for this lentil market class. The recently released cultivar Essex (Vandemark et al., 2011) had significantly higher yields than Eston (Table 5–2). Collectively, Essex and the three USDA-ARS breeding lines had yields that on average were 20% higher than Eston. Three entries produced seed that was significantly larger than Eston. The larger seed size of high-yielding lines compared to Eston is a concern, as it remains to be determined if these differences in seed size will be tolerated by traditional consumers of Eston lentils. The breeding line LC01602273E is currently being considered for release as a new cultivar on the basis of its high yield (15% greater than Eston) and the similarity of its seed size to that of Eston.

Considering selected traits of agronomic importance to lentil, no significant differences between entries were observed for days to maturity or plant height index (Table 5–2). This suggests that yield gains have been realized in newer lentil lines of the Eston market class without any appreciable increase in time needed to reach maturity or decrease in lodging tolerance. The canopy height was significantly greater in the two highest yielding entries than in all other entries.

The performances of newer cultivars and breeding lines of the Spanish Brown market class were compared with the cultivar Pardina (Table 5–2). Pardina is the only cultivar of the Spanish Brown market class that has been grown to date commercially in the United States. No significant differences in yield were detected between 'Pardina' and the four newer entries. The three highest yielding breeding lines had an average yield of approximately 1593 kg ha^{-1} (1423 lb ac^{-1}), which is 8% higher than the yield of 'Pardina.' However, the yield of Morena, a Spanish Brown lentil cultivar recently released by the USDA-ARS, was nearly 9% less than Pardina. 'Morena' was released on the basis of several years of evaluation in the PNW, where it consistently exhibited yields equivalent to or higher than 'Pardina,' along with having a higher canopy height that 'Pardina.' Recently, 'Morena' was found to be susceptible to Stemphylium blight, caused by the fungus *Stemphylium botryosum* Wallr. The disease has historically been a problem for lentil production in Canada and the U.S. Northern Plains (Banniza, 2011) but had not been detected in lentils in the PNW until 2012, when it was detected in seed increase fields of 'Morena' in Washington and Idaho. The low yields observed for 'Morena' are considered to be primarily due to its susceptibility to Stemphylium blight, which is becoming an emerging disease of importance for lentil production in the PNW.

Similar to comparisons among entries in the small yellow market class, no significant differences were observed between 'Pardina' and the newer entries for days to maturity or plant height index (Table 5–2). The three highest yielding breeding lines all produced seed that was significantly larger than 'Pardina.' The cultivar Morena has received very favorable reviews from growers of Spanish Brown lentils, largely due to its seed size, which is not significantly different than 'Pardina.'

The performances of newer cultivars and breeding lines of the large yellow market class were compared with the cultivar Brewer (Table 5–2). No significant differences in yield were detected between 'Brewer' and the four newer entries. The three highest yielding breeding lines had an average yield of 1359 kg ha^{-1} (1214 lb ac^{-1}), which is 14% higher than the yield of 'Brewer.' No significant differences were observed between 'Brewer' and the newer entries for days to maturity or plant height index. 'Brewer' produced significantly smaller seed than any other entry, and its canopy height was also significantly smaller than all other entries. Unlike the small yellow and Spanish Brown market classes of lentil, the medium and large yellow lentil cultivars cultivated in the United States, including Brewer, Merrit, and Riveland, span a range of seed sizes, and new breeding lines that have high yield coupled with larger seed size may be promising for commercial production.

It should be noted that although the advanced breeding lines included in this analysis appear to be promising candidates for release as new cultivars on the basis of their agronomic performance, they still must be examined for other important post-harvest traits that will impact their suitability for commercial production. These traits include how well the seeds can be split, seed water retention and cooking time, and palatability (Bressani and Elias, 1988). Other quality traits of importance to lentil cultivars include protein and carbohydrate digestibility and quantities of antinutritional factors such as trypsin inhibitors and tannins (Bressani and Elias, 1988).

Chickpea

Domesticated chickpea is thought to have been derived 8000 to 10,000 yr ago from initial selections in the progenitor species *Cicer reticulatum* Ladizinsky (Ladizinsky and Adler, 1976), which had a limited distributed throughout southern Turkey (Berger et al., 2003). During the early domestication of chickpea, it was converted from a fall-sown crop to a spring-sown crop in southwestern Asia and southeastern Europe, most likely in response to severe crop losses due to Ascochyta blight [*Ascochyta rabiei* (Pass.) Lab.], the most globally destructive disease of chickpea, both past and present (Abbo et al., 2003; Bayaa and Chen, 2011). Currently, there are 44 different species, 9 annual and 35 perennial, recognized in the genus *Cicer*, of which only one, *C. arientinum*, is cultivated (van der Maesen et al., 2007). Chickpeas can be divided into two major classes, the macrosperma, or "kabuli" class, and the microsperma, or "desi" class (Toker, 2009). Kabuli chickpea seeds are shaped like an "owl head" and are larger and lighter in color than desi type chickpeas, which have a "teardrop" seed shape. Kabuli type chickpeas are cooked and then used for salads, canned and eaten whole, or used to make hummus, whereas desi chickpeas are predominately split and then cooked (Newman et al., 1988).

Commercial chickpea production in the United States began in the 1930s in the San Joaquin Valley of southern California. The first chickpeas grown were large white kabuli chickpeas known as "Spanish White" or "Spanish Common" (Muehlbauer et al., 1982). Chickpea production in California was centered in the San Joaquin Valley, where they were sown in the fall, grown under irrigation, and harvested in March to April. In the 1980s, as demand for pea decreased and large increases in lentil production by Canada restricted lentil prices, chickpea production expanded in the dryland wheat production areas of Washington and Idaho. Currently, U.S. chickpeas are primarily grown in Washington and Idaho, with lesser amounts in Colorado, Montana, Nebraska, North Dakota, and California. Only kabuli chickpeas are produced commercially to any appreciable degree in the United States. Over the 25-yr period from 1987 through 2011, the area of chickpeas harvested annually in the United States has dramatically increased, with the majority of area being in Washington and Idaho (Fig. 5–13). During the

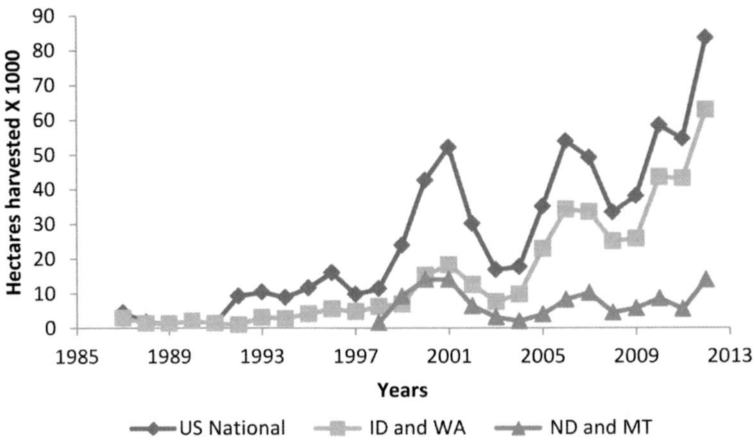

Fig. 5–13. Historical changes in the area (ha) of chickpeas harvested in the total United States, the U.S. Pacific Northwest (Idaho and Washington), and the U.S. Northern Plains (Montana and North Dakota) during 1987 to 2011.

most recent 5 yr (2007–2011) of the past 25-yr (1987–2011) period, the average area harvested in the United States was 46,612 ha, which is a 835% increase over the average of 5580 ha harvested in the United States during the first 5 (1987–1991) of the past 25 yr (USDA-NASS, 2013).

On-Farm Productivity Trends for Chickpeas

Although both the area harvested and total production of chickpeas in the United States have increased considerably over the recent 25-yr period from 1987 to 2011, chickpea yields in the United States on a per acre basis have increased by approximately 2.8% per year over a recent 25-yr period (1987–2011) (Fig. 5–14). This increase translates to a yield gain per year of 26 kg ha^{-1} (23.2 lb ac^{-1}). However, yields have not increased consistently over this time, with the greatest yield increases being observed during 1992 to 1996. These increases were likely due to the introduction of chickpea cultivars with partial resistance to Ascochyta blight and improved fungicides for blight control.

Basis for Yield Improvement in Chickpea

Crop yield in chickpea has several components, including the number of pods per plant, seeds per pod, and seed weight (Kumar and Arora, 1991; Güler et al., 2001). Early generation selection of individual plants has not resulted in improved chickpea yields, so commercial chickpea cultivars are pure lines predominately developed through modified bulk-pedigree descent breeding (Salimath et al., 2007). Large seeded chickpea lines tend to have lower yields than lines with smaller seeds, and these differences are attributed to larger seeded lines having fewer pods per plant and fewer seeds per pod than small seeded lines (Liu et al., 2003). A significant negative correlation was observed between yield and seed weight in kabuli chickpeas, while a significant positive correlation between yield and number of pods per plant was detected (Gowda et al., 2011). Similarly, Cobos et al. (2007) observed a significant negative correlation between seed weight and

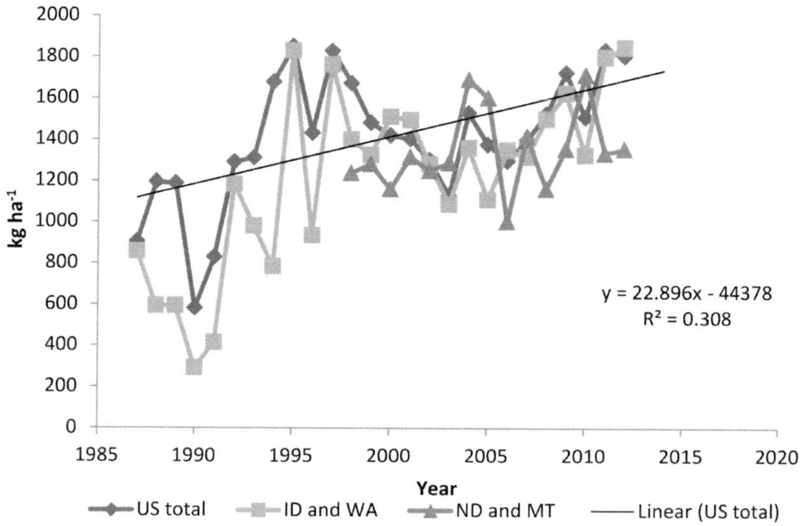

Fig. 5–14. Historical changes in on-farm chickpea yields (kg ha^{-1}) across the total United States, the U.S. Pacific Northwest (Idaho and Washington), and the U.S. Northern Plains (Montana and North Dakota) during 1987 to 2011.

yield in a population derived from a kabuli × desi cross, with a broad sense heritability estimate for yield of 0.14. Dehghani et al. (2010) determined that the total contributions to yield variance in chickpea of environmental effects, genotype effects, and genotype × environment interaction effects were approximately 60, 9, and 23%, respectively. Significant genetic and genetic × environment interaction effects were detected for both yield and seed weight in large seeded kabuli chickpea genotypes (Gowda et al., 2011).

Leaf type may also influence yield in chickpea. Chickpeas have two leaf types, unifoliate and fern type (Muehlbauer and Singh, 1987). Unifoliate leafs are characterized by a single lobate leaf attached to the petiole, while fern leaves are composed of several small serrated leaflets attached to a single petiole. Li et al. (2006) observed that fern leaf type chickpea cultivars had higher yields than unifoliate leaf type cultivars when grown in short season environments in Saskatoon. Higher yields in fern leaf types were attributed to a higher growth rate and greater partitioning of dry matter to seeds than in unifoliate leaf types.

Chickpea yield is also influenced by a range of diseases and abiotic stresses. Devastating crop losses occurred in the PNW in the 1980s because of epidemics of Ascochyta blight (Kaiser and Muehlbauer, 1988). Consequently, improving levels of resistance to Ascochyta blight remains a priority for chickpea breeding in the United States. Two different pathotypes of *A. rabiei* have been reported in the United States, and the evaluation of chickpea lines and cultivars for disease reaction suggests that resistance to each pathotype is inherited independently (Chen et al., 2004). Besides Ascochyta blight, chickpea production in the United States is also impacted by other diseases such as *Bean leaf roll virus* (Larsen and Schwinghamer, 2011) and wilt caused by *Fusarium oxysporum* f. sp. *cicieris* (Jiménez-Díaz et al., 2011).

Two other important traits for chickpea in the United States are early maturity and large seed size. Chickpeas are typically harvested in the United States during August to September to avoid autumn precipitation. However, chickpeas often have not sufficiently senesced by this time to be dry enough for harvesting. Consequently, most chickpea growers in the United States apply a desiccant herbicide, such as glyphosate, to the crop 7 to 10 d before harvesting (McKay et al., 2002). Earlier maturing chickpea cultivars might allow for harvest without using desiccant herbicide. Early maturing cultivars may also permit delayed sowing, which is part of an integrated strategy to manage ascochyta blight (Bayaa and Chen, 2011). Inheritance of maturity in chickpea can be conditioned by both epistatic (Anbessa et al., 2006) and quantitative (Lichtenzveig et al., 2006) gene action.

Large seed size is a critical trait for chickpeas intended for the fresh or canning markets. The USDA-NASS classifies chickpeas grown in the United States as "small" (<9 mm seed diameter) or "large" (≥9 mm) (USDA-NASS, 2013). This trait has considerable economic importance, as large chickpeas have a higher value than small chickpeas. For example, in 2011 the average value for large chickpeas was ≈\$39 cwt^{-1} compared with an average of ≈\$24 cwt^{-1} for small chickpeas. Previous reports suggest that inheritance of seed size in chickpea is conditioned by both genotype and genotype × environment effects (Gowda et al., 2011).

Chickpea is a self-pollinated diploid ($2N = 2X = 16$) crop that produces cleistogamous flowers. Current commercial cultivars are pure lines that have been developed through a modified bulk-pedigree descent breeding approach (Salimath et al., 2007). Initially in the United States, the landrace Spanish White was grown in California (Muehlbauer et al., 1982). Spanish White produces a large, very light colored seed that has a high value. Direct selections from Spanish White resulted in the cultivars UC-5 and Mission, and these three cultivars were grown during the early years of chickpea production in the PNW (Muehlbauer et al., 1982). However, the incidence of Ascochyta blight began to increase in Washington and Idaho in the early 1980s until an epidemic in 1987 resulted in nearly complete loss of the chickpea crop in this region (Bayaa and Chen, 2011). Chickpea accessions from ICARDA were screened for reaction to Ascochyta blight in the field in Washington, and resistant lines were identified (Kaiser, 1991). The resistant kabuli chickpea line FLIP 85–58 was crossed with Surutato 77, a large kabuli chickpea cultivar from Mexico, which had high yield, resistance to Fusarium wilt, and early flowering (Muehlbauer et al., 1982). The FLIP 85–58 × 'Surutato 77' cross resulted in two sibling line that were released as the cultivars Sanford (Muehlbauer et al., 1998a) and Dwelley (Muehlbauer et al., 1998b). These were the first two chickpea cultivars with appreciable levels of resistance to Ascochyta blight under field conditions. 'Surutato 77' also contributed early flowering and improved Fusarium wilt resistance to 'Dwelley' and 'Sanford.' In 2004, the kabuli chickpea cultivar Sierra was released (Muehlbauer et al., 2004). 'Sierra' was the result of the following three-way cross: Dwelley/FLIP 85–58/Spanish White. FLIP 85–58 provided additional resistance to Ascochyta blight, while Spanish White served as a source of large seed size and light seed color. Currently the majority of kabuli chickpeas produced in the United States are from 'Dwelley' and 'Sierra,' although small seeded chickpeas grown primarily for hummus production include cultivars developed by the University of Saskatchewan, such as 'CDC Frontier,' or local landraces that are commonly referred to as "Billybeans." In 2013 the USDA-ARS released two new large café kabuli chickpea cultivars,

'CA04900843C' and 'CA04900851C,' based on their high yields and large seed size demonstrated in advanced yield trials (Grain Legume Genetics and Physiology Research Unit, 2012).

Yield Gains Measured in Chickpea Cultivar Trials

The chickpea cultivars Dwelley (Muehlbauer et al., 1998b) and Sierra (Muehlbauer et al., 2004) represent the great majority of chickpeas produced in the United States. The performances of Dwelley and Sierra during 2011 and 2012 are compared of two new café kabuli chickpea cultivars approved for release by the USDA-ARS in 2013, CA04900843C and CA04900851C (Table 5–3). 'CA04900851C' has simple leaves like 'Sierra' and 'Dwelley,' while 'CA04900843C' has compound leaves. There were no significant differences among entries for yield, with 'Sierra' averaging 3% less than 'Dwelley,' while yields of 'CA04900843C' and 'CA04900851C' were 16 and 8% greater, respectively, than 'Dwelley.' 'CA04900843C' produced significantly larger seed than the other cultivars. 'CA04900843C' reached 50% flowering significantly earlier than the other cultivars. Differences between 'CA04900851C,' 'Sierra,' and 'Dwelley' were not significant for days to 50% flower or seed size.

Summary and Conclusions Regarding Yields in Cool Season Food Legumes

Over the past 25 yr, the production of cool season food legumes (peas, lentils, and chickpeas) in the United States has increased dramatically, from 240,000 t in 1987 to 570,000 t in 2011 (USDA-NASS, 2013). During this time the production area in the United States has risen considerably, from approximately 125,000 ha in 1987 to over 380,000 ha in 2011. Unfortunately, crop yield per unit of area (kg ha^{-1}) has not seen similar increases over the past 25 yr. Over the past 25 yr, dry pea yields in the United States have decreased by 0.3% per year, lentil yields in the United States have increased by only 0.1% per year, and chickpea yields have increased by approximately 2.8% per year.

Table 5–3. Yield and agronomic characteristics of newly released and older chickpea cultivars based on field trials conducted in Washington and Idaho in 2011 and 2012.

Entry	Release	Leaf type†	Yield‡	Seed size¶	Days to flower ¶	Reaction to Ascochyta blight#
			kg ha^{-1}	g 100 seed^{-1}		
CA04900843C	2013	C	2299	62.8 A	55.8 A	1.8
CA04900851C	2013	S	2148	56.5 B	57.5 B	2.0
Dwelley	1998	S	1981	53.2 B	56.3 AB	2.6
Sierra	2004	S	1923	54.3 B	56.0 AB	2.5
LSD (α = 0.05)			ns	3.9	1.6	ns

† C, compound; S, simple.

‡ Based on trials conducted at Pullman, WA; Genesee, ID; and Kendrick, ID in 2011 and 2012 (six location years).

¶ Based on trials conducted at Pullman, WA in 2011 and 2012.

Each plant within a plot was scored on a scale of 0–9 as follows: 1 = healthy plant; 3 = lesions clearly present but plant remains mostly green; 5 = lesions girdle stems, most leaves show lesions; 7 = plant dying but at least three green leaves present, and 9 = dead plant (Chen et al., 2004). The score of the most severely affected plant within a plot was used for the plot score.

Several factors are likely responsible for limited yield gains over the past 25 yr for these crops. Pea and lentil production has increased dramatically over this time, especially in the NGP. The cultivars grown in the NGP were originally selected on the basis of their performance in the PNW and were likely not well adapted for the higher disease pressure and more extreme climatic conditions encountered in the NGP. The continued efforts of the cool season food legume breeding program at North Dakota State University should develop new cultivars adapted for the NGP. The further evaluation in the NGP of plant materials from other breeding programs should also identify promising plant genotypes for the NGP.

Another likely factor restraining yield gains in cool season food legumes is the limited availability of new cultivars. Although private breeding efforts, in particular those of Progene Plant Research (Othello, WA), have produced several green pea cultivars, including Banner, Ariel, and Aragorn, there have been few new cultivars of cool season food legumes. 'Dwelley,' released in 1998 and 'Sierra,' released in 2004, dominate the U.S. chickpea market. The Spanish Brown lentil market has been dominated by the cultivar Pardina for over 30 yr. Important large green lentil cultivars including Laird and Brewer have been grown for at least the past 25 yr. However, the areas where these cultivars are cultivated have changed markedly in terms of pressure from diseases, pests, and weeds.

In conclusion, advances over the past 25 yr in the production of cool season food legumes in the United States has been the result of increasing the area of cultivation of these crops as opposed to being due to yield gains that can be attributed to genetic enhancement. Several cool season legume cultivars currently in production were released over 20 yr ago. New advanced breeding lines of peas, lentils, and chickpeas are currently being evaluated that have yields that are superior to commercial cultivars. Continuing efforts to develop pea and lentil lines that are especially adapted to the U.S. Northern Plains will result in improved yields in this emerging area of importance for U.S. cool season food legume production.

References

Abbo, S., J. Berger, and N.C. Turner. 2003. Evolution of cultivated chickpea: Four bottlenecks limit diversity and constrain adaptation. Funct. Plant Biol. 30:1081–1087. doi:10.1071/FP03084

Adams, M.W. 1967. Basis of yield component compensation in crop plants with special reference to the field bean, *Phaseolus vulgaris*. Crop Sci. 7:505–510. doi:10.2135/cropsci1967.0011183X000700050030x

Adams, M.W. 1982. Plant architecture and yield breeding. Iowa State J. Res. 56:225–254.

Anbessa, Y., T. Warkentin, A. Vandenberg, A., and R. Ball. 2006. Inheritance of time to flowering in chickpea in a short-season temperate environment. J. Hered. 97:55–61. doi:10.1093/jhered/esj009

Andersen, A.L., E.E. Down, and G. Whitford. 1960. The Sanilac pea bean—Its history, development, and characteristics. Mich., Agric. Exp. Stn., Q. Bull. 43(2):214–236.

Banniza, S. 2011. Stemphylium blight of lentil. In: W. Chen, H.C. Sharma, and F.J. Muehlbauer, editors, Compendium of chickpea and lentil diseases and pests. APS Press, St Paul, MN. p. 60–62.

Bayaa, B., and W. Chen. 2011. Ascochyta blight of chickpea. In: W. Chen, H.C. Sharma, and F.J. Muehlbauer, editors, Compendium of chickpea and lentil diseases and pests. APS Press, St Paul, MN. p. 34–40.

Beaver, J.S., and J.M. Osorno. 2009. Achievements and limitations of contemporary common bean breeding using conventional and molecular approaches. Euphytica 168:145–176. doi:10.1007/s10681-009-9911-x

Berger, J., S. Abbo, and N.C. Turner. 2003. Ecogeography of annual wild *Cicer* species: The poor state of the world collection. Crop Sci. 43:1076–1090. doi:10.2135/cropsci2003.1076

Biddle, A.J. 2000. Sclerotinia white mold. In: J.M. Kraft and F.L. Pfleger, editors, Compendium of pea diseases. 2nd ed. APS Press, St Paul, MN. p. 30–31.

Bingham, J. 1967. Breeding cereals for improved yielding capacity. Ann. Appl. Biol. 59:312–315. doi:10.1111/j.1744-7348.1967.tb04442.x

Bowen, S.C. 1898. History of the bean industry. In: Annual Report of the New York State Agricultural Society, Wynkoop Hallenbeck Crawford, Co., Albany, NY. p. 323–329.

Bressani, R., and L.G. Elias. 1988. Seed quality and nutritional goals in pea, lentil, faba bean and chickpea breeding. In: R.J. Summerfield, editor, World crops: Cool season food legumes. Kluwer, Dordrecht, The Netherlands. p. 381–404.

Bretag, T.W., and M. Ramsey. 2000. Ascochyta spp. In: J.M. Kraft and F.L. Pfleger, editors, Compendium of pea diseases. 2nd ed. APS Press, St. Paul, MN. p. 24–28.

Brick, M.A., P.F. Byrne, H.F. Schwartz, J.B. Ogg, K. Otto, A.L. Fall, and J. Gilbert. 2006. Reaction to three races of Fusarium wilt in the *Phaseolus vulgaris* core collection. Crop Sci. 46:1245–1252. doi:10.2135/cropsci2005.06-0102

Brick, M.A., and K.F. Grafton. 1999. Improvement of medium seeded race Durango cultivars. In: S. Singh, editor, Common bean improvement in the twenty-first century. Kluwer, Dordrecht, the Netherlands. p. 223–253.

Brick, M.A., J.B. Ogg, H.F. Schwartz, J.J. Johnson, and F. Judson. 2011. Registration of 'Croissant' pinto bean. J. Plant Reg. 5:299–303. doi:10.3198/jpr2010.07.0443crc

Broughton, W.J., G. Hernández, M. Blair, S. Beebe, P. Gepts, and J. Vanderleyden. 2003. Beans (*Phaseolus* spp.)—Model food legumes. Plant Soil 252:55–128. doi:10.1023/A:1024146710611

Buchwaldt, L. 2011. Anthracnose of lentil. In: W. Chen, H.C. Sharma, and F.J. Muehlbauer, editors, Compendium of chickpea and lentil diseases and pests. APS Press, St Paul, MN. p. 31–34.

Burke, D.W., and D.E. Miller. 1983. Control of *Fusarium* root rot with resistant beans and cultural management. Plant Dis. 67:1312–1317. doi:10.1094/PD-67-1312

Burkholder, W.H. 1918. The production of an anthracnose-resistant white marrow bean. Phytopathology 8:353–359.

Câmara, C.R.S., C.A. Urrea, and V. Schlegel. 2013. Pinto beans (*Phaseolus vulgaris* L.) as a functional food. Agriculture 3:90–111. doi:10.3390/agriculture3010090

Chen, W., C. Coyne, T. Peever, and F.J. Muehlbauer. 2004. Characterization of chickpea differentials for Ascochyta blight and identification of resistance sources for *Didymella rabiei*. Plant Pathol. 53:759–769. doi:10.1111/j.1365-3059.2004.01103.x

Cobos, M.J., J. Rubio, M.D. Fernández-Romero, R. Garza, M.T. Morena, T. Millán, and J. Gil. 2007. Genetic analysis of seed size, yield and days to flowering in a chickpea recombinant inbred line population derived from a Kabuli × Desi cross. Ann. Appl. Biol. 151:33–42. doi:10.1111/j.1744-7348.2007.00152.x

Cox, T.S., J.P. Shroyer, L. Ben-Hui, R.G. Sears, and T.J. Martin. 1988. Genetic improvement in agronomic traits of hard red winter wheat cultivars from 1919 to 1987. Crop Sci. 28:756–760.

Coyne, D.P. 1980. Modification of plant architecture and crop yield by breeding. HortScience 15:244–247.

Cubero, J.I., M. Pérez de la Vega, and R. Fratini. 2009. Origin, phylogeny, domestication and spread. In: W. Erskine, F.J. Muehlbauer, A. Sarker, and B. Sharma, editors, The lentil: Botany, production and uses. CABI, Oxfordshire, UK. p. 13–33.

Dehghani, H., S.H. Sabaghpour, and A. Ebadi. 2010. Study of genotype × environment interaction for chickpea yield in Iran. Agron. J. 102:1–8.

Down, E.E., and A.L. Andersen. 1956. Agronomic use of an X-ray induced mutant. Science 124:223–224.

Down, E.E., and J. W. Thayer, Jr. 1938. The Michelite bean. Spec. Bull. 295. Mich. Agric. Exp. Stn., East Lansing, MI.

Duvick, D.N., and K.G. Cassman. 1999. Post-green revolution trends in yield potential of temperate maize in the north-central United States. Crop Sci. 39:1622–1630.

Evans, L.T., and R.A. Fisher. 1999. Yield potential: Its definition, measurement, and significance. Crop Sci. 39:1544–1551.

FAOSTAT.2013. Production. Food and Agricultural Organization of the United Nations. http://faostat.fao.org/site/339/default.aspx (accessed on 1 Apr. 2013).

Fisher, A.G., M.A. Brick, D.R. Wood, M. Stack, H.F. Schwartz, J.B. Ogg, C.H. Pearson, J.F. Shanahan, and M. Ballarin. 1995. Registration of Fisher Pinto Bean. Crop Sci. 35:1511.

Fischer, R.A., D. Byerlee, and G.O. Edmeades. 2009. Can technology deliver on the yield challenge to 2050? Paper presented at: Expert Meeting on How to Feed the World in 2050. Food and Agricultural Organization, Rome. 24–26 Jun.

Galasso, I. 2003. Distribution of highly repeated DNA sequences in species of the genus *Lens* Miller. Genome 46:1118–1124.

Godfray, H.C.J., J.R. Beddington, I.R. Crute, L. Haddad, D. Lawrence, J.F. Muir, J. Pretty, S. Robinson, S.M. Thomas, and C. Toulmin. 2010. Food security: The challenge of feeding 9 billion people. Science 327:812–818.

Gowda, C.L., H.D. Upadhyaya, N. Dronavalli, and S. Singh. 2011. Identification of large-seeded high-yielding stable kabuli chickpea germplasm lines for use in crop improvement. Crop Sci. 51:198–209.

Grain Legume Genetics and Physiology Research Unit. 2012. Dry pea, lentil, chickpea, and winter legume breeding. USDA-ARS. http://www.ars.usda.gov/SP2UserFiles/Place/53482000/ProgressReport/Progress%20Report%202012%20V8.pdf (accessed Oct. 2013).

Gritton, E.T. 1986. Pea breeding. In: M.J. Bassett, editor, Breeding vegetable crops. AVI Publ., Westport, CT. p. 283–319.

Güler, M., S. Adak, and H. Ulukan. 2001. Determining relationships among yield and some yield components using path coefficient analysis in chickpea (*Cicer arientinum* L.). Eur. J. Agron. 14:161–166. doi:10.1016/S1161-0301(00)00086-1

Haglund, W.A. 2000. Fusarium wilt. In: J.M. Kraft and F.L. Pfleger, editors, Compendium of pea diseases. 2nd ed. APS Press, St. Paul, MN. p. 14–16.

Harlan, J.R. 1950. Collection of crop plants in Turkey. Agron. J. 42:258–259. doi:10.2134/agronj1950.00021962004200050012x

Harlan, J.R. 1992. Crops and man. CSSA, Madison, WI.

Hawtin, G.C., F.J. Muehlbauer, A.E. Slinkard, and K.B. Singh. 1988. Current status of cool season food legume crop improvement: An assessment of critical needs. In: R.J. Summerfield, editor, World crops: Cool season food legumes. Kluwer, Dordrecht, The Netherlands. p. 67–80.

Jiménez-Díaz, R.M., M. del Mar Jiménez-Gasco, B.B. Landa del Castillo, P. Castillo, and J.A. Navas-Cortés. 2011. Fusarium wilt of chickpea. In: W. Chen, H.C. Sharma, and F.J. Muehlbauer, editors, Compendium of chickpea and lentil diseases and pests. APS Press, St Paul, MN. p. 16–20.

Kaiser, W.J. 1991. Host range studies with the Ascochyta blight pathogen of chickpea. Int. Chickpea Newsl. 25:25–27.

Kaiser, W.J., and F.J. Muehlbauer. 1988. An outbreak of Ascochyta blight of chickpea in the Pacific Northwest, USA, in 1987. Int. Chickpea Newsl. 18:16–17.

Kaplan, L. 1965. Archeology and domestication in American Phaseolus (beans). Econ. Bot. 19(4):358–368. doi:10.1007/BF02904806

Kelly, J.D. 2000. Remaking bean plant architecture for efficient production. Adv. Agron. 71:109–143. doi:10.1016/S0065-2113(01)71013-9

Kelly, J.D. 2004. Advances in common bean improvement: Some case histories with broader applications. Acta Hortic. 637:99–122.

Kelly, J.D., M.W. Adams, A.W. Saettler, G.L. Hosfield, G.V. Varner, M.A. Uebersax, and J. Taylor. 1990. Registration of 'Sierra' pinto bean. Crop Sci. 30:745–746. doi:10.2135/cropsci1990.0011183X003000030062x

Kelly, J.D., and K.A. Cichy. 2012. Dry bean breeding and production technologies. In: M. Siddiq and M.A. Uebersax, editors, Dry beans and pulses: Production, processing, and nutrition. Wiley-Blackwell, Ames, IA, p. 23–54.

Kelly, J.D., J. Kolkman, and K. Schneider. 1998. Breeding for yield in dry bean (*Phaseolus vulgaris* L.). Euphytica 102:343–356. doi:10.1023/A:1018392901978

Kelly, J.D., and V.A. Vallejo. 2004. A comprehensive review of the major genes conditioning resistance to Anthracnose in common bean. HortScience 39:1196–1207.

Khodambashi, M., N. Bitaraf, and S. Hoshmand. 2012. Generation mean analysis for grain yield and its related traits in lentil. J. Agric. Sci. Technol. 14:609–616.

Kumar, L., and P.P. Arora. 1991. Basis of selection in chickpea. Int. Chickpea Newsl. 24:14–15.

Kraft, J.M. 1981. Registration of 792022 and 792024 pea germplasm. Crop Sci. 21:352–353.

Kraft, J.M. 1992. Registration of 90–2079, 90–2131, and 20–2322 pea germplasms. Crop Sci. 32:1076.

Kraft, J.M., M.J. Silbernagel, and F.J. Muehlbauer. 1972. Registration of PH-14–119 and PH-91–3-pea germplasm. Crop Sci. 12:399.

Ladizinsky, G., and A. Adler. 1976. The origin of chickpea *Cicer arientinum* L. Euphytica 25:211–217.

Larsen, R.C., and M.W. Schwinghamer. 2011. Diseases caused by viruses. In: W. Chen, H.C. Sharma, and F.J. Muehlbauer, editors, Compendium of chickpea and lentil diseases and pests. APS Press, St. Paul, MN. p. 72–77.

Li, L., R.A. Bueckert, Y. Gan, and T. Warkentin. 2006. Biomass and yield performance of kabuli chickpea cultivars with the fern or unifoliate leaf trait in the Northern Great Plains. Can. J. Plant Sci. 86:1089–1097.

Lichtenzveig, J., D. Bonfil, H.B. Zhang, D. Shtienberg, and S. Abbo. 2006. Mapping quantitative trait loci in chickpea associated with time to flowering and resistance to *Didymella rabiei* the causal agent of Ascochyta blight. Theor. Appl. Genet. 113:1357–1369.

Licker, R., M. Johnston, J.A. Foley, C. Barford, C.J. Kucharik, C. Monfreda, and N. Ramankutty. 2010. Mind the gap: How do climate and agricultural management explain the 'yield gap' of croplands around the world? Glob. Ecol. Biogeogr. 19:769–782.

Liu, P.Y., Y. Gan, T. Warkentin, and C. McDonald. 2003. Morphological plasticity of chickpea in a semiarid environment. Crop Sci. 43:426–429. doi:10.2135/cropsci2003.0426

Lopes, M.S., M.P. Reynolds, Y. Manes, R.P. Singh, J. Crossa, and H.J. Braun. 2012. Genetic yield gains and changes in associated traits of CIMMYT spring bread wheat in a 'Historic' set representing 30 years of breeding. Crop Sci. 52:1123–1131.

MacKey, J. 1979. Genetic potentials for improved yield. In: S. Rajki, editor, Proceedings of a Workshop on Agricultural Potentiality Directed by Nutritional Needs, Martonvásár, Hungary. 5–9 Jun. Akad. Kiado, Budapest. p. 121–143.

Marx, G.A., W.T. Schroeder, R. Provvidenti, and W. Mishanec. 1972. A genetic study of tolerance in pea (*Pisum sativum*) to Aphanomyces root rot. J. Am. Soc. Hortic. Sci. 97:619–621.

McGee, R.J., C.J. Coyne, M.-L. Pilet-Nayal, A. Moussart, B. Tivoli, A. Baranger, C. Hamon, G. Vandemark, and K. McPhee. 2012. Registration of pea germplasm lines partially resistant to Aphanomyces root rot for breeding fresh or freezer pea and dry pea types. J. Plant Reg. 6:203–207. doi:10.3198/jpr2011.03.0139crg

McKay, K., P. Miller, P. Jenks, B. Riesselman, K. Neill, D. Buschena, and A.J. Bussan. 2002. Growing chickpea in the Northern Great Plains. Bull. 1236. North Dakota State Univ. Ext. Serv., Fargo, ND.

McPhee, K.E., C.C. Chen, D.M. Wichman, and F.J. Muehlbauer. 2007. Registration of 'Windham' winter feed pea. J. Plant Reg. 1:117–118. doi:10.3198/jpr2006.12.0828crc

McPhee, K.E., and F.J. Muehlbauer. 2007. Registration of 'Specter' winter feed pea. J. Plant Reg. 1:118–119. doi:10.3198/jpr2006.12.0826crc

McPhee, K.E., and F.J. Muehlbauer. 2009. Registration of 'Riveland' lentil. J. Plant Reg. 3:5–9.

Michigan State University. 2009. One hundred years of bean breeding at Michigan State University: A chronology. Michigan State University. http://www.css.msu.edu/bean/100Years.cfm (accessed 1 Feb 2013).

Miklas, P.N., J.D. Kelly, S.E. Beebe, and M.W. Blair. 2006. Common bean breeding for resistance against biotic and abiotic stresses: From classical to MAS breeding. Euphytica 147:105–131.

Muehlbauer, F.J. 1987. Registration of 'Brewer' and 'Emerald' lentil. Crop Sci. 27:1087–1088.

Muehlbauer, F.J. 1992. Registration of 'Palouse' lentil. Crop Sci. 32:1070.

Muehlbauer, F.J., W. J. Kaiser, D. F. Bezdicek, K. J. Morrison, and D. G. Swan. 1982. Description and culture of chickpeas. Ext. Bull. 1112. Wash. State University, Pullman, WA.

Muehlbauer, F.J., W.J. Kaiser, and I. Kusmenoglu, I. 1998a. Registration of 'Sanford' chickpea. Crop Sci. 38:282.

Muehlbauer, F.J., W.J. Kaiser, and I. Kusmenoglu. 1998b. Registration of 'Dwelley' chickpea. Crop Sci. 38:282–283.

Muehlbauer, F.J., and K.E. McPhee. 2004. Registration of 'Merrit' lentil. Crop Sci. 44:1487–1488.

Muehlbauer, F.J., M. Mihov, A. Vandenberg, A. Tullu, and M. Materne. 2009. Improvement in developed countries. In: W. Erskine, F.J. Muehlbauer, A. Sarker, and B. Sharma, editors, The lentil: Botany, production and uses. CABI, Oxfordshire, UK. p. 137–154.

Muehlbauer, F.J., and K.B. Singh. 1987. Genetics of chickpea. In: M.C. Saxena and K.B. Singh, editors, The chickpea. CABI, Oxfordshire, UK. p. 99–125.

Muehlbauer, F.J., S.R. Temple, and W. Chen. 2004. Registration of 'Sierra' chickpea. Crop Sci. 44:1864.

Neumann, K., P.H. Verburg, E. Stehfest, and C. Müller. 2010. The yield gap of global grain production: A spatial analysis. Agric. Syst. 103:316–326.

Newman, C.W., R.K. Newman, and R.H. Lockerman. 1988. Utilization of food legumes in human nutrition. In: R.J. Summerfield, editor, World crops: Cool season food legumes. Kluwer, Dordrecht, The Netherlands. p. 405–412.

Nuland, D.S., and G. Taylor. 1996. Technological progress in dry bean yields. Annu. Rep. Bean Improv. Coop. 39:50–51.

Osorno, J.M., K.F. Grafton, G.A. Rojas-Cifuentes, J.R. Gelin, and A.J. Vander-Wal. 2010. Registration of 'Lariat' and 'Stampede' pinto beans. J. Plant Reg. 4:5–11.

Park, S.J. 1993. Response of bush and upright plant type selections to white mold and seed yield of common beans grown in various row widths in southern Ontario. Can. J. Plant Sci. 73:265–272.

Pfender, W. 2000. Aphanomyces root rot. In: J.M. Kraft and F.L. Pfleger, editors, Compendium of pea diseases. 2nd ed. APS Press, St Paul, MN. p. 9–13.

Pilet-Nayel, M., F.J. Muehlbauer, R. McGee, J. Kraft, A. Baranger, and C. Coyne. 2002. Quantitative trait loci for partial resistance to Aphanomyces root rot in pea. Theor. Appl. Genet. 106:28–39.

Porter, L.D., T. Paulitz, and W. Chen. 2011. Rhizoctonia seed, seedling, and wet root rot. In: W. Chen, H.C. Sharma, and F.J. Muehlbauer, editors, Compendium of chickpea and lentil diseases and pests. APS Press, St. Paul, MN. p. 28–29.

Progene Plant Research. 2013a. Ariel green field pea. http://www.progenellc.com (accessed Oct. 2013).

Progene Plant Research. 2013b. Aragorn green field pea. http://plantresearch.co.nz/images/pdf/Aragorn.pdf (accessed Oct. 2013).

Progene Plant Research. 2013c. Banner green field pea. http://www.progenellc.com (accessed Oct. 2013).

Ray, D.K., N. Ramankutty, N.D. Mueller, P.C. West, and J.A. Foley. 2012. Recent patterns of crop yield growth and stagnation. Nat. Comm. 3:1293 10.1038/ncomms2296.

Salimath, P.M., C. Toker, J.S. Sandhu, J. Kumar, B. Suma, S.S. Yadav, and P.N. Bahl. 2007. Conventional breeding methods. In: S.S. Yadav, R. Redden, W. Chen, and B. Sharma, editors, Chickpea breeding & management. CABI, Oxfordshire, UK. p. 369–390.

Sarawat, P., F.L. Stoddard, D.R. Marshall, and S.M. Ali. 1994. Heterosis for yield and related characters in pea. Euphytica 80:39–48. doi:10.1007/BF00039296

Schafer, E.G. 1926. Washington agriculture: Cereal crops. State Coll. Wash. Ext. Bull. 134:5–37.

Schwinghamer, M.W., and R.C. Larsen. 2011. Diseases caused by viruses. In: W. Chen, H.C. Sharma, and F.J. Muehlbauer, editors, Compendium of chickpea and lentil diseases and pests. APS Press, St. Paul, MN. p. 60–62.

Sharma, B. 2009. Genetics of economic traits. In: W. Erskine, F.J. Muehlbauer, A. Sarker, and B. Sharma, editors, The lentil: Botany, production and uses. CABI, Oxfordshire, UK. p. 76–101.

Shehata, M.A., C.R. Grau, D.W. Davis, and F.L. Pfleger. 1983. Breeding for resistance to *Aphanomyces euteiches* root rot and *Rhizoctonia solani* stem rot in peas. J. Am. Soc. Hortic. Sci. 108:1080–1085.

Shinn, J.R. 1916. Pea growing in Spokane county. State Coll. Wash. Ext. Bull. 19. State College of Washington, Pullman, WA.

Singh, S.P. 1982. A key for identification of different growth habits of *Phaseolus vulgaris* L. Annu. Rep. Bean Improv. Coop. 25:92–94.

Singh, S.P., and H.F. Schwartz. 2010. Breeding common beans for resistance to diseases: A review. Crop Sci. 50:2199–2223. doi:10.2135/cropsci2009.03.0163

Singh, S.P., A. Molina, C.A. Urrea, and J.A. Gutiérrez. 1993. Use of interracial hybridization in breeding the race Durango common bean. Can. J. Plant Sci. 73:785–793.

Singh, S.P., H. Terán, M. Lema, D.M.A. Strausbaugh, P.N. Miklas, H.F. Schwartz, and M.A. Brick. 2007. Seventy-five years of breeding dry bean of the Western USA. Crop Sci. 47:981–989.

Slinkard, A.E. 1981. Eston lentil. Can. J. Plant Sci. 61:733–734.

Sonnante, G., I. Galasso, and D. Pignone. 2003. ITS sequence analysis and phylogenetic inference in the genus *Lens* Mill. Ann. Bot. (London) 91:49–54. doi:10.1093/aob/mcg007

Specht, I.E., and J.H. Williams. 1984. Contribution of genetic technology to soybean productivity—Retrospect and prospect. In: W.R. Fehr, editor, Genetic contributions to yield gains of five major crop plants. CSSA Spec. Publ. 7. ASA, CSSA, and ASA, Madison, WI. p. 49–74.

Steadman, J.R. 1983. White mold—A serious yield limiting disease of bean. Plant Dis. 67:346–347. doi:10.1094/PD-67-346

Tar'an, B., T. Warkentin, D.J. Somers, D. Miranda, A. Vandenberg, S. Blade, and D. Bing. 2004. Identification of quantitative trait loci for grain yield, seed protein concentration and maturity in field pea (*Pisum sativum* L.). Euphytica 136:297–306. doi:10.1023/B:EUPH.0000032721.03075.a0

Terán, H., M. Lema, D. Webster, and S.P. Singh. 2009. 75 years of breeding pinto bean for resistance to diseases in the United States. Euphytica 167:341–351. doi:10.1007/s10681-009-9892-9

Timmerman-Vaughan, G.M., A. Mills, C. Whitfield, T. Frew, R. Butler, S. Murray, M. Lakeman, J. McCallum, A. Russell, and D. Wilson. 2005. Linkage mapping of QTL for seed yield, yield components, and developmental traits in pea. Crop Sci. 45:1336–1344. doi:10.2135/cropsci2004.0436

Toker, C. 2009. A note on the evolution of kabuli chickpeas as shown by indunced mutations in *Cicer reticulatum* Ladizinsky. Genet. Resour. Crop Evol. 56:7–12. doi:10.1007/s10722-008-9336-8

Tullu, A., B. Tar'an, T. Warkentin, and A. Vandenberg. 2008. Construction of an intraspecific linkage map and QTL analysis for earliness and plant height in lentil. Crop Sci. 48:2254–2264. doi:10.2135/cropsci2007.11.0628

Urrea, C.A., and S.P. Singh. 1994. Comparison of mass, F2-derived family, and single-seed-descent selection methods in an interracial population of common bean. Can. J. Plant Sci. 74:461–464. doi:10.4141/cjps94-085

Urrea, C.A., and S.P. Singh. 1995. Comparison of recurrent and congruity backcrossing for interracial hybridization in common bean. Euphytica 45:469–478.

Urrea, C.A., C.D. Yonts, D.J. Lyon, and A.E. Koehler. 2009. Selection for drought resistance in dry bean (*Phaseolus vulgaris* L.) derived from the Mesoamerican gene pool in western Nebraska. Crop Sci. 49:2005–2010. doi:10.2135/cropsci2008.12.0694

USDA-ERS. 2010. Vegetable and pulses: dry beans. USDA-ERS. http://www.ers.usda.gov/topics/crops/vegetables-pulses/dry-beans.aspx (accessed 15 Mar. 2013).

USDA-ERS. 2013. Dry edible beans. USDA-ERS. http://usda.mannlib.cornell.edu/MannUsda/viewDocumentInfo.do?documentID=1394 (accessed 29 Jan. 2013).

USDA-NASS. 2013. Statistics by subject. USDA-NASS http://www.nass.usda.gov/Statistics_by_Subject/index.php?sector=CROPS (accessed 1 Apr. 2013).

U.S. Dry Pea and Lentil Council. 2012. USDA/USADPLC 2012 Production Comparison Report. Pulse Pipeline 12(23):1.

Vandemark, G.J., K.E. McPhee, and F.J. Muehlbauer. 2011. Registration of 'Essex' lentil. J. Plant Reg. 5:19–21. doi:10.3198/jpr2010.03.0151crc

Vandemark, G.J., and L.D. Porter. 2010. First report of lentil root rot caused by *Aphanomyces euteiches* in Idaho. Plant Dis. 94:480. doi:10.1094/PDIS-94-4-0480B

van der Maesen, L.J.G., N. Maxted, F. Javadi, S. Coles, and A.M.R. Davies. 2007. Taxonomy of the genus *Cicer* revisited. In: S.S. Yadav, R. Redden, W. Chen, and B. Sharma, editors, Chickpea breeding & management. CABI, Oxfordshire, UK. p. 14–46.

Voldeng, H.D., E.R. Cober, D.J. Hume, C. Gillard, and M.J. Morrison. 1997. Fifty-eight years of genetic improvement of short-season soybean cultivars in Canada. Crop Sci. 37(2):428–431. doi:10.2135/cropsci1997.0011183X003700020020x

Wallace, D.H. 1985. Physiological genetics of plant maturity, adaptation, and yield. In: J. Janick, editor, Plant breeding reviews. Vol. 3. John Wiley & Sons, Hoboken NJ. p. 21–158.

Wallace, D.H., J.P. Baudoin, J.S. Beaver, D.P. Coyne, D.E. Halseth, P.N. Masaya, H.M. Munger, J.R. Myers, M. Silbernagel, K.S. Yourstone, and R.W. Zobel. 1993. Improving efficiency of breeding for higher crop yield. Theor. Appl. Genet. 86:27–40.

Wallace, D.H., and R.E. Wilkinson. 1965. Breeding for Fusarium root rot resistance in beans. Phytopathology 55:1227–1231.

White, J., and A. Gonzales. 1990. Characterization of the negative association between seed yield and seed size among genotypes of common bean. Field Crops Res. 23:159–175.

Wood, D.R., M. Ballarin, M.A. Brick, H.F. Schwartz, and C.H. Pearson. 1989. Registration of Bill Z pinto bean. Crop Sci. 29:488.

Wood, D.R., and J.G. Keenan. 1982. Registration of Olathe bean. Crop Sci. 22:1259–1260.

Wright, D.J. 1985. Combining peas for human consumption. In: P.D. Hebblethwaite, M.C. Heath, and T.C.K. Dawkins, editors, The pea crop. Buttersworth, London, UK. p. 441–451.

Youngman, V.E. 1967. Lentils—A pulse of the Palouse. Econ. Bot. 21:135–139.

Yu, K., S.J. Park, and V. Poysa. 2000. Marker-assisted selection of common beans for resistance to common bacterial blight: Efficacy and economics. Plant Breed. 119:411–415.

Zohary, D., and M. Hopf. 1973. Domestication of pulses in the old world. Science 182:887–894.

6

Maize

Stephen Smith,* Mark Cooper, Joseph Gogerty, Carlos Löffler, Delos Borcherding, and Kevin Wright

ABSTRACT

Studies measuring genetic gain with maize (*Zea mays* L.) hybrids released by DuPont Pioneer[1] were initiated in 1972. Previous reports of genetic gain at optimum planting density were 92 kg ha^{-1} yr^{-1} (1.5 bu ac^{-1} yr^{-1}) for 1930 to 1980 and 77 kg ha^{-1} yr^{-1} (1.2 bu ac^{-1} yr^{-1}) for 1930 to 2001. We report a genetic gain of 87.6 kg ha^{-1} yr^{-1} (1.4 bu ac^{-1} yr^{-1}) for hybrids released from 1930 to 2011 and planted at optimum density in the target production environment (TPE), 79.6 to 90.8 kg ha^{-1} yr^{-1} (1.3–1.5 bu ac^{-1} yr^{-1}) when grown in well-watered environments, and 49.9 to 59.1 kg ha^{-1} yr^{-1} (0.8–0.9 bu ac^{-1} yr^{-1}) when grown in drought stress environments. Genetic gain estimates were higher for a subset of single-cross hybrids when measured in the TPE (92.2 kg ha^{-1} yr^{-1} or 1.5 bu ac^{-1} yr^{-1}) and in Viluco, Chile for both well-watered (95.5 kg ha^{-1} yr^{-1} or 1.5 bu ac^{-1} yr^{-1}) and drought stressed (61.5 kg ha^{-1} or 1 bu ac^{-1} yr^{-1}) environments. Genetic gain contributions to yield increases during 1930 to 2011 were 79% (TPE) and 59 to 67% (well watered); for single-cross hybrids during 1963 to 2011, genetic gain contributions were 74% (TPE) and 66 to 77% (well watered). The effects of weather on yield trends were minimal and very dependent on annual fluctuations contributing an upward bias of 1.3 kg ha^{-1} yr^{-1} (0.02 bu ac^{-1} yr^{-1}) during 1950 to 2011. The mean contribution to yield from +Cry1Ab resistance to attack by the European corn borer (ECB) (*Ostrinia nubialis* Hubner) was 565 kg ha^{-1} (9 bu ac^{-1}). Tassel size continued to decrease, but erectness of the leaves plateaued. Trends toward shorter anther silk intervals and improved stalk and root lodging continued, while resistance to barrenness and improvement in stay-green peaked. Potential yields (PYs) showed no signs of plateauing. Increased rainfed yields are dependent on continued increases in genetic gain. Irrigated farm yields (FYs) may be less amenable to genetic gain and more dependent on improved farm management. We anticipate an increased rate of genetic gain of 126 kg ha^{-1} yr^{-1} (2 bu ac^{-1} yr^{-1}) for 2015 and subsequent years under rainfed conditions provided that investments in maize breeding continue and there is no substantial change in the economics of maize production. More effective use of genetic diversity and crop management will allow U.S. maize breeders and farmers to accommodate climate change for the foreseeable future.

[1] All products are trademarks of their manufacturer.

Abbreviations: AY, attainable yield; Bt, Bacillus thuringensis; CRW, corn rootworm; ECB, European corn borer; ECB2, second-generation ECB; FY, farm yield; G × E × M, genetic potential of the variety × environmental × management effects; HI, harvest index; NCGA, U.S. National Corn Grower Association; OPVs, open pollinated varieties; PY, potential yield; QTL, quantitative trait locus; RFLP, restriction fragment length polymorphism; TPE, target production environment.

Stephen Smith and **Kevin Wright** (Kevin.D.Wright@Pioneer.com), DuPont Pioneer, 7300 NW 62nd Ave., Johnston, IA 50131-1004. *Corresponding author (Stephen.Smith@Pioneer.com).
Mark Cooper (Mark.Cooper@Pioneer.com) and **Carlos Löffler** (Carlos.Loeffler@Pioneer.com), DuPont Pioneer, 7250 NW 62nd Ave., Johnston, IA 50131-0552. **Joseph Gogerty,** DuPont Pioneer, 1920 East McGregor Street, Algona, IA 50511-0557 (Joseph.Gogerty@Pioneer.com). **Delos Borcherding,** DuPont Pioneer, 205 Fair View Drive, Dallas Center, IA 50063 (Delos.Borcherding@Pioneer.com). **Kevin Wright,** DuPont Pioneer, 7300 NW 62nd Ave., Johnston, IA 50131-1004 (Kevin.D.Wright@Pioneer.com).

doi:10.2135/cssaspecpub33.c6

Yield Gains in Major U.S. Field Crops. CSSA Special Publication 33.
Stephen Smith, Brian Diers, James Specht, and Brett Carver, editors.

There are economic, health, and environmental imperatives to sustainably increase agricultural production (Borlaug, 2001; Royal Society, 2009; Tester and Langridge, 2010; Foley et al., 2011). Harvestable crop yield is a function of the genetic potential of the variety × environmental and management effects (G × E × M) and their interactions. As farms and breeding nurseries become increasingly more technological, scientific, capital, and information rich, it becomes even more important to apportion factors affecting crop yields. This information provides the basis for charting more effectively a course toward increased crop production through further improvements in plant breeding, farm management practices, and their interactions.

An objective approach to apportion yield improvement is to first measure genetic progress. This objective can be accomplished in various ways (de la Vega and Chapman, 2010): by measuring yields of cultivars that have different years of release with "progress evaluation trials" (Duvick, 1977, 2005a, 2005b; Duvick et al., 2004a, 2004b; Russell, 1984; Ustun et al., 2001) in a production system that best represents the targeted production environment (TPE), from analyses of commercial crop data (Ellis et al., 2001, 2004), or by using data from multi-environment trials (Frensham et al., 1998). Here we report on genetic gain estimated from progress evaluation trials or "era" studies. Major changes in farm management practice should be taken into account in such studies. For example, if varieties released over several decades were cultivated at significantly different planting densities, then those planting densities should be taken into account in the design of the performance trial. The genetic contribution of cultivars grown in other types of production environments can be measured with additional treatment split-plot trials in the same or other experiments. Genetic, physiological, and morphological data collected in such experiments provide opportunities for observing the collateral impact on various traits of selection for yield. These data can be used to formulate testable hypotheses about changes in the physiological, morphological, or genetic factors that contribute to yield improvement.

Recent Significant Advances in Maize Breeding Technologies

Improved Phenotyping

The goal of using replicated field trials coupled with appropriate statistical analyses is to maximize the statistical power to partition and thus to most accurately estimate genotypic contributions to complexly inherited and environmentally influenced traits such as yield. Precision phenotyping reduces the confounding effects of environment by providing a uniform screening environment for specific agronomic traits when screening in a natural setting is too imprecise. Successful application of precision phenotyping effectively increases the heritability of traits and therefore improves the rate of genetic gain following selection.

Precision phenotyping is not a new activity. For example, Ferde Dicke artificially infested maize with egg masses of the ECB (*Ostrinia nubialis* Hubner) to identify resistant corn genotypes in 1927. While precision phenotyping remains an intensive resource-demanding exercise, costs of generating genotypic or molecular marker data have dropped substantially. Consequently, greater emphasis is being placed on developing more manageable and standardized environments so that phenotypic assays more accurately reflect heritable differences, thereby enabling more widespread and effective use of marker-assisted selection. Phenotyping for stalk and root strength has improved. In the past, breeders had to depend on observations following natural storm events or resort to the use of mechanical "push" tests to simulate the effects of wind. Breeders now have access to mobile wind machines that screen hybrids for root lodging, stalk lodging, and brittle snap, which are traits that collectively cause significant yield losses, estimated at more than US$1 billion annually in North America. Genotyping rankings based on mechanical screening are compared with those obtained following natural weather events to ensure the former is a solid proxy for the latter when making hybrid advancement decisions. Specialized nurseries are used to research the phenotypic and genetic bases of tolerances to stresses imposed by heat, drought, and different levels of fertilizer. All these developments contribute to the deployment of increasingly knowledge-based breeding strategies.

Marker Assisted Breeding

Isozymes allowed the maize genome to be surveyed for up to 25 polymorphic loci and during the 1980s were deployed to monitor genetic purity, validate pedigrees, and demonstrate the potential to generate linkage maps and more efficiently select for genomic regions (quantitative trait loci or QTLs) associated with agronomic traits (Edwards et al., 1987; Stuber et al., 1987). However, for selection at a QTL to be operationally effective with adapted germplasm, a more discriminative and thorough sampling of the maize genome was required. Restriction fragment length polymorphisms (RFLPs) met these criteria (Burr et al., 1983). However, RFLPs were cumbersome to use, which limited the number that could be employed, plus multiband phenotypes of a given RFLP could arise and required care in their genetic interpretation. As a result, most RFLP-based studies were exploratory and produced results that were often inconsistently and poorly estimated (Beavis et al., 1991; Bubeck et al., 1993). Beavis (1998) reported significant bias in QTL identification and effect estimation, unless at least 500 and preferably 1000 segregating progeny were studied, thus necessitating the development of a new low cost marker system that would provide higher throughput and greater interpretational clarity.

Amplification of simple sequence repeats (Shattuck-Eidens et al., 1990) provided a multi-allelic mono-locus marker system that allowed numerous loci to be detected relatively inexpensively and with lesser amounts of sample DNA (Senior and Heun, 1993; Smith et al., 1997; Senior et al., 1998). New mapping approaches (association and linkage disequilibrium mapping) were developed that assessed correlations between genotypes and phenotypes in populations of individuals, covering a range of genetic relationships that took advantage of many more generations of recombination and thereby increased QTL resolution (Morrell et al., 2012). Previously, QTL mapping using individual F2 populations had been limited by the time required to establish segregating populations, narrow inferences to a broader base of germplasm, and limited precision due to relatively few recombinations.

During the last decade, maize breeders have integrated the use of several technologies that collectively increase both the number of progeny evaluated and the effectiveness of selection. Included among these technologies are: (i) ultrahigh throughput single nucleotide polymorphisms or other kinds of DNA sequence markers, (ii) genome wide association mapping, (iii) information management and database systems with each phenotypic datum linked (via computerized onboard combine phenotypic trait capture) to the GPS-tracked location of uniquely numbered field plots, (iv) additional data connections to molecular marker genotypes and generations of pedigree data tracing back to founder inbreds, (v) statistical methodologies to not only improve the estimation of genotypic means but also to improve genotypic mean-based imputation of the post-trial performance, (vi) inbred development via double-haploid technology (which cuts inbred creation from years to months and makes all of the observed between-progeny variance entirely ascribable to the additive genetic variance component), thereby making both additive variance and nonadditive variance based selections for genotypic potential much more efficient and effective, and (vii) use of off-season nurseries that at some sites include managed stress environments for progeny evaluation and selection.

Breeders focus much of their time determining how resources should be allocated in a breeding program to find a balance that maximizes the opportunity to achieve the highest possible rate of annualized genetic gain. Troyer and Wellin (2009) proposed a radical shift in traditional maize breeding to increase efficiency by "replacing yield trials of preliminary testcrosses with yield trials of new inbreds," and the integrated approach described above follows this path to some extent. Marker data are collected on inbred lines, with prediction tools then used to screen lines for both quantitative and qualitative traits before testcrosses are grown. However, predicted hybrid performance selection based on observed inbred performance is not sufficiently precise for traits under multigenic control as suggested by Troyer and Wellin (2009) (O.S. Smith, personal communication, 2013). Morrell et al. (2012) stated forthrightly that "genomic selection will revolutionize breeding in the next decade." This revolution is already underway in maize breeding.

Adding Exotic Genetic Diversity for Simply Inherited Traits via Genetic Engineering

Practical use of transgenic crops is fundamentally dependent on the identification and incorporation into a genome of a single gene or regulatory construct that makes a sufficiently positive economic impact (in addition to satisfying stewardship and regulatory requirements relating to human and environmental health). The potential for transgenic crops had its origins during the early 1980s when employees at a Monsanto Roundup herbicide manufacturing plant discovered that soil bacteria were breaking down the active ingredient, glyphosate. Researchers recognized that if the glyphosate resistance gene could be engineered into cultivars and maintain its function, then those transgenic crops would offer a novel form of weed control that did not require preplant tillage-mediated soil incorporation. This technological advance allowed producers to use minimum till and no-till options, thereby leading to improved soil tilthe, moisture availability, and conservation of both soil and fuel. The first glyphosate resistant varieties were soybean [*Glycine max* (L.) Merr] released in 1996; transgenic herbicide resistant maize varieties followed in 1999.

Another quickly recognized potential use of transgenics was to provide improved insect control. The ECB (*Ostrinia nubilalis* Hubner) is one of the most significant insect pests of U.S. maize (Jampatong et al., 2002; Dillehay et al., 2004).

Even by the 1950s, Loomis et al. (1957) reported that, "The development of adequate and economical measures of suppressing populations of this insect (*Ostrinia nubilalis*) is a problem which has concerned entomologists for nearly forty years." Historic annual economic losses and control costs exceeded US$1000 million in the 1990s (Mason et al., 1996). Most economic losses were from the second-generation ECB (ECB2) that, because of sheath feeding and tunneling into the stalk and ear shank, reduced the uptake of water and nutrients, as well as photosynthate production and transport. Yield losses were both direct and indirect because of feeding injury increasing the propensity for broken stalks, dropped ears, stalk rots, and other diseases (Mason et al., 1996). Steffey and Gray (2008) noted that, "It's not unreasonable to estimate that during outbreak years, such as 1949, 1989, and 1991, ECB took 15 to 20% of the yield for themselves."

It had been well established by 1957 that multigenetic sources of diversity existed among maize inbred lines to resist or withstand attack by the ECB (Loomis et al., 1957). Penny et al. (1967) developed five synthetic maize populations that were "essentially borer-resistant." Iowa State University released B68 in 1968, which provided sources of ECB resistance. Krakowsky et al. (2004) reported that inbreds B52, De811, and Mo47 exhibited resistance to stalk tunneling by the ECB. Although QTLs for ECB tolerance were detected, results were inconsistent because small environmental differences had significant confounding effects (Jampatong et al., 2002; Krakowsky et al., 2002, 2004). Nonetheless, Duvick (1984) reported breeders had achieved some success, noting that "newer hybrids were on the average significantly more tolerant to both natural and artificial infestations of second-generation ECB." Similarly, Duvick (1997) reported "linear gains in resistance to second-generation ECB, but not to first generation borer." However, a significant yield penalty (drag) was associated with many inbred lines, in contrast to their most closely related non-ECB resistant counterpart inbreds, and each resistant genotype gave only partial protection against the ECB2.

In 1901, Japanese biologist Shigetane Ishiwatari discovered that a protein produced by the bacterium *Bacillus thuringensis* (Bt) was associated with the death of silkworms (University of California, San Diego, 2013). Farmers began to use Bt as a pesticide in 1920, and commercialized spore formulations were available in France in 1938. Carson (1962) advocated the use of Bt toxin as an environmentally friendly naturally occurring insecticide. During the early 1980s, maize breeders became increasingly aware of the potential significance of genetic transformation coupled with radically exotic sources of genes to achieve insect resistance. However, it required years of research before this Bt gene could be isolated, linked with expression vectors, stably inserted into the parental lines of maize hybrids, and receive complete regulatory approvals in 1996. Hutchison et al. (2010) estimated the cumulative monetary benefits in suppressing corn borer populations over 14 yr in Illinois, Iowa, Minnesota, Nebraska, and Wisconsin at US$6 billion, with US$1.9 billion in benefits also accruing to non-Bt maize growers.

Significant losses to corn production in the central Corn Belt of the United States can also occur because of root feeding by larvae of the western rootworm (*Diabrotica virgifera virgifera* LeConte) and northern corn rootworm (*Diabrotica barberi* Smith and Lawrence) (Spencer et al., 2009). Sappington et al. (2006) estimated annual yield loss and management costs of US$1.7 billion. Larval feeding negatively affects plant height, root weight, photosynthesis, and yield (Owens et al., 1974; Godfrey et al., 1993). Root lodging can cause harvestability problems, reducing grain yield by 11 to 34% (Spike and Tollefson, 1991). Yield loss due to injury or lodging is highly variable and influenced by many biotic and abiotic factors (Gray and Steffey, 1998). Excessive rainfall, saturated soil, and soil compaction can ameliorate ill effects caused by damage to roots by the corn rootworm (CRW) (Spike and Tollefson, 1988; Ellsbury et al., 1994). The first transgenic maize varieties resistant to CRW were released in 2003. Previously, extensive and time-consuming scouting was required, while soil applied chemical control measures were not entirely adequate yet had significant environmental costs (Payne et al., 2003; Vaughn et al., 2005). Furthermore, certain rootworm species had evolved to overcome crop rotation strategies (Levine et al., 2002).

Adding Stress Resistance and Performance Attributes via Seed Treatments

Seed treatments have become increasingly important to U.S. maize agriculture during the past 20 yr (Munkvold, 2009). Seed-applied insecticides are used to control corn flea beetle (*Chaetocnema pulicaria* Melsheimer), insect vector of the bacterium *Pantoea sterwartii* (causal organism of Stewart's wilt), and so reduce bacterial infection in maize. Nematicidal seed treatments may also help reduce fungal infection via roots. New fungicides effective at lower application rates and with improved safety for nontarget organisms have been introduced, including those with improved efficacy against important seedling pathogens such as *Fusarium graminearum* and *Rhizoctonia* spp. Most recently, seed coatings with more enduring effects because of systemic incorporation have been introduced that improve seedling stands, delay senescence, increase tolerance to environmental stresses, improve CO_2 and nitrogen assimilation, and increase water use efficiency and yield (Munkvold, 2009).

U.S. Maize Yields and the Contribution of Genetic Improvement

U.S. Maize Production

The United States annually contributes approximately 40% of global maize production. The production of maize in the United States can be divided into three eras (Fig. 6–1). First, from 1866 to 1920 increased maize production was only possible by taking more land into cultivation; yields approached stagnation, increasing at 3 kg ha^{-1} yr^{-1} (0.05 bu ac^{-1} yr^{-1}) (Fig. 6–2) (USDA-NASS, 2013a). Annual commercial fertilizer use on U.S. farms was 5.5 Tg yr^{-1} (6.1 million tons yr^{-1}) in the 1920s, rising to 6.0 Tg yr^{-1} (6.6 million tons yr^{-1}) during the 1930s (Agriculture in the Classroom, 2013). Second, during 1940 to 1970, maize production further increased (Fig. 6–1) even as the land area used for maize cultivation shrank because of the successive introduction of higher yielding hybrids coupled with improved management practices, for example, commercial fertilizer use on U.S. farms doubled to 12.2 Tg yr^{-1} (13.5 million tons yr^{-1}) during the 1940s. From around 1942, U.S. maize yields rose from 1758 kg ha (28 bu ac^{-1}) to 9541 kg ha^{-1} (152 bu ac^{-1}) at an annual rate of 133 kg ha^{-1} (1.8 bu ac^{-1} yr^{-1}) (Fig. 6–2). Annual U.S. commercial fertilizer use rose to 20.3 Tg yr^{-1} (22.4 million tons yr^{-1}) during the 1950s and reached 29.4 Tg yr^{-1} (32.4 million tons yr^{-1}) during the 1960s. Third, during 1970 to 2012, U.S. maize production continued to increase; but beginning in the late 1970s, and again in the 1990s, a previous trend toward using less land to

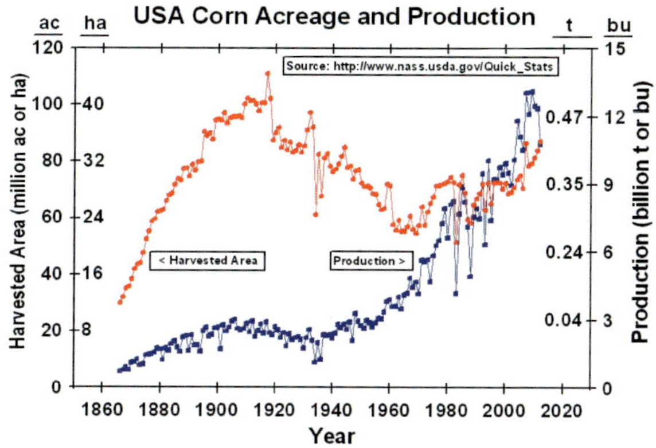

Fig. 6–1. U.S. maize acreage and production from 1865 to 2011 (USDA-NASS, 2013a).

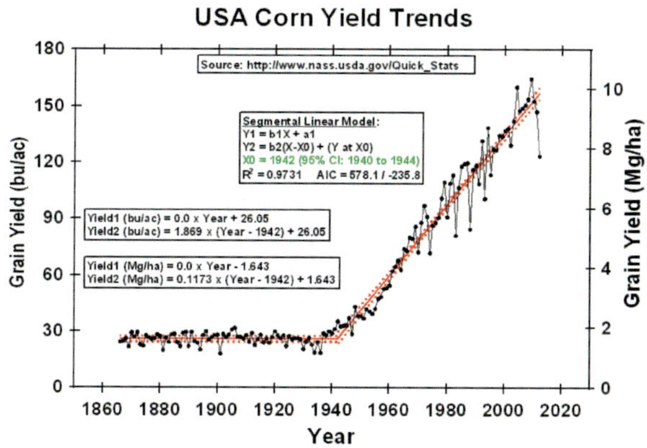

Fig. 6–2. Yields of U.S. maize from 1865 to 2011 (USDA-NASS, 2013a).

meet production needs was reversed. Usage of nitrogen fertilizer on Iowa maize area from 1964 to 2010 is shown in Fig. 6–3 (USDA-ERS, 2013a).

Maize growing regions of the United States are presented in Fig. 6–4a, 6–4b, and 6–4c. The top five maize producing states (percentage of annual U.S. production is shown in parentheses) include: Iowa (19%), Illinois (18.6%), Nebraska (11.5%), Minnesota (8.2%), and Indiana (8.1%). About 13% of U.S. maize is produced under irrigation, mostly in Nebraska (Grassini and Cassman, 2012). As of 10 May 2013, U.S. maize production was projected to be a record level of 359 Tg (14.14 billion bushels) (USDA, 2013) from a predicted harvested area of 39.4 million hectares (97.3 million acres). This estimate represents the highest planted U.S. acreage for maize since 1936 when an estimated 41.7 million hectares (103 million acres) were planted (USDA-NASS, 2013b).

During 1979 to 2011 maize cultivation has expanded north and west of the central Corn Belt (Fig. 6–4a and 6–4b). There has also been an increase in maize yields in the central Corn Belt and in irrigated areas in the midwestern and far

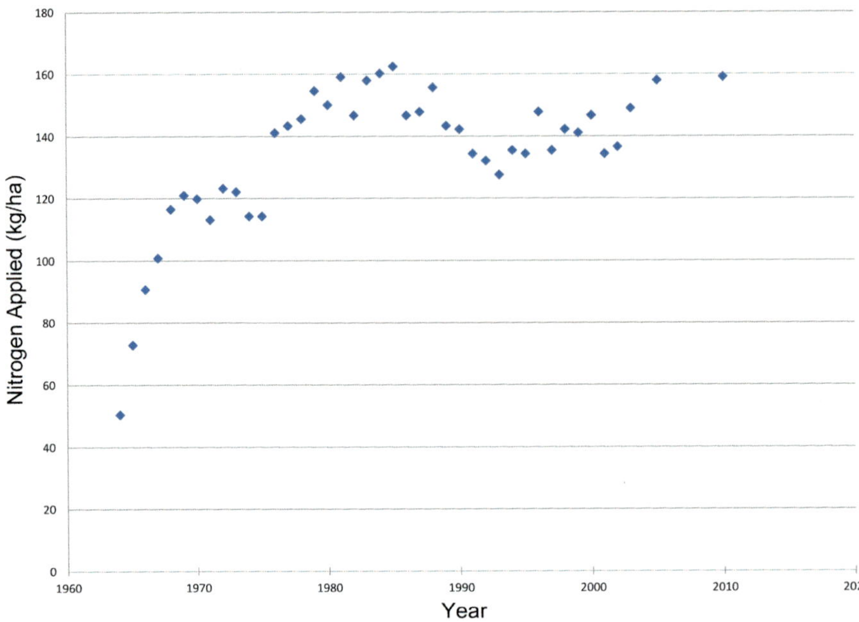

Fig. 6–3. Nitrogen use on Iowa maize acreage from 1964 to 2010 (USDA-NASS, 2010).

western United States (Fig. 6–4c). A major factor contributing to increased production of maize during 2000 to 2009 has been increased demand for ethanol, the production of which increased from 6.1 billion liters (1.6 billion gallons) to 40.9 billion liters (10.8 billion gallons) during this period (Wallander et al., 2011). Approximately 40% of the 2012 maize harvest was used to produce ethanol (USDA-ERS, 2013b). Increases in maize acreage during this period have come from shifting to more intensive maize crop rotations, double cropping, conversion of uncultivated hay, and reductions in cotton area but without a net decrease in soybean area (Wallander et al., 2011).

Recent Major Changes in U.S. Maize Farm Management

During 1989 to 2008, the percentage of conventionally tilled U.S. maize area fell, while conservation tillage percentage increased, with the percentage of no-till land substantively increasing (Fig. 6–5) (Conservation Technology Information Center, http://www.ctic.purdue.edu, accessed 14 Aug. 2013). Farmers adopted conservation tillage for numerous reasons, including management costs, limited availability of labor, soil and fuel conservation, and greater yield stability under water limiting environments (DeFelice et al., 2006). Use of nitrogen (N) in Iowa maize fields increased during the latter half of this period, rising by 23% from 123 kg ha^{-1} (110 lb ac^{-1}) to over 150 kg ha^{-1} (135 lb ac^{-1}) (Fig. 6–3) (USDA-NASS, 2010; USDA-ERS, 2013b). Increased mechanization, computerized data collection, and field mapping using global positioning satellites allowed improved management of soil, crop residue, and chemical and genetic (hybrid) inputs. The percentage of Iowa corn acres planted with transgenic maize hybrids rose from 30% (2000)

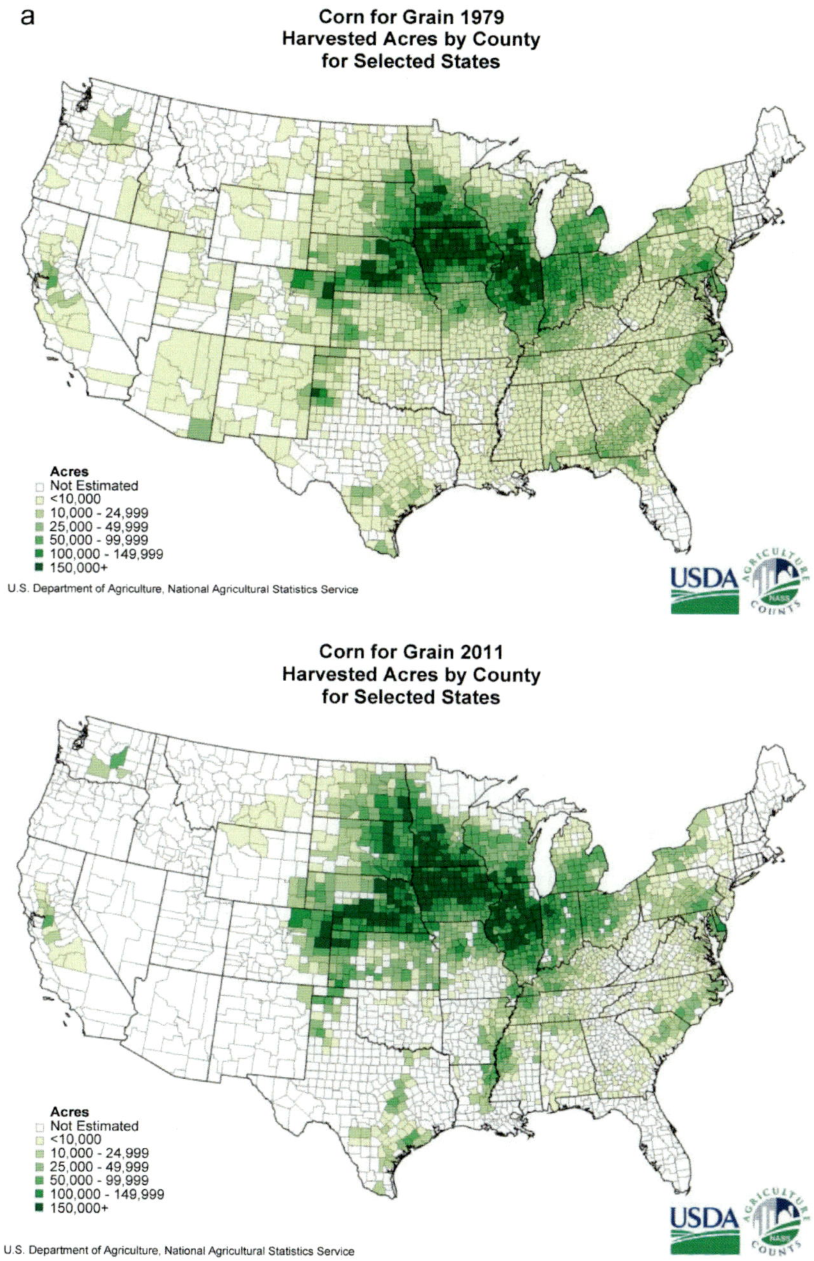

Fig. 6–4a. U.S. harvested areas of maize: 1979 compared with 2011. Kindly provided by Ray Roberts and Travis Thorson (USDA-NASS, 2013a).

b

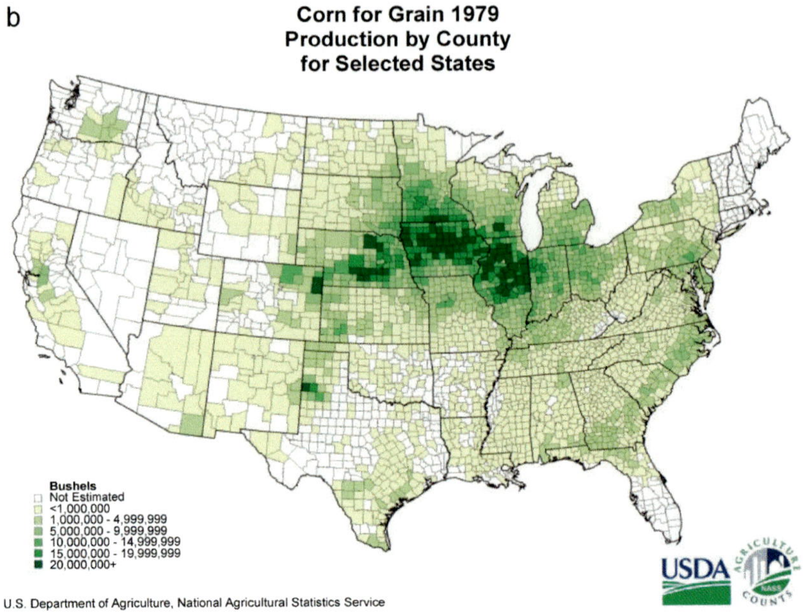

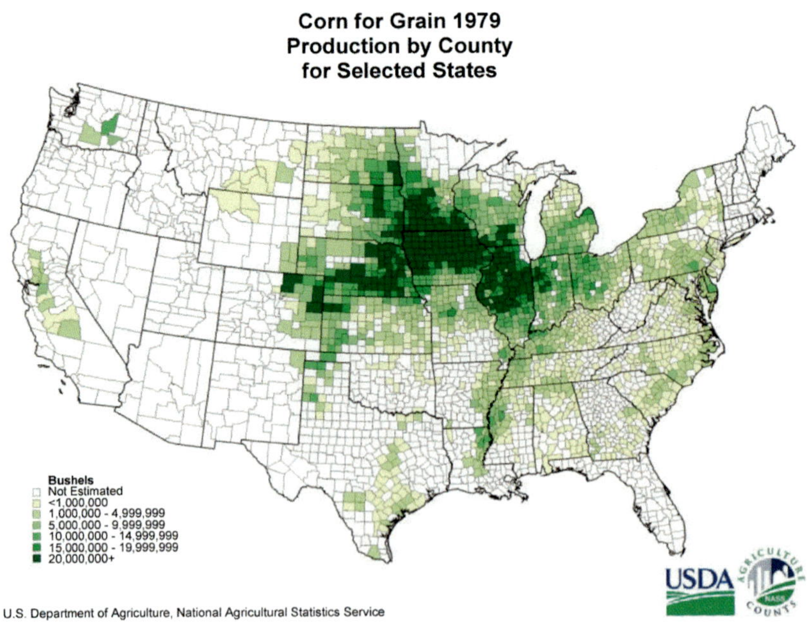

Fig. 6–4b. U.S. maize production by county: 1979 compared with 2011. Kindly provided by Ray Roberts and Travis Thorson (USDA-NASS, 2013a).

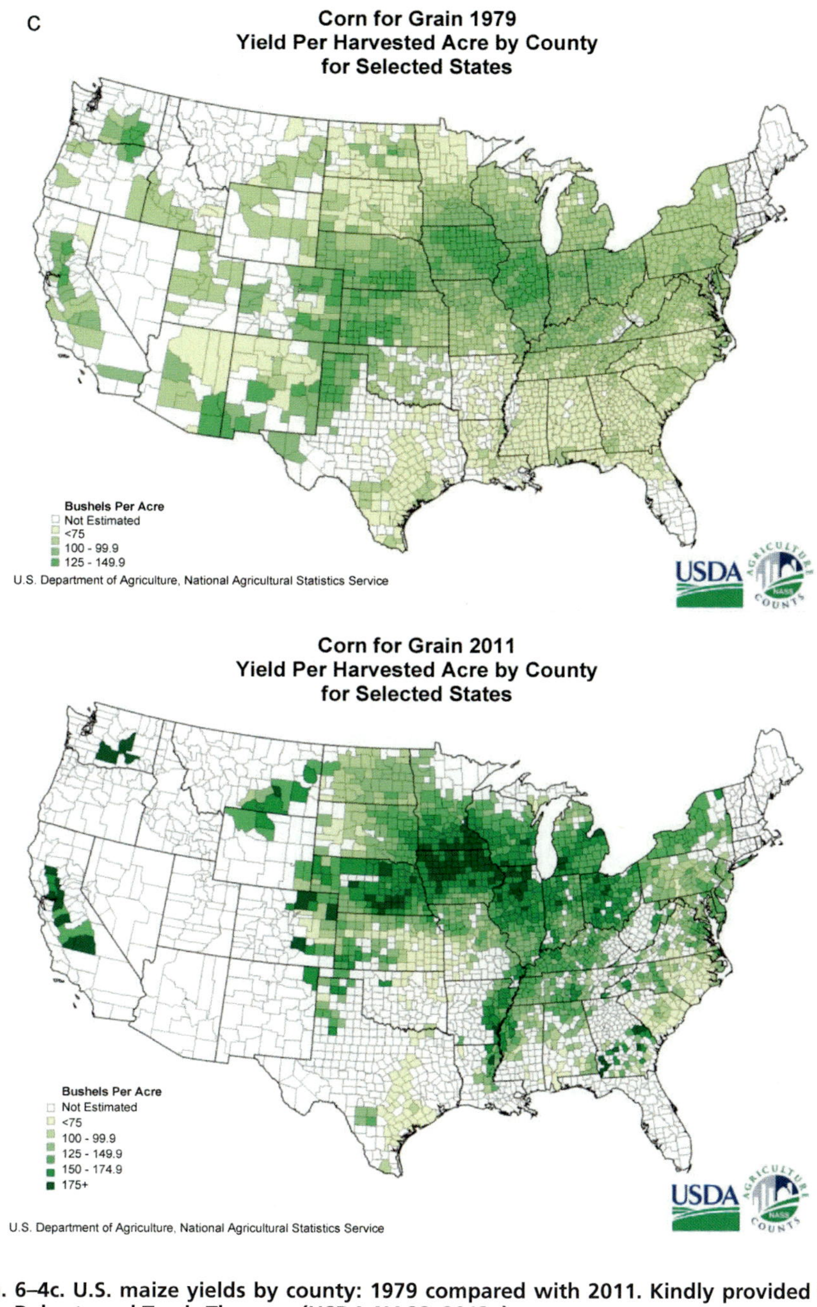

Fig. 6–4c. U.S. maize yields by county: 1979 compared with 2011. Kindly provided by Ray Roberts and Travis Thorson. (USDA-NASS, 2013a).

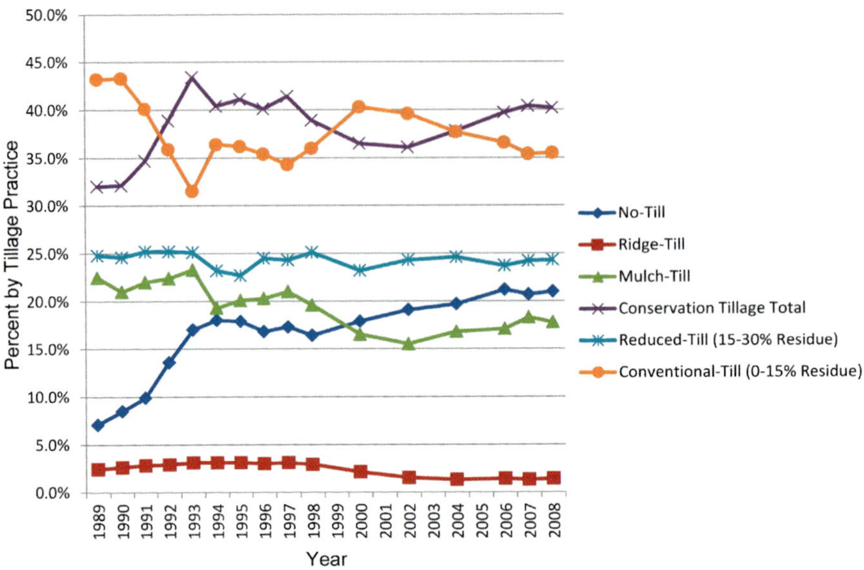

Fig. 6–5. Proportions of U.S. maize area according to tillage regimes (Conservation Technology Information Center, http://www.ctic.purdue.edu, accessed 14 Aug. 2013).

to over 90% (2012) (USDA-NASS, 2000–2012); the percentage of Iowa corn acres planted to stacked transgenic traits rose from 2 to 64%. The Center for Environmental Risk Assessment (2013) listed 30 transgenic trait combinations that have been approved for use in maize cultivation in the United States during the 2013 production season. These traits provide herbicide resistance (tolerances to either glufosinate herbicides, glyphosate herbicides, or both) and insect resistance (resistance to either lepidopteran insects, coleopteran insects, or both). During the past 20 yr, there has been increased use of seed treatments to control pests and diseases and to further enhance yields (Munkvold, 2009).

Maize planting dates have become earlier. During the late 1970s, 50% of Iowa maize area was planted by mid-May; during 2005 to 2009, 50% of maize was planted before the end of April. Several factors allow maize to be planted earlier, including wider planters, more cold stress tolerant hybrids, improved seed treatment, and reduced tillage systems (Abendroth and Elmore, 2010). Cardwell (1982) and Kucharik (2008) showed a significant association of early planting date with increased yield. However, increased yields are contributed by numerous factors, including germplasm improvement and management practices (Kucharik, 2008). Whether earlier planting to allow greater biomass accumulation by growing longer-season hybrids or to take advantage of stress avoidance during flowering or grain-fill have contributed to increased yields remains an open question. For Iowa, planting dates that are either too early (20 March–5 April) or too late (5–20 May) can yield significantly less than maize planted during mid-April to early May (Iowa State University, 2013). When planted too early, yields can be more variable as a result of cold stress or flooding leading to nonuniform stand establishment. Planting very early is a very risky proposition, not the least because replanting requires numerous additional costs (Iowa State University, 2001). Even if very early planting of a relatively long season hybrid should result in successful stand establishment, then additional costs to dry grain are likely to be incurred (Imad Saab personal communication, 2013.).

Examination of Weather Trends in the United States during 1866 to 2011 and Estimated Effects on Maize Yields

Leakey et al. (2006) found that in the absence of drought stress, enhanced levels of CO_2 (550 ppm vs. 376 ppm) did not stimulate photosynthesis biomass or yield of maize. Markelz et al. (2011) reported that an elevated level of CO_2 (550 ppm vs. 380 ppm) countered negative effects of drought stress on yields of maize regardless of nitrogen levels. We therefore anticipate that the net per se effects of rising CO_2 levels on maize yields in the central Corn Belt will, at least to date, have been negligible.

Weather, including include rainfall, solar radiation, and temperature, exerts an effect on measured trends in crop yields in two ways. Season-to-season variation in weather results in yield fluctuation (spikes or depressions), while interannual trends in any weather variable also might be present and exert a longer term impact (positive or negative) with respect to yield trends. Lobell and Asner (2003) estimated that 20% of increased U.S. maize yields that had previously been attributed to increased use of technology was more properly attributable to favorable weather. Kaufmann and Snell (1997) determined that 19% of maize yield variability during 1969 to 1987 was due to climate, with 74% attributable to social variables ("capitol, labor, fertilizer practices, etc."). Carlson et al. (1996) found strong correlations between July rainfall variability and maize yields for five midwestern states during 1931 to 1975. Kucharik and Ramankutty (2005) determined that during 1866 to 2002, U.S. maize yield variability fluctuated widely with an overall very slightly decreasing trend (slope −0.0136). Observation of U.S. maize yields (Fig. 6–2) supports their conclusions that variability was high during the 1890s, the 1930s, and generally post-1960s. Periods of relatively little yield variability were during the 1950s and early 1960s. Kucharik and Ramankutty (2005) also included the mid-1980s to 2002 as a period of low variability. However, observation of Fig. 6–2 would suggest that the mid-1990s to early-2000s better defined this period of lower variability.

We addressed both climate trend and interannual variability as contributing effects of weather on U.S. yields. We simulated annual maize yields that would have been possible on the basis of seasonal weather conditions in the absence of other environmental constraints such as diseases or pests, and we calculated a water stress index for each 30 by 30 km grid in the United States where maize was grown for each year during 1950 to 2011 using EnClass (285,839 simulations) (Löeffler et al., 2005; Messina et al., 2009). We then used these simulated yields as a covariate to adjust USDA annual average yields with an autoregressive model (Littell et al., 2006). Finally, to estimate simulated yields for the entire United States, we used the annual area harvested for each 30 by 30 km grid as a weighting factor. A comparison of actual USDA yield data from 1950 to 2011 and the annual simulated environmentally adjusted yields validates this approach (Fig. 6–6). For example, the relative overadjustments to yield in the 1950s indicate relatively poor responses in practice to favorable growing conditions (e.g., 1950 and 1951) and a low tolerance to drought (e.g., 1952–1955) of double-cross hybrids. The

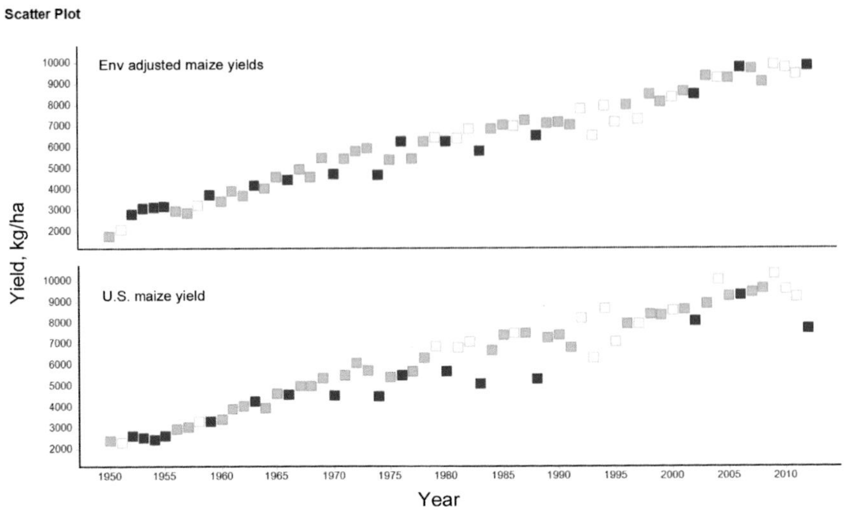

Fig. 6–6. Average U.S. maize yield (in kilograms per hectare) compared with adjusted maize yields for 1950–2011 using an autoregressive model adjustment based on EnClass simulated yield. The color intensity is based on value of water stress index (light = low stress, dark = high stress) (see text).

small yield adjustment for 1970 shows lower yields were due primarily to biotic factors (e.g., Southern corn leaf blight) rather than abiotic weather factors alone. Likewise, the yield dip in 1974 could be partially accounted for by drought in the central Corn Belt (mainly southern Minnesota, western Iowa, and Illinois); yields were also compromised that year by delayed planting due to a wet spring and a severe killing frost across much of the United States in early September, which we could not quantify, hence the relatively small positive yield adjustment. Low yields and large positive adjustments for 1983 and 1988 are due to drought. The dip in 1993 was due to low solar radiation accompanied by excessive rainfall and cool temperatures. During 1950 to 2011, the higher slope of actual yield (1.96) compared with the weather adjusted slope (1.93) reflects a trend in the annual rate of yield increase of 1.9 kg ha^{-1} yr^{-1} (0.03 bu ac^{-1} yr^{-1}), attributable to more favorable growing conditions for maize having occurred during these decades. These data are consistent with reports of a "benign climate" during 1954 to 1973, which favored crop production across the Corn Belt region (Baker et al., 1993; Carlson, 1990; Thompson, 1986; Andresen et al., 2001). However, we do not ascribe so great an effect on yield increase to the "benign climate" as do Kucharik and Ramankutty (2005), who correlate the highest annual percentage increase in yield (5%) with this period (1954–1973).

On this basis, maize era studies measured during 1950 to 2001 overestimate genetic gain by 1.9 kg ha^{-1} yr^{-1} (0.03 bu ac^{-1} yr^{-1}) (1.6%). However, evidence for a causal weather trend on yield still remains highly dependent on specific annual weather conditions. For example, when these same comparisons were made for 1950 to 2012, we saw no evidence of a weather trend on yield (2012 was a drought year not favorable to high maize yields). Lobell et al. (2011) concluded from a regression analysis of historical data and modeling that temperature and precipitation had little overall impact on maize yields in the United States during 1980 to 2008. We do not have sufficiently comprehensive data to extend use of this methodology of evaluating the effects of weather on U.S. maize yields before 1950. However, weather experienced during 1900 to 1957 in Iowa was described as "hostile" for maize yields (Carlson, 1990). Observations of temperature and precipitation for the State of Iowa during 1900 to 1957 indicate a trend of increasing temperatures and decreasing precipitation with superimposed annual fluctuations (Iowa State University, 2013).

Genetic Gain Contributing to U.S. Maize Yields from 1930 to 2011

Review of Previous Genetic Gain Studies

U.S. farmers began to test hybrid maize during the early 1930s. Droughts during 1934 and 1936 provided Iowa farmers first-hand evidence that hybrids were more resilient with yields that were double those of open pollinated varieties (OPVs) grown under the same conditions (Sutch, 2009). Adoption rates of hybrids reached 50% of maize area in Iowa during 1938 and 100% by 1942. Adoption rates of hybrid maize in the United States as a whole reached 50% during 1943 but only approached 100% by 1960 (Duvick, 2001). Frey (1971) estimated that by 1938 maize hybrids grown in Iowa had a genetic gain of 117% in comparison with open-pollinated varieties. However, subsequent hybrid releases seemed to provide less substantive yield gains. Consequently, during the 1940s and 1950s, concerns arose that a yield plateau had been reached (Russell, 1974). Robinson and Rhoades (1935) noted that improvements by breeding were dependent on G × E interaction. Frey (1971) noted that "Improvement in both production environment and in varieties must occur hand-in-hand to optimize production." To illustrate this point, Frey (1971) reported that only hybrids, not OPVs, produced higher yields per unit area as planting density increased.

Russell (1974) did not believe a plateau had been reached but nonetheless noted, "A plant breeder frequently wonders about the actual gains he has achieved for the crop species with which he works" and concluded, "Occasionally, a breeder needs to evaluate the progress in genetic improvement of their materials." Russell (1974) further remarked, "Maize breeders are confident that present-day hybrid genotypes are better able to respond to improved management, but we have no good evidence for the extent of the genetic contribution."

Similar questions arose in the proprietary breeding sector during 1970 to 1971, when William L. Brown, Vice President and Director of Corporate Research at Pioneer, approached the director of the corn breeding department, Donald N. Duvick, and asked [according to our (JSCS, MC) recollections of Don's later commentary]: "You corn breeders are always coming up with your favorite new hybrid, which you are sure is going to beat all the others, but then you throw many of them out the next year because they didn't perform. How do you know if you are really making progress?" Don Duvick responded by initiating yield trials of Pioneer hybrids in 1972 and 1973. Thus began an annual exercise to monitor genetic yield gain in maize using the so-called "era" that Pioneer hybrids had a record of wide usage by maize producers in the central U.S. Corn Belt.

Duvick (1984) reported a linear rate of genetic gain (at optimum planting density) of 92 kg ha^{-1} yr^{-1} (1.5 bu ac^{-1} yr^{-1}) for "era" hybrids released during 1930 to 1980. Duvick (1984) partitioned the rate of gain into 72 kg ha^{-1} yr^{-1} (1.1 bu ac^{-1} yr^{-1}) estimated for hybrids released during 1930 to 1955 compared with 112 kg ha^{-1} yr^{-1} (1.8 bu ac^{-1} yr^{-1}) estimated for hybrids released during 1956 to 1980. Duvick et al. (2004a, 2004b) and Duvick (2005a, 2005b) extended the estimation to include hybrids released during 1930 to 2001 and reported a linear rate of gain of 77 kg ha^{-1} yr^{-1} (1.2 bu ac^{-1} yr^{-1}). Haegele et al. (2013) measured genetic gain using 21 hybrids adapted to east-central Illinois that were commercially released by DeKalb during 1967 to 2006; none of the hybrids were transgenic. Measurements of genetic gain during 2009 and 2010 for these hybrids ranged from 56 kg ha^{-1} yr^{-1} (0.9 bu ac^{-1} yr^{-1}) to 86 kg ha^{-1} yr^{-1} (1.4 bu ac^{-1} yr^{-1}) for yield trials grown under low and high N regimes, respectively (Haegele et al., 2013).

Bubeck et al. (2006) examined genetic gain using topcross data to evaluate progress over cycles of maize inbred improvement. This approach provided a metric of breeding progress (empirical breeding values) among families of inbreds and also indicated progress on an individual trait basis. Yield gain per year among progeny topcrosses ranged from 33.9 to 111.7 kg ha^{-1} yr^{-1} (0.54–1.78 bu ac^{-1} yr^{-1}) for Stiff Stalk inbreds and 18.8 to 91.0 kg ha^{-1} yr^{-1} (0.30–1.45 bu ac^{-1} yr^{-1}) for non–Stiff Stalk inbreds. However, these data underestimated progress in a commercial program as hybrid topcrosses were made using a common tester chosen to maximize the general combining ability rather than the specific combining ability (these topcross hybrids yielded 6 to 17% less than the best commercial check hybrids).

Here we estimate genetic gain using Pioneer "era" hybrids (Duvick et al., 2004a, 2004b; Duvick, 2005a, 2005b), with an entry list extended to include Pioneer brand hybrids that were widely used in the TPE of the central U.S. Corn Belt during 2002 to 2011 (Table 6–1). In addition, we estimate genetic gain for yield in drought and well-watered environments. We estimated the contribution to genetic gain of the Cry1Ab gene. We also make comparisons with the FY using mean Iowa (rainfed) yields and mean Nebraska irrigated yields. We also compare with the PY using experimental data obtained under the auspices of the U.S. National Corn Grower Association (NCGA) yield contests.

Genetic Yield Gain for Pioneer "Era" Hybrids Released from 1930 to 2011

Yield (genetic gain) data for era hybrids released from 1930 to 2011 exhibited positive linearity when tested in any of three planting densities (Fig. 6–7) of 38.8 kg ha^{-1} yr^{-1} (0.6 bu ac^{-1} yr^{-1}) for a density of 30,000 plants ha^{-1} (12,000 plants ac^{-1}), 80.7 kg ha^{-1} yr^{-1} (1.3 bu ac^{-1} yr^{-1}) for 54,000 plants ha^{-1} (22,000 plants ac^{-1}), and 108 kg ha^{-1} yr^{-1} (1.7 bu ac^{-1} yr^{-1}) for 84,000 plants ha^{-1} (34,000 plants ac^{-1}). A transition to hybrids that yielded more at moderate planting densities occurred around 1940. Another subsequent transition occurred in the late 1980s to hybrids that yielded more at high plant densities.

It is of interest to consider the rate of yield gain extracting from the combined hybrid × density treatment means only *one* yield mean per hybrid, which is the highest of the three yield means generated by a given hybrid when tested in the three plant densities. The rate of gain for hybrids released from 1930 to 2011, when only the highest (optimum) G × D means were used, resulted in an

Table 6–1. Entry list of hybrids used in the era studies for hybrids released from 1930–2011. Transgenic hybrids are identified in respect to the specific trait they contain.

Hybrid	Release year	Number and combinations of transgenes	Herbicide resistance			Insect resistance	
BYD	1930	0	–	–	–	–	–
RYD1	1930	0	–	–	–	–	–
RYD2	1930	0	–	–	–	–	–
KYD	1930	0	–	–	–	–	–
351	1934	0	–	–	–	–	–
322	1936	0	–	–	–	–	–
307	1936	0	–	–	–	–	–
317	1937	0	–	–	–	–	–
330	1939	0	–	–	–	–	–
336	1940	0	–	–	–	–	–
340	1941	0	–	–	–	–	–
339	1942	0	–	–	–	–	–
344	1945	0	–	–	–	–	–
352	1946	0	–	–	–	–	–
350B	1948	0	–	–	–	–	–
347	1950	0	–	–	–	–	–
301B	1952	0	–	–	–	–	–
354	1953	0	–	–	–	–	–
329	1954	0	–	–	–	–	–
354A	1958	0	–	–	–	–	–
328	1959	0	–	–	–	–	–
3618	1961	0	–	–	–	–	–
3206	1962	0	–	–	–	–	–
3306	1963	0	–	–	–	–	–
3376	1965	0	–	–	–	–	–
3390	1967	0	–	–	–	–	–
3334	1968	0	–	–	–	–	–
3571	1968	0	–	–	–	–	–
3388	1970	0	–	–	–	–	–
3517	1971	0	–	–	–	–	–
3366	1972	0	–	–	–	–	–
3301A	1974	0	–	–	–	–	–
3529	1975	0	–	–	–	–	–
3541	1975	0	–	–	–	–	–
3382	1976	0	–	–	–	–	–
3377	1982	0	–	–	–	–	–
3378	1983	0	–	–	–	–	–
3475	1984	0	–	–	–	–	–
3379	1988	0	–	–	–	–	–
3362	1989	0	–	–	–	–	–
3417	1990	0	–	–	–	–	–
3394	1991	0	–	–	–	–	–
3563	1991	0	–	–	–	–	–
3279	1992	0	–	–	–	–	–
3489	1994	0	–	–	–	–	–
3335	1995	0	–	–	–	–	–
34G81	1997	0	–	–	–	–	–

Table continued.

Table 6–1. Continued.

Hybrid	Release year	Number and combinations of transgenes	Herbicide resistance			Insect resistance	
33A14	1997	1A	–	–	Cry1Ab	–	–
33G26	1998	0	–	–	–	–	–
33P66	1999	0	–	–	–	–	–
33P67	1999	1A	–	–	Cry1Ab	–	–
34B23	1999	0	–	–	–	–	–
33B50	1999	0	–	–	–	–	–
33B51	1999	1A	–	–	Cry1Ab	–	–
34G13	2000	0	–	–	–	–	–
34B24	2000	1A	–	–	Cry1Ab	–	–
33R77	2001	0	–	–	–	–	–
34M95	2001	1A	–	–	Cry1Ab	–	–
34N44	2002	1A	–	–	Cry1Ab	–	–
34H31	2002	0	–	–	–	–	–
33D31	2002	0	–	–	–	–	–
34N42	2003	2A	–	pat	–	Cry1F	–
34H32	2003	1A	–	–	Cry1Ab	–	–
34M93	2003	2A	–	pat	–	Cry1F	–
33N09	2004	0	–	–	–	–	–
33A84	2004	0	–	–	–	–	–
33D11	2005	0	–	–		–	–
34A16	2005	3A	–	pat	Cry1Ab	Cry1F	–
34P88	2005	0	–	–	–	–	–
33D13	2006	3A	–	pat	Cry1Ab	Cry1F	–
33D14	2007	4	CP4 EPSPS	pat	–	Cry1F	Cry34Ab1, Cry35Ab1
33T59	2007	3B	CP4 EPSPS	pat	–	Cry1F	Cry34Ab1, Cry35Ab1
34R67	2007	3C	CP4 EPSPS	pat	–	Cry1F	–
35H42	2007	3C	CP4 EPSPS	pat	–	Cry1F	–
34A20	2007	4	CP4 EPSPS	pat	–	Cry1F	Cry34Ab1, Cry35Ab1
33M16	2007	3C	CP4 EPSPS	pat	–	Cry1F	–
33W13	2007	3C	CP4 EPSPS	pat	–	Cry1F	–
32T84	2007	3C	CP4 EPSPS	pat	–	Cry1F	–
33F12	2007	0	CP4 EPSPS	–	–	–	–
31N30	2007	4	CP4 EPSPS	pat	–	Cry1F	Cry34Ab1, Cry35Ab1
35F40	2007	3C	CP4 EPSPS	pat	–	Cry1F	–
33W84	2008	4	CP4 EPSPS	pat	–	Cry1F	Cry34Ab1, Cry35Ab1
34F97	2008	4	CP4 EPSPS	pat	–	Cry1F	Cry34Ab1, Cry35Ab1
35K04	2008	4	CP4 EPSPS	pat	–	Cry1F	Cry34Ab1, Cry35Ab1
35K03	2008	3C	CP4 EPSPS	pat	–	Cry1F	–
34F29	2008	4	CP4 EPSPS	pat	–	Cry1F	Cry34Ab1, Cry35Ab1
33D49	2008	3C	CP4 EPSPS	pat	–	Cry1F	–

Table continued.

Table 6–1. Continued.

Hybrid	Release year	Number and combinations of transgenes	Herbicide resistance			Insect resistance	
33Y76	2008	4	CP4 EPSPS	pat	–	Cry1F	Cry34Ab1, Cry35Ab1
35A52	2008	0	CP4 EPSPS	–	–	–	–
32N73	2008	4	CP4 EPSPS	pat	–	Cry1F	Cry34Ab1, Cry35Ab1
33P83	2008	4	CP4 EPSPS	pat	–	Cry1F	Cry34Ab1, Cry35Ab1
32T16	2008	0	–	–	–	–	–
P0916HR	2009	3C	CP4 EPSPS	pat	–	Cry1F	–
P0916XR	2009	4	CP4 EPSPS	pat	–	Cry1F	Cry34Ab1, Cry35Ab1
P1395XR	2009	4	CP4 EPSPS	pat	–	Cry1F	Cry34Ab1, Cry35Ab1
32B34	2009	3C	CP4 EPSPS	pat	–	Cry1F	–
P1184HR	2009	3C	CP4 EPSPS	pat	–	Cry1F	–
32B09	2009	1B	CP4 EPSPS	–	–	–	–
P1162HR	2009	3C	CP4 EPSPS	pat	–	Cry1F	–
P1162XR	2009	4	CP4 EPSPS	pat	–	Cry1F	Cry34Ab1, Cry35Ab1
33W82	2009	0	–	–	–	–	–
P1151HR	2011	3C	CP4 EPSPS	pat	–	Cry1F	–

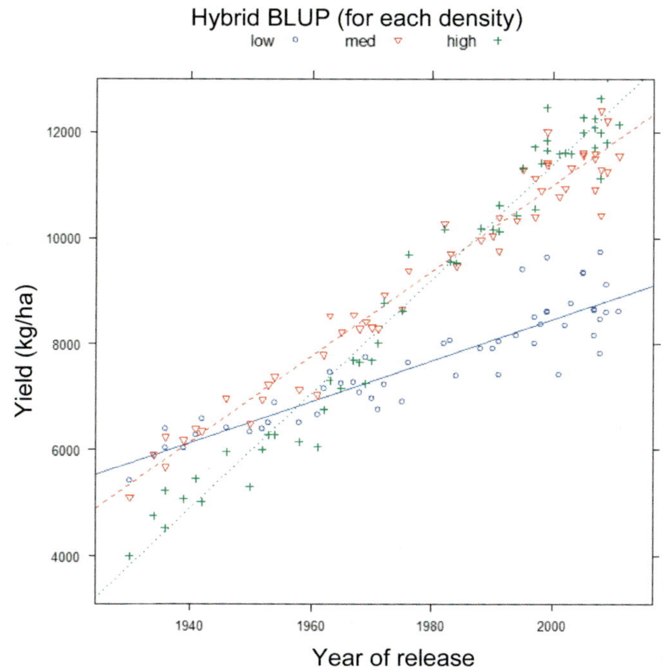

Fig. 6–7. Yield (genetic gain) for era hybrids released from 1930 to 2011 when planted at three planting densities.

estimated linear improvement of 87.6 kg ha^{-1} yr^{-1} (1.4 bu ac^{-1} yr^{-1}) (Fig. 6–8). This metric has been the norm for reporting genetic gain from these "era" studies and can also be interpreted as the rate of gain that would apply if G × D based improvement was substituted for pure G-based improvement. Rates of genetic gain measured at optimum planting density in well-watered conditions were 79.6 kg ha^{-1} yr^{-1} or 1.3 bu ac^{-1} yr^{-1} (Woodland, CA) (Fig. 6–9a) and 90.8 kg ha^{-1} yr^{-1} or 1.5 bu ac^{-1} yr^{-1} (Viluco, Chile) (Fig. 6–9b). Rates of gain measured at optimum planting density under drought stress imposed at grain fill and seed set were 59.1 kg ha^{-1} yr^{-1} or 1.5 bu ac^{-1} yr^{-1} (Woodland, CA) and 49.9 kg ha^{-1} yr^{-1} or 0.8 bu ac^{-1} yr^{-1} (Viluco, Chile). Double-cross hybrids released in the 1940s and 1950s expressed a relative inability to yield well in either of these environments. For comparison, Iowa farm maize yields during 1930 to 2011 increased at a rate of 111.6 kg ha^{-1} yr^{-1} (1.8 bu ac^{-1} yr^{-1}) (Fig. 6–8), while maize production on irrigated Nebraska farms (1947–2011) increased at an annual rate of 135.0 kg ha^{-1} yr^{-1} (1.8 bu ac^{-1} yr^{-1} (Fig. 6–9a and 6–9b). Percentage contributions of genetic gain to annual productivity gains during 1930 to 2011 were 79% for Iowa (rainfed) and 77% (Nebraska irrigated).

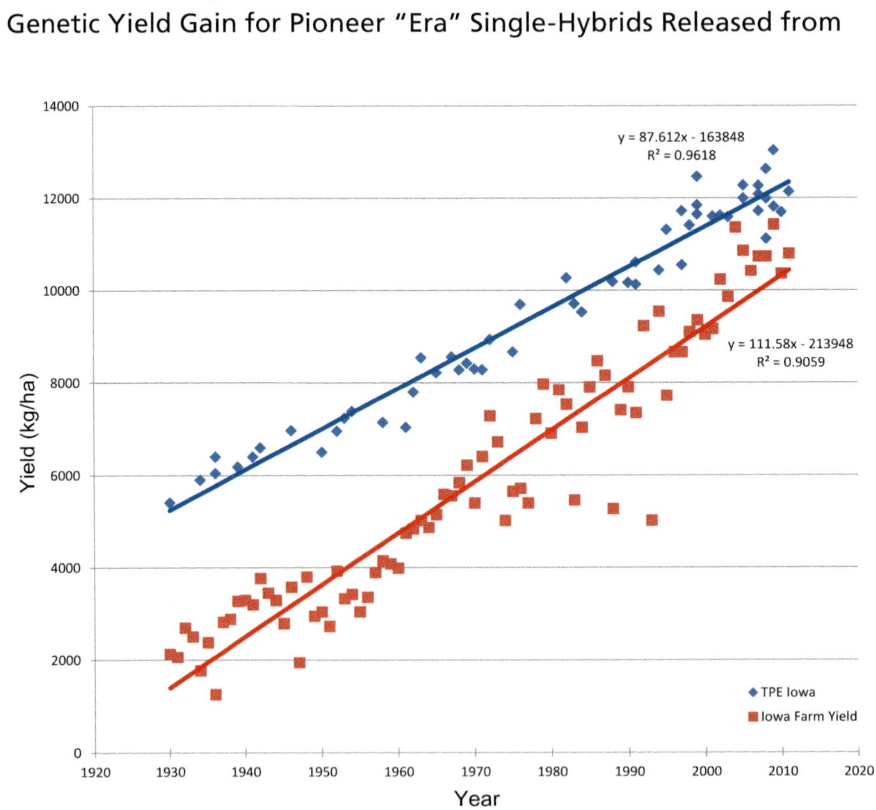

Fig. 6–8. Mean Best Linear Unbiased Predictor yields of Pioneer "era" maize hybrids released from 1930 to 2011 when grown at optimum planting densities in the TPE (target production environment). For comparison, the annual mean yields of maize in the Iowa (primarily rainfed) are also presented.

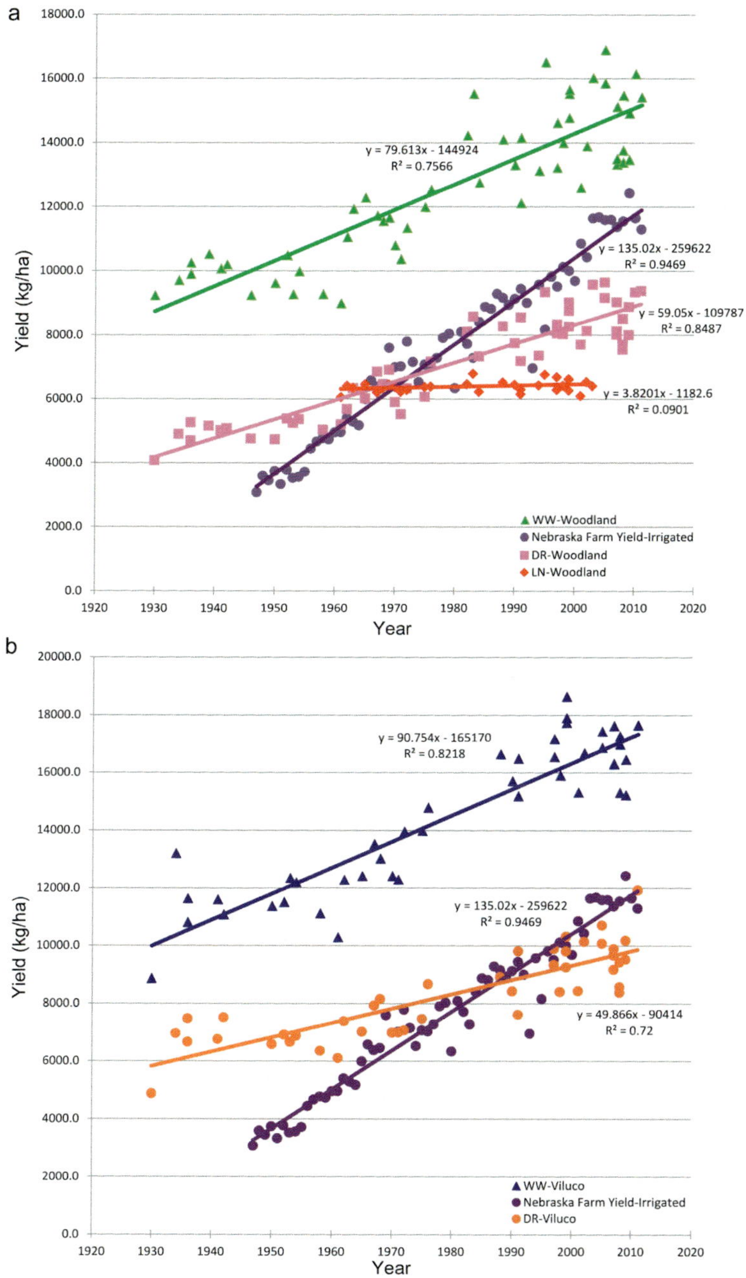

Fig. 6–9. (a) Mean Best Linear Unbiased Predictor yields of Pioneer "era" maize hybrids released from 1930 to 2011 when grown at optimum planting density in well-watered and drought environments (Woodland, CA). For comparison, annual mean yields of maize on irrigated farms in Nebraska are also presented. (b) Mean Best Linear Unbiased Predictor yields of Pioneer "era" maize hybrids released from 1930 to 2011 when grown at optimum planting density in well-watered and drought environments (Viluco, Chile). For comparison, annual mean yields of maize on irrigated farms in Nebraska are also presented.

1963 to 2011

We measured genetic gain from trialing "era" single-cross hybrids released from 1963 to 2011 grown in TPE plots as 92.2 kg ha^{-1} yr^{-1} (1.5 bu ac^{-1} yr^{-1}) (Fig. 6–10). Rates of gain for single-cross hybrids in well-watered conditions were 81.1 kg ha^{-1} yr^{-1} (1.3 bu ac^{-1} yr^{-1}) when grown in Woodland, CA (Fig. 6–11a) and 95.5 kg ha^{-1} yr^{-1} (1.5 bu ac^{-1} yr^{-1}) when grown in Viluco, Chile (Fig. 6–11b). However, the positive slope of genetic gain for a more recently released (1988–2011) subset of hybrids grown in well-watered conditions in Viluco, Chile was much reduced [17.8 kg ha^{-1} yr^{-1} (0.3 bu ac^{-1} yr^{-1})] (Fig. 6–11b). When grown under managed drought stress conditions, genetic gain was 58.0 kg ha^{-1} yr^{-1} (0.9 bu ac^{-1} yr^{-1}) when grown in Woodland, CA and 61.5 kg ha^{-1} yr^{-1} (1.0 bu ac^{-1} yr^{-1}) when grown in Viluco, Chile (Fig. 6–11a and 6–11b). For comparison, during this era of single-cross hybrids rate of yield increase on Iowa farms was 123.9 kg ha^{-1} yr^{-1} (2.0 bu ac^{-1} yr^{-1}) (Fig. 6–10), while the rate of yield increase on irrigated farms in Nebraska was 123.5 kg ha^{-1} yr^{-1} (2.0 bu ac^{-1} yr^{-1}) (Fig. 6–11a and 6–11b). Percentage contributions of genetic gain to annual productivity gains for single-cross hybrids were 74% for Iowa (rainfed) and 66 to 77% for Nebraska (irrigated).

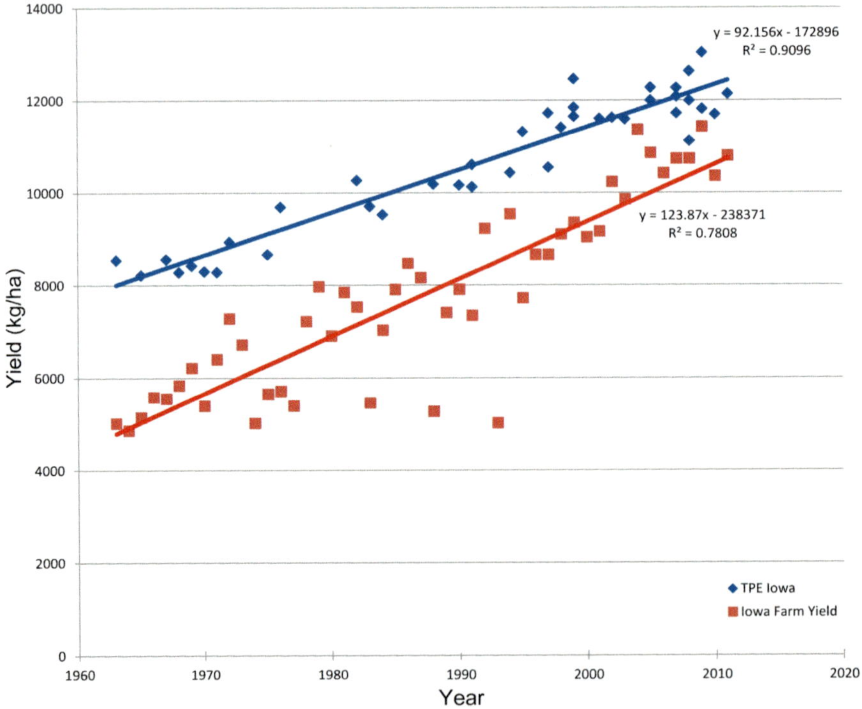

Fig. 6–10. Mean Best Linear Unbiased Predictor yields of Pioneer "era" maize hybrids released from 1963 to 2011 when grown at optimum planting densities the target production environment (TEP). For comparison, annual mean yields of maize in Iowa (primarily rainfed) are also presented.

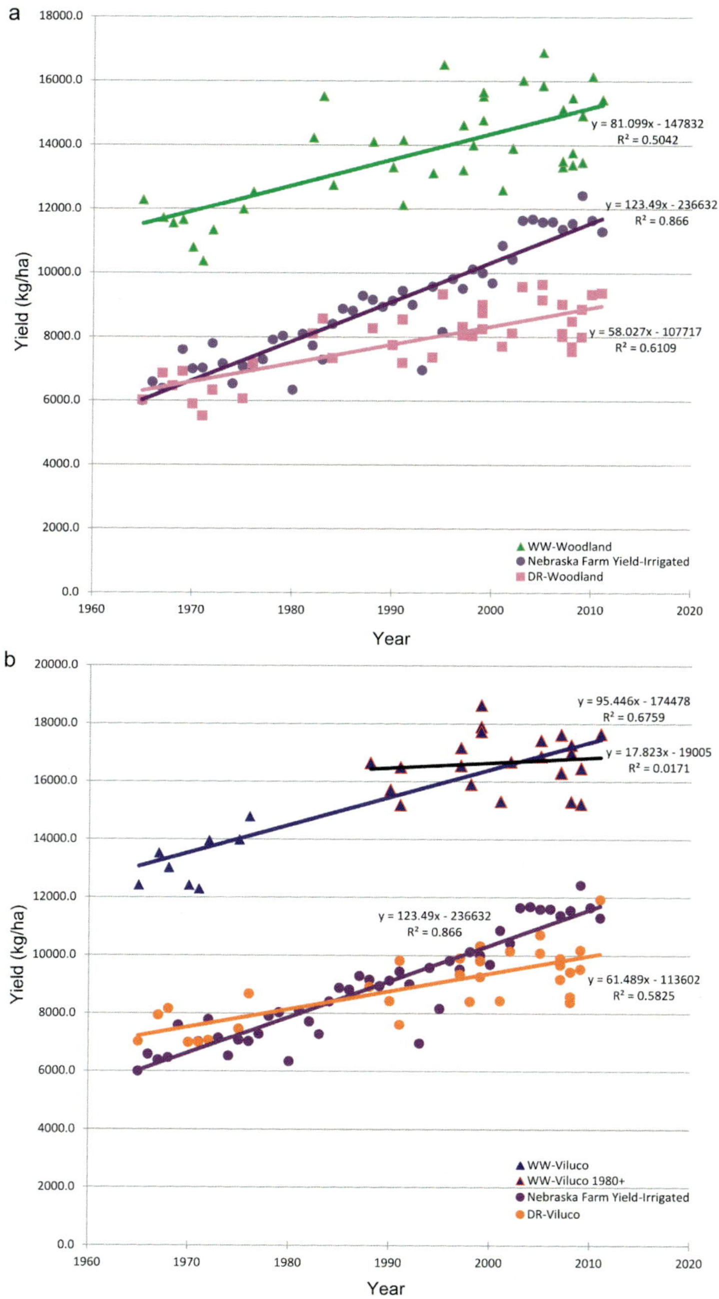

Fig. 6–11. (a) Mean Best Linear Unbiased Predictor yields of Pioneer singe-cross "era" maize hybrids released when grown at optimum planting density in well-watered and drought environments (Woodland, CA). For comparison, annual mean yields of maize achieved on irrigated farms in Nebraska are also presented. (b) Mean Best Linear Unbiased Predictor yields of Pioneer single-cross "era" maize hybrids when grown at optimum planting density in well-watered and drought environments (Viluco, Chile). For comparison, annual mean yields of maize achieved on irrigated farms in Nebraska are also presented.

Morphological Changes Associated with Genetic Gain

The mean harvest index (HI) increased from 0.46 for 1961 era hybrids to 0.49 for 2004 released hybrids (with maxima of 0.53). Of those traits categorized as contributing to increased HI (Duvick, 1984, 2005a, 2005b; Duvick et al., 2004a, 2004b), we found a continued trend toward reduced tassel size. In contrast, a trend toward more upright leaves plateaued in the 1990s (Fig. 6–12a). Duvick (1984, 2005a, 2005b) also categorized several morphological changes associated with higher planting densities (Fig. 6–12b). We detected a continuing trend toward shorter anther silk intervals and a lessening of stalk lodging, plus a slight but consistent improvement in root lodging. In contrast, resistance to plant barrenness (failure to produce grain) and greater stay-green plateaued in the 1990s.

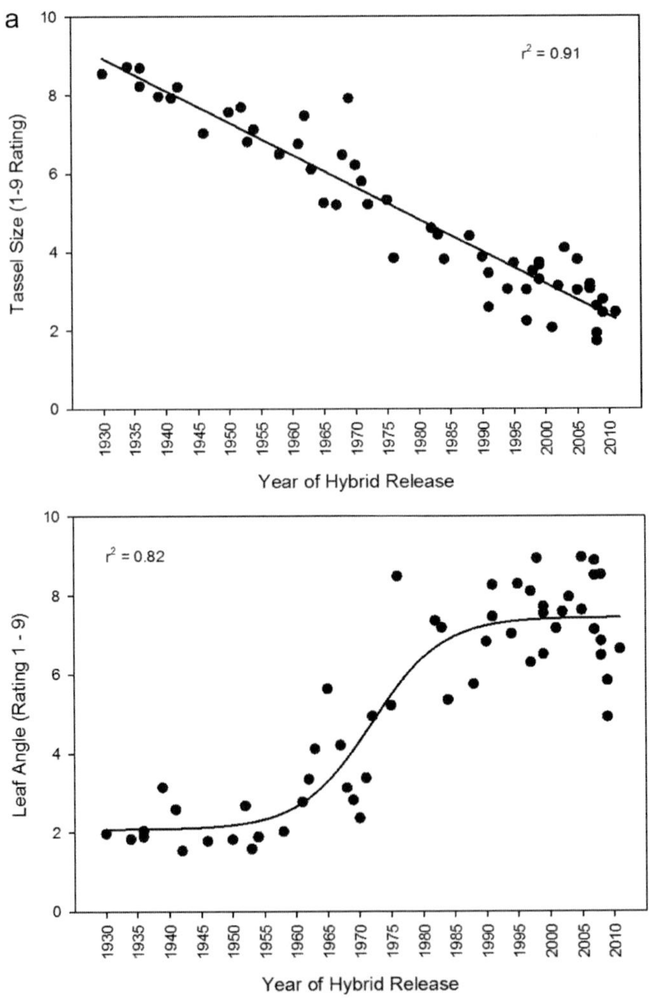

Fig. 6–12a. Changes in morphological attributes categorized as contributing to increased efficiency of grain production.

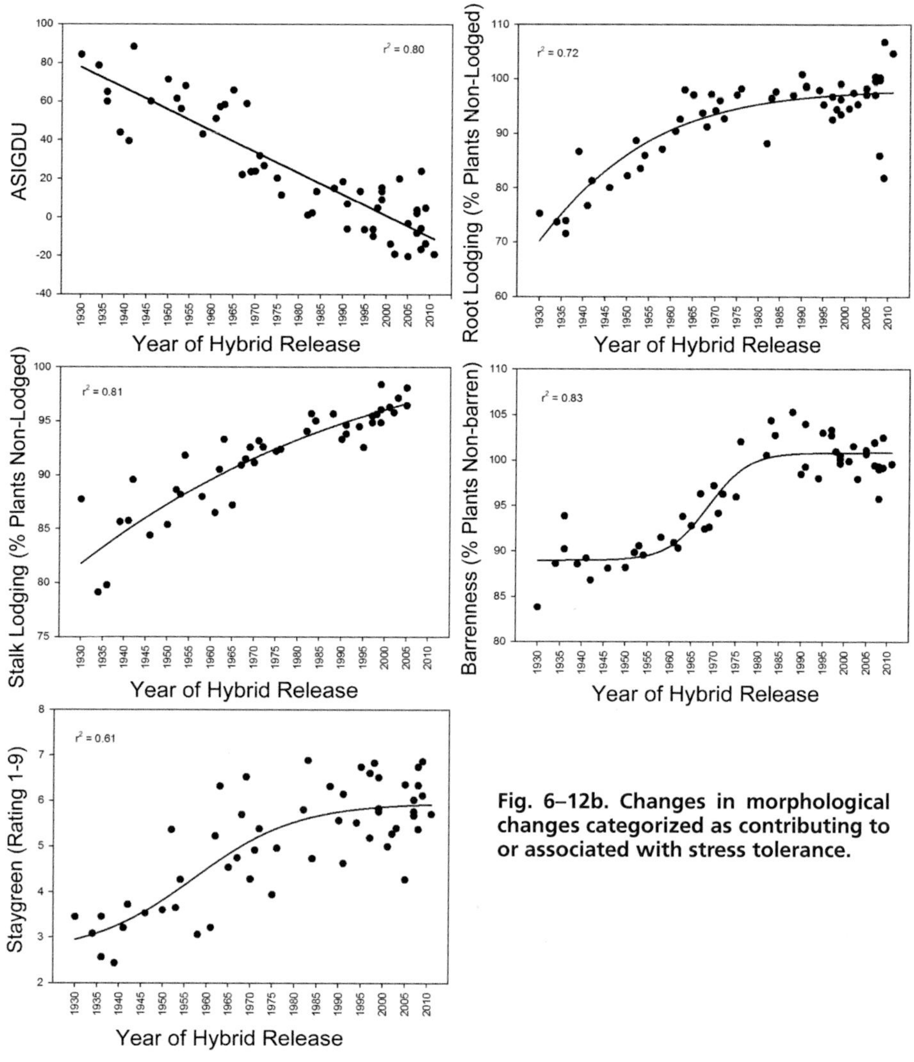

Fig. 6–12b. Changes in morphological changes categorized as contributing to or associated with stress tolerance.

Gains from Achieving Resistance to the European Corn Borer and Corn Rootworm

Hutchison et al. (2010) emphasized the multiplicity of factors affecting the spread of insect pests, including "insect biology, insect and crop phenology synchrony, reduced insecticide use, conservation tillage, soil type, glyphosate resistant crops, insect genetics, insect pathogens, preexisting insect population densities, and climate change." Consequently, determining the effect on yield of introducing resistance genes, whether they are native or transgenic, is fraught with difficulties because of the possible negative (drag) effects on yield in the absence of pests and complexities of the pest environment, including levels of pest infestation, pest biotype, soil nutrient level, incidence of disease, drought stress, and soil type, which can vary widely by location and by year.

The optimal method for measuring a trait effect is to evaluate isogenic pairs of genotypes differing in the contrasting alleles at a gene locus (or loci) governing the trait (Gurian-Sherman, 2009; Ubbenga, 2009). Here we report on yields of 15 isogenic pairs of Pioneer maize commercial hybrids with adaptations spanning the predominant maize maturity zones in North America. Each isogenic hybrid pair had either no transgene (non-GMO) or a +Cry1Ab transgene (i.e., a backcross-derived conversion of the former to the latter). Hybrids were planted in two rows of 5-m long plots arranged in randomized complete blocks at 3739 location–year combinations during 2000 to 2007. Experimental plots were distributed in North America among 24 states and 2 provinces (Fig. 6–13). Standard agronomic practices were used at all sites. Neither artificial infestations nor pesticides were applied to manage the ECB or any other Lepidopteran populations. Yield (in kilograms per square meter) was measured on a 150 g kg^{-1} moisture basis.

For each location–year combination, yield reduction due to infestation and feeding by corn borers was measured as:

$$\text{YLDRCB}\ (\%) = \sum_i \frac{A-B}{B} 100$$

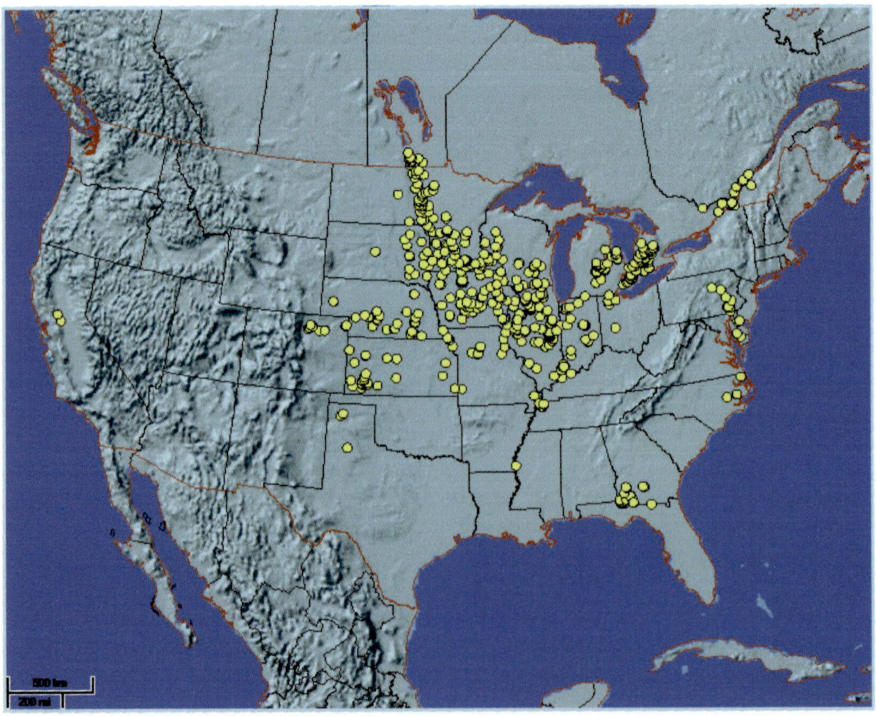

Fig. 6–13. Map of North America showing locations where isogenic pairs of Pioneer hybrids were planted to measure the effects of European corn borer resistance provides via Cry1Ab.

where *A* and *B* are observed yields at a given location for the isogenic hybrid pairs. Yield reduction due to infestation and feeding by corn borers is the average yield differential between the Cry1Ab gene plus and minus members of the of the isogenic hybrid pairs. Maps show examples of the extent and distribution of corn borer damage and characterize areas where low, medium, and high damage occurred (Fig. 6–14). Below mean infestation damage occurred in 2000 and 2004, mean infestation damage occurred in 2006 and 2007, and higher than mean infestation damage occurred in 2001, 2002, 2003, and 2005. The yield advantage for the

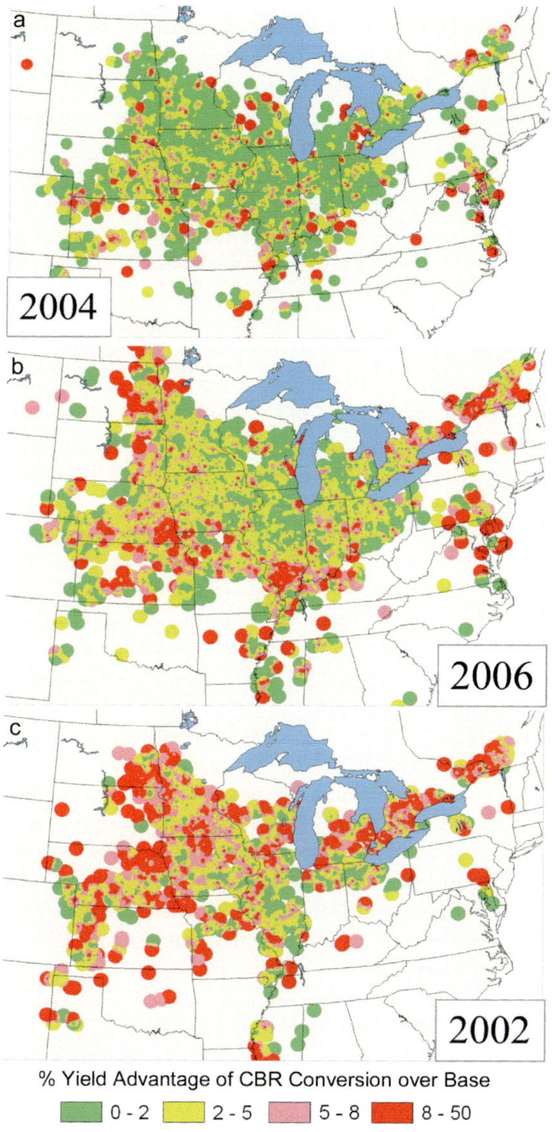

Fig. 6–14. Maps showing the distribution of corn borer damage representative of years with low (below mean) infestation damage (2000, 2004), years with medium (approximately mean) infestation damage (2006, 2007), and years with high (above mean) infestation damage (2001, 2002, 2003, 2005).

member of isogenic hybrid pairs that contained the Cry1Ab gene ranged from 2.5 to 8% (mean 5.3%) of the yield of the respective base non-Cry1Ab producing hybrids or 565 kg ha^{-1} (9 bu ac^{-1}).

Shi et al. (2013) reported a mean yield advantage for single event Bt transgenic hybrids of 411 kg ha (6.54 bu ac^{-1}). For ECB resistance incorporated into triple stacks, Edgerton et al. (2012) reported a yield advantage of 510 kg ha^{-1} (8.11 bu ac^{-1}), but Shi et al. (2013) reported a much lower yield advantage of 57 kg ha^{-1} (0.9 bu ac^{-1}). Since 2006, "it has been increasingly difficult to find ECB in most producers' fields and moth flights have been negligible" (Steffey and Gray, 2008). Consequently, as long as protection remains resilient, future studies in the United States of genetic protection against the ECB, if reliant on natural infestations alone, will underestimate the positive effects provided by insect resistant hybrids.

Across maize growing regions of the United States, Alston et al. (2002) estimated average yield increase factors resulting from the use of CRW resistant maize hybrids (in comparison with untreated conditions) ranging from 1.0 to 1.165 (low CRW pressure), 1.076 to 1.269 (moderate CRW pressure), and 1.164 to 1.393 (high CRW pressure). There are significant interactions with the soil type, water, and insect biotype. Corn rootworm infestation is more prevalent in wet clay soils, but if water and nutrient availability can be maintained, then yields might not suffer.

Genetic Gain for Maize Measured in Other Countries

We present the rates of genetic gain reported for maize grown in different countries (Table 6–2). Interpretative caution is advised because genetic gain data are context dependent. The contextual factors include germplasm, breeding strategies, crop management, cultivation practices, manual or mechanical harvesting, initial yield levels, and duration of study (i.e., persistence of genetic yield gain). Yield genetic gain data reported here for "era" hybrids, for example, show different yields and rates of gain depending on the location (estimates of genetic gain under irrigation and drought for trials conducted in Woodland, CA compared with Viluco, Chile), treatments (well watered, drought stress, levels of N), and chronological period of hybrid release studied. Different factors may be associated with genetic gain. For example, Luque et al. (2006) showed a positive yield response to increased planting density and increased yield on an individual plant basis for Argentinean maize hybrids. In contrast, for the United States, Duvick (1984, 2005a, 2005b) reported no increase in yield on an individual plant basis. In contrast, most maize breeders in China have yet to take full advantage of germplasm that is adapted to high yield potential at higher planting densities (Li et al., 2011). In France, Pierre Desbons (personal communication, 2005) demonstrated a large effect both on higher yield and rate of gain when single-cross hybrids were introduced replacing three- and four-way hybrids. It is prudent to conduct genetic gain studies in appropriate TPEs over a number of years or at least until consistent trends can be established.

Factors Contributing to Genetic Yield Gain

There has been much discussion during the preparation of this chapter regarding the validity of describing high plant density, high-production environments as stress environments. Traditional usage of the word "stress" describes factors that tend to reduce yields. In that context, how can stress environments also

Table 6–2. Rates of genetic gain reported for maize grown in different countries.

Country	Release period	Genetic gain	Comments	References
		kg ha^{-1} yr^{-1}		
Argentina	1965–1997	132–194		Luque et al. (2006)
	1972–1983	120		Löeffler et al. (1986)
	1979–1998	105–170		Eyherabide et al. (1994)
	1979–1988	170		Eyherabide and Damilano (2001)
Brazil	na	165–166		Alliprandini et al. (1998)
Canada	1950s–1980s	~85	hand harvested	Tollenaar (1989)
	1850s–1980s	~140	machine harvested	Tollenaar (1989)
China	1970–2000	50–57–95		Ci et al. (2011)
	1964–2001	52	spring corn, higher yielding	Wang et al. (2011)
	1964–2001	69	summer corn, lower yielding	Wang et al. (2011)
	1960s–2000s	62	not bred for high density planting	Li et al. (2011)
	1960s–2000s	81	bred for high density planting	Li et al. (2011)
France	1950–1985	100	very early and early maturing	Derieux et al. (1987)
	1950–1985	60	semi-early maturing	Derieux et al. (1987)
	1950–1985	80	late maturing	Derieux et al. (1987)
	1958–1978	30	three-way and four-way hybrids	Pierre Desbons (personal communication, 2005)
	1980–2002	140	single-cross hybrids	Pierre Desbons (personal communication, 2005)
Italy	1997–2012	70	on-farm trials, moving base of common checks	Alberto Verderio (personal communication, 2012)

Table continued.

Table 6–2. Continued.

Country	Release period	Genetic gain	Comments	References
		kg ha^{-1} yr^{-1}		
Romania	1975–2008	54–55	100 RM, nonstressed environments	Ion Ciocazanu (personal communication, 2012)
	1975–2008	75–85	103 RM, moderate nonstressed environments	Ion Ciocazanu (personal communication, 2012)
United States	1930–1980	92	target production environment, machine harvested	Duvick (1984)
	1930–1955	72	target production environment, machine harvested	Duvick (1984)
	1956–1980	112	target production environment, machine harvested	Duvick (1984)
	1930–2001	77	target production environment, machine harvested	Duvick et al. (2004a, 2004b)
	1930–2011	87	target production environment, machine harvested	this study
	1963–2011	92	target production environment, machine harvested	this study

be high-production environments? It is important to retain strict definitions to allow for clarity of discourse and understanding. Briefly, when U.S.-bred maize plants are grown at a high density, they individually yield less, but collectively, the population as a whole yields more. Thus, individual plants are subject to stress, including those growing in a high planting density, high-production environment. Older hybrids are particularly prone to reduced yields, including barrenness, particularly when subjected to high plant densities and especially when coupled with drought and heat stress. Newer hybrids are better adapted to high planting density, higher yield production environments. This is because they, amongst other attributes, have been selected to better mine soil for water and nutrients and collectively create a leaf canopy that maximizes the interception of solar radiation. As a result, some of these hybrids are less prone to drought-mediated yield reduction. Duvick (1984) ascribed the primary cause of genetic gain to the successive development of "increased stress resistance." Increased yields also resulted from the increased duration of photosynthesis (increased stay-green) and the greater partitioning of assimilates to seed, thereby resulting in greater HI. Duvick (1984) also noted a "close relationship between yielding ability and positive performance for defensive traits" (e.g., resistance to premature death, root and stalk lodging, barrenness, second-generation corn borers)—a relationship that was evident "especially at higher planting densities." Tollenaar and Wu (1999) also reported newer maize hybrids exhibiting an increased tolerance to competition from weeds. Studies of genetic gain using "era" hybrids in managed nurseries showed that newer hybrids yielded better in all stress environments, including in the severely depleted N environment, than hybrids released during or before the 1960s (S. Collinson, unpublished data, 2005, 2006). We, therefore, endorse and reinforce conclusions by Duvick (2005a, 2005b) that newer hybrids exhibit increased yield in both high-stress and low-stress environments. Nonetheless, an increased emphasis should also be placed on selecting for higher yields in relatively stress-free environments, including under irrigation.

We sought to compare the rates of genetic yield gain in U.S. maize contributed by simply inherited traits with the advances made by breeding with "base" germplasm. We adopted two approaches: First, we penalized yields of all single stack hybrids listed in Table 6–1 by 565 kg ha^{-1}, which is our estimated positive contribution to defending intrinsic genetic yield potential that is provided by Cry1Ab. We also penalized all hybrids containing Cry1F by 500 kg ha^{-1}, which was the estimated positive effect reported by Edgerton et al. (2012) for triple stacks that include ECB resistance. After revising the yield data accordingly for hybrids released during 1930 to 2011, the genetic gain was reduced to 82 kg ha^{-1} yr^{-1} (1.3 bu ac^{-1} yr^{-1}). This indicates a contribution of 6 kg ha^{-1} yr^{-1} (0.09 bu ac^{-1} yr^{-1}) or 7% of yield gain arising from monogenic sources of insect resistance and the remaining 93% arising from multigenic mediated yield gain (including, in this analysis, the net effect of herbicide resistance). We also compared regressions for "era" single-cross hybrids released between 1963 and 2011. We found a 31.4 kg ha^{-1} yr^{-1} (0.5 bu ac^{-1} yr^{-1}) positive advantage in genetic gain for yield for hybrids released that included either a single transgene or stack (hybrids released from 1997–2011) compared with those without GMOs (hybrids released from 1963–2005).

Second, we considered a one-time contribution of 500 kg ha^{-1} (8 bu yr^{-1}) made by protecting maize plants against the stresses imposed by both aboveground and belowground insects. We then computed the relative proportions of genetic gain that might be contributed during 20 yr (the life of a patent) by this genetic resistance and by breeding with native maize germplasm, conservatively assuming a genetic gain of 82 kg ha^{-1} yr^{-1} (1.3 bu ac^{-1} yr^{-1}). This computes as 500 kg ha^{-1} (8 bu ac^{-1}) (transgenic event) plus 20 $\times$ 82 or 1640 kg ha^{-1} (26 bu ac^{-1}) for the contribution of native germplasm. After 20 yr, the simply inherited trait contribution will be 23%, while that of the native germplasm will be 77%. These calculations are undeniably speculative. For example, they ignore any increase in the cycle time needed to incorporate transgenic traits, which reduces genetic gain per year, and they ignore yield improvement that could have been achieved by further selection of native ECB2 tolerance (Duvick, 2005a, 2005b; Duvick et al., 2004a, 2004b).

Potential Yields, Yield Achievement on Farms, Genetic Gain, and Future Prospects

Fischer and Edmeades (2010) defined potential yield as "the yield of an adapted cultivar when grown with the best management and without natural hazards such as hail, frost, or lodging, and without abiotic or biotoic stress limitations." The difference between PY and FY represents an exploitable yield gap (Cassman et al., 2003; Lobell et al., 2009), which remains to be closed through changes in management practice, including choice of cultivars (genetic gain). Attainable yield (AY) is "the yield a skillful farmer should reach when taking prudent account of economics and risk" (Fischer and Edmeades, 2010). In practice, average national yields tend to plateau at ~75 to 85% of PY, reflecting a yield limit that is logistically feasible and economically viable for most farmers (Cassman et al., 2003; Lobell et al., 2009).

Theoretical or simulated calculations of PY use assumptions that can become outdated as farm management practices change. In contrast, experimentally derived high yield data are available from annual yield contests such as those organized by the NCGA. Contest winners select particularly favorable growing environments and use management practices, some of which might not be economically viable or environmentally sustainable on a larger scale (Fischer and Edmeades, 2010; Grassini et al., 2011). Nonetheless, we drew on NCGA contest data as estimators of PY (under irrigated conditions) and of PYw (under rainfed conditions). These data were collated and kindly provided to us by Professor Jim Specht (University of Nebraska) from original data obtained from winning contest reports (NCGA). We took a conservative approach to the use of contest yield data by computing the annual mean yields of the top three placed contestants in each category, and we ignored data points that appeared as high-yielding outliers.

Rates of genetic gain previously presented in this chapter are summarized in Fig. 6–15. Rates of production gain for FYs achieved under irrigated and rainfed conditions are presented in Fig. 6–16. Mean 2011 yields, calculated from regression slopes (Fig. 6–15 and 6–16), percentage of genetic gain, and yields as percentage of PY and PYw are presented in Table 6–3. For PY, we present yield gains obtained under conventional compared with conservation (no-till, strip-till) tillage classes separately. In contrast, the similarity of NCGA contest yield data obtained in rainfed environments (PYw) using different tillage methods did not warrant their separation. For FY, we used annual means of irrigated maize production in Nebraska. For FYw, we used annual means of FYs in Iowa. Duvick and Cassman (1999) also used contest-winning yield data as surrogates for PY and PYw (see their Fig. 4b).

Our data for PY show large rate increases for irrigated environments ranging from 155 kg ha^{-1} yr^{-1} (2.5 bu ac^{-1} yr^{-1}) to 193.3 kg ha^{-1} yr^{-1} (3.1 bu ac^{-1} yr^{-1}) (Fig. 6–16). These data provide a chronological update to those presented by Duvick and Cassman (1999), who previously interpreted that contest-winning yield data had plateaued, because most of the yield gain for PY then occurred post-2000 (Fig. 6–16). Yield gains on irrigated Nebraska farms (126.3 kg ha^{-1} yr^{-1} or 2.0 bu ac^{-1} yr^{-1}) also showed continued progress but at a rate below those of contest winners. For yields under irrigation, we used two approaches to measuring the proportion of yield gain contributed by genetic gain: first, by using as the numerator the rate of genetic gain measured from yields of single-cross "era" hybrids grown in

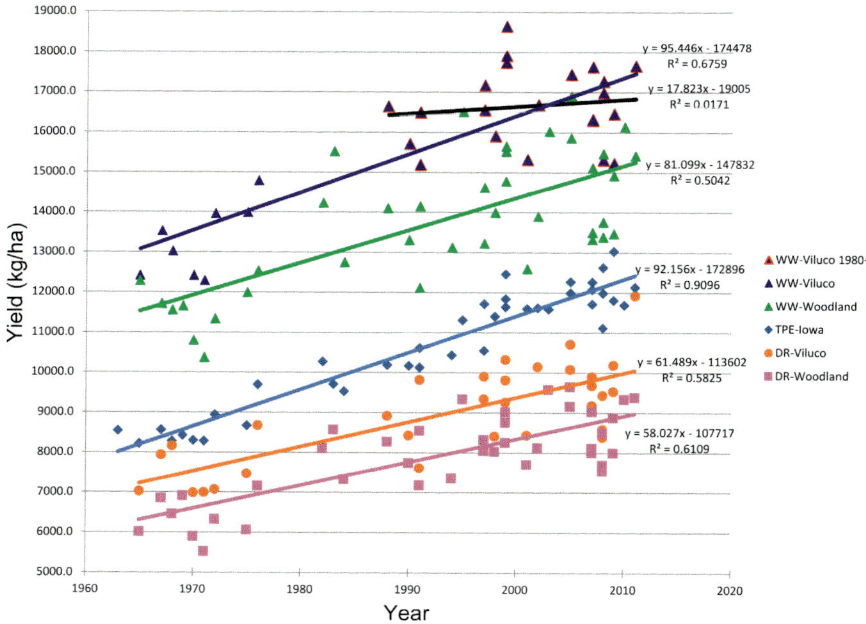

Fig. 6–15. Summary of genetic gain data for Pioneer era single-cross hybrids grown in target production environments, well-watered environments, and drought stress environments.

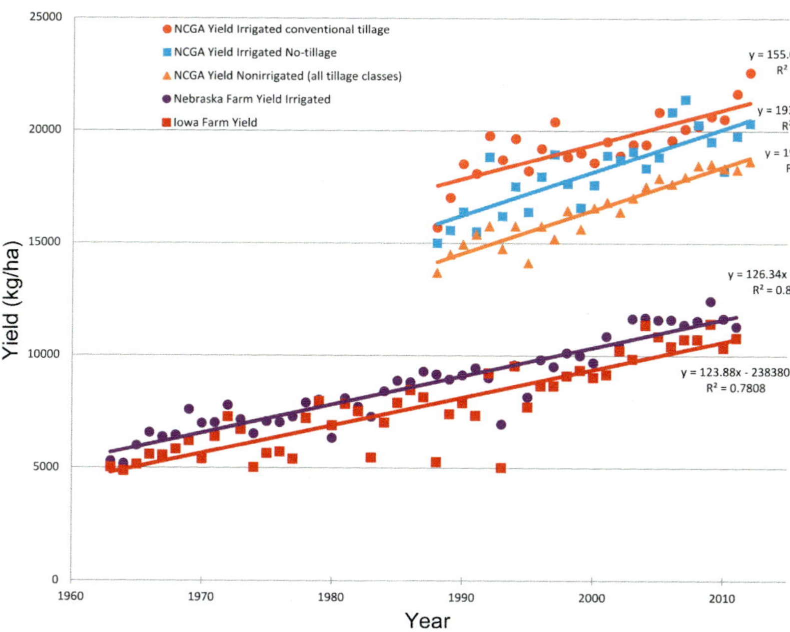

Fig. 6–16. Mean annual yields of the top three placed U.S. National Corn Grower Association (NCGA) contestants (outliers excluded) for rainfed (all tillage classes) and irrigated (conventional tillage and no tillage). U.S. annual maize production mean data for Iowa (primarily rainfed) and Nebraska irrigated farms are also provided.

Table 6–3. Single-cross hybrid era: 2011 yields, rates of yield gain, rates of genetic gain, percentage of genetic gain, and percentage of potential yield (PY) for maize grown under irrigated and rainfed conditions (see Fig. 6–15 and 6–16).

	2011 yield	Rate of yield or genetic gain	Genetic gain	Yield
	kg ha^{-1} (bu ac^{-1})	kg ha^{-1} yr^{-1} (bu ac^{-1} yr^{-1})	%	% PY
Irrigated				
Pioneer "era" hybrids	15,258.1–17,464.0 (244.0–279.2)	81.1–95.5 or 17.8† (1.3–1.5 or 0.3)	100%	72–83%‡
Nebraska irrigated	11,741 (187.7)	126.3 (2.0)	70 or 14%§	56–58%
PY, no-till and strip-till	20, 295 (324.5)	193.3 (3.1)	46 or 9%§	100%
PY, conventional tillage	21,108 (337.5)	155 (2.5)	57 or 12%§	100%
Rainfed				
Pioneer "era" target production environment	12429.7 (198.8)	92.2 (1.5)	100%	67%
Iowa farm yields	10,743 (171.8)	123 (2.0)	75%	58%
PY, all tillage classes	18,599 (297.4)	192.9 (3.1)	48%	100%
Drought				
Pioneer "era" hybrids	8975.3–10,052.4 (143.5–160.7)	58.0–61.5 (0.9–0.98)	100%	¶

† Mean genetic gain of 88.3 kg ha^{-1} yr^{-1} for single-cross hybrid era. Genetic gain of 17.8 kg ha^{-1} yr^{-1} (0.3 bu ac^{-1} yr^{-1}) for 1988–2011.

‡ Woodland–Viluco percentage of irrigated conventional till.

§ The lower percentage is computed using the rate of gain of 17.8 kg ha^{-1} yr^{-1} (0.3 bu ac^{-1} yr^{-1}) observed for era hybrids release for 1988–2011 (Fig. 15).

¶ No PY data for drought condition.

well-watered conditions (81.1–95.5 kg ha^{-1} yr^{-1} or 1.3–1.5 bu ac^{-1} yr^{-1}), and second, by using the rate of genetic gain (17.8 kg ha^{-1} yr^{-1} or 0.3 bu ac^{-1} yr^{-1}) measured in well-watered conditions in Viluco, Chile for hybrids released from 1988 to 2011. In 2011, mean Nebraska irrigated FYs were 56 to 58% of irrigated NCGA winning yields, approximately 20% below yields of Pioneer "era" hybrids, suggesting that opportunities remain to increase irrigated FYs through improved management including hybrid selection. However, there are several caveats: First, the extent to which managerial improvements can contribute via more effective use of inputs depends on a greater dissemination of knowledge, including from contest winners, to other farmers. Second, the extent to which greater nongenetic inputs can contribute to improved productivity depends on economic factors and potentially also on national or regional environmental regulations. Third, future PYs are dependent on future genetic gain. The highest ranking "era" hybrid yielded 18,643 kg ha^{-1} (297 bu ac^{-1}) when grown in well-watered conditions (Viluco, Chile), and the estimated rate of gain for more recently released "era" hybrids when grown in the well-watered environment of Viluco, Chile might lead to the cautioning of an impending genetic yield ceiling in relatively stress-free environments. However, there remained many era hybrids expressing yields well above the slope mean in both the Viluco and Woodland environments, indicating that useful genetic diversity to further increase yields in relatively stress-free, well-watered environments might not yet have been fully exploited. Duvick and Cassman (1999) recommended that maize breeders conduct hybrid performance trials at sites specifically selected for high yield potential. Today, maize breeders indeed conduct some of their breeding programs under low-stress environments, seeking to accentuate the positive effects of G × E interaction.

Yields obtained by NCGA contestants under rainfed conditions are increasing at a high rate (192.9 kg ha^{-1} yr^{-1} or 3.1 bu ac^{-1} yr^{-1}) and have reached very high levels (18,599 kg ha^{-1} or 297 bu ac^{-1} for 2011 mean yields) (Fig. 6–16), well above 2011 mean yields for (rainfed) Iowa farms (10,743 kg ha^{-1} or 172 bu ac^{-1}) and even above mean 2011 yields for Nebraska irrigated farms (11,741 kg ha^{-1} or 188 bu ac^{-1}) (Fig. 6–16). The mean 2011 Iowa FYs were 58% of PYw (identical to the percentage of PY for irrigated Nebraska yields) but much closer (86%) to the mean 2011 yield of the era rainfed plots than were mean 2011 irrigated Nebraska FYs to well-watered 2011 mean "era" yields (67–77%), indicating a potential PY ceiling for rainfed maize (Cassman et al., 2003; Lobell et al., 2009) unless (i) the current rate of genetic gain increases, (ii) input use increases (primarily N), or (iii) managerial skills continue to increase, thereby making more effective use of all available resources.

Persistent advances in yield made by NCGA contest winners in both irrigated and rainfed conditions indicate that there remains the potential to increase yields through improved farm management, most especially under irrigated conditions. However, the amount of realizable yield gain through improved management will depend on the dissemination of information, economic factors, and environmental regulations. Kip Cullers, a NGCA contest winner, noted that the primary value of contest plots was "so we can learn there and apply that to 11,000 acres" (Cullers, 2008). Extending such knowledge to 40 million hectares (~100 million acres) could significantly improve U.S. maize yields and production.

If current rates of yield gain on rainfed Iowa farms and the rate of genetic gain do not change, then FYs and era plot yields will coincide at approximately 14,161 kg ha^{-1} (226 bu ac^{-1}) during 2030 to 2035. For the foreseeable future, maize breeders are confident that not only can further genetic gain be achieved for rainfed maize agriculture, but recent increases in breeding efficiencies will further increase the rate of genetic gain. Estimates of future genetic gain from yields of early commercial development hybrids range from a conservative estimate of 126 kg ha^{-1} yr^{-1} (2 bu ac^{-1} yr^{-1}) (K. Boldman, personal communication, 2013) to 240 kg ha^{-1} yr^{-1} (3.8 bu ac^{-1} yr^{-1}) (Edgerton, 2009). Using this range of estimated future genetic gain Iowa FYs in 2025 of 12,491 to 13,621 kg ha^{-1} (199–217 bu ac^{-1}) and in 2050 of 15,630 to 19,584 kg ha^{-1} (249–312 bu ac^{-1}) are projected. For the United States as a whole, maize yields in 2025 of 11,487 to 12,617 kg ha^{-1} (183–201 bu ac^{-1}) and in 2050 of 14,625 to 18,580 kg ha^{-1} (233–296 bu ac^{-1}) would be projected. Achieving these projections will allow farmers in Iowa and in the United States to meet 75 to 98% of the goal to double 2008 yields by the year 2050 (Ray et al., 2013). Prospects for achieving these increases in yield will depend on continued research investments to secure an increased rate of genetic gain, further improvements in crop management, and no increase in the ratio of farm input costs relative to global grain prices.

Climate Change

Between 1999 and 2100, global temperatures and CO_2 levels are projected to increase (Intergovernmental Panel on Climate Change, 2007). North America and Europe are predicted to experience greater warming than Africa, Asia, or South America (Lobell and Gourdji, 2012). Relative to 1970 to 1999, summer surface temperatures in Iowa are projected to be between +1.5°C during the 2040s and +2.5°C by the 2080s (low emissions model) and +1.5°C (northern Iowa) and +2.5°C (southern Iowa) during the 2040s and +4°C (northern Iowa) and +4.5°C (southern Iowa) during the 2080s (high emissions model) (Walthall et al., 2012). To place these numbers in context, "Des Moines, in the heart of the U.S. Corn Belt, currently experiences fewer than 20 d above 32°C (89.6°F); this would double with a mean warming of 2°C (3.6°F)" (Rosenzweig et al., 2001). By 2100, summer temperatures in the midwestern United States are projected to reach from +3 to 4°C to +5 to 6°C (low and high emissions scenarios, respectively. Significant increases in frequencies of hot nights are also anticipated (Walthall et al., 2012). The growing season could increase by up to 1 mo. Trends in precipitation are harder to predict and have greater regional variability, although for the north-central U.S. rainfall could increase 5 to 15%. Historical data also show decadal fluctuations superimposed on longer-term trends, including both drought and wet intervals (Walthall et al., 2012). More extreme weather fluctuations are projected (Walthall et al., 2012), and such variability is more disruptive of cropping systems than changes in the mean (Katz and Brown, 1992).

Maize yields increase with temperatures up to 29°C, but higher temperatures are very harmful (Schlenker and Roberts, 2009). Increased precipitation can ameliorate some of the effects of increased temperature (Kucharik and Serbin, 2008). Air temperatures higher than 36°C (96.8°F) causes pollen to loose viability (Rosenzweig et al., 2001). High nighttime temperatures (21–27°C) contribute to yield loss due to wasteful respiration (Nafziger, 2011). The 63-yr mean (1950–2012) high temperature during the pollination period in central Iowa is currently 30°C. When temperatures >32°C have occurred during the pollination period in 13 of 63 (21%) years, then pollen tends to be released earlier in the day and successful pollinations occur both in breeding nurseries and on farms. Pollination failure has more often occurred because of poor silk receptivity caused by a combination of high heat and drought stress.

Projecting the effects of possible future climate scenarios is fraught with challenges. Hatfield et al. (2011) predict that provided adequate water is available, yields of maize would drop by 1.5% in 30 yr because of a rise in temperature alone. In contrast, unless there is amelioration of climate change effects through managerial adaptation, Schlenker and Roberts (2009) projected losses in U.S. maize yields of 20 to 30% by 2020 to 2049 and a 40 to 80% reduction by 2070 to 2099 compared with yields achieved earlier in the 21st century. Rosenzweig and Parry (1994) predicted crop productivity losses due to climate change in the midwestern United States of approximately 20%, whereas a recent report (World Bank, 2010) indicated a net gain in crop productivity for the central United States. Using a different model and taking a regional approach, Southworth et al. (2000) project that by 2050 to 2059 yields in the northern states (Wisconsin, Michigan, Ohio) will have generally increased, especially for longer season maize. In these regions, low temperatures currently limit grain fill, and increased temperatures will remove that limiting factor. In contrast, yields in the southern and central regions (Illinois, Indiana) will have declined because of increased surface air temperature contributing to earlier flowering and a shortening of the grain fill period. Long-season maize is projected to decrease in yield by 0 to −45% in southern and central locations of the Midwest, with midseason hybrids suffering dramatic decreases in yield from −15 to −40%. Tubiello et al. (2002) projected increases in growing season precipitation of 22 and 40%, respectively, for the central Corn Belt. As a further contrast, they project increased yields and reduced year-to-year variability for dryland maize grown in central Iowa (15–25%) and Illinois (15–38%), with greater proportional increases at northern locations Fargo, ND (25–50%) and Duluth and St Cloud, MN (30–50%), where previously both precipitation and relatively low temperatures had limited yields. In contrast, Tubiello et al. (2002) project reduced yields at current irrigated sites of −4 to −20%.

Kuleshov (1933) brought attention to the "extraordinary large area occupied in the world by *Zea mays* L." and "the large agro-ecological diversity encompassed; from 58° north to 40° south, across temperate, subtropical and tropical, from below sea level to 3000 m, and from arid regions (250 mm annual precipitation) to the tropics (5000–6,000 mm annual precipitation)." Kuleshov (1933) concluded: "genetic investigations have not yet exhausted the phenotypic richness in the maize plant." Bellon et al. (2011) also bring attention to the "remarkable diversity and climatic adaptability of maize races in Mexico." Nonetheless, they draw attention to the vulnerability of communities located in the highlands to climate change, where an estimated 18% of Mexican maize production occurs, because of the absence of locally adapted materials and the need to create new seed supply systems. Plant breeders create seed supply systems through their exchange of germplasm. However, maize breeders have been limited in their abilities to effectively source potentially useful new exotic genetic diversity because of uncertainties in how to prioritize choices from among thousands of accessions coupled with the subsequent high risk and resource intensive requirements to adapt and evaluate exotic germplasm (Tanksley and McCouch, 1997; McCouch, 2013). The vision and dogged determination expended over several decades by a small group of breeders coupled with a tenuous lifeline of support from the U.S. public and private sectors have ensured the availability of genetic resources diversity to the maize breeding community and demonstrated its potential utility (Hallauer and Sears, 1972; Brown, 1975; Goodman, 1985, 1990, 1999; Pollak and Salhuana, 2000). Finally, there are prospects that potentially useful maize diversity that has hitherto either been under-utilized in situ or that has resided on shelves in ex situ collections can now be more effectively accessed and evaluated. First, classification of the vast diversity of maize production environments has recently been updated (Hartkamp et al., 2000), thereby improving the efficiency of germplasm accession selection (Pollak and Corbett, 1993). Second, advances in the knowledge of physiology and use of DNA sequence data enable more effective mining of maize accessions for potentially useful genetic diversity (Romay et al., 2013). An internationally accessible informatics infrastructure is required to collate and cross reference these data to facilitate access by breeders to serve the needs of farmers (McCouch, 2013). There are many other managerial approaches that can be used to optimally adapt agriculture to changes in climate. These approaches include adjustment of planting and harvest dates, expansion of croplands, changing genotypes or species, altering fertilization rates, soil management and irrigation practices, and using climate forecasting to reduce production risks (Howden et al., 2007). Globally, Wheeler and von Braun (2013) argue that a "climate-smart agriculture" alone will be insufficient preparedness and a climate-smart food system is required.

Concluding Comments

We adhered to the original protocol for selection of hybrids that was established by Don Duvick when he initiated the "era" yield trials in the 1970s. We selected hybrids based solely on relative volume of sales regardless of whether they were solely "base" germplasm or included transgenic events. These genetic gain data therefore reflect any net yield drag caused by event insertion, time lag required to insert single transgenes or stacks, and net yield change caused by the action of transgenes (or stacks). We conducted an additional set of experiments to determine the contribution of the Cry1Ab gene to yield. In contrast, other recent studies of genetic gain in U.S. maize (Edgerton, 2009; Haegele et al., 2013) reported yield gains of hybrids before the insertion of transgenes; therefore, presumably for hybrids that would later be solely released commercially with transgenes, the hybrids actually yield trialed represented base germplasm at a precommercial stage of development.

We cannot address whether changes in tillage practice contributed to a reduction in genetic gain before 2001 because breeders might have then prioritized defensive traits as postulated by Duvick (1984). This is because our breeding and yield test plots, including "era" trials, are generally located on conventionally tilled land since uniform residue distribution is nearly impossible to attain in minimum or no-till fields, with consequent effects on the uniformity of seedling emergence, leading to less precision in phenotyping and thus lower heritability. Duvick and Cassman (1999) reported that performance issues for hybrids relating specifically to growth under conservation tillage conditions had not been a major reason for hybrid replacement on farms. Consequently, we see no strong evidence for the hypothesis that a focus by breeders on defensive traits contributed to a decline in the rate of genetic gain. Several other hypotheses could explain the reduced rate of genetic gain reported for "era" hybrids released from 1930 to 2001 (Duvick et al., 2004a, 2004b). There may have been either a reduction in the amount or efficiency of corn breeding and yield testing. During this period, breeding companies elected to introduce and strengthen their transgenic research and to move those products through the regulatory process and into variety development. In so doing, and because research budgets are never unlimited, some resources may have been invested in additional activities required to deliver the products of biotechnological research to farmers. In this context, it is useful to remember that the introduction of exotic germplasm or backcross breeding for defensive yield traits can also slow breeding purely for quantitative yield improvement whenever resources for the latter are diverted to the former.

Traits delivered via transgenics provided farmers with hitherto unavailable options in crop management and tillage practice and the ability to more effectively control insect pests. These achievements were possible through radically innovative uses of exotically sourced single-gene resistances that directly target either herbicide chemistry or insect physiology. None of these approaches purposely directly targets the genetics or physiology of the *Zea mays* L. germplasm itself. The deployment of additional single genes to further positively contribute to yield will be very challenging to achieve because of the inherent complexities contributed by numerous interactive genetic and physiological processes, including their interactions with the environment and management practices. Providing the best possible stewardship of existing major-gene resistance mechanisms with the ability to deploy additional resistance genes is of paramount importance (Moar et al., 2008; Heap, 2009; Gassmann et al., 2011; Onstad et al., 2011).

We recognize that estimates of the relative contributions of single genes compared with multigenic traits are subjective with a large degree of error. In addition, they are highly dependent on infestation levels of insect pests and abilities to reduce yield drag that may be associated with specific transgenic events. Nonetheless, we believe potential errors do not invalidate the conclusion that the vast majority of genetic yield gain has been contributed through quantitative genetic improvement using native maize germplasm. Duvick (2005a, 2005b) and Duvick et al. (2004a, 2004b) included the resistance to attack by the ECB2 in the list of stress responses that breeders had selected for by using native maize germplasm and that contributed genetic gain to maize. Therefore, we consider that the partitioning of yield as either "intrinsic" or "operational" (Gurian-Sherman, 2009) is a false dichotomy. Nonetheless, we do agree that scientific inquiry into the effectiveness and economic costs of different breeding approaches (Cox, 2002) must inform the development of breeding strategies. Likewise, entrepreneurs and policymakers must be informed of necessities to invest in research and product development to allow genetic gain to contribute to greater sustainable production of food, fiber, and fuel.

Photosynthesis is the main agent by which solar energy is transmuted to energy that sustains life on earth. We determined a small positive effect of climate contributing to improve U.S. maize yields during 1950 to 2011, corroborating other reports of benign, slightly favorable weather during at least part of that period. However, that trend was tenuous as shown by its negation with the addition of just a single year of weather and yield data confirming that while longer trends may exist, they have superimposed on them wide annual fluctuations. The degree of future annual fluctuations in weather may increase with a trend towards higher temperatures. We agree with Kucharik and Ramankutty (2005) that as mean yields rise then fluctuations in yield due to annual weather variability will tend to be more pronounced and may lead to the false perception of extreme weather occurrences. However, we do not agree with Kucharik and Ramankutty (2005) that reductions in the percentage of annual rate of gain are necessarily indicative of an imminently approaching yield ceiling. First, maintaining the same rate of yield gain will by definition mathematically cause the percent annual rate of gain to decline. Indeed, Frisvold et al. (2012) provide a very different perspective: They describe annual genetic improvements generated by plant breeders as "an annual stream of permanent yield improvements. Every year there has been a new incremental permanent increase in yields. One may properly think of the long-term benefits of genetic improvements as a stream of income streams." Second, yield data from contests show no indication of a decline in genetic or yield potential. Third, we anticipate an upturn in the genetic rate of genetic gain, at least under rainfed conditions, from observation of germplasm already in the commercial pipeline that is due to more efficient and effective breeding practices.

We purposely chose not to associate PYs achieved by contest winners with AYs on farms in general. This is in contrast to Duvick and Cassman (1999), who regarded contest-winning yields as "a reasonable estimate of the attainable yield level in a given year for a given region." Contest-winning yields evidence the genetic potential that resides in hybrids, but even with the best management practices, some of that potential is likely not to be practically achievable because of economic and possibly environmental realities. There is a lot of "yield space" between FY and PY, which is due to the conduct of different management practices including choice of hybrids, additional genetic potential that could be available in future hybrids, a wide range of differences in soil types, pests and diseases, and climates. If and how this "yield space" can be reduced will depend on achieving further genetic gain, management gain, and their positive interactions. Both breeding nurseries and farms are increasingly becoming more information rich environments. More detailed regional monitoring of weather trends could further improve optimum use of genotype × environmental interaction in both breeding and making crop management decisions. Grassini and Cassman (2012) demonstrated that improved management allowing high input-use efficiencies achieves both high yields and reduced potential to increase global warming. Prospects for continued increases in yield gains will ultimately depend on economic factors and environmental regulations, especially if they result in reduced inputs of water and nutrients from current levels. Continued research is required to develop maize hybrids and management practices that can yet more efficiently utilize available water, nutrient, and solar resources.

Duvick (1984) highlighted the rising cost of research relative to unit genetic gain achieved. We estimate from our experience of U.S. research and product development in maize breeding and biotechnology that this ratio has increased approximately 20-fold compared with 1980 (constant dollars). These investments are necessary to provide for increased knowledge of physiology and genetics, for the deployment of new breeding methods, and for the continual upgrading of skills by agronomists and plant breeders. Farmers and breeders have traditionally sourced varieties from a range of agro ecological zones, transporting them across national, regional, and continental boundaries to better adapt agriculture (Falcon and Fowler, 2002; Brush, 2005). A recent DNA sequence based comparison of U.S. maize germplasm developed by the public and private sectors showed that breeding programs, which develop the most productive hybrids, are also among the most genetically similar (Romay et al., 2013). Increased investments in germplasm development are vital to ensure a sufficiency of adapted genetic diversity so that genetic improvement can continue to contribute to yield gains on farms. Attaining sound stewardship and more effective use of genetic resources are prerequisites plant breeders must continually strive for to enable farmers to sustainably maximize efficiencies in converting solar energy to food, fiber, and fuel.

Acknowledgments

The authors wish to thank Ray Roberts and Travis Thorson (USDA-NASS) for their provision of maps showing maize usage, production, and yield in the United States and also thank Hua Mo for conducting the statistical analysis of U.S. maize yield data 1950 to 2011. S. Smith expresses personal thanks to Professor Jim Specht for his very thorough and insightful advice and critiques, which have significantly improved the chapter. Professor Specht is also to be acknowledged for his provision of several figures and for analytical data for NCGA contest yield data. Finally, neither this chapter nor this entire enterprise would have been possible without the expert coordination and dedicated hard work of Candace Madsen.

References

Abendroth, L., and R. Elmore. 2010. Planting date trends. Iowa State University, Agronomy Extension. http://www.agronext.iastate.edu/corn/production/management/planting/earlier.html (accessed 14 Aug. 2013).

Agriculture in the Classroom. 2013. Growing a nation: The story of American agriculture. Utah State University. http://www.agclassroom.org/gan/timeline/farm_tech.htm (accessed 19 Aug. 2013).

Alliprandini, L.F., A.P. Duarte, and R.A.D. Kanthack. 1998. [*Zea mays* L.] Genetic gain in commercial maize in summer and autumn–winter crops in the Paranapanema River Valley, Brazil, 1992 to 1997. Maydica 43:55–64.

Alston, J.M., J. Hyde, M.C. Marra, and P.D. Mitchell. 2002. An ex ante analysis of the benefits from the adoption of corn rootworm resistant transgenic corn technology. AgBioForum 5(3):71–84.

Andresen, J.A., G. Alagarswamy, C.A. Rotz, J.T. Richie, and A.W. LeBaron. 2001. Weather impacts on maize, soybean, and alfalfa production in the Great lakes region, 1895–1996. Agron. J. 93:1059–1070. doi:10.2134/agronj2001.9351059x

Baker, D.G., D.L. Ruschy, and R.H. Skaggs. 1993. Agriculture and the recent "benign climate" in Minnesota. Bull. Am. Meteorol. Soc. 74:1035–1040. doi:10.1175/1520-0477(1993)074<1035:AATRCI>2.0.CO;2

Beavis, W.D. 1998. QTL analyses power, precision and accuracy. In: A.H. Patterson, editor, Molecular dissection of complex traits. CRC Press, New York. p. 145–162.

Beavis, W.D., D. Grant, M. Albertsen, and R. Fincher. 1991. Quantitative trait loci for plant height in four maize populations and their association with qualitative genetic loci. Theor. Appl. Genet. 83:141–145. doi:10.1007/BF00226242

Bellon, M.R., D. Hodson, and J. Hellin. 2011. Assessing the vulnerability of traditional maize seed systems in Mexico to climate change. Proc. Natl. Acad. Sci. USA 108:13,432–13,437. doi:10.1073/pnas.1103373108

Borlaug, N.E. 2001. Feeding the world in the 21st century: The role of agricultural science and technology. Speech given at Tuskegee University, April 2001. http://www.agbioworld.org/biotech–info/topics/borlaug/borlaugspeech.html (accessed 14 Aug. 2013).

Brown, W.L. 1975. A broader germplasm base in corn and sorghum. Annu. Corn Sorghum Res. Conf. Proc. 30:81–89.

Brush, S.B. 2005. Protecting traditional agricultural knowledge. Wash. Univ. J. Law Policy. 17:58–109.

Bubeck, D.M., M.R. Carlone, R.L. Fox, M.D. Hoffbeck, R.L. Segebart, and D.S. Stucker. 2006. Breeding porgess measured in eight elite inbred families. Maydica 51:141–149.

Bubeck, D.M., M.M. Goodman, W.D. Beavis, and D. Grant. 1993. Quantitative trait loci controlling resistance to gray leaf spot in maize. Crop Sci. 33:838–847. doi:10.2135/cropsci1993.0011183X003300040041x

Burr, B., S.V. Evola, F.A. Burr, and J.S. Beckman. 1983. The application of restriction fragment length polymorphisms to plant breeding. In: J.K. Setlow and A. Hollander, editors, Genetic engineering principles and methods. Vol. 5. Plenum Press, London. p. 45–59.

Cardwell, V.B. 1982. Fifty years of Minnesota corn production: Sources of yield increase. Agron. J. 74:984–990. doi:10.2134/agronj1982.00021962007400060013x

Carlson, R.E. 1990. Heat stress, plant-available soil moisture, and corn yields in Iowa: A short- and long-term view. J. Prod. Agric. 3:293–297. doi:10.2134/jpa1990.0293

Carlson, R.E., D.P. Today, and S.E. Taylor. 1996. Midwestern corn yield and weather in relation to extremes of the Southern Oscillation. J. Prod. Agric. 9:347–352. doi:10.2134/jpa1996.0347

Carson, R. 1962. Silent spring. Houghton Mifflin, Boston, MA.

Cassman, K.G., A. Dobermann, D.T. Walters, and H.S. Yang. 2003. Meeting cereal demand while protecting natural resources and improving environmental quality. Annu. Rev. Environ. Resour. 28:315–358. doi:10.1146/annurev.energy.28.040202.122858

Center for Environmental Risk Assessment. 2012. GM Crop Database. ILSI Research Foundation, Washington DC. http://cera-gmc.org/index.php?action=gm_crop_database (accessed 19 Aug. 2013).

Ci, X., M. Li, X. Liang, Z. Xie, D. Zhang, X. Li, Z. Lu, G. Ru, L. Bai, C. Xie, Z. Hao, and S. Zhang. 2011. Genetic contribution to advanced yield for maize hybrids released from 1970 to 2000 in China. Crop Sci. 51:13–20. doi:10.2135/cropsci2010.04.0207

Cox, T.S. 2002. The mirage of genetic engineering. Am. J. Alternative Agric. 17:41–43.

Cullers, K. 2008. How I achieved a 154 bu yield. Paper presented at: Soy Super Session II: 38th Annual Corn and Sorghum Research Conference. Chicago, IL. 9–12 Dec.

DeFelice, M.S., P.R. Carter, and S.B. Mitchell. 2006. Influence of tillage on corn and soybean yield in the United States and Canada. Crop Manage. doi:10.1094/CM-2006-0626-01-RS

de la Vega, A.J., and S.C. Chapman. 2010. Mega-environment differences affecting genetic progress for yield and relative value of component traits. Crop Sci. 50:574–583. doi:10.2135/cropsci2009.04.0209

Derieux, M., M. Darrigrand, A. Gallais, Y. Barriere, D. Bloc, and Y. Montalant. 1987. Evaluation of the genetic gain in grain maize during the last thirty years in France. Agronomie 7:1–11. doi:10.1051/agro:19870101

Dillehay, B.L., G.W. Roth, D.D. Calvin, R.J. Kratochvil, G.A. Kuldau, and J.A. Hyde. 2004. Performance of Bt corn hybrids, their near isolines, and leading corn hybrids in Pennsylvania and Maryland. Agron. J. 96:818–824. doi:10.2134/agronj2004.0818

Duvick, D.N. 1977. Genetic rates of gain in hybrid maize yields during the past 40 years. Maydica 22:187–196.

Duvick, D.N. 1984. Genetic contributions to yield gains of U.S. hybrid maize, 1930 to 1980. In: W.R. Fehr, editor, Genetic contributions to yield gains of five major crop plants. CSSA Spec. Publ. 7. ASA and CSSA, Madison, WI. p. 1–47.

Duvick, D.N. 1997. What is yield? In: G.O. Edmeades et al., editors, Developing drought- and low N-tolerant maize. Proceedings of a Symposium, El Batan, Mexico. 25–29 Mar. 1996. CIMMYT, El Batan, Mexico. p. 332–335.

Duvick, D.N. 2001. Biotechnology in the 1930s: The development of hybrid maize. Nat. Rev. Genet. 2:69–74. doi:10.1038/35047587

Duvick, D.N. 2005a. Genetic progress in yield of United States maize (*Zea mays* L.). Maydica 50:193–202.

Duvick, D.N. 2005b. The contribution of breeding to yield advances in maize (*Zea mays* L.). Adv. Agron. 86:83–145. doi:10.1016/S0065-2113(05)86002-X

Duvick, D.N., and K.G. Cassman. 1999. Post-green revolution trends in yield potential of temperate maize in the north-central United States. Crop Sci. 39:1622–1630. doi:10.2135/cropsci1999.3961622x

Duvick, D.N., J.S.C. Smith, and M. Cooper. 2004a. Long-term selection in a commercial hybrid maize breeding program. In: J. Janick, editor, Plant breeding reviews: Long-term selection: Crops, animals, and bacteria. Vol. 24. Part 2. John Wiley & Sons, Hoboken, NJ. p. 109–151.

Duvick, D.N., J.S.C. Smith, and M. Cooper. 2004b. Changes in performance, parentage, and genetic diversity of successful corn hybrids, 1930–2000. In: C.W. Smith et al., editors, Corn: Origin, history, technology, and production. John Wiley & Sons, Hoboken, NJ. p. 65–97.

Edgerton, M.D. 2009. Increasing crop productivity to meet global needs for feed, food, and fuel Plant Physiol. 149:7–13.

Edgerton, M.D., J. Fridgen, J.R. Anderson, Jr., J. Ahlgrim, M. Criswell, P. Dhungana, T. Gocken, Z. Li, S. Mariappan, C.D. Pilcher, A. Rosielle, and S.B. Stark. 2012. Transgenic insect resistance traits increase corn yield and yield stability. Nat. Biotechnol. 30:493–496. doi:10.1038/nbt.2259

Edwards, M.D., C.W. Stuber, and J.F. Wendel. 1987. Molecular-marker-facilitated investigations for quantitative trait loci in maize. I. Numbers, genomic distribution and types of gene action. Genetics 116:113–125.

Ellis, R.N., K.E. Basford, M. Cooper, J.K. Leslie, and D.E. Byth. 2001. A methodology for analysis of sugarcane productivity trends. I. Analysis across districts. Aust. J. Agric. Res. 52:1001–1009. doi:10.1071/AR00130

Ellis, R.N., K.E. Basford, J.K. Leslie, D.M. Hogarth, and M. Cooper. 2004. A methodology for anlaysis of sugarcane productivity trends. 2. Comparing variety trials with commercial productivity. Aust. J. Agric. Res. 55:109–116. doi:10.1071/AR03074

Ellsbury, M.M., T.E. Schumacher, R.D. Gustin, and W.D. Woodson. 1994. Soil compaction effect on corn rootworm populations in maize artificially infested with eggs of western corn rootworm (Coleoptera: Chrysomelidae). Environ. Entomol. 23:943–948.

Eyherabide, G.H., and A.L. Damilano. 2001. Comparison of genetic gain for grain yield of maize between the 1980s and 1990s in Argentina. Maydica 46:277–281.

Eyherabide, G.H., A.L. Damilano, and J.C. Colazo. 1994. Genetic gain for grain yield of maize in Argentina. Maydica 39:207–211.

Falcon, W.P., and C. Fowler. 2002. Carving up the commons–emergence of a new international regime for germplasm development and transfer. Food Policy 27:197–222. doi:10.1016/S0306-9192(02)00013-1

Fischer, R.A., and G.O. Edmeades. 2010. Breeding and cereal yield progress. Crop Sci. 50:S85–S98. doi:10.2135/cropsci2009.10.0564

Foley, J.A., N. Ramankutty, K.A. Brauman, E.S. Cassidy, J.S. Gerber, M. Johnston, N.D. Mueller, C. O'Connell, D.K. Ray, P.C. West, C. Balzer, E.M. Bennett, S.R. Carpenter, J. Hill, C. Monfreda, S. Polasky, J. Rockström, J. Sheehan, S. Siebert, D.Tilman, and D.P.M. Zaks. 2011. Solutions for a cultivated planet. Nature 478:337–342. doi:10.1038/nature10452

Frensham, A.B., A.R. Barr, B.R. Cullis, and S.D. Pelham. 1998. A mixed model analysis of 10 years of oat evaluation data: Use of agronomic information to explain genotype by environment interaction. Euphytica 99:43–56. doi:10.1023/A:1018395731621

Frey, K.J. 1971. Improving crop yields through plant breeding. In: J.D. Eastin and R.D. Munson, editors, Moving off the yield plateau. ASA Spec. Publ. 20. ASA, CSSA, SSSA, Madison, WI. p. 15–58.

Frisvold, G., J. Sullivan, and A. Raneses. 2012. Who gains from genetic improvements in U.S. crops. AgBioForum 2:237–246.

Gassmann, A.J., J.L. Petzold-Maxwell, R.S. Keweshan, and M.W. Dunbar. 2011. Field-evolved resistance to Bt maize by western corn rootworm. PLoS ONE 6(7):E22629. doi:10.1371/journal.pone.0022629

Godfrey, L.D., L.J. Meinke, and R.J. Wright. 1993. Vegetative and reproductive biomass accumulation in field corn: Response to root injury by western corn rootworm (Coleoptera: Chrysomelidae). J. Econ. Entomol. 86:1557–1573.

Goodman, M.M. 1985. Exotic maize germplasm: Status, prospects, and remedies. Iowa State J. Res. 59:497–527.

Goodman, M.M. 1990. Genetic and germpolasm stocks worth conserving. J. Hered. 81:11–16.

Goodman, M.M. 1999. Broadening the genetic diversity in maize breeding by use of exotic germplasm. In: J.G. Coors and S. Pandey, editors, The genetics and exploitation of heterosis in crops. ASA, CSSA, SSSA, Madison, WI. p. 139–148.

Grassini, P., and K.G. Cassman. 2012. High-yield maize with large net energy yield and small global warming intensity. Proc. Natl. Acad. Sci. USA 109:1074–1079. doi:10.1073/pnas.1116364109

Grassini, P., J. Thorburn, C. Burr, and K. Cassman. 2011. High-yield irrigated maize in the western U.S. Corn Belt: I. On-farm yield, yield potential, and impact of agronomic practices. Agronomy and Horticulture University of Nebraska-Lincoln Faculty Publications. http://digitalcommons.unl.edu/agronomyfacpub/585 (accessed 14 Aug. 2013).

Gray, M.E., and K.L. Steffey. 1998. Corn rootworm (Coleoptera: Chrysomelidae) larval injury and root compensation of 12 maize hybrids: An assessment of the economic injury index. J. Econ. Entomol. 91:723–740.

Gurian-Sherman, D. 2009. Failure to yield: Evaluating the performance of genetically engineered crops. Union of Concerned Scientists. http://www.ucsusa.org/assets/documents/food_and_agriculture/failure–to–yield.pdf. (accessed 6 Mar 2013).

Haegele, J.W., K.A. Cook, D.M. Nichols, and F.E. Below. 2013. Changes in nitrogen use traits associated with genetic improvement for grain yield of maize hybrids released in different decades. Crop Sci. 53:1256–1268. doi:10.2135/cropsci2012.07.0429

Hallauer, A.R., and J.H. Sears. 1972. Integrating exotic germplasm into Corn Belt maize breeding programs. Crop Sci. 12:203–206. doi:10.2135/cropsci1972.0011183X001200020016x

Hartkamp, A.D., J.W. White, A. Rodriguez, M. Banziger, G. Srinivasan, G. Granados, and J. Crossa. 2000. Maize production environments revisited; A GIS-based approach. CIMMYT, El Batan, Mexico.

Hatfield, J.L., K.J. Boote, B.A. Kimball, L.H. Ziska, R.C. Izaurralde, D. Ort, A.M. Thomson, and D. Wolfe. 2011. Climate impacts on agriculture: Implications for crop production. Agron. J. 103:351–370.

Heap, I. 2009. The international survey of herbicide resistant weeds. http://weedscience.org (accessed 14 Aug. 2013).

Howden, S.M., J.F. Soussana, F.N. Tuniello, N. Chhetri, M. Dunlop, and H. Meinke. 2007. Adapting agriculture to climate change. Proc. Natl. Acad. Sci. USA 104:19,691–19,696. doi:10.1073/pnas.0701890104

Hutchison, W.D., E.C. Burkness, P.D. Mitchell, R.D. Moon, T.W. Leslie, S.J. Fleischer, M. Abrahamson, K.L. Hamilton, K.L. Steffey, M.E. Gray, R.L. Hellmich, L.V. Kaster, T.E. Hunt, R.J. Wright,

K. Pecinovsky, T.L. Rabaey, B.R. Flood, and E.S. Raun. 2010. Areawide suppression of European corn borer with Bt maize reaps savings to non-Bt maize growers. Science 330:222–225. doi:10.1126/science.1190242

Intergovernmental Panel on Climate Change. 2007. Climate change 2007: Synthesis report. Contribution of working groups I, II, and III to the fourth assessment report of the Intergovernmental Panel on Climate Change, edited by Core Writing Team, R.K. Pachauri, and A. Reissinger. IPCC, Geneva, Switzerland.

Iowa State University. 2001. Corn planting guide. ISU University Extension. http://www.extension.iastate.edu/publications/pm1885.pdf (accessed 19 Aug. 2013).

Iowa State University. 2013. Climate Science Program, data archive. ISU Climate Science Program. http://climate.engineering.iastate.edu/DataArchive.html (accessed 14 Aug. 2013).

Jampatong, C., M.D. McMullen, B.D. Barry, L.L. Darrah, P.F. Byrne, and H. Kross. 2002. Quantitative trait loci for first–and second–generation European corn borer resistance derived from the maize inbred Mo47. Crop Sci. 42:584–593. doi:10.2135/cropsci2002.0584

Katz, R.W., and B.G. Brown. 1992. Extreme events in a changing climate: Variability is more important than averages. Clim. Change 21:289–302. doi:10.1007/BF00139728

Kaufmann, R.K., and S.E. Snell. 1997. A biophysical model of corn yield: Integrating climatic and social determinants. Am. J. Agric. Econ. 79:178–190. doi:10.2307/1243952

Krakowsky, M.D., M.J. Brinkman, W.L. Woodman-Clikeman, and M. Lee. 2002. Genetic components of resistance to stalk tunneling by the European corn borer in maize. Crop Sci. 42:1309–1315. doi:10.2135/cropsci2002.1309

Krakowsky, M.D., M. Lee, W.L. Woodman-Clikeman, M.J. Long, and N. Sharopova. 2004. QTL mapping of resistance to stalk tunneling by the European corn borer in RILs of maize population B73 ´ De8. Crop Sci. 44:274–282. doi:10.2135/cropsci2004.0274

Kucharik, C.J. 2008. Contribution of planting date trends to increased maize yields in the central United States. Agron. J. 100:328–336. doi:10.2134/agrojnl2007.0145

Kucharik, C.J., and N. Ramankutty. 2005. Trends and variability in U.S. corn yields over the twentieth century. Earth Interact. 9:1–29. doi:10.1175/EI098.1

Kucharik, C.J., and S.P. Serbin. 2008. Impacts of recent climate change on Wisconsin corn and soybean yield trends. Environ. Res. Lett. 3:034003. doi:10.1088/1748-9326/3/3/034003

Kuleshov, N.N. 1933. World's diversity of phenotypes of maize. J. Am. Soc. Agron. 25:688–700. doi:10.2134/agronj1933.00021962002500100006x

Leakey, A.D.B., M. Uribelarrea, E.A. Ainsworth, S.L.O. Naidu, A. Rogers, D.R. Ort, and S.P. Long. 2006. Photosynthesis, productivity, and yield of maize are not affected by open-air elevation of CO_2 concentration in the absence of drought. Plant Physiol. 140:779–790. doi:10.1104/pp.105.073957

Levine, E., J.L. Spencer, S.A. Isard, D.W. Onstad, and M.E. Gray. 2002. Adaptation of the western corn rootworm to crop rotation: Evolution of a new strain in response to a management practice. Am. Entomol. 48:94–107.

Li, Y., X. Ma, T. Wang, Y. Li, C. Liu, Z. Liu, B. Sun, Y. Shi, Y. Song, M. Carlone, D. Bubeck, H. Bhardwaj, D. Whitaker, W. Wilson, E. Jones, K. Wright, S. Sun, W. Niebur, and S. Smith. 2011. Increasing maize productivity in China by planting hybrids with germplasm that responds favorably to higher planting densities. Crop Sci. 51:2391–2400. doi:10.2135/cropsci2011.03.0148

Littell, C., G.A. Milliken, W.W. Stroup, R.D. Wolfinger, and O. Schabenberger. 2006. SAS for mixed models. 2nd ed. SAS Inst., Cary, NC.

Lobell, D.B., and G.P. Asner. 2003. Climate and management contributions to recent trends in U.S. agricultural yields. Science 299:1032. doi:10.1126/science.1077838

Lobell, D.B., K.G. Cassman, and C.B. Field. 2009. Crop yield gaps: Their importance, magnitude and causes. Annu. Rev. Environ. Resour. 34:179–204. doi:10.1146/annurev.environ.041008.093740

Lobell, D.B., and S.M. Gourdji. 2012. The influence of climate change on global crop productivity. Plant Physiol. 160:1686–1697. doi:10.1104/pp.112.208298

Lobell, D.B., W. Schlenker, and J. Coste-Roberts. 2011. Climate trends and global crop production since 1980. Science 333:616–620. doi:10.1126/science.1204531

Löeffler, C.M., M.T. Salaberry, and J.C. Maggio. 1986. Stability and genetic improvement of maize yield in Argentina. Euphytica 35:449–458. doi:10.1007/BF00021853

Löeffler, C.M., J. Wei, T. Fast, J. Gogerty, S. Langton, M. Bergman, B. Merrill, and M. Cooper. 2005. Classification of maize environments using crop simulation and geographic information systems. Crop Sci. 45:1708–1716. doi:10.2135/cropsci2004.0370

Loomis, R.S., S.D. Beck, and J.F. Stauffer. 1957. The European corn borer, *Pyrausta nubilalis* (Hubn), and its principal host plant. V. A chemical study of host plant resistance. Plant Physiol. 32:379–385. doi:10.1104/pp.32.5.379

Luque, S.F., A.G. Cirilo, and M.E. Otegui. 2006. Genetic gains in grain yield and related physiological attributes in Argentine maize hybrids. Field Crops Res. 95:383–397. doi:10.1016/j.fcr.2005.04.007

Markelz, R.J., R.S. Strellner, and A.D.B. Leakey. 2011. Impairment of C4 photosynthesis by drought is exacerbated by limiting nitrogen and ameliorated by elevated [CO_2] in maize. J. Exp. Bot. 62:3235–3246. doi:10.1093/jxb/err056

Mason, C.E., M.E. Rice, D.D. Calvin, J.W. Van Duyn, W.B. Showers, W.D. Hutchison, J.F. Witkowski, R.A. Higgins, D.W. Onstad, and G.P. Dively. 1996. European corn borer: Ecology and management. North Regional Extension Publ. 327. Iowa State University, Ames.

McCouch, S. 2013. Feeding the future. Nature 499:23–24. doi:10.1038/499023a

Messina, C.D., G.L. Hammer, Z. Dong, D. Podlich, and M. Cooper. 2009. Modeling crop improvement in a G′E′M framework via gene–trait–phenotype relationships. In: V. Sadras and D. Calderini, editors, Crop physiology: Interfacing with genetic improvement and agronomy. Academic Press, San Diego, CA. p. 235–265.

Moar, W., R. Roush, A. Shelton, J. Ferre, S. MacIntosh, B.R. Leonard, and C. Abel. 2008. Field-evolved resistance to Bt toxins. Nat. Biotechnol. 26:1072–1074. doi:10.1038/nbt1008-1072

Morrell, P.L., E.S. Buckler, and J. Ross-Ibarra. 2012. Crop genomics: Advances and applications. Nat. Rev. Genet. 13:85–96.

Munkvold, G.P. 2009. Seed pathology progress in academia and industry. Annu. Rev. Phytopathol. 47:285–311. doi:10.1146/annurev-phyto-080508-081916

Nafziger, E. 2011. High temperatures and crop. The Bulletin: Pest management and crop development information for Illinois. University of Illinois Extension, Iowa City, Iowa. http://bulletin.ipm.illinois.edu/article.php?id = 1537 (accessed 14 Aug. 2013).

Onstad, D.W., P.D. Mitchell, T.M. Hurley, J.G. Lundgren, R.P. Porter, C.H. Krupke, J.L. Spencer, C.D. Difonzo, T.S. Baute, R.L. Hellmich, L.L. Buschman, W.D. Hutchison, and J.F. Tooker. 2011. Seeds of change: Corn seed mixtures for resistance management and integrated pest management. J. Econ. Entomol. 104:343–352. doi:10.1603/EC10388

Owens, J.C., J.F. Witkowski, J.J. Tollefson, R.R. Rogers, and D.C. Peters. 1974. Greenhouse evaluation of soil insecticides for western corn rootworm control. J. Econ. Entomol. 67:772–774.

Payne, J., J. Fernandez-Cornejo, and S. Daberkow. 2003. Factors affecting the likelihood of corn rootworm Bt seed adoption. AgBioForum 6:79–86.

Penny, L.H., G.E. Scott, and W.D. Guthrie. 1967. Recurrent selection for European corn borer resistance in maize. Crop Sci. 7:407–409. doi:10.2135/cropsci1967.0011183X000700050001x

Pollak, L.M., and J.D. Corbett. 1993. Using GIS datasets to classify maize-growing regions in Mexico and Central America. Agron. J. 85:1133–1139. doi:10.2134/agronj1993.00021962008500060007x

Pollak, L.M., and W. Salhuana. 2000. U.S. germplasm enhancement for maize project (U.S.–GEM). In: H.D. Cooper and C. Spillane, editors, Broadening the genetic bases of crop production. IPGRI/FAO, Rome. p. 319–329.

Ray, D.K., N.D. Mueller, P.C. West, and J.A. Foley. 2013. Yield trends are insufficient to double global crop production by 2050. PLoS ONE 8:1–8.

Robinson, J.L., and M.M. Rhoades. 1935. The 1935 Iowa corn yield test. Agric. Exp. Stn. Iowa State College of Agriculture and Mechanical Arts, Ames. p. 339–355.

Romay, M.C., M.J. Millard, J.C. Glaubitz, J.A. Pfeiffer, K.L. Swarts, T.M. Casstevens, R.J. Elshire, C.B. Acharya, S.E. Mitchell, S.A. Flint-Garcia, M.D. McMullen, J.B. Holland, E.S. Buckler, and C.A. Gardner. 2013. Comprehensive genotyping of the USA national maize seed bank. Genome Biol. 14:1–43.

Rosenzweig, C., A. Iglesius, X.B. Yang, P.R. Epstien, and E. Chivian. 2001. Climate change and extreme weather events–implications for food production, plant diseases, and pests. Global Change Hum. Health 24(2):90–104.

Rosenzweig, C., and M.L. Parry. 1994. Potential impact of climate change on world food supply. Nature 367:133–138. doi:10.1038/367133a0

Royal Society. 2009. Reaping the benefits: Science and the sustainable intensification of global agriculture. The Royal Society, London, UK. http://royalsociety.org/uploadedFiles/Royal_Society_Content/policy/publications/2009/4294967719.pdf (accessed 14 Aug. 2013).

Russell, W.A. 1974. Comparative performance for maize hybrids representing different eras of maize breeding. In: H.D. Loden and D. Wilkinson, editors, Proceedings of the 29th Annual Corn and Sorghum Research Conference, Chicago, IL. 10–12 Dec. 1974. Am. Seed Trade Assoc., Washington, DC. p. 81–101.

Russell, W.A. 1984. Agronomic performance of maize cultivars representing different eras of breeding. Maydica 29:375–390.

Sappington, T.W., B.D. Siegfried, and T. Guillemau. 2006. Coordinated *Diabrotica* genetics research: Accelerating progress on an urgent insect pest problem. Am. Entomol. 52:90–97.

Schlenker, W., and M.J. Roberts. 2009. Nonlinear temperature effects indicate severe damages to U.S. crop yields under climate change. Proc. Natl. Acad. Sci. USA 106:15,594–15,598. doi:10.1073/pnas.0906865106

Senior, M.L., and M. Heun. 1993. Mapping maize microsatellites and polymerase chain reaction confirmation of the targeted repeats using a CT primer. Genome 36:884–889. doi:10.1139/g93-116

Senior, M.L., J.P. Murphy, M.M. Goodman, and C.W. Stuber. 1998. Utility of SSRs for determining genetic similarities and relationships in maize using an agarose gel system. Crop Sci. 38:1088–1098. doi:10.2135/cropsci1998.0011183X003800040034x

Shattuck-Eidens, D.M., R.N. Bell, S. Neuhausen, and T. Helentjaris. 1990. DNA sequence variation within maize and melon: Observations from polymerase chain reaction amplification and direct sequencing. Genetics 126:207–212.

Shi, G., J.P. Chavas, and J. Lauer. 2013. Commercialized transgenic traits, maize productivity and yield risk. Nat. Biotechnol. 31:111–114. doi:10.1038/nbt.2496

Smith, J.S.C., E.C.L. Chin, H. Shu, O.S. Smith, S.J. Wall, M.L. Senior, S.E. Mitchell, S. Kresovich, and J. Ziegle. 1997. An evaluation of the utility of SSR loci as molecular markers in maize (*Zea mays* L.): Comparisons with data from RFLPs and pedigree. Theor. Appl. Genet. 95:163–173. doi:10.1007/s001220050544

Southworth, J., J.C. Randolph, M. Habeck, O.C. Doering, R.A. Pfeifer, D.G. Rao, and J. J. Johnston 2000. Consequences of future climate change and changing climate variability on maize yields in the midwestern United States. Agric. Ecosyst. Environ. 82:139–158. doi:10.1016/S0167-8809(00)00223-1

Spencer, J.L., B.E. Hibbard, J. Moeser, and D.W. Onstad. 2009. Behaviour and ecology of the western corn rootworm (*Diabrotica virgifera* LeConte). Agric. For. Entomol. 11:9–27. doi:10.1111/j.1461-9563.2008.00399.x

Spike, B.P., and J.J. Tollefson. 1988. Western corn rootworm (Coleoptera: Chrysomelidae) larval survival and damage potential to corn subjected to nitrogen and plant density treatments. J. Econ. Entomol. 81:1450–1455.

Spike, B.P., and J.J. Tollefson. 1991. Response of western corn rootworm-infested corn to nitrogen fertilization and plant density. Crop Sci. 31:776–785. doi:10.2135/cropsci1991.0011183X003100030048x

Steffey, K., and M. Gray. 2008. Survey for the second-generation European Corn Borer larvae, Illinois, 2008 (Revised). The Bulletin: Pest management and crop development information for Illinois. University of Iowa, Iowa City. http://ipm.illinois.edu/bulletin/article.php?id = 1051 (accessed 14 Aug. 2013).

Stuber, C.W., M.D. Edwards, and J.F. Wendel. 1987. Molecular marker-facilitated investigations of quantitative trait loci in maize. II. Factors influencing yield and its component traits. Crop Sci. 27:639–648. doi:10.2135/cropsci1987.0011183X002700040006x

Sutch, R. 2009. The impact of the 1936 corn–belt drought on American farmers' adoption of hybrid corn (draft of December 8, 2009). University of CA, Riverside. p. 1–42.

Tanksley, S.D., and S.R. McCouch. 1997. Seed banks and molecular maps: Unlocking genetic potential from the wild. Science 277:1063–1066. doi:10.1126/science.277.5329.1063

Tester, M., and P. Langridge. 2010. Breeding technologies to increase crop production in a changing world. Science 327:818–822. doi:10.1126/science.1183700

Thompson, L.M. 1986. Climate change, weather variability, climatic change, and grain production. Agron. J. 78:649–653. doi:10.2134/agronj1986.00021962007800040019x

Tollenaar, M. 1989. Genetic improvement in grain yield of commercial maize hybrids grown in Ontario from 1959–1988. Crop Sci. 29:1365–1371. doi:10.2135/cropsci1989.0011183X002900060007x

Tollenaar, M., and J. Wu. 1999. Yield improvement in temperate maize is attributable to greater stress tolerance. Crop Sci. 39:1597–1604. doi:10.2135/cropsci1999.3961597x

Troyer, A.F., and E.J. Wellin. 2009. Heterosis decreasing in hybrids: Yield test inbreds. Crop Sci. 49:1969–1976. doi:10.2135/cropsci2009.04.0170

Tubiello, F.N., C. Rosenzweig, R.A. Goldberg, S. Jagtap, and J.W. Jones. 2002. Effects of climate change on U.S. crop production: Simulation results using two different GCM scenarios. Part I: Wheat, potato, maize, and citrus. Clim. Res. 20:259–270. doi:10.3354/cr020259

Ubbenga, K.J. 2009. Farmers' expectations of yield differences between triple-stack and conventional corn hybrids. M.S. thesis, University of Illinois at Urbana-Champaign, Urbana.

University of California, San Diego. 2013. History of Bt. Aroian Laboratory, USCD. http://www.bt.ucsd.edu/bt_history.html (accessed 14 Aug. 2013).

USDA. 2013. World agricultural supply and demand estimates. USDA. http://www.usda.gov/wps/portal/usda/usdahome?contentid=2013/05/0090.xml (accessed 19 Aug. 2013).

USDA-ERS. 2013a. Fertilizer use and price. USDA-Economic Research Service. http://www.ers.usda.gov/data-products/fertilizer-use-and-price.aspx#26730 (accessed 14 Aug. 2013).

USDA-ERS. 2013b. Corn–background. USDA-Economic Research Service. http://www.ers.usda.gov/topics/crops/corn/background.aspx (accessed 14 Aug. 2013).

USDA-NASS. 2000–2012. Genetically engineered corn varieties by State and United States, 2000–2012. USDA-National Agricultural Statistics Service. www.ers.usda.gov/...Genetically_Engineered_Crops...US/alltables.xls (accessed 13 Jan. 2014).

USDA-NASS. 2010. Fertilizer use on corn by state (Iowa). USDA-National Agricultural Statistics Service. http://www.ers.usda.gov/data-products/fertilizer-use-and-price.aspx#26730 (accessed 13 Jan. 2014).

USDA-NASS. 2013a. Statistics by subject. USDA-National Agricultural Statistics Service. http://www.nass.usda.gov/Statistics_by_Subject/index.php?sector=CROPS (accessed 13 Jan. 2014).

USDA-NASS. 2013b. Prospective plantings March 28th. USDA-National Agricultural Statistics Service. http://www.usda.gov/nass/PUBS/TODAYRPT/pspl0313.pdf (accessed 14 Aug. 2013).

Ustun, A., F.L. Allen, and B.C. English. 2001. Genetic progress in soybean of the U.S. midsouth. Crop Sci. 41:993–998. doi:10.2135/cropsci2001.414993x

Vaughn, T., T. Cavato, G. Brar, T. Coombe, T. DeGooyer, S. Ford, M. Groth, A. Howe, S. Johnson, K. Kolacz, C. Pilcher, J. Purcell, C. Romano, L. English, and J. Pershing. 2005. A method of controlling corn rootworm feeding using a *Bacilllus thuringiensis* protein expressed in transgenic maize. Crop Sci. 45:931–938. doi:10.2135/cropsci2004.0304

Wallander, S., R. Claassen, and C. Nickerson. 2011. The ethanol decade: An expansion of U.S. corn production, 2000–09. Economic Information Bull. No. 79. Economic Research Service, USDA, Washington, DC.

Walthall, C.L., J. Hatfield, P. Backlund, L. Lengnick, E. Marshall, M. Walsh, S. Adkins, M. Aillery, E.A. Ainsworth, C. Ammann, C.J. Anderson, I. Bartomeus, L.H. Baumgard, F. Booker, B. Bradley, D.M. Blumenthal, J. Bunce, K. Burkey, S.M. Dabney, J.A. Delgado, J. Dukes, A. Funk, K. Garrett, M. Glenn, D.A. Grantz, D. Goodrich, S. Hu, R.C. Izaurralde, R.A.C. Jones, S-H. Kim, A.D.B. Leaky, K. Lewers, T.L. Mader, A. McClung, J. Morgan, D.J. Muth, M. Nearing, D.M. Oosterhuis, D. Ort, C. Parmesan, W.T. Pettigrew, W. Polley, R. Rader, C. Rice, M. Rivington, E. Rosskopf, W.A. Salas, L.E. Sollenberger, R. Srygley, C. Stöckle, E.S. Takle, D. Timlin, J.W. White, R. Winfree, L. Wright-Morton, and L.H. Ziska. 2012. Climate change and agriculture in the United States: Effects and adaptation. USDA Tech. Bull. 1935. Agriculture Research Service, USDA, Washington, DC.

Wang, T., X. Ma, Y. Li, D. Bai, C. Liu, Z. Liu, X. Tan, Y. Shi, Y. Song, M. Carlone, D. Bubeck, H. Bhardwaj, E. Jones, K. Wright, and S. Smith. 2011. Changes in yield and yield components of single-cross maize hybrids released in China between 1964 and 2001. Crop Sci. 51:512–525. doi:10.2135/cropsci2010.06.0383

Wheeler, T., and J. von Braun. 2013. Climate change impacts on global food security. Science 341:508–513. doi:10.1126/science.1239402

World Bank. 2010. World development report 2010: Development and climate change. World Bank, Washington, DC.

7

Peanut

C. Corley Holbrook,* Tim B. Brenneman, H. Thomas Stalker, W. Carroll Johnson III, Peggy Ozias-Akins, Ye Chu, George Vellidis, and Duncan McClusky

ABSTRACT

Average yields of peanut (*Arachis hypogaea* L.) in the United States set an all-time record of 4695 kg ha^{-1} in 2012. This far exceeded the previous record yield of 3837 kg ha^{-1} in 2008. Favorable weather conditions undoubtedly contributed to the record yields in 2012; however, these record yields would not have been achievable without numerous technological advances that have been made in peanut production. The cumulative effect of these technologies caused U.S. yields to increase sixfold from 739 kg ha^{-1} in 1909 to 4695 kg ha^{-1} in 2012. If we disregard the record yield of 2012, the average gain from 1909 to 2011 was 29.9 kg ha^{-1} yr^{-1}. These yield gains are due to improved cultivars, advances in agronomic practices, improvements in practices and chemistries for control of weeds and diseases, and increased use of precision agriculture, particularly for the digging and harvesting of the crop. Modern peanut cultivars have much higher yield potential; however, because of the synergism between production systems and plant breeding, it is difficult to precisely quantify the amount of the yield gains that are due to improved cultivars. In addition to yield per se, cultivar development has also resulted in improved resistance to important yield-limiting diseases. This has had important economic impacts on U.S. peanut production.

Abbreviations: GPS, global positioning system; HADSS, Herbicide Application Decision Support System; TSWV, *Tomato spotted wilt virus.*

C.C. Holbrook USDA-ARS, 115 Coastal Way, Tifton, GA 31793. *Corresponding author (Corley.holbrook@ars.usda.gov).
T.B. Brenneman, Univ. of Georgia, Plant Sci. Bldg., 2360 Rainwater Rd., Tifton, GA 31793-5766 (arachis@uga.edu).
H.T. Stalker, North Carolina State Univ., Dep. of Crop Sci., Greenhouse Unit 3, 214 Method Rd., Raleigh, NC 27695-7629 (tom_stalker@ncsu.edu).
W.C. Johnson III, USDA-ARS, 2747 Davis Rd., Tifton, GA 31793 (Carroll.johnson@ars.usda.gov).
P. Ozias-Akins (pozias@uga.edu) and **Y. Chu** (ye.chu.test@gmail.com), Univ. of Georgia, NESPAL, 2360 Rainwater Rd., Tifton, GA 31793-5766.
G. Vellidis, Univ. of Georgia, Engineering Bldg., 2329 Rainwater Rd., Tifton, GA 31793-5766 (yiorgos@uga.edu).
D. McClusky, Univ. of Georgia, Library, 2360 Rainwater Rd., Tifton, GA 31793-5766 (Mcclusky@uga.edu).

doi:10.2135/cssaspecpub33.c7

Yield Gains in Major U.S. Field Crops. CSSA Special Publication 33.
Stephen Smith, Brian Diers, James Specht, and Brett Carver, editors.

Average yields of peanut in the United States set an all-time record of 4695 kg ha^{-1} in 2012. This far exceeded the previous record yield of 3837 kg ha^{-1} in 2008. Favorable weather conditions undoubtedly contributed to the record yields in 2012; however, these record yields would not have been achievable without numerous technological advances that have been made in peanut production. Figure 7–1 illustrates the cumulative effect of these technological advances with average U.S. yields increasing sixfold from 739 kg ha^{-1} in 1909 to 4695 kg ha^{-1} in 2012. If we disregard the record yields of 2012, the average on farm gain from 1909 to 2011 was 29.9 kg ha^{-1} yr^{-1}. In this chapter we discuss some of the technological advances that have resulted in these remarkable yield gains.

Peanut is a legume that originated in South America. It is an allotetraploid species that evolved from a cross of two diploid progenitors. In evolutionary terms this was a relatively recent event, resulting in a low level of genetic polymorphism in *A. hypogaea*. Peanut is widely grown throughout the tropical, subtropical, and warm areas of the globe. Asia and Africa are important secondary centers of diversity for peanut.

China is the largest producer of peanut in terms of both acreage and total production. India ranks second, and the United States is a distant third. Peanut is an oilseed crop in most areas where it is produced. The majority of the peanuts produced in the United States are used to manufacture peanut butter, with much of the rest used to produce other food products. U.S. peanuts are grown on about 485,000 ha, concentrated in Alabama, Florida, Georgia, North Carolina, South Carolina, Mississippi, New Mexico, Oklahoma, Texas, and Virginia. Georgia leads all states with about 50% of U.S. production.

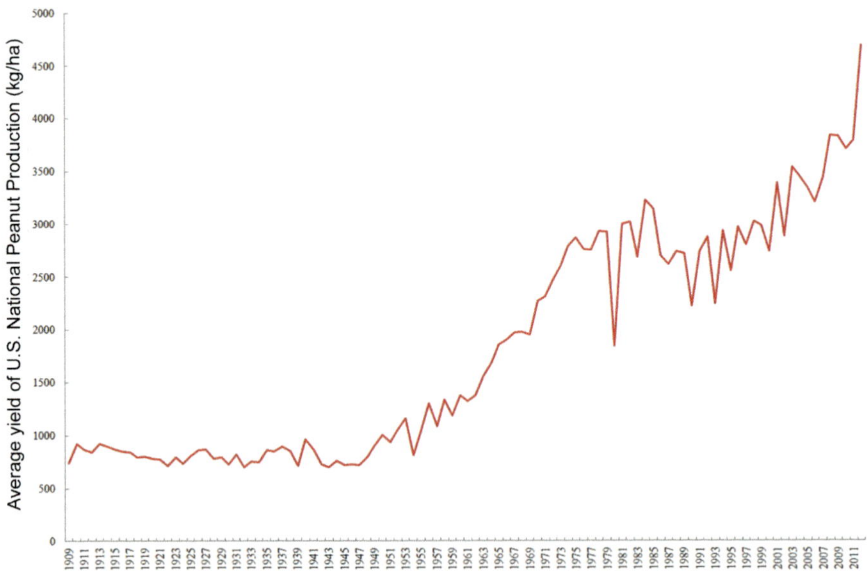

Fig. 7–1. Average yield for U.S. peanut production from 1909 to 2012. Source: USDA-NRCS (www.nass.usda.gov, accessed 27 Jan. 2014).

Worldwide average peanut yields have also increased because of technological advances (Fig. 7–2). Average yield was 849 kg ha^{-1} in 1961, increasing to 1774 kg ha^{-1} in 2011. Rates of yield increases for different regions of the world vary greatly and are primarily based on the level of adoption of improved technologies (Bantilan et al., 2003).

The remainder of this chapter will be presented in three sections. First will be an analysis of peanut yield gains in the United States, with individual evaluations of cultivar development, disease control, weed control, and precision agriculture. The next section will evaluate yield gains in other peanut-producing countries of the world. We conclude with a section on how advances in genomic technology may be used to provide future yield advances in peanut.

U.S. Peanut Yield Gains

Cultivar Development

Because of the synergism between production systems and plant breeding, the results from trials comparing old and new cultivars have to be interpreted with caution (Austin, 1994; Tracy et al., 2004). Older cultivars, bred for farming systems that prevailed when they were first grown, may be at a disadvantage under modern farming systems and may lack resistance to the prevalent strains of important pests and diseases. Nevertheless, cultivar development has undoubtedly contributed to increased peanut production. Peanut yields increased dramatically in the 1960s and 1970s. During the 1960s, production shifted from the lower-yielding Spanish to the higher-yielding runner market-type peanut. Since then, runners have been grown in the Southeast; large-seeded Virginia types in the

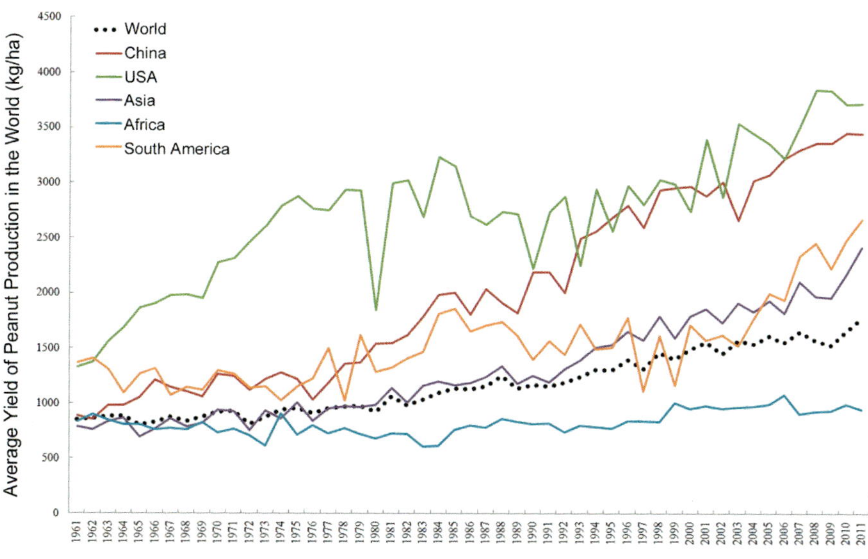

Fig. 7–2. Average yield of peanut production for the world and for some important production regions from 1961 to 2011. Source: Food and Agricultural Organization of the United Nations (http://faostat.fao.org, accessed 27 Jan. 2014).

Carolina–Virginia area; and runners, Spanish, Virginias, and Valencia types in the Southwest. Improvements in the cultivars of all market types occurred during the 1960s and 1970s, with the most significant being Florunner (runner type) (Norden et al., 1969) and Florigiant (Virginia type) (Carver, 1969), which were widely adopted by growers (McArthur et al., 1982). By 1981, average yields were 2996 kg ha^{-1} as compared with 1008 kg ha^{-1} in 1950. Warren (1998) summarized yield gains for nine crops in the United States during the 20th century. Yield gains ranged from twofold to sevenfold and were attributed to various crop specific factors. A threefold increase was documented for peanut, with most of that increase occurring in the 1950s, 1960s, and 1970s. The average yield of representative states of the three historical U.S. peanut-producing regions (Southwest, Southeast, and Virginia–Carolina regions) demonstrates parallel trends of improvement across the country (Fig. 7–3).

Duncan et al. (1978) reported significant improvement in runner-type cultivars in Florida since the start of their breeding program in 1928. They reported as much as a 100% increase in yield potential due to improved cultivars in the 1970s when compared with the first cultivar released from the Florida program.

Peanut yields increased from an average of 1120 kg ha^{-1} in the 1940s to over 3360 kg ha^{-1} in the 1980s in the Virginia–North Carolina production area (Mozingo et al., 1987). This increase can be attributed to genetic improvement as well as improved cultural and management practices. Cultivar development programs were not started until the mid-1940s in North Carolina and the mid-1950s in Virginia (Mozingo et al., 1987). It was estimated that improved cultivars accounted for 18.5% of the yield increase from the 1940s to the 1980s, with a yearly increase of 14.7 kg ha^{-1} from genetic improvement (Mozingo et al., 1987).

Studies have indicated that improved reproductive efficiency has contributed to the breeding advances for higher yields. Reproductive efficiency is the

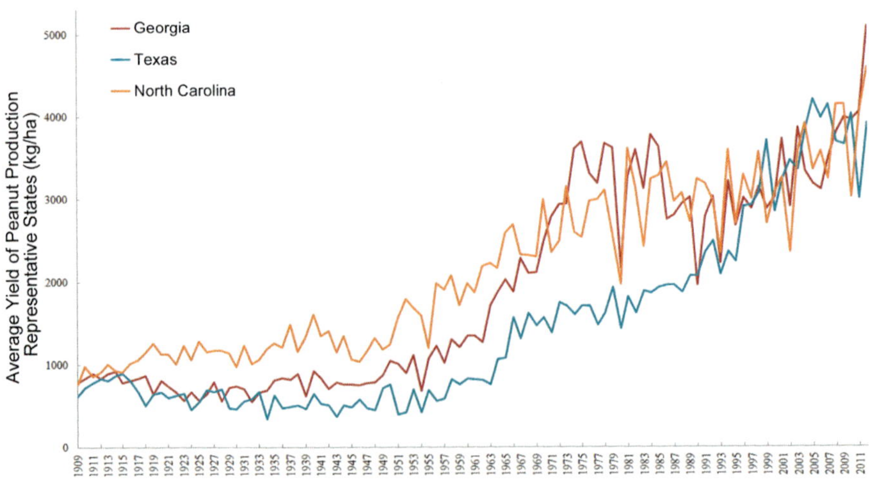

Fig. 7–3. Average yield of peanut production in the major state representing each of the three historical U.S. peanut-producing regions (Southwest, Southeast, and Virginia–Carolina) from 1909 to 2012 Source: USDA-NRCS (www.nass.usda.gov, accessed 27 Jan. 2014).

measure of a plant's ability to produce mature fruit relative to total plant growth. Smith (1954) reported low reproductive efficiency expressed by peanut cultivars available in 1944. Only 15% of the plant's flowers produced mature fruit. Later studies indicated that reproductive efficiency could be improved (Emery et al., 1973). The cultivar NC 17 exhibited a 57% greater fruit to plant weight ratio than Florigiant (the leading commercial Virginia-type cultivar at the time). Coffelt et al. (1989) also found that more recent cultivars were earlier maturing and had higher reproductive efficiency than older cultivars. Work with cultivars developed by the Florida breeding program showed an increase in allocation of photosynthate to reproductive structures during pod filling from 41% for the first cultivar released in 1943 to 98% for a cultivar released in 1977 (Duncan et al., 1978). Mozingo et al. (1987) also documented an increase in peanut yields in response to breeding efforts.

Wells et al. (1991) examined growth and dry matter partitioning for 10 Virginia-type peanut genotypes bred for increased yield in North Carolina. In addition to increasing yield, greater breeding efforts resulted in genotypes with less vegetative mass, shorter main stem lengths, and greater reproductive allocation than genotypes from earlier breeding cycles. Furthermore, an earlier transition from vegetative to reproductive growth was related to the observed alterations and resulted in greater dry matter allocation into reproductive structures at the apparent expense of vegetative growth.

Boote and Tollenaar (1994) used a crop growth simulation modeling approach to evaluate the importance of what they termed the five "P's" of yield potential: photosynthesis, prior events, pod filling period, partitioning, and prior accumulation and remobilization. For indeterminate crops such as peanut, yield improvement has been achieved from large increases in partitioning (Duncan et al., 1978). A comparison of peanut cultivars released in Florida illustrated the yield improvement associated with increased intensity of partitioning to pods and more rapid transition to a full pod load (Duncan et al., 1978). The oldest improved cultivar, Dixie Runner, was grown in the 1940s and 1950s and produced a considerable number of leaves even while filling pods. Breeding efforts resulted in high-yielding cultivars such as Florunner that added pods more rapidly, and once into active pod fill, the plant partitioned less dry matter to vegetation and more to pods, resulting in higher yields. Boote and Tollenaar (1994) concluded that the pod filling period duration is the most likely factor to account for future yield increases; however, yield improvement also may come from increased stress tolerance.

Large increases in peanut yields did not occur during the 1980s and 1990s. Warren (1998) proposed that this may have been due to two negative factors. First, most of the effective fumigant nematicides were no longer registered, resulting in yield reductions from root-knot nematodes (*Meloidogyne arenaria* Chitwood). Second, the increase of areas planted with peanut in the 1980s and early 1990s disrupted established crop rotations and increased losses from soilborne diseases and nematodes. The emergence of the *Tomato spotted wilt virus* (TSWV) also inhibited yield advances during this period of time because all widely grown cultivars were highly susceptible. Fortunately, sources of resistance were identified in the peanut germplasm collection (Isleib et al., 2001); however, it took many years before highly resistant cultivars were developed and released. The presence of the virus and the lack of highly resistant varieties forced farmers into later planting dates to avoid thrips (*Thrips palmi* Karny) feeding on the plants and transmitting the virus, which also reduced yields (J. Beasley, personal communication, 2013). Since 1992, peanut hectarage has declined, which allows the reestablishment of better crop rotations aiding disease and nematode management (Warren, 1998). Perhaps even more importantly for recent yield increases have been several cultivars released with excellent resistance to TSWV (Culbreath and Srinivasan, 2011).

In addition to yield per se, cultivar development has resulted in improved resistance to important yield-limiting diseases. This has had important economic impact on U.S. peanut production. The United States maintains a large (>8000 accession) and genetically diverse collection of peanut germplasm. Isleib et al. (2001) conducted a study to: (i) summarize and document the use of genetic resources in cultivar development and (ii) estimate the resulting economic impact from use of the germplasm collection. Tables showing pedigrees for all U.S. cultivars were developed, which allowed the average germplasm ancestry to be calculated. This study showed that the use of this germplasm has had a significant economic impact on U.S. peanut farmers. The largest impact has been through the development of cultivars with resistance to Sclerotinia blight (*Sclerotinia minor* Jagger), root-knot nematode, and TSWV, which has had an economic benefit of more than US$200 million annually for U.S. peanut producers.

Disease Control

Peanut is highly susceptible to many foliar and soilborne diseases. Since the plant fruits belowground, the pods are exposed to direct attack from pathogens in the soil, which can decrease quality as well as yields. Diseases have had a major influence on peanut yields and production practices from the earliest days the crop was grown, although USDA bulletins around the turn of the century had little to say about diseases. In the 1940s, disease management became more of a concern, and sulfur or copper–sulfur compounds were applied either by hand or by two-row, mule-drawn applicators for partial control of foliar diseases (Woodruff, 1966). Thus began a series of improvements in fungicide chemistry and application methods that continues to the present time. Simultaneously, there were concerted efforts by researchers to improve the levels of genetic resistance to the most damaging diseases. The combined effects of fungicides and disease resistance have played a major role in mitigating the effects of disease and increasing yields of peanut.

One chemical input that had a profound effect on peanut yields is seed treatment, a prophylaxis that was widely accepted because of its obvious benefits. Early treatments relied on mercury as the active ingredient, which was very effective and greatly increased plant stands. Seed treatment was particularly important when mechanical threshing was employed because damage decreased germination by 25 to 75% compared with hand shelling (Woodruff, 1966). Mechanical shelling and handling of seed has improved tremendously over the years, but even in recent years a good seed treatment will greatly increase peanut yields, often by more than 1000 kg ha^{-1} on the basis of multiple years of testing in Georgia (Brenneman, unpublished data, 2010–2013). The industry currently uses wettable powder mixtures (vs. liquid treatments that often damage the fragile seed coat) containing multiple fungicide classes to provide control of the wide range of potential seed and seedling pathogens known to attack peanut.

Early leaf spot (*Cercospora arachidicola* Hori) and late leaf spot (*Cercosporidium personatum* Berk. & M.A. Curtis) have caused serious yield losses of peanut for many years. Early trials in Georgia indicated that controlling leaf spot could increase yields by 20 to 25% (Woodroof et al., 1944), a return that much more than paid for the cost of the fungicides. Later estimates for losses to foliar diseases have been as high as 50% (Smith and Littrell, 1980). Tractor applicators improved the ease of application, but sulfur dust had limited efficacy and was easily washed off. The introduction of chlorothalonil (Bravo) and benomyl (Benlate) ushered in a new era of spraying versus dusting and led to dramatic improvements in the control of foliar diseases during the 1970s. This allowed utilization of the cultivar Florunner, which had great yield potential but was highly susceptible to the leaf spot diseases and peanut rust (*Puccinia arachidis* Speg.). Rust is one of the most damaging diseases of peanut in more tropical production areas, and in some years can causes significant losses in the more southern production areas of the United States. Newer chemistries such as the triazoles and strobilurins provided even stronger tools for managing these foliar diseases and are no doubt one reason they are not more prevalent. Growers rapidly adopted good spray programs, and recent research has shown that modern cultivars also have a high level of tolerance for defoliation from leaf spot (Navia Gine et al., 2011). Losses to foliar diseases in recent years have been minimal, although disease control practices are a significant part of production costs.

Soilborne diseases such as stem rot (*Sclerotium rolfsii* Sacc.) have caused severe losses in peanut for many years, with nearly 50% plant mortality in one early report from North Carolina (Prince, 1944). Stem rot is currently the most damaging disease in Georgia with loss estimates ranging from 5 to 10% of statewide yields (Georgia Cooperative Extension Service estimates). Other diseases such as Sclerotinia blight (*S. minor*) and Cylindrocladium black rot (*Cylindrocladium parasiticum* Crous, Wingfield, & Alfenas) were discovered on peanut in the 1960s and 1970s. Both of these diseases are significant problems in the Virginia–Carolina region, and by the early 1980s they were causing statewide yield losses in Virginia of 5 to 10% and 1 to 3%, respectively (Phipps, 1981, 1982, 1983, 1984). Sclerotinia blight is also a major problem in parts of the Southwest. Losses to both diseases have been reduced in recent years due to new fungicides and more resistant cultivars. Rhizoctonia limb rot (*Rhizoctonia solani* Kühn) is not a new disease but emerged to cause greater losses when irrigation and more intensive crop management practices were employed in the 1970s and 1980s.

The new fungicides developed for leaf spot in the 1970s provided no control of soilborne fungi. Chlorpyrifos, which is primarily an insecticide, and pentachloronitrobenzene (PCNB) both provided some suppression of stem rot. However, they had to be applied as granules and control was marginal at best. Tebuconazole was registered for use on peanut in 1994. This was the first effective fungicide spray that provided significant control of soilborne diseases (Brenneman et al., 1991), apparently by washoff and redistribution down the plant to the tissues near the soil surface (Woodward et al., 2012). Other effective fungicides such as flutolanil and azoxystrobin were subsequently labeled. However, losses to these diseases have been difficult to manage with fungicides because the targeted initial infection sites are near or even below the soil surface. Losses to stem rot in Georgia in recent years often approach 10% of the total state crop but would undoubtedly be higher without the use of fungicides.

Although some cultivars such as York (Gorbet and Tillman, 2011) and Georgia-07W (Branch and Brenneman, 2008) have some resistance to stem rot, they are not widely accepted by growers; thus, most currently grown varieties are highly susceptible to the disease. Historically, peanut growers have chosen to grow the highest yielding cultivars available regardless of disease resistance, as long as fungicides are cost effective. Fungicide trials with cultivars such as GA-06G (Branch, 2007) demonstrate that stem rot alone has the potential to reduce yields by as much as 2000 kg ha^{-1} (Brenneman, unpublished data, 2013). However, the commercial impact of improved disease resistance is demonstrated by the cultivar Bailey (Isleib et al., 2011), which has excellent resistance to stem rot, late leaf spot, and TSWV, allowing the use of reduced-cost fungicide programs without sacrificing yield (Chapin et al., 2009; Chapin and Thomas, 2011). Growers in the Virginia–Carolina production region have rapidly adopted Bailey as the predominant Virginia-type variety because of the combined economic advantage of lower input costs and a greatly reduced risk of catastrophic loss from erratic fungicide performance against soil diseases.

Other diseases causing serious yield losses are caused by viruses, primarily TSWV, and nematodes, primarily peanut root-knot nematode. TSWV caused huge losses in the United States for about 10 yr starting in the mid-1990s, and continued increase of the disease could have seriously threatened the peanut industry. A concerted research and extension effort led to the development of a TSWV Risk Index (Brown et al., 2005) that combined multiple management inputs. Use of the index has reduced TSWV losses to minimal levels (Culbreath and Srinivasan, 2011). Nematode damage continues to be significant, especially in shorter rotations and sandy soils. The recent withdrawal of aldicarb from the market left Telone (1,3-dichloropropene) as the primary means of chemical control, but supply and application issues have limited its use. However, excellent genetic resistance (near immunity) is now available, which offers growers a powerful new tool for control and has greatly decreased yield losses and production costs in runner peanut when resistant cultivars are used (Holbrook et al., 2008). Newer cultivars with nematode resistance and greater yield potential are being developed that may have even greater impact in the future.

Future progress for reducing yield losses from diseases in peanut will be by incorporating genetic resistance into high-yielding cultivars. Rapid progress is being made in sequencing the peanut genome and identifying potential resistance genes. Marker-assisted selection has already streamlined phenotyping efforts for several traits and will be expanded to others as good markers are identified. Currently, there are no transgenic peanut cultivars in production, but efforts are being made to register a cultivar with an oxalate oxidase gene from barley (*Hordeum vulgare* L. ssp. *vulgare*) conferring resistance to Sclerotinia blight (Livingstone et al., 2005).

In summary, different diseases are present in various areas of the peanut belt, making it difficult to generalize regarding their effects on pod yield. Clearly, diseases are a huge constraint that must be dealt with to consistently produce the crop wherever it is grown, and the integrated management programs described above have played a major role in achieving the record yields recently documented in the United States.

Weed Control

Weeds were not effectively controlled in peanut before the use of herbicides. Starting in the 1950s, and with continuing improvement to the present time, many herbicides for use on peanut have been developed. An excellent selection is now on the market that controls most weeds in peanut fields.

Row patterns and seeding rates are cultural practices that affect weed management in peanut. Historically, peanut is seeded in rows 91 to 97 cm apart. Anecdotally, this row spacing is attributed to the width needed for mule-drawn implements to operate, and this row spacing remains common in many crops throughout peanut-producing regions. Peanut seeded in multiple, narrow-spaced rows have been shown to be slightly higher yielding than peanut seeded in wide rows, although weed control was not necessarily improved by quick canopy closure (Buchanan and Hauser, 1980; Hauser and Buchanan, 1981; Johnson et al., 2005; Wehtje et al., 1984). In narrow row patterns, seeds are widely spaced in each of the multiple rows, giving the same overall seeding rate (kg ha^{-1}) as peanut seeded in wide rows. In 1984, county agent surveys indicated that narrow-spaced rows (<91-cm spacing) accounted for only 8% of the peanut acreage in Georgia. Two developments in the mid-1990s quickly changed this trend. Vacuum planters became readily accepted by growers and allowed precise metering of seed. Older planter technologies did not uniformly meter seed, potentially causing wide gaps in the stand of narrow-spaced rows. Fewer skips lessen weed competition and thus enhance peanut yields. In addition, TSWV became endemic in the region, and multiple narrow-spaced rows reduced disease risk (Brown et al., 2003), presumably because the thrips vector preferred feeding on isolated rather than densely spaced plants. This was included as part of a disease-reducing strategy, and peanut growers quickly adopted the practice of seeding peanut in narrow row patterns. By 2003, 41% of the peanut acreage in Georgia was seeded in narrow row patterns. It should be emphasized that in addition to increasing yield (Tillman et al., 2006), widespread use of vacuum planters and increased acceptance of multiple narrow row patterns enhance weed control by lessening bare ground caused by skips and promoting quick canopy closure.

Cultivation was traditionally an integral component in peanut weed management (Bridges et al., 1984; Hauser and Parham, 1969). In 1993, county agent surveys indicated that 73% of the acreage in Georgia was cultivated for weed control. Since then, new herbicide developments improved overall weed control, and cultivation was no longer needed to control escapes. In addition, conservation tillage peanut production systems are now common, and cultivation is rarely feasible in these systems. As a result of these factors, only 19% of the peanut acreage was cultivated for weed control in 2003.

This remarkable reduction in cultivation over a 10-yr period directly addresses one of the inherent vulnerabilities in peanut production—susceptibility of peanut to infection by soil-inhabiting fungi. Boyle (1952, 1956) reported the relationship between incidences of stem rot and displaced soil thrown on peanut plants from cultivation. Disease incidence can be reduced by using cultivation techniques that lessens soil movement (Garren and Duke, 1957). Not needing to cultivate for control of weed escapes is obviously a preferable solution to lessen disease epidemics and protect peanut yield.

Many of the newly registered herbicides are more versatile, consistent, and have a broader spectrum than herbicides available in the 1980s. Of these, four active ingredients (diclosulam, flumioxazin, imazapic, and imazethapyr) have superior residual weed control properties that were previously unavailable (Grey et al., 2003; Wilcut et al., 1991).

Despite the broad spectrum of weeds controlled by the recently developed herbicides, no single herbicide adequately controls all weed species in peanut. Knowledge of weed history and correct herbicide selection improves overall weed control and protects peanut yield. The Herbicide Application Decision Support System (HADSS) is an expert system developed by North Carolina State University to help peanut growers correctly choose herbicides and combinations based on weed species present (http://www.webhadss.ncsu.edu). HADSS has been validated and modified for use in other peanut-producing states. The combination of improved herbicides and readily accessible databases have contributed to better weed control and improved peanut yield.

There were no selective postemergence herbicides in 1980 registered on any crop (including peanut) that controlled emerged annual grasses. In 2013, there were three postemergence herbicides registered for use on peanut to control escaped annual and perennial grasses in peanut: sethoxydim, clethodim, and fluazifop. Annual and perennial grasses are among the most competitive weeds in peanut and significantly reduce yields even when present at low densities, along with causing substantial harvest losses (Chamblee et al., 1982; Johnson and Mullinix, 2005; McCarty, 1983; Wilcut, 1991; York and Coble, 1977). These herbicides provide excellent postemergence control of grasses in peanut with minimal phytotoxicity (Grichar, 1991; Grichar and Boswell, 1986; Johnson et al., 2002; Prostko et al., 2001; Wilcut, 1991). Registration of these herbicides were major weed control milestones in peanut production and have largely eliminated yield losses from annual grasses that escaped earlier control efforts.

In 2013, 21 herbicide active ingredients were registered in the United States for weed control in peanut. In contrast, there were 12 herbicide active ingredients registered for use on peanut in 1980. Overall, weed management in peanut has improved in recent years, and weeds cause fewer losses than 30 yr ago. This improvement is largely due to herbicides that are more consistent, versatile, and have a broader spectrum than earlier herbicides.

Precision Agriculture

Global positioning system (GPS) guidance of farm machinery has been adopted by increasingly larger segments of the farming community over the past decade because of the inherent gains in efficiency that it provides. It is now common to find farmers who own multiple vehicles (tractors, sprayers, and harvesters) equipped with GPS guidance. In many areas of the United States, the most advanced and expensive form of GPS guidance, real-time kinematic (RTK)-based automated steering (auto-steer), is also being quickly adopted by farmers. Many peanut farmers have adopted auto-steer because it can theoretically result in large yield gains when used to plant and invert peanuts—one of the region's most important crops (Leidner, 2012).

The yield gains are realized by reducing digging losses because auto-steer offers peanut farmers the potential of being able to follow the planting centerline with both accuracy and precision when inverting their peanuts. Ortiz et al. (2013)

used auto-steer to intentionally invert peanuts at increasing distances from the planting centerline and showed that peanut yields decreased with distance. Digging losses may range from 15 to 30% of the peanut crop's potential yield (Ortiz et al., 2013). Jackson et al. (2011a) measured digging losses of 26% in strip tillage peanuts planted and inverted without auto-steer at a location in southern Georgia.

A study was conducted for 2 yr on a working farm in Georgia to assess the benefits of using auto-steer to plant and invert peanut on field rows with varying degrees of curvature (Vellidis, unpublished data, 2010, 2011). When all data were grouped together, auto-steer outperformed conventional driving by 579 kg ha^{-1} in 2010 and 451 kg ha^{-1} in 2011.

International Peanut Yield Gains

Peanut is grown in more than 100 countries. China has the largest production with 13.42 million tonnes, followed by India (7.7 million tonnes), and the United States (1.88 million tonnes). Nigeria and Indonesia each produce more than 1 million tonnes of peanut (http://www.soyatech.com/peanut_facts.htm). Peanut production systems range from very primitive with only hand labor and few inputs of fertilizer or chemical controls for weeds or diseases to other systems that are highly technical and mechanized. A large percentage of the crop is crushed for oil, but there is an increasing trend for small farmers to supply local markets or for family subsistence. Thus, one strategy for increasing yields will not work in all cropping systems because of constraints on labor; funds to purchase fertilizers and chemicals for weed, disease and pest controls; and the inability to acquire improved seeds by many farmers in lesser developed countries.

Most countries that produce peanut have research programs to improve production, with the majority being conducted by governmental agencies. The resources available to many of these programs are minimal, and progress toward improved yields and quality are difficult to obtain without adequate funding or access to agricultural supplies. In many countries, the research and extension efforts are still not adequate enough to have significant impacts on peanut yields, and even when excellent programs are in place, farming conditions reduce their effectiveness for significant yield increases. For example, many lesser developed countries have disorganized agriculture segments of their economies that lead to seed quality issues, nonavailability of high quality cultivars, inadequate access to chemicals for timely applications of weed, disease, or insect control, insufficient extension capacities, and lack of incentives to farmers to grow a high quality crop (Sylla, 2010). On the other hand, excellent research programs have evolved in Argentina, Australia, Brazil, China, and other countries with corresponding yield and quality improvements.

To address production and yield problems in the semiarid tropics, which to a large extent consist of the large land areas in Asia and Africa, The International Crops Research Institute for the Semi-Arid Tropics (ICRISAT) was established in 1976. The institute is located in India, which in the 1970s had both the largest area and total peanut production of any country. Since then, ICRISAT established several satellite stations in Africa to better serve this part of the world. Yield loss due to diseases and unfavorable weather (mostly lack of water) in peanut production areas are the greatest detriments to high yields throughout much of Africa, Asia, and parts of South and Central America (Vara Prasad et al., 2010). A large share of the efforts at ICRISAT is to breed peanuts with high levels of disease resistance, in particular to early and late leaf spots, peanut rust, and viruses. While significant progress has been made, yield potentials can only be obtained if chemicals are also applied to the peanut crop. The Peanut Collaborative Research Support Program (Peanut CRSP) was established in the early 1980s by the U.S. Agency for International Development as an international effort to enhance peanut production, quality, and safety. Scientists in the Peanut CRSP have worked closely with ICRISAT scientists to address specific problems in the countries where they have cooperative agreements to enhance peanut research and extension to farmers. Perhaps the greatest single step towards increasing yields has been training national scientists so that both research and extension will have application in local farming situations.

Cummins and Jackson (1982) surveyed peanut researchers and producers from around the world to catalog important areas of needed research for peanut improvement. The highest priorities were to breed peanuts for disease resistance, higher yield potential, drought tolerance, insect resistance, aflatoxin resistance, and early maturity. Leaf spots and rust were the most important disease problems, followed by foliage insects, root, stem, and pod rots, and rosette virus. Soil fertility, seeding date and plant populations, tillage, soil pH, and intercropping also contribute to low yields. In the semiarid tropics, the lack of mechanization is a major constraint to producing higher yields because about 2000 labor hr ha^{-1} are needed to produce a crop (Byerlee, 1979). Unfortunately, all of the above constraints remain important today in most of the semiarid tropics.

The first International Workshop on Groundnuts was held at the ICRISAT Center, India, during 1980. This was the first time that scientists from around the world had met to exchange ideas and lay plans to address the problems restricting peanut production in lesser developed countries. It quickly became apparent that many of the yield constraints were widespread across most of the world's production regions, in particular, widespread occurrence of early and late leaf spot diseases and peanut rust; a complex of insects; few cultivars with high yielding potential; lack of seed technologies; and primitive farming systems and cultural practices (ICRISAT, 1980). Other yield constraints such as bud necrosis virus and soilborne diseases were more localized but were also major yield constraints in many areas. The conference led to more directed efforts to address these problems through research and extension efforts. In addition, programs were put in place to better educate farmers in production technologies that would lead to higher yields and seed quality. A second conference was held in 1992 at ICRISAT to address international approaches to production and peanut improvement. This conference was attended by scientists from 44 countries. It was obvious from the reports that significant progress had been made in many areas to increase yields and production, but many of the constraints to peanut yields remained (Nigam, 1992). Twenty years later, yields have almost universally increased across production regions (Fig. 7–2), but diseases, insects, agricultural infrastructure, and lack of water remain major constraints to production in most production areas.

When the ICRISAT research program was initiated, it had as one of its primary objectives the creation of high yielding breeding lines or cultivars with resistance to the main factors (i.e., diseases, insects, drought) limiting production. At the time, sources of resistance in *A. hypogaea* lines to many of the peanut pathogens were published in the literature, but resistance levels were generally moderate to low (Nigam et al., 1980). Thus, scientists at ICRISAT screened large parts of the germplasm collection (more than 10,000 entries) for late leaf spot, rust, several viruses and other diseases, insects, drought resistance, and physiological traits (Dwivedi et al., 2007a). Many superior lines were subsequently used in breeding programs, and their germplasm lines and cultivars are widely distributed throughout Asia and Africa for use in national breeding programs and/or directly by farmers. To increase peanut yields, there also must be a system put in place to grow cultivars with high yielding potential, weeds must be controlled, diseases and insects must be suppressed, and there must be good growing conditions with adequate moisture to sustain the crop. Research at ICRISAT and in national programs have addressed these problems, and the knowledge base is available to significantly increase yields.

The global yield of peanut has steadily increased during the past half century. Average world yield in the 1970s was 930 kg ha^{-1}, which increased to 1080 kg ha^{-1} in the 1980s (16.1% increase) (Fletcher et al., 1992). A large part of these yield increases were in China, Myanmar, and to a lesser degree in Indonesia, Senegal, and the United States. Average yields decreased 17% in African countries during the same period (Talawar, 2004). Beginning in the 1970s there has been a 34% increase in food use (vs. oil), which indicates that peanut has become an important food source for families. The average global yields in 2011 rose to 1550 kg ha^{-1}, with the United States (3440 kg ha^{-1}), China (3300 kg ha^{-1}), and Egypt (3170 kg ha^{-1}) leading in yield per hectare (USDA, 2013). A closer look at several individual countries may shed some light on the reasons for higher peanut yields on a global scale.

India

Until the 1990s, India was the largest producer of peanut (or groundnuts) globally and remains near the top of the list in cultivated hectares. Yields have historically been very poor with an average of 736 kg ha^{-1} in 1980, rising to 930 kg ha^{-1} in 1990 and to a little more than 1000 kg ha^{-1} in 2011 (USDA, 2013). Although yields remain very low, a 35.9% yield increase (1.3% yr^{-1}) was realized in India. Peanut production in 1980 was mostly on small farms with oxen as the source of cultivation, and this method of cultivation is still used by many farmers, but mechanization has been an important factor in yield improvements. During the 1970s, there were no cultivars available with resistance to the leaf spots, peanut rust, or insects. The Indian government established the National Research Centre for Groundnut in 1979 with a mandate to improve peanut production at all levels in the country. Scientists soon discovered that minimal inputs (e.g., 150 Rs ha^{-1}) resulted in significant yield increases. For example, controlling white grubs in infested areas more than doubled yields and significantly increased net profits (Singh, 1980). Nigam et al. (1994) reported that most increases throughout the 1980s and early 1990s were due to more productive cultivars and improved technologies in India. Even with higher yields under managed systems, low levels of disease resistance to leaf spots, peanut rust, and other diseases caused significant yield losses. Prolonged periods of drought greatly restricts yield potentials, but when irrigation is used, yields can more than double (Talawar, 2004). Many cultivars are recommended for different production regions in India, some of which yield 2500 to 3500 kg ha^{-1} and include Spanish, Virginia runners, and Virginia bunch types (Talawar, 2004). Yield potentials have greatly increased in recent years. Planting date is the single most important nonmonetary input that affects yield in India (Talawar, 2004), where delayed planting significantly decreases yields. In addition to breeding, cropping systems research has indicated that peanut–cereal intercropping can give yield advantages to peanut (Reddy et al., 1980).

China

Peanut production in China has had a long history, and the country is currently the largest peanut producer with more than 40% of the global production even though it constitutes only 20% of the total world area in peanut production. Yield rose from under 1500 kg ha^{-1} in the 1980s to about 3300 kg ha^{-1} in 2011 (USDA, 2013). The Chinese government encouraged farmers by providing incentives to intensify production, and since the 1980s production has significantly increased because of a combination of improved cultivars and cultivation practices. Most notable is the use of polyethylene mulching to suppress weeds and retain soil moisture, and this procedure has led to yield increases from 18 to 49% (Yao, 2004). The downside of using the polyethylene is the labor required to place it in the field, disposal of the plastic, and residues left in the field after harvest. Mulching also can extend the growing season, particularly in cooler parts of the country (Yao, 2004).

Egypt

Although there is a perception that growing peanut (and other crops) on small farms is a significant restriction to yield, this is not always the case. An example is peanut production in Egypt where about half of the peanut crop is produced on large farms and the other half on small land holdings of only a few hectares. In the 1980s, Egypt's average yield was about 1800 kg ha^{-1}, whereas in 2011 it is among the highest of all countries at 3170 kg ha^{-1} (76% increase) (USDA, 2013). To raise exports, the Egyptian government decided to make peanut production a priority and initiated incentives for greater production (El Nockrashy et al., 1986). A production package was developed that included fertilizer applications, control of diseases, and development of improved cultivars. Implementation of these practices by both the large and small farmers in Egypt resulted in the significant yield increases.

Thailand

Comparisons of kilograms per hectare yields among countries is an imperfect indicator of the success of programs to increase production. An example is Thailand where the average yield in 2011 was 1770 kg ha^{-1}, which is approximately half the yield of countries such as the United States, China, Egypt, and Argentina (USDA, 2013). During the 1980s and early 1990s, the yield was only 1350 kg ha^{-1}, which means that during the past 20 yr they have realized a 31% yield increase or 1.6% per year. Furthermore, much of the crop in Thailand is used for the boiling market, which means that peanuts are harvested while immature and with greatly reduced yields as compared with a fully mature crop. Eighty percent of the crop is grown on nonirrigated land in the upland area where soils have low fertility. The other 20% is grown on residual water from rice paddies. The wet and dry seasons are significantly different in terms of pests and disease pressures, and breeding programs have split resources to address the specific problems in different cropping systems. Major production constraints in Thailand are drought in the rainfed areas, low soil fertility, weeds, diseases, and pests. Major pests are leaf miners (*Aproaerema modicella* Deventer), thrips, subterranean ants (*Dorylus orientalis* Westwood), and leafhoppers (*Empoasca fabae* Harris). In Thailand, 97% of peanuts are used domestically for direct consumption. The use of peanut oil is not significant, and there are few small oil mills with low capacity (Peanut Collaborative Research Support Program, 1994).

In summary, global yields of peanut have steadily increased during the past half century at about 1.5% yr^{-1}, in large part because of improved cultivars and production technologies. However, yields across sub-Sahara Africa remain extremely low because of lack of an agricultural infrastructure, water deficit, and cultivars that are highly susceptible to diseases. While the peanut plant can survive extreme drought conditions, yields are greatly suppressed. Thus, the two greatest needs for increasing yields in the low-input systems of most small farmers are for cultivars to be developed with higher drought tolerance (or shorter maturity to avoid droughts) and incorporation of high levels of disease resistance. With the marginal levels of resistance to early and late leaf spot in *A. hypogaea*, this will be a difficult task without utilizing other *Arachis* species to enhance resistance levels. Government incentives to increase production has been a key ingredient for increasing yields and export markets in several countries; however, a large percentage of the crop is still produced on small farms with primitive technologies, and the seed is either consumed by producers or sold in local markets. Aflatoxin is a severe problem in all areas where there is drought, and the infrastructure for testing and eliminating the problem has not been developed in many areas, especially in Africa. Peanut has great potential for additional yield increases at the global level, but a comprehensive production system must be put in place at the local levels to suppress constraints on yield potential.

Future Yield Gains for Peanut

Yield gains in peanut due to genetics up to now have been achieved through conventional breeding, although the application of molecular marker-assisted breeding for this goal is on the horizon. Peanut genomics (genome sequencing) has trailed many other legumes such as soybean [*Glycine max* (L.) Merr.] (Schmutz et al., 2010) and even common bean (*Phaseolus vulgaris* L.) and pigeonpea [*Cajanus cajan* (L.) Millsp.] (http://www.phytozome.net/commonbean.php#A; Varshney et al., 2012). Peanut genome sequencing was initiated in 2012 with a plan to sequence not only tetraploid cultivated peanut but also its diploid progenitor species, *Arachis duranensis* Krapov. & W.C. Greg (A genome) and *Arachis ipaensis* Krapov. & W.C. Greg (B genome) (Kochert et al., 1996). Independently sequencing and assembling the diploid progenitors is part of a strategy to improve assembly of the large (2.8 Gb) and complex tetraploid genome whose subgenomes were estimated to have diverged from one another only around 3 million years ago (Moretzsohn et al., 2012). Further complicating peanut genomics is the recent evolutionary origin of cultivated peanut only several thousands of years ago (Singh and Simpson, 1994) and the low molecular diversity that has been observed in genic regions of the genome. Even typically highly variable regions of a genome such as those containing simple sequence repeats (SSRs) show only 5 to 10% polymorphism between parents in divergent biparental crosses (Guo et al., 2012; Pandey et al., 2012; Qin et al., 2012). A second highly variable marker type based on miniature inverted-repeat transposable elements (MITEs) recently has shown promise for genetic mapping in peanut (Shirasawa et al., 2012). While PCR-based variable repeat or insertion–deletion markers such as these are likely to be used for genetic mapping in peanut over the short term, single nucleotide polymorphism (SNP)-based markers will be more broadly employed in the future, particularly when whole genome sequence for peanut, which may harbor greater variability than coding sequence, is generated, and SNP-detection methods and computational tools advance to handle complex, polyploid genomes (Edwards et al., 2013; Poland and Rife, 2012).

One could envision marker-assisted breeding in peanut progressing along two trajectories: (i) focus on specific traits with relatively simple inheritance for which markers accelerate selection and (ii) genome-wide selection from which information from related training populations is applied during multiple rounds of selection from breeding populations. Feasibility of the first trajectory already has been demonstrated for peanut (Holbrook et al., 2011). Two simply inherited traits, dominant nematode resistance introgressed from a wild relative, *Arachis cardenasii* Krapov. & W.C. Greg, and recessive high-oleic, low-linolenic acid resulting from known mutations have been selected using molecular markers (Barkley et al., 2011; Chu et al., 2011; Church et al., 2000). Phenotypic assays for each of these traits imposes constraints of time, cost, sample destruction, and/or additional generations, whereas marker-assisted selection enables sampling of seedling tissues, screening with a fast, economical DNA extraction and PCR-based assay, scoring of zygosity, recovery of plants and seeds from tested lines, and their immediate use in the crossing program, even as heterozygotes. Marker-assisted backcrossing and the identification of materials likely to carry the desired traits in later generations are obvious targets for markers linked with a locus that is highly heritable and has a major contribution to the phenotypic variation. Traits with more complex inheritance have not been as amenable to marker-assisted breeding (Bernardo, 2008; Dwivedi et al., 2007a, and 2007b). For example, the complex trait of yield in rice is associated with hundreds of quantitative trait loci (QTLs) that may influence yield component traits such as numbers of tillers, inflorescence characteristics, and grain size (Xing and Zhang, 2010). In peanut, physiological traits such as rate of photosynthesis and stomatal conductance, as well as response to vapor pressure deficit, show genetic variation, contribute to variation in yield under stressed and nonstressed environments, and can be genetically mapped (Devi et al., 2010; Nautiyal et al., 2012). A molecular breeding approach shifting towards genomic selection is likely to occur over the next several years as the cost of DNA sequencing continues to decline and whole genome sequences become available for most crops, including peanut (Jackson et al., 2011b). Genomic selection operates with information from hundreds to thousands of markers, each used to calculate a contribution to phenotypic effects and estimate breeding values that can be used to predict performance (Jannink et al., 2010; Meuwissen et al., 2001).

Traditional breeding in peanut has relied primarily on crosses within the tetraploid germplasm, and breeders have recognized the value of exotic tetraploid sources of alleles for disease–pest resistance. Accessing the diploid gene pool has been more challenging, but not insurmountable, and key traits such as nematode resistance have been introgressed from diploid sources (Simpson and Starr, 2001). The potential to more widely access novel alleles from diploids is emerging from the intense development of synthetic tetraploids based on combinations of A- and B-genome species in crosses and artificial chromosome doubling of hybrids (Favero et al., 2006). Synthetic tetraploids created from the progenitor species of peanut cross readily with cultivated forms and hybrids show little segregation distortion (Fonceka et al., 2009; Stalker et al., 2014). This reproductive behavior expands opportunities to search for gene combinations that impact not only disease–pest resistance but also even potentially more complex traits such as yield, as demonstrated by Fonceka et al. (2012).

Transgenic peanuts have not yet reached commercialization, in part because the associated costs of deregulation are too steep for minor and specialty crops (Wang and Brummer, 2012). Methods for transformation of peanut have been developed (Ozias-Akins and Gill, 2001), and while efficiencies are not as high as for some other crops, a large body of research on transgene efficacies has accumulated (Guo et al., 2011; Qin et al., 2011). Introduced traits up to now would only indirectly affect yield by reducing losses to diseases and pests or increasing resistance to abiotic stresses. These losses can be significant, however, and the difference between actual yields and yield potential can be reduced by genetic means in growing regions that lack access to disease or other stress control measures. Economic and social barriers to the deployment of transgenic peanut probably will prevent the use of this option to enhance yield for some time to come, although use of transgenics as tools for research continues to be beneficial.

Acknowledgments

Acknowledgement is given to Dr. A.K. Culbreath for his critical review and comments of a previous draft of this manuscript.

References

Austin, R.B. 1994. Plant breeding opportunities. In: K.J. Boote, J.M. Bennett, T.R. Sinclair, and G.M. Paulsen, editors, Physiology and determination of crop yield. ASA, CSSA, and SSSA, Madison, WI. p. 567–586.

Bantilan, M.C.S., U.K. Deb, and S.N. Nigam. 2003. Impacts of genetic improvement in groundnut. In: R.E. Evenson and D. Gollin, editors, Crop variety improvement and its effects on productivity: The impact of international agricultural research. CABI Publ., Wallingford, UK. p. 293–313.

Barkley, N.A., K.D. Chenault Chamberlain, M.L. Wang, and R.N. Pittman. 2011. Genotyping and fatty acid composition analysis in segregating peanut (*Arachis hypogaea* L.) populations. Peanut Sci. 38:11–19. doi:10.3146/PS10-17.1

Bernardo, R. 2008. Molecular markers and selection for complex traits in plants: Learning from the last 20 years. Crop Sci. 48:1649–1664. doi:10.2135/cropsci2008.03.0131

Boote, K.J., and M. Tollenaar. 1994. Modeling genetic yield potential. In: K.J. Boote, J.M. Bennett, T.R. Sinclair, and G.M. Paulsen, editors, Physiology and determination of crop yield. ASA, CSSA, and SSSA, Madison, WI. p. 533–565.

Boyle, L.W. 1952. Factors to be integrated in the control of southern blight on peanut. Phytopathology 42:282.

Boyle, L.W. 1956. Fundamental concepts in the development of control measures for southern blight and root rot on peanut. Plant Dis. Rep. 40:661–665.

Branch, W.D. 2007. Registration of 'Georgia-06G' peanut. J. Plant Regist. 1:120. doi:10.3198/jpr2006.12.0812crc

Branch, W.D., and T.B. Brenneman. 2008. Registration of 'Georgia-07W' peanut. J. Plant Regist. 2:88–91.

Brenneman, T.B., A.P. Murphy, and A.S. Csinos. 1991. Activity of tebuconazole on *Sclerotium rolfsii* and *Rhizoctonia solani*, two soilborne pathogens of peanut. Plant Dis. 75:744–747.

Bridges, D.C., R.H. Walker, J.A. McGuire, and N.R. Martin. 1984. Efficiency of chemical and mechanical methods for controlling weeds in peanuts (*Arachis hypogaea*). Weed Sci. 32:584–591.

Brown, S.L., A.K. Culbreath, J.W. Todd, D.W. Gorbet, J.A. Baldwin, and J.P. Beasley, Jr. 2005. Development of a method of risk assessment to facilitate integrated management of spotted wilt of peanut. Plant Dis. 89:348–352.

Brown, S., J. Todd, A. Culbreath, J. Baldwin, and J. Beasley. 2003. *Tomato spotted wilt virus* of peanut: Identifying and avoiding high risk situations. Bull. 1165. Univ. of Georgia Coop. Ext. Service, Athens, GA.

Buchanan, G.A., and E.W. Hauser. 1980. Influence of row spacing on competitiveness and yield of peanuts (*Arachis hypogaea*). Weed Sci. 28:401–409.

Byerlee, D. 1979. Rural labor markets in West Africa with emphasis on the semi-arid tropics. In: J. G. Ryan, editor, Proceedings of the International Workshop on Socioeconomic Constraints to Development of Semi-Arid Tropical Agriculture, Hyderabad, India. 19–23 Feb. 1979. ICRISAT, Hyderabad, India.

Carver, W.A. 1969. Registration of Florigiant peanuts1 (Reg. No. 1). Crop Sci. 9:849–850. doi:10.2135/cropsci1969.0011183X000900060069x

Chamblee, R.W., L. Thompson, Jr., and H.D. Coble. 1982. Interference of broadleaf signalgrass (*Brachiaria platyphylla*) in peanuts (*Arachis hypogaea*). Weed Sci. 30:45–49.

Chapin, J.W., and J.S. Thomas. 2011. Improved disease resistance to Virginia-type peanuts–Developing appropriate management programs for S.C. production conditions. Proc. Am. Peanut Res. Educ. Soc. 43:66–67 (Abstr.).

Chapin, J.W., J.S. Thomas, T.G. Isleib, F.M. Shokes, W.D. Branch, and B.L. Tillman. 2009. Field evaluation of Virginia-type peanut cultivars for resistance to *Tomato spotted wilt virus*, late leaf spot, and stem rot. Peanut Sci. 37:1–7.

Chu, Y., C.L. Wu, C.C. Holbrook, B.L. Tillman, G. Person, and P. Ozias-Akins. 2011. Marker-assisted selection to pyramid nematode resistance and the high oleic trait in peanut. Plant Gen. 4:110–117. doi:10.3835/plantgenome2011.01.0001

Church, G.T., C.E. Simpson, M.D. Burow, A.H. Paterson, and J.L. Starr. 2000. Use of RFLP markers for identification of individuals homozygous for resistance to *Meloidogyne arenaria* in peanut. Nematology 2:575–580. doi:10.1163/156854100509367

Coffelt, T.A., M.L. Seaton, and S.W. VanScoyoc. 1989. Reproductive efficiency of 14 Virginia-type peanut cultivars. Crop Sci. 29:1217–1220. doi:10.2135/cropsci1989.0011183X002900050026x

Culbreath, A.K. and R. Srinivasan. 2011. Epidemiology of spotted wilt disease of peanut caused by *Tomato spotted wilt virus* in the southeastern U.S. Virus Res. 159:101–109.

Cummins, D.G., and C.R. Jackson, editors. 1982. World peanut production, utilization and research. Spec. Publ. 16. Univ. of Georgia, Athens, GA.

Devi, M.J., T.R. Sinclair, and V. Vadez. 2010. Genotypic variation in peanut for transcription response to vapor pressure deficit. Crop Sci. 50:191–196. doi:10.2135/cropsci2009.04.0220

Duncan, W.G., D.E. McCloud, R.L. McGraw, and K.J. Boote. 1978. Physiological aspects of peanut yield improvement. Crop Sci. 18:1015–1020. doi:10.2135/cropsci1978.0011183X0018000 60028x

Dwivedi, S.L., D.J. Bertioli, J.H. Crouch, J.F. Valls, H.D. Upadhyaya, A. Favero, M. Moretzsohn, and A.H. Paterson. 2007a. Peanut. In: C. Kole, editor, Oilseeds. Genome mapping and molecular breeding in plants. Vol. 2. Springer, New York. p. 115–151.

Dwivedi, S.L., J.L. Crouch, D.J. Mackill, Y. Xu, M.W. Blair, M. Ragot, H.D. Upadhyaya, and R. Ortiz. 2007b. The molecularization of public sector crop breeding: Progress, problems, and prospects. Adv. Agron. 95:163–318. doi:10.1016/S0065-2113(07)95003-8

Edwards, D., J. Batley, and R.J. Snowdon. 2013. Accessing complex crop genomes with next-generation sequencing. Theor. Appl. Genet. 126:1–11. doi:10.1007/s00122-012-1964-x

El Nockrashy, A.S., O. Galal, and J. Davenport. 1986. Chapter IV. The more and better food intervention projects. In: More and better food—An Egyptian demonstration project (BOSTID) project on improvement of peanut productivity. Board on Science and Technol. for Intern. Dev., Washington, DC. p. 71–124.

Emery, D.A., J.C. Wynne, and P.W. Rice. 1973. Can reproductive efficiency in cultivated peanuts be improved? Oleagineux 28:399–403.

Favero, A.P., C.E. Simpson, J.F.M. Valls, and N.A. Vello. 2006. Study of the evolution of cultivated peanut through crossability studies among *Arachis ipaensis, A. duranensis*, and *A. hypogaea*. Crop Sci. 46:1546–1552. doi:10.2135/cropsci2005.09-0331

Fletcher, S.M., P. Zhang, and D.H. Carley. 1992. Groundnuts: Production, utilization, and trade in the 1980s. In: S.N. Nigam, editor, Groundnut: A global perspective. Proceedings of an International Workshop, Hyderabad, India. 25–29 Nov. 1991. ICRISAT. Hyderabad, India. p. 17–32.

Fonceka, D., T. Hodo-Abalo, R. Rivallan, I. Faye, M.N. Sall, O. Ndoye, A.P. Favero, D.J. Bertioli, J.C. Glaszmann, B. Courtois, and J.F. Rami. 2009. Genetic mapping of wild introgressions into cultivated peanut: A way toward enlarging the genetic basis of a recent allotetraploid. BMC Plant Biol. 9:103. doi:10.1186/1471-2229-9-103

Fonceka, D., H.A. Tossim, R. Rivallan, H. Vignes, I. Faye, O. Ndoye, M.C. Moretzsohn, D.J. Bertioli, J.C. Glaszmann, B. Courtois, and J.F. Rami. 2012. Fostered and left behind alleles in peanut: Interspecific QTL mapping reveals footprints of domestication and useful natural variation for breeding. BMC Plant Biol. 12:26. doi:10.1186/1471-2229-12-26

Garren, K.H., and G.B. Duke. 1957. The peanut stem rot problem and a preliminary report on interrelations of "non-dirting" weed control and other practices to stem rot and yield of peanuts. Plant Dis. Rep. 41:424–431.

Gorbet, D.W., and B.L. Tillman. 2011. Registration of 'York' peanut. J. Plant Registr. 5:289–294. doi:10.3198/jpr2010.11.0644crc

Grey, T.L., E.P. Prostko, D.C. Bridges, W.C. Johnson III, E.F. Eastin, W.K. Vencill, B.J. Brecke, G.E. MacDonald, J.A. Tredaway, J.W. Everest, G.R. Wehtje, and J.W. Wilcut. 2003. Residual weed control with imazapic, diclosulam, and flumioxazin in southeastern peanut (*Arachis hypogaea*). Peanut Sci. 30:22–28. doi:10.3146/pnut.30.1.0005

Grichar, W.J. 1991. Control of Texas panicum (*Panicum texanum*) and southern crabgrass (*Digitaria ciliaris*) in peanuts (*Arachis hypogaea*) with postemergence herbicides. Peanut Sci. 18:6–9. doi:10.3146/i0095-3679-18-1-3

Grichar, W.J., and T.E. Boswell. 1986. Postemergence grass control in peanut (*Arachis hypogaea*). Weed Sci. 34:587–590.

Guo, B.Z., C.Y. Chen, Y. Chu, C.C. Holbrook, P. Ozias-Akins, and H.T. Stalker. 2011. Advances in genetics and genomics for sustainable peanut production. In: N. Benkeblia, editor, Sustainable agriculture and new biotechnologies. CRC Press, Boca Raton, FL. p. 341–368.

Guo, Y., S. Khanal, S. Tang, J.E. Bowers, A.F. Heesacker, N. Khalilian, E.D. Nagy, D. Zhang, C.A. Taylor, H.T. Stalker, P. Ozias-Akins, and S.J. Knapp. 2012. Comparative mapping in intraspecific populations uncovers a high degree of macrosynteny between A- and B-genome diploid species of peanut. BMC Genomics 13:608. doi:10.1186/1471-2164-13-608

Hauser, E.W., and G.A. Buchanan. 1981. Influence of row spacing, seeding rates, and herbicide systems on the competitiveness and yield of peanut. Peanut Sci. 8:74–81. doi:10.3146/i0095-3679-8-1-18

Hauser, E.W., and S.A. Parham. 1969. Effects of annual weeds and cultivation on the yield of peanuts. Weed Res. 9:192–197. doi:10.1111/j.1365-3180.1969.tb01473.x

Holbrook, C.C., P. Ozias-Akins, Y. Chu, and B. Guo. 2011. Impact of molecular genetic research on peanut cultivar development. Agronomy 1:3–17. doi:10.3390/agronomy1010003

Holbrook, C.C., P. Timper, A.K. Culbreath, and C.K. Kvien. 2008. Registration of 'Tifguard' peanut. J. Plant Regist. 2:92–94. doi:10.3198/jpr2007.12.0662crc

ICRISAT. 1980. Proceedings of an International Workshop on Groundnuts, Hyderabad, India. 13–17 Oct. 1980. ICRISAT Center, Hyderabad, India.

Isleib, T.G., C.C. Holbrook, and D.W. Gorbet. 2001. Use of plant introductions in peanut cultivar development. Peanut Sci. 28:96–113. doi:10.3146/i0095-3679-28-2-11

Isleib, T.G., S.R. Milla-Lewis, H.E. Pattee, S.C. Copeland, M.C. Zuleta, B.B. Shew, J.E. Hollowell, T.H. Sanders, L.O. Dean, K.W. Hendrix, M. Balota, and J.W. Chapin. 2011. Registration of 'Bailey' peanut. J. Plant Regist. 5:27–39. doi:10.3198/jpr2009.12.0742crc

Jackson, J.L., J.P. Beasley, R.S. Tubbs, R.D. Lee, and T.L. Grey. 2011a. Fall-bedding for reduced digging losses and improved yield in strip-till peanut. Peanut Sci. 38:31–40. doi:10.3146/0095-3679-38.1.31

Jackson, S.A., A. Iwata, S.H. Lee, J. Schmutz, and R. Shoemaker. 2011b. Sequencing crop genomes: Approaches and applications. New Phytol. 191:915–925. doi:10.1111/j.1469-8137.2011.03804.x

Jannink, J.L., A.J. Lorenz, and H. Iwata. 2010. Genomic selection in plant breeding: From theory to practice. Briefings Funct. Genomics 9:166–177. doi:10.1093/bfgp/elq001

Johnson, W.C. III, and B.G. Mullinix, Jr. 2005. Texas panicum (*Panicum texanum*) interference in peanut (*Arachis hypogaea*) and implications for treatment decisions. Peanut Sci. 32:68–72. doi:10.3146/0095-3679(2005)32[68:TPPTII]2.0.CO;2

Johnson, W.C. III, E.P. Prostko, and B.G. Mullinix, Jr. 2002. Texas panicum (*Panicum texanum*) control in strip-tillage peanut (*Arachis hypogaea*) production. Peanut Sci. 29:141–145. doi:10.3146/pnut.29.2.0012

Johnson, W.C. III, E.P. Prostko, and B.G. Mullinix, Jr. 2005. Improving the management of dicot weeds in peanut with narrow row spacings and residual herbicides. Agron. J. 97:85–88.

Kochert, G., H.T. Stalker, M. Gimenes, L. Galgaro, C.R. Lopes, and K. Moore. 1996. RFLP and cytogenetic evidence on the origin and evolution of allotetraploid domesticated peanut, *Arachis hypogaea* (Leguminosae). Am. J. Bot. 83:1282–1291. doi:10.2307/2446112

Leidner, J. 2012. Precision farming payoff in peanuts. Southeast. Peanut Farmer Mag. 50(5):10–12.

Livingstone, D.M., J.L. Hampton, P.M. Phipps, and E.A. Grabau. 2005. Enhancing resistance to *Sclerotinia minor* in peanut by expressing a barley oxalate oxidase gene. Plant Physiol. 137:1354–1362. doi:10.1104/pp.104.057232

McArthur, W.C., V.N. Grise, H.O. Doty, Jr., and D. Hacklander. 1982. U.S. peanut industry. Agric. Econ. Report No. 493. Natl. Econ. Div., Econ. Res. Serv., USDA, Washington, DC.

McCarty, M.T. 1983. Economic thresholds of annual grasses in agronomic crops. Ph.D. diss., North Carolina State University, Raleigh, NC.

Meuwissen, T.H.E., B.J. Hayes, and M.E. Goddard. 2001. Prediction of total genetic value using genome-wide dense marker maps. Genetics 157:1819–1829.

Moretzsohn, M.C., E.G. Gouvea, P.W. Inglis, S.C.M. Leal-Bertioli, J.F.M. Valls, and D.J. Bertioli. 2012. A study of the relationships of cultivated peanut (*Arachis hypogaea*) and its most closely related wild species using intron sequences and microsatellite markers. Ann. Bot. (Lond.) 111:113–126.

Mozingo, R.W., T.A. Coffelt, and J.C. Wynne. 1987. Genetic improvement in large-seeded Virginia-type peanut cultivars since 1944. Crop Sci. 27:228–231.

Nautiyal, P.C., V. Ravindra, A.L. Rathnakumar, B.C. Ajay, and P.V. Zala. 2012. Genetic variations in photosynthetic rate, pod yield and yield components in Spanish groundnut cultivars during three cropping seasons. Field Crops Res. 125:83–91.

Navia Gine, P.A., A.K. Culbreath, B.L. Tillman, C.C. Holbrook, W.D. Branch, and N.B. Smith. 2011. Relationships between defoliation by late leaf spot and yield in new runner-type peanut cultivars. Proc. Am. Peanut Res. Educ. Soc. 43:38–39.

Nigam, S.N., editor. 1992. Groundnut: A global perspective. Proceedings of an International Workshop, Hyderabad, India. 25–29 Nov. 1991. ICRISAT, Hyderabad, India.

Nigam, S.N., S.L. Dwivedi, and R.W. Gibbons. 1980. Groundnut breeding at ICRISAT. In: Proceedings of an International Workshop on Groundnuts, Hyderabad, India. 13–17 Oct. 1980. ICRISAT Center, Hyderabad, India. p. 62–68.

Nigam, S.N., L.J. Reddy, H.D. Upadhyaya, and H.D. Dwivedi. 1994. Genetic enhancement in groundnut. In: M.V.R. Prasad, R.K. Shastry, C.V. Raghuvaiah, and T. Damodaram, editors, Sustainability in oilseeds. National Seminar on Oilseeds Research and Development in India: Status and strategies, Hyderabad, India. 2–5 Aug. 1993. Indian Society of Oilseeds Research, Hyderabad, India.

Norden, A.J., R.W. Lipscomb, and W.A. Carver. 1969. Registration of Florunner peanuts (Reg. No. 2). Crop Sci. 9:850. doi:10.2135/cropsci1969.0011183X000900060070x

Ortiz, B.V., K.B. Balkcom, L. Duzy, E. van Santen, and D.L. Hartzog. 2013. Evaluation of agronomic and economic benefits of using RTK-based autoguidance systems on peanut production. J. Precis. Agric. doi:10.1007/s11119–012–9297-y

Ozias-Akins, P., and R. Gill. 2001. Progress in the development of tissue culture and transformation methods applicable to the production of transgenic peanut. Peanut Sci. 28:123–131. doi:10.3146/i0095-3679-28-2-14

Pandey, M.K., E. Monyo, P. Ozias-Akins, X. Liang, P. Guimarães, S.N. Nigam, H.D. Upadhyaya, P. Janila, X. Zhang, B. Guo, D.R. Cook, D.J. Bertioli, R. Michelmore, and R.K. Varshney. 2012. Advances in *Arachis* genomics for peanut improvement. Biotechnol. Adv. 30:639–651. doi:10.1016/j.biotechadv.2011.11.001

Peanut Collaborative Research Support Program. 1994. The external evaluation panel report for the Peanut CRSP. Univ. of Georgia. http://pdf.usaid.gov/pdf_docs/PDABK164.pdf (accessed 22 Jan. 2013).

Phipps, P.M. 1981. Applied research on field crop disease control. Annual report. Information Series 72. Virginia Polytech. Inst. and State Univ., Suffolk.

Phipps, P.M. 1982. Applied research on field crop disease control. Annual report. Information Series 82. Virginia Polytech. Inst. and State Univ., Suffolk.

Phipps, P.M. 1983. Applied research on field crop disease control. Annual report. Information Series 101. Virginia Polytech. Inst. and State Univ., Suffolk.

Phipps, P.M. 1984. Applied research on field crop disease control. Annual report. Information Series 123. Virginia Polytech. Inst. and State Univ., Suffolk.

Poland, J.A., and T.W. Rife. 2012. Genotyping-by-sequencing for plant breeding and genetics. Plant Gen. 5:92–102. doi:10.3835/plantgenome2012.05.0005

Prince, A.E. 1944. Diseases of peanuts in North Carolina. Plant Dis. Rep. 28:1080–1081.

Prostko, E.P., W.C. Johnson III, and B.G. Mullinix, Jr. 2001. Annual grass control with preplant incorporated and preemergence applications of ethalfluralin and pendimethalin in peanut (*Arachis hypogaea*). Weed Technol. 15:36–41. doi:10.1614/0890-037X(2001)015[0036:AGCWPI]2.0.CO;2

Qin, H., S. Feng, C. Chen, Y. Guo, S. Knapp, A. Culbreath, G. He, M. Wang, X. Zhang, C.C. Holbrook, P. Ozias-Akins, and B. Guo. 2012. An integrated genetic linkage map of cultivated peanut (*Arachis hypogaea* L.) constructed from two RIL populations. Theor. Appl. Genet. 124:653–664. doi:10.1007/s00122-011-1737-y

Qin, H., Q. Gu, J.L. Zhang, L. Sun, S. Kuppu, Y.Z. Zhang, M. Burow, P. Payton, E. Blumwald, and H. Zhang. 2011. Regulated expression of an isopentenyltransferase gene (IPT) in peanut significantly improves drought tolerance and increases yield under field conditions. Plant Cell Physiol. 52:1904–1914. doi:10.1093/pcp/pcr125

Reddy, M.S., C.N. Floyd, and R.W. Willey. 1980. Groundnut in intercropping systems. In: Proceedings of an International Workshop on Groundnuts, Hyderabad, India. 13–17 Oct. 1980. ICRISAT Center, Hyderabad, India. p. 133–142.

Schmutz, J., S.B. Cannon, J. Schlueter, J.X. Ma, T. Mitros, W. Nelson, D.L. Hyten, Q.J. Song, J.J. Thelen, J.L. Cheng, D. Xu, U. Hellsten, G.D. May, Y. Yu, T. Sakurai, T. Umezawa, M.K. Bhattacharyya, D. Sandhu, B. Valliyodan, E. Lindquist, M. Peto, D. Grant, S.Q. Shu, D. Goodstein, K. Barry, M. Futrell-Griggs, B. Abernathy, J.C. Du, Z.X. Tian, L.C. Zhu, N. Gill, T. Joshi, M. Libault, A. Sethuraman, X.C. Zhang, K. Shinozaki, H.T. Nguyen, T.A. Wing, P. Cregan, J. Specht, J. Grimwood, D. Rokhsar, G. Stacey, R.C. Shoemaker, and S.A. Jackson. 2010. Genome sequence of the palaeopolyploid soybean. Nature 463:178–183. doi:10.1038/nature08670

Shirasawa, K., P. Koilkonda, K. Aoki, H. Hirakawa, S. Tabata, M. Watanabe, M. Hasegawa, H. Kiyoshima, S. Suzuki, C. Kuwata, Y. Naito, T. Kuboyama, A. Nakaya, S. Sasamoto, A. Watanabe, M. Kato, K. Kawashima, Y. Kishida, M. Kohara, A. Kurabayashi, C. Takahashi, H. Tsuruoka, T. Wada, and S. Isobe. 2012. *In silico* polymorphism analysis for the development of simple sequence repeat and transposon markers and construction of linkage map in cultivated peanut. BMC Plant Biol. 12:80. doi:10.1186/1471-2229-12-80

Simpson, C.E., and J.L. Starr. 2001. Registration of 'COAN' peanut. Crop Sci. 41:918. doi:10.2135/cropsci2001.413918x

Singh, A.K., and C.E. Simpson. 1994. Biosystematics and genetic resources. In: J. Smart, editor, The groundnut crop: A scientific basis for improvement. Chapman and Hall, London. p. 96–137.

Singh, V. 1980. Indian coordinated research project on oilseeds with special reference to groundnut. In: R.W. Gibbons, editor, Proceedings of the International Workshop on Groundnuts, Hyderabad, India. 13–17 Oct. 1980. ICRISAT, Hyderabad, India. p. 19–24.

Smith, B.W. 1954. *Arachis hypogaea*. Reproductive efficiency. Am. J. Bot. 41:607–616. doi:10.2307/2438286

Smith, D.H., and R.H. Littrell. 1980. Management of peanut foliar diseases with fungicides. Plant Dis. 64:356–361. doi:10.1094/PD-64-356

Stalker, H.T., S.P. Tallury, P. Ozias-Akins, S.C. Leal-Bertioli, and D. Bertioli. 2014. The value of diploid peanut relatives for breeding and genomics. Peanut Sci. (in press).

Sylla, F. 2010. Revitalization of groundnut sector in West Africa (Gambia, Guinea Bissau, and Senegal). Global Agricultural Information Network (GAIN Report). USDA, Washington, DC. http://gain.fas.usda.gov/Recent%20GAIN%20Publications/Revitalization%20of%20the%20Groundnut%20sector%20in%20West%20Africa_Dakar_Senegal_12-3-2010.pdf (accessed 22 Jan. 2013).

Talawar, S. 2004. Peanut in India: History, production, and utilization. Peanut in Local and Global Food Syst. Ser. Rep. 5. Univ. of Georgia, Athens, GA. http://www.worldpeanutinfo.com/documents/IndiaProduction.pdf (accessed 22 Jan. 2013).

Tillman, B.L., D.W. Gorbet, A.K. Culbreath, and J.W. Todd. 2006. Response of peanut cultivars to seeding density and row pattern. Crop Manage. doi:10.1094/CM-2006-0711-01-RS

Tracy, W.F., I.L. Goldman, A.E. Tiefenthaler, and M.A. Schaber. 2004. Trends in productivity of U.S. crops and long-term selection. Plant Breed. Rev. 24:89–108.

USDA. 2013. Table 13. Peanut area, yield, and production. Foreign Agriculture Service, USDA. http://www.fas.usda.gov/psdonline/psdreport.aspx?hidReportRetrievalName=BVS&hidReportRetrievalID=918&hidReportRetrievalTemplateID=1 (accessed 22 Jan. 2013).

Vara Prasad, P.V., V.G. Kakani, and H.D. Upadhyaya. 2010. Growth and production of groundnut. In: Soils, plant growth and crop production. Encyclopedia of Life Support Systems (EOLSS). Vol. 2. Oxford, UK. p. 1–26.

Varshney, R.K., W.B. Chen, Y.P. Li, A.K. Bharti, T.K. Saxena, J.A. Schlueter, M.T.A. Donoghue, S. Azam, G.Y. Fan, A.M. Whaley, A.D. Farmer, J. Sheridan, A. Iwata, R. Tuteja, R.V. Penmetsa, W. Wu, H.D. Upadhyaya, S.P. Yang, T. Shah, K.B. Saxena, T. Michael, W.R. McCombie, B.C. Yang, G.Y. Zhang, H.M. Yang, J. Wang, C. Spillane, D.R. Cook, G.D. May, X. Xu, and S.A. Jackson. 2012. Draft genome sequence of pigeonpea (*Cajanus cajan*), an orphan legume crop of resource-poor farmers. Nat. Biotechnol. 30:83–89. doi:10.1038/nbt.2022

Wang, Z.Y., and E.C. Brummer. 2012. Is genetic engineering ever going to take off in forage, turf and bioenergy crop breeding? Ann. Bot. (Lond.) 110:1317–1325. doi:10.1093/aob/mcs027

Warren, G.F. 1998. Spectacular increases in crop yields in the United States in the twentieth century. Weed Technol. 12:752–760.

Wehtje, G., R.H. Walker, M.G. Patterson, and J.A. McGuire. 1984. Influence of twin rows on yield and weed control in peanuts. Peanut Sci. 11:88–91. doi:10.3146/i0095-3679-11-2-10

Wells, R., T. Bi, W.F. Anderson, and J.C. Wynne. 1991. Peanut yield as a result of fifty years of breeding. Agron. J. 83:957–961. doi:10.2134/agronj1991.00021962008300060007x

Wilcut, J.W. 1991. Efficacy and economics of common bermudagrass (*Cynodon dactylon*) control in peanuts (*Arachis hypogaea*). Peanut Sci. 18:106–109. doi:10.3146/i0095-3679-18-2-11

Wilcut, J.W., F.R. Walls, Jr., and D.N. Horton. 1991. Imazethapyr for broadleaf weed control in peanuts (*Arachis hypogaea*). Peanut Sci. 18:26–30. doi:10.3146/i0095-3679-18-1-9

Woodroof, N.C., J.R. Cole, and J.H. Hunter. 1944. Leafspot control for increased peanut yields. Cir. 145. Georgia Agric. Exp. Stn., Athens, GA.

Woodruff, J.G. 1966. Peanuts: Production, processing, products. Avi Publishing Company, Westport, CT.

Woodward, J.E., T.B. Brenneman, and B.G. Mullinix. 2012. Irrigation timing impacts the efficacy of foliar-applied fungicides toward foliar and soilborne pathogens of peanut. Plant Dis. 96:1785–1790. doi:10.1094/PDIS-01-12-0042-RE

Xing, Y.Z., and Q.F. Zhang. 2010. Genetic and molecular bases of rice yield. Annu. Rev. Plant Biol. 61:421–442. doi:10.1146/annurev-arplant-042809-112209

Yao, G. 2004. Peanut production and utilization in the People's Republic of China. Peanut in Local and Global Food Syst. Ser. Rep. 4. Univ. of Georgia, Athens. 26 pp. http://www.caes.uga.edu/commodities/fieldcrops/peanuts/pins/documents/ChinaProduction.pdf (accessed 22 Jan. 2013).

York, A.C., and H.D. Coble. 1977. Fall panicum interference in peanuts. Weed Sci. 25:43–47.

8

Potato

Shelley H. Jansky,* David M. Spooner, and Paul C. Bethke

ABSTRACT

For more than half a century, yield gains for potato (*Solanum tuberosum* L.) have outpaced those for most other crops. Virtually all of this improvement in productivity has come from improved management of the crop and a shift to production in geographic regions with higher yield potential. Genetic gains for total yield have been negligible in comparison, although newer potato varieties have produced significant economic benefits in terms of increased marketable yield and improved tuber quality. Continuing the trend of increasing yield per unit area will have significant benefits for society and for producers during the upcoming decades, when competition for land and water resources is expected to be fierce, and costs of production are likely to increase. It seems likely that genetic improvement will make a greater contribution to productivity increases, but only if customers are willing to accept newer varieties. Improvements in our understanding of potato genetics and genomics are likely to contribute to the goal of increasing yield by making the breeding process more efficient. Biotechnology has the potential to make a substantial contribution to yield increases as well, particularly though increased resistance to insect pests and diseases including late blight [caused by *Phytophthora infestans* (Mont.) de Bary], *Potato virus Y*, and Verticillium wilt (caused by *Verticillium dahliae* Kleb.). Refinements in the how the crop is grown and stored are continuing to increase yields, and this trend shows no sign of plateauing. Thus, it seems likely that the remarkable improvements in potato yield that have occurred in the past several decades will continue into the future.

Abbreviations: PVMI, Potato Variety Management Institute; PVPA, Plant Variety Protection Act; UPGA, United Potato Growers of America; UPGI, United Fresh Potato Growers of Idaho.

Shelley H. Jansky, David M. Spooner (david.spooner@ars.usda.gov), and **Paul C. Bethke** (paul.bethke@ars.usda.gov), USDA-ARS, Vegetable Crops Research Unit, 1575 Linden Dr., Madison, WI 53706-1590. *Corresponding author (shelley.jansky@ars.usda.gov; shjansky@wisc.edu).

doi:10.2135/cssaspecpub33.c8

Yield Gains in Major U.S. Field Crops. CSSA Special Publication 33.
Stephen Smith, Brian Diers, James Specht, and Brett Carver, editors.

Potato is the most important vegetable crop in the United States. It is produced throughout the country and across all seasons. Total production in 2012 (21,182,800 t) was 57% higher than that in 1980 (13,466,840 t), although production area decreased by 10.8% (from 516,357 ha to 459,982 ha).[1] More potatoes are now produced on less land because productivity has improved at a rate of 1 to 2% yr^{-1}.

Most potatoes are produced in the western states (Fig. 8–1) and harvested in the fall (Fig. 8–2). The highest production is in (listed in order of total production, based on 2012 data): Idaho, Washington, Wisconsin, North Dakota, Oregon, and Colorado. The fall crop in these six states accounted for 73% of the total U.S. production in 2012. Yield per hectare varies widely among growing seasons (Fig. 8–3), states (Fig. 8–4), and production regions (Fig. 8–5). Yields are highest in Washington, Oregon, and southwestern Idaho, where longer growing seasons, irrigated production, and abundant solar radiation combine to produce high yields. In Fig. 8–4, counties in southwestern Idaho are presented separately from the remainder

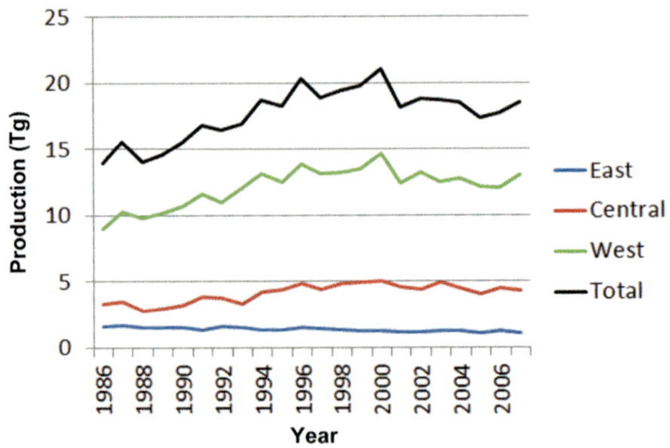

Fig. 8–1. Average U.S. potato production by region, 1986 through 2008.

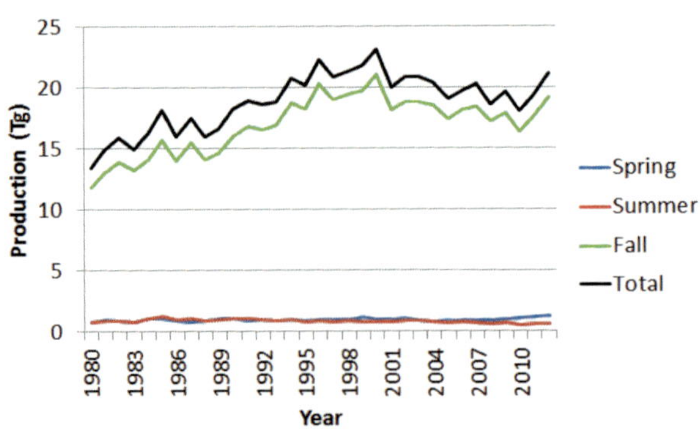

Fig. 8–2. Average U.S. potato production by season, 1980 through 2012.

[1] Unless stated otherwise, all data in this chapter were provided by the U.S. Department of Agriculture National Agricultural Statistics Service (USDA-NASS, http://www.nass.usda.gov/, accessed 23 Jan. 2014).

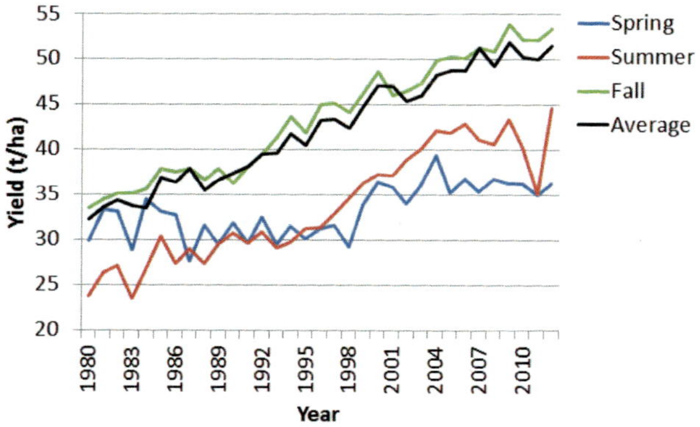

Fig. 8–3. Average U.S. potato yield by season, 1980 through 2012.

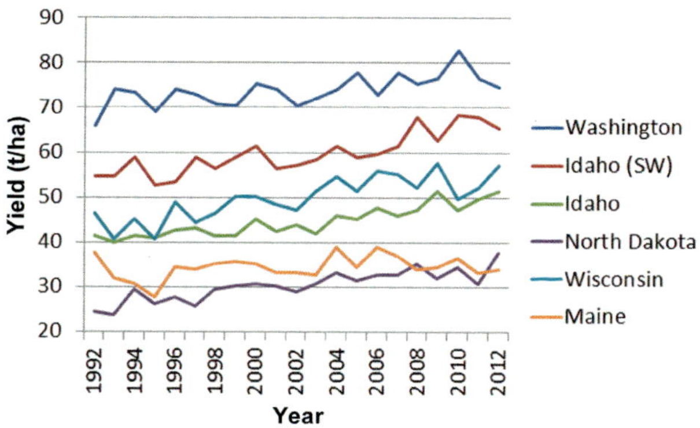

Fig. 8–4. Average U.S. potato yield by state, 1992 through 2012.

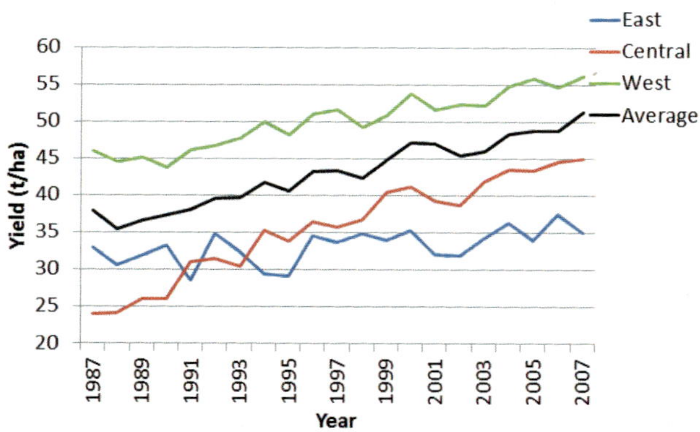

Fig. 8–5. Average U.S. potato yield by region, 1987 through 2007.

of Idaho. The southwestern counties are at a lower elevation and consequently have a longer growing season and higher yields relative to other growing regions of Idaho. Yields in Maine and North Dakota are lower because of a shorter growing season and a larger percentage of nonirrigated production.

Potato yields per unit of land have increased more dramatically than those of any other vegetable crop during the past seven decades (Tiefenthaler et al., 2003). Average U.S. potato yields increased linearly at a rate of 460 kg ha^{-1} yr^{-1} from 1935 to 1998. Between 1987 and 2007, average yield in the United States increased at an annual rate of 720 kg ha^{-1} (Fig. 8–5). The central states realized the largest gain, producing on average an extra tonne of potatoes on each hectare of land every year. Gains in the United States are credited to grower specialization, improved production and storage technology, improved cultural practices, a shift in production to the high-yielding environments of the Pacific Northwest, and the introduction of new cultivars, mainly through public breeding programs (Douches et al., 1996; Johnson et al., 2010). Water and nutrient availability are critical determinants of potato productivity (Fig. 8–6). High potato yields reflect management practices that match water and nutrient supply with crop needs throughout the growing season. From 1980 through 2000, U.S. potato production increased steadily at a rate of 416,000 t yr^{-1} (Fig. 8–2, $R^2 = 0.90$). Since the turn of the 21st century, however, overall production has declined slightly with time. Because a large portion of total production results from the fall crop, the trend lines for the fall crop are similar to those for overall potato production (Fig. 8–2). The spring and summer potato crops supply the fresh and chip markets. However, these crops are relatively small, and production has not changed substantially in recent decades. Raw product from the fall crop has displaced some of the spring crop as a result

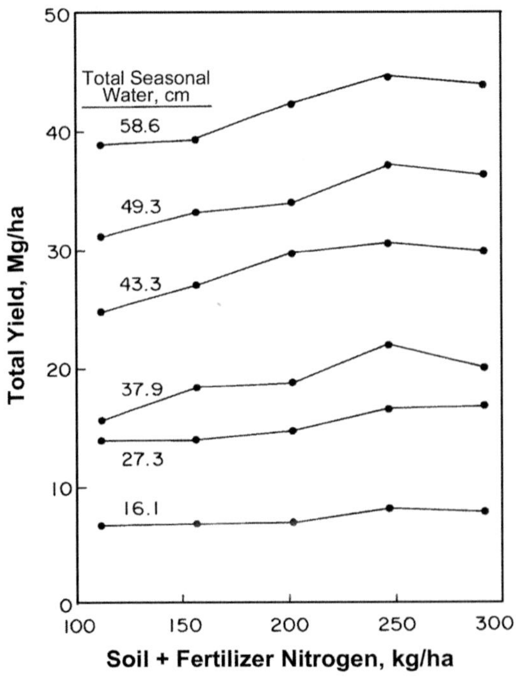

Fig. 8–6. Effect of nitrogen and water inputs on total yield of Russet Burbank potatoes (Ojala et al., 1990).

of improvements in storage technology and the development of cultivars that can be stored for extended lengths of time while still maintaining acceptable quality. The winter crop is very small (1%) and is not separated from the spring crop in recent USDA-NASS reports, so the winter crop is not considered in this chapter.

The total area of U.S. potatoes harvested increased slightly from 1980 until 2000, when it began to decrease (Fig. 8–7 and 8–8). A similar trend was observed with fall potatoes; spring and summer harvested area has declined slowly throughout the past 30 yr (Fig. 8–7). The decline in harvested area is attributed in part to continued improvements in yield per land area, which have outpaced potato demand. These improvements have resulted from refinements in agronomic practices as well as a slow shift from the predominant cultivar Russet Burbank to other cultivars with similar or better yields and a higher percentage of marketable tubers. The formation of the United Fresh Potato Growers of Idaho

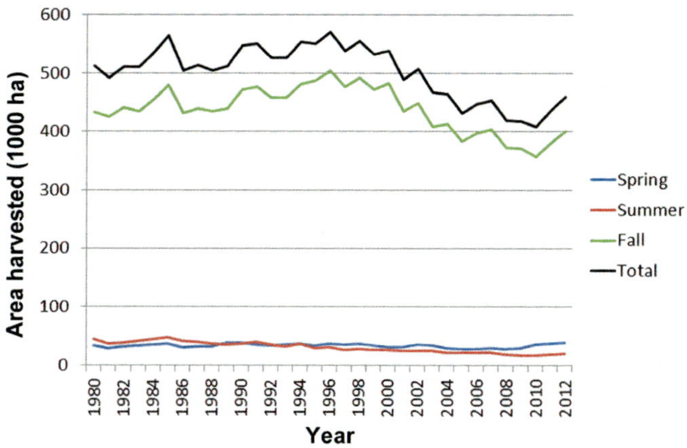

Fig. 8–7. Area of potatoes harvested in the United States by season, 1980 through 2012.

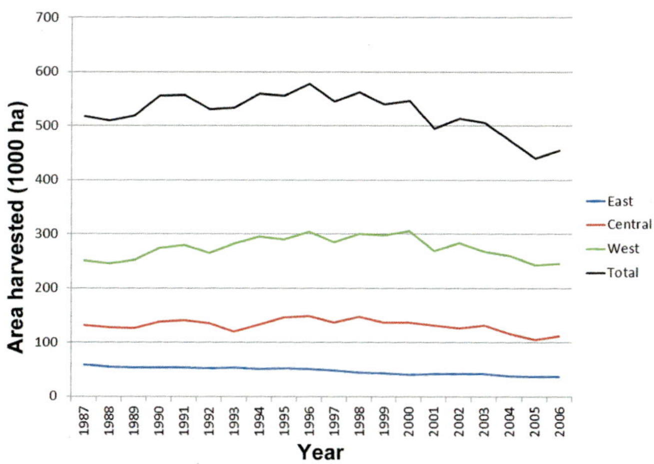

Fig. 8–8. Area of potatoes harvested in the United States by region, 1987 through 2006.

(UPGI) in 2004, followed by the United Potato Growers of America (UPGA) in 2005 is a major factor in the stabilization of production in the 21st century. The goal of these farmer cooperatives is to maintain profitability by managing potato supply to keep it in line with demand. By limiting production and consequently supply, UPGI and UPGA have been successful in reducing market volatility of fresh potatoes, while increasing prices (McGary, 2012).

Potato is a high-input crop. The total cost of production typically exceeds \$1,500 ha^{-1}, with costs of up to \$7,500 ha^{-1}, or more, in regions of high disease and pest pressure (Thornton, 2001; Johnson et al., 2010). For example, estimated 2012 costs of production in southwestern Idaho for the cultivar Russet Burbank stored for 4.5 mo were \$3,700 for ownership costs and \$7,400 for operating costs, resulting in a total of over \$11,000 ha^{-1} (Patterson, 2012). Chemical control agents are applied to over 80% of potato acreage in the United States (Wiese, 1998). Most of the crop is harvested in the fall from production areas across the northern tier of states. Since processing facilities cannot handle the entire crop at harvest, most potatoes are stored. Storage costs are expensive—estimated at \$850 to \$1,100 ha^{-1} in the Northwest (Johnson et al., 2010; Patterson, 2012).

Processing facilities for frozen potato products are concentrated near the major production areas in the Pacific Northwest, the upper Midwest, and the Northeast (Johnson et al., 2010). Potato chip processing plants, however, are built around human population centers. Chips are fragile to transport and it is more cost-effective to transport dense tubers than light bags of chips. In the East, growers of fresh potatoes are typically also packers (Johnson et al., 2010). In contrast, in the West, the production and packing businesses have more commonly been separate entities. That is changing, though, and more growers in the West are developing packing businesses (Johnson et al., 2010).

Genetic Contributions to Productivity Increases

Commercial potato is a clonally propagated tetraploid crop. The predominant breeding strategy is a phenotypic recurrent selection scheme that consists of crossing two parents, selecting a very small percentage of hybrid offspring, and then selection of clonally propagated clones for several years under progressively larger and more rigorous field trials. This strategy has not changed in the past century (Douches et al., 1996; Carputo and Frusciante, 2011). Based on cultivar releases published in the *American Journal of Potato Research*, U.S. Plant Variety Protection applications, and the *World Catalogue of Potato Varieties* (Peiterse and Hils, 2009), 182 potato varieties were released in the United States between 1980 and 2012 (Table 8–1).

Cultivar development in potato focuses on market classes, each with different target properties. Russet potatoes are used for both the fresh market and the frozen processing market, which is dominated by frozen french fry production. Round white potatoes are consumed fresh or processed into chips. Round reds are used almost exclusively in the fresh market. Specialty potato varieties, including fingerlings and colored flesh potatoes, have a small but growing share of the fresh market.

Historically, most potatoes were grown for fresh consumption, but in 1989 the use of the fall potato crop for processing surpassed its use for the fresh market for the first time (Johnson et al., 2010). Since then, the majority of the potato

Table 8–1. Potato cultivars released in the United States between 1980 and 2012.

Year	Name	Market class	Use
1980	Allagash Russet	Russet	Dual purpose
1980	Delta Gold	Specialty	Dual purpose
1980	Crystal	White	Fresh
1980	Dakchip	White	Chip
1980	All Red	Specialty	Fresh
1980	Lemhi Russet	Russet	Dual purpose
1980	Belchip	White	Chip
1981	Highlat Russet	Russet	Fresh
1981	Rosa	White	Dual purpose
1981	Belrus	Russet	Dual purpose
1981	ChipBelle	White	Chip
1981	RhineRed	Red	Fresh
1982	Sangre	Red	Fresh
1983	Alasclear	White	Fresh
1983	Yankee Chipper	White	Chip
1983	Agassiz	Russet	Fresh
1983	Erik	Red	Fresh
1983	Redsen	Red	Fresh
1983	Candy Stripe	Specialty	Fresh
1984	Islander	White	Dual purpose
1984	Yankee Supreme	White	Dual purpose
1984	Reddale	Red	Fresh
1984	Tolaas	Russet	Dual purpose
1984	Russette	Russet	Dual purpose
1985	Campbell 14	White	Fresh
1985	Sunrise	White	Dual purpose
1985	Elba	White	Fresh
1985	Hampton	White	Fresh
1985	NorKing Russet	Russet	Dual purpose
1985	Langlade	White	Dual purpose
1986	FL 657	White	Chip
1987	CalGold	Specialty	Fresh
1987	Ute Russet	Russet	Fresh
1987	Russet Norkotah	Russet	Fresh
1987	Tejon	White	Fresh
1987	HiLite Russet	Russet	Dual purpose
1988	Krantz	Russet	Fresh
1988	Coastal Russet	Russet	Fresh
1989	Iditared	Red	Fresh
1989	Saginaw Gold	Specialty	Dual purpose
1989	Kanona	White	Chip
1989	NemaRus	Russet	Fresh
1990	Norwis	White	Chip
1990	Somerset	White	Chip
1990	Eide Russet	White	Fresh
1990	Allegany	White	Dual purpose
1990	Steuben	White	Dual purpose
1990	Snowden	White	Chip
1991	LaBelle	White	Dual purpose
1991	Castile	White	Dual purpose

Table continued.

Table 8–1. Continued.

Year	Name	Market class	Use
1991	Frontier Russet	Russet	Dual purpose
1991	Gemchip	White	Chip
1992	Russet Nugget	Russet	Dual purpose
1992	Fontenot	Red	Fresh
1992	Michigold	Specialty	Fresh
1992	Spartan Pearl	White	Dual purpose
1992	Goldrush	Russet	Fresh
1992	Norqueen Russet	Russet	Fresh
1992	Ranger Russet	Russet	Dual purpose
1992	Coastal Chip	White	Chip
1993	Red Cloud	Red	Fresh
1993	Genesee	White	Fresh
1993	Chipeta	White	Chip
1993	Suncrisp	White	Dual purpose
1994	Itasca	White	Dual purpose
1994	Mainechip	White	Chip
1994	Prestile	White	Fresh
1994	Red Ruby	Red	Fresh
1994	Norgold Russet	Russet	Fresh
1994	Catalina	White	Fresh
1994	Gilroy	White	Dual purpose
1994	Mindy	White	Dual purpose
1995	Portage	White	Fresh
1995	NorDonna	Red	Fresh
1996	Bonaparte	Specialty	Fresh
1996	St. Johns	White	Fresh
1996	NorValley	White	Chip
1997	CalWhite	White	Dual purpose
1997	Mainestay	White	Fresh
1998	Andover	White	Dual purpose
1998	Pike	White	Chip
1999	Cherry Red	Red	Fresh
1999	Quaggy Joe	White	Fresh
1999	Keuka Gold	Specialty	Fresh
1999	Reba	White	Dual purpose
1999	Dakota Pearl	White	Chip
2000	Calred	Red	Fresh
2000	Century Russet	Russet	Fresh
2000	Umatilla Russet	Russet	Dual purpose
2000	Winema	Red	Fresh
2000	Russet Legend	Russet	Dual purpose
2000	Silverton Russet	Russet	Processing
2001	Jacqueline Lee	White	Fresh
2001	Liberator	White	Chip
2001	Michigan Purple	Specialty	Fresh
2001	Eva	White	Dual purpose
2001	Klamath Russet	Russet	Fresh
2001	Mazama	Red	Fresh
2001	Amey	Russet	Dual purpose
2002	FL 1889	White	Chip

Table continued.

Table 8–1. Continued.

Year	Name	Market class	Use
2002	FL 1930	White	Chip
2002	FL 1944	White	Chip
2002	Gem Russet	Russet	Dual purpose
2002	IdaRose	Red	Fresh
2002	Ivory Crisp	White	Chip
2002	Bannock Russet	Russet	Dual purpose
2002	Red Companion	Red	Fresh
2003	Boulder	White	Dual purpose
2003	Modoc	Red	Fresh
2003	Wallowa Russet	Russet	Dual purpose
2003	Alturas	Russet	Dual purpose
2004	FL 1867	White	Chip
2004	FL 1879	White	Chip
2004	FL 1901	White	Chip
2004	FL 1922	White	Chip
2004	Monticello	White	Chip
2004	Dakota Jewel	Red	Fresh
2004	Red Pearl	Red	Fresh
2005	FL 1900	White	Chip
2005	FL 2006	White	Chip
2005	FL 2027	White	Chip
2005	Summit Russet	Russet	Dual purpose
2005	Dakota Crisp	White	Chip
2005	Dakota Diamond	White	Chip
2005	Sierra Gold	Specialty	Fresh
2005	Purple Majesty	Specialty	Fresh
2005	Millennium Russet	Russet	Dual purpose
2006	FL 2048	White	Chip
2006	FL 2049	White	Chip
2006	FL 2053	White	Chip
2006	FL 2072	White	Chip
2006	GemStar Russet	Russet	Fresh
2006	Western Russet	Russet	Dual purpose
2006	Reeves Kingpin	Russet	Processing
2006	Beacon Chipper	White	Chip
2006	Marcy	White	Chip
2006	Dakota Rose	Red	Fresh
2006	Defender	Russet	Processing
2006	Villetta Rose	Red	Fresh
2006	White Pearl	White	Chip
2007	Blazer Russet	Russet	Dual purpose
2007	Adirondack Blue	Specialty	Fresh
2007	Adirondack Red	Specialty	Fresh
2007	MegaChip	White	Chip
2007	Canela Russet	Russet	Fresh
2008	FL 2086	White	Chip
2008	Willamette	White	Dual purpose
2008	Premier Russet	Russet	Dual purpose
2009	FL 2085	White	Chip
2009	FL 2101	White	Chip

Table continued.

Table 8–1. Continued.

Year	Name	Market class	Use
2009	Highland Russet	Russet	Processing
2009	Kalkaska	White	Chip
2009	Missaukee	White	Chip
2009	Purple Haze	Specialty	Fresh
2009	Dakota Trailblazer	Russet	Dual purpose
2009	Crimson Red	Red	Fresh
2009	Purple Pelisse	Specialty	Fresh
2009	Sunset Red	Red	Fresh
2009	TerraRosa	Specialty	Fresh
2009	Freedom Russet	Russet	Dual purpose
2010	Classic Russet	Russet	Fresh
2010	Red Maria	Red	Fresh
2010	Clearwater Russet	Russet	Dual purpose
2010	Yukon Gem	Specialty	Fresh
2010	Sierra Rose	Red	Fresh
2011	Alpine Russet	Russet	Processing
2011	Palisade Russet	Russet	Dual purpose
2011	Teton Russet	Russet	Dual purpose
2011	Huckleberry Gold	Specialty	Fresh
2012	FL 2126	White	Chip
2012	FL 2137	White	Chip
2012	FL 2215	White	Chip
2012	Lamoka	White	Chip
2012	Waneta	White	Chip
2012	Dakota Russet	Russet	Dual purpose
2012	Owyhee Russet	Russet	Dual purpose
2012	FF x RG 387	Specialty	Fresh
2012	Amarosa	Specialty	Fresh
2012	Accumulator	White	Chip
2012	Lelah	White	Chip
2012	Nicolet	White	Chip
2012	Tundra	White	Chip

crop has been used for processing, mainly frozen, chip, and dehydrated products (Fig. 8–9). Processors pay a premium for specific gravity (dry matter content) within a desirable range (Johnson et al., 2010). Other important factors are tuber size, shape and size distribution, a lack of bruising and internal defects, sensory qualities (taste, aroma, and texture of processed product), and low reducing sugar levels, both at harvest and after storage. High amounts of the reducing sugars glucose and fructose result in dark colored products when fried (Malone et al., 2006). Dark chips and fries are undesirable to consumers.

Fresh market potatoes include round tubers with white flesh and white or red skin, specialty types including fingerings and tubers with yellow, red, or blue flesh, and long russets. Russet potatoes are typically sold to institutional buyers in "count carton" packages. The "count" refers to the number of potatoes of similar size in the standard size carton (Johnson et al., 2010). Consequently, varieties that produce tubers of a consistent size are desirable. For example, the cultivar Russet Norkotah can be a challenge to grow, mainly due to its susceptibility to diseases such as Verticillium wilt and *Potato virus Y*. However, it is popular because of

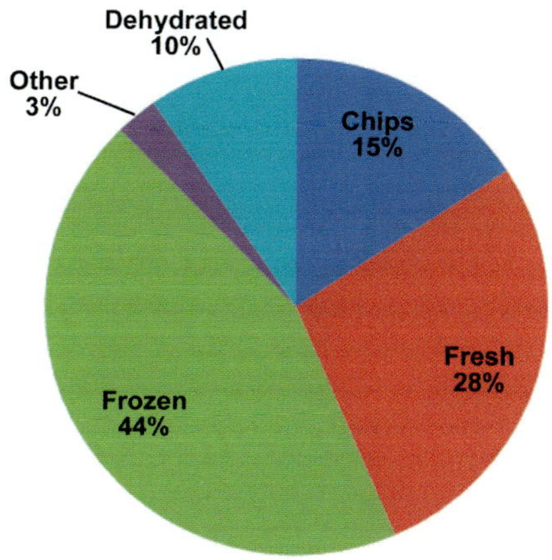

Fig. 8–9. Major uses of the U.S. potato crop in 2010 (U.S. Potato Board, www.uspotatoes.com, accessed 1 Apr. 2014).

its exceptional tuber appearance and uniformity. Red, white, and specialty potatoes are more often sold in bags containing tubers of variable size. While specific gravity and sugar levels are not major fresh market considerations, appearance is critical. Cultivars must be resistant to diseases that cause blemishes, such as common scab (caused by *Streptomyces scabies* Thaxter), black scurf (caused by *Rhizoctonia solani* Kühn) and silver scurf (caused by *Helminthosporium solani* Durieu & Mont.). In addition, red-skinned cultivars that retain their color during storage are rare, but highly desirable.

A study comparing modern with historical cultivars was not able to detect genetic improvements in yield or specific gravity during the 20th century (Douches et al., 1996). The authors concluded that a century of potato breeding had not resulted in genetic advances for these traits. This study used seven modern varieties released between 1976 and 1991, and it was performed at a single site in Michigan. Consequently, it may not have detected yield advances in other cultivars released during that time period, in more recent cultivars, or in the same cultivars grown in other parts of the country, especially the West where the growing season is longer and average yields are higher. In addition, while yield and dry matter content are critical for the success of new processing cultivars, a number of additional agronomic, disease resistance, storage, and quality traits are also considered. Consequently, a discussion of genetic gain for yield is only a part of the potato-breeding story. For example, early vine maturity can increase profitability by capturing a high value market, escaping disease, and extending the harvest season. Many new cultivars are earlier to mature than older varieties (Douches et al., 1996). It is difficult to combine early maturity with high yield, however, because the tuber-bulking window is narrow. Consequently, yield stasis may be expected when breeding for earlier maturing cultivars. Tubers of fresh market cultivars released in recent decades are more uniform in size and smooth in appearance than those grown in the early part of the 20th century, although yields are not necessarily higher. Percent marketable yield provides a better measure of breeding progress than total yield, since it determines the farmer's profitability. Most modern cultivars produce a high proportion of tubers that meet grading standards for both the fresh and processing markets, but this characteristic is not exclusive to recent releases (Douches et al., 1996).

Despite an emphasis on disease resistance by U.S. breeding programs, this trait is not a predominant feature in commercially significant new cultivars. Several new cultivars have been released with disease or pest resistances as a distinctive feature (Novy et al., 2006; Groza et al., 2009; Douches et al., 2001; Goth et al., 1989; Plaisted et al., 1999; Holm et al.,1987), but none have become major cultivars. Cultural practices are still generally economical and reliable enough for disease control, so traits other than disease resistance determine the success of new cultivars. Popular new cultivars, such as Umatilla Russet, Dakota Pearl, and Dakota Rose exhibit no exceptional disease or pest resistance (Mosley et al., 2000; Thompson et al., 2005, 2006). They have found a niche due to exceptional quality traits. Russet Norkotah is highly susceptible to *Potato virus Y* and Verticillium wilt, but it is grown widely because it has a desirable tuber appearance. Clonal line selections of Russet Norkotah with increased resistance to *Potato virus Y* and Verticillium wilt have replaced standard Russet Norkotah in many growing areas. These line selections are higher yielding than standard Russet Norkotah, and increased disease resistance may contribute to yield increases under some circumstances (Miller et al., 1999; Jansky and Miller, 2010; Whitworth et al., 2010). Major quick service restaurants and food processors are placing increasing emphasis on the sustainable production of their products. This may result in disease resistance becoming more important to the potato industry if there is increased financial pressure to decrease pesticide and fungicide application.

The predominant market class in most major production states is russets for french fry processing. The dominant cultivar in the United States is Russet Burbank. The french fry processing industry has been built around this cultivar, and the shift to new varieties has been slow. This is in part because tubers of Russet Burbank, unlike those of most other varieties, can be stored for up to 12 mo and used to make high-quality french fries. However, Russet Burbank has a number of problems (Pavek et al., 1992). It has a high requirement for nitrogen fertilizer and is susceptible to major diseases such as Verticillium wilt and virus. In addition, heat and water stress commonly cause malformed tubers and serious internal defects. Across many growing locations, only 55 to 85% of total harvested yield for Russet Burbank is graded as U.S. No. 1 (Whitworth et al., 2011; Novy et al., 2002, 2008). Consequently, Russet Burbank is being displaced by new cultivars.

Current potato production has a much more diverse cultivar base than it did 30 yr ago. This is especially apparent when fall production by cultivar is considered on a state by state basis. The major potato production states Idaho and Washington in the West, Wisconsin and North Dakota in the Midwest, and Maine in the East. The 1986 NASS report lists only the top four cultivars, but in 2012 it provides names of all cultivars that occupy 1% or more of land planted. Consequently, it is not possible to use these data to compare diversity of cultivars in 1986 with that in 2012. However, in many states in 1986, the top four cultivars were responsible for most of the production, indicating little diversity in cultivars grown at that time.

In the West, the land area planted to Russet Burbank has decreased dramatically since 1986. New varieties now grown in place of Russet Burbank include Russet Norkotah, Ranger Russet, and Umatilla Russet (Fig. 8–10 and 8–11). Russet Norkotah was released by the North Dakota breeding program in 1987 (Johansen et al., 1988). In yield trials performed in North Dakota, it out-yielded Russet

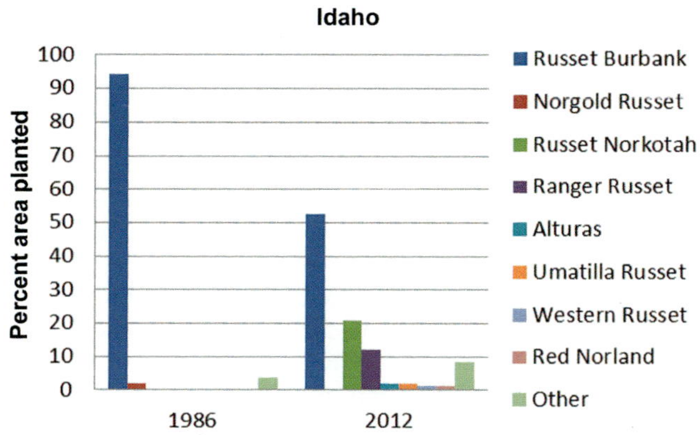

Fig. 8–10. Percent of production area planted to major cultivars in Idaho, 1986 and 2012.

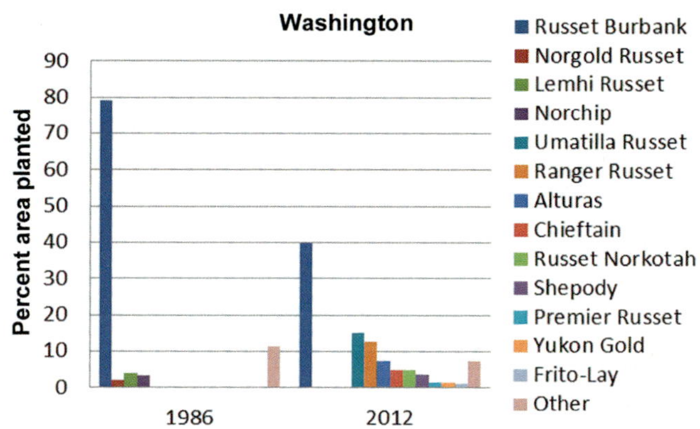

Fig. 8–11. Percent of production area planted to major cultivars in Washington, 1986 and 2012.

Burbank by an average of 7 t ha^{-1}. Russet Norkotah is also much earlier to mature than Russet Burbank. It can be processed out of the field or after a short storage period, but is typically used for fresh consumption. Russet Norkotah has become the standard for early, fresh market russets, producing a high proportion of U.S. No. 1 tubers. It supplies much of the count carton market for russet baking potatoes.

Ranger Russet was released by the USDA and agricultural experiment stations in the West in 1991 (Pavek et al., 1992). Russet Burbank is in the pedigree of two of its great-grandparents. In trials performed in the West, Ranger Russet consistently produced higher total yields, U.S. No. 1 yields, and specific gravity than Russet Burbank (Table 8–2). Solids accumulate faster during tuber bulking, and marketable tubers are produced about 2 wk earlier than Russet Burbank tubers. Ranger Russet tubers also have fewer internal defects, such as hollow heart and heat necrosis, produce lower levels of reducing sugars, and are less susceptible to sugar-end defects than Russet Burbank.

Table 8–2. Mean yield (t ha^{-1}) and specific gravity (a measure of dry matter) of Ranger Russet compared to Russet Burbank in trials from Idaho, Colorado, Oregon and Washington (adapted from Pavek et al., 1992).

	Ranger Russet	Russet Burbank
Total yield	55	53
U.S. No. 1 yield	44	34
Specific gravity	1.089	1.082

Umatilla Russet was also released by the USDA and agricultural experiment stations in the West in 1998 (Mosley et al., 2000). It produces a high proportion of high-quality marketable tubers for the french fry industry. In yield trials performed in the West, it out-yielded Russet Burbank by an average of 21%, based on marketable yield. It has fewer internal and external defects than Russet Burbank. Both Ranger Russet and Umatilla Russet produce French fries that meet quality specifications set by several large quick service restaurant chains. This has generated a strong demand for these cultivars.

In the Midwest, Russet Burbank production in Wisconsin in 2012 was one-third of what it was in 1986, but Russet Burbank production in North Dakota increased threefold during the same period (Fig. 8–12 and 8–13). However, the 2012 production data reveal that several new varieties have expanded acreage in the Midwest, including Russet Norkotah, Ranger Russet, Umatilla Russet, Goldrush (Johansen et al., 1994), and Silverton Russet. New chip processing cultivars have made an impact in the Midwest as well. They include Prospect (a cultivar developed in Canada), Ivory Crisp (Love et al., 2003), Dakota Pearl (Thompson et al., 2005), and Snowden. Large gains have been made in breeding for processing quality traits in these cultivars, especially low reducing sugar content (Douches et al., 1996; Love et al., 1998), but similar gains in yield have not occurred (Table 8–3).

In Maine, as in North Dakota, Russet Burbank production has increased dramatically in recent decades (Fig. 8–14). This is mainly the result of a need to supply a major french fry processing plant in New Brunswick. The 2012 production pattern by cultivar in Maine is similar to that in North Dakota, with Russet

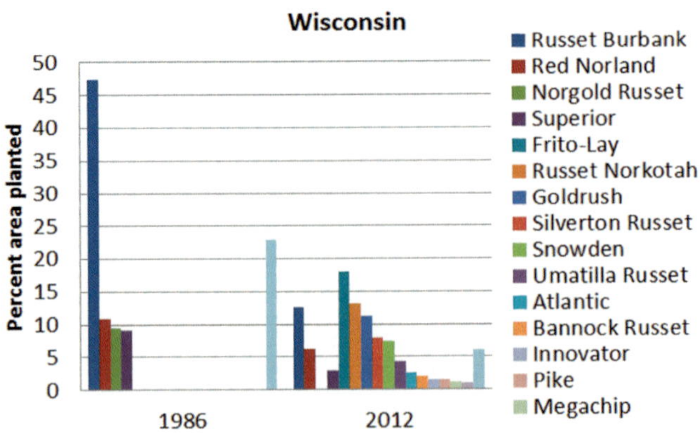

Fig. 8–12. Percent of production area planted to major cultivars in Wisconsin, 1986 and 2012.

Burbank dominating the area planted, and a large number of additional cultivars occupying small niches. New cultivars include Russet Norkotah (Johansen et al., 1988), Snowden, Goldrush, Blazer Russet (Stark et al., 2007), and Innovator.

As seen in Fig. 8–3, 8–4, and 8–5, yield per unit area has increased over time. It is impossible to determine precisely the relative contributions of cultivar, environment, and the cultivar × environment interaction to these gains. However,

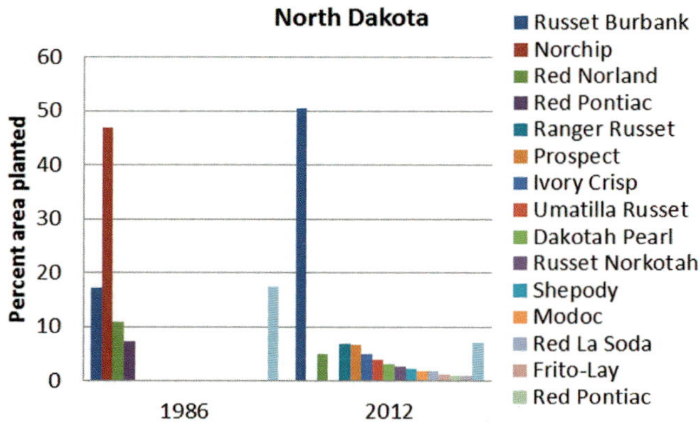

Fig. 8–13. Percent of production area planted to major cultivars in North Dakota, 1986 and 2012.

Table 8–3. Mean yield and specific gravity (a measure of dry matter) of potato cultivars released in four time periods (adapted from Douches et al., 1996).

	1861–1893	1930–1949	1950–1969	1970–1991
Total yield, t ha^{-1}	52.8	60	43.25	45.6
U.S. No. 1 yield, t ha^{-1}	38	53	33.75	38.8
Specific gravity	1.076	1.069	1.069	1.075

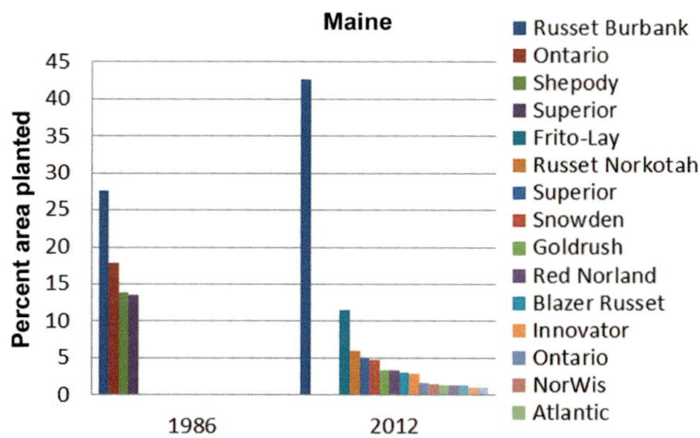

Fig. 8–14. Percent of production area planted to major cultivars in Maine, 1986 and 2012.

one way to dissect the genetic component out of total yield gain data is to determine the relative proportions of land area occupied by new cultivars over time and compare them to yield increases. In 1992, USDA-NASS began to report, by state, all cultivars that occupy at least 1% of land area planted. Using these data, we determined the proportion of total planted area occupied by cultivars released in 1980 or later (Fig. 8–15). In four of the five major potato producing states, the proportion of land area planted to new cultivars has increased in the past two decades. This proportion has actually decreased in Maine with time. The fastest rate of gain in land planted to new varieties is in the state of Washington (1.7% yr^{-1}). Idaho is similar, with a gain of 1.6% yr^{-1}. Gains in Wisconsin and North Dakota are similar, at just under 1% yr^{-1}. This graph is similar to a graph of yield per acre in these states for the same time period (Fig. 8–4). High adoption rates of new cultivars in Idaho and Washington are reflected by higher yield gains across the same time period. In contrast, Maine has decreased its production of new varieties in the past two decades and has realized the slowest yield gain by far. These data suggest that even at the very coarse level of total yield, adoption of new varieties in recent decades has been accompanied by yield gains.

The 182 potato varieties that have been released in the United States since 1980 include 88 white, 49 russet, 25 red, and 20 specialty cultivars (Table 8–1). Most of the white cultivars were developed for the chip industry, and a major feature of many of them is the ability to produce light-colored chips following storage at temperatures of 9°C or less. The most successful new chip cultivars developed by public breeding programs are Snowden and Dakota Pearl, for which 4471 and 2439 ha, respectively, were planted in 2012. Dakota Pearl is similar to older chip cultivars in yield and specific gravity, but produces light chips after up to 7 mo in cold storage (Thompson et al., 2005). This provides producers with flexibility in the dates for which the crop is sent to the processing plant. Most new russet cultivars are dual purpose, but often better suited for either the fresh or french fry market. The most successful new russet cultivars are Ranger Russet, Russet Norkotah and line selections made from it, Umatilla Russet and Alturas. Ranger Russet and Umatilla Russet have established themselves in the french fry market due to agronomic qualities and the ability to produce fries that

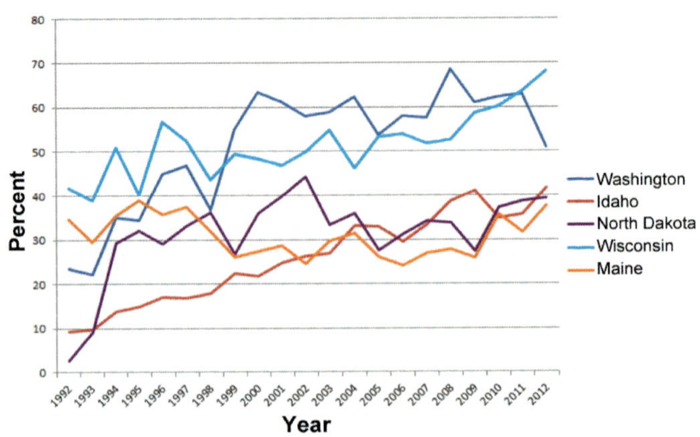

Fig. 8–15. Percent of production area in five states planted to new cultivars (1980 or later), 1992 through 2012.

have the taste and texture demanded by several large restaurant chains. Alturas is a late-maturing cultivar but is superior to Russet Burbank for yield and specific gravity (Novy et al., 2003). It addition, it is resistant to cold-induced sweetening, allowing tubers to be processed after low temperature storage. The most successful new red-skinned cultivars are Dakota Rose, Rosara, and Modoc. The success of these cultivars is likely due in large part to their ability retain bright red skin color during storage, extending marketable shelf life (Thompson et al., 2006; Rykbost et al., 2003).

Most cultivars have been released from public (federal and state) programs and are freely available to growers. The only major private potato breeding program is operated by Frito-Lay, which focuses on round white cultivars to supply its chip processing needs (FL clones in Table 8–1). The Plant Variety Protection Act (PVPA) was amended in 1994 to add protection to potatoes and other tuber crops. The PVPA allows breeders to recover the cost of developing new varieties through licensing fees and royalties. It has changed the cultivar release landscape. In 2005, the Potato Variety Management Institute (PVMI) was created by the potato commissions in Washington, Oregon, and Idaho to promote varieties developed by joint breeding efforts in those states. License fees and royalties from PVMI cultivars support the breeding programs and marketing efforts. Another likely effect of PVPA is that the rate of cultivar release has increased, from 42 cultivars in the 1980s, to 44 cultivars in the 1990s and finally 96 cultivars since 2000 (Table 8–1).

Transgenic Potatoes

Potato is easily amenable to in vitro manipulation, so it is a good candidate for transgenic modification (Vayda and Belknap, 1992). Genetically modified potato varieties with improved resistance to abiotic stress, fungal and bacterial diseases, or pests could potentially increase tuber yields. However, transgenic potatoes are not currently being grown in the United States. In the late 1990s, NatureMark, a subsidiary of the Monsanto company, developed transgenic potatoes under the NewLeaf brand name (Kaniewski and Thomas, 2004). The initial goal was to introduce *Potato leaf roll virus* resistance into Russet Burbank. That work was successful and produced highly resistant lines by expressing the *Potato leaf roll virus* replicase gene in potato (Thomas et al., 2000). Parallel research incorporated the *Bt* gene conferring resistance to Colorado potato beetle [*Leptinotarsa decemlineata* (Say)] into the varieties Russet Burbank, Snowden, Atlantic, and Superior (Kaniewski and Thomas, 2004; Perlak et al., 1993). NewLeaf potato lines with Colorado potato beetle resistance were commercially available in 1995. Potato lines that combined Colorado potato beetle and *Potato leaf roll virus* resistance (NewLeaf Plus) and Colorado potato beetle and *Potato virus Y* resistance (NewLeaf Y) were also developed and were commercialized in 1998 (Duncan et al., 2002; Kaniewski and Thomas, 2004). NewLeaf potatoes were grown in the Pacific Northwest, while commercial production of NewLeaf Y was concentrated in the central and eastern United States. Growers of NewLeaf potatoes benefited from lower insecticide costs for controlling Colorado potato beetle and aphids, which carry viruses. However, in 1999, a strong anti-biotechnology campaign was launched against NewLeaf potatoes. Under consumer pressure, major buyers of potatoes for processing, including McDonald's, Wendy's, McCain, and Frito-Lay, prohibited the use of genetically modified potatoes in their products. In addition, antibiotechnology sentiments in Europe and Asia affected the export market. By the end of 2001, the NatureMark Company was dissolved.

Recently, the J.R. Simplot company has developed a research program that does not use genes from outside of the potato gene pool for the construction of transgenic lines (Rommens et al., 2007). This technology strategy, trademarked Innate, is focusing initially on consumer-oriented traits, including reduced bruising and light-colored fry products (Rommens et al., 2008). By focusing on traits for which consumers can see a benefit, and educating the public about the value of transgenic crops, these scientists hope to re-establish transgenic potatoes in the United States.

Environmental Factors Interacting with Genetic Gain

The USDA-NASS crop production reports describe major environmental factors that impact fall potato production each year. Those comments are summarized in Table 8–4, beginning with 1985, when they were first reported. Water problems are the most prevalent environmental production constraint. Yield reductions of 20% or more were common due to drought in the late 1980s and flooding in the 1990s and 2000s. Drought obviously affects dryland production, for example in the Red River Valley of North Dakota and Minnesota. However, a lack of irrigation water in some regions also impacts production. For example, drought in the West at the turn of the 21st century led to a shortage of irrigation water. In 2001, this lack of water for irrigated production reduced potato yields in California by

Table 8–4. Examples of major constraints for U.S. fall potato production since 1985. Data obtained from USDA NASS reports.

Year	Production constraint
1985	Frost damage in MT, WY, and eastern ID
1988	Drought in the Midwest; water shortage and frost in OR
1989	Drought in the Red River Valley of ND and MN
1990	Drought in the Red River Valley of ND and MN
1991	Heat in the East
1992	Warm harvest temperatures in ME
1993	Flooding in the Midwest
1994	Field and storage breakdown in the East; considerable acreage in ME not harvested
1995	Late blight losses in the Pacific Northwest; heat in the East
1996	Flooding in MI
1997	Flooding in the Red River Valley of ND and MN
1998	Summer heat in the West
1999	Flooding in the Red River Valley; drought in ME, PA, OH and IN
2000	Flooding in the Midwest
2001	Drought in the West
2003	Heat in CA
2005	Flooding in the Midwest
2007	Potato virus Y in the West

70% and in Oregon by 32%. The past two decades have been marked by a significant number of serious floods in the Midwest. Flooded fields are often abandoned because when tubers are harvested from flooded fields, disease pressure in storage increases and storage quality suffers. In 1993, more than 20,000 ha of potatoes in the Red River Valley were abandoned due to flooding.

Temperature appears to be the second most significant environmental production constraint. Spring frost damage in the West in 1985 led to the loss of 4,000 ha. Excessive heat during the production season impacts both yield and quality. Heat at harvest is a problem because tubers cannot be sufficiently cooled in storage facilities. In 1992, warm fall temperatures resulted in large reductions in marketable yield in Maine due to storage losses. Finally, severe disease causes large yield reductions, as seen by tuber breakdown in the East in 1994, late blight in the West in 1995, and *Potato virus Y* in the West in 2007.

As a tuber crop, potato is vulnerable to large losses due to both disease and physiological stresses. Losses in marketable yield as a result of spoilage, dehydration, and respiration in storage are commonly 5% or more (National Potato Council, 2013). In addition, potatoes that do not meet market standards due to bruising, greening, sprouting, and tuber disease are counted as losses. Average harvest loss, based on data available from 2008 through 2012, ranged from 32.6% in North Dakota to 10.8% in Wisconsin.

Future Prospects

Global climate change models predict a decrease in potato yields throughout much of the United States (Hijmans, 2003). In most regions, yield losses are expected, mainly due to increases in temperature throughout the growing season. Yield losses in the United States in the middle decades of the 21st century are predicted to be 33% if production practices and cultivars do not change. Predicted losses decrease to 6% if planting dates and cultivar maturity class are changed to minimize exposure to high temperatures. Heat tolerance is expected to increase yields by 5% in most regions. Consequently, potato farmers will likely benefit if breeders add heat tolerance to their selection criteria. If irrigation water becomes less available and rainfall becomes more sporadic, it will also be important to improve water use efficiency in potato. Finally, it will become imperative for potato breeders to develop cultivars with higher nitrogen use efficiency to offset the increasing cost of nitrogen fertilizer and minimize groundwater contamination with nitrates.

The potato crop is well positioned to utilize diverse germplasm resources to meet future production demands. Abundant, accessible genetic diversity is readily available to breeders through a worldwide network of gene banks (Jansky et al., 2013). Germplasm is available to contribute to breeding efforts to improve heat tolerance, drought tolerance, and nitrogen use efficiency (Reynolds and Ewing, 1989; Errebhi et al., 1999; Cabello et al., 2012). In addition, while disease resistance is currently not a major determinant of cultivar success, that scenario is changing in the face of tightening environmental regulations, emergence of new pests and pathogens, and increasing demand for sustainable production practices. Systematic efforts are needed to survey genetic diversity in potato relatives to efficiently identify sources of valuable traits or alleles (Jansky et al., 2009; Chung et al., 2011; Spooner et al., 2009; Jansky et al., 2006, 2008; Cai et al., 2011). This germplasm can then be introgressed into adapted parental clones for use in cultivar improvement (Jansky et al., 2012; De Jong et al., 2001; Jansky et al., 2011; Domanski et al., 2004). Breeding programs in the United States share populations of early generation clones to allow evaluation in multiple environments. This effort will become increasingly important if state and federal funding resources continue to shrink. Innovative breeding strategies such as sexual polyploidization, parental line testing, diploid breeding, and inbred line development have the potential to move potato breeding forward at a faster rate in the 21st century (Buso et al., 1999; Alberino et al., 2004; Bradshaw, 2009; Lindhout et al., 2011).

The sequencing of the potato genome in 2011 opened new opportunities to utilize genomics for potato improvement (Xu et al., 2011). The USDA-funded SolCAP project was developed to provide genomic resources accessible to potato breeders. It has provided insights into the genetic composition of potato cultivars and is leading to the development of mapping resources for marker development (Hirsch et al., 2013). A major effort of the SolCAP project was to genotype and phenotype a so-called diversity panel, which is a collection of about 300 clones relevant to the breeding community. This resulted in data on a common set of clones that continues to grow as researchers utilize this resource. The development of community resources such as the diversity panel will benefit all researchers and should serve as a model for the future.

References

Alberino, S., D. Carputo, G. Caruso, M.R. Ercolano, and L. Frusciante. 2004. Field performance of families and clones obtained through unilateral sexual polyploidization in potato (*Solanum tuberosum*). Adv. Hortic. Sci. 18:47–52.

Bradshaw, J.E. 2009. Potato breeding at the Scottish Plant Breeding Station and the Scottish Crop Research Institute: 1920–2008. Potato Res. 52:141–172. doi:10.1007/s11540-009-9126-5

Buso, J.A., L.S. Boiteux, and S.J. Peloquin. 1999. Multitrait selection system using populations with a small number of interploid (4x-2x) hybrid seedlings in potato: Degree of high-parent heterosis for yield and frequency of clones combining quantitative agronomic traits. Theor. Appl. Genet. 99:81–91. doi:10.1007/s001220051211

Cabello, R., F. Mendiburu, M. Bonierbale, P. Monneveux, W. Roca, and E. Chujoy. 2012. Large-scale evaluation of potato improved varieties, genetic stocks and landraces for drought tolerance. Am. J. Potato Res. 89:400–410. doi:10.1007/s12230-012-9260-5

Cai, X., D. Spooner, and S. Jansky. 2011. A test of taxonomic and biogeographic predictivity: Resistance to *Potato virus Y* in wild relatives of the cultivated potato. Phytopathology 101:1074–1080. doi:10.1094/PHYTO-02-11-0060

Carputo, D., and L. Frusciante. 2011. Classical genetics and traditional breeding. In: J.M. Bradeen and C. Kole, editors, Genetics, Genomics, and Breeding of Potato. CRC Press, Enfield, NH. p. 20–40.

Chung, Y.S., K. Holmquist, D.M. Spooner, and S.H. Jansky. 2011. A test of taxonomic and biogeographic predictivity: Resistance to soft rot in wild relatives of cultivated potato. Phytopathology 101:205–212. doi:10.1094/PHYTO-05-10-0139

De Jong, H., K.G. Proudfoot, and A.M. Murphy. 2001. The germplasm release of F87084, a fertile, adapted clone with multiple disease resistances. Am. J. Potato Res. 78:141–149. doi:10.1007/BF02874769

Domanski, L., E. Zimnoch-Guzowska, M. Domanska, K. Zgorska, and M. Paczowska. 2004. The germplasm release of M-62774 and M-62805, two potato clones with cold-sweetening resistance. Folia Hortic. 16(2):33–40.

Douches, D., K. Jastrzebski, J. Coombs, W. Kirk, K. Felcher, R. Hammerschmidt, and R. Chase. 2001. Jacqueline Lee: A late-blight-resistant tablestock variety. Am. J. Potato Res. 78:413–419. doi:10.1007/BF02896372

Douches, D.S.S., D. Maas, K.R. Jartrzebski, R.W. Chase, and K. Jastrzebski. 1996. Assessment of potato breeding progress in the USA over the last century. Crop Sci. 36:1544–1552. doi:10.2135/cropsci1996.0011183X003600060024x

Duncan, D.R., D. Hammond, J. Zalewski, J. Cudnohufsky, W. Kaniewski, M. Thornton, J.T. Bookout, P. Lavrik, G.J. Rogan, and J. Feldman-riebe. 2002. Field performance of transgenic potato, with resistance to Colorado potato beetle and viruses. HortScience 37:275–276.

Errebhi, M., C. Rosen, and F. Lauer. 1999. Evaluation of tuber-bearing Solanum species for nitrogen use efficiency and biomass partitioning. Am. J. Potato Res. 76:143–151. doi:10.1007/BF02853579

Goth, R., J. Shumaker, R. Webb, D. Wilson, R. Moore, L.L. Sanford, and K.G. Haynes. 1989. Nemarus: A golden nematode resistant, russet-skin potato variety. Am. Potato J. 66:703–707. doi:10.1007/BF02896826

Groza, H.I., B.D. Bowen, A.J. Bussan, F.M. Navarro, W.R. Stevenson, J.P. Palta, and J. Jiang. 2009. Freedom Russet—A dual purpose russet potato cultivar with resistance to common scab and good fry quality. Am. J. Potato Res. 86:406–414. doi:10.1007/s12230-009-9096-9

Hijmans, R.J. 2003. The effect of climate change on global potato production. Am. J. Potato Res. 80:271–280. doi:10.1007/BF02855363

Hirsch, C., C. Hisrch, K. Felcher, J. Coombs, D. Zarka, A. VanDeynze, W. DeJong, R. Veilleux, S. Jansky, P. Bethke, D. Douches, and C. Buell. 2013. Retrospective view of North American potato (*Solanum tuberosum* L.) breeding in the 20th and 21st centuries. G3 3(6):1003–1013. doi:10.1534/g3.113.005595.

Holm, D., J. Twomey, and M. Workman. 1987. Ute Russet: An oblong russet potato cultivar with resistance to leafroll net necrosis and *Sclerotium rolfsii*. Am. J. Potato Res. 64:559–562. doi:10.1007/BF02853755

Jansky, S., H. Dempewolf, E. Camadro, M. Bonierbale, E. Zimnoch-Guzowska, D. Bisnognin, and S. Reinhard. 2013. A case for crop wild relative preservation and use in potato (*Solanum tuberosum* L.). Crop Sci. 53: 746–754. doi:10.2135/cropsci2012.11.0627

Jansky, S.H., A. Hamernik, and P.C. Bethke. 2011. Germplasm release: Tetraploid clones with resistance to cold-induced sweetening. Am. J. Potato Res. 88:218–225. doi:10.1007/s12230-011-9186-3

Jansky, S., A. Hamernik, and Y.S. Chung. 2012. M7 germplasm release: A tetraploid clone derived from *Solanum infundibuliforme* for use in expanding the germplasm base for french fry processing. Am. J. Potato Res. 89:448–452. doi:10.1007/s12230-012-9266-z

Jansky, S.H., and J.C. Miller, Jr. 2010. Evaluation of Verticillium wilt resistance in Russet Norkotah and six strains selections. Am. J. Potato Res. 87:492–496.

Jansky, S.H., R. Simon, and D.M. Spooner. 2006. A test of taxonomic predictivity: Resistance to white mold in wild relatives of cultivated potato. Crop Sci. 46:2561–2570. doi:10.2135/cropsci2005.12.0461

Jansky, S.H., R. Simon, and D.M. Spooner. 2008. A test of taxonomic predictivity: Resistance to early blight in wild relatives of cultivated potato. Phytopathology 98:680–687. doi:10.1094/PHYTO-98-6-0680

Jansky, S.H., R. Simon, and D.M. Spooner. 2009. A test of taxonomic predictivity: Resistance to the Colorado potato beetle in wild relatives of cultivated potato. J. Econ. Entomol. 102:422–431. doi:10.1603/029.102.0155

Johansen, R.H., B. Farnsworth, D.C. Nelson, G.A. Secor, N. Gudmestad, and P.H. Orr. 1988. Russet Norkotah: A new russet-skinned potato cultivar with wide adaptation. Am. Potato J. 65:597–604. doi:10.1007/BF02908344

Johansen, R., B. Farnsworth, G. Secor, N. Gudmestad, A. Thompson-Johns, and E. Holm. 1994. Goldrush: A new high quality russet-skinned potato cultivar. Am. Potato J. 71:809–815. doi:10.1007/BF02849376

Johnson, S.B., N. Olsen, C. Rosen, and D.M. Spooner. 2010. Commercial potato production in North America. Am. J. Potato Res. 87(suppl):1–90.

Kaniewski, W.K., and P.E. Thomas. 2004. The potato story. AgBioForum 7:41–46.

Lindhout, P., D. Meijer, T. Schotte, R.C.B. Hutten, R.G.F. Visser, and H.J. Eck. 2011. Towards F1 hybrid seed potato breeding. Potato Res. 54:301–312. doi:10.1007/s11540-011-9196-z

Love, S., A. Mosley, R. Novy, D. Corsini, R. Thornton, S. James, and D. Hane. 2003. Ivory crisp: A potato variety with high tuber solids and cold chipping ability. Am. J. Potato Res. 80:207–213. doi:10.1007/BF02855692

Love, S.L., J.J. Pavek, A. Thompson-Johns, and W. Bohl. 1998. Breeding progress for potato chip quality in North American cultivars. Am. J. Potato Res. 75:27–36. doi:10.1007/BF02883514

Malone, J.G., V. Mittova, R.G. Ratcliffe, and N.J. Kruger. 2006. The response of carbohydrate metabolism in potato tubers to low temperature. Plant Cell Physiol. 47:1309–1322. doi:10.1093/pcp/pcj101

McGary, S.D. 2012. Fresh potato price volatility and efforts of a potato growers association to reduce the volatility through production controls. In: 22nd Annual World Forum and Symposium of the International Food and Agribusiness Management Association. p. 1–27.

Miller, J.C. Jr., D.C. Scheuring, and G.C.J. Fernandez. 1999. Selection, evaluation, and identification of improved Russet Norkotah strains. Am. J. Potato Res. 76:161–167.

Mosley, A.R., S.R. James, D.C. Hane, K.A. Rykbost, C.C. Shock, B.A. Charlton, J.J. Pavek, S.L. Love, D.L. Corsini, and R.E. Thornton. 2000. Umatilla Russet: A full season long Russet for processing and fresh market use. Am. J. Potato Res. 77:83–87. doi:10.1007/BF02853734

National Potato Council. 2013. 2013 potato statistical yearbook. NPC, Washington, DC.

Novy, R., D. Corsini, S. Love, J. Pavek, A. Mosley, S. James, D. Hane, C. Shock, K. Kykbost, C. Brown, and R. Thornton. 2003. Alturas: A multi-purpose, russet potato cultivar with high yield and tuber specific gravity. Am. J. Potato Res. 80:295–301. doi:10.1007/BF02854312

Novy, R., D. Corsini, S. Love, J. Pavek, A. Mosley, S. James, D. Hane, C. Shock, K. Rykbost, C. Brown, and R. Thornton. 2002. Bannock russet: A dual-purpose, russet potato cultivar with high U.S. No. 1 yield and multiple disease resistances. Am. J. Potato Res. 79:147–153. doi:10.1007/BF02881524

Novy, R., S. Love, D. Corsini, J. Pavek, J. Whitworth, A. Mosley, S. James, D. Hane, C. Shock, K. Rykbost, C. Brown, R. Thornton, N. Knowles, M. Pavek, N. Olsen, and D. Inglis. 2006. Defender: A high-yielding, processing potato cultivar with foliar and tuber resistance to late blight. Am. J. Potato Res. 83:9–19. doi:10.1007/BF02869605

Novy, R.G., J.L. Whitworth, J.C. Stark, S.L. Love, D.L. Corsini, J.J. Pavek, M.I. Vales, S.R. James, D.C. Hane, C.C. Shock, B.A. Charlton, C.R. Brown, N.R. Knowles, M.J. Pavek, T.L. Brandt, and N. Olsen. 2008. Premier Russet: A dual-purpose, potato cultivar with significant resistance to low temperature sweetening during long-term storage. Am. J. Potato Res. 85:198–209. doi:10.1007/s12230-008-9013-7

Ojala, J., J. Stark, and G. Kleinkopf. 1990. Influence of irrigation and nitrogen management on potato yield and quality. Am. J. Potato Res. 67:29–43. doi:10.1007/BF02986910

Patterson, P.E. 2012. 2012 cost of potato production in Idaho: 5-year trend. Agric. Econ. Ext. Ser. Series 12-04. University of Idaho Agricultural Economics Extension, Moscow, ID.

Pavek, J., D. Corsini, S. Love, D. Hane, D. Holm, W. Iritani, S. James, M. Martin, A. Mosley, J. Ojala, C. Stanger, and R. Thornton. 1992. Ranger Russet: A long Russet potato variety for processing and fresh market with improved quality, disease resistance, and yield. Am. J. Potato Res. 69:483–488. doi:10.1007/BF02853837

Peiterse, L., and U. Hils. 2009. World catalogue of potato varieties. Agrimedia GmbH, Bergen/Dumme, Germany.

Perlak, F., T. Stone, Y. Muskopf, L. Peterson, G. Parker, S. McPherson, J. Wyman, S. Love, G. Reed, D. Biever, and D. Fischoff. 1993. Genetically improved potatoes: Protection from damage by Colorado potato beetles. Plant Mol. Biol. 22:313–321. doi:10.1007/BF00014938

Plaisted, R., D. Halseth, B. Brodie, S. Slack, J. Sieczka, B. Christ, K. Paddock, and M. Peck. 1999. Reba : A mid to late season golden nematode resistant variety for use as tablestock or chipstock. Am. J. Potato Res. 76:1–4. doi:10.1007/BF02853550

Reynolds, M., and E. Ewing. 1989. Heat tolerance in tuber bearing *Solanum* species: A protocol for screening. Am. Potato J. 66:63–74. doi:10.1007/BF02854425

Rommens, C.M., M.A. Haring, K. Swords, H.V. Davies, and W.R. Belknap. 2007. The intragenic approach as a new extension to traditional plant breeding. Trends Plant Sci. 12:397–403. doi:10.1016/j.tplants.2007.08.001

Rommens, C.M., H. Yan, K. Swords, C. Richael, and J. Ye. 2008. Low-acrylamide french fries and potato chips. Plant Biotechnol. J. 6:843–853. doi:10.1111/j.1467-7652.2008.00363.x

Rykbost, K., S. James, A. Mosley, B. Charlton, D. Hane, E. Eldredge, R. Voss, R. Johansen, S. Love, and R. Thornton. 2003. Modoc: A potato variety with bright red skin and early maturity for fresh market. Am. J. Potato Res. 80:235–240. doi:10.1007/BF02855359

Spooner, D.M., S.H. Jansky, and R. Simon. 2009. Tests of taxonomic and biogeographic predictivity: Resistance to disease and insect pests in wild relatives of cultivated potato. Crop Sci. 49:1367–1376. doi:10.2135/cropsci2008.04.0211

Stark, J., R. Novy, S. Love, J. Whitworth, D. Corsini, J. Pavek, A. Mosley, M. Pavek, N. Knowles, R. Thornton, S. James, D. Hane, N. Olsen, M. Vales, and C. Brown. 2007. Blazer russet: An early to mid-season potato cultivar with high U.S. No. 1 yields and good processing and culinary qualities. Am. J. Potato Res. 84:447–457. doi:10.1007/BF02987881

Thomas, P., C. Lawson, J. Zalewski, G. Reed, and W. Kaniewski. 2000. Extreme resistance to *Potato leafroll virus* in potato cv. Russet Burbank mediated by the viral replicase gene. Virus Res. 71:49–62. doi:10.1016/S0168-1702(00)00187-8

Thompson, A., R. Novy, B. Farnsworth, G. Secor, N. Gudmestad, J. Sowokinos, E. Holm, J. Lorenzen, and D. Preston. 2005. Dakota Pearl: An attractive, bright white-skinned, cold-chipping cultivar with tablestock potential. Am. J. Potato Res. 82:481–488. doi:10.1007/BF02872226

Thompson, A., G. Secor, J. Lorenzen, B. Farnsworth, R. Novy, N. Gudmestad, E. Holm, and D. Preston. 2006. Dakota Rose: A bright red tablestock potato cultivar that retains its skin color in storage. Am. J. Potato Res. 83:317–323. doi:10.1007/BF02871592

Thornton, M. 2001. The rise and fall of NewLeaf potatoes. In: Biotechnology: Science and society at a crossroad. NABC Rep. 15. p. 235–243. http://nabc.cals.cornell.edu/pubs/nabc_15/chapters/Thornton.pdf (accessed 24 Jan. 2014).

Tiefenthaler, A., I. Goldman, and W. Tracy. 2003. Vegetable and corn yields in the United States, 1900-present. HortScience 38:1080–1082.

Vayda, M., and W. Belknap. 1992. The emergence of transgenic potatoes as commercial products and tools for basic science. Transgenic Res. 1:149–163. doi:10.1007/BF02522533

Whitworth, J., P. Hamm, and C. McIntosh. 2010. Effect of potato virus Y on yield of a clonal selection of Russet Norkotah. Am. J. Potato Res. 87:310–314.

Whitworth, J.L., R.G. Novy, J.C. Stark, J.J. Pavek, D.L. Corsini, S.L. Love, N. Olsen, S.K. Gupta, T. Brandt, M.I. Vales, A.R. Mosley, S. Yilma, S.R. James, D.C. Hane, B.a. Charlton, C.C. Shock, N.R. Knowles, M.J. Pavek, J.S. Miller, and C.R. Brown. 2011. Alpine Russet: A potato cultivar having long tuber dormancy making it suitable for processing from long-term storage. Am. J. Potato Res. 88:256–268. doi:10.1007/s12230-011-9190-7

Wiese, M. 1998. Use, target pests, and economic impact of pesticides applied to potatoes in the United States. USDA-NAPIAP Report 2-CA-98.

Xu, X., S. Pan, S. Cheng, B. Zhang, D. Mu, et al. 2011. Genome sequence and analysis of the tuber crop potato. Nature 475:189–195. doi:10.1038/nature10158

9

Rangeland and Warm-Season Forage Grasses

Kevin B. Jensen* and William F. Anderson

ABSTRACT

The development of rangeland grasses, particularly the wheatgrasses and wildryes, and warm-season grasses was initiated in the late 1800s to early 1900s and was associated with the development of the midwestern and western United States. During this period, undisturbed rangeland was replaced with cultivated crops; however, subsequent droughts left this land void of vegetation and subject to soil erosion and dunning. In an attempt to mitigate the problem, plant collectors were sent throughout the world with the charge to secure seeds and plants valuable for a variety of purposes. Many of the wheatgrasses and wildryes were introduced into the United States at this time. However, associated with these introductions was a lack of knowledge about how these grasses reproduce, chromosome numbers, and complex allpolyploidy and autopolyploidy. Hence, most of the initial cultivar development was limited to direct seed increase after a collection had been evaluated for establishment, persistence, and seed and forage yield. Occasionally, several cycles of mass selection would be conducted to eliminate off types in the seed production field. Even today, under the current directives to restore disturbed rangelands to their native habitat, native range grass development programs are centered on the release of locally adapted ecotypes with limited to no genetic selection. As funding increased for range grass breeding in the 1950s, basic research questions regarding different modes of reproduction and genetic relationships within and among the different species and genera allowed breeders to explore the use of intraspecific and interspecific hybrids and interploidy breeding through chromosome doubling. The introgression of new and unique germplasm into range grass breeding programs resulted in increased seedling establishment, stand persistence, and seed and forage yields. In addition, forage digestibility was enhanced. Molecular tools have been used to expand genetic variation by identifying geographical regions with the greatest genetic diversity for either inclusion into a breeding program or development of ecotypes with a unique genetic background.

Abbreviations: ADG, average daily animal gain; AFLP, amplified fragment length polymorphisms; DM, dry matter; DMY, dry matter yield; FRRL, Forage and Range Research Laboratory; FSU, former Soviet Union; IVDMD, in vitro dry matter digestibility; RRPS, restricted recurrent phenotypic selection; VIR, Vavilov Institute of Plant Industry.

Kevin B. Jensen, USDA-ARS, Forage and Range Research Lab, 695 North 1100 East, Utah State Univ., Logan, UT 84332-6300. *Corresponding author (Kevin.Jensen@ars.usda.gov).
William F. Anderson, USDA-ARS, Crop Genetics and Breeding Research Unit, 115 Coastal Way, Tifton, GA 31793 (Bill.anderson@ars.usda.gov).

doi:10.2135/cssaspecpub33.c9

Yield Gains in Major U.S. Field Crops. CSSA Special Publication 33.
Stephen Smith, Brian Diers, James Specht, and Brett Carver, editors.

Livestock ranchers depend on grassland grazing for a substantial part of their livestock management systems. Grassland forages compose up to 85% of the feed supply for ruminant animal products, especially in warm climates (Coleman et al., 2004a). However, very different approaches to managing grasslands have emerged over the past two centuries in the United States. Much of the eastern portion of the United States relies on smaller confined parcels of land, referred to as pastures, which are planted with a particular species and genotype. In this case, native or introduced forage species have been evaluated and selected for yield or high animal performance. For the southeastern United States, the predominant species for this purpose has been perennial warm-season (C_4) grasses that can take advantage of favorable temperatures and rainfall. Because of confinement of most ruminants, supplemental feed is necessary and includes silage, haylage, or hay from some of these same grass species. In contrast, livestock managed in western portions of the United States are often allowed to graze on wide expanses of rangeland that normally consist of multiple native grass species. Because of the large fluctuations in climate and soil types in the west, the complex of grasses species can be quite different from one area to another.

To understand the genetic improvements in dryland ranges grasses and their importance to rangeland ecosystems and agriculture, it is important to understand rangeland history. As the development in the western United States moved forward, the Carey Act of 1894, Newlands Act in 1909, and the Enlarged Homestead Act in 1909 made it possible for large tracks of native grasslands to be removed and placed under cultivation for dryland wheat and other cereal crops between 1905 and 1920. These events have been referred to by some as the greatest ecological disaster in America. During the 1930s, a prolonged drought led to the failure of annual crops in these areas. Land that was previously in tall grass prairie and big sagebrush communities was now abandoned crop land subject to soil erosion. Such dust storms were felt across the country; for example, in May 1934, one storm deposited 12 million tons of dust in Chicago (West, 2013).

Early revegetation efforts on rangeland in the early 1900s involved attempts to reestablish the native grasses found on the ranges by early settlers. However, grasses such as timothy (*Phleum pratense* L.), redtop (*Agrostis gigantea* Roth), smooth brome (*Bromus inermis* Leyss.), orchardgrass (*Dactylis glomerata* L.), and Kentucky bluegrass (*Poa pratensis* L.) were often used because seeds of these species were available. Unfortunately, most of the early seeding attempts on western rangelands failed because of improper species selections (Asay et al., 2001), primarily because they were not adapted to big sagebrush (*Artemisia tridentata* Nutt.) communities, typically composed of an understory of cool-season grasses.

The lack of perennial vegetation and the inability to reestablish perennial vegetation subsequently led to increasing annual plant invasion by cheatgrass (*Bromus tectorum* L.) and medusahead [*Taeniatherum caput-medusae* (L.) Nevski], creating unfavorable conditions for perennial vegetation by changing the soil structure and properties (Norton et al., 2004, 2007) and increasing fire frequency and disturbance (Robins et al., 2013). As a result of these and other disturbances, much of the U.S. arid and semiarid rangelands are currently classified as severely disturbed (Abella, 2010). Furthermore, forecasts based on predicted climate change models suggest trends toward hotter and drier conditions, thus potentially increasing the already high rate of rangeland degradation (D'Antonio and

Vitousek, 1992; Baldwin et al., 2003; Diffenbaugh et al., 2005; Chambers, 2008) and expanding severely disturbed rangelands (Christensen et al., 2007; Karl et al., 2009; Munson et al., 2011). It was not until the mid-1930s, when there was a need to reseed and stabilize these sites, that dryland grass breeding began in earnest. The mission of these breeding programs was to develop grasses that could be used to revegetate highly erodible, disturbed, and abandoned crop lands (Vogel and Moore, 1993).

Some of the earliest reports of range grass development originate from the late 1800s or early 1900s when N.E. Hansen, a special agent for the USDA Division of Botany from 1897 to 1898, was charged with the mission to secure seeds and plants valuable for a variety of purposes (Asay and Jensen, 1996a). As a result of his collections trips, Hansen was the first to introduce crested wheatgrass [*Agropyron cristatum* (L.) Gaertn.] into the United States in 1898. Five seed lots, designated as PIs 835, 837, 838, 1010, and 1012, from the Valuiki Experiment Station near what is now Volgograd, Russia were evaluated at experiment stations in Alabama, Indiana, Michigan, Colorado, and Washington (Dillman, 1946; Lorenz, 1986). In 1906, Hansen introduced a second set of five seed lots (PIs 19537–19541) labeled as *A. desertorum* (Fisch. ex Link) Schult. and one seed lot labeled *A. cristatum* (PI 19536) from the Valuiki Experiment Station. These introductions were distributed to 15 experimental stations throughout the United States. The first cultivar of crested wheatgrass was the diploid ($2n = 2x = 14$) Fairway released in 1932 (Kirk, 1932; Lorenz, 1986), which was selected from PI 19536 for its leafiness, fine texture, and dryland turf quality. 'Fairway' along with common crested wheatgrass, believed to be of the tetraploid ($2n = 4x = 28$) form, were seeded extensively throughout the 1930s. 'Fairway' is still used today in many grass mixtures for revegetation.

Range grasses that are of particular interest are the cool-season wheatgrasses and wildryes within the Triticeae tribe. These are some of the most valuable range grasses in the temperate regions of western and southern North America. These diverse groups of grasses are planted on disturbed, dry environments where they stabilize soils, provide forage for livestock and wildlife during the relatively cool periods of spring, early summer, fall, and winter, and reduce weed invasion and the frequency and intensity of wildfires.

Challenges

Range grasses have unique breeding challenges, including these key ones: diversity in pollination systems (self- and cross-pollination, self-incompatibility, and apomixis); complex genetic systems that may include high chromosome numbers and differing ploidy levels (autopolyploids, allopolyploids, and autoallopolyploids); heterozygosity within and between populations; poor seed production; and a long-lived perennial habit that requires many years to evaluate persistence and productivity.

Another confounding factor associated with range grass breeding is the occurrence of significant genotype × environment interactions. U.S. rangeland has highly variable soil types and intra- and inter-variation in annual precipitation and temperatures, thus making it difficult to estimate on-farm annual yields and then allocate those gains into genetics, management, and their interactions over time. A comprehensive study examining the establishment and persistence of 18 perennial cool-season range grasses at 34 field sites in the Intermountain West and Great Plains regions of the United States concluded that stand establishment and persistence were positively correlated with mean annual precipitation (Robins et al., 2013). 'Ree' intermediate wheatgrass [*Thinopyrum intermedium* (Host) Barkworth & D.R. Dewey] was evaluated over eight different environments in Canada and North Dakota, and the eight mean dry matter yield (DMY) values ranged from 2.7 to 7.2 Mg ha^{-1} (Wilson and Smoliak, 1977). In the same study, seed yields ranged from 160 to 585 kg ha^{-1}. For both traits, there was a significant cultivar by location interaction.

In most cross-pollinating grasses, inbreeding leads to a depression of vigor and loss of fertility; hence, an understanding of the level of self-fertility is important in developing populations for future selection. In wheatgrasses and wildryes, the full spectrum from complete self-fertility to complete self-sterility can be encountered (Jensen et al., 1990), along with sexual reproduction to apomixis (Murphy, 2003). Within this group, there are many more self-fertile species than self-sterile species. About two-thirds of the approximately 250 wheatgrasses and wildryes are self-fertilizing. The greatest majority of self-pollinating species are in the genus *Elymus*, with a few in *Thinopyrum* and *Leymus*. Species within *Agropyron, Pseudoroegneria, Psathyrostachys, Pascopyrum,* and the *Epinephelus lanceolatus* (Scribner & Smith) Gould complex are predominately cross-pollinating (Jensen et al., 1990). The long-awned *Elymus rectisetus* (Nees in Lehm.) Á Löve is apomictic, and *E. scaber* (R. Br.) Á Löve varies from facultative apomixis to sexual reproduction (Wang and Jensen, 2009).

The inflorescence structure of warm-season grasses may determine self- or cross-pollination. For example, buffalograss [*Buchloë dactyloides* (Nutt.) Engelm.] is dioecious with phenotypically different male and female plants, which ensures cross-pollination (Shearman et al., 2004). Eastern gamagrass (*Tripsicum dactyloides* L.), which has staminate and pistillate flowers on the same plant (Springer and Dewald, 2004), is also cross-pollinated. However, most C_4 grasses are monoecious (self-incompatible), which is controlled by at least two segregating genes (Vogel and Burson, 2004). Genetic improvement of C_4 grasses through breeding is most often accomplished via selection and hybridization of sexually propagated self-incompatible parents; however, a significant number of species are apomictic, which requires the search and discovery of naturally occurring sexual ecotypes for genetic improvement (Vogel and Burson, 2004).

Genetic Diversity

Various molecular markers, such as amplified fragment length polymorphisms (AFLP), restriction fragment length polymorphisms (RFLP), and random amplified polymorphic DNA (RAPD), have been employed to assess genetic diversity and structure in rangeland grasses, inferring evolutionary history and gene flow among and within populations. Of particular interest is the difference between cross- and self-pollinated species in the distribution of genetic variation within and between populations (collections). As a general rule, cross-pollinating grasses have 70% of their genetic variation within a population, with 30% between populations, suggesting that selection within an individual population may result in satisfactory gains given the presence of genetic variation for the traits of interest.

Jensen et al. (2009a) reported that among Siberian wheatgrass [*A. fragile* (Roth) P. Candargy] cultivars, 88% of the total genetic variation was ascribable to variants existing within cultivars, with the remaining 12% arising from differences among cultivars. In a similar study, Jensen et al. (2006a) reported that 17% of the detected AFLP was due to differences among seven crested wheatgrass [*A. cristatum* and *A. desertorum*] cultivars and their hybrid populations. The remaining 83% was detected within cultivars and hybrid populations. Mellish et al. (2002) reported 12 and 88% of the AFLP variation partitionable, respectively, among and within two diploid and four tetraploid crested wheatgrass populations. Similar percentages (12–27) of AFLP variation were found among Russian wildrye [*Psathyrostachys juncea* (Fisch.) Nevski] hybrid populations and cultivars (Jensen et al., 2005a). Using AFLP and the optimum Bayesian cluster model, Larson et al. (2004) identified 21 bluebunch wheatgrass [*Pseudoroegneria spicata* (Pursh) Á. Löve] groups with 18% of the DNA polymorphism partitioned among the 21 regional groups, 15% among localities within groups, and 67% within accessions. On the basis of the proportion of shared DNA fragments, Larson et al. (2001) reported in California brome [*Bromus carinatus* Hook. & Arn.], a self-pollinating grass, that 5% of the variation was within a population and 95% was between populations, the opposite of that observed in cross-pollinating grasses. Larson et al. (2001) reported that self-pollinating purple needlegrass [*Nassella pulchra* (Hitchc.) Barkworth] populations from the western United States had 14% of their total variation within populations and 86% between populations.

Breeding

The great drought of the 1930s and the ensuing soil erosion led to the establishment of grass breeding and evaluation programs across the United States. Because of the lack of available plant materials capable of meeting the needs from the 1930s to 1950s, many of these programs began by collecting and evaluating large numbers of both native and introduced grasses for a targeted environment and selecting superior collections for release. Even today, given the desire by many to restore the rangelands of the western United States to their original state, much of the work done on native grasses has emphasized pre-variety germplasm or wildland collected materials with little or no genetic selection or testing (Asay et al., 2001). Most of the breeding and selection for improved seedling establishment and persistence with a lesser emphasis on DMY has been within the introduced grasses such as crested and Siberian wheatgrass, Russian wildrye, and intermediate and tall wheatgrass [*Thinopyrum ponticum* (Podp.) Z.-W. Liu & R.-C. Wang], with limited selection taking place in native grasses such as Snake River wheatgrass [*Elymus wawawaiensis* J. Carlson & Barkworth], bluebunch wheatgrass, and slender wheatgrass [*Elymus trachycaulus* (Link) Gould ex Shinners]. The majority of breeding and improvement of warm-season grass DMY for the southeastern United States has also been done with introduced species such as bermudagrass (*Cynodon* sp.) and bahiagrass (*Paspalum notatum* Flügge). Selection and release of lines selected from ecotypes have been done for some native species with limited success. In this section we will review significant dryland grasses documenting the gains in seed traits, seedling establishment, plant persistence, and DMY for native and non-native grasses.

Because of sterility problems in the interploidy hybrids, most breeding programs have used recurrent phenotypic selection to increase frequencies of favorable genes for the above traits within ploidy levels, primarily the diploids and tetraploids. Others methods include half-sib family performance for evaluation of combining ability for desired traits to select superior clonal materials for synthetic recombination.

Wheatgrasses: Agropyron

Crested Wheatgrass

Since its introduction from Asia in the early 1900s, crested wheatgrass has become the major cool-season grass used to improve semiarid rangelands of western North America. As noted previously, most early seedings of crested wheatgrass were made with unimproved introductions, or mixtures consisting of both 'Fairway' and 'Standard' types. Intensive breeding work with crested wheatgrass began in the United States in 1936, when the USDA received increased funding for grass breeding. Research stations in Mandan, ND and Bozeman, MT were the earliest breeding centers, and improved germplasm has since been developed in breeding programs conducted by the Agricultural Research Service in Lincoln, NE and Logan, UT.

All *Agropyron* (crested and Siberian wheatgrasses) species are composed of the basic (**P**) genome modified by structural rearrangements as an autoploid series with three ploidy levels: diploid (**PP**), tetraploid (**PPPP**), and hexaploid ($2n = 6x = 42$; **PPPPPP**) (Dewey and Asay, 1982). Asay and Dewey (1983) made all possible crosses among diploid, tetraploid, and hexaploid levels and concluded, on the basis of chromosome pairing and fertility, that the crested wheatgrasses should be treated as a single gene pool. However, because of *Agropyron's* unique genomes, there is little potential to utilize secondary and tertiary gene pools for improvement (Jensen et al., 2006a).

Early diploid crested wheatgrass cultivars include Parkway, which was developed from Fairway germplasm on the basis of improved leafiness, plant height, and vigor (Alderson and Sharp, 1994), and Ruff, which was selected for its broad bunch growth habit (USDA Extension Service, 1978). The earliest tetraploid cultivars released in 1953 were Nordan and Summit. 'Nordan' was selected for its upright growth form, relatively large seeds, and good seedling vigor (Rogler, 1954; Wolfe and Morrison, 1957). Because of seed processing problems, 'Summit' is not widely used (Alderson and Sharp, 1994). Another tetraploid cultivar Ephraim (Stevens et al., 1983) was selected for its ability to produce rhizomes by the third year after establishment. The cultivar Kirk (Knowles, 1990), a tetraploid *A. cristatum*, was selected for vigor, fertility, seed retention, and reduced awn development (Table 9–1).

The greatest progress to date in crested wheatgrass breeding has been achieved at the tetraploid levels. Hybridization schemes involving $6x$-$2x$, $6x$-$4x$, and $4x$-$2x$ and hybrids between colchicine-induced tetraploids ($C4x$) and natural tetraploids ($N4\ x$) have shown potential for expanding the genetic resources of 4x breeding programs (Asay et al., 1986). Of particular interest were the $C4x$-$N4x$ crosses where Tai and Dewey (1966) found colchicine-induced tetraploids ($C4x$) to be relatively fertile. The $C4x$ clones readily crossed with natural N4x plants, and fertility and general vigor of the $C4x$-$N4x$ hybrids were comparable to the parents.

Table 9–1. List of cultivars, year released, and breeding history for the wheatgrasses and wildryes.†

Cultivar	Year of release	Originated from	Selection cycles/type
		Wheatgrasses	
Crested wheatgrass [*Agropyron cristatum* (L.) Gaertn. & *A. desertorum* (Fisch. ex Link) Schult.]			
Fairway	1932	PI 19536.	Direct seed increase.
Nordan	1953	PI 469225.	Single plant selections for plant vigor made under open pollination for two generations.
Summit	1953	Western Siberian Experiment Station, Omsk, former Soviet Union.	Field rouging of seed increase field for uniformity.
Parkway	1969	Fairway.	Several generations of recurrent selection (RS) for vigor, height, and leafiness followed by polycross progeny tests.
Ruff	1974	Fairway.	Selected from seven lots of fairway and carried for three generations.
Ephraim	1983	PI 109012.	Direct seed increase.
Hycrest	1984	Fairway (doubled) × Standard tetraploid.	Five cycles of RS for fertility, forage and seed yield, and seedling establishment. (18-clone synthetic.)
Kirk	1987	Collections from Turku, Finland crossed with nine local strains of European origin.	Two cycles of phenotypic selection for vigor, fertility, reduced seed shattering, and reduced awn development.
Douglas	1994	PI 406442—Soviet Union (exceptionally wide leaves); PIs 401076, 401080, and 401085—Iran; and PI 173622—Turkey.	Two cycles of RS for broad-leaf types, vegetative vigor, seed yield potential, and response to pests and drought. (10-clone synthetic.)

Table continued.

Table 9–1. Continued.

Cultivar	Year of release	Originated from	Selection cycles/type
CD-II	1996	Hycrest.	One cycle of RS for vegetative vigor, absence of purple leaves during the spring, and leafiness. Two subsequent cycles of RS for leafiness, vegetative vigor, and seedling vigor were conducted. (10-clone synthetic.)
Goliath	2003	Parkway (doubled) × plants from the cultivar Kirk.	Several generations of RS for seed size, floret fertility, regrowth, and plant height. (Syn-2 population.)
Hycrest II	2008	Fairway (PI 279802; doubled)—parent to Hycrest.	Six cycles of RS for plant vigor, seedling vigor, and persistence under drought. (8-plant synthetic.)
Siberian wheatgrass [*A. fragile* (Roth) P. Candargy]			
P-27	1953	Collection from Kazakhstan in 1934.	Included in row nurseries and field evaluations for narrow spikes, fine leaves, and adaptation to sandy soils. Individual clones selected in 1949.
Vavilov	1994	Collections received from Vavilov Institute of Plant Industry (VIR) in Russia, Stavropol Botanical Garden in Russia, Eskisehir Plant Breeding Station in Turkey, and selections from P-27.	Three cycles of RS for vegetative vigor, response to drought, resistance to diseases and insects, seedling vigor, seed yield, and plant type. (14-clone synthetic.)
Vavilov II	2008	PIs from Kazakhstan (22 collections) intercrossed with plants from Vavilov II.	One cycle of RS for plant vigor and persistence under extreme drought. (50-clone synthetic.)
Stabilizer	2012	Originated form PIs 598669, 598670, 598671, 598674, 598675, and 598714 from the dry desert regions of Kazakhstan.	Two cycles of selection for blue-gray color, pubescence, and low growth habit. (23-plant synthetic.)

Cultivar	Year of release	Originated from	Selection cycles/type
Intermediate wheatgrass [*Thinopyrum intermedium* (Host) Barkworth & D.R. Dewey]			
Ree	1945	PI 68568.	Direct seed increase.
Greenar	1945	Originated from PI 98568 from the former Soviet Union in 1937.	Open-pollinated selections made from planting one generation after introduction.
Amur	1952	PI 131532 from Manchuria, China.	Increase under isolation removing awned and off type plants.
Topar	1953	PI 107330 from Tashkent, Turkistan, former Soviet Union.	Seed from selected plants were bulked and increased under isolation. Nonpubescent, slow sod forming, and weak plants were removed.
Chief	1961	Collections from the former Soviet Union and cultivar Ree.	One cycle of mass selection for high seed yield and seed quality. Five lines bulked to produce breeder seed.
Oahe	1961	Cultivar Ree.	One cycle of selection in self- and open-pollinated progeny for percent seed set followed by one cycle of polycross testing and selection for seed set, forage yield, and rust resistance. (4-clone synthetic.)
Luna	1963	PI 106831 from Turkestan.	Direct seed increase after extensive evaluation trials for seedling vigor, fast establishment, and forage yield.
Greenleaf	1966	Commercial seed lots from Davenport, WA and Bismark, ND—unknown origin.	57 plants selected for forage. Open-pollinated progenies evaluated in greenhouse for seedling vigor and spread. (12-line synthetic.)
Tegmar	1968	PI 109219—Bolu, Turkey.	Open-pollinated selections made from planting one generation after introduction.

Table continued.

Table 9–1. Continued.

Cultivar	Year of release	Originated from	Selection cycles/type
Slate	1969	PIs 98568 (cultivar Ree) and 131532 (cultivar Amur).	Breeder seed was constituted by blending equal quantities of seed from the two PIs and selecting for erect growth habit and green color.
Clarke	1980	Collections from the Moscow Botanical Garden and R.P Knowles breeding lines.	Four to eight cycles of RS for winter-hardiness, drought tolerance, establishment vigor, resistance to aphis-virus infection, seed quality and yield, and forage production. (20-clone synthetic.)
Reliant	1991	Intermating of 24 cultivars and experimental lines adapted to the northern Great Plains region.	One cycle of selection for resistance to leaf spot followed by an additional cycle of selection within clonal plots and open-pollinated progeny for forage and seed yield, winter survival, forage quality, and leaf spot resistance. (6-clone synthetic.)
Manska	1992	PI 116252—former Soviet Union.	One cycle of selection based on polycross progeny performance for leaf-spot disease, spring recovery, nutritional quality, and forage and seed yields. (116-clone synthetic.)
Rush	1994	German Botanical Garden.	Direct increase after several evaluation trials.
Beefmaker	2003	PIs 345586 (Rostov), 273733 (Kursk), 273732 (Rostov), 315353 (Voronezh), 315067 (St. Petersburg Botanical Garden), and 315355 (Stavropol) from the former Soviet Union.	Two cycles of RS for forage yield, leafiness, erectness, disease resistance, and digestibility.
Haymaker	2003	PIs 440015, 440008, and 440011 from Svetlograd and Stavropol in the Caucasian region of Russia, and cultivar Slate.	One cycle of RS for increased DMY and digestibility.
Manifest	2007	10 collections near Stavropol and Svetlograd, in the Caucasian region of Russia.	One cycle of selection with the parent clones selected on the basis of polycross progeny performance for forage yield, seed yield, spring recovery, and resistance to leaf-spot disease.
Tall wheatgrass [*T. ponticum* (Podp.) Z.-W. Liu & R.-C. Wang]			

Cultivar	Year of release	Originated from	Selection cycles/type
Largo	1937	Collection near Bandirma, Turkey.	Direct seed increase of collection.
Alkar	1951	PI 98526—VIR.	Mass selection within spaced plants of initial collection with seed bulked.
Jose	1965	PI 150123—Eurasia.	Direct seed increase with off type plants removed.
Orbit	1966	PI 98526—VIR, Alkar, and seed from locally selected strains.	Composite of nine open-pollinated lines and one 3-clone synthetic, which originated from one cycle of selection for winter survival, seed, and forage yield.
Platte	1972	PI 98526 (VIR).	Several generations of mass selection for erect bunch type with good seed set released as Nebraska 98526 germplasm. Nebraska 98526 was screened for survival on saline-alkali soils. Surviving plants were subject to two cycles of selection on saline soils and designated as Nebraska 98. Platte was synthesized by growing Nebraska 98526 and Nebraska 98 in alternate rows and bulk harvesting the field.
Tyrrell	1981	Cultivar Largo.	Multiple cycles of RS for tolerance to high salinity.
Bluebunch wheatgrass [*Pseudoroegneria spicata* (Pursh) Á. Löve]			
Whitmar	1946	Collection from Palouse prairie grassland, Colton, Whitman County, WA.	Superior collection in observational tests involving more than 500 bluebunch wheatgrass collections. Direct seed increase from open-pollinated seed collected from the evaluation trial.
Goldar	1986	Collection from Malley Ridge, Umatilla National Forest, Asotin, WA.	Superior collection in observational tests involving more than 500 bluebunch wheatgrass collections. Direct seed increase from open-pollinated seed collected from the evaluation trial.
P-7	2002	Collections from Washington; Oregon; Nevada; Utah; Idaho; Montana; and British Columbia, Canada.	One generation of intermating the different populations followed by seed increase.
Anatone	2004	Collection near Anatone, Asotin County, WA.	Selected from a series of comparative field trials evaluating more than 80 bluebunch wheatgrass collections from the western United States. Direct seed increase.

Table continued.

Table 9–1. Continued.

Cultivar	Year of release	Originated from	Selection cycles/type
Snake River wheatgrass [*Elymus wawawaiensis* J. Carlson & Barkworth]			
Secar	1980	Collection from the Snake River gorge, Lewiston, ID.	Superior collection in observational tests involving more than 500 bluebunch wheatgrass collections. Direct seed increase from open-pollinated seed collected from the evaluation trial.
Discovery	2008	Collections from Whitman and Asotin counties in WA and ID.	Collections were crossed and seed harvested in bulk to generate Discovery breeder seed.
Thickspike *wheatgrass* [*E. lanceolatus (Scribn.* & J.G. Sm.) Gould]			
Sodar	1954	Collection from Canyon City, Grant County, OR.	Several cycles of mass selection to remove off types from the population.
Critana	1971	Collection from Havre, Hill County, MT.	Direct seed increase.
Elbee	1980	Collections from southern Alberta and southwest Saskatchewan.	8-clone synthetic based on polycross performance for forage and seed yields, plant height, aggressive rhizomes, and days to maturity.
Schwendimar	1994	Collection from the Columbia River, east of The Dalles, OR.	Several generations of mass selection to remove off types from the population.
Bannock	1995	Collections from Pocatello, ID; The Dalles, OR; and Quincy, WA.	Six individual plants that were more vigorous and productive were selected. Increased under isolation and seed bulked.
Slender wheatgrass [*E. trachycaulus* (Link) Gould ex Shinners]			
Primar	1946	Collection near Beebe, MT.	Direct seed increase after evaluation.
Revenue	1970	Seed of a single plant collected near Revenue, SK.	Superior collection in observational tests involving more than 750 slender wheatgrass collection. Direct seed increase from open-pollinated seed collected from the evaluation trial.

Cultivar	Year of release	Originated from	Selection cycles/type
San Luis	1984	Collection from San Luis Valley, Rio Grande County, CO.	Direct seed increase of collection.
Pryor	1988	Seed collected from several plants in Carbon County, MT.	Direct seed increase of collection.
Adanac	1990	Seed from a single plant collected near Climax, SK.	Direct seed increase of the collection.
Hillcrest	1994	Seed collected from a single plant near Crowsnest Pass of southwestern AB.	Direct seed increase of this collection.
Highlander	1995	Seed collected from a single plant near Crowsnest Pass of southwestern AB.	Direct seed increase of this collection.
Mountaineer	1995	Seed from a single plant collected in the Rocky Mountains south of Kananaskis Country, AB.	Direct seed increase of this collection.
FirstStrike	2006	Collections from Colorado and Wyoming.	One cycle of RS for plant persistence and visual plant vigor.
Copperhead	2007	Collection from the upper Clark Fork River basin, western MT on soil impacted by acid and heavy metal contamination.	Superior collection in observational tests on acid and heavy metal sites. Direct seed increase of the collection.
Western wheatgrass [*Pascopyrum smithii* (Rydb.) Á. Löve]			
Barton	1970	Collection from Walnut Creek near Heizer, Barton County, KS.	Direct seed increase of this collection.
Rosana	1972	Collection from native meadows along Porcupine Creek, northwest of Forsyth, MT.	Direct increase of seed collection after evaluation of 60 western wheatgrass accessions.

Table continued.

Table 9–1. Continued.

Cultivar	Year of release	Originated from	Selection cycles/type
Arriba	1973	Collection near Flagler, CO.	Bulk increase from initial planting followed by five seed increase generations.
Flintlock	1975	30 collections from central and southwestern New Brunswick and northwestern Kansas.	Two cycles of RS from the base population for seed yield and rhizome spread. (100-clone synthetic.)
Walsh	1982	Traces back to 468 western wheatgrass ecotypes collected in southern Alberta and southwestern Saskatchewan.	On the basis of clonal and polycross progeny performance for forage and seed yield, disease resistance, rhizome spread, and plant height 20 clones were selected and polycrossed.
Rodan	1983	Commercial seed field of unknown origin in Missouri River Bottoms, Mandan, ND.	13-clone synthetic resulting from one cycle of polycross progeny performance for cover, leafiness, and resistance to stem rust followed by 7 open-pollinated seed increase generations.
Recovery	2012	Cultivars Rosana and Rodan, and a collection from Fort Carson, CO.	Two cycles of RS for seedling vigor.
		Wildryes	
Russian wildrye [*Psathyrostachys juncea* (Fisch.) Nevski]			
Vinall	1960	PI 75737 (former Soviet Union), 108493 (eastern Kazakhstan, and 111549 (unknown origin).	5-clone synthetic based on four generations of RS on a single plant basis with progeny testing for seed yield and resistance to lodging.
Sawki	1963	Collections from Western Siberia Experiment Station in Omsk, former Soviet Union.	10-clone synthetic resulting from two cycles of RS for seed and forage yield, and erectness.
Mayak	1971	Breeding lines 1546, 1495, 2355, and PI 75737 (former Soviet Union) obtained from Mandan, ND and Swift Current, SK.	20-clone synthetic resulting from open-pollinated and polycross progeny tests for seed and forage yield, and resistance to leaf spot diseases.

Cultivar	Year of release	Originated from	Selection cycles/type
Cabree	1976	Commercial seed field of unknown origin in Manyberries, AB.	10 plants with excellent seed retention were selfed, and 6 of the S1 plants superior for seed retention, seedling vigor, and seed and forage yield were combined to form a 6-clone synthetic.
Swift	1978	Cultivar Swaki and Mandan breeding line 1546.	Four cycles of RS were conducted for seedling emergence, spring vigor, leafiness, seed and forage yield, and disease resistance.
Bozoisky-Select	1984	PI 406468 (VIR).	23-clone synthetic resulting from two cycles of RS for improved vigor, leafiness, seed yield, coleoptiles length, and seedling vigor.
Tetracan	1988	Chromosome doubled diploids of unknown origin.	One cycle of selection for seedling vigor and spring vigor, resistance to leaf spot diseases, leafiness, and seed production.
Mankota	1991	PIs 314675 and 272136 (Alma-Ata, former Soviet Union) and a bulk breeding line Mandan 34.	6-clone synthetic resulting from one cycle of polycross progeny testing for seed and forage yield, spring vigor, and resistance to leaf spot diseases.
Bozoisky II	2006	Two base population synthetics A and B. Synthetic A traces to cultivar Vinall, P210 (NRCS), PI 314675, cultivar Bozoisky, and PIs 406471 and 314082 (former Soviet Union). Synthetic B traces to cultivars Vinall and Bozoisky, P210, and PIs 272136, 314675, 369234 (Novosibirsk, Siberia), 406469 (former Soviet Union), and 314082.	Three cycles of RS was conducted in each of the two base population synthetics for plant vigor, increased seed weight and seedling vigor. Base synthetics were recombined in a crossing block followed by one cycle of RS for vegetative vigor, and forage production, and seedling vigor. (15-clone synthetic.)
Basin wildrye [*Leymus cinereus* (Scribn. & Merr.) Á. Löve]			
Magnar	1979	University of Saskatchewan (unknown origin).	Several selection of mass selection for the most vigorous types.
Trailhead	1991	Collection made near Roundup, MT.	Direct increase of seed collection after evaluation of 125 basin wildrye collections.

Table continued.

Table 9–1. Continued.

Cultivar	Year of release	Originated from	Selection cycles/type
Washue	2003	Collection made within the Anaconda Smelter Superfund site, Deer Lodge, County, MT.	Direct seed increase of collection.
Continental	2009	Cultivars Magnar and Trailhead (doubled).	Cross between Magnar and doubled Trailhead followed by two cycles of RS in the hybrid population for visual plant vigor.
Beardless wildrye [*L. triticoides* (Buckley) Pilger]			
Shoshone	1980	Collection from the Riverton, WY fairgrounds.	Direct seed increase of collection.
Rio	1991	Collection made in Stratford, Kings County, CA.	Direct seed increase after initial evaluation with 12 other beardless wildrye collections.
Beach wildrye [L. mollis (Trin.) Pilger]			
Benson	1991	Collection made near Narrow Cape, Kodiak Island, AK.	Direct increase via vegetative sprigs after initial evaluation with 18 other collections.
Reeve	1991	PI 345978 (Norway).	Direct seed increase of collection after initial evaluation with 18 other collections.
Altai wildrye [*L. angustus* (Trin.) Pilger]			
Prairieland	1976	Collections from the Steppes of Kustany and Voronezh of the former Soviet Union.	22-clone synthetic resulting from multiple cycles of RS for seed and forage yields, seed quality, resistance to leaf spot diseases, and seedling vigor.
Eejay	1989	Collections from the Steppes of Kustany and Voronezh of the former Soviet Union.	12-clone synthetic resulting from three cycles of RS for high seed and forage yield, resistance to leaf spot diseases, and seed quality within a green (nonglaucous) population.
Pearl	1989	Collections from the Steppes of Kustany and Voronezh of the former Soviet Union.	4-clone synthetic resulting from three cycles of RS for high seed and forage yield, resistance to leaf spot diseases, and seed quality within a blue (glaucous) population.
Mustang	2004	Cultivars Prairieland, Eejay, and Pearl and PI 499650 (Urumqi, Xinjiang Province).	Three cycles of RS selection for total seed yield, 100-seed weight, and visual plant vigor, resulted in a 14-clone synthetic with increased seedling establishment.

† Abbreviations: RS, recurrent selection; VIR, Vavilov Institute of Plant Industry

Moreover, the hybrid populations responded favorably to selection for improved fertility and became the basis for future crested wheatgrass breeding programs (Dewey and Pendse, 1968).

'Hycrest' was the first interspecific hybrid of crested wheatgrass to be released (Asay et al., 1985a). The initial hybrids between C4*x* of *A. cristatum* and N4*x* of *A. desertorum* were made in 1962 with additional hybrids made in 1967. From 1967 to 1974, three cycles (C_3) of recurrent selection were conducted for fertility and seed yield. Subsequently, 18 clones were identified and intercrossed to develop the first synthetic generation of 'Hycrest.' 'Hycrest' is a tetraploid; however, on occasion aneuploids of 30 and 32 chromosomes have been observed, with the extra chromosomes attributed to B-chromosomes previously observed in crested wheatgrass (Dewey and Pendse, 1968). 'Hycrest' consistently produced 28 seeds spike^{-1} compared with 14 seeds spike^{-1} in 'Nordan' (Asay et al., 1986).

The most noteworthy attributes of 'Hycrest' are its increased seedling establishment and ability to compete with invasive annual weeds. When averaged over four western locations, 'Hycrest' had significantly higher seedling establishment (94%) than 'Nordan' (76%), 'Fairway' (76%), and 'Ephraim' (73%) (Asay et al., 2001). Improvements in plant persistence were less than for seedling establishment; however, 'Hycrest' averaged 86% stand density after 7 yr compared with 80, 82, and 77% for 'Nordan,' 'Fairway,' and 'Ephraim,' respectively (Asay et al., 2001). When averaged over five locations in Idaho, Utah, and Montana, forage yield of 'Hycrest' averaged 2.7 Mg ha^{-1} compared with 1.9 and 2.0 Mg ha^{-1} for 'Nordan' and 'Fairway' crested wheatgrass, respectively (Asay et al., 1986). In a subsequent comparison, Asay et al. (2001) found that forage yields of 'Hycrest' were 4.2 Mg ha^{-1} compared with 'Nordan' at 3.9 Mg ha^{-1} and 'Fairway' at 3.3 Mg ha^{-1} on semi-arid range sites. In an attempt to increase the leafiness and absence of purple leaves during the early spring growth in 'Hycrest,' two cycles of recurrent selection were conducted, resulting in the release of the cultivar CD-II (1996) crested wheatgrass (Asay et al., 1997).

Douglas (1994) is the only hexaploid ($2n = 6x = 42$) crested wheatgrass cultivar. 'Douglas' was derived from hybrids between PI 406422 (former Soviet Union [FSU]) and four other hexaploid collections, three from Iran (PIs 401076, 401080, and 401085), and one from Turkey (PI 173622). PI 406422, which expressed exceptionally broad leaves, was used as the female parent to retain the cytoplasm of this PI in the breeding population. Two cycles of recurrent selection for broad-leaf types, vegetative vigor, seed yield, and seedling emergence from a deep planting depth resulted in significant increases in leaf width and seedling vigor in 'Douglas' over 'Nordan,' 'Fairway,' and 'Ephraim' but equivalent to 'Hycrest' (Asay et al., 1995a).

In 2003, another interploidy hybrid cultivar Goliath (Coulman, 2006) was released. 'Goliath' originated from colchicine-doubled plants of the diploid cultivar Parkway crossed with plants from the natural tetraploid cultivar Kirk, followed by two cycles of selection with emphasis on seed size, floret fertility, vigor, regrowth, and plant height. Relative to its parents, 'Goliath' is taller, produces fewer, heavier tillers, has greater seed weight, and exhibits a 3% increase in DMY.

In 2008 'Hycrest II' was released with improved seedling establishment (Jensen et al., 2009b). 'Hycrest II' originated by intercrossing 10 C4x plants of 'Fairway' crested wheatgrass (PI 279802) in 1960 to 1961(Tai and Dewey, 1966; Dewey and Pendse, 1968). This C4x population was one of the parents to the cultivar Hycrest (Asay et al., 1985a). From 1976 to 2002, the C4x population went through five cycles of recurrent phenotypic selection for seedling establishment, seed yield, and visual plant vigor (as an estimate of DMY). Within each cycle, the top 5% were selected from approximately 2000 spaced plants. Selection within the C4x population for seedling emergence from a deep planting depth (7.6 cm) resulted in significant gains in seedling establishment in 'Hycrest II' (71%) compared with 55 and 63% in 'Hycrest' and 'Nordan,' respectively (Jensen et al., 2009b). Given the origin of 'Hycrest II,' the reduction in AFLP bands from 617 in 'Hycrest' to 603 in 'Hycrest II' was not unexpected. On the basis of similarity coefficients, the genetic diversity of 'Hycrest II' (0.795) was lower than 'Nordan' (0.770) but similar to 'Hycrest' (0.793) (Jensen et al., 2006a, 2009b).

Current crested wheatgrass development involves interspecific hybrids between hexaploid accessions of crested wheatgrass from the FSU and tetraploid cultivars Hycrest and Hycrest II. The FSU accessions are characterized by exceptionally broad leaves, excellent forage quality, and color retention 2 to 3 wk longer in the growing season. The goal is to transfer the broad-leaf, high forage quality, and color retention traits into the genetic background of 'Hycrest II,' while maintaining drought tolerance and persistence. Currently, this interspecific population is meiotically stable at 28 chromosomes, fertile, and segregating for the broad-leaf character and seedling establishment (Jensen et al., 2006a).

Siberian Wheatgrass

The Siberian form of crested wheatgrass has the same **P** genome as crested wheatgrass, is more drought resistant, and is better adapted to sandy soils than other crested wheatgrass types. Siberian wheatgrass has narrow awnless inflorescences, finer leaves, and thinner stems than standard crested wheatgrass (Jensen et al., 2009a). The original cultivar P-27 was released in 1953, where it was derived directly from a 1934 collection from the Institute of Plant Industry in St. Petersburg, Russia (Alderson and Sharp, 1994). 'P-27' remained the only Siberian wheatgrass commercially available until 'Vavilov,' a 14-clone synthetic, was released in 1995 (Asay et al., 1995b). The parental germplasm for 'Vavilov' was derived from collections received from the Vavilov Institute of Plant Industry (VIR) in St. Peterburg, Russia; the Stavropol Botanical Garden in Stavropol, Russia; the Eskisehir Plant Breeding Station in Eskisehir Turkey; and selections within 'P-27.' During the development of 'Vavilov,' the breeding population underwent three cycles of recurrent selection for vegetative vigor, response to drought, diseases and insects, seedling vigor (emergence from deep seedings), seed yield, and upright plant type. In response to selection, 'Vavilov' had a 56% increase in seedling establishment compared with 'P-27'; in addition, persistence increased by 47% over 'P-27,' averaged over six locations in Utah, Idaho, and Wyoming. Dry matter yields were also 14% higher than 'P-27' (Asay et al., 1995b) (Table 9–1).

The most recently released Siberian wheatgrass cultivar Vavilov II is a broad-based 50-clone synthetic selected for seedling establishment and persistence under drought (Jensen et al., 2009a). Parental lines for 'Vavilov II' included plants from 'Vavilov' and collections made from the Steppes of Kazakhstan in 1988 (Jensen et al., 2000). Seedling establishment in 'Vavilov II' across six locations throughout the western United States were significantly higher at 64% compared with 44% for 'Vavilov.' Across those same locations, plant stands of 'Vavilov II' were more persistant after initial seedling establishment at 77% compared with 60% for 'Vavilov.' 'Vavilov II' had a 26% increase in DMY over 'Vavilov' when combined across Washington and Wyoming sites (Jensen et al., 2009a). Thus, in crested and Siberian wheatgrass, selection for improved seedling establishment and persistence has resulted in an increase in DMY, likely resulting from denser stands of grass rather than a correlated response to selection.

Wheatgrasses: Thinopyrum

Intermediate Wheatgrass

Intermediate wheatgrass has probably had the most direct selection for agronomic traits of any of the range grasses. Intermediate wheatgrass is a complex autoallohexaploid ($2n = 6x = 42$; **EEEEStSt**) composed of the **E** genome originating from *Thinopyrum* and the **St** genome from *Pseudoroegneria* (Liu and Wang, 1992; Kishii et al., 2005). At metaphase, chromosomes generally associate as bivalents, suggesting that it may be a segmental autoallohexaploid. Although several cultivars have been released, most of the early germplasm programs utilized PI 98568, from Maikop, FSU, as the base germplasm. This collection contributed to the development of cultivars Ree, Chief, Greenar, Oahe, and Slate (Asay and Jensen, 1996a). Chief was selected directly out of Ree followed by one cycle of recurrent selection for improved yield and seed quality and is still the most widely used cultivar in Canada (Alderson and Sharp, 1994). The pubescent cultivar Luna originated from PI 106831 (Ashkabad, Turkmenistan) and 116252 (FSU), respectively (Asay and Jensen, 1996a) (Table 9–1).

'Greenar' was selected directly out of PI 98568 after 10 yr of evaluation for late maturity, dark green leaves, and good seedling vigor. This cultivar is still being used today and will be the baseline for assessing increased DMY gains through selection (Asay and Jensen, 1996a). The cultivar Oahe (1961) is a 4-clone synthetic and is still used extensively in the Northern Great Plains and Intermountain regions as a dryland range grass (Asay and Jensen, 1996a). It resulted from two cycles of selection from the cultivar Ree on the basis of polycross testing for high seed set, DMY, and rust resistance (Ross, 1963). In an evaluation trial across the Northern Great Plains (North Dakota, Montana, Nebraska, and Utah) from 2001 to 2003, 'Oahe' averaged 4.9 Mg ha^{-1}, an 8% increase over the cultivar Greenar [Forage and Range Research Laboratory, (FRRL), Logan, UT, unpublished data, 2013].

In 1966, the cultivar Greenleaf, a 12-line synthetic, originated from seed collections taken from a commercial pubescent intermediate wheatgrass seed field (origin unknown) in 1956. 'Greenleaf' was selected for increased seedling establishment, earliness of spring growth, DMY, winter hardiness, predominance of bright green foliage, and pubescence of spikelets (Wilson and Smoliak, 1978). Dry matter yields for 'Greenleaf' across Alberta, Saskatchewan, Manitoba, and North Dakota averaged 5.2 Mg ha^{-1} compared with 5.1, 5.2, and 5.6 Mg ha^{-1} for 'Oahe,' 'Ree,' and 'Chief,' respectively. Seed yields of 'Greenleaf' were greater than 'Topar' and 'Ree' but similar to 'Oahe' and 'Chief' (Wilson and Smoliak, 1977).

In 1980, the Research Station in Swift Current, SK released the cultivar Clarke, a 20-clone synthetic with increased drought tolerance, winter hardiness, and high seed yield. After four to eight cycles of selection, DMY in 'Clarke' was similar to 'Chief' and 7% higher than 'Greenleaf,' which in the previous study yielded less than 'Chief' by 7%. 'Clarke' yielded 28% more seed than 'Chief' and 45% more than 'Greenleaf' (Lawrence, 1982).

'Reliant,' a 6-clone synthetic, traces to a source population of 2500 individually spaced plants. The source population was derived by intermating plants from 24 different intermediate wheatgrass cultivars and experimental strains and bulking equal quantities of seed from each parent plant (Berdahl et al., 1992a). One cycle of selection for resistance to leaf spot and root rot diseases, spring recovery, heading date, nutritive quality, lodging resistance, and DMY was conducted. Four years after establishment, DMY were 100, 96, 88, and 82%, respectively, for 'Reliant,' 'Chief,' 'Clarke,' and 'Greenleaf.' In North Dakota, DMY of 'Reliant' averaged 4.5 Mg ha^{-1} compared with 4.2 Mg ha^{-1} for 'Oahe.' In vitro dry matter digestibility (IVDMD) at anthesis averaged 618, 563, and 560 g kg^{-1} for 'Reliant,' 'Oahe,' and 'Chief,' respectively (Berdahl et al., 1992a).

'Manska' was derived from a source population consisting of 5160 individually spaced plants from 11 different seed lots of PI 116252. After one cycle of selection, which included nutritional quality and DMY, 'Manska' had a 6% increase in in vitro digestible organic matter at anthesis compared with 'Oahe' over several locations. Dry matter yields of 'Manska' were similar to or slightly less than 'Oahe' and 'Reliant'; however, the increased digestibility resulted in an increase in average daily animal gain (ADG) of 1.22 kg in 'Manska' compared with 1.06 kg for 'Oahe' (Berdahl et al., 1993). Seed yields were 13% higher for 'Reliant' than 'Manska.' At Mead, NE, 'Manska' averaged 7.2 Mg ha^{-1} compared with 6.8 Mg ha^{-1} for cultivars Oahe and Slate. Dry matter yields averaged over four test sites and multiple years in North Dakota were 4.2, 4.2, and 4.5 Mg ha^{-1}, respectively, for 'Manska,' 'Oahe,' and 'Reliant' (Berdahl et al., 1993).

'Rush' intermediate wheatgrass, released in 1996 (USDA–NRCS, 1996), is a selection originating from PI 281863 and obtained from the German Botanical Garden in Berlin in 1962. 'Rush' was selected for its superior seedling emergence and DMY on sites with average annual precipitation of 30 cm or above. Across an irrigation gradient that ranged from 20 to 80 cm of water a week, 'Rush' DMY ranged from 1.9 to 5.5 Mg ha^{-1} (Robins, 2010).

Cultivar Haymaker originated from PIs 440015, 440008, and 440011 from the FSU and the cultivar Slate. One cycle of selection within this population for increased DMY resulted in Haymaker having the greatest average DMY compared with released cultivars (Vogel et al., 2005b) with nutritional quality less than Beefmaker (Vogel et al., 2005a), but similar to the other released cultivars.

'Beefmaker' intermediate wheatgrass originated from two cycles of selection within PIs 345586, 273733, 273732, 315353, 315067, and 315355 for DMY and IVDMD by using a modified form of restricted recurrent phenotypic selection. Tested across Nebraska, Kansas, Nebraska, and Wyoming, Beefmaker averaged 7 to 20 g kg^{-1} greater IVDMD than did forage of other released cultivars (Vogel et al., 2005a). Dry matter yields were lower in Beefmaker than Haymaker but similar to other released cultivars.

The most recent cultivar Manifest, released in 2007 (USDA–NRCS, 2007), originated from 10 collections near Stavropol and Svetlograd, in the Caucasian region of Russia. Parent clones of 'Manifest' were selected on the basis of the performance of their polycross progeny for dry matter (DM) and seed yield, spring recovery, and resistance to leaf spot. In cooperative regional trials from 2001 to 2003, Manifest produced 5.2 Mg ha^{-1} DMY compared with 4.9, 4.7, 4.7, 4.6, 4.4, and 4.4 Mg ha^{-1} for Oahe, Reliant, Haymaker, Greenar, Manska, and Beefmaker (FRRL, unpublished data, 2013), which represents a 12% increase in DMY from the earlier cultivar Greenar.

Tall Wheatgrass

Tall wheatgrass is one of the most saline or alkali tolerant cultivated grasses and is particularly noted for its capacity to produce forage and persist in areas that are too alkaline or saline for other productive crops (Asay and Jensen, 1996a). Tall wheatgrass is an autodecaploid ($2n = 10x = 70$; **EEEEEEEEEE**) composed of the **E** genome that originates from the diploid *Thinopyrum elongatum* (Host) D.R. Dewey from the Mediterranean region (Dewey, 1984). The gene base of tall wheatgrass included in North America breeding programs is relatively narrow, with most cultivars tracing to one or two plant introductions. 'Largo' released in 1937, was developed by bulk increase of PI 109452, an accession collected by the Westover–Enlow expedition in Turkey (Alderson and Sharp, 1994). 'Alkar,' which is the widest used cultivar to date, was developed by mass selection within space plants from a large population of PI 98526 from the FSU in 1934. Initial DMY reported for 'Alkar' was 15.0 Mg ha^{-1} with seed yields of 675 kg ha^{-1} (Schwendiman, 1972). 'Alkar' tall wheatgrass ranked among the four highest entries for overall biomass yield, averaging 22.8 Mg ha^{-1} per year over five cuttings, in a comparison of 21 warm-season and six cool-season grasses grown under an irrigation gradient (Robins, 2010).

The cultivar Jose, released in 1965, was developed from PI 150123 (Central Europe) by mass increase, accompanied by removal of atypical plants (Asay and Jensen, 1996a). On an upland site near Hays, KS, cultivars Alkar and Jose averaged 5.6 and 4.3 Mg ha^{-1} over 3 yr, respectively (Harmoney, 2007). 'Orbit,' released in 1966, is a composite of nine open-pollinated lines and one 3-clone synthetic that was derived from PI 98526 and locally selected strains (Lawrence, 1977a). 'Orbit' was selected on the basis of winter hardiness and high seed and forage yield. Over 17 station years across western Canada, 'Orbit' averaged 5.9 Mg ha^{-1}, a 5% DMY gain over 'Alkar' (Lawrence, 1967a). 'Orbit' averaged 538 kg ha^{-1} of seed, a 2% increase over 'Alkar.' The cultivar Platte released in 1972 originated from selections within PI 98526 (VIR) and another breeding line of unknown origin obtained from Cheyenne, WY. The cultivar is noted for its winter hardiness and improved DM and seed production (USDA Extension Service, 1978). In 1981, the cultivar Tyrrell, essentially derived from Largo, was released for its high salinity tolerance (Oram, 1981).

Selection pressure within tall wheatgrass has been minimal. Vogel and Moore (1998) reported that sufficient variation exists in NPGS collections, particularly PIs 98526, 264770, 283163, and 401006, to improve DMY and nutritional quality through selection. This, combined with significant heritability estimates for seed weight (0.86) and DMY (0.55) reported in tall wheatgrass (Pistorale et al., 2008), suggest that gains in DMY, seed weight and yield, and increased nutritional quality are possible.

Wheatgrasses: Pseudoroegneria

Bluebunch Wheatgrass

Bluebunch wheatgrass is an important native grass of the Palouse Prairie and Intermountain sage brush sites. This species is composed of the **St** genome, or some variation of it, in diploid ($2n = 14$; **StSt**) and tetraploid ($2n = 28$; **StStStSt**) taxa. As with *Agropyron*, the polyploid races of *Pseudoroegneria* are autoploids or near-autoploids (Stebbins and Pun, 1953; Dewey, 1970). In addition to being the foundation of the genus *Pseudoroegneria*, the **St** genome is a component of all *Elymus* species.

Improvement of bluebunch wheatgrass to date has been largely at the pre-variety germplasm level with limited to no selection practiced. The first bluebunch wheatgrass cultivar Whitmar (subspecies *inermis* Á. Löve) was released in 1946 (Alderson and Sharp, 1994), originating from a collection made on the Palouse prairie grassland of Washington State. 'Whitmar' demonstrated superior forage and seed yields under drought in a common garden nursery of over 500 bluebunch wheatgrass collections, representing six ecotypes from the Pacific Northwest (Alderson and Sharp, 1994). 'Goldar' bluebunch wheatgrass (PI 5399873), a collection from the Malley Ridge, Umatilla National Forest in Asotin, WA, was subsequently selected from the same evaluation nursery followed by several cycles of mass selection for establishment and persistence (Gibbs et al., 1991) (Table 9–1).

To date, all bluebunch wheatgrass cultivars originated in the Palouse area of Washington State. In an attempt to broaden the genetic diversity in bluebunch wheatgrass, Jones et al. (2002) released 'P-7' bluebunch wheatgrass as a selected class pre-variety germplasm derived as a multiple-origin polycross generated by intermating collections from Washington (13 collections), Oregon (3), Nevada (1), Utah (2), Idaho (3), Montana (1), and British Columbia (1) and cultivars Whitmar and Goldar. It also combined diploid and tetraploid germplasm. Overall nucleotide-sequence within-population variation was greater for 'P-7' (38.7 ± 1.6) than for 'Whitmar' (34.2 ± 1.5) or 'Goldar' (33.9 ± 1.5) (Larson et al., 2000). 'Anatone,' a selected class pre-variety germplasm released in 2004 by the Rocky Mountain Research Station, was selected from an evaluation nursery of approximately 80 collections of bluebunch and Snake River wheatgrass, including 'Goldar' and 'Whitmar,' grown under arid conditions. Collections were compared for stand, vigor, seedling establishment (under cold temperatures), adaptability, and seed production. The original collections came from east of Anatone, in Asotin County, WA, similar to 'Goldar.' Since 2004, numerous pre-variety germplasm and source identified releases have been made. Development efforts in bluebunch wheatgrass have been centered on ecological adaptation and genetic diversity rather than selection for increased seedling establishment, plant persistence, and DMY; hence, on the basis of range trails across the west (FRRL, unpublished data, 2013), there has been no improvement in the above traits compared with initial collections, but improvement in these traits may be possible under selection.

Wheatgrasses: Elymus

Elymus is by far the largest genus of the wheatgrasses (Dewey, 1984) and is the most complex genomically with chromosome numbers ranging from $2n = 28$ (allotetraploids) to $2n = 56$ (allooctaploids). Within this genus, all species contain at least one basic **St** genome from *Pseudoroegneria*, in combination with one or more genomes **H** (small seeded *Hordeum* sp.), **Y** (unknown origin), **W** (*Australopyrum* sp.) or **P** (crested wheatgrass sp.) (Wang et al., 1995) in multiple polyploid series.

Snake River Wheatgrass

Before 1986, Snake River wheatgrass was treated taxonomically as bluebunch wheatgrass. However, genetic relationships described by Carlson and Barkworth (1997) demonstrated that Snake River wheatgrass is more closely related to thickspike wheatgrass in that both are allotetraploids ($2n = 28$; **StStHH**) composed of the **St** genome originating from the genus *Pseudoroegneria* and the **H** genome from the genus *Hordeum.* Snake River wheatgrass differs genomically from bluebunch wheatgrass in that the latter is composed exclusively of the **St** genome.

'Secar' Snake River wheatgrass, released in 1980, was selected from the same 500-collection evaluation nursery as 'Whitmar' and 'Goldar' for its divergent awns, early maturity, and drought resistance. The collection originated from the Lewiston Grade in the Snake River gorge, near Lewiston, ID in 1938. 'Discovery' was released in 2008 and originated from four collections that trace to Whitman and Asotin counties in southeastern Washington and Idaho County. These collections were crossed and seed was harvested in bulk to generate 'Discovery.' The stand frequency of 'Discovery' was 49% compared with 27% for 'Secar' on dryland range sites in the Great Basin receiving between 20 and 38 cm of annual precipitation. The DMY of 'Discovery' was 108% greater than 'Secar' (Jones, 2008). On the basis of variation for 100-seed weight and DM and seed yield among PIs and half-sib families of Snake River wheatgrass, and narrow-sense heritability estimates (h^2) of 0.57 ± 0.17, 0.57 ± 0.17, and 0.63 ± 0.10, respectively (Jensen et al., 2012), it seems reasonable to expect that further selection within this species would be effective (Table 9–1).

Thickspike Wheatgrass

Thickspike wheatgrass, called northern wheatgrass in Canada, is a native allotetraploid species ($2n = 28$) with a genome constitution of **StStHH**. Genetic introgression between thickspike and bluebunch wheatgrass frequently occurs in nature. Montana wheatgrass (Syn. *Agropyron albicans* Scrib. and Smith) and Wyoming wheatgrass (Syn. *A. griffithsii* Scribn. and Smith. ex Piper) are such examples. Introgression between thickspike and slender wheatgrass has also been reported (Dewey, 1975). Thickspike wheatgrass is so closely related, both biologically and morphologically, to streambank wheatgrass (*A. riparium* Scribner and J.G. Smith) that Dewey (1984) treated streambank wheatgrass as a glabrous form of thickspike wheatgrass.

'Sodar' thickspike (streambank) wheatgrass, released in 1954, originated from a collection made near Canyon City, OR and was the best of 11 collections evaluated for low-growth habit and narrow and smooth leaves (Asay and Jensen, 1996a). 'Sodar' was improved through several generations of mass selection to reduce off types. 'Critana' thickspike (streambank) wheatgrass, released in 1971, originated from collections along roadside cuts in north central Montana near Havre in 1960. It was selected from approximately 60 collections representing native materials from Montana and Wyoming (Stroh et al., 1972). At Decker, MT, there were no differences between 'Sodar' and 'Critana' for establishment; however, 'Sodar' persisted longer than 'Critana' (Asay et al., 2001) (Table 9–1).

'Elbee,' released in 1980, is an 8-clone synthetic selected from 49 ecotypes that originated from southern Alberta and southwestern Saskatchewan. The eight clones were selected on the basis of the performance of polycross progeny for DM and seed yield, uniformity in plant height, and aggressive creeping roots. Under dryland conditions, 'Elbee' produced 7% more DMY than 'Critana' at 3136 kg ha^{-1}. Seed yields were 20% higher in "Elbee' than 'Critana' (Smoliak and Johnson, 1982).

The most recent cultivars include Schwendimar (1994) and Bannock (1995) thickspike wheatgrass. 'Schwendimar' originated from a collection made on windblown sands along the banks of the Columbia River, east of The Dalles, OR, in 1934. Mass selection over several generations was conducted to reduce the frequency of aberrant plants, and then the seed was bulked. 'Bannock' was a composite of six collections made from Pocatello, ID; The Dalles, OR; and Quincy, WA before 1948. Subsequently, in the 1950s, six individual plants were selected on the basis of visual vigor and DMY, and seed was increased. On the basis of evaluation trials in Yakima, WA and Blue Creek, UT, seedling establishment was not significantly different between 'Bannock' (68%) and 'Critana' (62%); however, 'Bannock' exhibited greater persistence (60%) than 44% for 'Critana.' At Blue Creek, UT, 'Bannock' produced 12% more DMY than 'Critana' (Asay et al., 2001). In trials at Beaver, UT during 2008 to 2012, there were no significant differences between 'Sodar,' 'Critana,' 'Bannock,' and 'Schwendimar' in their ability to establish on the basis of stand frequency. Forage yields of 'Bannock' at 830 kg ha^{-1} were 49, 41, and 30% greater than for 'Sodar,' 'Critana,' and 'Schwendimar,' respectively (FRRL data, unpublished data, 2013).

Slender Wheatgrass

Slender wheatgrass has been one of the most widely used self-pollinating native grasses in revegetation programs on the rangelands of the western United States and Canada (Asay and Jensen, 1996a). With the emphasis on providing native grasses for restoration (i.e., revegetating with plant materials found only in the local area), many of the recent slender wheatgrass releases have been derived from direct seed increases of a single collection or plant on the basis of a unique trait or ecosystem. Such released germplasm requires no comparative data to support the release. The first commercial cultivar Primar was released in 1946 and originated from a collection near Beebe, MT in 1933. This collection was compared with over 120 collections of slender wheatgrass and tested over 10 yr across the Pacific Northwest (Myers, 1951). The cultivar Revenue released in 1970, originated from a single plant selected near Revenue, SK in 1961. It was evaluated and selected from over 750 collections from 1959 to 1969 (Crowle, 1970). 'Revenue' showed a 5% yield advantage over 'Primar'; however, 'Primar' had a 13% advantage in digestible dry matter and a higher leaf-to-stem ratio (Crowle, 1970). Seed yield in 'Revenue' was similar to 'Primar' (Table 9–1).

'San Luis' slender wheatgrass released in 1984, originating from the direct seed increase of a collection near the Upper Colorado Environmental Plant Center in the San Luis Valley in Rio Grande County, CO (http://plant-materials.nrcs.usda.gov/copmc/, accessed 12 Feb. 2014). 'Pryor' slender wheatgrass, released in 1988, resulted from the direct seed increase without selection of a collection made on a dry drainage way within a saline upland range site within the Clarks Fork of the Yellowstone River Valley between the Pryor and Beartooth Mountains (Majerus et al., 1991).

'Adanac,' released in 1990, originated from a single plant collected near Climax, SK in 1961 with subsequent seed increase. 'Adanac' has 7% higher DMY than 'Revenue.' 'Hillcrest' and 'Highlander' released in 1994 and 1995, respectively, were derived from direct seed increase from a single plant collected in 1985 near Crowspass, Hillcrest, AB. They were evaluated with 100 slender wheatgrass collections for seed production and vigor. 'Hillcrest' had a mean seed yield of 594 kg ha^{-1} compared with 1034 kg ha^{-1} for 'Highlander' and 892 kg ha^{-1} for 'Revenue' (Darroch and Acharya, 1996). 'Mountaineer,' referred to as broadglumed wheatgrass [*E. trachycaulus* subsp. *violaceus* (Horne.) Á & D Löve], was derived from a single plant collected in 1984 from a high elevation site south of Kananaskis Country, AB (Darroch et al., 2001). 'Copperhead' selected class germplasm originated directly from a collection near Wisdom, MT where the site was severely impacted by smelter fallout, surface wind, and water transported contaminates. Copperhead germplasm has exhibited superior emergence, survival, and biomass production on amended acid and heavy metal impacted soil.

'FirstStrike' slender wheatgrass, released in 2006, is a multi-origin composite of four collections from Colorado and Wyoming. After one cycle of selection for seedling establishment (i.e., emergence from a deep planting depth), persistence, and plant vigor, 'FirstStrike' averaged 45 and 8% denser stands than 'Pryor' and 'San Luis,' respectively, over nine trials across the western United States (Jensen et al., 2007). 'FirstStrike' germinated 5 d earlier than 'Pryor' on three different soil types (sandy loam, loam, and sandy). As expected, with better initial stands, persistence was greater in 'FirstStrike' by 51% over plots of 'Pryor.' Dry matter yield increased in 'FirstStrike' by 27% over that of 'Pryor' (Jensen et al., 2007).

Wheatgrasses: Pascopyrum

Western Wheatgrass

Western wheatgrass [*Pascopyrum smithii* (Rydb.) Á. Löve] is native to North America and is distributed throughout the western two-thirds of the United States and Canada (Hitchcock, 1971). It is an important rangeland and pasture grass and frequently occurs as the dominant species in native plant associations because of its aggressive rhizomes (Asay and Jensen, 1996a). It is a true allooctaploid with a genomic formula of **StStHHNsNsXmXm** (Dewey, 1984; Wang et al., 1995). Intermediate parents came from *Elymus* (*E. lanceolatus;* **StStHH**) and *Leymus* (*L. triticoides;* **NsNsXmXm**), and its diploid ancestors came from *Pseudoroegneria* (**StSt**), *Hordeum* (**HH**), *Psathyrostacys* (**NsNs**), and unknown diploid (**XmXm**). On the basis of chloroplast DNA, its maternal lineage traced through *E. lanceolatus* to a diploid *Pseudoroegneria* species (Jones et al., 2000).

Early seedings were made with seed harvested from native stands. 'Barton' (1970) was a direct seed increase from a collection along Walnut Creek near Heizer, KS (Alderson and Sharp, 1994). Seed harvested commercially from native meadows along Porcupine Creek, northwest of Forsyth, MT (1959), with subsequent seed increase, resulted in the cultivar Rosana (1972). This seed lot was evaluated with approximately 60 other western wheatgrass collections representing native germplasm from Montana and Wyoming (Alderson and Sharp, 1994). 'Arriba' (1973) was derived from a collection near Flager, CO in 1957, followed by five seed increase generations. (Alderson and Sharp, 1994). 'Flintlock' (1975) resulted from two cycles of recurrent selection for DMY, seed production, and rhizome spread, which originated from 30 collections across central and southwestern Nebraska and northwestern Kansas (Alderson and Sharp, 1994) (Table 9–1).

The cultivar Walsh (1982), a 20-clone synthetic, traces back to 468 different western wheatgrass ecotypes collected in 1968 and 1969 from indigenous stands in southern Alberta and southwestern Saskatchewan. The clones were selected on the basis of polycross performance for DM and seed yield, leafiness, rhizome spread, and uniformity in plant height and maturity (Smoliak and Johnson, 1984). Parental materials for 'Rodan' (1983) came from a commercial seed field of unknown origin grown in the Missouri River valley near Mandan, ND. On the basis of progeny-row performance for density, foliage cover, leafiness, and stem rust, 13 half-sib families were intercrossed and bulked to form this cultivar. Over the next seven generations, low intensity selection was conducted to remove off types (Barker et al., 1984).

The most recent western wheatgrass cultivar, Recovery (2011), combined germplasm from Rosana (28%), D2945 (50%), a different seed lot of the same population that gave rise to Rodan, and WW117FC (22%) a native collection made near Fort Carson, CO. Selection for seedling emergence from a deep planting depth resulted in 'Recovery' having significantly higher frequency of emerged seedlings at 60% compared with 'Rosana' (42%), 'Rodan' (45%), 'Arriba' (45%), 'Barton' (42%), and 'Flintlock' (53%) (Waldron et al., 2011). There was no increase in DMY in Recovery compared with the other cultivars.

Wildryes

The wildrye grasses are a diverse group of perennial grasses primarily included in the genera *Psathyrostachys, Elymus,* and *Leymus*. The wildryes are valuable sources of forage and habitat for livestock and wildlife. The wildryes are adapted throughout the temperate regions of the world, primarily in Asia, Europe, and North America (Asay and Jensen, 1996a).

Wildryes: Psathyrostachys

Russian Wildrye

Psathyrostachys species are predominately diploid (**NsNs**); however, autotetraploid (**NsNsNsNs**), and autohexaploid (**NsNsNsNsNsNs**) collections have recently been developed in Russia, Kazakhstan, and China, where this genus is native to the steppe and desert regions. The basic **Ns** genome of this genus is widely distributed among North American and Asian species of *Leymus* and *Pascopyrum* (Dewey, 1984; Jensen et al., 2005a). Only one species in the genus, Russian wildrye, has achieved recognition as an important forage grass used on rangelands of western North America, providing forage for grazing during the fall and winter (Rogler and Schaaf, 1963).

A Russian wildrye breeding program was initiated in 1936 by the USDA-ARS at the Northern Great Plains Field Station in Mandan, ND. Improved strains were released from this program in 1941 and 1942 for general use in North and South Dakota (Rogler and Schaaf, 1963). Selected from the improved strains, the first cultivar of Russian wildrye, Vinall, was released in 1960 as a 5-clone synthetic selected for increased seed yield and resistance to lodging (Rogler and Schaaf, 1963). The cultivar Sawki was released in 1963 as a 10-clone synthetic selected from germplasm obtained from the Western Siberian Experiment Station in Omsk, FSU for improved DM and seed yield (Lawrence, 1967b, 1977b) (Table 9–1).

Parent selections for the cultivar Mayak, a 20-clone synthetic released in 1971, were made based on open-pollinated and polycross progeny tests with emphasis on increased DM and seed yield and resistance to leaf diseases. This cultivar combined germplasm from breeding programs in Swift Current, SK and Mandan, ND. At nine locations in western Canada and in Mandan, ND, Mayak had DMY of 2834 kg ha^{-1}, a 6 to 7% increase over the earlier cultivars Sawki and Vinall, respectively (Lawrence, 1972). There was also an increase in seed production of 16% or more in 'Mayak' over 'Sawki' and 'Vinall.' Up to this point, relatively slow seedling growth and development hampered the overall use and success of Russian wildrye as a range grass. Breeding programs began to select Russian wildrye seedlings on their rate of emergence from a deep planting depth (Maguire, 1962) in attempts to improve seedling establishment in Russian wildrye (Table 9–1).

'Cabree,' released in 1976 as a 6-clone synthetic, was selected for enhanced seedling establishment. The original clones were selected on the basis of increased seed retention from a Russian wildrye field of unknown origin at the Agricultural Canada Research Substation in Manyberries, AB in 1955. Those clones were selfed and their S_1 progeny evaluated in Lethbridge, AB under dryland conditions for seed retention, seedling establishment, DM and seed yield, and culm length (Smoliak, 1976). On the basis of selfed progeny performance, six clones were selected and intercrossed. In Lethbridge, AB and Swift Current, SK, 'Cabree' exhibited 37 and 31% seed shattering compared with 'Vinall' at 59 and 50%, 'Sawki' at 69 and 58%, and 'Mayak' at 65 and 59%, respectively (Smoliak, 1976). Dry matter yields increased in 'Cabree' by 1% over 'Mayak,' 9% over 'Sawki,' and 24% over 'Vinall' (Smoliak, 1976).

The cultivar Swift is a 26-clone synthetic released in 1978 after four cycles of recurrent selection for seedling emergence from a deep planting depth, disease resistance, seed quality, and seed and forage yield (Lawrence, 1979). Selection for seedling emergence from a deep planting depth resulted in 'Swift' having significantly more seedling emerge after planting at 63% compared with 'Mayak' (43%), 'Swaki' (37%), and 'Cabree' (30%) in Indian Head, SK. Mean DMY in Swift were equal to Sawki, Mayak, and Cabree in cooperative tests across western Canada for 50 station years (Lawrence, 1979).

'Bozoisky-Select' (Asay et al., 1985b) was developed from PI 406468 (Bozoisky), obtained from the FSU, and has since become one of the prominent Russian wildrye cultivar in rangeland seedings. The breeding population was subjected to two cycles of recurrent phenotypic selection for improved vigor, leafiness, seed yield, coleoptile length (i.e., correlated to seedling vigor), and seedling establishment. At eight semiarid range locations in Utah, Idaho, and Wyoming, 'Bozoisky-Select' yielded 23% more forage than 'Vinall' during the first two production years (Asay et al., 1985b). Seedling establishment of 'Bozoisky-Select' was equal to or superior to 'Vinall' in over 20 trials representing the sagebrush (*Artemisia* spp.), juniper (*Juniperus* spp.), shadscale [*Atriplex confertifolia* (Torr. & Frem.) S. Wats], greasewood [*Sarcobatus vermiculatus* (Hook.) Torr.] and Indian ricegrass [*Oryzopsis hymendoides* (Roem. & Schult.) Ricker] ecosystems.

The next phase of Russian wildrye breeding involved the development of induced tetraploids, which exhibited increased seedling establishment and DMY over the diploid cultivars (Berdahl and Barker, 1991). These increases may in part be due to greater complementary gene interaction at the tetraploid level (Bingham et al., 1994) and increased cell size in tetraploids (Dewey, 1979). The first tetraploid

Russian wildrye cultivar Tetracan was released in 1988 and was derived from colchicine-induced tetraploids from several diploid strains (Dhindsa and Slinkard, 1963). Chromosome doubling resulted in an increase in seed weight with 'Tetracan' averaging 456 mg 100-seed^{-1} compared with 276, 243, and 289 mg 100-seed^{-1} for 'Swift,' 'Mayak,' and 'Cabree,' respectively (Lawrence et al., 1990). Seedling emergence of 'Tetracan' from a 4.5-cm planting depth at 25 d was 55% compared with 41, 18, and 30% for 'Swift,' 'Mayak,' and 'Cabree,' respectively. The effect of chromosome doubling, however, had little effect on DM and seed yield when compared with the diploid cultivars (Lawrence et al., 1990).

In 1998, RWR-Tetra-1 germplasm (Jensen et al., 1998) was the first Russian wildrye release to include natural occurring Russian wildrye tetraploid germplasm tracing back to four collections made in 1988 from Kazakhstan. The four natural tetraploid collections were combined with six induced tetraploids followed by two cycles of selection for vegetative vigor, seed yield, 100-seed weight, and seedling emergence from a 7.6-cm depth in the greenhouse. Dry matter yield in RWR-Tetra-1 did not differ significantly from Tetracan and the diploid cultivars Bozoisky-Select, Cabree, and Vinall; however, the 100-seed weight and rate of emergence were significantly greater than the other diploid cultivars.

The Russian wildrye cultivar Mankota (Berdahl et al., 1992b) is a 6-clone synthetic that traces back to PI 314675 from near Alma-Ata, FSU and Mandan bulk populations. One cycle of selection for coleoptile length, emergence from a 5-cm planting depth in the greenhouse, stand establishment in the field, and DM and seed yield was conducted. 'Mankota' averaged 17% higher DMY than did 'Vinall' across five locations (Berdahl et al., 1992b). Under drought conditions, DMY averaged 1.9 and 1.4 Mg ha^{-1} for 'Mankota' and 'Swift,' respectively. Seed yield and in vitro digestible organic matter was lower in 'Mankota' than in 'Swift,' suggesting that in Russian wildrye, as DMY increases there is a corresponding decrease in seed yield and digestibility.

The most recent Russian wildrye cultivar Bozoisky-II (Jensen et al., 2006b) included parental germplasm from Bozoisky-Select, Vinall, and plant introductions from Kazakhstan. 'Bozoisky-II' contains more DNA polymorphisms and thus has a broader genetic background than 'Bozoisky-Select,' 'Cabree,' 'Mankota,' 'Swift,' and 'Tetracan' (Jensen et al., 2006b). Selection emphasis on 'Bozoisky-II' was for increased seedling establishment and DMY. Initial stands and persistence of 'Bozoisky-II' were similar to 'Bozoisky-Select' and 'Mankota' but greater than 'Tetracan' in trials across Utah, Nebraska, North Dakota, and Montana. In the same trails, 'Bozoisky II' produced 6, 9, and 23% more DMY than 'Bozoisky-Select,' 'Mankota,' and 'Tetracan,' respectively (Jensen et al., 2006b).

Wildryes: Leymus

Species of the genus *Leymus* are long-lived perennials that are distributed from the coastal regions of the North Sea (Beach wildrye) across Central Asia [Altai wildrye, *Leymus angustus* (Trin.) Pilger] and western North America [Great Basin wildrye, *L. cinerenus* (Scribn. and Merr.) Á. Löve and beardless wildrye, *L. triticoides* (Buckl.) Pilger]. *Leymus* grasses often occupy saline and alkaline sites (Asay and Jensen, 1996b). More than half of *Leymus* species are allotetraploids ($2n = 28$), represented genomically as **NsNsXmXm**. The higher polyploid species ($2n$ = 42–84) are complex autoallopolyploids (Wang et al., 1995; Jensen and Wang, 1997), with some data suggesting they might be segmental autopolyploids composed of the $\mathbf{Ns_1Ns_2}$ genomes (Zhang and Dvořák, 1991; Anamthawat-Jónsson and Bödvarsdóttir, 2001). At the higher ploidy levels in *Leymus*, unreduced gametes have been observed (Jensen et al., 2002). With the exception of *Thinopyrum*, the genus *Leymus* is of most interest to cereal breeders as it represents a valuable source of resistance to drought, salinity, and disease through induced amphiploidy between *Leymus* and *Triticum* (Dewey, 1984; Asay and Jensen, 1996b).

Basin Wildrye

Early cultivars of basin wildrye were selections made from superior collections evaluated extensively followed by seed increase. 'Magnar' is a native octaploid basin wildrye, blue-green in color, tall growing, with broad course leaves, and is frequently used for winter forage and wind barriers. It was released in 1979 after several generations of selection for vigorous types. The parent material of 'Magnar' was originally received from the University of Saskatchewan in Saskatoon, Canada in 1939; however, detailed collection site information is not known. 'Trailhead' basin wildrye, released in 1991, originated from a seed collection near Roundup, MT. It was evaluated along with 125 basin wildrye collections from Montana and Wyoming. 'Trailhead' was superior in DMY and stand longevity. The original collection of 'Trailhead' was directly increased without selection (Asay and Jensen, 1996b). 'Washoe' germplasm basin wildrye is a pre-varietal selected class release from a collection originating near Anaconda, MT. The collection site is within the Anaconda Smelter Superfund Site where the pH ranged from 4.6 to 5.6 and arsenic, cadmium, copper, lead, and zinc concentrations ranged from moderate to excessive (Marty, 2003). 'Washoe' was directly increased without selection (Table 9–1).

Continental is a cultivar (Jones et al., 2009) derived from chromosome-doubled 'Trailhead' plants pollinated by the natural octoploid, 'Magnar.' Two cycles of recurrent selection were completed for secondary tiller production (one cycle), visual vegetative vigor, and upright growth habit. Dry matter yield production in 'Continental' was intermediate to both parents, averaging 23.6 g plant^{-1}. As expected, on the basis of the doubling of the chromosome number, 'Continental' seed was significantly heavier at 4.8 mg seed^{-1} compared with 'Magnar' at 4.1 and 'Trailhead' at 3.3. The stand density of 'Continental' was significantly greater, at 52% across four locations in Utah and Colorado, than 'Magnar' and 'Trailhead' at 21 and 42%, respectively (Jones et al., 2009). There were no significant differences in DMY between 'Continental,' 'Magnar,' and 'Trailhead' basin wildrye (FRRL, unpublished data, 2013).

Beardless Wildrye

Cultivar development in beardless wildrye has been restricted to the evaluation of ecotypes. The first beardless wildrye cultivar, Shoshone, originated in 1958 from a stand in Riverton, WY. On the basis of morphological characters (Asay and Jensen, 1996b) and chloroplast DNA sequences (Culumber, 2007), 'Shoshone' was shown to be similar to Eurasian *L. multicaulis* (Kar. and Kar.) Tzvelev (Young-Mathews and Winslow, 2010). The most recent cultivar, Rio, was collected in 1973 from a native stand in Stratford, Kings County, CA. It demonstrated superior seed viability and initial sod establishment in comparison with 12 other beardless wildrye collections from California (Benes et al., 2012; Young-Mathews and Winslow, 2010) (Table 9–1).

Hybrids between beardless and basin wildrye are partially fertile, and it has been suggested that it may be possible to introgress simply inherited traits from one species to another (Dewey, 1984). In particular, Dewey (1984) suggested that the seed germination of beardless wildrye could be improved by introgression of genes from basin wildrye, and the growth habit of caespitose basin wildrye could be improved by introgression of rhizome genes for rhizome production from beardless wildrye. Recent F_1 hybrids have displayed increased DMY, and progeny of these hybrids expressed transgressive segregation for DMY, forage quality, and many other traits including rhizome production and seed retention (Larson et al., 2006; Larson and Mayland, 2007). These observations indicate the potential of improving DMY, seed retention, and nutritional quality through hybridization.

Beach Wildrye

Beach wildrye is a species complex consisting of North American species *L. mollis* (Trin.) Pilger ($2n = 28$) and its Eurasian counterpart *L. arenarius* (L.) Hochst ($2n = 56$). Although beach wildrye is not considered to be an outstanding forage grass, it has a long history as a forage crop in Alaska (Klebesadel, 1985) and is used on coastal reclamation and dune stabilization. Two cultivars, Benson and Reeve, were developed by the Alaska Plant Materials Center (Wright, 1994a, 1994b). 'Benson' was derived from a native collection from one plant of *L. mollis* collected near Narrow Cape, Kodiak, AK. Reeve originated from a Eurasian collection of *L. arenarius* (PI 345978), where it outperformed 18 other beach wildrye ecotypes in hardiness and seed production from 1979 to 1983 (Table 9–1).

Altai Wildrye

Altai wildrye ($2n = 12x = 84$) forage maintains its nutritional value better during the late summer and early fall than many cool-season grasses. These qualities, combined with its erect culms and leaves that stand above the snow, make Altai wildrye a potentially valuable species for extending the grazing season (Asay, 1987; Lawrence, 1976). The most serious limitations of Altai wildrye are poor seedling establishment and low seed yields (Voigt and MacLauchlan, 1985). The first commercial cultivar was Prairieland, released in 1976. It is a 22-clone synthetic that traces back to two plant introductions, Sc 15011 collected on the Steppe of Kustanay, FSU in 1937 and Sc 15012 from Voronezh, FSU in 1934. Plants of 'Prairieland' segregated for light green (non-glaucous) to blue-green (glaucous). Dry matter and seed yield of 'Prairieland' over 23 station years was 4.4 Mg ha^{-1} and 208 kg ha^{-1}, respectively (Lawrence, 1976). 'Pearl,' released in 1989, is a 4-clone synthetic derived from three cycles of recurrent selection for DM and seed yield, increased seedling emergence from a deep planting depth in the greenhouse, freedom from leaf spot, and seed quality in a glaucous population that traces back to the collections used to develop 'Prairieland.' Selection resulted in 'Pearl' having significantly higher seedling emergence than 'Prairieland' (70 vs. 41%), and it produced 22% more seed than 'Prairieland'; however, 'Pearl' produced 6% less DMY than 'Prairieland' over six locations (Lawrence et al., 1991a) (Table 9–1).

The ancestry of 'Eejay,' released in 1989, traces back to seed lots originating from Sc 15011 and Sc 15012. 'Eejay' is a 12-clone synthetic arising from three cycles of recurrent selection for increased DM and seed yield, emergence from deep seeding, and freedom from leaf spot diseases within a non-glaucous population. 'Eejay' yielded 18% more seed than 'Prairieland' over 43 station years and was similar to 'Pearl' (Lawrence et al., 1991b). As with 'Pearl,' there were no observed increases in DMY when compared with 'Prairieland.'

'Mustang' Altai wildrye, released in 2004 was derived from four cycles of selection for seedling establishment (emergence from deep planting depth) and DM and seed yield. The parental germplasm from 'Mustang' was derived from selections within PI 499650 ('DT-3185'; 79%), 'Prairieland' (7%), 'Eejay' (7%), and 'Pearl' (7%). PI 499650 originated about 50 km southwest of Urumqi in Xinjiang Province, People's Republic of China.

'Mustang' was evaluated in the Northern Plains Regional Trials in Utah, Nebraska, North Dakota, and Montana for DMY, seedling establishment, and persistence from 2001 to 2003. Combined over six locations and years, 'Mustang' produced 21 and 26% more DMY than 'Prairieland' and 'Pearl,' respectively. Selection for seedling establishment resulted in a 23 and 18% increase in 'Mustang' over 'Prairieland' and 'Pearl,' respectively. Overall persistence was measured by stand density in 2003, and 'Mustang' had a 45 and 30% increase in stand density compared with 'Prairieland' and 'Pearl,' respectively (Jensen et al., 2005b). Across 34 field studies, 'Mustang' had at least a 27% or greater increase in stand density than either 'Pearl' (48%) or 'Prairieland' (32%) (Robins et al., 2013).

Interspecific Hybridization

With the identification of genome relationships in the Triticeae grasses (Dewey, 1984), our understanding and ability to transfer and or combine traits utilizing parental materials from primary, secondary, and tertiary gene pools from different species or genera have been enhanced. Because of the complex genome relationships among wheatgrass and wildrye grasses, each genome combination has its own gene pool, yet there are large secondary and tertiary gene pools that can be used to enhance germplasm development within the range grasses. The best example of using grasses from the secondary gene pool for germplasm enhancement is the release of RS hybrids (Asay and Dewey, 1981), which are hexaploid **StStStStHH** derivatives of the cross between hexaploid quackgrass [*Elymus repens* (L.) Gould; **StStStStHH**] and tetraploid bluebunch wheatgrass (*P. spicata*; **StStStSt**). This 42-chromosome hybrid combined the grazing and salt tolerance of quackgrass with the drought tolerance and palatability of bluebunch wheatgrass into populations that eventually led to the release of the cultivar NewHy (Asay et al., 1991).

Native C_4 Forages

Indiangrass (*Sorghastrum* spp.), eastern gamagrass [*Tripsacum dactyloides* (L.) L.], big bluestem (*Andropogon gerardii* Vitman), and gramas (*Bouteloua* sp.) are among the C_4 grasses that are native to North America and have undergone minimal selection for DMY. Apomictic tetraploids ($2n = 72$) of eastern gamagrass as well as a sexual diploid ($2n = 36$) were released in 1988 (USDA-NRCS, 2006; Springer and Dewald, 2004). One of the major restrictions with eastern gamagrass is the low seed production. The latest cultivar Verl, released in 2005 by the USDA-ARS with the Oklahoma Agricultural Experiment Station and USDA-NRCS (Springer et al., 2006), has excellent seed production but is susceptible to maize billbug (*Sphenoph-*

orus maidis Chittenden) and southern cornstalk borer (*Diatraea crambidoides* Grote) (USDA-NRCS 2006; Springer and Dewald, 2004). Although native grasses other than switchgrass (*Panicum virgatum* L.) do not compete in yield with most of the introduced bunchgrasses, eastern gamagrass and big bluestem are favorable for conversion to ethanol (Anderson et al., 2008). With renewed interest in bio-fuel production from cellulosic material, breeding efforts have been renewed.

Big bluestem is another native species that has lacked effort for genetic improvement. Of the limited number of improved cultivars available on the market, most are synthetic composites (Vogel et al., 2006a, 2006b). Among the traits that need improvement in big bluestem are reduced hard seededness, fast establishment, and resistance to late season lodging (Anderson et al., 2008). However, big bluestem produces twice the biomass per unit of applied nitrogen than either switchgrass or indiangrass [*Sorghastrum nutans* (L.) Nash] (Perry and Baltensperger, 1979) and can be converted to ethanol more efficiently than switchgrass (Weimer and Springer 2007), making it a favorable bioenergy feedstock.

Tropical C_4 Grasses

Tropical and subtropical C_4 grass species are in the panicoid tribe, which optimizes plant growth at temperatures of 30 to 40°C and high light intensities (Burton, 1993). A number of grasses from tropical origins have been introduced for use as forage in the United States. To date, the three with the longest breeding history are bermudagrass [*Cynodon dactylon* (L.) Pers.], stargrass [*Cynodon nelmfuensis* (K. Schum.) Pilg.], and bahiagrass (*Paspalum notatum* Flueggé).

A number of other tropical grasses have been introduced and evaluated. Buffelgrass (*Cenchrus ciliaris* L.) is drought tolerant and appropriate for dry areas of Texas. Most cultivars are a result of ecotype selection (Hanselka et al., 2004), but two obligate apomictic hybrids ('Nueces' and 'Llano') from sexual mutants and apomictic males have been developed (Bashaw, 1980). Cultivars have been produced from mostly apomicitic ecotype selections for kleingrass (*Panicum coloratum* L.) (Tischler and Ocumpaugh, 2004), dallisgrass (*Paspalum dilatatum* Poir.) (Evers and Burson, 2004), digitgrass (*Digitaria* sp.) (Pitman et al., 2004), braciariagrass (*Brachiaria* sp.) (Miles et al., 2004; Vendramini et al., 2012), limpograss [*Hemarthria altissima* (Poir.) Stapf and C. E. Hubbard] (Vendramini et al., 2008; Inyang et al., 2010), old world bluestems (Coleman et al., 2004b), and lovegrass (*Eragrostis* sp.) (Voight et al., 2004).

Hybridization and selection has resulted in improved cultivars of napiergrass (*Pennisetum purpureum* Schumach). There are a few breeding programs that are developing and evaluating interspecific hybrids of napiergrass with pearl millet (*Pennisetum glaucum* [L.] R. Br.) in the United States. Interest in the tall bunchgrass genus has been increased recently as a potential bioenergy feedstock because of DMY of up to 60 Mg ha^{-1} in southern Florida (Prine et al., 1997) and over 20 Mg ha^{-1} in North Florida and Georgia (Anderson et al., 2008). Though the vegetatively propagated napiergrasses are used extensively throughout the tropics as forage, only a few cultivars have been bred and released in the United States. The tall biomass cultivar Merkeron, selected for improved DMY and disease resistance, originated from a cross between a high yielding clone and a dwarf leafy clone of napiergrass (Burton, 1989b). A semi-dwarf genotype was selected from progeny of 'Merkeron' and was released as 'Mott' (Sollenberger et al., 1988). The napiergrass nursery has also been assessed for genetic diversity by AFLP, and of the 240 scorable bands, approximately 43% were polymorphic (Harris et al., 2010). This information will assist breeders in determining parental lines for breeding that can maximize genetic variability.

Bermudagrass

One of the most productive warm-season pasture grasses is bermudagrass. Bermudagrass will generally produce more hay than other C_4 grasses in Southern United States with sufficient nutrients and water. The *Cynodon* species are native to Africa and were brought to the United States in precolonial times (Taliaferro et al., 2004). It became a prominent forage across the South by the early 1800s, and though it is not a prolific seed producer, fecundity is sufficient for naturalization to occur across a large geographic range. Unimproved pasture of naturalized bermudagrass is generally given the name 'common' bermudagrass. Similar to many perennial grasses, bermudagrass is self-incompatible, so superior clones are vegetatively propagated. Pastures are propagated by sprigging rhizomes, stolons, and root sections through adapted machinery (Hanna and Anderson, 2008). The expense of vegetative establishment is offset by pasture persistence. Many pastures have undergone rotational or continuous grazing for over 50 yr (Rouquette et al., 2011), without reestablishment.

Bermudagrass and stargrass land races are composed of a number of species in the *Cynodon* genus and are generally either diploid ($2n$ = 18) or tetraploid ($2n$ = 36) (Taliaferro et al., 2004). Germplasm collections from Africa and Asia were made in the early 20th century (Hanna and Anderson, 2008; Harlon, 1970) and large nurseries are maintained in Tifton, GA (USDA-ARS) and in Stillwater, OK (Oklahoma State University). Most of the accessions have been phenotypically characterized, a core collection has been developed (Anderson, 2005), and diversity has been assessed (Anderson et al., 2009b).

Some selections from naturalized stands of *C. dactylon* were made in the late 19th century and early 20th century on the basis of biotypes (Tracy, 1917). Others were selected from fairground parks ('Hardy'), agricultural stations ('Greenfield') and in farmers' fields ('Tift') (Taliaferro et al., 2004). The first cultivar from a controlled breeding program, Coastal, was released in 1943 from a cross between Tift and South African accessions (Burton, 1943). Hybrid vigor can be fixed since Coastal and many subsequent cultivars are vegetatively propagated. Breeding consists of crossing, evaluating a large number of progeny, selection, increase, and testing (Burton, 1947). Subsequent releases of vegetatively propagated hybrids from Burton's breeding program included 'Suwanee' (drought tolerant) (Burton, 1962) 'Midland'(Hein, 1953), 'Tifton 44' (cold hardiness) (Burton and Monson, 1978), 'Coastcross 1'(Burton, 1972), 'Tifton 68' (good forage quality) (Burton and Monson, 1984), 'Tifton 78' (forage quality and yield) (Burton and Monson, 1988), and 'Tifton 85' (superior in forage quality and yield) (Burton et al., 1993). Oklahoma State University has released 'Hardie' (high yield) (Taliaferro and Richardson, 1980), 'Brazos' (good forage quality) (Eichhorn et al., 1984), and 'Midland 99' (high yield and superior persistence) (Taliaferro et al., 2002). 'Callie' (Watson, 1974) was released from Mississippi, 'Grazer' from Louisiana (Eichhorn et al., 1996), 'Russell' (Ball et al., 1996) from Alabama, and 'Florakirk' (Mislevy et al., 1999) from Florida. 'Florico' (Mislevy et al., 1989a) and 'Florona' (Mislevy et al., 1989b) are stargrass cultivars (*Cynodon nlemfluënsis*) released from Florida experiment stations. 'Alicia' was released by a private seed company. An extensive list of released bermudagrass and stargrass cultivars was presented by Taliaferro et al. (2004).

Yield has significantly increased for sprigged bermudagrass over the past 70 yr. Much of the increase can be attributed to genetics. Most evaluations are performed through small-plot clipping trials that remove the biomass as hay at harvest. Using basic breeding techniques and field trials, Burton (1943) developed 'Coastal,' which produced twice as much DMY as common bermudagrass. Since its release, 'Coastal' has been propagated to an estimated 4 million hectares (Hanna and Anderson, 2008) and is still considered the standard for comparison in all yield trials. Cultivars such as Tifton 44 and Midland 99 were released because of greater persistence to cold temperatures. Midland 99 is more adapted to the colder climates of Oklahoma, Arkansas, and Missouri and has produced greater DMY for the Southern Plains (Table 9–2). Cultivars developed after Coastal have shown moderate yield increases until the release of Tifton 85 in 1993. Tifton 85 has had consistently higher yields than other cultivars in Tifton, GA (Table 9–3) as well as other locations in the South (Table 9–4). 'Tifton 85' is a pentaploid ($2n = 45$) from a cross between 'Tifton 68' and plant introduction 290884 from South Africa. 'Tifton 68' is a hexaploid ($2n = 54$) resulting from a cross between diploid and tetraploid stargrass germplasm and is highly digestible but cold sensitive. The pollen parent (PI 290884) is a cold-hardy tetraploid *C. dactylon* with resistance to fall armyworm [*Spodoptera frugiperda* (J.E. Smith)]. Two other cultivars that have done well in recent tests are Russell and Coastcross II (Table 9–3). Russell has sustained yields that are close to Tifton 85 and has the advantage of requiring less time to cure because of being finer stemmed, but it is lower in digestibility. 'Coastcross II' is derivative of 'Coastcross I' but has not been officially released.

Seeded cultivars of bermudagrass are desirable to reduce the cost and time for establishment and for reseeding in northern transitional zones that periodically receive killing frosts. However, it has been a challenge to select for both high forage yields and sufficient seed yields. Seeded cultivars have been developed by private seed companies and agricultural experiment stations. Released cultivars

Table 9–2. Dry matter yield of bermudagrass [Cynodon dactylon (L.) Pers.)] cultivars grown in Oklahoma, Arkansas, and Missouri from 1990 to 1998.†

Cultivar	Year released	Haskell, OK 1995–1998	Chickasha, OK 1995–1995	Ardmore, OK 1990–1997	Batesville, AR 1995–1998	Mt. Vernon, MO 1996–1998
		Mg ha^{-1}				
Midland	1953	18.0b	18.3b	7.4b	13.6bc	
Greenfield	1954	15.3c	12.5c		14.8ab	
Hardie	1974	20.4a	22.4a	7.6b		8.6b
Tifton 44	1978	20.6a	23.2a	7.5b	16.4a	10.6a
World Feeder	1991	14.7c	12.5c		12.0c	
Quickstand	1993	16.0c	14.5c			
Midland 99	1999	21.8a	21.2a	8.9a	16.7a	11.9a

† Adapted from Taliaferro et al. (2008). Means within a column not followed by the same letter are significantly different ($P < 0.05$).

such as Wrangler, Cheyenne, and Laredo have been lower yielding than sprigged cultivars (Table 9–3. In the transition areas for bermudagrass, seeded cultivars are of greater importance because of the need for reestablishment after cold winters that kill bermudagrass. Seeded cultivars such as Pasto Rico, Ranchero Frio, and Cheyenne were similar to Tifton 44 for yield and for risk assessment (Clapham et al., 2009). Experimental lines such as P2 (developed at Tifton, GA) have performed well in tests compared with other seeded lines (Table 9–3) but do not produce

Table 9–3. Dry matter yield of bermudagrass [*Cynodon dactylon* (L.) Pers.] cultivars grown at Tifton, GA from 2006 to 2011.

Cultivar	Year released	2006	2007	2008	2009	2010	2011	Mean
		Mg ha^{-1}						
		Sprigged types						
Coastal	1943	16.3c†	14.7d	17.1def	17.5cd	15.4cd	19.9cd	16.8e
Coastcross 1	1972	22.2b	19.2c	19.3cde	19.1cd	16.7c	17.8d	19.0d
Tifton 44	1978	15.3c	12.6de	13.6fg	14.2ef	13.5de	18.4cd	14.6f
Tifton 85	1991	29.4a	27.3a	26.5a	28.2a	22,8a	24.8a	26.5a
Russell	1994	22.8b	22.7b	20.3cde	23.1b	20,.1b	23.2ab	22.0c
Coastcross 2	2004	28.3a	26.5a	24.5ab	23.1b	21.9ab	22.9ab	24.5b
		Seeded types						
Cheyenne	1989	16.5c	10.8e	16.7def	11.8fg	9.8ef	13.6e	13.2f
Wrangler	1999	7.6d	3.7f	5.4h	6.0h	3.5g	4.5f	5.1h
P2 Exp*	Not released	24.7b	17.7c	20.9bc	19.2c	16.0c	22.1bc	20.1d

† Means within a column not followed by the same letter are significantly different ($P < 0.05$).

Table 9–4. Mean dry matter yield of bermudagrass [*Cynodon dactylon* (L.) Pers.] cultivars grown at Alabama, Georgia, Texas, and Mississippi between 2002 and 2010.

Cultivar	Year released	Headland, AL† 2005–2007	Calhoun, GA‡ 2003–2005	Griffin, GA‡ 2003–2005	Tifton, GA‡ 2003–2005	Overton, TX§ 2002–2004	Starkville, MS¶ 2003–2010
		Mg ha^{-1}					
		Sprigged types					
Coastal	1943	14.7b#	16.0d	14.6b	16.1b	12.3b	10.0
Alicia	1965		18.1bc	14.5b	19.2a		12.9
Tifton 44	1978	13.9c	17.2cd	11.4c			9.6
Tifton 85	1991	22.4a	22.3a	16.4a	19.4a	14.0a	12.8
Russell	1994	14.4b	19.6b	16.6a	19.3a		12.1
		Seeded type					
Cheyenne	1989	12.0d	11.7e	10.2ab	10.3c	11.2c	8.4

† Data from Wiregrass Research and Extension Center, Headland, AL.

‡ Data from University of Georgia.

§ Data from Texas A&M University.

¶ Data from Mississippi State University.

Means within a column not followed by the same letter are significantly different ($P < 0.05$).

high yields of seed. Research is being conducted to determine optimum methods for seed production from this population. Another possible approach would be to employ the restricted recurrent phenotypic selection (RRPS) technique used for bahiagrass improvement (Burton, 1982) to develop a seeded population. Current research using this technique is in the preliminary stages with the primary obstacle being the ability to select progeny with concurrent anthesis times for the production of adequate seed for the next round of selection (Anderson, unpublished data, 2013).

Forage value is also often expressed through animal performance in grazing trials measured by the production of meat, milk, or wool. Average daily animal gain is measured for grazing cattle on particular cultivars for an extended period of time (Coleman et al., 2004a). Cattle grazing 'Tifton 85' pastures produced higher ADG and had greater stocking rates than those grazing Coastal (Corriher et al., 2007; Rouquette et al., 2004). However, grazing trials are land and labor intensive and not practical for evaluating a large number of breeding lines. A more practical selection technique called in vitro dry matter digestibility or disappearance (IVDMD) is useful in breeding programs (Tilley and Terry, 1963; Monson et al., 1969). This technique was used in breeding for higher quality bermudagrass (Burton and Monson, 1972) and resulted in the release of 'Coastcross I' (Burton, 1972) and 'Tifton 85' (Burton et al., 1993). Three-year grazing trials have confirmed the effectiveness of the IVDMD method. In one trial, the ADG of 'Tifton 78' (bred using the IVDMD technique) was 25% greater than 'Coastal' (Hill et al., 1990). In trials of 'Tifton 85,' the steer ADG ha^{-1} increase was 46% higher than 'Tifton 78' (Hill et al., 1995). 'Tifton 85,' which also has improved DMY, has increased intake and digestion in part because of a lower percentage of ester-linked ferulic acid linkages to lignin (Mandebvu et al., 1999; Hill et al., 2001).

Dry matter yield of bermudagrass is affected by temperature, day length, and solar radiation (Burton et al., 1988) and yields typically decline from mid to late summer. Clipping frequency for hay is generally recommended every 4 to 5 wk. More frequent harvests will result in lower overall yields but higher crude protein and digestibility (Mandebvu et al., 1999). Many studies have been performed over the past half century, primarily on 'Coastal,' to determine the best management practices for bermudagrass. Studies demonstrated that addition of nitrogen (N), phosphorous (P_2O_5), and potassium (K_2O) fertilizer on 'Coastal' significantly increased yields (Adams and Stelly, 1962; Burton, 1954). Although yields were shown to increase with nitrogen fertilizer applications up to 1120 kg ha^{-1} (Fisher and Caldwell, 1959), the current recommendations are to apply approximately 440 kg ha^{-1} N; 110 kg ha^{-1} P; and 330 kg ha^{-1} K on split applications when bermudagrass is cut for hay (Taliaferro et al., 2004). However, these recommended fertilizer rates are highly dependent on expected yields, fertilizer costs, and hay prices.

Bahiagrass

Bahiagrass is native to South America and was introduced to the United States in 1913 (Scott, 1920). It was first tested for forage and turf purposes in Florida and Georgia during the 1930s (Gates et al., 2004). Bahiagrass is often preferred over bermudagrass in soils that are less fertile and have less water holding capacity. Bahiagrass has greater persistence than bermudagrass under low fertility, drought, flooding, and continuous stocking. Over one million hectares of bahiagrass pasture is present in Florida, and it is a major forage in Georgia, Alabama, and Mississippi for beef cattle production with low-input requirements (Gates et al., 2004). Bahiagrass also has the advantage of being established by seed. There are two types of domesticated bahiagrass used for pasture or hay, which can be differentiated by ploidy and reproduction.

The 'Argentine' type is tetraploid ($2n = 40$) and reproduces through apomixis. Landraces of the tetraploids are either facultative or obligate apomicts. 'Paraguay 22' is the only germplasm developed through selection (Gates et al., 2004). However, recent attempts have been made to develop sexual tetraploids and cross these as females with apomictic tetraploids. Sexual tetraploids have been induced through seed or tissue culture treatments (Quesenberry et al., 2010). These crosses produced 80% sexual progeny with high genetic variability. High broad sense heritabilities were observed for growth habit and freezing resistance (Acuña et al., 2009). Selected sexual hybrids were then backcrossed with apomictic pollen to produce 12 backcross populations. Considerable heterosis for desirable traits was found within these populations; however, the percentage of highly apomictic progeny was reduced to 3% (Acuña et al., 2011).

The Pensacola type bahiagrass germplasm lines are sexual diploids ($2n = 20$) and self-incompatible (Burton, 1955). Pensacola derived its name from a collection made near the docks at Pensacola, FL (Burton, 1967). Though bahiagrass yields were improved through the development of hybrids, lack of seed production prohibited the use of the cultivars Tifhi-1 and Tifhi-2 (Gates et al., 2004). The most successful method of breeding was through RRPS, which Burton used to develop 'Tifton 9' (Burton, 1989a). Selection was based on visual appearance of individual plants lacking competition. In each cycle, 20% of the plants were selected from an original population of 1000 plants. A cycle could be completed in 12 mo (Burton, 1974, 1982). Burton reported a 16.4% increase in DMY after six cycles of selection compared with the original Pensacola population (Burton, 1982). After three subsequent cycles, seed was increased, tested, and released as 'Tifton 9.' Selection was continued to produce lines T18 and T23 (Burton et al., 2008). However, the later selections were more erect and resulted in reduced persistence in grazing trials (Werner and Burton, 1991). The RRPS system has been employed to further improve yield (Blount et al., 2008) and to reduce hard-seededness which results in faster germination and improved plant stands (Anderson et al., 2009a). 'UF Riata' was released as a higher yielding and cold tolerant bahiagrass (A.R. Blount unpublished data, 2013). 'TifQuik' has quicker emergence and similar yields to 'Tifton 9' (Anderson et al., 2011). Recent yield trials in Georgia, Florida, and Mississippi indicate that 'Tifton 9' has retained the yield increases over 'Pensacola' in Georgia and Florida. 'UF Riata' has outperformed 'Tifton 9' bahiagrass (Table 9–5). As in all grasses, fertilization is necessary for optimal growth for bahiagrass. A three-fold increase in yield was reported when N was increased from 0 to 270 kg ha^{-1} (Beaty et al., 1960). Tillers increase with N application and the large rhizomes and roots of bahiagrass supply a pool of nutrients (Beaty et al., 1977; Impithuksa and Blue, 1978). In general, bahiagrass is less dependent on P and K fertilization than bermudagrass (Gates et al., 2004).

Table 9–5. Mean dry matter yield of bahaigrass (*Paspalum notatum* Flueggé) cultivars grown in Georgia, Florida, and Mississippi between 2004 and 2012.

Cultivar	Year released	Tifton, GA 2006–2011	Tifton, GA 2005–2010	Ona, FL† 2005–2006	Marianna, FL† 2006	Starkville, MS‡ 2004–2007	Starkville, MS‡ 2008–2012
		Mg ha^{-1}					
Pensacola	1944	12.9c§	12.3b	8.7b	8.3b	7.6b	8.2ab
Argentine	1950		11.5b	8.8b	7.2b	8.1b	7.9b
Tifton 9	1987	15.5b	14.5a	9.9ab	9.7a	9.9a	9.2a
Sand Mt.	1999		12.1b	8.3b	8.3b	8.3b	8.8ab
UF Riata	2008	17.3a	15.4a	11.4a	10.6a		
TifQuik	2008	16.4ab	14.1a	9.7ab	10.0a		

† Data from University of Florida.

‡ Data from Mississippi State University.

§ Means within a column not followed by the same letter are significantly different ($P < 0.05$).

Summary

Documenting gains in seedling establishment, stand persistence, and forage yield in range pasture grasses will continue to be a challenge given the strong entry × environment interaction that exists. The present contribution of improved cool- and warm-season grass varieties with increased seedling establishment, stand persistence, and in some cases DMY increases is particularly notable in the reseeding efforts on disturbed rangelands in the western United States and pastures of the southern United States. The utilization of hybrids and interploidy breeding resulted in the biggest gains in seedling establishment, persistence, and forage yield in the wheatgrasses and wildryes. Using basic breeding techniques and field trials, yield gains in bermudagrass have significantly increased over time.

At present, the primary economic driving force for development of high yielding grasses is the emerging cellulosic bio-energy and bio-based product industry, which can use many of these cool- and warm-season grasses adapted to the southern and Great Plains regions of the United States. However, as land currently under feedstock cultivation for livestock is converted to cellulosic bio-energy, there will be a net decrease in forage available for livestock, hence necessitating a need to develop high yielding grasses adapted to marginal rangelands that frequently experience drought and hot temperatures. In addition, given predicted climate change models that suggest trends toward hotter and drier conditions, there is a need to develop high yielding grasses adapted to these changing environments. Also, if rainfall becomes more erratic, the need for drought and saline tolerant cultivars will be become more apparent.

Efforts to develop improved grasses for these changing environments will still need to focus on seedling establishment, stand persistence, and then on increased forage yields. The ability of a seed to germinate under hot, dry conditions on soils ranging from shallow to saline is generally the limiting factor in subsequently obtaining high forage yields. The development of high water-use efficient plants (i.e., increased biomass production given the same amount of water) will be needed.

Efforts to expand the establishment of semi-tropical warm-season grasses toward the transition zone through increased winter hardiness will be needed. With a plethora of new genetic materials and genomic tools available to range and pasture grass breeders, with adequate funding, increases in establishment, persistence, disease resistance, and forage traits should be expected.

References

Abella, S.R. 2010. Disturbance and plant succession in the Mojave and Sonoran Deserts in the American Southwest. Int. J. Environ. Res. Public Health 7:1248–1284. doi:10.3390/ijerph7041248

Acuña, C.A., A.R. Blount, K.H. Quesenberry, K.E. Kenworthy, and W.W. Hanna. 2009. Bahiagrass tetraploid germplasm: Reproductive and agronomic characterization of segregating progeny. Crop Sci. 49:581–588. doi:10.2135/cropsci2008.07.0402

Acuña, C.A., A.R. Blount, K.H. Quesenberry, K.E. Kenworthy, and W.W. Hanna. 2011. Tetraploid bahiagrass hybrids: Breeding technique, genetic variability and proportion of heterotic hybrids. Euphytica 179:227–235. doi:10.1007/s10681-010-0276-y

Adams, W.E., and M. Stelly. 1962. Fertility requirements of Coastal bermudagrass and crimson clover grown on Cecil sandy loam I. Yield response to fertilization. J. Range Manage. 15:84–87. doi:10.2307/3894867

Alderson, J., and W.C. Sharp. 1994. Grass varieties in the United States. Agric. Handb. 170. U.S. Gov. Print. Office, Washington, DC.

Anamthawat-Jónsson, K., and Bödvarsdóttir. 2001. Genomic and genetic relationships among species of *Leymus* (Poaceae: Triticeae) inferred from 18S-26S ribosomal genes. Am. J. Bot. 88:553–559.

Anderson, W.F. 2005. Development of a forage bermudagrass (*Cynodon* sp.) core collection. Grassl. Sci. 51:305–308. doi:10.1111/j.1744-697X.2005.00040.x

Anderson, W.F., M.D. Casler, and B.S. Baldwin. 2008. Improvement of perennial forage species as feedstock for bio-energy. In: W. Vermerris, editor, Genetic improvement of bioenergy crops. Springer, New York. p. 309–346.

Anderson, W.F., W.W. Hanna, and R.N. Gates. 2011. Registration of 'TifQuik' bahiagrass. J. Plant Reg. 5:147–150. doi:10.3198/jpr2010.07.0427crc

Anderson, W.F., R.N. Gates, W.W. Hanna, A.R. Blount, P. Mislevy, and G. Evers. 2009a. Recurrent restricted phenotypic selection for improving stand establishment of bahiagrass. Crop Sci. 49:1322–1327. doi:10.2135/cropsci2008.09.0556

Anderson, W.F., A. Maas, and P. Ozias-Akins. 2009b. Genetic variability of a forage bermudagrass core collection. Crop Sci. 49:1347–1358. doi:10.2135/cropsci2008.06.0330

Asay, K.H. 1987. Revegetation in sagebrush ecosystem. In: J.A. Onsager, editor, Integrated pest management: State-of-the-art in the sagebrush ecosystem. USDA-ARS Publ. 50. U.S. Gov. Print. Office, Washington, DC. p. 19–27.

Asay, K.H., N.J. Chatterton, K.B. Jensen, T.A. Jones, B.L. Waldron, and W.H. Horton. 2003. Breeding improved grasses for semiarid rangelands. Arid Land Res. Manage. 17:469–478. doi:10.1080/713936115

Asay, K.H., N.J. Chatterton, K.B. Jensen, R.R.-C. Wang, D.A. Johnson, W.H. Orton, A.J. Palazzo, and S.A. Young. 1997. Registration of 'CD-II' crested wheatgrass. Crop Sci. 37:1023. doi:10.2135/cropsci1997.0011183X003700030083x

Asay, K.H., and D.R. Dewey. 1981. Registration of Agropyron repens × A. spicatum germplasms RS-1 and RS-2. Crop Sci. 21:351. doi:10.2135/cropsci1981.0011183X002100020044x

Asay, K.H., and D.R. Dewey. 1983. Pooling the genetic resources of the crested wheatgrass species-complex. In: J.A. Smith and V.W. Hays, editors, Proceedings of the XIV International Grassland Congress, Lexington, KY. 15–24 June 1981. Westview Press, Boulder, CO. p. 124–127.

Asay, K.H., D.R. Dewey, F.B. Gomm, W.H. Horton, and K.B. Jensen. 1986. Genetic progress through hybridization of induced and natural tetraploids in crested wheatgrass. J. Range Manage. 39:261–263. doi:10.2307/3899064

Asay, K.H., D.R. Dewey, F.R. Gomm, F.B. Johnson, and J.R. Carlson. 1985a. Registration of 'Hycrest' crested wheatgrass (Reg. No. 16). Crop Sci. 25:368–369.

Asay, K.H., D.R. Dewey, F.R. Gomm, F.B. Johnson, and J.R. Carlson. 1985b. Registration of 'Bozoisky-Select' Russian wildrye. Crop Sci. 25:575–576. doi:10.2135/cropsci1985.0011183X002500030046x

Asay, K.H., D.R. Dewey, W.H. Horton, K.B. Jensen, P.O. Currie, N.J. Chatterton, W.T. Hansen, II, and J.R. Carlson. 1991. Registration of NewHy RS wheatgrass. Crop Sci. 31:1384–1385. doi:10.2135/cropsci1991.0011183X003100050075x

Asay, K.H., W.H. Horton, K.B. Jensen, and A.J. Palazzo. 2001. Merits of native and introduced triticeae grasses on semiarid rangelands. Can. J. Plant Sci. 81:45–52. doi:10.4141/P99-131

Asay, K.H., and K.B. Jensen. 1996a. Wheatgrasses. In: L.E. Moser, D.R. Buxton, and M.D. Casler, editors, Cool-season forage grasses. Agron. Monogr. 34. ASA, CSSA, and SSSA, Madison, WI. p. 691–724.

Asay, K.H., and K.B. Jensen. 1996b. Wildryes. In: L.E. Moser, D.R. Buxton, and M.D. Casler, editors, Cool-season forage grasses. Agron. Monogr. 34. ASA, CSSA, SSSA, Madison, WI. p. 725–748.

Asay, K.H., K.B. Jensen, D.A. Johnson, N.J. Chatterton, W.T. Hansen, W.H. Horton, and S.A. Young. 1995a. Registration of 'Douglas' crested wheatgrass. Crop Sci. 35:1510–1511. doi:10.2135/cropsci1995.0011183X003500050057x

Asay, K.H., D.A. Johnson, K.B. Jensen, N.J. Chatterton, W.H. Horton, W.T. Hansen, and S.A. Young. 1995b. Registration of Vavilov Siberian crested wheatgrass. Crop Sci. 35:1510. doi:10.2135/cropsci1995.0011183X003500050056x

Baldwin, C.K., F.H. Wagner, and U. Lall. 2003. Water resources. In: F.H. Wagner, editor, Rocky Mountain/Great Basin regional climate-change assessment. A report to the Rocky Mountain/Great Basin Regional Team for the US Global Change Research Program. Utah State University, Logan. p. 79–112.

Ball, D.M., M.M. Eichhorn, Jr., R.A. Burdett, Jr., and D.M. Bice. 1996. Registration of Russell bermudagrass. Crop Sci. 36:467. doi:10.2135/cropsci1996.0011183X003600020046x

Barker, R.E., J.D. Berdahl, and E.T. Jacobson. 1984. Registration of Rodan western wheatgrass. Crop Sci. 24:1215–1216. doi:10.2135/cropsci1984.0011183X002400060060x

Barker, R.W., L.K. Holzworth, and K.H. Asay. 1985. Genetic resources of wheatgrasses and wildrye species native to the rangelands of western North America. In: Range plant improvement in western North America. Proceedings of a Symposium at the Annual Meeting of the Society for Range Management, Salt Lake City, UT. 14 Feb. 1985. Society for Range Management, Denver, CO. p. 9–13.

Bashaw, E.C. 1980. Registration of Nueces and Llano buffelgrass. Crop Sci. 20:112. doi:10.2135/cropsci1980.0011183X002000010029x

Beaty, E.R., J.L. Engel, and J.D. Powell. 1977. Yield, leaf growth, and tillering in bahiagrass by N rate and season. Agron. J. 69:308–311. doi:10.2134/agronj1977.00021962006900020026x

Beaty, E.R., R.A. McCreery, and J.D. Powell. 1960. Response of Pensacola bahiagrass to nitrogen fertilization. Agron. J. 52:453–455. doi:10.2134/agronj1960.00021962005200080009x

Benes, S.E., D.D. Adhikari, S.R. Gratten, and R.L. Snyder. 2012. Evapotranspiration potential of forages irrigated with saline-sodic drainage water. Agric. Water Manage. 105:1–7. doi:10.1016/j.agwat.2011.11.019

Berdahl, J.D., and R.E. Barker. 1991. Characterization of autotetraploid Russian wildrye produced with nitrous oxide. Crop Sci. 31:1153–1155. doi:10.2135/cropsci1991.0011183X003100050014x

Berdahl, J.D., R.E. Barker, J.F. Karn, J.M. Krupinsky, R.J. Haas, D.A. Tober, and I.M. Ray. 1992a. Registration of 'Reliant' intermediate wheatgrass. Crop Sci. 32:1072. doi:10.2135/cropsci1992.0011183X003200040058x

Berdahl, J.D., R.E. Barker, J.F. Karn, J.M. Krupinsky, R.J. Naas, D.A. Tober, and I.M. Ray. 1992b. Registration of 'Mankota' Russian wildrye. Crop Sci. 32:1073. doi:10.2135/cropsci1992.0011183X003200040059x

Berdahl, J.D., R.E. Barker, J.F. Karn, J.M. Krupinsky, I.M. Ray, K.P. Vogel, K.J. Moore, T.J. Klopfenstein, B.E. Anderson, R.J. Haas, and D.A. Tober. 1993. Registration of 'Manska' intermediate wheatgrass. Crop Sci. 33:881. doi:10.2135/cropsci1993.0011183X003300040063x

Bingham, E.T., R.W. Groose, D.R. Woodfield, and K.K. Kidwell. 1994. Complementary gene interactions in alfalfa are greater in autotetraploids than diploids. Crop Sci. 34:823–829. doi:10.2135/cropsci1994.0011183X003400040001x

Blount, A.R., C.L. Mackowiak, K.H. Quesenberry, T.R. Sinclair, and P. Mislevy. 2008. Bahiagrass breeding for the U.S. southern Coastal Plain. In: 2008 annual meetings abstracts. [CD-ROM]. ASA, CSSA, and SSSA, Madison, WI.

Burton, G.W. 1943. Coastal bermudagrass. Circ. 10. Georgia Coastal Plain Exp. Stn., Tifton.

Burton, G.W. 1947. Breeding bermuda grass for the southeastern United States. J. Am. Soc. Agron. 39:551–569. doi:10.2134/agronj1947.00021962003900070001x

Burton, G.W. 1954. Coastal bermuda grass. Bull. NS2. Georgia Agric. Exp. Stn., Athens.

Burton, G.W. 1955. Breeding Pensacola bahiagrass, *Paspalum notatum*: I. Method of reproduction. Agron. J. 47:311–314. doi:10.2134/agronj1955.00021962004700070008x

Burton, G.W. 1962. Registration of varieties of bermudagrass. Crop Sci. 2:352–353. doi:10.2135/cropsci1962.0011183X000200040023x

Burton, G.W. 1967. A search for the origin of Pensacola bahiagrass. Econ. Bot. 21:379–382. doi:10.1007/BF02863165

Burton, G.W. 1972. Registration of Coastcross-1 bermudagrass. Crop Sci. 12:125. doi:10.2135/cropsci1972.0011183X001200010049x

Burton, G.W. 1974. Recurrent restricted phenotypic selection increases forage yields of Pensacola bahiagrass. Crop Sci. 14:831–835. doi:10.2135/cropsci1974.0011183X001400060016x

Burton, G.W. 1982. Improved recurrent restricted phenotypic selection increases bahiagrass forage yields. Crop Sci. 22:1058–1061. doi:10.2135/cropsci1982.0011183X002200050040x

Burton, G.W. 1989a. Registration of 'Tifton-9' Pensacola bahiagrass. Crop Sci. 29:1326. doi:10.2135/cropsci1989.0011183X002900050049x

Burton, G.W. 1989b. Registration of 'Merkeron'napiergrass. Crop Sci. 29:1327. doi:10.2135/cropsci1989.0011183X002900050050x

Burton, G.W. 1993. African grasses. In: J. Janick and J.E. Simon, editors, New crops. John Wiley & Sons, New York. p. 294–298.

Burton, G.W., W.F. Anderson, and R.N. Gates. 2008. Registration of higher herbage Pensacola bahiagrass germplasm lines T18 and T23. J. Plant Reg. 2:51–52. doi:10.3198/jpr2007.03.0168crg

Burton, G.W., R.N. Gates, and G.M. Hill. 1993. Registration of Tifton 85 bermudagrass. Crop Sci. 33:644–645. doi:10.2135/cropsci1993.0011183X003300030045x

Burton, G.W., J.E. Hook, J.L. Butler, and R.E. Hellwig. 1988. Effect of temperature, daylength and solar radiation on production of Coastal bermudagrass. Agron. J. 80:557–560. doi:10.2134/agronj1988.00021962008000040002x

Burton, G.W., and W.G. Monson. 1972. Inheritance of dry matter digestibility in bermudagrass, *Cynodon dactylon* (L.). Pers. Crop Sci. 12:375–378. doi:10.2135/cropsci1972.0011183X001200030036x

Burton, G.W., and W.G. Monson. 1978. Registration of Tifton 44 bermudagrass (Reg. No. 10). Crop Sci. 18:911. doi:10.2135/cropsci1978.0011183X001800050065x

Burton, G.W., and W.G. Monson. 1984. Registration of Tifton 68 bermudagrass. Crop Sci. 24:1211. doi:10.2135/cropsci1984.0011183X002400060051x

Burton, G.W., and W.G. Monson. 1988. Registration of Tifton 78 bermudagrass. Crop Sci. 28:187. doi:10.2135/cropsci1988.0011183X002800010047x

Carlson, J.R., and M.E. Barkworth. 1997. *Elymus wawawaiensis*: A species hitherto confused with *Pseudoroegneria spicata* (Triticeae, Poaceae). Phytologia 83:312–330.

Chambers, J.C. 2008. Invasive plant species and the Great Basin. In: J.C. Chambers, N. Devoe, and A. Evenden, editors, Collaborative management and research in the Great Basin—Examining the issues and developing a framework for action. Gen. Tech. Rep. RMRS-GTR-204. USDA Forest Service, Rocky Mountain Research Station, Fort Collins, CO. p. 38–41.

Christensen, J.H., B. Hewitson, A. Busuioc, A. Chen, X. Gao, I. Held, R. Jones, R.K. Kolli, W.-T. Kwon, R. Laprise, V. Magaña Rueda, L. Mearns, C.G. Menéndez, J. Räisänen, A. Rinke, A. Sarr and P. Whetton. 2007. Regional climate projections. In: S. Solomon, D. Qin, M. Manning, Z. Chen, M. Marquis, K.B. Averyt, M. Tignor and H.L. Miller, editors, Climate change 2007: The physical science basis. Contribution of Working Group I to the Fourth Assessment Report of the Intergovernmental Panel on Climate Change. Cambridge Univ. Press, Cambridge, UK. p. 847–849.

Clapham, W.M., J.M. Fedders, and C.D. Teutsch. 2009. Risk analysis, analysis of variance. Forage Grazinglands doi:10.1094/FG-2009-0331-01-RS

Coleman, S.W., J.E. Moore, and J.R. Wilson. 2004a. Quality and utilization. In: L.E. Moser, B.L. Burson, and L.E. Sollenberger, editors, Warm-season (C_4) grasses. Agron. Monogr. 45. ASA, CSSA, and SSSA, Madison, WI. p. 267–308.

Coleman, S.W., C.M. Taliaferro, and R.J. Tyrl. 2004b. Old world bluestems. In: L.E. Moser, B.L. Burson, and L.E. Sollenberger, editors, Warm-season (C_4) grasses. Agron. Monogr. 45. ASA, CSSA, and SSSA, Madison, WI. p. 909–936.

Corriher, V.A., G.M. Hill, J.G. Andrae, M.A. Froetschel, and B.G. Mullinix, Jr. 2007. Cow and calf performance on Coastal or Tifton 85 pastures with aeschynomene creep grazing paddocks. J. Anim. Sci. 85:2762–2771. doi:10.2527/jas.2007-0015

Coulman, B. 2006. Goliath crested wheatgrass. Can. J. Plant Sci. 86:743–744. doi:10.4141/P05-199

Crowle, W.L. 1970. Revenue slender wheatgrass. Can. J. Plant Sci. 50:748–749. doi:10.4141/cjps70-141

Culumber, C.M. 2007. DNA barcoding of western North American taxa: *Leymus* (Poaceae) and *Lepidium* (Brassicaceae). Thesis, Utah State University, Logan.

D'Antonio, C.M., and P.M. Vitousek. 1992. Biological invasions by exotic grasses, the grass/fire cycle, and global change. Annu. Rev. Ecol. Syst. 23:63–87.

Darroch, B.A., and S.N. Acharya. 1996. AEC Hillcrest awned slender wheatgrass. Can. J. Plant Sci. 76:345–347. doi:10.4141/cjps96-062

Darroch, B.A., S.N. Acharya, and J. Woosaree. 2001. AEC Mountaineer broadglumed wheatgrass. Can. J. Plant Sci. 81:745–747. doi:10.4141/P00-163

Dewey, D.R. 1970. Hybrids of South American *Elymus agropyroides* with *Agropyron caespitosum, Agropyron subsecundum*, and *Sitanion hystrix*. Bot. Gaz. 131:210–216. doi:10.1086/336534

Dewey, D.R. 1975. Introgression between *Agropyron dasystachyum* and *A. trachycaulum*. Bot. Gaz. 136:122–128. doi:10.1086/336792

Dewey, D.R. 1979. Some applications and misapplications of induced polyploidy to plant breeding. In: W.H. Lewis, editor, Polyploidy. Plenum Press, New York. p. 445–470.

Dewey, D.R. 1984. The genomic system of classification as a guide to intergeneric hybridization with the perennial Triticeae. In: J.P. Gustafson, editor, Gene manipulation in plant improvement. Plenum Press, New York. p. 209–279.

Dewey, D.R., and K.H. Asay. 1982. Cytogenetic and taxonomic relationships among three diploid crested wheatgrasses. Crop Sci. 22:645–650. doi:10.2135/cropsci1982.0011183X002200030052x

Dewey, D.R., and P.C. Pendse. 1968. Hybrids between *Agropyron desertorum* and induced-tetraploid *Agropyron cristatum*. Crop Sci. 8:607–611. doi:10.2135/cropsci1968.0011183X000800050030x

Dhindsa, S.S., and A.E. Slinkard. 1963. Variability in Russian wildrye (*Elymus junceus* Fisch.). Crop Sci. 3:405–406. doi:10.2135/cropsci1963.0011183X000300050010x

Diffenbaugh, N., J. Pal, R. Trapp, and F. Giorgi. 2005. Fine scale processes regulate the response of extreme events to global climate change. Proc. Natl. Acad. Sci. USA 102:15,774–15,778. doi:10.1073/pnas.0506042102

Dillman, A.C. 1946. The beginnings of crested wheatgrass in North America. J. Am. Soc. Agron. 38:237–250. doi:10.2134/agronj1946.00021962003800030004x

Eichhorn, M.M., Jr., B.D. Nelson, C.R. Montgomery, A.V. Davis, and W.B. Hallmark. 1984. Brazos—A new bermudagrass for pastures and hay in Louisiana. La. Agric. 27(3)18–19.

Eichhorn, M.M., Jr., W.M. Oliver, W.B. Hallmark, W.A. Young, A.V. Davis, and B.D. Nelson. 1986. Registration of 'Grazer' bermudagrass. Crop Sci. 26:835. doi:10.2135/cropsci1986.0011183X002600040046x

Evers, G.E., and B.L. Burson. 2004. Dallisgrass and other *Paspalum* species. In: L.E. Moser, B.L. Burson, and L.E. Sollenberger, editors, Warm-season (C_4) grasses. Agron. Monogr. 45. ASA, CSSA, and SSSA, Madison, WI. p. 681–713.

Fisher, F.L., and A.G. Caldwell. 1959. The effects of continued use of heavy rates of fertilizers on forage production and quality of Coastal bermudagrass. Agron. J. 51:99–102. doi:10.2134/agronj1959.00021962005100020012x

Gates, R.N., C.L. Quarin, and C.G.S. Pedreira. 2004. Bahiagrass. In: L.E. Moser, B.L. Burson, and L.E. Sollenberger, editors, Warm-season (C_4) grasses. Agron. Monogr. 45. ASA, CSSA, and SSSA, Madison, WI. p. 651–680.

Gibbs, J.L., G. Young, and J.R. Carlson. 1991. Registration of 'Goldar' bluebunch wheatgrass. Crop Sci. 31:1708. doi:10.2135/cropsci1991.0011183X003100060083x

Hanna, W.W., and W.F. Anderson. 2008. Development and impact of vegetative propagation in forage and turf bermudagrasses. Agron. J. 100:103–107.

Hanselka, C.W., M.A. Hussey, and F.F. Ibarra. 2004. Buffelgrass. In: L.E. Moser, B.L. Burson, and L.E. Sollenberger, editors, Warm-season (C_4) grasses. Agron. Monogr. 45. ASA, CSSA, and SSSA, Madison, WI. p. 477–502.

Harlon, J.R. 1970. *Cynodon* species and their value for grazing and hay. Herb. Abstr. 40:233–238.

Harmoney, K.R. 2007. Persistence of heavily-grazed cool-season grasses in the central Great Plains. Forage Grazinglands doi:10.1094/FG-2007-0625-01-RS

Harris, K., W. Anderson, and R. Malik. 2010. Genetic relationships among napiergrass (*Pennisetum purpureum* Schum.) nursery accessions using AFLP markers. Plant Genet. Resour. 8:63–70. doi:10.1017/S1479262109990165

Hein, M.A. 1953. [*Cynodon dactylon* (L.) Pers.] Registration of varieties and strains of bermudagrass, II. Agron. J. 45:572–573. doi:10.2134/agronj1953.00021962004500110016x

Hill, G.M., G.W. Burton, P.R. Utley, and R.N. Gates. 1990. Yield and grazing comparisons of Tifton 79–16, Tifton 78 and Coastal bermudagrasses. J. Anim. Sci. 68:56.

Hill, G.M., R.N. Gates, and J.W. West. 2001. Advances in bermudagrass research involving new cultivars for beef and dairy production. J. Anim. Sci. 79. E48–E58.

Hill, G.M., R.N. Gates, J.W. West, and G.W. Burton. 1995. Tifton 85 bermudagrass utilization in beef, dairy and hay production. In: Proceedings of the Workshop on the Foraging Potential of the Genus Cynodon, Juiz de Fora, Brazil. 7–9 Nov. 1995. EMBRAPA/CNPGL, Juiz de Fora, Brazil. p. 139–150.

Hitchcock, A.S. 1971. Manual of grasses of the United States. 2nd ed. Dover Publ. Inc., New York.

Impithuksa, V., and W.G. Blue. 1978. The fate of fertilizer nitrogen applied to Pensacola bahiagrass on sandy soils as indicated by nitrogen-15. Proc. Soil Crop Sci. Soc. Fla. 37:213–217.

Inyang, U., J.M.B. Vendramini, L.E. Sollenberger, B. Sellers, A. Adesogan, L. Paiva, and A. Lunpha. 2010. Forage species and stocking rate effects on animal performance and herbage responses of 'Mulato' and bahiagrass pastures. Crop Sci. 50:1079–1085. doi:10.2135/cropsci2009.05.0267

Jensen, K.B., K.H. Asay, N.J. Chatterton, and D.R. Dewey. 2002. Registration of *Leymus* hybrid-1 wildrye germplasm. Crop Sci. 42:675–676.

Jensen, K.B., K.H. Asay, D.A. Johnson, W.H. Horton, A.J. Palazzo, and N.J. Chatterton. 1998. Registration of tetraploid Russian wildrye [*Psathyrostachys juncea* (Fisch.) Nevski] RWR-Tetra-1 germplasm. Crop Sci. 38:1405. doi:10.2135/cropsci1998.0011183X003800050057x

Jensen, K.B., K.H. Asay, D.A. Johnson, S.R. Larson, B.L. Waldron, and A.J. Palazzo. 2006b. Registration of 'Bozoisky-II' Russian wildrye. Crop Sci. 46:986–987. doi:10.2135/cropsci2005.06-0163

Jensen, K.B., K.H. Asay, D.A. Johnson, and B.J. Li. 2000. Characterization of Siberian wheatgrass germplasm from Kazakhstan (Poaceae: Triticeae). J. Range Manage. 53:347–352. doi:10.2307/4003444

Jensen, K.B., S.R. Larson, and B.L. Waldron. 2005b. Registration of 'Mustang' Altai wildrye. Crop Sci. 45:1168–1169. doi:10.2135/cropsci2004.0165CV

Jensen, K.B., S.L. Larson, B.L. Waldron, and K.H. Asay. 2006a. Cytogenetic and molecular characterization of hybrids between 6x, 4x, and 2x ploidy levels in crested wheatgrass. Crop Sci. 46:105–112. doi:10.2135/cropsci2005.0148

Jensen, K.B., S.L. Larson, B.L. Waldron, and D.A. Johnson. 2005a. Characterization of hybrids from induced × natural tetraploids of Russian wildrye. Crop Sci. 45:1305–1311. doi:10.2135/cropsci2004.0455

Jensen, K.B., S.R. Larson, B.L. Waldron, and J.G. Robins. 2009b. 'Hycrest II,' a new crested wheatgrass cultivar with improved seedling establishment. J. Plant Reg. 3:57–60. doi:10.3198/jpr2008.04.0230crc

Jensen, K.B., I. Mott, J.G. Robins, and B.L. Waldron. 2012. Genetic improvement and diversity in Snake River wheatgrass (*Elymus wawawaiensis*) (Poaceae: Triticeae). Rangeland Ecol. Manag. 65:76–84. doi:10.2111/REM-D-10-00159.1

Jensen, K.B., A.J. Palazzo, B.L. Waldron, and B.S. Bushman. 2007. Registration of 'FirstStrike' slender wheatgrass. J. Plant Reg. 1:24–25. doi:10.3198/jpr2006.11.0694crc

Jensen, K.B., A. Palazzo, B.L. Waldron, J.G. Robins, B.S. Bushman, D.A. Johnson, and D. Ogle. 2009a. Improved establishment characteristics of 'Vavilov II' Siberian wheatgrass. J. Plant Reg. 3:61–64. doi:10.3198/jpr2008.04.0215crc

Jensen, K.B., and R.R.-C. Wang. 1997. Cytological and molecular evidence for transferring *Elymus coreanus* and *Elymus californicus* from the genus *Elymus* to *Leymus* (Poaceae:Triticeae). Int. J. Plant Sci. 158:872–877. doi:10.1086/297502

Jensen, K.B., Y.F. Zhang, and D.R. Dewey. 1990. Mode of pollination of perennial species of the Triticeae in relation to genomically defined genera. Can. J. Plant Sci. 70:215–225. doi:10.4141/cjps90-024

Jones, T.A. 2008. Notice of release of Discovery Snake River wheatgrass. Native Plants J. 9:99–101. doi:10.2979/NPJ.2008.9.2.99

Jones, T.A., S.R. Larson, D.C. Nielson, S.A. Young, N.J. Chatterton, and A.J. Palazzo. 2002. Registration of P-7 bluebunch wheatgrass germplasm. Crop Sci. 42:1754–1755. doi:10.2135/cropsci2002.1754

Jones, T.A., S.D. Parr, S.R. Winslow, and M.A. Rosales. 2009. Notice of release of 'Continental' basin wildrye. Native Plants J. 10:57–60. doi:10.2979/NPJ.2009.10.1.57

Jones, T.A., M.G. Redinbaugh, and Y. Zhang. 2000. The western wheatgrass chloroplast genome originates in *Pseudoroegneria*. Crop Sci. 40:43–47. doi:10.2135/cropsci2000.40143x

Karl, T.R., J.M. Melillo, and T.C. Peterson, editors. 2009. Global climate change impacts in the United States. Cambridge Univ. Press, New York.

Kirk, L.E. 1932. Crested wheatgrass. Univ. Saskatchewan Agric. Ext. Bull. 54. Regina, Saskatchewan, Canada.

Kishii, M., R.R.-C. Wang, and H. Tsujimoto. 2005. GISH analysis revealed new aspect of genomic constitution of *Thinopyrum intermedium*. Czech J. Genet. Plant Breed. 41:92–95.

Klebesadel, L.J. 1985. Beach wildrye: Characteristics and uses of a native Alaskian grass of uniquely costal distribution. Agroborealis 17:31–38.

Knowles, R.P. 1990. Registration of Kirk crested wheatgrass. Crop Sci. 30:749. doi:10.2135/cropsci1990.0011183X003000030069x

Larson, S.R., E. Carter, C.L. McCracken, and D. Dyer. 2001. Mode of reproduction and amplified fragment length polymorphism variation in purple needlegrass (*Nassella pulchra*): Utilization of natural germplasm sources. Mol. Ecol. 10:1165–1177. doi:10.1046/j.1365-294X.2001.01267.x

Larson, S.R., T.A. Jones, Z.-M. Hu, C.L. McCracken, and A. Palazzo. 2000. Genetic diversity of bluebunch wheatgrass cultivars and a multiple-origin polycross. Crop Sci. 40:1142–1147. doi:10.2135/cropsci2000.4041142x

Larson, S.R., T.A. Jones, and K.B. Jensen. 2004. Population structure of *Pseudoroegneria spicata* (Poaceae: Triticeae) modeled by Bayesian clustering of AFLP genotypes. Am. J. Bot. 91:1789–1800. doi:10.3732/ajb.91.11.1789

Larson, S.R., and H.F. Mayland. 2007. Comparative mapping of fiber, protein, and mineral content QTLs in two inter-specific *Leymus* wildrye full-sib families. Mol. Breed. 20:331–347. doi:10.1007/s11032-007-9095-9

Larson, S.R., X. Wu, T.A. Jones, K.B. Jensen, N.J. Chatterton, and B.L. Waldron. 2006. Comparative mapping growth habit, plant height, and flowering QTLs in two inter-specific families of *Leymus*. Crop Sci. 46:2526–2539. doi:10.2135/cropsci2005.12.0472

Lawrence, T. 1967a. Orbit, tall wheatgrass. Can. J. Plant Sci. 47:611–612. doi:10.4141/cjps67-105

Lawrence, T. 1967b. Sawki, Russian ryegrass. Can. J. Plant Sci. 47:612–613. doi:10.4141/cjps67-106

Lawrence, T. 1972. Mayak, Russian ryegrass. Can. J. Plant Sci. 52:121–122. doi:10.4141/cjps72-019

Lawrence, T. 1976. Prairieland, Altai wild ryegrass. Can. J. Plant Sci. 56:991–992. doi:10.4141/cjps76-164

Lawrence, T. 1977a. Registration of Orbit tall wheatgrass. (Reg. No. 11). Crop Sci. 17:980. doi:10.2135/cropsci1977.0011183X001700060054x

Lawrence, T. 1977b. Registration of Sawki Russian wildrye (Reg. No. 42). Crop Sci. 17:978–979. doi:10.2135/cropsci1977.0011183X001700060048x

Lawrence, T. 1979. Swift, Russian wild ryegrass. Can. J. Plant Sci. 59:515–518. doi:10.4141/cjps79-082

Lawrence, T. 1982. Registration of Clarke intermediate wheatgrass. Crop Sci. 22:898. doi:10.2135/cropsci1982.0011183X002200040047x

Lawrence, T., P.G. Jefferson, and C.D. Ratzlaff. 1991a. Eejay, Altai wild ryegrass. Can. J. Plant Sci. 71:551–553. doi:10.4141/cjps91-081

Lawrence, T., P.G. Jefferson, and C.D. Ratzlaff. 1991a. Pearl, Altai wild ryegrass. Can. J. Plant Sci. 71:547–549. doi:10.4141/cjps91-080

Lawrence, T., A.E. Slinkard, C.D. Ratzlaff, N.W. Holt, and P.G. Jefferson. 1990. Tetracan, Russian wild ryegrass. Can. J. Plant Sci. 70:311–313. doi:10.4141/cjps90-037

Liu, Z.-W., and R.R.-C. Wang. 1992. Genome analysis of *Thinopyrum junceiforme* and *T. sartorii*. Genome 35:758–764. doi:10.1139/g92-116

Lorenz, R.J. 1986. Introduction and early use of crested wheatgrass in the Northern Great Plains. In: K.L. Johnson, editor, Crested wheatgrass: Its values, problems and myths. Symposium proceedings of the Crested Wheatgrass Symposium, Logan, UT. 3–7 Oct. 1983. Utah State Univ., Logan. p. 9–20.

Maguire, J.D. 1962. Speed of germination—Aid in selection and evaluation for seedling emergence and vigor. Crop Sci. 2:176–177. doi:10.2135/cropsci1962.0011183X000200020033x

Majerus, M.E., J.G. Scheetz, and L.K. Holzworth. 1991. Pryor slender wheatgrass, a conservation plant for Montana and Wyoming. Leaflet. USDA-SCS, Plant Materials Center, Bridger, MT.

Mandebvu, P., J.W. West, G.M. Hill, R.N. Gates, R.D. Hatfield, B.G. Mullinix, A.H. Parks, and A.B. Caudle. 1999. Comparison of Tifton 85 and Coastal bermudagrass for yield, nutrient traits, intake, and digestion by growing beef steers. J. Anim. Sci. 77:1572–1586.

Marty, L. 2003. Washoe germplasm basin wildrye. Leaflet. USDA–NRCS, Plant Materials Center, Bridger, MT.

Mellish, A., B. Coulman, and Y. Ferdinandez. 2002. Genetic relationships among selected crested wheatgrass cultivars and species determined on the basis of AFLP markers. Crop Sci. 42:1662–1668. doi:10.2135/cropsci2002.1662

Miles, J.W., C.B. DoValle, I.M. Rao, and V.P.B. Euclides. 2004. Brachiariagrasses. In: L.E. Moser, B.L. Burson, and L.E. Sollenberger, editors, Warm-season (C_4) grasses. Agron. Monogr. 45. ASA, CSSA, and SSSA, Madison, WI. p. 745–783.

Mislevy, P., W.F. Brown, R. Caro-Caster, J. Vicente-Chandler, L.S. Dunavin, D.W. Hall, R.S. Kalmbacher, A.J. Overman, O.C. Ruelke, R.M. Sonada, A. Sotomayor-Rios, R.L. Stanley, Jr., and M.J. Williams. 1989a. 'Florico' stargrass. Circ. S-361. Florida Agric. Exp. Stn., Gainesville.

Mislevy, P., W.F. Brown, L.S. Dunavin, D.W. Hall, R.S. Kalmbacher, A.J. Overman, O.C. Ruelke, R.M. Sonada, R.L. Stanley, Jr., and M.J. Williams. 1989b. 'Florona' stargrass. Circ. S-362. Florida Agric. Exp. Stn., Gainesville.

Mislevy, P., W.F. Brown, R.S. Kalmbacher, L.S. Dunavin, W.S. Judd, T.A. Kucharek, O.C. Ruelke, J.W. Noling, R.M. Sonoda, and R.L. Stanley, Jr. 1999. Registration of Florakirk bermudagrass. Crop Sci. 39:587. doi:10.2135/cropsci1999.0011183X003900020055x

Monson, W.G., R.S. Lowery, and I. Forbes, Jr. 1969. In vitro nylon bag vs. two-stage in vitro digestion: Comparison of two techniques for estimating dry-matter digestibility of forages. Agron. J. 61:587–589. doi:10.2134/agronj1969.00021962006100040032x

Munson, S.M., J. Belnap, C.D. Schelz, M. Moran, and T.W. Carolin. 2011. On the brink of change: Plant responses to climate on the Colorado Plateau. Ecosphere 2:1–15. doi:10.1890/ES11-00059.1

Murphy, M.A. 2003. Relationships among taxa of *Elymus* (Poaceae: Triticeae) in Australia: Reproductive biology. Aust. Syst. Bot. 16:633–642.

Myers, W.M. 1951. Registration of varieties and strains of wheatgrass (Agropyon spp.), Primar slender wheatgrass (Reg. No. 1). Agron. J. 43:239. doi:10.2134/agronj1951.00021962004300050010x

Norton, J.B., T.A. Monaco, and U. Norton. 2007. Mediterranean annual grasses in western North America: Kids in a candy store. Plant Soil 298:1–5. doi:10.1007/s11104-007-9364-8

Norton, J.B., T.A. Monaco, J.M. Norton, D.A. Johnson, and T.A. Jones. 2004. Soil morphology and organic matter dynamics under cheatgrass and sagebrush-steppe plant communities. J. Arid Environ. 57:445–466. doi:10.1016/S0140-1963(03)00104-6

Oram, R.N. 1981. Register of Australian herbage plant cultivars: A. Grasses, 18. Wheatgrass *Agropyron elongatum* (Host.) Beauv. (Tall wheatgrass) cv. Tyrrell. Reg. no. A-18a-1. J. Aust. Inst. Agric. Sci. 47:179–180.

Perry, L.J., and D.D. Baltensperger. 1979. Leaf and stem yields and forage quality of three N fertilized warm season grasses. Agron. J. 71:355–358. doi:10.2134/agronj1979.00021962007100020031x

Pistorale, S.M., L.A. Abbott, and A. Andres. 2008. Genetic diversity and broad sense heritability in tall wheatgrass (*Thinopyrum ponticum*). Cienc. Invest. Agrar. 35:259–264.

Pitman, W.D., C.G. Chambliss, and J.B. Hacker. 2004. Digitgrass and other species of *Digitaria*. In: L.E. Moser, B.L. Burson, and L.E. Sollenberger, editors, Warm-season (C_4) grasses. Agron. Monogr. 45. ASA, CSSA, and SSSA, Madison, WI. p. 715–743.

Prine, G.M., J.A. Stricker, and W.V. McConnell. 1997. Opportunities for bioenergy development in lower south USA. In: Making a business from biomass in energy, Environment, Chemicals, Fibers and Materials. Proceedings of the Third Biomass Conference of the America, Montreal, Quebec, Canada. 24–29 Aug. 1997. Pergamon, Oxford, UK. p. 227–235.

Quesenberry, K.H., J.M. Dampier, Y.Y. Lee, R.L. Smith, and C.A. Acuña. 2010. Doubling the chromosome number of bahiagrass via tissue culture. Euphytica 175:43–50. doi:10.1007/s10681-010-0165-4

Robins, J.G. 2010. Cool-season grasses produce more total biomass across the growing season than do warm-season grasses when managed with an applied irrigation gradient. Biomass Bioenergy 34:500–505. doi:10.1016/j.biombioe.2009.12.015

Robins, J.G., K.B. Jensen, T.A. Jones, B.L. Waldron, M.D. Peel, C.W. Rigby, K.P. Vogel, R.B. Mitchell, A.J. Palazzo, and T.J. Cary. 2013. Stand establishment and persistence of perennial cool-season grasses in Intermountain West and northern Great Plains. Rangeland Ecol. Manag. 66(2):181–190.

Rogler, G.A. 1954. Nordan crested wheatgrass. N. D. Agric. Exp. Stn. Bull. 16:150–152.

Rogler, G.A., and H.M. Schaaf. 1963. Growing Russian wildrye in the western states. USDA Leaflet 313. U.S. Gov. Print Office, Washington, DC.

Ross, J.G. 1963. Registration of Oahe intermediate wheatgrass (Reg. No. 5). Crop Sci. 3:373. doi:10.2135/cropsci1963.0011183X000300040046x

Rouquette, F.M., Jr., W.F. Anderson, K.R. Harris-Shultz, and G.R. Smith. 2011. Stand maintenance and genetic diversity of bermudagrass pastures under different stocking intensities during a 38-year period. Crop Sci. 51:2886–2894. doi:10.2135/cropsci2011.03.0123

Rouquette, F.M., Jr., J.L. Kerby, G.H. Nimr, and W.C. Ellis. 2004. Protein supplementation of stocker calves grazing Tifton 85 and Coastal bermudagrass. Rep. 2004-1. Texas Agric. Exp. Stn., Overton.

Schwendiman, J.L. 1972. Registration of Alkar tall wheatgrass. (Reg. No. 7). Crop Sci. 12:260.

Scott, J.M. 1920. Bahia grass. J. Am. Soc. Agron. 12:112–113. doi:10.2134/agronj1920.0002196200 1200030004x

Shearman, R.C., T.P. Riordan, and P.G. Johnson. 2004. Buffalograss. In: L.E. Moser, B.L. Burson, and L.E. Sollenberger, editors, Warm-season (C_4) grasses. Agron. Monogr. 45. ASA, CSSA, and SSSA, Madison, WI. p. 1003–1026.

Smoliak, S. 1976. Cabree Russian ryegrass. Can. J. Plant Sci. 56:993–996. doi:10.4141/cjps76-165

Smoliak, S., and A. Johnson. 1982. Registration of Elbee thickspike wheatgrass (Reg. No. 13). Crop Sci. 22:162–163. doi:10.2135/cropsci1982.0011183X002200010051x

Smoliak, S., and A. Johnson. 1984. Registration of 'Walsh' western wheatgrass. Crop Sci. 24:1216. doi:10.2135/cropsci1984.0011183X002400060061x

Sollenberger, L.E., G.M. Prine, W.R. Ocumpaugh, W.W. Hanna, C.S. Jones, Jr., S.C. Schank, and R.S. Kalmbacher. 1988. 'Mott' dwarf elephantgrass—A high quality forage for the subtropics and tropics. Circ. S-356. Florida Agric. Exp. Stn., Gainesville.

Springer, T.L., and C.L. Dewald. 2004. Eastern gamagrass and other *Tripsacum* species. In: L.E. Moser, B.L. Burson, and L.E. Sollenberger, editors, Warm-season (C_4) grasses. Agron. Monogr. 45. ASA, CSSA, and SSSA, Madison, WI. p. 955–973.

Springer, T.L., C.L. Dewald, P.L. Sims, R.L. Gillen, V.H. Louthan, W.J. Cooper, C.M. Taliaferro, C. Maura, Jr., S. Pfafe, R.L. Wynia, J. Henry, S.B. Brukerhoff, M. Van de Grinten, P.R. Salon, M. Houck, Jr., and R.G. Esquivel. 2006. Registration of 'Verl' eastern gamagrass. Crop Sci. 46:477–478. doi:10.2135/cropsci2005.05-0043

Stebbins, G.L., Jr., and F.T. Pun. 1953. Artificial and natural hybrids in the Gramineae, tribe Hordeae. V. Diploid hybrids of *Agropyron*. Am. J. Bot. 40:444–449. doi:10.2307/2438528

Stevens, R., S.B. Monsen, N. Shaw, E.D. McArthur, G. James, G. Davis, K.R. Jorgensen, and J.N. Davis. 1983. Naming and release of Ephraim crested wheatgrass. Release notice. USDA Forest Service, Provo, UT.

Stroh, J.R., A.A. Thornburg, and D.E. Ryerson. 1972. Registration of Critana thickspike wheatgrass. Crop Sci. 12:394. doi:10.2135/cropsci1972.0011183X001200030045x

Tai, W., and D.R. Dewey. 1966. Morphology, cytology, and fertility of diploid and colchicine-induced tetraploid crested wheatgrass. Crop Sci. 6:223–226. doi:10.2135/cropsci1966.00111 83X000600030002x

Taliaferro, C.M., J.A. Anderson, J.L. Baker, A.A. Hopkins, S.W. Coleman, B.A. Phillips, L.J. Sandage, J.L. Moyer, T.L. Hanson, and R. Crawford. 2008. 'Midland 99.' A new forage bermudagrass. Oklahoma State University, Department of Plant and Soil Sciences. http://pss.okstate.edu/publications/publications-master-list/copy_of_publications/forages/midland99.pdf (accessed 12 Feb. 2014).

Taliaferro, C.M., J.A. Anderson, W.L. Richardson, J.L. Baker, S.W. Coleman, E.A. Phillips, L.J. Sandage, J.L. Moyer, T.L. Hanson, R.L. Kallenbach, and R.J. Crawford. 2002. Registration of Midland 99 bermudagrass. Crop Sci. 42:2212–2213. doi:10.2135/cropsci2002.2212

Taliaferro, C.M., and W.L. Richardson. 1980. Registration of 'Hardie' bermudagrass. Crop Sci. 20:413. doi:10.2135/cropsci1980.0011183X002000030039x

Taliaferro, C.M., F.M. Rouquette, Jr., and P. Mislevy. 2004. Bermudagrass and stargrass. In: L.E. Moser, B.L. Burson, and L.E. Sollenberger, editors, Warm-season (C_4) grasses. Agron. Monogr. 45. ASA, CSSA, and SSSA, Madison, WI. p. 417–475.

Tilley, J.M.A., and R.A. Terry. 1963. A two-stage technique for the *in vitro* digestion of forage crops. J. Br. Grassl. Soc. 18:104–111. doi:10.1111/j.1365-2494.1963.tb00335.x

Tischler, C.R., and W.R. Ocumpaugh. 2004. Kleingrass, blue panic, and vine mesquite. In: L.E. Moser, B.L. Burson, and L.E. Sollenberger, editors, Warm-season (C_4) grasses. Agron. Monogr. 45. ASA, CSSA, and SSSA, Madison, WI. p. 623–649.

Tracy, S.M. 1917. Bermuda grass. USDA Farmers Bull. 814. U.S. Gov. Print. Office, Washington, DC.

USDA Extension Service. 1978. New crops cultivars. Publ. ESC 584(13):209–211.

USDA-NRCS. 1996. Release of 'Rush' intermediate wheatgrass. USDA-NRCS, Aberdeen Plant Materials Center, Aberdeen, ID.

USDA-NRCS. 2006. Eastern gamagrass (*Tripsicum dactyloides)*: A plant for forage, conservation, and bioenergy. USDA-NRCS. http://directives.sc.egov.usda.gov/media/pdf/tn_b_74_a.pdf (accessed 12 Feb. 2014).

USDA-NRCS. 2007. Release of 'Manifest' intermediate wheatgrass. USDA-NRCS, Bismarck Plant Materials Center, Bismarck, ND.

Vendramini, J.M.B., J.D. Arthington, and W.F. Brown. 2008. Use of limpograss in grazing systems in Florida. Forage Grazinglands doi:10.1094/FG-2008-1212-01-RV

Vendramini, J.M.B., L.E. Sollenberger, G.C. Lamb, J.L. Foster, K. Liu, and M.K. Maddox. 2012. Forage accumulation, nutritive value, and persistence of 'Mulato II' brachiariagrass in northern Florida. Crop Sci. 52:914–922.

Vogel, K.E., and B.L. Burson. 2004. Breeding and genetics. In: L.E. Moser, B.L. Burson, and L.E. Sollenberger, editors, Warm-season (C_4) grasses. Agron. Monogr. 45. ASA, CSSA, and SSSA, Madison, WI. p. 51–94.

Vogel, K.E., R.B. Mitchell, T.J. Klopfenstein, and B.E. Anderson. 2006a. Registration of 'Bonanza' big bluestem. Crop Sci. 46:2313–2314. doi:10.2135/cropsci2006.02.0068

Vogel, K.E., R.B. Mitchell, T.J. Klopfenstein, and B.E. Anderson. 2006b. Registration of 'Goldmine' big bluestem. Crop Sci. 46:2314–2315. doi:10.2135/cropsci2006.02.0070

Vogel, K.P., and K.J. Moore. 1993. Native North American grasses. In: J. Janick and J.E. Simon, editors, New crops. John Wiley & Sons, New York. p. 284–293.

Vogel, K.P., and K.J. Moore. 1998. Forage yield and quality of tall wheatgrass accessions in the USDA Germplasm collection. Crop Sci. 38:509–512. doi:10.2135/cropsci1998.0011183X00380 0020039x

Vogel, K.P., P.E. Reece, D.D. Baltensperger, G. Schuman, and R.A. Nicholson. 2005a. Registration of 'Beefmaker' intermediate wheatgrass. Crop Sci. 45:414–415. doi:10.2135/cropsci2005.0414

Vogel, K.P., D. Tober, P.E. Reece, D.D. Baltsensperger, G. Schuman, and R.A. Nicholson. 2005b. Registration of 'Haymaker' intermediate wheatgrass. Crop Sci. 45:415–416. doi:10.2135/ cropsci2005.0415

Voight, P.W., and R.S. MacLauchlan. 1985. Native and other western grasses. In: R.F. Barnes et al., editors, Forages: The science of grassland agriculture, Iowa State Univ. Press, Ames. p. 177–187.

Voight, P.W., N.F.G. Rethman, and M.M. Poverene. 2004. Lovegrasses. In: L.E. Moser, B.L. Burson, and L.E. Sollenberger, editors, Warm-season (C_4) grasses. Agron. Monogr. 45. ASA, CSSA, and SSSA, Madison, WI. p. 1027–1056.

Waldron, B.L., K.B. Jensen, A.J. Palazzo, T.J. Cary, J.G. Robins, M.D. Peel, D.G. Ogle, and L. St. John. 2011. 'Recovery,' a new western wheatgrass cultivar with improved seedling establishment on rangelands. J. Plant Reg. 5:367–373. doi:10.3198/jpr2010.09.0527crc

Wang, R.R.-C., and K.B. Jensen. 2009. Wheatgrass and wildrye grasses (Triticeae). In: R.J. Singh, editor, Genetic resources, chromosome engineering, and crop improvement. Vol. 5. Forage crops. CRC Press, New York. p. 41–79.

Wang, R.R.-C., R. Von Bothmer, J. Dvorak, G. Fedak, I. Linde-Laursen, and M. Muramatsu. 1995. Genome symbols in the Triticeae (Poaceae). In: R.R.-C. Wang, K.B. Jensen, and C. Jaussi, editors, Proceedings of the 2nd International Triticeae Symposium, Logan, UT. 20–24 June 1993. Utah State Univ., Logan. p. 29–34.

Watson, V.H. 1974. A chronological review of the selection, research, and distribution of a bermudagrass called Callie. Report. Mississippi Agric. For. Exp. Stn., Starkville.

Weimer, P.J., and T.L. Springer. 2007. Fermentability of eastern gamagrass, big bluestem and sand bluestem grown across a wide variety of environments. Bioresour. Technol. 98:1615–1621. doi:10.1016/j.biortech.2006.06.003

Werner, B.K., and G.W. Burton. 1991. Recurrent restricted phenotypic selection for yield alters morphology and yield of Pensacola bahiagrass. Crop Sci. 31:48–50. doi:10.2135/cropsci1991.0011183X003100010011x

West, L. 2013. What was the worst environmental disaster in the United States? About.com. http://environment.about.com/od/environmentalevents/f/dustbowl.htm (accessed 12 Feb. 2014).

Wilson, D.B., and S. Smoliak. 1977. Greenleaf pubescent wheatgrass. Can. J. Plant Sci. 57:289–291. doi:10.4141/cjps77-041

Wilson, D.B., and S. Smoliak. 1978. Registration of Greenleaf pubescent wheatgrass Reg. no. 12). Crop Sci. 18:918.

Wolfe, H.H., and K.J. Morrison. 1957. Nordan crested wheatgrass. Ext. Circ. 274. State College of Washington, Pullman.

Wright, S.J. 1994a. Registration of 'Benson' Beach wildrye. Crop Sci. 34:537.

Wright, S.J. 1994b. Registration of 'Reeve' Beach wildrye. Crop Sci. 34:537.

Young-Mathews, A., and S.R. Winslow. 2010. Plant guide for beardless wildrye (*Leymus triticoides*). USDA-NRCS, Lockeford Plant Materials Center, Lockeford, CA.

Zhang, H.-B., and J. Dvořák. 1991. The genome origin of tetraploid species of *Leymus* (Poaceae: Triticeae) inferred from variation in repeated nucleotide sequences. Am. J. Bot. 78:871–884.

Zhang, G.-Y., and J. Dvořák. 1995. Chromosomal control of the tolerance of gradually and suddenly imposed salt stress in the *Lophopyrum elongatum* and wheat, *Triticum aestivum* L., genomes. Theor. Appl. Genet. 90:229–236.

10

Rice

K.S. McKenzie,* X. Sha, K.A.K. Moldenhauer, S.D. Linscombe, N.B. Lyman, and L.L. Nalley

ABSTRACT

As one of the world's most important food crops, rice (*Oryza sativa* L.) is produced in a mechanized direct-seeded production system in the United States, in contrast to a transplanted production system in Asia, where most of the world's rice is grown and consumed. The United States is a major rice exporting country but accounts for less than 1% of the world's production. From 1981 to 2011, U.S. rice yields increased 86 kg ha^{-1} yr^{-1} to a 3-yr national average of 7.8 t ha^{-1}. U.S. rice is produced in a southern region including Arkansas, Louisiana, Mississippi, Texas, and Missouri and in a second region in the central valley of California. Yield levels, and the rate and time period of yield increases, have varied among the U.S. rice-growing regions. To ascertain the genetic yield gain and realized maximum yields, regional data from rice breeding programs in Louisiana, Arkansas, and California were analyzed and values were estimated. Results confirmed the difference in the genetic yield gain, with Louisiana achieving 87 kg ha^{-1} yr^{-1} and reaching 7.1 t ha^{-1} and California achieving 25 kg ha^{-1} yr^{-1} and reaching 9.3 t ha^{-1}. Future prospects for U.S. rice yield enhancement, including the role that hybrid or transgenic rice might play, and obstacles to continued progress like grain quality and environmental and regulatory issues are discussed.

Abbreviations: ARPT, Arkansas Rice Performance Trials; CL, Clearfield; GHG, greenhouse gas; RES, Rice Experiment Station; RREC, Rice Research and Extension Center; UA, University of Arkansas; UCCE, University of California Cooperative Extension; URRN, Uniform Regional Rice Nursery.

K.S. McKenzie, Rice Experiment Station, California Cooperative Rice Research Foundation, Inc., P.O. Box 306, 955 Butte City Highway, Biggs, CA 95917 *Corresponding author (KMcKenzie@crrf.org).
X. Sha (xsha@uark.edu) and **K.A.K. Moldenhauer** (kmolden@uark.edu) Rice Research and Extension Center, Univ. of Arkansas, Div. of Agriculture, 2900 Hwy 130 E, Stuttgart, AR 72160.
S.D. Linscombe, Rice Research Station, Louisiana State Univ. Agricultural Center, 1373 Caffey Rd., Crowley, LA 70578 (SLinscombe@agcenter.lsu.edu).
N.B. Lyman (nblyman@uark.edu) and **L.L. Nalley** (llnalley@uark.edu), Dep. of Agricultural Economics and Agribusiness, Univ. of Arkansas, Fayetteville, AR 72701.

doi:10.2135/cssaspecpub33.c10

Yield Gains in Major U.S. Field Crops. CSSA Special Publication 33.
Stephen Smith, Brian Diers, James Specht, and Brett Carver, editors.

Rice production is fundamental to global food security. Rice provides 19% of the caloric intake and 13% of the protein for the world's population, with Southeast Asia receiving 48 and 38% of its caloric and protein intake from rice, respectively (World Rice Statistics, 2012). Although grown on five continents, rice is primarily produced in a tropical climate, predominately in eastern, southeastern, and southern Asia, and is generally consumed domestically. Eighty percent of the world's rice belongs to the tropical subspecies (indica), with the remaining 20% being the temperate subspecies (japonica) that is produced in temperate and subtropical regions or at higher elevations where cooler temperatures are prevalent (Zeigler, 2012). U.S. rice production falls into the japonica group with plantings from 2002 to 2012 averaging 1.2 million hectares (USDA-NASS, 2012). The United States accounts for less than 1% of the world's 158 million hectares of rice, but it is among the top five rice exporters.

Rice in the United States is direct seeded under a highly mechanized lowland flooded production system. The average size of rice farms in the United States is large (183 ha) relative to those in Asia (0.5–3 ha), where the majority of rice is transplanted (Baldwin et al., 2011). Rice yield trends from 1981 to 2011 for Japan, Australia, Uruguay, and the United States are plotted in Fig. 10–1 to illustrate some differing yield growth patterns among temperate rice-growing countries. Unlike the United States, Japan uses a transplanted production system in small paddies to produce rice on 1.65 million hectares. Japan has a long history of rice production with many disease, pest, and weather factors impacting rice production. With an advanced rice research infrastructure and advanced technology available to rice growers, they are self-sufficient in rice production and do not export rice. There has not been a major emphasis on increasing production in Japan because per capita consumption of rice is declining, and adherence to grain quality market requirements often takes priority over advancements in yield potential. Figure 10–1 and Table 10–1 show that Japan's rice yields have increased approximately 1 t ha^{-1} over this 30-yr period.

Australia, Uruguay, and the United States are, in contrast, young rice-producing countries, all exporters of rice, and have experienced significant yield increases. Rice production in Australia occurs in New South Wales under high solar radiation, low humidity, and less pest pressure relative to global rice production. A direct-seeded lowland production system is used in rotation with other crops, and rice farmers are supported by an advanced research infrastructure and advanced technology. Despite these favorable conditions that promote higher productivity, Australian rice producers face low-temperature events at the sensitive reproductive stage, causing floret sterility and thus high yield variability as shown in Fig. 10–1. Rice production in Australia has also been subject to water restrictions due to droughts, and the annual plantings from 2002 to 2012 have ranged from a low of 2000 ha to a high of 108,000 ha as a function of water constraints. Yields have grown significantly since the 1980s from approximately 7 to over 9 t ha^{-1} (Table 10–1). Reinke et al. (2012) reviewed the challenges to improving rice production and identified major research targets for Australia.

Uruguay has a small production area (171,000 ha) in a crop rotation, with 85% of the rice planted to indica cultivars (Deambrosi et al., 2012). Uruguay is a net exporting country, quickly adopting the latest production research and technology. National rice yields have increased rapidly in 30 yr (Fig. 10–1). Much of

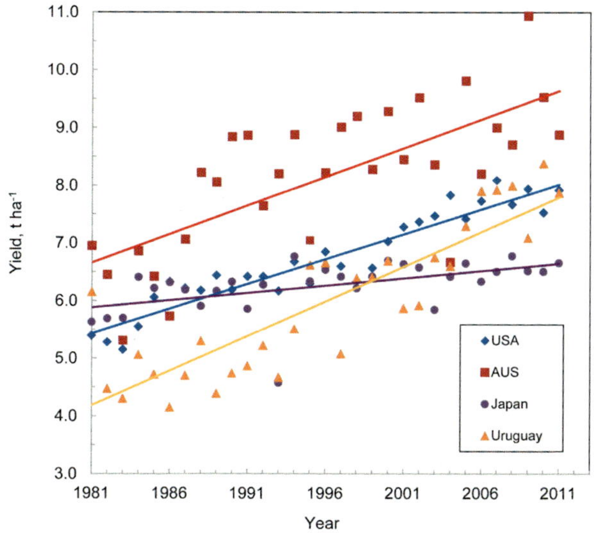

Fig. 10–1. Annual rice yields and linear trend lines from 1981 to 2011 for the United States, Australia, Japan, and Uruguay. Yield gains for this period in each country are reported in Table 1. Source: http://www.fas.usda.gov/psdonline/psdQuery.aspx.

Table 10–1. Yield gains in commercial rice production from 1981 to 2011 in four major rice-producing countries.

Country	Yield†		Yield gain‡	R^2
	Start	End		
	t ha⁻¹		kg ha⁻¹ yr⁻¹	
United States	5.3	7.8	86	0.89
Australia	6.2	9.8	99	0.51
Japan	5.7	6.6	25	0.26
Uruguay	5.0	7.8	120	0.75

† Mean of three consecutive years for both start and end periods.

‡ Slope of linear regression line shown in Fig. 10–1.

that yield increase was initially due to a shift to more modern semidwarf cultivars. During the second half of this period, 1996 to 2010, 'El Paso 144' was grown on approximately 70% of the national area, and Uruguay's average rice yield increased almost 2 t ha^{-1} (G. Zorrilla, personal communication, 2012). Zorrilla suggests that little of the yield improvement since 1996 can be attributed to genetic gain since one variety accounted for most of the production during that period.

From 1981 to 2011, on-farm U.S. rice yields increased 47% (2.5 t ha^{-1}), or 86 kg ha^{-1} yr^{-1} (Fig. 10–1 and Table 10–1). Reilly and Fuglie (1998), in their review of future yield growth in crops, produced historical trends for the period 1939 to 1994 and trend projections to 2019 based on linear and exponential models for U.S. rice yields. They projected yields for 2011 to be approximately 8 and 11 t ha^{-1} for the linear and exponential growth rates, respectively. Average U.S. rice yield for 2009 to 2011 was 7.8 t ha^{-1} (Table 10–1), close to the linear growth prediction but well below the prediction from the exponential model. Their predicted rate of gain for rice yield was 85 kg ha^{-1} yr^{-1}, almost identical to the calculated value of 86 kg ha^{-1} yr^{-1} in Table 10–1.

Regional Yield Trends in the United States

Rice is grown in the United States in two distinct regions. The largest area is the southern United States that includes Arkansas, Louisiana, Mississippi, Texas, and Missouri, where commercial production is primarily U.S. long grains from tropical japonica cultivars and, more recently, F_1 hybrids with indica and tropical japonica parentage. The Sacramento Valley in Northern California is the other rice-producing area that produces temperate japonicas and primarily markets U.S. medium grains (Calrose market type). The two U.S. rice production regions have developed different germplasm pools to address biotic and abiotic factors, as well as regional market quality requirements. Cultivars developed by breeding programs in the southern United States are grown throughout the region but are not adapted and produced in California. Likewise, California cultivars are not adapted or produced in the southern United States.

Improving yield potential remains a primary objective in both areas. There has been significant investment in rice cultivar improvement efforts in the United States since the 1970s, made possible by support from state rice "check-off programs," in which growers are assessed voluntarily per hundred weight of production. Funds from the check-off support breeding and other research activities through research grants that are awarded by grower research boards. In California, this funding expanded breeding research at a grower-owned rice experiment station, and in the southern United States, it has supported the land grant universities that have assumed cultivar development responsibility from the USDA-ARS. In the southern region, there has also been a large investment by the private international rice breeding company, RiceTec, to introduce hybrid rice cultivars that account for an estimated one-third of that production area. In the following sections, rice yields will be examined using regional research data from the three major U.S. public rice breeding programs to estimate genetic gain in U.S. rice yields in recent decades.

Louisiana

Though U.S. rice cultivation began in 1685 when the first-ever rice seeds were introduced from Madagascar to the Carolinas, the modern U.S. rice industry started on the coastal prairies of southwest Louisiana in the 1880s when a unique combination of factors allowed the establishment of commercial rice production in the region. These included the completion of the Southern Pacific Railroad from Houston, TX to New Orleans, LA, the migration of northern and midwestern farmers into the prairie, the availability of cheap land, and the mechanization of rice farming with the mechanical binder and stationary thresher. This produced a tremendous growth in rice cultivation, milling, and marketing in the region (Babineaux, 1967).

Figure 10–2 shows the history of Louisiana rice yields and also identifies some of the notable cultivars and technology that are believed to have contributed to significant yield increases. During the last 10 yr, an average of 0.19 million hectares of rice was grown annually in the state, accounting for 15% of the U.S.

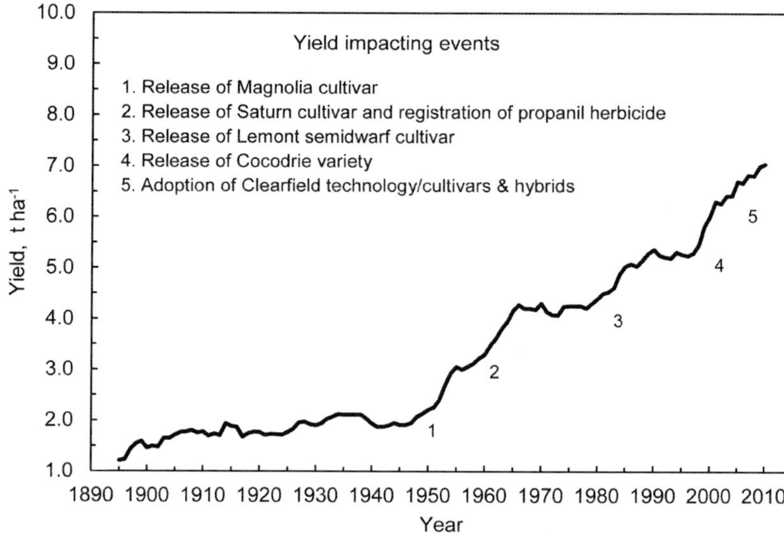

Fig. 10–2. History of Louisiana statewide rice yields (3-yr moving average) and major yield impacting events. Source: USDA-NASS (2012).

rice acreage (USDA-NASS, 2012), and consisted of long- and medium-grain rice from the tropical japonica subspecies. Like other southern rice-growing states, a steady yield increase has been achieved in commercial production in Louisiana during the last 30 yr, estimated at an annual rate of 1.5% or 85 kg ha^{-1} yr^{-1} (Table 10–2). The adoption of higher yielding cultivars during this timeframe, as opposed to improved production technology, is generally given most of the credit for this substantial yield increase. The adoption of non-transgenic imidazolinone herbicide resistant cultivars (Clearfield [CL]) to control red rice has been an important recent yield contributor that combines a new genetic trait with production technology. Red rice is a competitive weedy form of *Oryza sativa* L. that shatters, has strong grain dormancy so it persists in the soil and reseeds itself, and has a colored bran (aleurone layer) making it an unacceptable contaminant in packaged rice. In addition to controlling this weed, the Clearfield system allowed for more efficient use of applied nitrogen fertilizer by the cultivated crop. The following discussion provides support to the significant role that improved cultivars have made to yield increases in Louisiana.

The most direct way to estimate genetic gain is the evaluation of cultivars from different eras in common environments. However, this method is usually applied to a small genetic sample and can be confounded by the coadaptation of newer cultivars with evolving cultural practices and environmental conditions (Lillemo et al., 2009; Graybosch and Peterson, 2010). Historical regional yield trial data can also provide a reliable estimate of genetic gain achieved by breeding activities (Grandillo et al., 1999; Mackay et al., 2011).

A historical set of replicated yield trial data from the Uniform Regional Rice Nursery (URRN) was analyzed to estimate the contribution of improved cultivars to Louisiana rice production during the last 30 yr. The URRN is conducted

Table 10–2. Within-state yield gains in commercial rice production from 1981 to 2011.

Region	Yield†		Yield gain	Attributable genetic gain	Years
	Start	End			
	—— t ha^{-1} ——		—— kg ha^{-1} yr^{-1} ——		
United States	5.3	7.8	86	–	1981–2011
Louisiana	4.6	7.1	85	72	1982–2011
	4.6	5.2	58	41	1982–1995
	5.2	7.1	116	87	1996–2011
Arkansas	5.9	7.4	85	30	1986–2010
California‡	7.7	9.3	42	25	1981–2011
	8.5	9.3	74	50	1996–2011

† Average of three consecutive years from USDA-NASS (2012).

‡ Medium grain only.

by the USDA-ARS in cooperation with state breeding programs and typically contains 200 entries, including advanced experimental lines and commercial check cultivars that are tested in Arkansas, Louisiana, Mississippi, Missouri, and Texas. General standard production practices are followed, although no fungicides are applied. Only a small subset of the 200 entries is released to commercial growers each year, and some of those lines are primarily selected for traits other than higher yield potential, such as aroma, disease resistance, and earliness. Therefore, a subset of 24 entries that included popular cultivars grown in Louisiana during that time period, as well as experimental lines that were eventually released as cultivars from the Rice Research Station, Louisiana State University Agricultural Center, near Crowley, LA, were included in the analysis. The popularity of rice cultivars was determined according to the Annual Rice Variety Acreage Survey, which was published or compiled by the Rice Technical Working Group, The Rice Journal, and the Rice Miller's Association (Saichuk, 2012).

All yield, milling, maturity, and plant height data were obtained from the URRN conducted at the Rice Research Station. Of the selected cultivars, five were early cultivars developed by public breeding programs in Arkansas and Texas, while the rest were developed by the Rice Research Station and widely grown in other southern states. Similar to the procedure described by Breseghello et al. (1998), cultivar least-squares means obtained via the GLM procedure of SAS (SAS 9.3, SAS Institute) were regressed on the year of release to obtain an estimate of genetic gain for a specific trait. Regression analyses were conducted for the time period 1982 to 1996 (first half), 1997 to 2011 (second half), and for the entire time period from 1982 to 2011. The mid-1990s marked a transition to higher yielding cultivars such as Cypress and Cocodrie as well as the introduction of Clearfield technology a few years later, thus providing a meaningful milestone for monitoring genetic progress.

Average rough rice yield increased linearly for 24 rice cultivars released from 1977 to 2011 (Fig. 10–3) at the rate of 72 kg ha^{-1} ($P < 0.001$) or 0.8% per year. Genetic gains for grain yield of cultivars developed before and after 1996 were 41 ($P = 0.09$) and 87 kg ha^{-1} yr^{-1} ($P = 0.01$), respectively, indicating that yield gains have accelerated during the last 15 yr (Table 10–2). The genetic gain accounts for over 50% of the on-farm yield increase experienced during the same period. Similar genetic

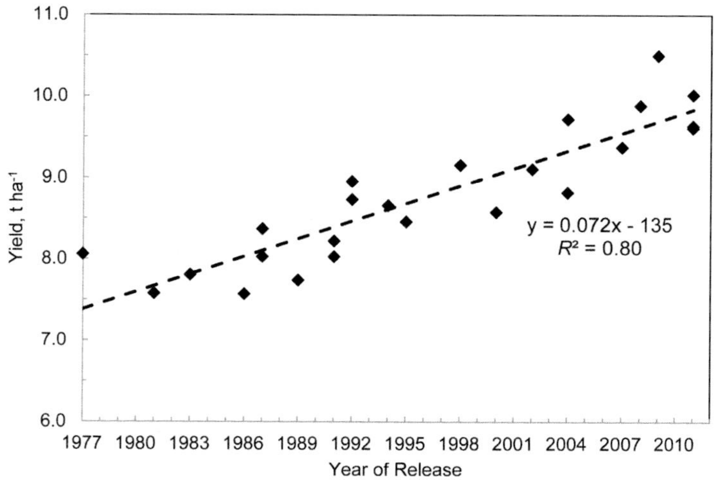

Fig. 10–3. Change in rice yield of the 24 most popular and/or released Louisiana rice cultivars tested in the Uniform Regional Rice Nursery at Crowley, LA from 1977 to 2011.

gains were reported in rice and other cereals (Anzoua et al., 2010; Peng et al., 2000; Tabien et al., 2008; Cox et al., 1988; Lillemo et al., 2009). By growing rice cultivars released from 1944 to 1992 by the Texas rice breeding program, Tabien et al. (2008) estimated genetic gain for grain yield to be 42 kg ha^{-1} yr^{-1}.

The development of landmark cultivars played a critical role in the yield increase witnessed in the last three decades. Lemont, the first commercially successful U.S. semidwarf long-grain cultivar released in 1983 (Bollich et al., 1985), dominated the southern rice landscape from 1985 until it was replaced by Cypress (Linscombe et al., 1993b) in the mid-1990s. Because of its high yield potential, early maturity, high and stable milling yields, and good grain appearance, Cypress and its derived imidazolinone-resistant (CL) cultivar CL161 (Sha et al., 2007) were widely grown in Louisiana and adjacent rice-growing states until the late 2000s. Meanwhile, another Louisiana release, Cocodrie (Linscombe et al., 2000), became the leading long-grain cultivar in the southern United States in 2001. 'Cypress' and 'Cocodrie' typified the trait complexes central to the Rice Research Station rice breeding program, which include semidwarf and compact plant type, very early maturity, good milling yields, and wide adaptation. Because of its effective control of noxious weedy red rice, CL rice rapidly gained popularity after its inception in the early 2000s. Even though early CL cultivars were inferior to the non-CL conventional cultivars in yield potential (Sha et al., 2007), the latest releases such as CL151 (Blanche et al., 2011) and CL111 possess higher yield potential than leading conventional cultivars, which probably contributed to the accelerated yield increase in the last decade. Although declining in production during the last 15 yr, medium-grain rice represents a sizable acreage in Louisiana and Arkansas. The semidwarf cultivar Bengal (Linscombe et al., 1993a) released in 1992 rapidly replaced the older tall cultivar Mars. 'Bengal' accounted for almost all the medium-grain rice produced in Louisiana from the mid-1990s to the late 2000s when it was finally replaced by 'Jupiter' (Sha et al., 2006).

The significant genetic gain in rice yields, especially for newer cultivars released after 1996, cannot be attributed to a broadening of the inherently narrow genetic base in U.S. rice. All leading Louisiana rice cultivars released after 1996 were selected from crosses among older U.S. cultivars or breeding lines. Increased investment of both financial and human resources into rice genetic improvement is a more likely cause. Nevertheless, germplasm introductions, such as 'IR8' and 'TN1,' did play an important role in the development of semidwarf cultivars. Broadening genetic diversity of U.S. rice by incorporating potentially important traits from foreign introductions should be actively pursued to achieve further yield gains.

Arkansas

Arkansas has a long history of rice production. In 1844, Captain Robertson of the steamer Sallie reported a number of farmers growing rice. One farmer, Mr. Halley, produced rice near Thompson Creek, Craighead County, AR (Spicer, 1964). In the spring of 1897, the first hectare of rice was planted on the Grand Prairie of Arkansas (Spicer, 1964). By 1905, there were 405 ha of rice grown (Slaton, 2001b), and the first USDA rice crop report for Arkansas was published. Today, Arkansas farmers produce more than 45% of the rice grown in the United States under variable production conditions across the state. Arkansas rice farmers depend on the public variety development programs that provide improved cultivars suited to changing conditions in the field and marketplace.

The Arkansas rice producers provide check-off funds that are administered by the Arkansas Rice Research and Promotion Board and help support the rice breeding program directed by University of Arkansas (UA) scientists in cooperation with researchers in other states and the USDA-ARS. Check-off funding for the breeding program began in 1980 and has since increased tenfold. Twenty-five cultivars have been released from the UA breeding program during this period.

The Rice Research and Extension Center (RREC), formerly the Rice Branch Experiment Station, opened in 1926 with G.H. Banks as director. The rice breeding and cultivar development program has been continuously active at the UA Division of Agriculture's RREC since 1931. First directed by USDA-ARS personnel and more recently by UA scientists, the rice breeding program has released 41 cultivars adapted to the Arkansas rice-growing regions. Zenith (Johnston, 1958), released in 1936 (Fig. 10–4), was the first cultivar released. Breeders at RREC work in close cooperation with the other faculty and numerous support scientists located at the center and at the UA campus in Fayetteville, AR to develop production packages specific to new cultivars with information on herbicide usage, disease and insect management, and fertility requirements. These production packages tailored to individual cultivars became available to producers with the release of new cultivars in the 1960s.

Rice production in Arkansas increased dramatically after the rice allotment restrictions established by the U.S. Secretary of Agriculture were lifted in 1974 (Fig. 10–4). In 1978, there were over 440,000 ha of rice grown in Arkansas. From 2001 to 2010, the rice area in Arkansas grew to approximately 610,000 ha.

The average yield per hectare also increased over time, even with an increase in the state's rice plantings. Yields remained in the 2.0 to 2.5 t ha^{-1} range from 1906 until 1955, when medium-grain rice was the primary type grown in Arkansas. There was a dramatic increase to 3.5 t ha^{-1} with the addition of 'Bluebonnet 50,' a

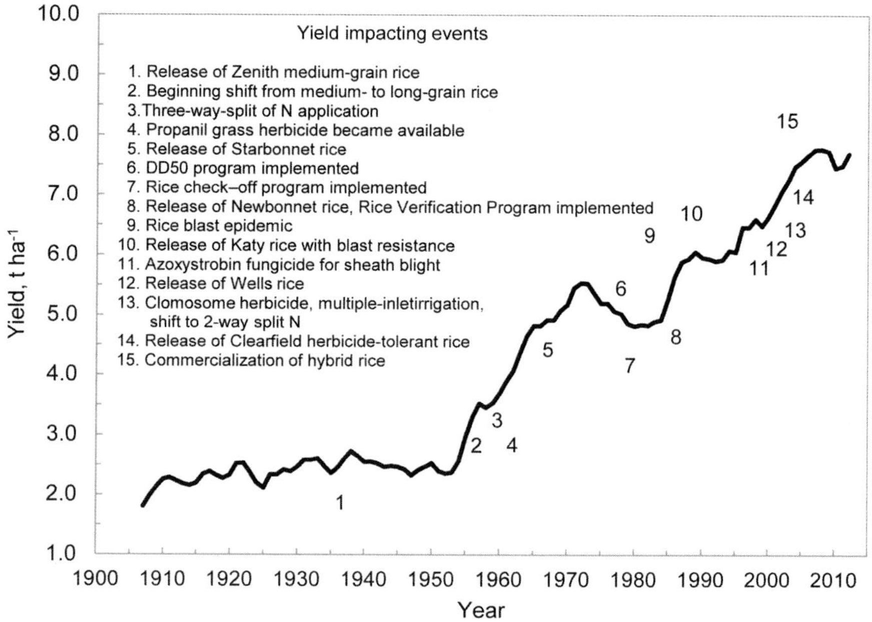

Fig. 10–4. History of Arkansas statewide rice yields (3-yr moving average) and major yield impacting events. Source: USDA-NASS (2012).

long-grain rice (Johnston, 1958) introduced from Texas in 1955 (Reid 1953, 1954, 1956; USDA-NASS, 2012). During this time, long-grains began their rise to prominence in Arkansas. 'Nato' (Johnston, 1958), released by the Rice Research Station, Crowley, LA in 1956, gave a resurgence to medium-grain rice, but long-grain cultivars eventually became dominant in Arkansas. Over the next 12 yr, there was a slow but steady increase in yield up to 5.0 t ha^{-1} in 1967 (USDA-NASS, 2012). Between 1970 and 1973, the average yields fluctuated around 5.5 t ha^{-1} with producer adoption of 'Starbonnet' (Johnston et al., 1968; Slaton, 2001b). When acreage controls were removed, yields declined to approximately 5.0 t ha^{-1} in 1974 and remained there until 1985. This drop may have been due to management constraints associated with the rapidly increasing planted area, including limited availability of equipment and irrigation, planting on land that may not have been ideally suited for rice, and the addition of new inexperienced rice producers.

In 1985, yields jumped to 5.8 t ha^{-1} with the availability of the higher yielding 'Newbonnet' (Johnston et al., 1984), which accounted for 61% of the rice grown in Arkansas (Fig. 10–4). Newbonnet remained on more than 35% of the acreage until 1992, when it was replaced by Katy (Moldenhauer et al., 1990), the first of the blast-resistant cultivars released by UA. From 1985 to 1987, when 'Newbonnet' was planted on more than 50% of the Arkansas rice acreage, rice blast (*Magnaporthe oryzae*) became a significant threat in the state because of Newbonnet's susceptibility. By 1989, 'Katy' was released with resistance to the common blast races in Arkansas. Another important cultivar during this time period was Lemont (Bollich et al., 1985) (high of 28% of the acreage in 1988), the first widely planted semidwarf variety in the southern United States. In 1996, with the advent of Kaybonnet, another of the blast-resistant cultivars (Gravois et al., 1995), the average yields jumped to an all-time high of 6.9 t ha^{-1}. 'Kaybonnet' was planted on 33% of Arkansas' acreage in 1996 and 1997.

The average yield in Arkansas remained between 6.3 and 6.8 t ha^{-1} until 2001 when 'Wells' (Moldenhauer et al., 2007) and 'Cocodrie' each occupied 30% of the acreage, and the state average yield again reached a high of 7.1 t ha^{-1} (Fig. 10–4) (Wilson and Branson, 2003; Wilson and Runsick, 2008; Hardke and Wilson, 2013). 'Wells' remained the prominent variety of rice grown in Arkansas until 2010. By 2003, 'Wells' was planted on 45% of the Arkansas acreage and 'Cocodrie' had dropped to 22%, but the yields continued to climb to a record high of 7.3 t ha^{-1}. The next significant yield increase was to 7.8 t ha^{-1} in 2004, when 'Wells,' 'Francis,' and 'CL161' occupied 40, 11, and 13% of the acreage, respectively. Part of this increase may be attributed to red rice control by the Clearfield system. In 2007, the state average yield was a record again at 8.1 t ha^{-1}, when 'Wells' and 'Francis' occupied 35 and 11% of the Arkansas acreage, respectively. In 2007, conventional hybrids, CL hybrids, and CL purelines occupied 8, 11, and 11% of the acreage, respectively. Clearfield cultivars were first released in 2001 (Fig. 10–4), and by 2008 they were grown on 39% of the acreage, climbing to nearly 60% by 2010, and have remained there since (Slaton, 2001a). Hybrids were first introduced in 2002, and 2006 was the first year they occupied more than 5% of the acreage. In 2012, hybrids including the CL hybrids accounted for approximately 35% of all rice planted in Arkansas.

The Arkansas Rice Performance Trials (ARPT), sponsored by the UA Division of Agriculture, has provided annual yield data on cultivars commercially available in Arkansas since 1971. Tests are conducted on experiment stations and in producer fields by using a randomized block design. Analysis of the ARPT has shown substantial gains from genetic improvement in cultivars over time. Using a Just–Pope production function (Just and Pope, 1979) holding non-genetic yield determinants such as weather, soil type, and technological advances constant over time, the UA breeding program released improved cultivars that increased producer yields by 30 kg ha^{-1} yr^{-1} over the past three decades (Nalley et al., 2011). Similarly, the observed genetic gain has not been found to increase yield variance or risks that are associated with experimental yield tests. The constant yield variance throughout the test period can be viewed as a successful "maintenance" function of a breeding program. That is, without the breeding program, rice yields may otherwise have deteriorated and become more volatile as pathogens like blast reduced yield and increased yield variability as earlier sources of resistance were defeated. Kropff et al. (1994) showed similar results.

Statewide yields, UA variety yields, and the proportion of UA cultivars harvested in Arkansas are presented in Fig. 10–5. ARPT yields from UA cultivars were generally higher than the statewide average, and both exhibited statistically significant ($P < 0.001$) positive trends over time. Yields increased experimentally and commercially by roughly 85 kg ha^{-1} yr^{-1} between 1986 and 2010. This annual rate is roughly 50 kg ha^{-1} yr^{-1} greater than the genetic gain attributable to the UA rice breeding program reported by Nalley et al. (2011). The difference is likely due to improved management practices and inputs. The downward trend in the area harvested from UA cultivars has resulted from greater availability of commercially released hybrid rice cultivars developed by RiceTec and the recent proliferation of CL rice cultivars, both hybrid and pureline. The recently started

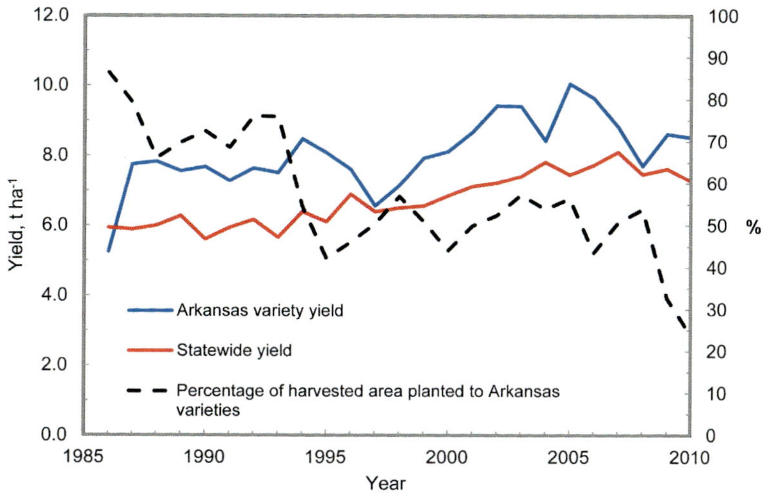

Fig. 10–5. Arkansas statewide rice yield, Arkansas variety trial yield, and University of Arkansas variety share from 1986 to 2010. Sources: USDA-NASS (2012), Arkansas Rice Performance Tests, and University of Arkansas DD50 Program.

UA hybrid breeding program seeks to offer alternatives to hybrid cultivars developed by private companies that are largely responsible for the sharp decline in rice area planted to UA cultivars from 54% in 2008 to 23% in 2010 (Fig. 10–5).

Yield improvements attributable to the UA breeding program have led to substantial economic gains for rice producers in Arkansas and neighboring states, as would be expected from either public or private breeding investments. Nalley et al. (2011) estimated the value of the genetic gains to be on an overall upward trend since 1986 and to approach nearly 90 million dollars in 2007 (Fig. 10–6). Estimates indicate that yield improvements have come at an increasing rate, which suggests that the breeding program has become more effective over the study period. In combination with the increased adoption of cultivars developed privately and by breeding programs from surrounding states, the effectiveness of the UA rice breeding program suggests a bountiful future for Arkansas rice production.

Through the years many important management practices have also affected rice yields. In 1914, tractors were first utilized in rice production in Arkansas, and by the late 1920s, changes were incorporated that improved irrigation (Spicer, 1964). These changes included land leveling in the Grand Prairie of Arkansas using homemade wooden levels, and by 1928, a levee puller had been designed that would pull both sides of the levee at once (Spicer, 1964). By the 1960s, herbicides were improving and having an impact on rice production in Arkansas, and propanil was the first in 1961 to give grass control in the rice crop (Slaton, 2001b). Before the availability of effective grass herbicides, the only noteworthy herbicides were phenoxy herbicides for the control of broadleaf weeds.

Improved herbicides allowed the development of rice cultivars that have higher harvest indices, yielded more grain per hectare, and partitioned more of their nitrogen into grain across a range of N fertilizer rates (Bufogle et al., 1997). Consequently, the advent of selective grass weed control allowed for improved

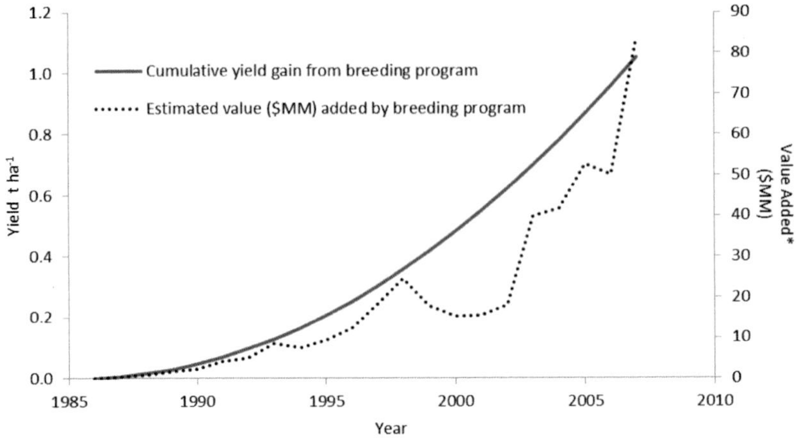

Fig. 10–6. Cumulative rice yield gain and value gains associated with University of Arkansas rice breeding program. The estimated value added by breeding program represents the product of yield in t ha^{-1} added by the breeding program in year t, the market price of rough rice in year *t*, and the Arkansas rice area harvested in year *t*. Source: Nalley et al. (2011).

nitrogen fertilization practices, increased nitrogen partitioning to the grain, and enhanced selection for yield by the elimination of constraints on yield. In 1968, molinate was first used in commercial rice fields in Arkansas and with propanil dominated chemical weed control measures until 1990, when the first propanil-resistant barnyard grass was discovered in Poinsett County, Arkansas. Molinate was still used until the early 2000s.

Quinclorac was released in 1992 as a replacement for propanil, and by 1992, quinclorac-resistant barnyard grass was also documented in Arkansas. In 2000, clomazone was labeled for rice in Arkansas, and it is still used today. These herbicides allowed breeders to select rice cultivars that did not have to compete with weeds, paving the way for shorter, less competitive plant types. The reduced competition between rice plants also allowed for higher yields. Cultivars with improved plant type (erect leaves and culms and shorter stature) and higher yield could now be released to producers. Azoxystrobin fungicide also allowed the expansion of semidwarf cultivars that were very susceptible to sheath blight (*Rhizoctonia solani* Kühn). As with any yield limiting factor, cultivars are selected for a given set of environmental and cultural conditions. With time, the prevailing disease organisms and weeds adapt genetically to the changing conditions. Research and technology has kept up with these pressures, and a resulting decrease in yield due to disease and weed infestations has not occurred in Arkansas.

Soil fertility improvements also played a major role in the yield increase in Arkansas rice cultivars. A three-way split method of nitrogen application was developed in Arkansas and implemented in the mid-1960s, which allowed the management of plant height to optimize yield. Over the next two decades, cultivars were selected that would perform well in this environment. This practice of selecting lines for a three-way-split nitrogen application was in effect until the mid-1980s, when a single pre-flood application was applied to the breeding nurseries for selection of cultivars that could utilize higher rates of nitrogen fertilizer in early development. The advent of the three-way-split fertilizer application was part of the impetus for the DD50 program to manage the rice crop. The DD50 program developed by Arkansas scientists was tested in 1975 with 43 producers and implemented in 1978 with 540 producers participating (Slaton, 2001b). The DD50 is still utilized today so that producers can stay abreast of management decisions and get the best yields possible in their situation.

California

Commercial rice production was successfully established in California in 1912. In 1908, USDA scientists began evaluating a number of plant introductions from Asia to identify suitable rice germplasm for rice production on heavy clay soils of the Central Valley of California. Two short-grain selections—'Caloro,' an early-maturing selection from the Japanese introduction 'Wataribune,' and 'Colusa,' a selection from a Chinese introduction developed by the USDA in Louisiana—were released by 1918 and constituted virtually all of California rice production until the 1950s. In 1948, a medium-grain variety developed through pedigree breeding, 'Calrose' (Johnston, 1958), was released by the California Rice Experiment Station (RES). It showed improved adaptation, cold tolerance, higher yield potential, and good grain quality and by 1975 occupied 70% of the California rice acreage. In 1969 California rice growers established a grower-funded research assessment, the model for check-off programs previously mentioned, which allowed the establishment of an accelerated rice improvement program (McKenzie et al., 1994). By the late 1970s, adapted first-generation semidwarf cultivars became available to California growers and were rapidly adopted. An aggressive rice breeding program has continued to date, releasing 20 improved "Calrose" medium grains and a similar number of other cultivars, including short grains, long grains, and special-purpose types. These improved RES cultivars have been grown on more than 90% of California's rice acreage from 1980 through 2013.

A number of improvements have contributed to increased California rice yields (Fig. 10–7). Those considered most significant include inputs like application of nitrogen fertilizer, chemical weed control, and precision farming. The first significant genetic improvement in yield was seen with the production of medium-grain rice variety 'Calrose' in the 1950s. From 1978 to 1988, in what may be characterized as the "green revolution" period, yield increased at the linear rate of 180 kg ha^{-1} yr^{-1}($R^2 = 0.63$). This was the result of the release of a series of new rice cultivars with the semidwarf plant type and earlier maturity, augmented by research-based agronomic management information, advances in herbicides, and an innovative-progressive grower clientele. McKenzie et al. (1994) estimated that 60% of this increase was attributed to improved rice cultivars. From 1989 to 1994 there were additional yield increases (168 kg ha^{-1} yr^{-1}, $R^2 = 0.42$) in California, supported by a second generation of semidwarf cultivars. This was a yield high point with a 9.5 t ha^{-1} average for the period 1991 to 1994. Yields declined and several reasons have been suggested by researchers and growers for this decline. Weed control problems developed because of the appearance of herbicide-resistant weeds, herbicide availability and application restrictions, and a phase-out of straw burning that all impacted productivity and crop management. Late spring rains that delayed planting, below-average temperatures, and fall rainfall impacting harvest are considered the major weather contributors to reduced

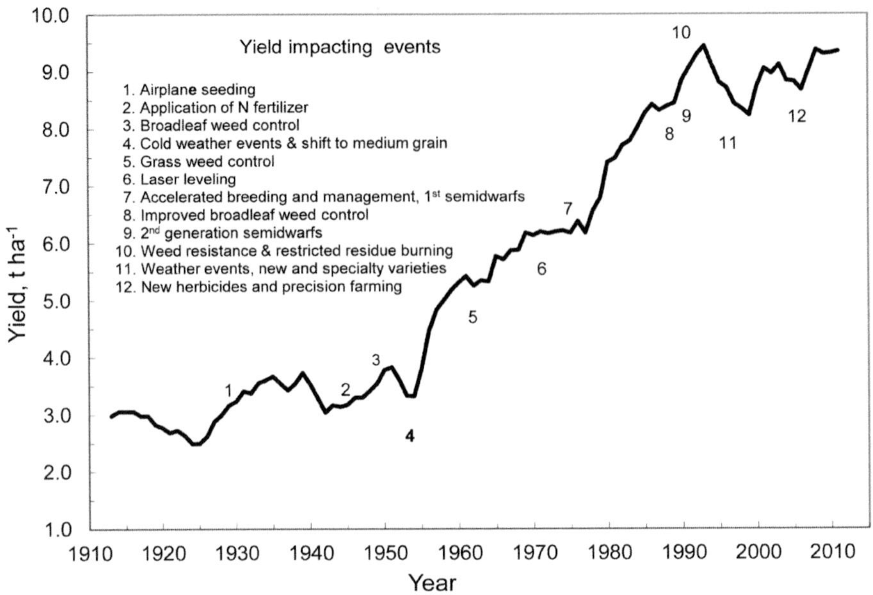

Fig. 10–7. History of California statewide rice yields (3-yr moving average) and major yield impacting events. Sources: USDA-NASS (2012) and J.E. Hill (personal communication, 2013).

yields levels. There was also an increase in acreage devoted to specialty cultivars that have significantly lower yield potential.

Over time, rice growers have adapted to the straw and weed management challenges while adding precision farming practices such as sensor-based fertilizer and pesticide application, improved laser leveling, and higher capacity equipment. Third-generation improved Calrose rice cultivars became available and were also adopted. All of these factors have contributed to returning the statewide commercial medium-grain rice yield to a 9.5 t ha^{-1} average for the period 2007 to 2011. Yield increased at the linear rate of 72 kg ha^{-1} yr^{-1} ($R^2 = 0.38$) for the period from 1995 to 2011. Yields did fluctuate during this period as can be seen in Fig. 10–7.

California's harvested rice area from 2001 to 2011 has averaged 145,400 ha, with a maximum of 145,790 ha in 2004 and a minimum of 116,390 ha in 2001 due to water shortages. Production area by cultivars is not collected in California by the USDA, the California Department of Agriculture, University of California Cooperative Extension (UCCE), seed producers, or marketing organizations in any complete, accurate, and/or publicly available form. Calrose medium grains have similar quality and marketing and are frequently comingled when delivered to the drying facility at harvest or later when shipped for milling. Thus, varietal yield comparisons are not available at this level; however, the variety testing program does provide an opportunity to examine recent yield gains.

In 1985, 'M-202' (Johnson et al., 1986) was released to growers. This early maturing medium grain proved to be very broadly adapted, becoming the predominant California rice cultivar for the next two decades. It still remains in the top three in California rice acreage on the basis of foundation seed sales (McKenzie, unpublished data, 2013). Two subsequent releases, 'M-205' (Johnson, 2002) and 'M-206' (Johnson, 2005), now hold the top two positions. Because of the longevity of M-202, its serves as a good baseline cultivar to examine yield progress in California rice.

The grower-funded cultivar testing program in California is conducted by the UCCE with funding from the California Rice Research Board. These tests are conducted annually in grower fields and also at the RES. Experimental entries and check cultivars are tested in one of three maturity groups—very early (four locations), early (four locations), or intermediate to late (three locations)—which include standard check cultivars. 'M-202' has been a continuing check entry since 1988 in at least two maturity groups. Using seven or more replicated experiments per year, we found that mean yields for M-202, for the highest yielding medium-grain entry, and for the commercial California medium-grain cultivars tracked in a parallel pattern during the period 1988 to 2011, indicating that commercial and experimental yields were impacted similarly in each growing season (Fig. 10–8a). The similar trends also exemplify the value of the testing program in reflecting commercial rice production in California. The top experimental entry in most years produced superior yields to 'M-202,' underscoring the presence of breeding material with higher yield potential. Not surprisingly, test plot yield means tended to exceed the commercial yield means, although this gap was less than 10%. Average yields over this period were 9.1, 10.2, and 10.8 t ha^{-1} for the state medium-grain average, 'M-202,' and the top medium-grain entry, respectively. The trend line for commercial yield is advancing upward, and a t test for $b > 0$ was significant ($P = 0.02$). The California medium-grain yields grew by 0.4 t ha^{-1} over the 23-yr period or about 4%. Trend lines for 'M-202' and the top experimental lines were not significantly different from $b = 0$ at the 0.05 level.

M-205 and M-206 were released to growers in 2000 and 2003, respectively, and have grown in acreage, displacing M-202 as the top cultivar. Figures 10–8b and 10–8c track the experimental plot yield for those two cultivars, the check standard 'M-202,' and the commercial medium-grain yield averages. Again, both plots showed a parallel pattern reflecting the yield potential of that particular growing season. M-205 and M-206 both showed a yield advantage over their predecessor cultivar M-202. The average yield increase over 'M-202' for both 'M-205' and 'M-206' was 0.5 t ha^{-1} or 5%. Over the time period in these figures, t tests of the trend lines were not significantly different from 0 at the 0.05 probability level.

Table 10–3 contains summary data from 68,000 ha of medium-grain rice processed from 2004 to 2008 by the Butte County Rice Growers Association (BUCRA) (C. Hoff, personal communication, 2012). It provides commercial-level validation of the yield genetic gain achieved by 'M-205' and 'M-206,' averaging 0.5 and 0.4 t ha^{-1} more than 'M-202,' respectively, essentially identical to the values presented previously from the statewide variety tests (Fig. 10–8b and 10–8c). In addition, 'M-205' and 'M-206' have improved milling yield and stability compared with 'M-202' (Mutters and Thompson, 2011). Table 10–3 documents, at the commercial level, genetic gains for rice milling yields achieved by 'M-205' and 'M-206' that generate additional financial returns to growers and millers. Rice must be processed before it is used. The hulls are removed from the "paddy" rice, which leaves brown rice and in turn is milled by friction or abrasion to yield white milled rice. Kernels may fissure and break in the field and/or during processing. Rice is

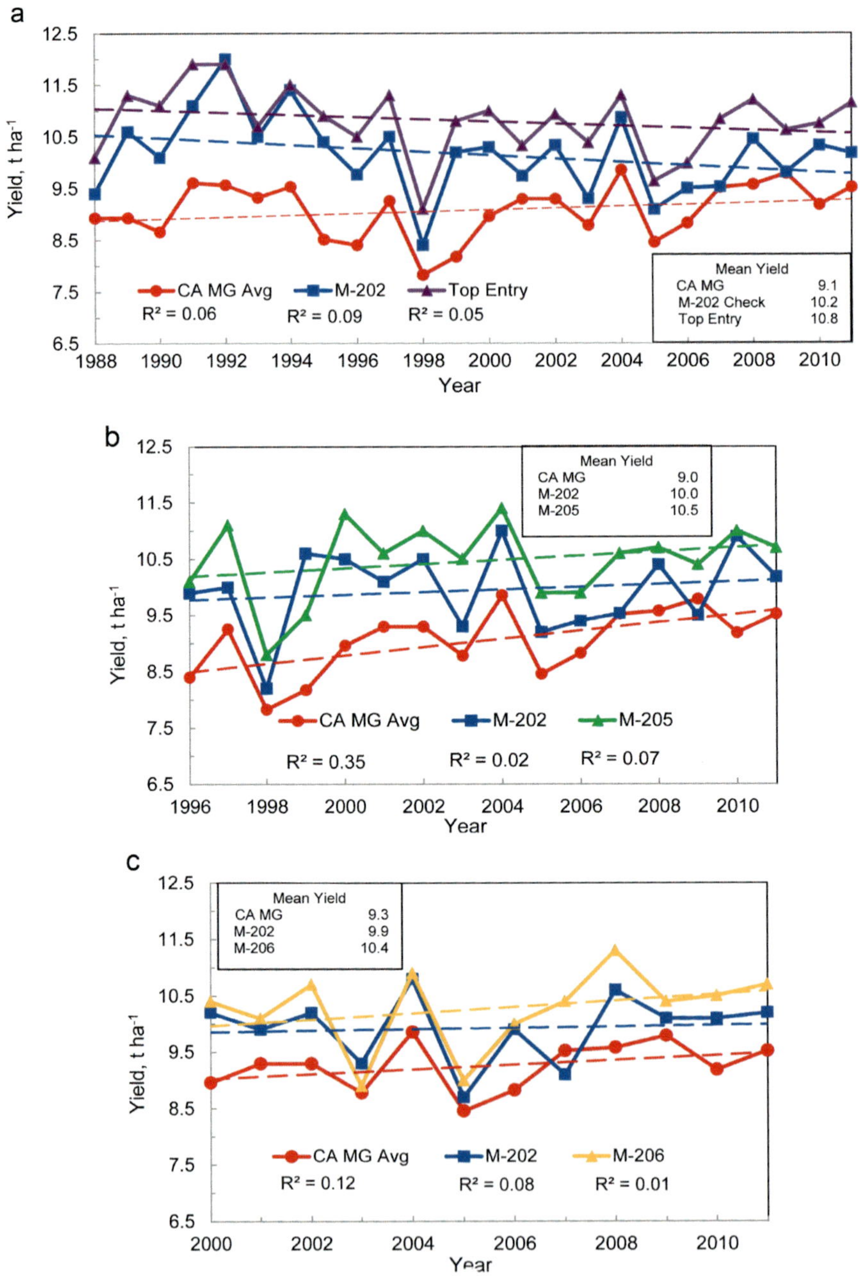

Fig. 10–8. Yield comparisons for commercially produced medium grain (MG) rice in California versus key cultivars tested in replicated research trials: (a) California MG rice yields, multi-location yield means for M-202, and the top experimental MG entry, 1998–2010; (b) California MG rice yields and multi-location yield means for M-202 and M-205, 1996–2010; and (c) California MG rice yields, multi-location means for M-202 and M-206, 2000–2010. Sources: USDA-NASS (2012) and University of California Cooperative Extension Statewide Yield Tests.

Table 10–3. Production and milling appraisal data for medium grain cultivars M-202, M-205, and M-206. Source: Butte County Rice Growers Association, 2004–2008.

Year	Cultivar			Milled rice: Whole			Milled rice: Total			Yield			Area			Production		
	M-202	M-205	M-206	M-202	M-205	M-206	M-202	M-205	M-206	M-202	M-205	M-206	M-202	M-205	M-206	M-202	M-205	M-206
	%									t ha^{-1}			ha × 1000			t × 1000		
2004	20.8	67.6	7.5	59.9	62.4	60.8	68.9	69.9	68.2	10.8	10.7	11.1	2.8	9.1	1.0	29.8	96.9	10.8
2005	6.3	46.3	43.3	61.5	64.1	63.3	68.5	69.7	68.9	8.7	9.0	8.9	0.8	5.9	5.6	7.2	53.2	49.7
2006	5.5	58.3	33.6	59.6	65.7	65.7	69.7	71.1	70.6	9.1	9.7	9.3	0.8	8.2	4.9	7.5	79.5	45.9
2007	1.7	65.0	31.6	66.9	67.8	66.4	69.8	71.3	70.2	9.6	10.3	10.1	0.3	9.5	4.7	2.5	97.3	47.3
2008	0.4	67.7	31.0	66.4	65.7	64.8	71.1	70.7	70.0	9.4	10.5	10.2	0.1	9.7	4.6	0.6	102.2	46.8
Avg.	6.9	61.0	29.4	62.9	65.1	64.2	69.6	70.5	69.6	9.5	10.0	9.9	0.9	8.5	4.2	9.5	85.8	40.1

appraised for market, and grower payments are based in part on the percent milling yield. If the milling yield—especially the whole kernel percentage—is low, the value of the rice is greatly reduced, so selection for high and stable milling yield is an important quality component in rice breeding. No genetic gain in milling yield was reported in Arkansas (Moldenhauer et al., 2009) and Texas (Tabien et al., 2008) and as part of the Louisiana yield analysis presented earlier in this chapter. However, the long-grain cultivar Cypress discussed earlier is recognized as having achieved high and stable milling yields through improved resistance to kernel fissuring (Jodari and Linscombe, 1996).

The improved yield potential of 'M-205' is partially attributed to improved lodging resistance, although it is more susceptible to cool temperature-induced sterility than 'M-202'. 'M-206' is not as lodging resistant as 'M-205,' but it has a sufficient level of cold tolerance for planting in cool production areas or in delayed planting situations where 'M-205' is not recommended. Greater kernel weights for 'M-205' and 'M-206' may contribute to their yield superiority over 'M-202'.

Medium-grain experimental entries yielding above 'M-205' have been generated in the RES rice breeding program (Fig. 10–8a). High yielding lines also are often seen in long-grain and short-grain materials. Of note is that some of these high yielding lines were from crosses with *Oryza rufipogon*, used as a source of disease resistance. However, after several crosses with adapted germplasm, they still contain significant deficiencies, especially in the area of grain quality, and thus are unsuitable for commercial markets. Rice is primarily appraised and marketed in the whole kernel form, making appearance, milling, and grain quality very important varietal requirements for U.S. rice. Improvement in yield potential in rice is achieved with larger panicles, increased kernel size, and/or increased tillering. These traits typically are not positively associated with milling yield or visual grain quality characteristics, such as kernel translucency, shape, and uniformity. Larger panicles reduce uniformity in kernel ripening, larger kernels tend to be chalkier, and increased tillering introduces more variability in maturity. Requirements for disease resistance and stress tolerance are further obstacles. The genetic challenge remains to correct these deficiencies and recombine positive traits, which makes further yield progress a slow and daunting task for rice breeders.

Future Prospects

Table 10–2 summarizes recent yield gains for U.S. rice, reported in this chapter to be associated with semidwarf habit, herbicide tolerance, disease resistance, specific cultivars, and F_1 hybrid cultivar development. Using the values presented and the proportion of acreage from Louisiana, Arkansas, and California, we estimate the annual U.S genetic rice yield gain to be about 40 kg ha^{-1} yr^{-1} from 1981 to 2011. Yield advancement and rate of gain have varied by production region and time period. Yield advancement has been more notable in Louisiana and Arkansas since the 1980s (Fig. 10–2 and 10–4), whereas California yield levels, although higher, have leveled off, indicating a yield plateau may have been reached (Fig. 10–7).

Introgression of new germplasm into future cultivars is an identified approach for increasing yield, as U.S. rice cultivars have an extremely narrow genetic base. After studying the pedigrees of 140 released rice cultivars or germplasm lines, Dilday (1990) revealed that all rice cultivars developed in the southern United States could be traced to 22 initial introductions from the early 1900s. Later, Lu et al. (2005) conducted single sequence repeat (SSR) marker analysis on the same set of cultivars and experimental lines and found that the population structure in U.S. rice was established before 1930 and remained essentially intact after over 70 yr of controlled crossing and artificial selection by various breeding programs. Only in rare cases has exotic diverse germplasm been incorporated into crossing programs, because such germplasm usually upsets the balance of desired quality traits such as amylose content and gelatinization temperature (Rutger et al., 1998), both of which contribute to the cooking and eating quality of rice. Nevertheless, a steady farm yield increase has been achieved in U.S. rice since the inception of the industry, as previously mentioned. New germplasm has been introduced to the United States in the form of hybrid rice.

The most successful yield-enhancing technology in recent decades has been the development and successful commercial production of hybrid rice by Chinese scientists (Yuan, 2003). This technology has spread worldwide and has been successfully commercialized in the southern United States and South America by RiceTec (Chu et al., 2008; Walton, 2003). In recent years, there has been a steadily increasing area in the southern United States planted to hybrid rice, estimated to be about 25% of the production area in 2012. Hybrid rice has yet to be commercialized in California although proprietary research by RiceTec was initiated in 2012. There is widespread interest and investment at the international level in hybrid rice breeding, not only because of the importance of rice as a world food crop but also because of the revenue generation potential unique to hybrid cultivars.

Sha et al. (2009) reported a 28% yield advantage for commercial hybrids over conventional long-grain check cultivars in 2006 to 2007 multi-location variety tests. In Arkansas in the 2010 to 2012 ARPT, the best hybrid was 6% better than the best cultivar for both the 3-yr average and for 2012 data (Hardke et al., 2012). In limited tests at the California RES, the yield advantage of introduced indica hybrids did not exceed 11% over the top purelines and in many cases showed no yield advantage. Determining the yield advantage of hybrid rice in the United States at the commercial level is more difficult to document, but a 10% advantage might be expected on the basis of our experiences. An observed advantage of hybrids is their ability to yield better than pureline cultivars on marginal rice soils or under adverse conditions caused by weather, disease, or management problems.

The land grant university based rice breeding programs in Arkansas, Louisiana, and Texas have initiated hybrid research and breeding programs as a demonstrated avenue to higher yields. Louisiana and Arkansas have hybrid lines under evaluation in multi-location yield tests. In contrast, the RES has not initiated a hybrid rice breeding program. Challenges facing hybrid rice development in California include temperate japonica cultivars with low heterosis, already higher commercial yields, required cold tolerance, limited acreage and resources to support a hybrid breeding effort, a water-seeded production system requiring higher seeding rates, grain quality and market acceptance, and seed production and seed import restrictions.

Breeding and commercial hybrid seed production pose significant challenges for a self-pollinated inbred plant like rice in a mechanized direct-seeded production system (i.e., not transplanted). A three-line system (cytoplasmic genetic male sterility with restorer genes) has been developed, and a two-line system (environment-sensitive genetic male sterility) is also being used. Yields in seed production fields are very low, <3 t ha^{-1} (Wanjari et al., 2006), in comparison with conventional pureline seed yields of 9 t ha^{-1}, making hybrid seed much more expensive and requiring much lower commercial seeding rates.

In recent years, the U.S. rice industry has encountered some marketing problems because of poor long-grain appearance (chalkiness, kernel size, and uniformity) (Ott Mayer, 2012). Environmental factors like high night temperatures at grain filling have been experienced in the southern United States during this period and have been shown to increase percent chalk in rice. Genetics is also a factor, with some cultivars having higher levels of chalkiness. Hybrid rice in the United States has performed below lower-yielding pureline cultivars in terms of milling yield and grain appearance. However, recently released hybrids show improved milling yield compared with their predecessors but still show some weaknesses with chalkiness and grain quality, as also observed in some higher yielding pureline cultivars. With rice primarily marketed in a whole-kernel form, millers, marketers, growers, and breeders must simultaneously address yield and quality to satisfy and maintain market share. Hybrid breeders are addressing specific quality deficiencies in commercial hybrids, and pureline breeders will continue to address yield and further enhancement of grain quality. Growers and millers must determine the trade-offs between yield and quality, and marketers will have to consider discounts or premiums for rice grain quality.

Achieving additional increments of yield in rice may become increasingly difficult and illusive as the maximum yield potential of the crop is approached. Grain yields of 16.5 t ha^{-1} were reported in replicated plot experiments in the Yunnan province using the hybrid 'Liangyoupeijiu' (Katsura et al., 2008). Environmental factors contributing to the high yield were high incident solar radiation, low nighttime temperature during the midseason growth stage, and long growth duration (157 d) that enhanced biomass accumulation. Growth duration of U.S. cultivars is considerably shorter, ranging from 100 to 140 d, and about 100 kg d^{-1} can be used to calculate potential yield (A. Dobermann, personal communication, 2009). Yields of 16 t ha^{-1} in research plots were also observed in New South Wales, Australia (P. Snell, personal communication, 2013). Yields of 22 t ha^{-1} have been reported in the popular press but have not received firm acceptance and documentation in the scientific community (Dobermann, 2013). Yield levels in the range of 13 to 14 t ha^{-1} have been recorded in single-location replicated large plot experiments by the authors.

Weed control presents a formidable challenge and cost for growers in U.S. direct-seeded rice production. Clearfield cultivars, exclusive to the southern United States, provide non-transgenic herbicide tolerance and have become very popular and successful. BASF Agrochemicals has indicated they have developed a new non-transgenic herbicide-tolerant trait for rice that they plan to bring to the U.S. rice market by about 2018. This will provide an additional herbicide-resistant trait for rice under a different mode of action (acetyl-Coenzyme A carboxylase) for control of red rice, the development of herbicide-tolerant weeds, and possibly to extend the life of the Clearfield rice system. Transgenic rice has not achieved acceptance commercially or politically in the United States or anywhere in the world to date, in contrast to crops like maize (*Zea mays* L.) and soybean [*Glycine max* (L.) Merr]. With its small genome, rice is considered a model system for genetic research and is relatively easy to engineer and regenerate. The private sector organizations like Monsanto, Bayer, and Japan Tobacco were very actively working in rice in the late 1990s, developing herbicide and insect-tolerant materials as well as researching yield traits. Those programs were apparently postponed, but interest still remains. Recently, at the 2013 U.S. Rice Outlook Conference in San Diego, CA, a presentation by BASF indicated that they intend to bring genetically modified rice to the United States and the world market in the mid-2020s, amounting to "10% yield increases."

There is transgenic research underway in rice. Arcadia Biosciences, for example, reports it is developing rice with improved nitrogen use efficiency, drought tolerance, and salt tolerance. Golden rice, which enhances β carotene content as a precursor for vitamin A, is scheduled to be released in the near future in some countries in Southeast Asia. China has yet to decide if it will approve the release of transgenic insect-resistant rice. This would clearly have some major health and safety benefits regarding insecticide use and exposure for the Chinese rice field worker. International rice organizations and universities, although continuing genetic engineering with rice, have restricted that activity to genetic research and not rice cultivar development. Rice breeders recognize that this technology would introduce production traits and yield traits that may enhance yield, but its market acceptance and possible intellectual property restrictions keep this an area of uncertainty. While transgenic rice may eventually become a commercial reality, there is currently much disagreement in the U.S. rice industry on its arrival and acceptance.

The use of genetic markers has received considerable emphasis in rice research and breeding. Markers have now become routine in U.S. rice breeding programs for selecting and confirming genes for blast resistance and several grain quality traits, and mapping efforts for many other traits are widespread (Boyett et al., 2013; Andaya, 2012). Markers for yield enhancement or stress tolerance represent active areas of research but do not appear to have progressed to the breeding program level at this date.

Although not a new concept, changing rice from a C_3 photosynthesis pathway to a C_4, like maize, has been identified as a means to bring about yield increases of 30% or more. The C_4 Rice Project (http://c4rice.irri.org/) was initiated with the goal of "bringing C_4 rice to breeders in the developing world in 15 yr through a global dynamic integrated research effort." An International C_4 Rice Consortium has achieved significant international funding, and research is underway (von Caemmerer et al., 2012). This project presents very formidable challenges that include both the development of rice with a different biochemical photosynthetic pathway and accompanying leaf and stem changes to Krantz anatomy (Sheehy et al., 2007). Packaging this together into a commercially acceptable plant that produces marketable grain will create challenges to rice breeders.

Environmental and regulatory trends cause concern that yield increases may be slowed or rice yields may decline. Declines in yields of the Green Revolution's IR-8 at the International Rice Research Institute are one example (Kropff et al., 1994). Climate change is also a cited concern. Lowland rice production with its flooded anaerobic soil emits the greenhouse gas (GHG) methane (CH_4). Altering production practices (aerobic production) offers one area to potentially reduce CH_4 emissions (van Groenigen et al., 2012). However, van Groenigen et al. do note that "it is unclear whether or how the GHG intensity of cropping systems will be affected by future temperature conditions." Nighttime air temperatures appear to be the reason for reduced yields in response to higher temperatures as higher day temperatures had little or no effect on yield (Peng et al., 2004). Negative yield effects from high temperatures include reduced pollination and fertilization (Matsui et al., 2000, 2001; Jagadish et al., 2007); whole kernel milling yield, chalk, amylopectin, and amylose formation (Counce et al., 2005; Cooper et al., 2006, 2008, Ambardekar et al., 2011; Lanning et al., 2011, 2012); and increased respiration (Mohammed and Tarpley, 2009). An increase in frequency or strength of crop damaging weather events like hurricanes would also have yield impacts. Even a reduction in light intensity because of cloud cover would limit canopy photosynthesis and thus yield (Dingkuhn et al., 1990; Peng, 2000). For U.S. rice, if water was not available, or if aerobic production was used as in upland or rainfed systems, much lower yields would be expected.

Regulatory restrictions on water, fertilizers, agrochemicals for pest control, crop residue management, tillage and air quality, and GHG already exist or are anticipated in the United States. With its aquatic production system, rice receives extensive regulatory oversight by state and federal governments. Most regulations are not yield enhancing but rather the opposite. Growers adapt their production system and management to regulatory constraints. Some regulations may not have a direct yield impact; however, they can raise production costs and restrict management options to achieve higher yields.

Recently, the U.S. rice industry has come under scrutiny because of reported "high" detectable arsenic levels in rice. Currently, no public health impact has been identified and no FDA standards for rice or other crops have been established. Typically, U.S. samples have tracked below international Codex thresholds, but media-driven concerns have initiated calls for research and production changes (i.e., to aerobic rice) to lower arsenic levels, and regulatory federal legislation has been proposed. Manipulating an established and effective production system or adding difficult new traits to breeding efforts to meet regulatory goals would only compromise yield increases for the future.

Reilly and Fuglie (1998) noted that "the trend in the next century may be to improve 'environmental' productivity, that is, reducing the environmental impact of agriculture production." Agronomic rice research in the United States is giving emphasis to "sustainability" and complex environmental issues including GHG. This is driven by granting agencies and by the rice industry to address anticipated demands from retailers and regulators. The U.S. rice industry, with its flooded habitat, has successfully documented and promoted the habitat value of its aquatic production system for waterfowl and wildlife. Rice growers are now being encouraged to expand the habitat aspects of their rice production through programs offering some financial benefit, as well as for providing public and political goodwill. Addressing these environmental issues in production practices and breeding would seemingly divert needed attention to increasing rice yields. Thus, there appears to be growing conflict among mitigating environmental impact from rice production and maintaining economic sustainability of U.S. rice production, versus meeting the anticipated global demand for rice consumption in the future. Reilly and Fuglie (1998) pointed out that yield improvement per hectare could be considered an environmental benefit if it allows the land area for rice production to not expand or even possibly contract, with the released area returned to natural ecosystems. Van Groenigen et al. (2012) state that GHG emissions from croplands need to be based on yield rather than on land area to take into consideration the consequences of any GHG mitigation on food production.

At what level these arguments are accepted in the United States by the environmental community and ultimately the government is uncertain.

Rice has evolved to be produced across a global environmental range. The genes that provide such environmental fitness, including possibly genes from other organisms, will provide the genetic base for future yield increases. Breeders continuously design crosses, evaluate and select the materials, and adopt new technologies to develop cultivars to perform in their region. Thus, adapting to market requirements, regulatory and resource restrictions, and climatic conditions are integral in the breeding process. It is, however, expected that rice researchers in the United States will be committed to increasing yield per hectare as a primary breeding objective.

References

Ambardekar, A.A., T.J. Siebenmorgen, P.A. Counce, S.B. Lanning, and A. Mauromoustakos. 2011. Impact of field-scale nighttime air temperatures during kernel development on rice milling quality. Field Crops Res. 122:179–185. doi:10.1016/j.fcr.2011.03.012

Andaya, C.B. 2012. DNA marker lab, p. 30–34. 2012 Rice breeding progress report. California Coop. Rice Res. Foundation. http://www.crrf.org/linked/2012annualreport.pdf (accessed 11 Feb. 2014).

Anzoua, K.G., K. Junichi, H. Toshihiro, I. Kazuto, and J. Yutaka. 2010. Genetic improvements for high yield and low soil nitrogen tolerance in rice (*Oryza sativa* L.) under a cold environment. Field Crops Res. 116:38–45. doi:10.1016/j.fcr.2009.11.006

Babineaux, L.P., Jr. 1967. A history of the rice industry of southwestern Louisiana. M.A. thesis. Univ. of Southwestern Louisiana, Lafayette. http://ereserves.mcneese.edu/depts/archive/FTBooks/babineaux.htm (5 Feb. 2014).

Baldwin, K., E. Dohlman, N. Childs, and L. Forman. 2011. Consolidation and structural change in the U.S. rice sector. Rep. RCS-11d-01. USDA-ERS. http://www.ers.usda.gov/media/111364/rcs11d01_1_.pdf (11 Feb. 2014).

Blanche, S.B., X. Sha, D.L. Harrell, D.E. Groth, K.F. Bearb, L.M. White, and S.D. Linscombe. 2011. Registration of 'CL151' rice. J. Plant Reg. 5:177–180. doi:10.3198/jpr2010.05.0245crc

Bollich, C.N., B.D. Webb, M.A. Marchetti, and J.E. Scott. 1985. Registration of 'Lemont' rice. Crop Sci. 25:883–885. doi:10.2135/cropsci1985.0011183X002500050038x

Boyett, V.A., V. Booth, V. Thompson, K.A.K. Moldenhauer. D, Ahrent, J. Bulloch, H. Sater, and S. Pinson. 2013. Molecular genetics at the University of Arkansas Rice Research and Extension Center. In: R.J. Norman and K.A.K. Moldenhauer, editors, B.R. Wells rice research studies 2012. Arkansas Agric. Exp. Stn. Res. Ser. 609. Univ. of Arkansas, Fayetteville. p. 57–60.

Breseghello, F., O.P. de Morais, and P.H.N. Rangel. 1998. A new method to estimate genetic gain in annual crops. Genet. Mol. Biol. 21:551–555. doi:10.1590/S1415-47571998000400024

Bufogle, A., Jr., P.K. Bollich, J.L. Kovar, R.E. Macchiavelli, and C.W. Lindau. 1997. Rice variety differences in dry matter and nitrogen accumulation as related to plant stature and maturity group 1. J. Plant Nutr. 20:1203–1224. doi:10.1080/01904169709365328

Chu, Q.R., F. Cuevas, and J. Nelsen. 2008. Hybrid rice research and development in the Americas. In: Proceedings of the 32nd Rice Technical Working Group, San Diego, CA. 18–21 Feb. 2008. Louisiana State Univ. Agric. Ctr., Baton Rouge. p. 45.

Cooper, N.T.W., T.J. Siebenmorgen, and P.A. Counce. 2008. Effects of nighttime temperature during kernel development on rice physicochemical properties. Cereal Chem. 85:276–282. doi:10.1094/CCHEM-85-3-0276

Cooper, N.T.W., T.J. Siebenmorgen, P.A. Counce, and J.F. Meullenet. 2006. Explaining rice milling quality variation using historical weather data analysis. Cereal Chem. 83:447–450. doi:10.1094/CC-83-0447

Counce, P.A., R.J. Bryant, C.J. Bergman, R.C. Bautista, Y.J. Wang, T.J. Siebbenmorgen, K.A.K. Moldenhauer, and J.F. Meullenet. 2005. Rice milling quality, grain dimensions, and starch branching as affected by high night temperatures. Cereal Chem. 82:645–648. doi:10.1094/CC-82-0645

Cox, T.S., R.G. Sears, T.J. Martin, J.P. Shroyer, and L. Ben-Hui. 1988. Genetic improvement in agronomic traits of hard red winter wheat cultivars from 1919 to 1987. Crop Sci. 28:756–760. doi:10.2135/cropsci1988.0011183X002800050006x

Deambrosi, E., F. Perez de Vida, and A. Roel. 2012. Temperate rice in Uruguay. In: K.K. Jena and B. Hardy, editors, Advances in temperate rice research. Int. Rice Res. Inst., Los Baños, Philippines. p. 103–105.

Dilday, R.H. 1990. Contribution of ancestral lines in the development of new cultivars of rice. Crop Sci. 30:905–911. doi:10.2135/cropsci1990.0011183X003000040030x

Dingkuhn, M., H.F. Schnier, S.K. De Datta, K. Dorffling, C. Javellana, and R. Pamplona. 1990. Nitrogen fertilization of direct-seeded vs. transplanted rice: II. Interactions among canopy properties. Crop Sci. 30:1284–1292. doi:10.2135/cropsci1990.0011183X003000060025x

Dobermann, A. 2013. Another new yield record? Let's move beyond it. Int. Rice Res. Inst., Los Baños, Philippines. http://irri.org/index.php?option com_k2&view item&id 12487:another-new-rice-yield-record?-let%E2%80%99s-move-beyond-it&lang en (accessed 24 Feb. 2013).

Grandillo, S., D. Zamir, and S.D. Tanksley. 1999. Genetic improvement of processing tomatoes: A 20-year perspective. Euphytica 110:85–97. doi:10.1023/A:1003760015485

Gravois, K.A., K.A.K. Moldenhauer, F.N. Lee, R.J. Norman, R.S. Helms, J.L. Bernhardt, B.R. Wells, R.H. Dilday, P.C. Rohman, and M.M. Blocker. 1995. Registration of 'Kaybonnet' rice. Crop Sci. 35:587–588. doi:10.2135/cropsci1995.0011183X003500020052x

Graybosch, R.A., and C.J. Peterson. 2010. Genetic improvement in winter wheat yields in the Great Plains of North America, 1959–2008. Crop Sci. 50:1882–1890. doi:10.2135/cropsci2009.11.0685

Hardke, J.T., and C.E. Wilson, Jr. 2013. Trends in Arkansas rice production. In: R.J. Norman and K.A.K. Moldenhauer, editors, B.R. Wells rice research studies 2012. Arkansas Agric. Exp. Stn. Res. Ser. 609. Univ. of Arkansas, Fayetteville. p. 38–47.

Hardke, J.T., C.E. Wilson Jr., K. Moldenhauer, Y. Wamishe, R. Carrtwright, R, Norman, D. Frizzell, J. Branson, M. Blocker, J. Bulloch, E. Castaneda, L. Shmidt, and R. Mazzanti. 2012. Arkansas rice performance trials, 2010–2012. Rice Info. Sheet 171. Univ. of Arkansas Coop. Ext. Serv., Fayetteville.

Jagadish, S., P.Q. Craufurd, and T.R. Wheeler. 2007. High temperature stress and spikelet fertility in rice (*Oryza sativa* L.). J. Exp. Bot. 58:1627–1635. doi:10.1093/jxb/erm003

Jodari, F., and S.D. Linscombe. 1996. Grain fissuring and milling yields of rice cultivars as influenced by environmental conditions. Crop Sci. 36:1496–1502. doi:10.2135/cropsci1996.0011183X003600060014x

Johnson, C.W. 2002. Rice cultivar 'M-205'. U.S. Patent 6,492,582. Date issued: 10 December.

Johnson, C.W. 2005. Rice cultivar 'M-206'. U.S. Patent 6,911,589. Date issued: 28 June.

Johnson, C.W., S.T. Tseng, and J.J. Mantenbreok. 1986. Registration of 'M-202' rice. Crop Sci. 26:198. doi:10.2135/cropsci1986.0011183X002600010052x

Johnston, T.H. 1958. Registration of rice varieties. Agron. J. 50:694–700. doi:10.2134/agronj1958.00021962005000110021x

Johnston, T.H., K.A. Kuenzel, F.N. Lee, B.R. Wells, S.E. Henry, and R.H. Dilday. 1984. Registration of 'Newbonnet' rice. Crop Sci. 24:209–210. doi:10.2135/cropsci1984.0011183X002400010053x

Johnston, T.H., B.D. Webb, and K.O. Evans. 1968. Registration of 'Starbonnet' rice. Crop Sci. 8:400–401. doi:10.2135/cropsci1968.0011183X000800030056x

Just, R., and R. Pope. 1979. Production function estimation and related risk considerations. Am. J. Agric. Econ. 61:276–284. doi:10.2307/1239732

Katsura, K., S. Maeda, I. Lubis, T. Horie, W. Cao, and T. Shiraiwa. 2008. The high yield of irrigated rice in Yunnan, China— 'A cross location analysis'. Field Crops Res. 107:1–11. doi:10.1016/j.fcr.2007.12.007

Kropff, M.J., K.G. Cassman, S. Peng, R.B. Matthews, and T.L. Setter. 1994. Quantitative understanding of yield potential. In: K.G. Cassman, editor, Breaking the yield barrier. Proceedings of a Workshop on Rice Yield Potential in Favorable Environments, Los Baños, Philippines. 29 Nov. to 4 Dec. 1993. Int. Rice Res. Inst., Manila, Philippines. p. 21–38.

Lanning, S.B., T.J. Siebenmorgen, A.A. Ambardekar, P.A. Counce, and R.J. Bryant. 2012. Effects of nighttime air temperature during kernel development of field-grown rice on physicochemical and functional properties. Cereal Chem. 89:168–175. doi:10.1094/CCHEM-12-11-0146

Lanning, S.B., T.J. Siebenmorgen, P.A. Counce, A.A. Ambardekar, and A. Mauromoustakos. 2011. Extreme nighttime air temperatures in 2010 impact rice chalkiness and milling quality. Field Crops Res. 124:132–136. doi:10.1016/j.fcr.2011.06.012

Lillemo, M., L. Reitan, and A. Bjørnstad. 2009. Increasing impact of plant breeding on barley yields in central Norway from 1946 to 2008. Plant Breed. 129:484–490.

Linscombe, S.D., F. Jodari, P.K. Bollich, D.E. Groth, L.M. White, Q.R. Chu, R.T. Dunand, and D.E. Sanders. 2000. Registration of 'Cocodrie' rice. Crop Sci. 40:294.

Linscombe, S.D., F. Jodari, K.S. McKenzie, P.K. Bollich, L.M. White, D.E. Groth, and R.T. Dunand. 1993a. Registration of 'Bengal' rice. Crop Sci. 33:645–646. doi:10.2135/cropsci1993.0011183X003300030046x

Linscombe, S.D., F. Jodari, K.S. McKenzie, P.K. Bollich, L.M. White, D.E. Groth, and R.T. Dunand. 1993b. Registration of 'Cypress' rice. Crop Sci. 33:355. doi:10.2135/cropsci1993.0011183X003300020039x

Lu, H., M.A. Redus, J.R. Coburn, J.N. Rutger, S.R. McCouch, and T.H. Tai. 2005. Population structure and breeding pattern of 145 U.S. rice cultivars based on SSR marker analysis. Crop Sci. 45:66–76. doi:10.2135/cropsci2005.0066

Mackay, I., A. Horwell, J. Garner, J. White, J. McKee, and H. Philpott. 2011. Reanalyses of the historical series of UK variety trials to quantify the contributions of genetic and environmental factors to trends and variability in yield over time. Theor. Appl. Genet. 122:225–238. doi:10.1007/s00122-010-1438-y

Matsui, T., K. Omasa, and T. Horie. 2000. High temperature at flowering inhibits swelling of pollen grains, a driving force for thecae dehiscence in rice (*Oryza sativa* L.). Plant Prod. Sci. 3:430–434. doi:10.1626/pps.3.430

Matsui, T., K. Omasa, and T. Horie. 2001. The difference in sterility due to high temperatures during the flowering period among japonica-rice varieties. Plant Prod. Sci. 4:90–93. doi:10.1626/pps.4.90

McKenzie, K.S., C.W. Johnson, S.T. Tseng, J.J. Oster, and D.M. Brandon. 1994. Breeding improved rice cultivars for temperate regions: A case study. Aust. J. Exp. Agric. 34:897–905. doi:10.1071/EA9940897

Mohammed, A.-R., and L. Tarpley. 2009. Impact of high nighttime temperature on respiration, membrane stability, antioxidant capacity, and yield of rice plants. Crop Sci. 49:313–322. doi:10.2135/cropsci2008.03.0161

Moldenhauer, K.A.K., F.N. Lee, J.L. Bernhardt, R.J. Norman, N.A. Slaton, C.E. Wilson, M.M. Anders, R.D. Cartwright, and M.M. Blocker. 2007. Registration of 'Wells' rice. Crop Sci. 47:442–443. doi:10.2135/cropsci2006.06.0419

Moldenhauer, K.A.K., F.N. Lee, R.J. Norman, R.S. Helms, B.R. Wells, R.H. Dilday, P.C. Rohman, and M.A. Marchetti. 1990. Registration of 'Katy' rice. Crop Sci. 30:747–748. doi:10.2135/cropsci1990.0011183X003000030065x

Moldenhauer, K.A.K., L. Nally, B. Watkins, J. Gibbons, and K. Gravois. 2009. Impact of rice variety development—Changes, challenges, and future. Paper presented at: Footprints in the landscape: Sustainability throughout plant and soil sciences. ASA, CSSA, and SSSA Annual Meetings, Pittsburgh, PA. 1–5 Nov. 2009.

Mutters, R.G., and J.F. Thompson. 2011. Crop management and environmental condition effects on rice yield and quality. Rep. RP13. California Rice Res. Board, Yuba City.

Nalley, L.L., K.A.K. Moldenhauer, and N. Lyman. 2011. The genetic and economic impact of the UA's Breeding Program: 1983–2007. J. Agric. Appl. Econ. 43:131–142.

Ott Mayer, K.S. 2012. U.S. long-grain rice industry: At a crucial crossroads. Delta Farm Press. http://deltafarmpress.com/site-files/deltafarmpress.com/files/uploads/2012/09/Rice-Quality-WP-FPO.pdf (accessed 11 Feb. 2014).

Peng, S. 2000. Single-leaf and canopy photosynthesis of rice. Stud. Plant Sci. 7:213–228. doi:10.1016/S0928-3420(00)80017-8

Peng, S., J. Huang, J.E. Sheehy, R.C. Laza, R.M. Visperas, X. Zhong, G.S. Centeno, G.S. Khush, and K. G. Cassman. 2004. Rice yields decline with higher night temperature from global warming. Proc. Natl. Acad. Sci. 101:9971–9975.

Peng, S., R.C. Laza, R.M. Visperas, A.L. Sanico, K.G. Cassman, and G.S. Khush. 2000. Grain yield of rice cultivars and lines developed in the Philippines since 1966. Crop Sci. 40:307–314. doi:10.2135/cropsci2000.402307x

Reid, W.M. 1953. Rice acreage in the United States 1953. The Rice Millers' Assoc., New Orleans, LA.

Reid, W.M. 1954. Rice acreage in the United States 1954. The Rice Millers' Assoc., New Orleans, LA.

Reid, W.M. 1956. Rice production in the United States 1956. The Rice Millers' Assoc., New Orleans, LA.

Reilly, J.M., and K.O. Fuglie. 1998. Future yield growth in field crops: What evidence exists? Soil Tillage Res. 47:275–290. doi:10.1016/S0167-1987(98)00116-0

Reinke, R., G. Beecher, B. Dunn, and P. Snell. 2012. Temperate rice in Australia. In: K.K. Jena and B. Hardy, editors, Advances in temperate rice research. Int. Rice Res. Inst., Los Baños, Philippines. p. 1–14.

Rutger, J.N., K.S. McKenzie, K.A.K. Moldenhauer, A.M. McClung, K.A. Gravois, S.D. Linscombe, and D.G. Kanter. 1998. Rice quality breeding efforts in the US. In: J. Chataigner, editor, Rice quality: A pluridisciplinary approach. CIHEAM-IAMM, Montpellier, France. p. 311–315.

Saichuk, J. 2012. Rice variety acreage tables. In: Proceedings of the 34th Rice Technical Working Group Meeting, Hot Springs, AR. 27 Feb to 1 Mar 2012. Louisiana State Univ. Agric. Cent., Baton Rouge. p. 17–26.

Sha, X.Y., S.D. Linscombe, S.B. Blanche, and D.E. Groth. 2009. Yield advantage of hybrid rice over conventional and Clearfield long-grain rice in the southern United States. In: F. Xie and B. Hardy, editors, Accelerating hybrid rice development. Int. Rice Res. Inst., Los Baños, Philippines. p. 687–693.

Sha, X., S.D. Linscombe, and D.E. Groth. 2007. Field evaluation of imidazilinone-tolerant Clearfield rice (*Oryza sativa*) at nine Louisiana locations. Crop Sci. 47:1177–1185. doi:10.2135/cropsci2006.09.0592

Sha, X., S.D. Linscombe, D.E. Groth, J.A. Bond, L.M. White, Q.R. Chu, H.S. Utomo, and R.T. Dunand. 2006. Registration of 'Jupiter' rice. Crop Sci. 46:1811–1812. doi:10.2135/cropsci2005.10-0393

Sheehy, J.E., P.L. Mitchell, and B. Hardy, editors. 2007. Charting a new pathway to C_4 rice. International Rice Research Institute, Los Baños, Philippines.

Slaton, N. 2001a. DD50 computerized rice management program. In: Rice production handbook. Misc. Publ. 192. Univ. of Arkansas Coop. Ext. Serv., Fayetteville. p. 31–36.

Slaton, N. 2001b. Introduction. In: Rice production handbook. Misc. Publ. 192. Univ. of Arkansas Coop. Ext. Serv., Fayetteville.

Spicer, J.M. 1964. Beginnings of the rice industry in Arkansas. In: Museum of the Arkansas Grand Prairie. Arkansas Cty. Agric. Mus., Stuttgart. p. 122.

Tabien, R.E., S.O. Samonte, and A.M. McClung. 2008. Forty-eight years of rice improvement in Texas since the release of cultivar Bluebonnet in 1944. Crop Sci. 48:2097–2106. doi:10.2135/cropsci2007.12.0680

USDA-NASS. 2012. Quick stats. USDA, Washington, DC. http://www.nass.usda.gov/Quick_Stats/ (accessed 13 Nov. 2012).

van Groenigen, K.J., C. van Kessel, and B.A. Hungate. 2012. Increased greenhouse-gas intensity of rice production under future atmospheric conditions. Nat. Clim. Change 3:288–291. doi:10.1038/nclimate1712

von Caemmerer, S., W.P. Quick, and R.T. Furbank. 2012. The development of C4 rice: Current progress and future challenges. Science 336:1671–1672. doi:10.1126/science.1220177

Walton, M. 2003. Hybrid rice for mechanized agriculture. In: S.S. Virmani, C.X. Mao, and B. Hardy, editors, Hybrid rice for food security, poverty alleviation, and environmental protection. Proceedings of the 4th International Symposium on Hybrid Rice, Hanoi, Vietnam. 14–17 May 2002. Int. Rice Res. Inst., Los Baños, Philippines. p. 97–102.

Wanjari, R.H., K.G. Mandal, P.K. Ghosh, T. Adhikari, and N.H. Rao. 2006. Rice in India: Present status and strategies to boost its production through hybrids. J. Sustain. Agric. 28:19–39. doi:10.1300/J064v28n01_04

Wilson, C.E., Jr., and J.W. Branson. 2003. Trends in Arkansas rice production. In: R.J. Norman and J.F. Meullenet, editors, B.R. Wells rice research studies 2002. Arkansas Agric. Exp. Stn. Res. Ser. 504. Univ. of Arkansas, Fayetteville. p. 15–20.

Wilson, C.E., Jr., and S. Runsick. 2008. Trends in Arkansas rice production. In: R.J. Norman, J.F. Meullenet, and K.A.K. Moldenhauer, editors, B.R. Wells rice research studies 2007. Arkansas Agric. Exp. Stn. Res. Ser. 560. Univ. of Arkansas, Fayetteville. p. 11–20.

World Rice Statistics. 2012. National and sub-national rice data. Int. Rice Res. Inst., Los Baños, Philippines. http://ricestat.irri.org:8080/wrs/ (accessed 13 Nov. 2012).

Yuan, L.P. 2003. Recent progress in breeding super hybrid rice in China. In: S.S. Virmani, C.X. Mao, and B. Hardy, editors, Hybrid rice for food security, poverty alleviation, and environmental protection. Proceedings of the 4th International Symposium on Hybrid Rice, Hanoi, Vietnam. 14–17 May 2002. Int. Rice Res. Inst., Los Baños, Philippines. p. 3–6.

Zeigler, R.S. 2012. Foreword. In: K.K. Jena and B. Hardy, editors, Advances in temperate rice research. Int. Rice Res. Inst., Los Baños, Philippines. p. v.

11

Sorghum

Roger Monk,* Cleve Franks, and Jeff Dahlberg

ABSTRACT

Sorghum [*Sorghum bicolor* (L.) Moench], a vital cereal grain in the semiarid areas of the tropics and subtropics, has shown continual productivity gains in response to both changing agronomic practices and genetic improvement. The development and deployment of hybrid sorghum dramatically increased yields in the United States, Mexico, Australia, Argentina, and several other countries, but improved cultivars and landraces have contributed only modest gains across much of Africa and India. Since sorghum is primarily grown under marginal conditions, genetic improvement has focused on tolerance to biotic and abiotic stresses, with host plant resistance to disease pathogens, insects, and drought tolerance being the key mechanism of improvement. Agronomic practices, including minimum tillage and effective herbicides, fungicides, and insecticides, have all contributed to improved stand establishment and production stability. Close cooperation between public and private research organizations has also contributed to sorghum yield gains. The genetic diversity of sorghum has been the primary source of new traits and increased heterosis in hybrids, but with the sequencing and publication of the sorghum genome, sorghum researchers are using new technologies, such as molecular markers, to improve the efficiency of breeding programs.

R. Monk, DuPont Pioneer, 115 Meyer St., P.O. Box 97, Taft, TX 78390. *Corresponding author (roger.monk@pioneer.com).
C. Franks, DuPont Pioneer, 2260 CR 60, Plainview, TX 79072 (cleve.franks@pioneer.com).
J. Dahlberg, Kearney Agricultural Research and Extension Center, 9240 S. Riverbend Ave., Parlier, CA 93648 (jadahlberg@ucanr.edu).

doi:10.2135/cssaspecpub33.c11

Yield Gains in Major U.S. Field Crops. CSSA Special Publication 33.
Stephen Smith, Brian Diers, James Specht, and Brett Carver, editors.

Sorghum remains one of the most important cereal crops in semiarid environments and has ranked as one of the top 25 major commodities in the world since 1961 according to FAO statistics (http://www.fao.org/statistics/en/, accessed 28 Jan. 2014). Sorghum is grown primarily in the semiarid areas of the tropics and subtropics and is a vital source of income and sustenance to people and livestock in many stressful, variable, and unpredictable environments. In 2010, sorghum ranked fifth in global production among the cereal crops, preceded by maize (*Zea mays* L.), paddy rice (*Oryza sativa* L.), wheat (*Triticum aestivum* L.), and barley (*Hordeum vulgare* L.) (FAOSTAT—Crops. http://faostat3.fao.org/faostat-gateway/go/to/home/E, accessed 28 Jan. 2014). Averaged across all years since 1961 and on a global basis, 45 million ha were harvested with a production of 59.5 Mt per year (Fig. 11–1). Yields averaged 1308 kg ha^{-1} (20.8 bu ac^{-1}) (Fig. 11–2). Sorghum acreage and yield peaked worldwide around 1985.

Yield Trends in Countries that Predominantly Cultivate Sorghum Hybrids

The first significant public U.S. investment in sorghum occurred in the late 1800s when the USDA introduced several landraces of sorghum (see Smith and Frederiksen, 2000 for a review of the introduction of sorghum to the United States). The first significant yield increases in sorghum occurred through rather simple selection for adaptation by farmers and research scientists in the early 1900s. Yields of landrace cultivars of sorghums were approximately 500 kg ha^{-1} (8 bu ac^{-1}) in the early 1900s; however, national yields had increased 300% by the mid 1950s to more than 1500 kg ha^{-1} (24 bu ac^{-1}) (Fig. 11–3). This production gain for yield was accomplished through improved farming practices and plant breeding with selection by both farmers and research scientists that allowed them to expand on the inherent diversity within these early sorghum introductions. Among the 30 or so initial plant introductions were milo, kafir, feterita, hegari, sudangrass and sweet sorghum types, which formed the basis of pre-hybrid varieties as well

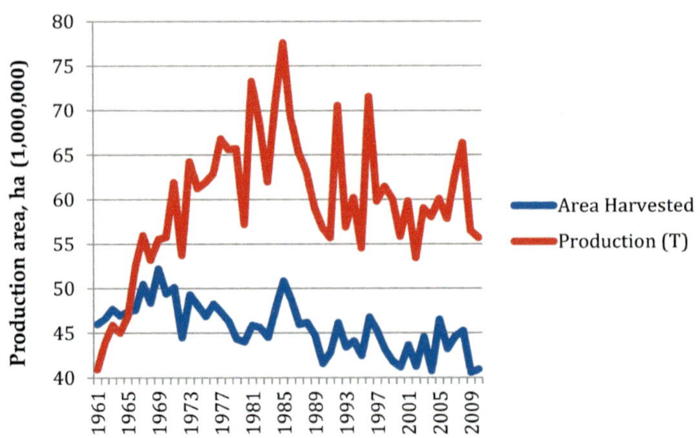

Fig. 11–1. World production and area harvested for sorghum from 1961 through 2010 (FAOSTAT—Crops. http://faostat3.fao.org/faostat-gateway/go/to/home/E, accessed 28 Jan. 2014).

as the mechanism for hybridization through cytoplasmic sterility. The demand for cereal grains and sorghum's consistency in drought-prone areas resulted in increased sorghum acreage to approximately 10.1 million ha by 1957.

Before the late 1950s, hybrid seed was not available for sorghum production. With the discovery of cytoplasmic male sterility in sorghum (Stephens and Holland, 1954), a system was developed that could efficiently produce hybrids on a commercial scale. The sharp increase in production and yields seen in the early

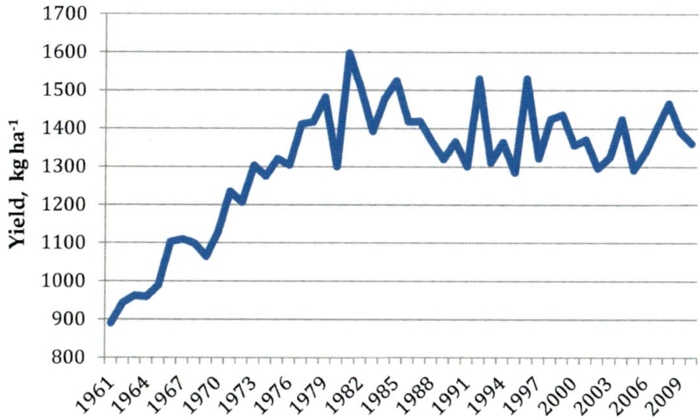

Fig. 11–2. Average world sorghum yields from 1961 through 2010 (FAOSTAT—Crops. http://faostat3.fao.org/faostat-gateway/go/to/home/E, accessed 28 Jan. 2014).

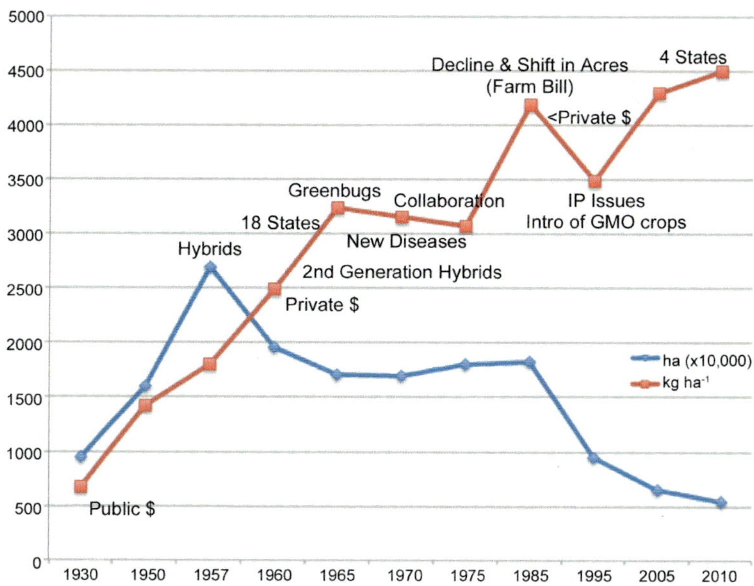

Fig. 11–3. United States sorghum grain yield plotted with area planted (×10,000) to sorghum from 1930 through 2010 (USDA-NASS, http://www.nass.usda.gov/Quick_Stats/, accessed 28 Jan. 2014).

to late 1960s can be directly attributed to the widespread adoption of hybrids by sorghum farmers, primarily in developed countries. Within the United States, movement away from cultivars to hybrids was quite rapid. In 1956, less than 1% of the total sorghum acreage in the United States was planted to hybrids, but by 1959 it was estimated that 75% of the total acreage was planted to hybrids. By the early 1960s, more than 90% of sorghum acreage was planted to hybrids (Bennett et al., 1990). Maunder (Smith and Frederiksen, 2000) suggested this rapid adoption to hybrid grain sorghums was due in part to growers' familiarity with the success of corn hybrids and their yield gains. He attributed much of this change to heterosis, which resulted in increased grain number per inflorescence and earlier flowering, which in turn improved water use efficiency, escape from insect pressure and tolerance to temperature extremes (Maunder, 1998). Improvements in farm equipment, irrigation, and farming practices also contributed to these production gains.

The second significant increase in U.S. sorghum yields occurred between the time hybrid grain sorghum was introduced in 1957 and 1965, just before the first reports of greenbug [*Schizaphis graminum* (Rondani)] becoming a major insect pest of sorghum. Despite a national decrease of acreage by 38%, during this 8-yr span, national production increased by 13.8%. Average yields increased almost 79%. These yield increases were the result of second-generation hybrids, developed by the private seed industry, and increased research infrastructure and investment in sorghum breeding by18 public universities. Second-generation hybrids based on novel yellow endosperm germplasm utilized parental lines that offered improved agronomic traits and diverse heterotic patterns that contributed to yield increases, especially in dryland situations (Bennett et al., 1990).

Between 1965 and 1975, national sorghum yields decreased by 5%. There are several reasons for this reduction, including the backcrossing of greenbug resistance into commercial hybrids which focused plant breeding efforts on the incorporation of a single trait at the expense of total plant improvement (Bennett et al., 1990). Outbreaks of downy mildew (*Peronosclerospora sorghi* W. Weston & Uppal), *Maize dwarf mosaic virus*, and small seed malady also contributed to the stagnation of yield gains during this time frame, although breeders successfully corrected these biotic stresses genetically.

Between 1975 and 1985, U.S. sorghum yields increased by 36%. These increases can be directly attributed to greater investment in sorghum research and the strong collaborative efforts between public and private sectors during the previous 10 yr. The stabilization of yield with the introduction of greenbug-tolerant hybrids, greater stalk strength and standability, and improvements in tolerance to abiotic stresses, such as the introduction of the stay-green characteristic, all contributed to the increase in yields seen during this time.

Another drop in U.S. sorghum yields occurred from 1985 through 1995. Several factors unrelated to genetic gain can explain much of this decline, which represented a drop in yield of roughly 16%. In 1985, the U.S. passed major Farm Bill legislation that had a net negative impact on sorghum acres, and, perhaps more importantly, on the makeup of the acres on which sorghum was planted. Farm Bill payments and insurance policy pushed sorghum acres out of traditionally productive sorghum acres and into more marginal agricultural lands. These lands were more prone to drought and heat stress, resulting in wide swings in national yields for sorghum from year to year, primarily corresponding with rain and high temperature patterns. The sorghum acreage in Missouri is an example of this shift. In 1985, Missourian farmers planted more than 560,000 ha (1.38 million acres) of sorghum; this had reduced to 26,000 ha (64,220 ac) in 2012. A similar pattern occurred across much of the South and in Nebraska, where 930,000 ha (2.30 million acres) planted to sorghum in 1985 became <60,000 ha (148,200 acres) planted in 2012. Previously cultivated sorghum areas were lost primarily to soybean [*Glycine max* (L.) Merr.] and corn and were often higher yielding areas with either abundant rainfall or supplementary irrigation. In this sense, sorghum was a victim of the success of maize breeders, whose success in increasing hybrid yields and introducing ground-breaking traits like herbicide resistance and Bt encouraged many growers to plant their more productive land to maize. This downturn in sorghum cultivated area has in turn led to less interest and fewer research dollars invested in sorghum by both the private and public sectors.

Despite significant reductions in research dollars from both the private and public sectors, the most recent trends in sorghum yield gains from 1995 through 2010 have shown increases. In addition, the impact of the biotechnology industry has affected sorghum in terms of freedom to operate issues, germplasm exchange, and graduate student training. For most of sorghum's history following hybridization, the close relationship between public (state and federal) and private researchers meant that genetic sources of biotic and abiotic resistance and new technologies were rapidly and widely disseminated across the industry. During the past 20 yr, with the emphasis on molecular aspects, the development of materials has entered an era of intellectual property protection and the quantity and diversity of germplasm released by public institutions has dramatically decreased. The exchange of germplasm even between public institutions has greatly diminished. In addition, many graduate student programs have become less field oriented and increasingly expensive with the need for advanced molecular training. During the 1995 to 2010 time frame, only four major universities, Kansas State University, Texas A&M University, University of Nebraska, and Purdue University, had active sorghum research programs, and, through consolidation and restructuring, the number of private sector seed companies declined dramatically as well. Even with these trends, national yield statistics showed a slight increase in yield from 1995 through 2010. However, the shift towards sorghum being grown on less productive and more risky acreages is dramatically illustrated by comparing the percentage of harvested to planted acres, that is, the rate of crop abandonment. Between 1982 and 1997, harvested area averaged 89% and was less than 85% of the planted crop harvested in only 1 yr out of fifteen (Fig. 11–4). In the 15 yr since, there have been 9 yr when less than 85% of the crop was harvested and only 1 yr in which the proportion harvested exceeded 90% (Fig. 11–4). The fact that sorghum yields have stayed relatively stable in this time frame is testimony to the crop's inherent resilience, and gives credence to the claim that sorghum breeders have indeed made genetic gains in recent years. These gains, however, are masked by a trend towards sowing sorghum under increasingly harsh and challenging conditions.

Of the other four major countries that predominately plant sorghum hybrids, some interesting trends can be noted (Fig. 11–5). Both Mexico and Argentina have experienced yield trends similar to the United States. These countries developed robust national and private sector sorghum research programs and strong relationships with U.S. grain sorghum programs. First-generation hybrids were key

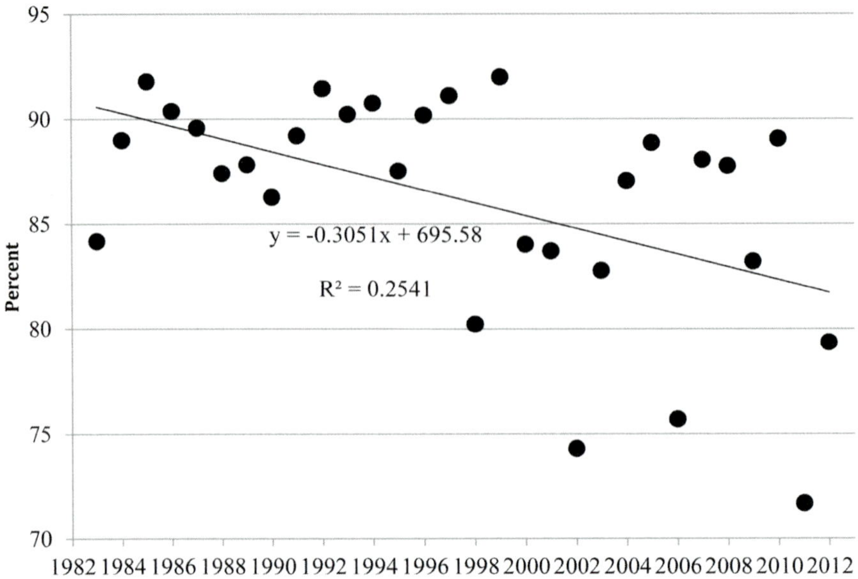

Fig. 11–4. Grain sorghum acres harvested, as a percentage of acres planted.

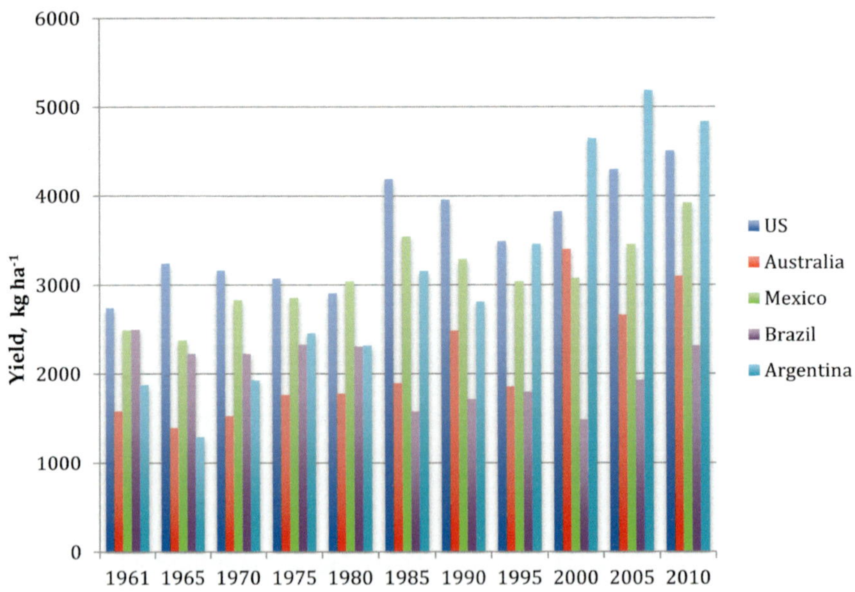

Fig. 11–5. Yield trends in the five largest sorghum producing countries that plant primarily sorghum hybrids (FAOSTAT—Crops. http://faostat3.fao.org/faostat-gateway/go/to/home/E, accessed 28 Jan. 2014).

to increasing national sorghum yields in these countries, and later increases in yield by 1985 point to the importance of second-generation hybrids in further expanding these yield gains. From 1985 through 2010, third- and fourth-generation hybrids were introduced and produced additional increases in yields. These yield gains were reflected in the increased diversity of germplasm sources and introduction of new traits into these hybrids.

Australian sorghum production is almost exclusively grown as a rainfed crop on clay soils with moderate to high water holding capacity and subject to major cyclical weather events that impact yield. Farming systems in Australia rely wholly on stored soil moisture to ensure profitable yields. Two major limiting factors have impacted research and breeding needs over the years, sorghum midge [*Contarinia sorghicola* (Coquillett)] and postanthesis water deficit stress. From 1960 to 1987, major breeding efforts concentrated on defensive traits for these two major issues (Jordan et al., 1998). Similar to the situation in the United States during this timeframe, there was continued expansion of acres into poorer environments and a loss of high yielding acres to cotton (*Gossypium hirsutum* L.) that has contributed to increasingly variable average yields (David Jordan, 2013, personal communication).

From the early 1960s to the mid 1980s, sorghum yields averaged 1774 kg ha^{-1} (28 bu ac^{-1}), and trends (Fig. 11–5) were fairly flat. These yields may have been reflective of both environmental and insect pressures. In 1993, hybrids with good levels of midge resistance and stay-green were introduced, and average yields rose to 2753 kg ha^{-1}, (43.8 bu ac^{-1}) by 2010, an increase of 55%. It can be argued that once hybrids with midge resistance were introduced, the breeding programs began to pursue yield improvements aggressively through use of exotic germplasm and integration of crop improvement technologies for rapid genetic advances (Jordan et al., 2009).

The story of the success of hybrid sorghum is impressive in terms of national statistics, but these figures mask the actual potential of the crop. Fischer and Edmeades (2010) discussed the difficulty of assessing farm yield compared to the attainable or potential yields. They defined potential yield as the yield achieved by growing well-adapted varieties under optimum conditions, unlimited by insect, disease, nutrient, water or other abiotic or biotic pressures. Attainable yield, however, refers to the yield of a prudent farmer while factoring in economics and risk. The potential yield of current grain sorghum hybrids can be demonstrated in various state variety performance trials, where yields above 11,200 kg ha^{-1} (180 bu ac^{-1}) are routinely achieved (Mascagni, et al., 2013). Annual yield contests sponsored by the National Sorghum Producers often report yields in excess of 12,100 kg ha^{-1} (190 bu ac^{-1}) (see www.sorghumgrowers.com for yield contest results, accessed 28 Jan. 2014). Such yields are three to four times higher than the average farm yield, so one of the goals of breeding programs is to address genetic methods to close these gaps.

Yield Trends in Countries that Predominantly Use Local Landraces or Cultivars

National breeding programs and several international programs (see INTSORMIL reports at http://intsormil.org/smabout.htm, accessed 28 Jan. 2014; Rai et al., 1999; Kumar et al., 2011) have worked to establish research centers and train both researchers and growers throughout Africa and India to assist in improving yields in sorghum; however, with the exception of a few countries, yield gains in sorghum have remained relatively stagnant for the last 40 yr and reflect increased area under production as well as yield per area. The FAO reported that in 1961 average yields in Least Developed Countries was 717 kg ha^{-1}. Yields were 769 kg ha^{-1}in 2010, an increase of just over 7% in 40 yr.

Several African countries have made significant progress in sorghum yield gains over these past 40 yr (Fig. 11–6). South Africa has shown very positive yield gains, while Ethiopia has seen a steady increase in yields during the same time frame. Nigeria, Mali, and India have seen relatively modest growth, while Sudan and Niger have seen a decline in yields. South Africa has been aggressively investing in improved landraces for many areas. In addition an emergent commercial seed industry has resulted in adoption of hybrids by South African farmers in more developed areas. Ethiopia's strong sorghum research program has also been successfully breeding for improved cultivars. Nigeria, Mali, and India have seen slower adoption of improved cultivars for various reasons, including poor agricultural infrastructure to deliver improved cultivars to farmers, low rates of adoption by rural farmers, movement of productive agricultural land away from sorghum to other crops, and civil unrest and disruption. It can be argued that in Africa and India there are greater disease and insect pressures on sorghum that are not as relevant in the developed world, but it is quite clear that sorghum's genetic diversity should hold the potential to be able to be used by breeders to overcome much of the disease and insect pressures that face these farmers.

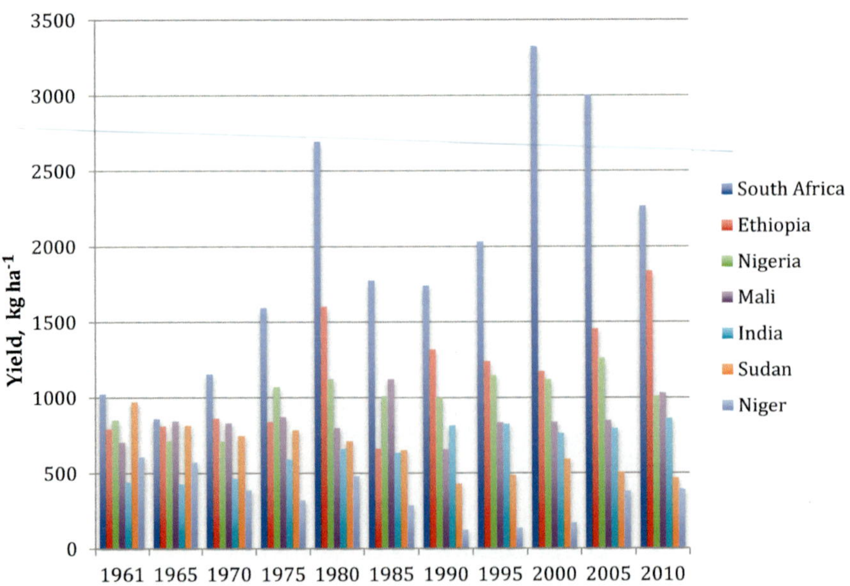

Fig. 11–6. Average sorghum yields from 1961 through 2010 (FAOSTAT—Crops. http://faostat3.fao.org/faostat-gateway/go/to/home/E, accessed 28 Jan. 2014) for top producing African countries and India.

Sorghum Yield Gains

The yield gains demonstrated in sorghum reflect changes in both genetics and agronomic practices. Determining the individual contribution of each is not easy, and a strong interaction component has been shown in maize (Duvick, 1986). Sorghum production benefited greatly in the High Plains of the United States during the 1960s and 1970s from the availability and relative low costs of nitrogen fertilizer and the low price of natural gas, which held irrigation costs low (Miller and Kebede, 1984). Beginning in the 1980s, a strong conservation movement began to influence crop production practices across the central plains of the United States, and minimum tillage helped stabilize production in dryland areas (Assefa and Staggenborg, 2010).

Water management in dryland systems has also benefited from practices such as furrow diking, terrace benching, and adjusting row spacing and plant populations based on available water at planting. A significant increase in stand establishment success was aided by the development of broad spectrum fungicides and, more recently, by insecticidal seed treatments that protect the germinating seed and seedling. Weed control in sorghum was simplified by the advent of pre-emergent herbicides and the effectiveness of a range of herbicides against broadleaf weeds, but weed control is as yet not as effective, economical, or simple as the over-the-top glyphosate [*N*-(phosphonomethyl)glycine] or glufosinate (DL-phosphinothricin) systems used in maize, soybean, and cotton.

Estimates vary widely, but Duvick (2005) summarized a number of studies to estimate that 50 to 60% of the gain in maize yields during a 70-yr period from 1930 to 2000 could be attributed to improved genetics. Miller and Kebede (1984) estimated that the genetic contribution to yield increases in sorghum was almost 40%. Unger and Baumhardt (1999), while specifically addressing yield increases under dryland conditions on the High Plains of Texas, showed a yield increase of 139% from 1956 to 1997, with 46% of the gain due to improved hybrids. The rest of the increase was attributed to better water management to increase the soil water available at planting, the factor most responsible for the increase. More recent studies indicate the contribution of improved genetics may be as high as 60 to 65% (Assefa and Staggenborg, 2010). These studies all indicate a yield gain of between 1 and 2% per year. Mason et al. (2008) compared the rate of gain in maize and sorghum from a small set of hybrids in eastern Nebraska from 1950 to 1989 in rainfed and irrigated environments. They found that under these environments maize outyielded sorghum and that maize yields had increased at a faster rate during the 40-yr span.

Improvements in genetic gain for yield have taken place within the framework of anatomical, physiological and phenological changes. Duvick (2005) summarized a number of maize studies examining such changes and determined that total biomass and kernels per square meter have consistently increased with time, but that the most dramatic increases occur when plants are planted in high densities or abiotic stress, where decreased barrenness, better synchronization of anthesis–silking, improved lodging resistance, and more erect leaves all play a role in increasing stability and yield. There are far fewer studies comparing the traits associated with increased genetic gain in sorghum. Miller and Kebede (1984) compared characteristics of hybrids produced from inbreds developed in the 1950s with hybrids produced from 1980s-era inbreds and hybrids produced by crossing the old by the new inbreds. The newer hybrids showed a 48% increase in yield and a substantial increase in kernels per panicle for the newer hybrids as well as increases in total plant weight, height, and leaf area at four different growth stages. The hybrids made from old by new parents were intermediate in response. They did not find an increase in harvest index or kernel weight, indicating that the increase in yield was a function of the increase in total production rather than a shift in partitioning. The new hybrids showed a stronger stay-green response, retaining green leaf area until maturity, while the old hybrids began to lose leaf area shortly after anthesis. Assefa and Staggenborg (2011) compared commercial hybrids spanning five decades, under differential water regimes in a greenhouse study. They found that total water use was similar for all hybrids, but that the total biomass produced was greater for newer hybrids, indicating a net increase in water use efficiency in sorghum hybrids with time.

Factors Influencing Genetic Gain

In general, sorghum is produced in stressful environments and is grown by farmers as a lower input cost, stable, lower margin crop. Although commercial companies place emphasis on combining ability and extensive performance testing, these grower expectations have led sorghum scientists, and particularly academic scientists, to focus on the improvement of host plant resistance as the primary defense against plant diseases, insects, and abiotic stress, and the general goal has been to protect yield rather than to maximize yield potential. To address these realities, research trial sites are intentionally established in regions where plants are typically exposed to high levels of biotic or environmental stress. New pathotypes of diseases continue to evolve in sorghum and require continuous breeding attention to combat these threats effectively. New races of head smut [*Sporisorium reilianum* (J.G. Kühn) Langdon & Full.] have been identified on previously resistant hybrids in Tamaulipas, Mexico (Prom et al., 2011). In addition, the downy mildew pathogen *Peronosclerospara sorghi* Weston and Uppal. has developed a new pathotype that overcomes both readily available genetic sources of resistance and Metalaxyl fungicide [methyl *N*-(methoxyacetyl)-*N*-(2,6-xylyl)-DL-alaninate] (Isakeit and Jaster, 2005) and has been identified in both the United States and the Bajio region of Mexico. *Colletotrichum sublineolum* Henn. ex Sacc. & Trotter, the pathogen responsible for anthracnose, has been shown to have numerous pathotypes and haplotypes, which makes developing host plant resistance an ongoing and continual challenge (Moore et al., 2008). In more temperate climates, *Fusarium* spp. cause significant lodging and grain deterioration. In tropical climates, foliar pathogens can cause severe yield losses. Since few of these pathogens are economically or effectively controlled by pesticide applications, host plant resistance is the primary means of control beyond stewardship practices of crop rotation, residue management, and good agronomic practices. Identifying sources of resistance, introgressing resistance genes into adapted breeding populations and testing the resulting progeny has historically required planting nurseries in naturally occurring disease locations or the use of controlled environments, inoculation, and screening large numbers of entries. The development of molecular markers holds the promise of new tools for screening large populations of sorghum in the future for specific defensive traits.

In the United States, the effectiveness of insecticidal seed treatments has reduced the impact of greenbug, yellow sugarcane aphid [*Sipha flava* (Forbes)], chinch bugs [*Blissus leucopterus* (Say)], and fire ants [*Solenopsis invicta* (Buren)] on the germination and early growth of seedlings. Commercial breeding efforts to incorporate host plant resistance to these pests have been reduced, although several public programs continue to monitor and identify new sources of resistance. Other countries continue to work on a number of insect pests, with midge tolerance a component of most hybrids in Australia, and shoot fly [*Atherigona soccata* (Rondani)] tolerance an important goal of ICRISAT programs (Kumar et al., 2011).

To offer sorghum producers better weed control options, several sorghum programs are beginning to develop herbicide tolerance in sorghum. In 2007, Kansas State University released a set of materials that had genetic tolerance to ALS chemistry, allowing a method of post emergence weed control (Tuinstra and Al-Khatib, 2006). These sources were developed by introgressing the tolerance identified in naturally occurring shattercane (*S. bicolor* ssp. *bicolor*) populations. DuPont Crop Protection licensed the trait from Kansas State University to develop and register the herbicide, while offering a number of sorghum seed companies the opportunity to license the trait. Incorporation of these traits has been arduous, with fertility restoration, yellowing, and several agronomic traits all presenting challenges. Chemical registration, stewardship practices, and hybrid availability are targeted for 2015–2016.

GMO solutions to weed and insect pests have not been pursued in the United States due to a number of factors. In the late 1990s when glyphosate-tolerant corn, soybean, and cotton were being developed and distributed, sorghum was not included because of the fear of the rapid escape of the trait into a related aggressive weedy species, particularly Johnsongrass (*S. halapense* L. Pers., $2N = 40$). The other overarching limitation has been the costs associated with registration of a GMO trait in sorghum, especially given the low-input production systems sorghum is grown under. There are no insurmountable technical issues with the development of GMO sorghum, and a number of projects have been registered with the USDA-APHIS. The technology is currently being used to develop biofortified sorghum for Africa, in a project sponsored by the Bill and Melinda Gates Foundation, Africa Harvest, DuPont Pioneer, and others. The project is focused on developing a more nutritious sorghum with increased levels of lysine, vitamin A, zinc, and iron to improve the health of people who depend on this crop for food.

The inherent drought tolerance of sorghum makes it the grain of choice in many agricultural production areas, and selecting and enhancing the traits responsible for drought tolerance is the focus of a number of programs. In contrast to the conclusions of Mason et al. (2008), Staggenborg et al. (2008) maintained that in areas of limited potential for maize, sorghum is quite well adapted and an economically sound choice. The traits responsible for the drought tolerance of sorghum have been well documented (Jordan and Monk, 1980) and include avoidance mechanisms such as leaf rolling, epicuticular wax, and reduced stomatal conductance, as well as adaptive mechanisms such as osmotic adjustment and deeper rooting. The progress in genetic gain has been attributed to increased drought tolerance mechanisms, notably nonsenescence or stay-green and increased root biomass (Assefa and Staggenborg, 2011). Recent work (Mace et al., 2012) has also explored root angle as an adaptive mechanism. Nevertheless, sorghum yields remain keenly affected by the timing of drought and or high-temperature stress. Rosenow and Clark (1995) described drought tolerance as either preflowering or postflowering drought tolerance. Developing lines with improved postflowering stress tolerance via stay-green has been the more successful approach, with concomitant improvement in charcoal rot resistance and reduced preharvest lodging. Sorghum yield is still directly related to available water, so the movement of sorghum out of the mid southern regions and further west into Kansas and Nebraska has pushed sorghum into highly variable climates.

Phenotypic Variation Still Key in Sorghum Genetic Gain

The sorghum genus contains rich diversity in both the cultivated races as well as in related genera. The World Collection of sorghum at ICRISAT contains more than 37,000 accessions, and the GRIN collections in the United States at Griffin, GA and Fort Collins, CO have more than 42,000. However, the identification and incorporation of useful traits remains difficult (Jordan et al., 2011). Since the crop originated in tropical environments, many accessions in the collections are photoperiod sensitive, meaning they will not flower under the short-night conditions of temperate climates. To utilize this material in a temperate environment involves either converting the material to a temperate type by substituting photoperiod genes in a backcrossing–selection system or conducting the introgression in a tropical environment. Early pioneers in the sorghum industry, recognizing the limited genetic diversity available within the founder lines, established the Sorghum Conversion Program, a joint project of the USDA and Texas A&M University (Stephens et al., 1967). This program released 840 converted lines and provided sources of disease tolerance, drought tolerance, insect tolerance, and other yield enhancing traits (Table 11–1). The traits made accessible from the conversion program and introduced into elite breeding programs have made significant contributions to the continued genetic gain around the world. Additionally, scientists involved in germplasm introduction and identification have made astute observations of significant descriptors that could be related to plant introductions, enhancing their value to all breeders.

Table 11–1. Partial list of important traits identified in partially converted and converted lines released through the sorghum conversion program.

Key traits	Cited sources (Compilations)
Plant pathogen tolerance Downy mildew Anthracnose Head smut Fusarium head blight Rust Grain mold Charcoal rot *Maize dwarf mosaic virus*	Frederiksen and Rosenow, 1979; Duncan et al., 1991; Castor and Frederiksen, 1980
Insect tolerance Midge Greenbug Banks grass mite	Johnson and Teetes, 1979; Foster et al., 1977
Agronomic traits Stay-green Preflowering stress tolerance Postflowering stress tolerance Acid soil tolerance	Rosenow et al., 1997; Duncan et al., 1991

Germplasm and phenotypic research were strengthened with the launch of INTSORMIL, a USAID Collaborative Research Support Program in 1980. This program was instrumental in coordinating sorghum research in developing countries and training sorghum scientists in all areas of breeding, food science, agronomy, and other related disciplines. INTSORMIL fostered strong collaborative partnerships between public and private sorghum researchers in the United States and internationally. In Australia, similar close links between the public and private sectors exist (Jordan et al., 2006). Australian sorghum researchers in a range of disciplines act collaboratively to establish long-term research goals and yield targets. The industry currently supports four commercial seed companies and a public breeding program jointly funded by a grower levy and government investment. For the last 35 yr, the public program focused primarily on producing germplasm with resistance to sorghum midge and stay-green drought tolerance, while simultaneously selecting for yield. Germplasm from the program was widely used by the commercial programs to the extent that during the 1990s, the majority of hybrid seed sold in Australia had one or both parents from the public programs. Since 2000, the program has focused on grain yield per se and increasing the genetic diversity while maintaining stay-green and midge-tolerance traits (Jordan et al., 2012).

The separation of maize germplasm into heterotic groups has been a guiding principal in inbred and hybrid development. In sorghum, the principle division has been between lines that either maintain or restore cytoplasmic male sterility. The first reported source of A1 cytoplasmic sterility in sorghum was identified in the F_2 progeny of a cross between Double Dwarf Yellow Sooner Milo and Texas Blackhull Kafir (Stephens and Holland, 1954). Since that time, five additional cytoplasmic male sterility (CMS) systems have been identified (Reddy et al., 2008). In commercial breeding programs, the characteristics of shriveled anthers, complete sterility/restoration, the prevalence of restorers across a range of germplasm, and environmental stability have led to almost universal adaptation of the A1 system in the United States, Australia, and Argentina. Sorghum has not had an epidemic of disease or insects related to the use of a single cytoplasm, as occurred in corn with the Southern leaf blight [*Bipolaris maydis* (Y. Nisik. & C. Miyake) Shoemaker] outbreak associated with T cytoplasm in the early 1970s.

Future Prospects

Molecular Markers

It is close to a universal truth that sorghum is planted and grown in conditions in which yield expression is less than maximal. Most growers see sorghum as a means of reliably producing a crop where few other species could be cultivated. Consequently, while top-end potential is important, the primary focus in a larger sense is on yield stability. Sorghum hybrids must be able to withstand both biotic and abiotic stresses, and there are few sorghum fields in the United States that do not suffer from extreme temperatures or water deficits in the course of a growing season. Many of the molecular tools now being developed serve to address the very sort of defensive issues from which sorghum suffers. One of the principle uses of molecular markers is in the selection of traits that are difficult or impossible to select for under most environmental conditions. The stay-green trait in sorghum, for example, can only be directly selected for under growing conditions that impose severe postflowering drought stress. Other defensive traits, such as disease or insect resistance, rely on either the incidental manifestation of the pest in nurseries or yield trials, or the use of designated research facilities with pathology or entomology support. Since most sorghum breeding programs do not have in-house personnel for greenhouse or laboratory screening, the customary means of facing this challenge is to plant yield trials in locations apt to suffer from biotic or abiotic stress, as the needs of the program dictate. As a result, many sorghum improvement programs have focused on defensive traits. While this strategy has proven effective at improving the overall levels of tolerance to these challenges in sorghum, it has also diverted resources that may have otherwise been used in improving maximum yields of sorghum.

With the remarkable reduction in the cost of molecular markers in recent years (Bernardo, 2008; Eathington et al., 2007), even breeders in relatively minor crops like sorghum are better able to avail themselves of these tools. Breeding programs within larger corporations or organizations often have the opportunity to leverage the investments in infrastructure made for the major crops, and marker data points are now a primary consideration in terms of resource allocation within these programs. The overarching requirement becomes precision phenotyping capabilities through managed stress environments, disease and insect inoculations, or machinery capable of inducing lodging. Such tools are expensive to develop, and collecting consistent data requires significant commitment and skill. Statistical tools and information management also become key factors in the ability to associate marker data with precise data for traits successfully.

Many of the traits that have proven so challenging to breed for in sorghum are excellent candidates for the utilization of marker-assisted selection. Head smut resistance (Oh et al., 1994; Li et al., 2012), downy mildew resistance (Jeger et al., 1998), anthracnose resistance (Biruma et al., 2012), leaf blight resistance (Martin et al., 2011), and stay-green (Harris et al., 2007) all show mono- or oligogenic inheritance, and QTL maps for each are available in the public domain. Mace and Jordan (2010) published a sorghum consensus map showing the location of 35 major effect genes in sorghum, many of agronomic importance.

The existence of these maps, along with the expansion of marker utilization in breeding programs, points to the practical use of marker assisted selection and marker assisted backcrossing on a large scale in sorghum. Using markers to select for important defensive traits will allow breeders to invest their perpetually limited field resources on the direct improvement of yield. Whether through forward breeding or targeted introgression of specific traits, the application of molecular markers in sorghum breeding promises to change its focus from one of yield protection to one of yield enhancement.

Many of the recent developments in sorghum have come as a result of the sequencing and publication of the sorghum genome (Paterson et al., 2009). Using this tool, sorghum researchers can now go from conventional quantitative trait loci mapping directly to candidate genes, often capitalizing on the synteny that exists across major grass genera (Mace et al., 2013; Satish et al., 2012). The online reference sequence has also been of tremendous utility in the development and deployment of more and higher quality molecular markers in sorghum, especially when used in conjunction with high-throughput sequencing (Morris et al.,

2013; Nelson et al., 2011). Targeting induced local lesions in genomes (TILLing) populations have been created and have proven useful in sorghum; their use in sorghum relies on the existent reference sequence for the elucidation of genic association to mutant phenotypes (Xin et al., 2008; Blomstedt et al., 2012).

Heterotic Patterns in Sorghum

Sorghum heterotic pools are essentially defined by whether or not they restore fertility to cytoplasmic male sterile lines, a relationship that has shaped sorghum breeding and production since the late 1950s. While sorghum breeders have made strides in increasing the heterosis observed in crosses between A1 cytoplasmic sterile lines and their restorers, the resultant hybrid performance of any experimental cross is not predictable with any degree of accuracy. Surprisingly little research has been conducted on this topic, perhaps due to the restrictive cost of molecular marker work to date (Jordan et al., 2003; Gabriel, 2005; Ben-Israel et al., 2012). As with marker-assisted selection, the sequencing of the sorghum genome and the ongoing advancements in marker technology offer sorghum breeders the potential to perform genome-wide selection, hybrid performance predictions, and begin to define heterotic patterns in a much more informed manner. A partnership led by Texas A&M University and funded by the U.S. Sorghum Checkoff Board has been formed, involving both public and private sorghum institutions, to address this challenge. This effort seeks to characterize major grain sorghum genetic pools within the United States through high-throughput sequencing, distribute the sequence information, and use the information in the prediction of hybrid performance.

New Uses

Several new uses of sorghum have the potential to enhance yield demands in things other than traditional grain or biomass yield. Sorghum is unique in that it is one of the few agronomic crops that can be used in all of the renewable fuel technologies being explored. Grain can be used interchangeably in the starch-to-ethanol industry, while sweet sorghums provide a sugar stream similar to sugarcane. More importantly, Dahlberg et al. (2011) showed that variability in sorghum compositional characteristics can be developed to create, through breeding, sorghum biomass feedstocks that could meet the demands of a wide range of feedstock requirements for renewable fuels. Fuel equivalents per acre may become the driving force for yield enhancement in some of these specialty crops. This is also true with some of the current research pointing to the nutraceutical potential of various sorghums (Awika and Rooney, 2004; Dykes and Rooney, 2006).

There is also renewed interest in sorghum from researchers in other species for its superior adaptation to dryland cropping systems. Just as maize has historically served as a model species for the study of heterosis and classical genetics, sorghum holds great potential as a model drought-tolerant crop species. Many of the drought and heat tolerance mechanisms that sorghum already possesses are the very traits that are being aggressively pursued in major research efforts within the private sector. While some of these mechanisms and pathways will undoubtedly assist those seeking to enhance drought tolerance in maize and other species, the concomitant exploration of sorghum physiology and genetics will also help sorghum breeders better understand their crop.

References

Assefa, Y., and S.A. Staggenborg. 2010. Grain sorghum yield with hybrid advancement and changes in agronomic pactices from 1957 through 2008. Agron. J. 102:703–706. doi:10.2134/agronj2009.0314

Assefa, Y., and S.A. Staggenborg. 2011. Phenotypic changes in grain sorghum over the last five decades. J. Agron. Crop Sci. 197:249–257. doi:10.1111/j.1439-037X.2010.00462.x

Awika, J.M., and L.W. Rooney. 2004. Sorghum phytochemicals and their potential impact on human health. Phytochemistry 65:1199–1221. doi:10.1016/j.phytochem.2004.04.001

Ben-Israel, I., B. Kilian, H. Nida, and E. Fridman. 2012. Heterotic trait locus (HTL) mapping identifies intra-locus interactions that underlie reproductive hybrid vigor in sorghum bicolor. PLoS ONE 7(6):E38993. doi:10.1371/journal.pone.0038993

Bennett, W.F., B.B. Tucker, and A.B. Maunder. 1990. Modern grain sorghum production. Iowa State Univ. Press, Ames.

Bernardo, R. 2008. Molecular markers and selection for complex traits in plants: Learning from the last 20 years. Crop Sci. 48:1649–1664. doi:10.2135/cropsci2008.03.0131

Biruma, M., T. Martin, I. Fridborg, P. Okori, and C. Dixelius. 2012. Two loci in sorghum with NB-LRR encoding genes confer resistance to Colletotrichum sublineolum. Theor. Appl. Genet. 124:1005–1015. doi:10.1007/s00122-011-1764-8

Blomstedt, C.K., R.M. Gleadow, N. O'Donnell, P. Naur, K. Jensen, T. Laursen, C.E. Olsen, P. Stuart, J.D. Hamill, B.L. Møller, and A.D. Neale. 2012. A combined biochemical screen and TILLING approach identifies mutations in *Sorghum bicolor* L. Moench resulting in acyanogenic forage production. Plant Biotechnol. J. 10:54–66. doi:10.1111/j.1467-7652.2011.00646.x

Castor, L.L., and R.A. Frederiksen. 1980. *Fusarium* and *Curvularia* grain molds in Texas. In: Sorghum Diseases, a World Review. Proceedings of the International Workshop on Sorghum Diseases, Hyderabad, India, 11–15 Dec. 1978. ICRISAT, Andhra Pradesh, India. p. 93–102.

Dahlberg, J., E. Wolfrum, B. Bean, and W.L. Rooney. 2011. Compositional and agronomic evaluation of sorghum biomass as a potential feedstock for renewable fuels. J. Biobased Mater. Bioenergy 5:507–513. doi:10.1166/jbmb.2011.1171

Duncan, R.R., P.J. Bramel-Cox, and F.R. Miller. 1991. Contributions of introduced sorghum germplasm to hybrid development in the USA. In: H.L. Shands and L.E. Wiesner, editors, Use of plant introductions in cultivar development. Part 1. CSSA Spec. Publ. 17. CSSA, Madison, WI. p. 69–102. doi:10.2135/cssaspecpub17.c5

Duvick, D.N. 1986. Plant breeding: Past achievements and expectations for the future. Econ. Bot. 40:289–297. doi:10.1007/BF02858986

Duvick, D.N. 2005. The contribution of breeding to yield advances in maize (*Zea mays*). Adv. Agron. 86:83–145. doi:10.1016/S0065-2113(05)86002-X

Dykes, L., and L.W. Rooney. 2006. Sorghum and millet phenols and antioxidants. J. Cereal Sci. 44:236–251. doi:10.1016/j.jcs.2006.06.007

Eathington, S.R., T.M. Crosbie, M.D. Edwards, R.S. Reiter, and J.K. Bull. 2007. Molecular markers in a commercial breeding program. Crop Sci. 47:S154–S163. doi:10.2135/cropsci2007.04.0015IPBS

Fischer, R.A., and G.O. Edmeades. 2010. Breeding and cereal yield progress. Crop Sci. 50: S-85–S-98. doi:10.2135/cropsci2009.10.0564

Foster, D.G., G.L. Teetes, and J.W. Johnson. 1977. Resistance in sorghum to the Banks grass mite. J. Econ. Entomol. 70:259–262.

Frederiksen, R.A., and D.T. Rosenow. 1979. Breeding for disease resistance in sorghum. In: Biology and breeding for resistance to arthropods and pathogens in agricultural plants. MP-1451. Proceedings of a short course, 22 July–4 Aug. 1979. Texas A&M Univ., College Station. p. 137–167.

Gabriel, K. 2005. A study of heterotic relationships in sorghum. Ph.D. diss. Texas A&M Univ., College Station.

Harris, K., P.K. Subudhi, A. Borrell, D. Jordan, D. Rosenow, H. Nguyen, P. Klein, R. Klein, and J. Mullet. 2007. Sorghum stay-green QTL individually reduce post-flowering drought-induced leaf senescence. J. Exp. Bot. 58:327–338. doi:10.1093/jxb/erl225

Isakeit, T., and J. Jaster. 2005. Texas has a new pathotype of *Peronosclerospora sorghi*, the cause of sorghum downy mildew. Plant Disease 89:529. doi:10.1094/PD-89-0529A

Jeger, M.J., E. Gilijamse, C.H. Bock, and H.D. Frinking. 1998. The epidemiology, variability, and control of the downy mildews of pearl millet and sorghum, with particular reference to Africa. Plant Pathol. 47:544–569.

Johnson, J.W., and G.L. Teetes. 1979. Breeding for arthropod resistance in sorghum. In: Biology and breeding for resistance to arthropods and pathogens in agricultural plants. MP-1451. Proceedings of a short course, 22July–4 Aug. 1979. Texas A&M Univ., College Station. p. 168–180.

Jordan, D.R., G.L. Hammer, and R.G. Henzell. 2006. Breeding for yield in the DPI&F breeding program. In: A. K. Borrel, R. G. Henzell, and D. R. Jordan, editors, Proceedings 5th Australian Sorghum Conference, Gold Coast, Australia, 30 Jan.–2 Feb. 2006. CD-ROM. Range Media Pty Ltd., QLD, Australia.

Jordan, D.R., R.G. Henzell, and A.W. Cruickshank. 2009. Can public and private sector plant breeding coexist? Lessons from 50 years of synergy in sorghum. ABRAO J. Breeding and Genetics Vol. 41 Special Supplement August 2009. Proc. of the 14th Australasian Plant Breeding Conference and 11th SABRAO Congress, 2009, Cairns, Australia.

Jordan, D.R., C.H. Hunt, A.W. Cruickshank, A.K. Borrel, and R.G. Henzell. 2012. The Relationship between the stay-green trait and grain yield in elite sorghum hybrid grain sorghum in a range of environments. Crop Sci. 52:1153–1161. 1–9. doi:10.2135/cropsci2011.06.0326

Jordan, D.R., E.S. Mace, A.W. Cruickshank, C.H. Hunt, and R.G. Henzell. 2011. Exploring and exploiting genetic variation from unadapted sorghum germplasm in a breeding program. Crop Sci. 51:1444–1457. doi:10.2135/cropsci2010.06.0326

Jordan, W.R., and R.L. Monk. 1980. Enhancement of drought resistance of sorghum: Progress and limitations. In: Proc. 35th Annual Corn and Sorghum Res. Conf., Am. Seed Trade Assoc. Chicago, IL. p. 185–204.

Jordan, D.R., Y.Z. Tao, I.D. Godwin, R.G. Henzell, M. Cooper, and C.L. McIntyre. 1998. Loss of genetic diversity associated with selection for resistance to sorghum midge in Australian sorghum. Euphytica 102:1–17. doi:10.1023/A:1018311908636

Jordan, D.R., Y. Tao, D. Godwin, R.G. Henzell, M. Cooper, and C.L. Mclityre. 2003. Prediction of hybrid performance in grain sorghum using RFLP markers. Theor. Appl. Genet. 106:559–567.

Kumar, A.S., B.V.S. Reddy, H.C. Sharma, C.T. Hash, P.S. Rao, B. Ramaiah, and P.S. Reddy. 2011. Recent advances in sorghum genetic enhancement research at ICRISAT. Am. J. Plant Sci. 2:589–600. doi:10.4236/ajps.2011.24070

Li, Y., J. Zou, L. Ma, C. Ma, X. Li, D. Lu, S. Lu, and L. Hao. 2012. Development of head smut resistance-linked sequence characterized amplified regions markers in sorghum. Int. J. Agric. Biol. 14:613–616.

Mace, E.S., and D.R. Jordan. 2010. Location of major effect genes in sorghum (*Sorghum bicolor* (L.) Moench). Theor. Appl. Genet. 121:1339–1356. doi:10.1007/s00122-010-1392-8

Mace, E.S., C.H. Hunt, and D.R. Jordan. 2013. Supermodels: Sorghum and maize provide mutual insight into the genetics of flowering time. Theor. Appl. Genet. 126(5):1377–1395. doi:10.1007/s00122-013-2059-z.

Mace, E.S., V. Singh, E.J. Oosterom, G.L. Hammer, C.H. Hunt, and D.R. Jordan. 2012. QTL for nodal root angle in sorghum (*Sorghum bicolor* L. Moench) co-locate with QTL for traits associated with drought adaptation. Theor. Appl. Genet. 124(1):97–109. doi:10.1007/s00122-011-1690-9.

Martin, T., M. Biruma, I. Fridborg, P. Okori, and C. Dixelius. 2011. A highly conserved NB-LRR encoding gene cluster effective against *Setosphaeria turcica* in sorghum. BMC Plant Biol. 11(3):151. doi:10.1186/1471-2229-11-151

Mason, S.C., D. Kathol, K. Eskridge, and T.D. Galwha. 2008. Yield increase has been more rapid for maize than grain sorghum. Crop Sci. 48:1560–1568. doi:10.2135/cropsci2007.09.0529

Mascagni, H.J., Jr., K. Arceneaux, S. Brown, M. Deloach, J. Fluitt, D. Harrell, S. Harrison, C. Hollier, D. Kerns, J. Kruse, J. Leonards, J. Lofton, B. Padgett, R. Regan, and S. Sterling. 2013. Performance of grain sorghum hybrids in Louisiana 2012. LAES Research Summary 195.

Moore, J.W., M. Ditmore, and D.O. TeBeest. 2008. Pathotypes of *Colletotrichum sublineolum* in Arkansas. Plant Dis. 92:1415–1420. doi:10.1094/PDIS-92-10-1415

Morris, G.P., P. Ramu, S.P. Deshpande, C.T. Hash, T. Shah, H.D. Upadhyaya, O. Riera-Lizarazu, P.J. Brown, C.B. Acharya, S.E. Mitchell, J. Harriman, J.C. Glaubitz, E.S. Buckler, and S. Kresovich. 2013. Population genomic and genome-wide association studies of agroclimatic traits in sorghum. Proc. Natl. Acad. Sci. USA 110:453–458. doi:10.1073/pnas.1215985110

Maunder, A.B. 1998. Why hybrids? In: J. Axtel, editor, Proc. Hybrid Sorghum Workshop, INTSORMIL, Lincoln, NE.

Miller, F.R., and Y. Kebede. 1984. Genetic contributions to yield gains in sorghum, 1950 to 1980. In: W.R. Fehr, editor, Genetic contributions to yield gains in five major crop plants. CSSA Spec. Publ. 7. ASA and CSSA, Madison, WI. doi:10.2135/cssaspecpub7.c1

Nelson, J.C., S. Wang, Y. Wu, X. Li, G. Antony, F.F. White, and J. Yu. 2011. Single-nucleotide polymorphism discovery by high-throughput sequencing in sorghum. BMC Genomics 12:352. doi:10.1186/1471-2164-12-352

Oh, B.J., R.A. Frederiksen, and C.W. Magill. 1994. Identification of molecular markers linked to head smut resistance gene (Shs) in sorghum by RFLP and RAPD analyses. Phytopathology 84:830–833. doi:10.1094/Phyto-84-830

Paterson, A. H., J. E. Bowers, R. Bruggmann, I. Dubchak, J. Grimwood, H. Gundlach, G. Haberer, U. Hellsten, T. Mitros, A. Poliakov, J. Schmutz, M. Spannagl, H. Tang, X. Wang, T. Wicker, A. K. Bharti, J. Chapman, F. Alex Feltus, U. Gowik, I. V. Grigoriev, E. Lyons, C. A. Maher, M. Martis, A. Narechania, R. P. Otillar, B. W. Penning, A. A. Salamov, Y. Wang, L. Zhang, N. C. Carpita, M. Freeling, A. R. Gingle, C. T. Hash, B. Keller, P. Klein, S. Kresovich, M. C. McCann, R. Ming, D. G. Peterson, M. ur-Rahman, D. Ware, P. Westhoff, K. F. X. Mayer, J. Messing, and D. S. Rokhsaret. 2009. The *Sorghum bicolor* genome and the diversification of grasses. Nature 457:551–556. doi:10.1038/nature07723

Prom, L.K., R. Perumal, S.R. Erattaimuthu, J.E. Erpelding, N. Montes, G.N. Odvody, C. Greenwald, Z. Jin, R. Frederiksen, and C.W. Magill. 2011. Virulence and molecular genotyping studies of *Sporisorium reilianum* isolates in sorghum. Plant Disease 95:523–529. doi:10.1094/PDIS-10-10-0720

Rai, K.N., D.S. Murty, D.J. Andrews, and P.J. Bramel-Cox. 1999. Genetic enhancement of pearl millet and sorghum for the semi-arid tropics of Asia and Africa. Genome 42:617–628. doi:10.1139/g99-040

Reddy, B.V.S., A. Kumar, and S.L. Kaul. 2008. Alternative cytoplasmic male sterility systems in sorghum and their utilization. In: Sorghum improvement in the new millennium. ICRISAT, Patancheru, India, p. 132–144.

Rosenow, D.T., and L.E. Clark. 1995. Drought and lodging resistance for a quality sorghum crop. In: Proc. 50th Annual Corn and Sorghum Res. Conf., Am Seed Trade Assoc., Chicago, IL. p. 82–96.

Rosenow, D.T., G. Ejeta, L.E. Clark, M.L. Gilbert, R.G. Henzell, A.K. Borrell, and R.C. Muchow. 1997. Breeding for pre- and post-flowering drought stress resistance in sorghum. In: Proceedings of the International Conference on Genetic Improvement of Sorghum and Pearl Millet, Lubbock, TX, 22–27 Sept. 1996. Publ. 97-5. ICRISAT, Patancheru, India. p. 400–411.

Satish, K., R. Madhusudhana, P.G. Padmaja, N. Seetharama, and J.V. Patil. 2012. [*Sorghum bicolor* (L.) Moench] Development, genetic mapping of candidate gene-based markers and their significant association with the shoot fly resistance quantitative trait loci in sorghum. Mol. Breed. 30:1573–1591. doi:10.1007/s11032-012-9740-9

Smith, C.W., and R.A. Frederiksen. 2000. History of cultivar development in the United States: From "Memoirs of A. B. Maunder—Sorghum breeder.". In: C.W. Smith and R.A. Frederiksen, editors, Sorghum: Origin, history, technology, and production. John Wiley & Sons, New York. p. 191–223.

Staggenborg, S.A., K.C. Dhuyvetter, and W.B. Gordon. 2008. Grain sorghum and corn comparisons: Yield, economic and environmental response. Agron. J. 100:1600–1604. doi:10.2134/agronj2008.0129

Stephens, J.C., and R.F. Holland. 1954. Cytoplasmic male-sterility for hybrid sorghum seed production. Agron. J. 46:20–23. doi:10.2134/agronj1954.00021962004600010006x

Stephens, J.C., F.R. Miller, and D.T. Rosenow. 1967. Conversion of alien sorghums to early combine genotypes. Crop Sci. 7:396. doi:10.2135/cropsci1967.0011183X000700040036x

Tuinstra, M.R., and K. Al-Khatib. 2006. Acetolactate synthase herbicide resistant sorghum. U.S. Provisional Patent Application No. 60/873,529. Filed 7 Dec. 2006.

Unger, P.W., and R.L. Baumhardt. 1999. Factors related to dryland grain sorghum yield increases: 1939 through 1997. Agron. J. 91:870–875. doi:10.2134/agronj1999.915870x

Xin, Z., M.L. Wang, N.A. Barkley, G. Burow, C. Franks, G. Pederson, and J. Burke. 2008. Applying genotyping (TILLING) and phenotyping analyses to elucidate gene function in a chemically induced sorghum mutant population. BMC Plant Biol. 8:103. doi:10.1186/1471-2229-8-103

12

Soybean

James E. Specht,* Brian W. Diers, Randall L. Nelson, Jose Francisco Ferraz de Toledo, Jessica A. Torrion, and Patricio Grassini

ABSTRACT

Soybean [*Glycine max* (L.) Merr.] yield improvement arises from continual producer adoption of genetic technology—the annual releases of new cultivars that express ever-greater genetic yield potential, and agronomic technology—the periodic transformations in crop and soil management practices that enhance environmental yield potential. The contribution of both factors, and their interaction, to soybean yield is assessed in this chapter. In the United States, the rate of on-farm yield improvement from 1924 to 2012 is 23.3 kg ha^{-1} yr^{-1}, whereas the Brazilian rate from 1977 to 2013 is 43.5 kg ha^{-1} yr^{-1}. When a segmental regression model was applied, the post-1983 U.S. rate increased 25% to 29.4 kg ha^{-1} yr^{-1}. The genetic rates of yield improvement in the United States for Maturity Group (MG) II, III, and IV cultivars, estimated by growing cultivars from different eras in common environments, were 23.1, 22.8, and 19.5 kg ha^{-1} yr^{-1}. In the states in which these MG cultivars are adapted, the on-farm rates are 27.3 (MG II & III) and 21.4 (MG IV) kg ha^{-1} yr^{-1}. Statistical modeling indicates that the more recent genetic improvement rates for the three MGs are 31.1, 29.4, and 26.5 kg ha^{-1} yr^{-1}. Comparison of on-farm yield improvement and genetic yield improvement in high-yield irrigated production environments suggests that about two-thirds of the on-farm yield improvement is likely arising from the contribution of continual release of ever-higher yielding soybean cultivars that are quickly adopted by soybean producers. Periodic advances in agronomic practices adopted by producers likely account for the other one-third, but recent publications have documented another contributor—synergistic genetic × agronomic interactions, wherein the difference between the yields of modern and obsolete cultivars widens when the agronomic productivity is enhanced by better crop management practices (e.g., earlier planting date). The impact of rising atmospheric CO_2 levels on soybean yield was judged to be small, but it will be cumulative over time.

Abbreviations: AIC, Akaike information criterion; AYI, agronomic yield improvement; DOY, day of the year; GYI, genetic yield improvement; MAS, marker-assisted selection; MG, maturity group; PVP, Plant Variety Protection [Act]; QTL, quantitative trait loci; RUE, radiation use efficiency; SCN, soybean cyst nematode; WUE, water-use efficiency.

J.E. Specht and **P. Grassini** (pgrassini2@unl.edu), Dep. of Agronomy and Horticulture, 322 Keim Hall, East Campus, 1875 N. 38th St., Univ. of Nebraska, Lincoln, NE 68583-0915. *Corresponding author (jspecht1@unl.edu).

B.W. Diers, Dep. of Crop Sciences, 268 National Soybean Research Center, 1101 W. Peabody Drive, Univ. of Illinois, Urbana, IL 61801 (bdiers@illinois.edu).

R.L. Nelson, Soybean/Maize Germplasm, Pathology & Genetics Research Unit, USDA-ARS, 1101 W. Peabody Drive, Urbana, IL 61801 (randall.nelson@ars.usda.gov).

J.F.F. de Toledo, Avenida dos Imigrantes, 2715, Apt. 104, Bairro Centro, 78890-000, Sorriso, MT, Brazil (JToledo@nidera.com.br).

J.A. Torrion, Crop Physiology, Northwestern Agricultural Research Center (NWARC), Dep. of Research Centers, Montana State Univ., 4570 MT Hwy 35, Kalispell, MT 59901 (jessica.torrion@montana.edu).

doi:10.2135/cssaspecpub33.c12

Yield Gains in Major U.S. Field Crops. CSSA Special Publication 33.
Stephen Smith, Brian Diers, James Specht, and Brett Carver, editors.

Soybean production has undergone enormous growth in the past century. During this period in the United States, soybean went from a minor crop to the second most important based on value. This chapter focuses on the genetic and agronomic improvements in soybean in the United States and reviews the literature on what genetic and physiological changes led to these improvements. We also have included sections on soybean improvement in Brazil because of the increasing importance of production in that country. In fact, Brazil's 2013 soybean production (May harvest) exceeded U.S. 2013 soybean production (November harvest). Before discussing soybean crop improvement, let us first provide a historical perspective.

U.S. Soybean History

Soybean is a relatively new crop in the United States. The first recorded production of soybean in the United States was near Savannah, GA in 1765 (Hymowitz and Harlan, 1983). There is no record of additional soybean production until the 20th century. The USDA began its now annual survey of area, production, and seed yield statistics in 1924 (Fig. 12–1, 12–2, 12–3). In the decade from 1929 to 1939, soybean production area quadrupled (Fig. 12–1). However, soybean was initially grown in the United States primarily as a forage crop until 1941, when the area harvested for seed finally exceeded the area grown for all other purposes (Probst and Judd, 1973). Between 1942 and 1949, the area planted to soybean did not change appreciably, although total seed production more than doubled as producers began to harvest soybean for its seed instead of forage. In 1950, the area harvested for seed began a steady increase that peaked in 1979 at a then record of 28.9 million ha (71.4 million acres) that was not exceeded again until 1998. Thereafter, the harvested area held steady in a range of 29 to 31 million hectares, except for a notable dip in 2007, which was due to a shift from soybean to corn (*Zea mays* L.) production by producer expectations in the spring of that year that corn-derived ethanol production would drive up corn prices.

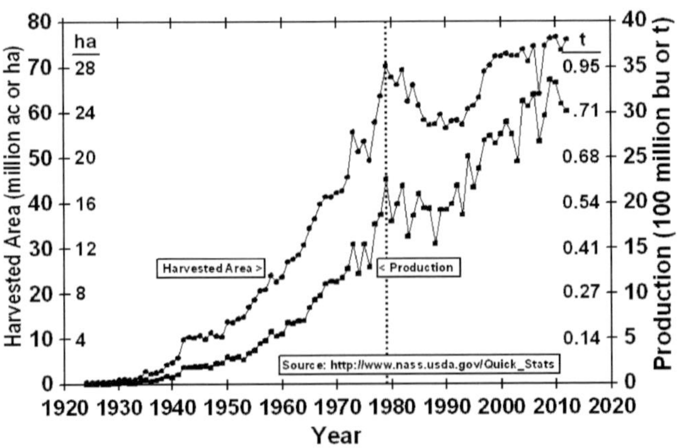

Fig. 12–1. U.S. soybean statistics from 1924 to 2012 for harvested area (left *y* axis; acres or hectares) and production (right *y* axis; tons or bushels). The vertical dotted line denotes the record peak (since exceeded) in soybean area and production that occurred in 1979.

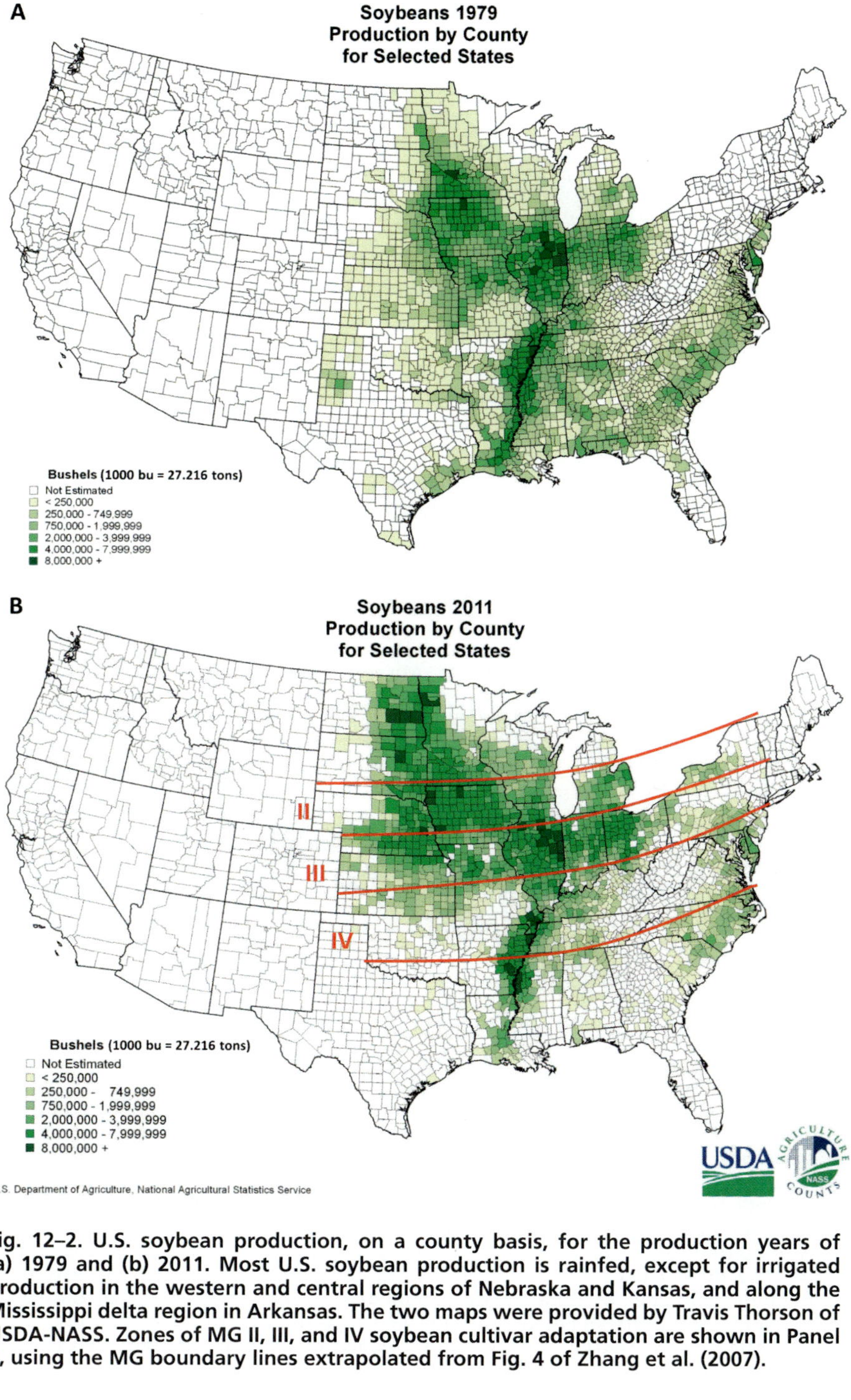

Fig. 12–2. U.S. soybean production, on a county basis, for the production years of (a) 1979 and (b) 2011. Most U.S. soybean production is rainfed, except for irrigated production in the western and central regions of Nebraska and Kansas, and along the Mississippi delta region in Arkansas. The two maps were provided by Travis Thorson of USDA-NASS. Zones of MG II, III, and IV soybean cultivar adaptation are shown in Panel B, using the MG boundary lines extrapolated from Fig. 4 of Zhang et al. (2007).

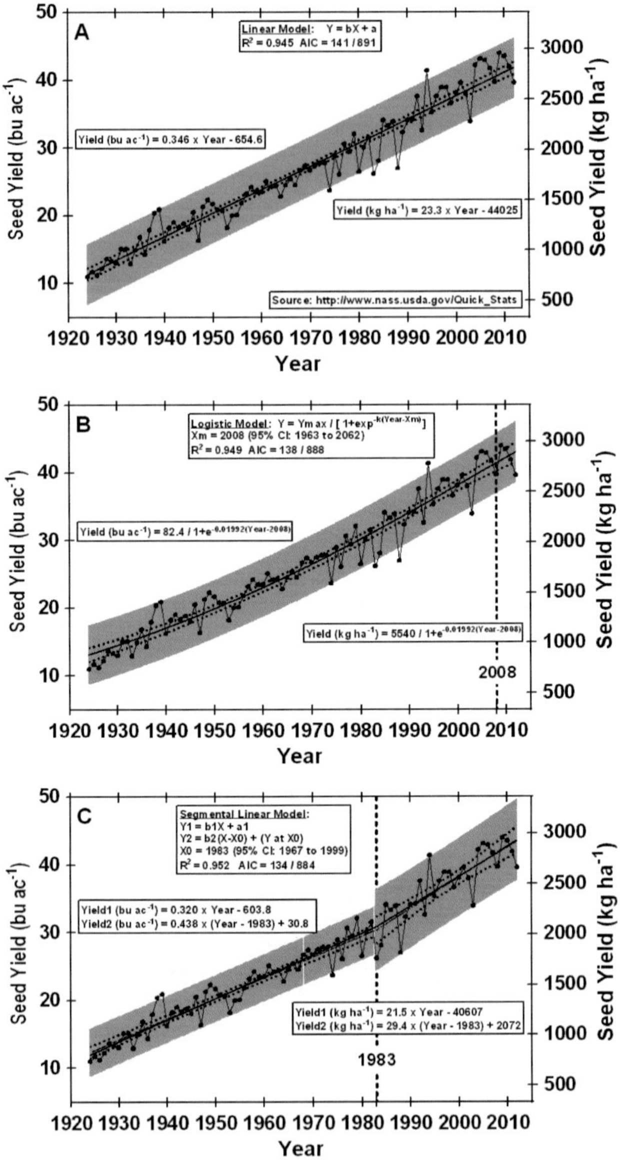

Fig. 12–3. U.S. on-farm soybean yield statistics from 1924 to 2012 plotted as 89 solid symbols with connecting lines (left *y* axis in bu ac^{-1}; right *y* axis in kg ha^{-1}). Panels A, B, and C display the respective linear, logistic, and segmental linear models fit to these data, with regression statistics provided in each panel text box. For an objective determination of the most probable model, an Akaike information content (AIC) value was computed for the fit of each model in bushels per acre or kilograms per hectare units. The dotted lines depict the 95% confidence interval around the estimated regression trend line, whereas the grey highlighted zone reflects the 95% prediction interval. A dashed vertical line denotes (B) the estimated year of the inflection point in the logistic (sigmoid) curve fit or (C) the estimated breakpoint year between the two regression line trends in the segmental linear fit.

The increase in U.S. total soybean production during the past three decades (Fig. 12–1) was coupled with a remarkable geographic shift in the location of that production. Compare, for example, U.S. county production statistics in 1979 (Fig. 12–2A) with those in 2011 (Fig. 12–2B). During this 33-yr time frame, soybean production shifted northward to become more concentrated in the north-central United States. The most noteworthy aspect of this shift was the movement of soybean production into the northern Great Plains states of South and North Dakota. In North Dakota, for example, soybean harvested area increased from 83,400 ha in 1979 to 1.6 million hectares in 2011 and then to 1.9 million hectares in 2012 (USDA-NASS, http://www.nass.usda.gov/, accessed 3 Feb. 2014).

On-farm U.S. soybean yield quadrupled from the USDA-NASS-estimated national yield of 740 kg ha^{-1} (11 bu ac^{-1}) in 1924 to a recent record yield of 2960 kg ha^{-1} (44 bu ac^{-1}) in 2009 (Fig. 12–3A). A simple linear regression model fit to the on-farm U.S. soybean yield data from 1924 to 2012 indicated a 23.3 kg ha^{-1} yr^{-1} rate of increase. Although the linear fit accounted for most of the variation (i.e., R^2 value of 0.95), the linear regression residuals failed a normality test (P = 0.025), indicating a violation of the assumption of homoscedasticity. To account for a slight curvature pattern that is evident in the data point deviation from linearity (Fig. 12–3A), a logistic model was fit to the data (Fig. 12–3B). A logistic model invokes the need to estimate three parameters (k, X_m, Y_{max}), one more than a simple linear model (b, a). The sigmoid curvature in a logistic model is indicative of an initially slow, but accelerating growth rate up to the inflection point, but thereafter a decelerating rate on the approach to a zero-rate plateau. A logistic model is more intellectually satisfying than a linear model from a physiological point of view, given that there is a biological limit to yield improvement. Specht et al. (1999) had made note of that biophysical limitation and for that reason used a logistic model in their analysis of historic soybean yield from 1924 up to 1999. The maximum thermodynamic efficiency at which the energy of solar radiation can be converted into the chemical energy of crop shoot, root, and seed dry matter ultimately sets a yield limit for a given crop. Sinclair and Horrie (1989), in their review of the literature, estimated a radiation use efficiency (RUE) maximum of 1.2 g MJ^{-1} for soybean. Sinclair and Rufty (2012) noted that a functional practical limit for soybean yield, based on their estimate for soybean transpiration efficiency, was about 6 t ha^{-1} for a seasonal transpiration amount of 450 mm.

To objectively test two alternative models to identify the one that better fits the data set, the Akaike information criterion (AIC) value can be used (Sakamoto et al., 1986; Posada and Buckley, 2004). The AIC is a well-known statistic that rewards the model with a better R^2 fit, but penalizes the model that requires estimation of more parameters to achieve that better fit. A lower AIC value was found for the logistic model, which translated into an 81% probability that it was more probable (i.e., preferred) than the linear model for these 1924 to 2012 U.S. on-farm soybean yield data. Moreover, the logistic model residuals passed a normality test ($P = 0.09$).

The estimated logistic model X_m parameter of 2008 (SE: ± 22.4 yr) is of interest because it denotes the time point at which the rate of on-farm yield improvement transitions from the end of an accelerating phase to the start of a deceleration phase. Though actual on-farm yield continues to rise in both phases, a logistic model implicitly projects a yield plateau (i.e., Y_{max}, the yield limit parameter) in some distant year. The Y_{max} estimate was ~5543 ± 1284 kg ha^{-1} and may seem to be in conflict with the estimated theoretical (absolute max) yield potential of approximately 8100 kg ha^{-1} for soybean grown in a "perfect" environment (Specht et al., 1999). Keep in mind, however, that each annual U.S. soybean yield estimate originates from USDA-NASS-enumerator yield estimation from fields that vary in productivity. Highly fertile and well-managed individual soybean fields may approach the theoretical max in the future. However, field-to-field productivity variability will invariably keep future yield means at county, state, and national levels much lower than that, perhaps at only 80 to 85% thereof, as Grassini et al. (2009, 2011) and Van Wart et al. (2013) have documented. In a recent analysis, Sinclair and Rufty (2012) used crop simulation, based on the currently known limits for soybean RUE, water-use efficiency (WUE), and nitrogen-use efficiency, to estimate that even in years of above-normal, well-distributed rainfall, on-farm rainfed soybean yield was not likely to be much greater than ~6000 kg ha^{-1} in the north-central United States. Thus, a logistic Y_{max} parameter estimate of 5543 kg ha^{-1} for a collective set of farms across the nation is not an unreasonable value. Note that while state soybean yield contest winners in the last 6 yr currently generate yields in the 6400 kg ha^{-1} range, their neighbors likely generate far lower yields. A significant yield contest winner outlier has been the soybean producer who has won yield contests in Missouri, with a yield of 10,000 kg ha^{-1} in 2006, 10,400 kg ha^{-1} in 2007, and 10,800 kg ha^{-1} in 2010. Although those yields were certified by the sponsor of the contest, further scientific analysis of these yields is warranted.

Although the logistic model was objectively preferred over a simple linear model, on the basis of an AIC comparison, an alternative model that could also be applied is a trisegmental linear model. In the trisegmental model, the sigmoid curvature in the logistic model is replaced with three linear segments circumscribed by two breakpoints (see Bastidas et al., 2008, and references therein for more details). Such a model is appropriate for evaluating data sets in which there is incipient evidence of a yield plateau. Egli (2008b), who examined U.S. county yield data for U.S. production years up to 2005, did not detect any significant evidence of nascent yield plateaus in that data set. Although we now have seven more years of data, a trisegmental linear model still could not be fit to the data, or more precisely, that model could not detect a second breakpoint year in the current yield data. That failure is not inconsistent with the estimated logistic-based inflection point year of 2008. Rather, only 5 yr of a logistic model-based deceleration rate of yield may not be enough time for any yield plateau to become evident in the yield data, at least to the degree that a trisegmental model could detect it.

One can still attempt to fit a two-segment linear model that, in effect, replaces the first curved phase of the fitted logistic model (Fig. 12–3B) with two linear lines demarked by a breakpoint year (Fig. 12–3C). In fact, the fit of this model to the 1924 to 2012 yield data generated the best AIC value of the three competing models, with the calculations indicating that it was 96% more probable than either of the other two models, despite the need to estimate more model parameters. Segmented linear models are widely used when trend line transitions appear to be more blunt than smooth. Still, even though the logistic model projects an eventual yield plateau, the lack of a trisegmental fit implies with some certainty that no yield plateau is statistically discernible in the current historic soybean yield data.

The two-segment model does suggest that the rate of U.S. on-farm soybean yield improvement transitioned to a 50% faster rate after 1983 (i.e., the X0 parameter year estimate in Fig. 12–3C). That inference would be consistent with the report of Twine and Kucharik (2009), who observed that the net productivity of U.S. corn and soybean crop systems trended upward at a much faster rate during the period from 1982 to 2002 than it did during the prior period of 1950 to 1981. These authors asserted that, although genetic and agronomic technologies played a dominant role in these more rapid productivity increases, climate change also played a role. They concluded that a positive summer precipitation trend during this period, coupled with a concurrent negative temperature trend, likely contributed about 20 to 25% of the observed yield increases in that 20-yr period. In Chapter 6 in this book, Smith et al. (2014) used a scientific statistical modeling approach to estimate any positive (or negative) long-term weather trend in this same time frame on maize yield. They detected a favorable weather trend that, when annualized, amounted to a 1.9 kg ha^{-1} contribution to the observed upward trend in on-farm maize yields. Given the documented constant corn/soybean yield ratio of 3 to 1 (in kg ha^{-1} units) in productive crop systems (Specht et al., 1999; Egli, 2008a), one might speculatively propose that one-third (i.e., 0.63 kg ha^{-1} yr^{-1}) of the foregoing estimated weather contribution to maize might also be contributing to the observed upward trend in on-farm soybean yields. Finally, it is also worth noting the greater year-to-year variability about the soybean yield trend line after the 1983 breakpoint year as seen in Fig. 12–3C, and such post-1983 variability is also evident in maize (see Fig. 6–2 in Chapter 6 of this volume, Smith et al., 2014).

The previous analyses provide objectively modeled on-farm U.S. yield improvement rates that can be used as a baseline for estimating genetic and agronomic contributions, plus the contribution of any synergistic interactions between modern cultivars and modern agronomic practices on yield improvements. Also worthy of some consideration is the impact on soybean productivity of the ever-persistent annual increase in atmospheric CO_2 and its consequence– future climate change. Predictions of future yields assume, of course, a continuation of genetic and agronomic improvements based on current basic technologies, and do not take into account the potential of proposed breakthrough technologies, such as genetically engineering soybean plants to improve photosynthetic systems (Zhu et al., 2010). We now turn to the issue of soybean genetic improvement.

History of Genetic Improvement in the United States

The first soybean cultivars grown in the United States were introductions imported from Asia. Morse and Cartter (1939) described 108 soybean cultivars then available in the United States, and all were introductions, selections from introductions, or the progeny of natural crosses within introductions. Of the 108 cultivars, 43 were forage types, with only 14 of the 108 grown on significant U.S. acreage.

These introduced cultivars provided the genetic basis for U.S. soybean production until the 1940s when cultivars that were the products of scientific plant breeding within the United States were released. These cultivars were released based on data collected in regional tests (i.e., the Uniform Soybean Tests) that were coordinated by the USDA. Those tests were initiated in the northern United States in 1939 and in the southern United States in 1943. During the first 15 yr of cooperative regional testing, 86 introductions were parents of the many experimental lines entered into the USDA tests. Thirty-seven of these 86 introductions eventually were among the more than 80 introductions that provided the genetic base of North American soybean production. Seventeen of the 37 are included in the 35 ancestral lines that contributed 95% of the genes to the current U.S. genetic base, as calculated from pedigree analysis. Most other members of the genetic base were selected from the collection as sources of disease resistance (Gizlice et al., 1994).

Most of the introductions used as parents before 1955 were almost certainly chosen based on their performance in yield trials. Some were entries in the Uniform Soybean Tests, but many, if not most, were evaluated in state tests for which records are not available. Testing procedures were not well developed in those early years, and the lack of mechanization kept entry numbers in the trials small. When the USDA Soybean Germplasm Collection was established in 1949, only 1677 (i.e., 20%) of 8225 documented soybean introductions into the United States were still extant (Bernard et al., 1987, 1989). How many of these introductions would have been available to breeders for local testing when organized plant breeding began in the 1930s is unknown. Given the limitations of breeding programs during this period, it is certainly possible that the available introductions with the highest yield potential may not have been selected for use as parents.

The first major cultivar developed from intentional cross-pollination was Lincoln. It was derived from unknown parents (most likely from China) and released in 1943. In the Group III tests from 1942 to 1945, Lincoln yielded 7% more and was 8 d earlier than the best domestically developed cultivar, Chief. Lincoln yielded 11% more and was 10 d earlier than the best introduced cultivar, Patoka. Lincoln also yielded 16% more than the plant introduction Dunfield, which was of similar maturity. The yield advantage of Lincoln made it a very attractive parent. The first lines with Lincoln as a parent were entered in the Uniform Soybean Tests in 1948, and between 1948 and 1955, 65% of all of the lines tested in MG 0 to IV had Lincoln as a parent. It was common at that time to share partially inbred segregating populations among states. As a result, many of the lines from different states that were tested during this period had the same pedigree. Seven states had Uniform Test entries in MG 0 to IV derived from the Lincoln(2) × Richland backcross. With the availability of much improved cultivars such as Lincoln, the direct use of plant introductions as parents decreased rapidly. The release of the southern cultivar Lee in 1953 had a similar effect on breeding in the southern United States. Lee and its sister lines from the cross of S-100 × CNS dominated the pedigrees of entries in the Southern Uniform Soybean Tests for a decade after the release of Lee. As calculated from pedigree analysis (Gizlice et al., 1994), Lincoln provided 24% of the genetic base of cultivars in the northern United States, and Lee provided 46% of the genetic base of cultivars in the southern United States, based on cultivar releases up to 1994. Together these two cultivars account for approximately 35% of the U.S. genetic base.

The actual pool from which the major ancestors of U.S. cultivars were selected, and the procedures by which those lines were selected are not precisely known. However imperfect that process was, the North American ancestral base has apparently been sufficient to maintain continual genetic yield improvement for nearly 80 yr. The diversity of these ancestral lines has been documented by several studies using DNA markers (Diwan and Cregan, 1997; Keim et al., 1992; Kisha et al., 1998; Skorupska et al., 1993; Thompson and Nelson, 1998). DNA marker analyses show that the diversity of major North American ancestors is equivalent to that of a relatively large set of introductions used in recent population development for yield improvement, and to the major ancestors of modern Chinese cultivars (Brown-Guedira et al., 2000; Kisha et al., 1998; Li et al., 2001).

The major ancestors of North American soybean also have considerable variation for morphological and biochemical traits (Gizlice et al., 1993).

Based on DNA marker analyses (Brown-Guedira et al., 2000; Kisha et al., 1998; Thompson and Nelson, 1998), the major ancestral lines can be placed into four groups: (i) Lincoln, S-100, A.K. (Harrow), Illini, and Dunfield; (ii) Arksoy, Ralsoy, and Haberlandt; (iii) Mandarin (Ottawa), Richland, and Capital; and (iv) Jackson, Ogden, Roanoke, and Mukden. Korean, Perry and CNS, the remaining ancestral lines that contributed at least 1% of genes to the current gene pool were diverse from each other and were outliers relative to the four other groups. These four major ancestral groupings are generally consistent with what is known about pedigrees and geographical origin of these lines. In the first group, A.K. (Harrow) and Illini trace to a single introduction (A.K.) from China, and S-100 was selected from Illini (Bernard et al., 1988). All three members of the second group have a putative origin in Korea, and Ralsoy was selected from Arksoy (Bernard et al., 1988). In the fourth group, Tokyo, the only major plant introduction from Japan used in North American breeding, is in the pedigree of both Jackson and Ogden. The listed pedigree of Capital is Strain 171 × A.K. (Harrow). Strain 171 was lost, and DNA marker analysis does not support a relationship between A.K. (Harrow) and Capital. Based on a coefficient of pedigree analysis, the first group of ancestors above contributed more than 35% of the parentage to recent U.S. cultivars, the second contributed 4%, the third group 22%, and the fourth group 14%. Many successful crosses had parents that were from different genetic groups. As was already noted, Lincoln(2) × Richland was an eminently successful backcross for northern soybean breeders. Blackhawk and Hawkeye are the progeny of Mukden × Richland. Mandarin (Ottawa)(2) × A.K. (Harrow) produced Harosoy. In the south, Lee was the product of S-100 × CNS. Bragg was selected from Jackson × D49–2491 (a sister line of Lee). Roanoke × N45–745 (Ogden × CNS) were the parents of Hood.

Cultivars from the southern United States are not as genetically diverse as cultivars from the north, primarily because more than 40% of the parentage of southern cultivars traces to just two ancestors, CNS and S-100 (Gizlice et al., 1994; Sneller, 1994; 2003). A single F_2 plant from the cross of CNS × S-100 was the progenitor of Lee and its three sib lines, which together account for 37% of the parentage in southern soybean cultivars (Carter et al., 1993).

As activity expanded in the public breeding sector, cultivar development programs generated 94 soybean cultivars that were registered in the journal *Crop Science* between 1942 and 1972 (Hartwig, 1973). A few of the released cultivars became popular producer choices, and this resulted in much of the U.S. production being dominated by a relatively small number of public cultivars, at least until the early 1980s. In 1957, for example, 12 cultivars accounted for 87% of all U.S. land planted to soybean, with 55% of that land planted to just four of those 12 cultivars. In 1965, five cultivars were grown on 67% of the U.S. harvested soybean land. Six years later, the 10 leading cultivars were grown on 80% of the soybean land (Hartwig, 1973). The cultivar Williams and its phytophthora root rot resistant successors, Williams 79 and Williams 82, were dominant cultivars popular with producers until about 1982, when the acreage of proprietary cultivars began to markedly increase.

The signing into law of the Plant Variety Protection (PVP) Act in 1970 gave a strong incentive for proprietary seed companies to invest in soybean breeding because it gave them the right to control the marketing and distribution of the cultivars they developed (Fehr, 1987). In 1985, patent protection for seed-propagated plants was allowed through utility patents, and such patents provided a more stringent protection of cultivars than did PVP. The first soybean cultivar was patented in 1994 (Mikel et al., 2010). On the basis of U.S. PVPs, U.S. utility patents, and cultivar journal registrations, Mikel et al. (2010) reported that from 1970 to 2008, 2242 soybean cultivars were registered, with about 80% created by breeders associated with proprietary companies, and the remainder created by breeders in public institutions.

There is very little genetic divergence between recently released public and proprietary cultivars in terms of their genetic base (Sneller, 1994, 2003; Mikel et al., 2010). Only 3% of the total variation in coefficient of parentage values among 312 MG 00 to VII cultivars from 1999 to 2001 was detectable in an analytical contrast of public vs. proprietary lines when sampled from the same MG (Sneller, 2003). Proprietary lines have a larger contribution from Lincoln, and less from Mandarin (Ottawa), CNS, and S-100, compared to public lines, primarily because of the extensive use in proprietary breeding programs of Asgrow A3127 and its progeny, which have considerable ancestry from Lincoln (Sneller, 1994, 2003).

The transition from public to proprietary cultivars as the primary source of new cultivar releases for U.S. soybean producers occurred relatively quickly and is illustrated when the number of acres of certified seed of publicly developed cultivars grown each year is graphed (Fig. 12–4). Certified seed is the route (i.e., seed class) by which public cultivars are made available for retail sale to soybean producers. Thus, tracking the annual acreage of this seed class serves as a proxy for acreage of public cultivars grown by U.S. producers in the following year (assuming that all certified seed is sold and planted). The planted area of certified seed of public cultivars dropped substantially, from more than 672,200 ha in 1978 to 46,900 ha in 2011 (data are courtesy of Chet Boruff, CEO of AOSCA, 2013, personal communication), despite a U.S. soybean harvested area of 24 to 32 million ha annually during this same time frame. The transition from public to proprietary cultivars is illustrated in an independent manner by the percentage of Nebraska soybean acres occupied by named soybean cultivars, which was tracked annually by the Nebraska NASS office from 1958 to 1983 (Fig. 12–4, inset). Comparable cultivar area reports from other states were not found, and this Nebraska cultivar survey was unfortunately discontinued after proprietary cultivars gained a majority share (63%) of the Nebraska producer seed sales market in 1983. In that year, the popular proprietary cultivar A3127 was grown on 23% of the Nebraska state soybean area; the next most popular was the public cultivar Williams and its 79 and 82 derivatives, which combined were grown on 11% of the area (USDA-NASS, http://www.nass.usda.gov/, accessed 3 Feb. 2014) (Fig. 12–4).

The active investment of the proprietary sector in cultivar development paralleled an increase in all soybean breeding activity. Carter et al. (2004) showed that during the period from 1945 to 1990, the number of public soybean breeders in the United States increased from 13 to about 50, although that number subsequently declined to about 40 in 2000, and to about 30 in 2013. In contrast, the number of U.S. proprietary soybean breeders, which was 10 in 1971, substantively increased to about 70 in 2000. Based on attendance at the recent annual Soybean Breeders' Workshops, it would appear that since the turn of the century, the number of soybean breeders at proprietary organizations has continued to increase, whereas the number of public soybean breeders has not appreciably changed. Proprietary breeders

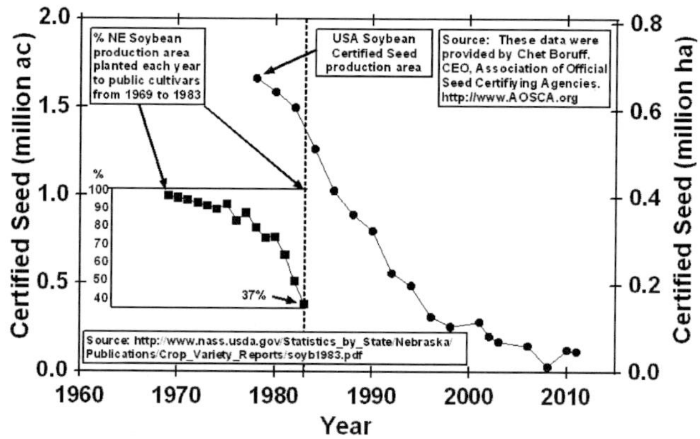

Fig. 12–4. Historic data on U.S. soybean certified seed production area from 1978 to 2011. The inset graph provides historical data on the estimated percentage of soybean production area in Nebraska planted to public cultivars from 1969 until 1983, when the reporting was discontinued. Reports of cultivar percentage area data in other states are not available (i.e., either not collected, or the records were not retained).

have now taken over the primary role of developing cultivars, with public sector breeders focusing their efforts on more basic breeding and genomics research.

In addition to the increased number of breeders, the scale of breeding activities now possible for an individual soybean breeder has increased substantially with time. During the past 40 yr, the development of sophisticated multi-row plot planters and self-propelled, data-collection-computerized plot combines have allowed breeders to dramatically increase the number of plots that they can manage. This increase has been coupled with improved statistical analytic techniques for partitioning out more of the systematic and random errors in the ever-larger experimental data sets. These advances have enhanced the opportunity for individual breeders to achieve greater input and output efficiencies as they scale up the size of their cultivar development programs. Eathington et al. (2007) reported that Monsanto's use of specialized breeders and investments in equipment had resulted in an 80% increase in yield plot capacity in the 4 to 5 yr before the date of this publication.

Breeders also have increased their annual rates of gain by reducing the number of years needed to develop new cultivars through the use of off-season winter nurseries and the release of cultivars at an earlier generation of inbreeding. Breeders in North America grow two generations of single-seed descent in tropical winter nurseries, and most grow plant rows, seed increases, and yield tests in South America. Large proprietary seed companies have developed their own off-season nurseries that have been seamlessly incorporated into their breeding systems.

History of Genetic Improvement in Brazil

Although the first references of soybean cultivation in Brazil date from as early as 1882, the crop became economically important in the country during the 1970s.

Table 12–1 documents the main cultivars grown in each 5-yr period from 1960 through 2012. This time frame can be divided into three important phases:

1. Introduction of germplasm from the southern United States for cultivation in regions in Brazil south of 23° S latitude.
2. The migration of soybean cropping towards the tropics.
3. The introduction of transgenic soybean cultivars, indeterminate growth habit germplasm, and the no-till system.

In the 1960s and 1970s, soybean cropping depended on the introduction of southern U.S. cultivars with determinate growth habit, such as Hardee, Bossier, and Davis. Other cultivars, such as Santa Rosa and Parana, were experimental lines developed in the United States but were given Brazilian names for marketing purposes. It was not until the 1980s that locally bred cultivars developed from U.S. germplasm became important to supply the needs of soybean farmers. At the time, however, production was concentrated in latitudes south of the Tropic of Capricorn.

Proper understanding of the soybean reaction to short day conditions of latitudes less than 23°S was the main challenge to overcome in the 1970s and 80s to allow soybean cultivation to move northward towards the tropical areas of Brazil.

Table 12–1. Mean soybean yields in Brazil for each of a succession of 5-yr periods in the 1961to 2010 time frame, plus a 2-yr period after 2010. Also listed are key cultivars grown during those periods. Data adapted from Arantes and Souza (1993). The 5-yr periods are denoted by the vertical grid lines in Fig. 12–5.

5-yr period	Phase	Period mean yield	Cumulative yield increase	Key cultivars grown during the 5-yr period
		kg ha^{-1}	%	
1961–1965	I	441	(base period)	Abura, Aliança, Arksoy, Mammoth Yellow, Mojana, Pelican
1966–1970	I	999	126.5	IAC-2, Industrial, Mineira, Santa Rosa, Viçoja
1971–1975	I	1218	21.9	Hardee, UFV-1
1976–1980	I	1452	19.2	Bossier, Davis, IAC-5, Parana, UFV-2
1981–1985	II	1739	19.8	Nova Bragg, BR-Savanna, FTCristalina, Doko, FT-11, IAC-8, UFV-5
1986–1990	II	1867	7.4	FT-Estrela, MGBR-Garimpo, OCEPAR-Primavera, UFV-10
1991–1995	II	2144	14.8	BR-16, CD 201, CD 202, BMGOPA 301, BMGOPA 313
1996–2000	II	2328	8.6	BR-16, Embrapa 48, BRS 232, CD 202, CD 215, MGBR-46 (Conquista)
2001–2005	III	2544	9.3	Embrapa 48, CD 202, BRMT-Pintado, M-Soy 8866, BRS ValiosaRR
2006–2010	III	2723	7.0	Vmax, M-Soy 8866, M-9144RR, BMX Apollo, BMX Potencia, BRS Valiosa, Anta 8.2
2010–2011	III	2883	5.9	NA 5909RG, BMX ApolloRR, BMX PotenciaRR, VMax, NA 7337RR, TMG 132, M-9144RR

Several individual genes or combinations of a few genes that expressed late flowering under short day conditions (i.e., long juvenile trait) were identified in many of the available cultivars from southern Brazil, or in some plant introductions originating from east Asian countries (Kiihl and Garcia, 1989). Breeding this material allowed soybean to move northward, initially to 16° S, and later to latitudes of 10°S or less. In the 1990s, genetic studies showed that soybean response to photoperiod was controlled by several genes and that a continuous response to photoperiod could be obtained in the newly developed cultivars (Toledo et al., 1993). This allowed the development of cultivars with a range of sensitivity to day length, such that early, medium, and late maturity cultivars were available for good adaptation to diverse Brazilian growing areas, thereby resulting in consequent increases in yield.

With the photoperiod response issue solved, soybean breeding in Brazil next focused on increasing yields. However, in the early 1990s disease and pest problems, caused by the spread of soybean cyst nematode (SCN) (*Heterodera glycines* Ichinohe) stem canker (*Diaporthe phaseolorum* var. *meridionalis* F.A. Fern), and highly pathogenic races of *Cercospora sojina* Hara, the pathogen that causes the disease frog-eye leaf spot, resulted in a large increase in disease-induced yield losses. The challenge posed by stem canker was swiftly defeated by breeders who replaced susceptible with resistant cultivars. However, the SCN problem spread to most production areas in Brazil in about 5 yr, and several pathogenic races emerged (Dias et al., 2005). Resistance to SCN is now necessary in many areas in central Brazil to maintain a highly productive soybean crop. Despite these challenges, yield gains were maintained throughout this period during the movement towards the low latitude areas (Table 12–1; Fig. 12–5). These yield gains were likely supported by the introduction of new genotypes into breeding programs to provide the resistance genes necessary for defeating the stem canker and SCN problems. This hypothesis was corroborated by a study on the genetic basis of the Brazilian soybean genotypes, which demonstrated that all available cultivars up to 1988 were derived from just a few parents, resulting in a narrow

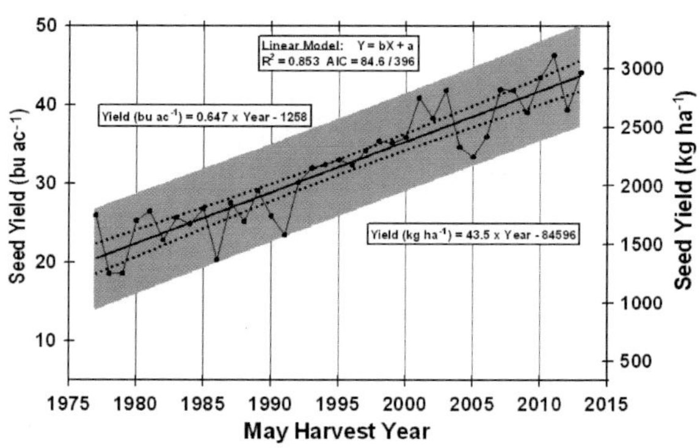

Fig. 12–5. Brazil soybean yield statistics from 1977 to 2013 plotted as 37 solid symbols with connecting lines. A linear model provided a more probable AIC fit to the data than did the logistic and segmented regression models (so the latter two are not shown here). The 95% confidence interval around the estimated regression trend line is indicated with dotted lines, whereas the gray highlighted zone reflects the 95% prediction interval.

genetic basis (Hiromoto and Vello, 1986; Miranda et al., 2007). Hiromoto and Vello (1986) analyzed the ancestry of 69 out of 74 cultivars (the remaining 5 cultivars did not have a known ancestry) that had been used for cultivation in Brazil in the 1983–1984 season. These 69 cultivars had 11 ancestors that together accounted for approximately 89% of their pedigrees.

Cultivars such as Davis, Parana, and Santa Rosa predominated as sources of frog-eye leaf spot resistance; Tracy-M, Sharkey, IAC-12 and FT-Estrela as sources of resistance to stem-canker; and Braxton, Sharkey, Peking, and a few plant introductions as sources of root knot [*Meloidogyne javanica* (Treub) Chitwood and *M. incognita* (Kofoid and White) Chitwood] and SCN resistance for Brazilian breeding programs. As the crop moved northward, these sources provided needed resistance and most likely also provided additional genetic diversity to the common Brazilian gene pool, thereby allowing genetic progress towards higher yields. Until the year 2000, the soybean crop in Brazil relied mostly on soybean cultivars with a determinate growth habit. However, Brazilian breeders soon began to search for solutions to newly encountered problems within the determinate habit germplasm pool.

The third phase of soybean breeding in Brazil began in 2000 (Table 12–1), and during this phase, the Brazilian cropping system evolved, and cultivars adapted to this system were developed. Historically, the best cultivars adapted to either southern or central regions of Brazil were those that matured between 120 and 125 d. These cultivars usually had high yield and were least affected by genotype × environment interaction, which made them preferred by farmers (de Toledo et al., 2006). On the other hand, the search for earlier cultivars that allowed double cropping of soybean with either corn or cotton (*Gossypium hirsutum* L.) in a single season was reinforced by the introduction of Asian soybean rust (*Phakospora pachyrhizi* Sydow) to Brazil. These earlier cultivars either had a better chance of escaping the disease or required fewer fungicide applications. However, high yielding cultivars earlier than 115 to 120 d were difficult to obtain from determinate type soybeans because they needed about 45 d from emergence to flowering to achieve good growth. This meant that early maturity could only be achieved in the determinate types by shortening the reproductive period, and this resulted in a yield reduction. The introduction of indeterminate growth habit soybean was a logical solution in the search for earliness since they continue to grow after flowering.

The introduction of indeterminate cultivars started in the southernmost growing states in Brazil with cultivars originating from Argentina and North America. However, indeterminate cultivars rapidly moved northward towards central Brazil. The impact of these early indeterminate types in the Brazilian soybean cropping system was greater than expected. The short-cycle indeterminate soybean cultivars fit well into a no-till system that significantly improves soil fertility by preventing soil erosion and increasing organic matter. These indeterminate cultivars have shorter internodes that lessen plant lodging and have smaller and erect leaves that facilitate canopy penetration with multiple sprays that were needed to mitigate soybean rust. Above all, however, these cultivars introduced a new set of yield genes that were previously untapped in Brazil, and these genes have resulted in greater yields in most regions of southern and central Brazil (Godoi et al., 2013). It must be mentioned that cultivation of early types was also made possible by the introduction of the Roundup Ready (Monsanto) soybean, which allowed weed control without the soybean leaf damage caused by other post-emergence herbicides. The early-season determinate genotypes also had less time to recover from injury induced by other herbicides, resulting in more yield loss compared to long-season determinate types with a conventional herbicide program. Seed of these early season indeterminate cultivars are sold today at a premium to Brazilian farmers since the high yields coupled with a second crop have increased farmers returns.

Breeding efforts in Brazil were performed primarily by the public sector until the mid to late 1980s. Before this time, efforts from a few private companies were short-lived and had little impact in the Brazilian seed market. State and national research institutions and universities that had invested in postgraduate education of professionals, many of whom were sent to institutions outside Brazil, soon obtained returns from their investments. The researchers returned to the country with state-of-the-art agronomic, genetics, and breeding knowledge and were able to advance successful soybean breeding programs. Campinas Agronomic Institute (IAC), Agronomic Institute of Paraná (IAPAR), Federal University of Viçosa (UFV), Minas Gerais State Agriculture Corporation (EPAMIG), Goias State Research Corporation (EMGOPA), and Brazilian Agriculture Research Corporation (EMBRAPA) are the best-known examples of those institutions, and all developed soybean breeding programs. These programs were responsible for the movement of soybean towards the tropics and, despite limited coordination by EMBRAPA, remained basically independent, at least until the early 1990s.

From 1993 onward, EMBRAPA, some state-owned research institutions, and several Brazilian private companies dedicated to soybean seed production developed innovative public-private associations to breed soybean cultivars. The seed producers organized into five or six private foundations in the most important soybean production states and helped to finance the EMBRAPA and state soybean breeding programs. These associations coupled the creativity of the well-trained public scientists with the market capacity and financial strength of the private companies to produce and sell soybean cultivars. These public–private arrangements were successful and reached an approximate seed market share of 75% throughout Brazil in the late 1990s (de Toledo et al., 2000).

Starting in the early 1990s, a few local private breeding programs and farmer associations (called cooperatives) that had invested in soybean breeding programs began to have an impact in the seed market. The contribution of soybean cultivars released by the Research Department of the Organization of Cooperatives of Parana State (OCEPAR), succeeded in 1995 by Central Cooperative for Agricultural Research (COODETEC) in the South, and by Terasawa Sementes and associates in Southern and Central Brazil, became important in this decade.

Up to the mid 1990s, the large multinational players did not show interest in the Brazilian soybean cultivar market mainly due to the absence of regulatory legislation. However, the intellectual property landscape changed with the approval of the Variety Protection Law on 24 Apr. 1997 by the Brazilian Congress along with other industrial patent legislation. This resulted in strong growth in investment by multinationals, which brought the latest biotechnology advances and the establishment of breeding programs to Brazil. These investments, coupled with the wide acceptance of the Roundup Ready soybean and the maturing of the no-till system, helped the Brazilian soybean crop to enter its third phase as previously mentioned. Presently, soybean cultivars released by the proprietary companies are planted on approximately 90% of the growing areas (The Kleffmann Group, unpublished data, 2012–2013).

Rate of Genetic Improvement in the United States

A number of researchers have estimated the rate of genetic yield gain for soybean in North America by simultaneously evaluating in one or more production environments cultivars released during different decades. Specht and Williams (1984) evaluated 240 MG 00 to IV cultivars released from 1902 to 1977 in a Nebraska irrigated production system and observed an average genetic-based yield increase of 18.8 kg ha^{-1} yr^{-1}. Boerma (1979) evaluated yield gains in MG VI to VIII soybean cultivars in Georgia and reported genetic improvement of 13.7 kg ha^{-1} yr^{-1} from 1942 to 1973. These genetic yield gain estimates were similar to the 15.9 kg ha^{-1} yr^{-1} and 18.7 kg ha^{-1} yr^{-1} rates later reported by Salado-Navarro et al. (1993), who generated these estimates in four Florida environments. Voldeng et al. (1997) estimated an 11 kg ha^{-1} yr^{-1} rate of genetic yield improvement in MG 000, 00, and 0 cultivars released from 1934 to 1992. In tests conducted in Tennessee, Ustun et al. (2001) estimated the rate of genetic gain from the 1940s to the late 1980s to be 14 kg ha^{-1} yr^{-1}. De Bruin and Pedersen (2008) reported genetic yield gains of 25.4 kg ha^{-1} yr^{-1} for a few MG II and III cultivars released during a period spanning 1938 to 2004. Wilcox (2001) evaluated progress in public soybean breeding programs by evaluating performance data from cooperative tests across the northern United States and southern Canada. He reported the rates of yield improvements as follows: MG 00, 21.6 kg ha^{-1} yr^{-1}; MG 0, 25.8 kg ha^{-1} yr^{-1}; MG I, 30.4 kg ha^{-1} yr^{-1}; MG II, 29.3 kg ha^{-1} yr^{-1}; MG (III), 30.6 kg ha^{-1} yr^{-1}; and MGIV, 29.5 kg ha^{-1} yr^{-1}.

Genetic yield improvement over time was recently estimated in a collaborative genetic gain experiment that included MG II, III, and IV cultivars. Detailed information about these yield tests can be found in Rincker et al. (2014). In brief, 60 MG II, 59 MG III, and 49 MG IV cultivars, with release years spanning the years 1923 to 2008, were evaluated for yield, agronomic traits, and seed composition in replicated trials conducted in 15 states of the north-central and eastern United States, plus one Canadian province during 2010 and 2011. These trials were conducted at sites in or near the zones of matching cultivar MG adaption. The U.S. soybean MG II, III, and IV zone boundaries recently described by Zhang et al. (2007) are depicted in Fig. 12–2B. These authors used 1998 through 2003 maturity data from U.S. experimental research and extension soybean trials to update a prior maturity zone map published by Scott and Aldrich (1970).

In the 2-yr project, there were a total of 27 combined test site–year environments for the MG II and for the MG IV, but 26 for the MG III cultivars. The data from the tests were analyzed with linear and nonlinear statistical techniques to detect the degree of change in yield (or other traits) that occurred over time as new cultivars replaced predecessors. This objective was accomplished by discerning which regression model provided the best AIC-based value for cultivar yield vs. cultivar year of release.

Averaged over the trials in which a given MG set of genotypes was evaluated, it was apparent that seed yield improved as old cultivars were replaced with newer ones (Fig. 12–6). The linear model estimates of annualized genetic yield improvement over time (i.e., cultivar release year) were 23.1 kg ha^{-1} yr^{-1} for MG II (Fig. 12–6A), 22.8 kg ha^{-1} yr^{-1} for MG III (Fig. 12–6B), and 19.5 kg ha^{-1} yr^{-1} for MG IV

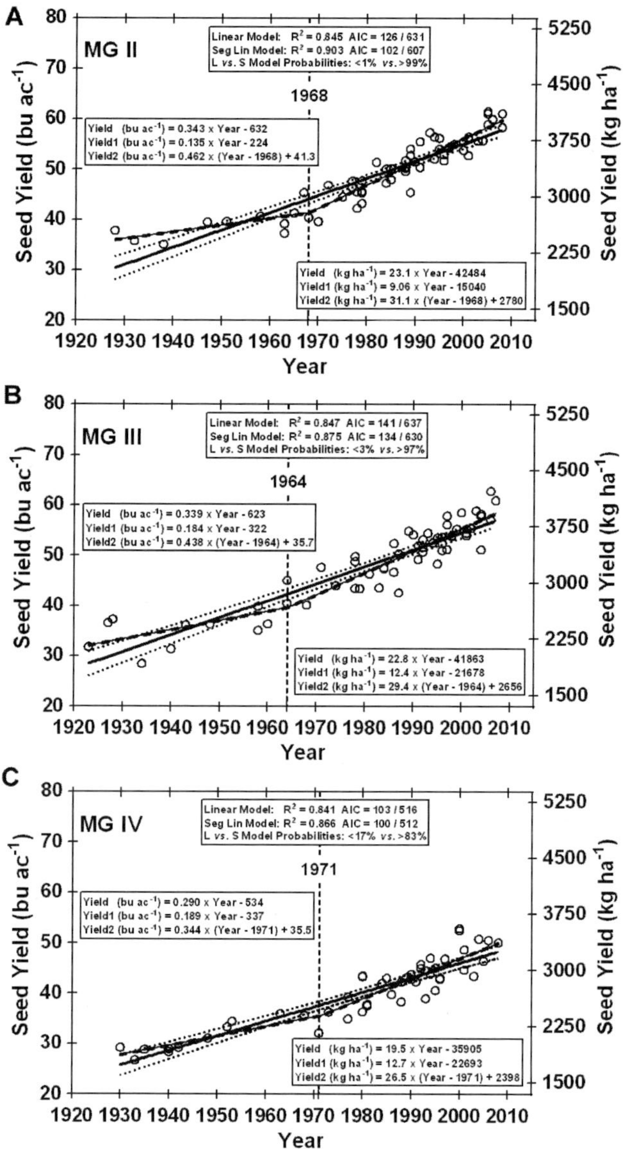

Fig. 12–6. Estimated annual rates of soybean genetic yield improvement in (A) MG II, (B) MG III, and (C) MG IV, based on multi-site performance trials conducted in 2010–2011 with a respective set of 60, 59, and 49 public and proprietary cultivars. The annual rates were obtained by regressing cultivar yield on year of cultivar release (Rincker et al., 2014). The data were first fit to a linear model (dotted lines are the standard error values) and then to a segmental linear model. The metrics for the two models are shown in the text boxes of each panel. The vertical dashed line in each panel denotes the breakpoint year derived from a reiterative convergent fit of the nonlinear two-regression model to the data. For an objective comparison of the two models, the Akaike information content (AIC) value was computed for each model (in bu ac^{-1} units and kg ha^{-1} units) to determine which of the two models was more probable.

(Fig. 12–6C). These estimates are, with but two exceptions, greater than the rates reported in the publications cited at the start of this chapter section. Note that the estimated linear rate of 23.3 kg ha^{-1} yr^{-1} for on-farm U.S. yield improvement from 1924 to 2012 (Fig. 12–3A) was not substantially different from the linear rates estimated from the MG II and III yield tests (Fig. 12–6A,B). A null hypothesis that the three MG linear regressions could be globally fit to one common coefficient and one common intercept was rejected based on an F test ($P < 0.001$). Additional F tests (not presented here) showed that the MG II and III regression coefficients and y intercepts were not significantly different; however, the MG IV parameter values were significantly lower. Hypothetically, the lower linear yield trend line of the MG IV set may reflect either an intrinsically lower basal genetic yield potential achieved to date in MG IV per se, or a lower environmental yield potential at the locations where MG IV cultivar releases were tested, or both. To evaluate these two hypotheses, Rincker et al. (2014) restricted the data analysis to just the 15 (of the 27) environments at which performance trials were conducted for both MG III and MG IV. Although the reduction in site number lowered the statistical power, this reanalysis showed that the coefficients and intercepts of the two MGs were not significantly different in these 15 environments, providing evidence that the MG IV set was not, on average, lower yielding than the MG III set. This result also suggested that environmental yield potential influences the magnitude of the genetic improvement regression coefficients and intercepts. In fact, mean yields were less than 2000 kg ha^{-1} in five of the 12 environments involving only MG IV trials compared to one such environment for MG III. Means of all MG II environments exceeded 2000 kg ha^{-1}.

As was the case with U.S. on-farm yields (Fig. 12–3C), an AIC-based comparison of the simple vs. the segmental linear models indicated that the latter was statistically more probable than the former for genetic yield gain in the three MGs (Fig. 12–6). Using the estimated breakpoint years of 1968 for MG II, 1964 for MG III, and 1971 for MG IV, it is apparent that genetic yield improvement in each MG was enhanced between 2.1 and 3.4 times after those breakpoint years (Fig. 12–6). However, one should probably view the low and flatter pre-breakpoint year genetic yield gain rates with some caution, simply because of the paucity of cultivar releases during the early stages of U.S. soybean breeding. The post-breakpoint year greater genetic yield improvement is most likely the result of increases in breeding activity, especially by proprietary seed companies, that was stimulated by the passage of the 1970 PVP act.

Note that the respective MG II, III, and IV post-breakpoint (slope2) yield gain rates of 31.1, 29.4, and 26.5 kg ha^{-1} yr^{-1} (Fig. 12–6) also are concordant with the slope2 rate of 29.4 kg ha^{-1} yr^{-1} of on-farm U.S. yield improvement rate after 1983 (Fig. 12–3C). This concordance would suggest that genetic improvement accounts for nearly all of the U.S. on-farm soybean yield improvement. However, it must be kept in mind that the genetic yield gain estimates reported here apply only to MG II, III, and IV soybean cultivars, although the 20 states growing those three MGs (as shown in Fig. 12–2B) actually accounted for 73% of the U.S. harvested area during the 1979 to 2011 time frame (Fig. 12–2A,B), with the MG II and III states accounting for 50% (USDA-NASS, http://www.nass.usda.gov/, accessed 3 Feb. 2014). Still, a more appropriate comparison would use the on-farm yield improvement rates in just the states or U.S. regions where cultivars of those MGs are adapted and grown. In this regard, the U.S. on-farm yield data for individual

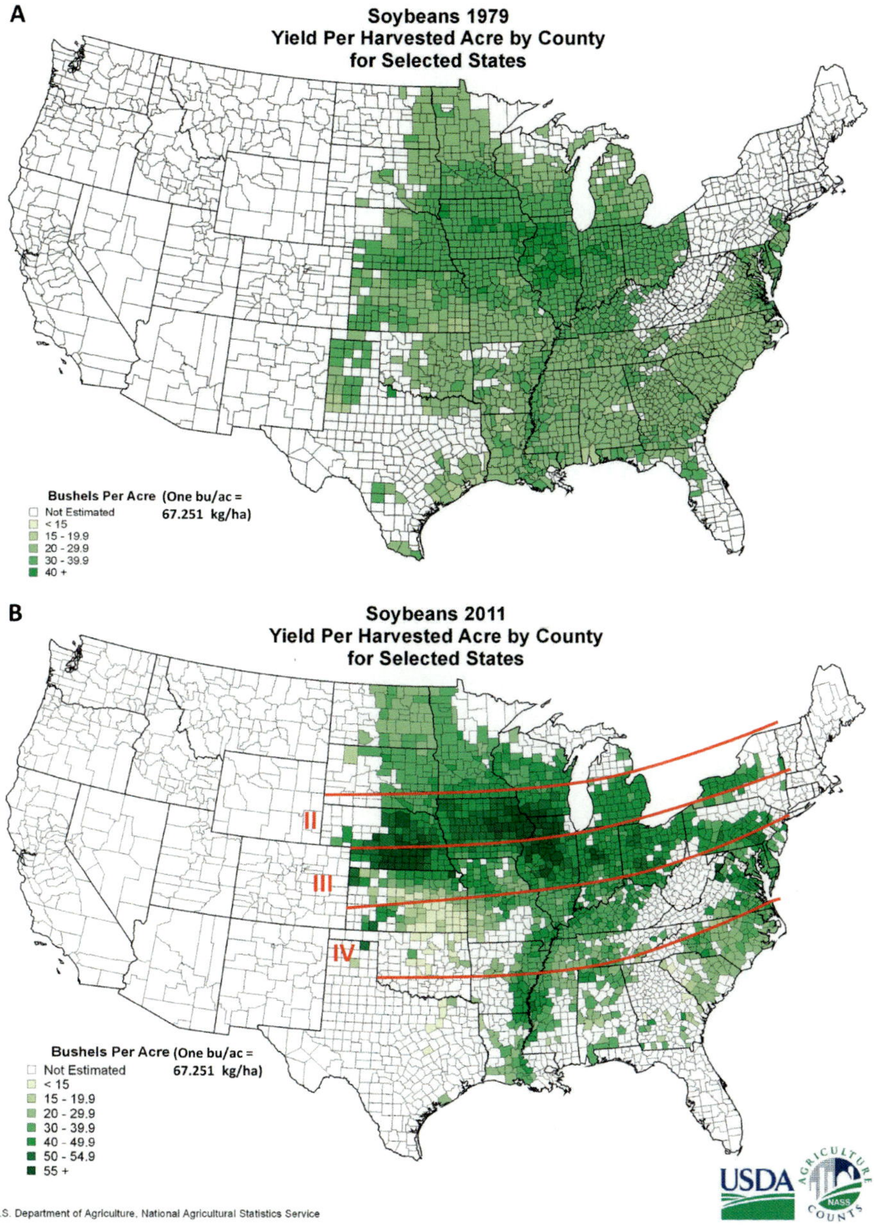

Fig. 12–7. U.S. soybean yields, on a county basis, for the production years of (A) 1979 and (B) 2011. Most U.S. soybean production is rainfed, except for irrigated production in central and western regions of Nebraska and Kansas, and along the Mississippi delta region in Arkansas. The two maps were provided by Travis Thorson of NASS. Zones of MG II, III, and IV soybean cultivar adaptation are shown in Panel B, with boundary lines extrapolated from Fig. 4 of Zhang et al. (2007).

U.S. counties for the production years of 1979 and 2011 (Fig. 12–7A,B) indicate that soybean yields have moved to much higher levels in the north-central states of Nebraska, Iowa, Illinois, Indiana, and Ohio (mostly MG II and III zones) than the yield levels still prevailing in other states.

On-farm yield improvement linear regression line trends in the 20 individual states for which MG II, III, and IV cultivars are adapted and grown (as shown in the Fig. 12–7B MG zone map) are presented in Table 12–2. In addition, a composite regression trend line was computed for those states in which mostly MG II and III cultivars are grown vs. a composite regression trend line for those states in which mostly MG IV cultivars are grown (Fig. 12–8). The separate aggregation into MG II–III vs. IV is not necessarily perfect, because some states span more than one MG zone, notably Illinois, which spans all three MG zones, and Missouri and Kansas, which span the MG III and IV zones (Fig. 12–7B). So, for these three states, state soybean county production data was parsed into a MG III (or earlier in the case of IL) group of counties and into a MG IV group of counties to enable the calculation of separate estimates for the rate of on-farm yield improvement in the MG III and IV sections of these three states (Table 12–2).

The estimated linear genetic rates of yield improvement in each of the three MGs (Fig. 12–6A,B,C), which were estimated by performance trials sited in MG II, III, and IV locations of those north-central states (Rincker et al., 2014), are just 8 to 11% points short of matching the 20-state composite rate of on-farm yield improvement of 24.7 kg ha^{-1} yr^{-1} (Fig. 12–8). However, more specificity is provided by noting that the MG II and III linear genetic yield gain rates of 23.1 and 22.8 kg ha^{-1} yr^{-1} (Fig. 12–6A,B), when averaged (i.e., 23.0), represent about 84% of the linear on-farm yield gain rate of 27.3 kg ha^{-1} yr^{-1} computed for the collective group of MG II and III states (Fig. 12–8). Thus, by that measure, even if producers are adopting the best genetic technology (i.e., new cultivars) as soon as it becomes available, nongenetic factors are likely contributing the remaining 16% of observed on-farm yield improvement in the MG II and III states. The linear MG IV genetic rate of 19.5 kg ha^{-1} yr^{-1} is about 91% of the lower linear on-farm yield gain rate of 21.4 kg ha^{-1} yr^{-1} computed for the MG IV states (Fig. 12–8).

What about genetic rates of yield gain in the context of on-farm rates of yield gain in individual states? The estimated linear rates of genetic yield gain in the MG II and III cultivar groups (e.g., 23.1 and 22.8 kg ha^{-1} yr^{-1}, respectively) account for just 70 to 82% of the linear rates of on-farm yield improvement observed for individual states of Nebraska (32.5 kg ha^{-1} yr^{-1}), Wisconsin (29.3 kg ha^{-1} yr^{-1}), Iowa (29.3 kg ha^{-1} yr^{-1}), Indiana (28.5 kg ha^{-1} yr^{-1}), and the MG II and III zones of Illinois (IL23) (27.8 kg ha^{-1} yr^{-1}) (Table 12–2). The percentage weakens further when one considers the estimated linear rate of on-farm yield improvement (42.1 kg ha^{-1} yr^{-1}) in just the Nebraska irrigated soybean production system (Fig. 12–9) in MG II and III, which accounts for about half of that state's total 2.02 million hectares. One would assume that the increased genetic yield potential that has been bred into modern cultivars could be readily captured and realized in production systems where seasonal water stress is no longer a yield-limiting factor. A comparison of the average MG II–III genetic improvement rate of 23.0 kg ha^{-1} yr^{-1} with the rate of Nebraska irrigated on-farm yield improvement indicates that the former accounts for just 55% of the latter (i.e., 23.0/42.1). However, by using the estimated mean genetic improvement rate of 26.6 kg ha^{-1} yr^{-1} derived for the MG II and III historic cultivar sets that were grown at Nebraska irrigated sites in the

Table 12–2. Equations and R^2 values derived from the regression of on-farm soybean yield on production years (1924–2012) for the north-central states in which cultivars of maturity group (MG) II (or earlier), III, and/or IV are grown (see MG adaptation zone map in Fig. 12–7B). In some cases, fewer than 89 yr of data are available (source: USDA-NASS). Regression was performed for each state, the USA, for only the MG II, III, and IV states (MG234), for only the MG II and III states (MG23), and for only the MG IV states (MG4). In addition, regression was performed for just the Illinois counties in MG II and III zones (IL23) and just those in the MG IV zone (IL4), and for counties in Kansas and Missouri in their MG III zone (KS3 and MO3) and in their MG IV zone (KS4 and MO4). All regression coefficients were significantly different from zero ($P = 0.05$). Also shown, for comparative purposes, are the 2013 yield predictions of the equations.

State(s) or USA	Data points	Regression equations		R^2	Predicted 2013 yields	
	yr	bu ac^{-1} yr^{-1} X – bu ac^{-1}	kg ha^{-1} yr^{-1} X – kg ha^{-1}		bu ac^{-1}	kg ha^{-1}
AR	89	$Y = 0.311X - 591$	$Y = 20.9X - 39{,}800$	0.845	35.4	2380
DE	89	$Y = 0.292X - 553$	$Y = 19.7X - 37{,}200$	0.726	35.7	2400
IA	89	$Y = 0.436X - 827$	$Y = 29.3X - 55{,}600$	0.917	50.5	3390
IL	89	$Y = 0.386X - 730$	$Y = 25.0X - 49{,}100$	0.917	48.1	3240
IL23	86	$Y = 0.414X - 782$	$Y = 27.8X - 52{,}600$	0.918	51.4	3460
IL4	86	$Y = 0.365X - 694$	$Y = 24.5X - 46{,}700$	0.865	40.3	2710
IN	89	$Y = 0.424X - 806$	$Y = 28.5X - 54{,}200$	0.934	48.8	3280
KS	89	$Y = 0.323X - 616$	$Y = 21.7X - 41{,}400$	0.737	34.4	2310
KS3	73	$Y = 0.393X - 751$	$Y = 26.4X - 50{,}500$	0.652	40.3	2710
KS4	73	$Y = 0.230X - 435$	$Y = 15.5X - 29{,}300$	0.409	28.3	1900
KY	89	$Y = 0.370X - 704$	$Y = 24.9X - 47{,}400$	0.868	41.0	2760
MD	89	$Y = 0.321X - 609$	$Y = 21.6X - 40{,}900$	0.806	38.4	2580
MI	89	$Y = 0.358X - 678$	$Y = 24.0X - 45{,}600$	0.898	41.4	2790
MO	89	$Y = 0.366X - 697$	$Y = 24.6X - 46{,}900$	0.872	40.1	2690
MO3	69	$Y = 0.326X - 617$	$Y = 21.9X - 41{,}500$	0.677	39.8	2680
MO4	69	$Y = 0.326X - 617$	$Y = 21.9X - 41{,}500$	0.758	38.1	2560
NE	75	$Y = 0.483X - 924$	$Y = 32.5X - 62{,}100$	0.840	48.8	3280
NJ	75	$Y = 0.277X - 522$	$Y = 18.6X - 35{,}100$	0.652	36.0	2420
NY	62	$Y = 0.410X - 784$	$Y = 27.6X - 52{,}700$	0.886	41.7	2810
OH	89	$Y = 0.374X - 706$	$Y = 25.1X - 47{,}500$	0.898	45.7	3070
OK	89	$Y = 0.226X - 428$	$Y = 15.2X - 28{,}800$	0.631	26.6	1790
PA	79	$Y = 0.404X - 770$	$Y = 27.2X - 51{,}700$	0.865	43.7	2940
TN	89	$Y = 0.335X - 638$	$Y = 22.5X - 42{,}900$	0.809	36.7	2470
VA	89	$Y = 0.266X - 503$	$Y = 17.9X - 33{,}800$	0.767	33.8	2270
WI	89	$Y = 0.436X - 834$	$Y = 29.3X - 56{,}100$	0.870	44.4	2980
WV	42	$Y = 0.389X - 741$	$Y = 26.2X - 49{,}800$	0.911	42.5	2860
USA	89	$Y = 0.346X - 655$	$Y = 23.3X - 44{,}000$	0.945	42.1	2830
MG234	89	$Y = 0.367X - 695$	$Y = 24.7X - 46{,}700$	0.939	40.8	2740
MG23	89	$Y = 0.406X - 769$	$Y = 27.3X - 51{,}700$	0.944	45.1	3040
MG4	89	$Y = 0.319X - 605$	$Y = 21.4X - 40{,}700$	0.895	35.9	2420

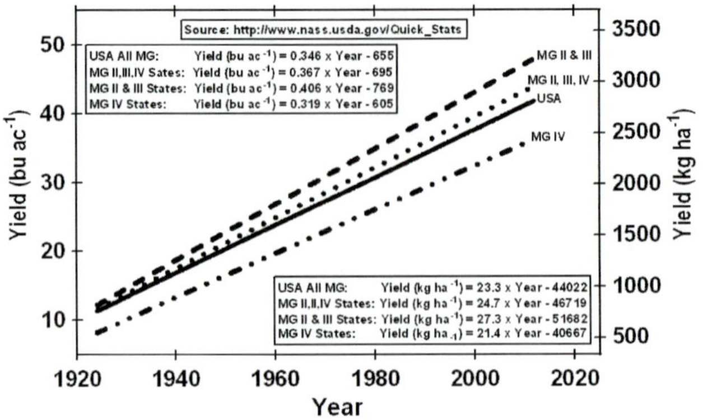

Fig. 12–8. Soybean linear yield trends from 1924 to 2012 for (1) the entire USA (solid line), (2) a composite of all 20 individual states in which MG II (or earlier), III, and/or IV (or later) cultivars are grown (dotted line), (3) a composite of just those states in which cultivars of just MG II (or earlier) plus MG III are grown (dashed line), and (4) a composite of just those states in which MG IV (or later are grown (dotted-dashed line). County data for the states of Illinois, Missouri, and Kansas were parsed into a northern MG III (and earlier) region and into a southern MG IV (and later) region, so that these regions could be appropriately assembled into the foregoing two composites. See Table 12–2 for individual state regressions. Metrics for the linear regression equations are presented in the text boxes in the graph.

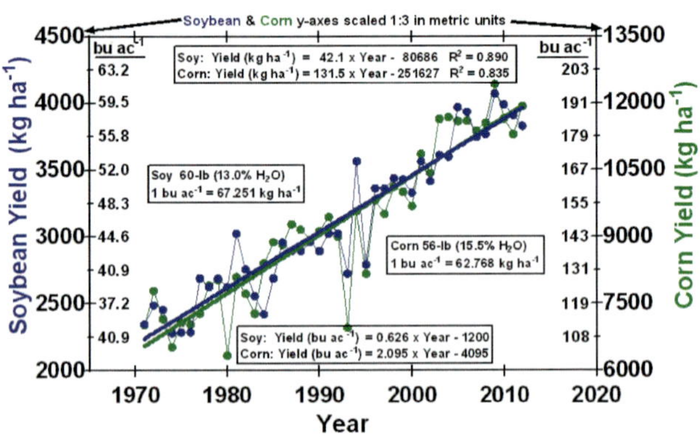

Fig. 12–9. A graph depicting the rates of on-farm yield improvement for soybean (left axis) and for corn (right axis) in Nebraska's irrigated production systems from 1971 to 2012. Trend line equations derived from the linear regression of crop yield on production year are presented in the top (metric units) and bottom (English units) text boxes. Because the 42 annual soybean/corn yield ratio data points have averaged approximately 1:3 (in kg ha^{-1} units), the left and right metric unit *y* axes in the graph were scaled 1:3 to make more visually evident the near-constancy over time of this yield ratio overtime (i.e., the top left *y* axis number of 4500 is 1/3 of the top right *y* axis number of 13500, as are the bottom numbers of 2000 and 6000). The constancy of this corn/soybean yield ratio was first reported by Specht et al. (1999) and then documented more extensively by Egli (2008a). Note that, because soybean and corn bushels have different volume weights (60-lb vs. 56-lb per bushel, respectively), the 42 yield ratio data points, when expressed in bushels per acre yield values (shown inside the left and right *y* axes of the graph) and then averaged, translate into a irrigated soybean/corn yield ratio of 1:3.25 (e.g., 52/167). Source: http://www.nass.usda.gov/Statistics_by_State/Nebraska/index.asp.

2010–2011 study reported by Rincker et al. (2014), that percentage can be adjusted upward to 63% (i.e., 26.6/42.1). All of the foregoing comparisons clearly indicate that genetic improvement cannot be the sole foundation for all on-farm yield improvement. Still, one could reasonably infer that genetic improvement in cultivar genetic yield potential has likely accounted for about two-thirds of observed on-farm yield improvement, particularly when both are measured in an environmental context of optimal opportunity for genetic yield potential expression.

It is of interest to soybean researchers to compare rates of on-farm soybean vs. corn yield improvement, particularly given the perspective of many that the improvement in corn is "lapping" that of soybean (Specht, 2013). This comparison is best accomplished in a scenario where corn and soybean are annually rotated in an irrigated production system. Why? Because (i) the two crops are produced (even if in alternate years) on the same land, (ii) any yield differential between the two crops arising from seasonal rainfall deficit differentials between July (a corn critical period of silking and tasseling) and August (a soybean critical stage of seed-filling) is mitigated by irrigation; and finally (iii) the greater environmental yield potential in an irrigated system (minimized water stress) likely permits a near-optimal expression of the intrinsic genetic yield potential of both the soybean cultivars and corn hybrids.

In Fig. 12–9, Nebraska irrigated soybean and corn on-farm yield trends are presented for the 1971 to 2012 time frame. When making a corn–soybean yield comparison, one must be cognizant of the intrinsic physiological differences between a C-4 cereal and a C-3 legume that inherently makes corn about three times more productive than soybean in an absolute sense. The predominant physiological differences between corn and soybean are, respectively: (i) C-4 vs. C-3 photosynthesis, (ii) the energy input consequences inherent in the production of a mostly carbohydrate maize seed vs. a mostly protein and lipid soybean seed, and (iii) the energy consequences of generating amino N obtained mostly from the uptake of nitrate from the soil vs. generating amino N obtained from both soil nitrate uptake and N_2 fixation (Specht et al., 1999; Salvagiotti et al., 2008; Sinclair and Rufty, 2012). In Nebraska's irrigated system, this physiological productivity difference routinely translates into a 3:1 ratio of corn/soybean yield in kg ha^{-1} units (3.25/1.00 in bu ac^{-1} units; the difference between these two ratios is the result of the standard bushel of soybean weighing 60 pounds, whereas a bushel of corn weighs 56 pounds). This ratio has long been a commonly used yardstick for assessing producer management skills when using irrigated corn and soybean rotations in Nebraska (Specht et al., 1999). With this in mind, we intentionally scaled the tick-marked metric scale of the right *y* axis for corn yield in the Fig. 12–9 graph to be a 3.0× match to the tick-marked metric scale of left *y* axis for soybean yield. When displayed in this manner, the near-parallel soybean and corn trend lines reveal that the relative rates of on-farm yield improvement are nearly identical. That is, each 3 kg ha^{-1} (i.e., 3.25 bu ac^{-1}) increment in irrigated on-farm maize yield improvement has, with time, been invariably matched with at least a 1 kg ha^{-1} (i.e., 1.00 bu ac^{-1}) increment of irrigated on-farm soybean yield improvement. This might seem surprising, but it is not necessarily specific to Nebraska irrigated systems, given that corn/soybean ratios of 3.0, 2.9, and 2.9 (in metric units) were also documented as constant in Illinois, Indiana, and Iowa by Egli (2008a) for the time frames of data he examined (i.e., from 1973, 1966, 1972, respectively, up to 2005, the last year of his analysis). In any event, it is quite evident in the Fig. 12–9 graph that corn yield improvement is not meaningfully "lapping" soybean yield improvement from a crop physiological point of view.

One unexpected discovery in the results reported by Rincker et al. (2014) was the detection of a strong relationship of cultivar year of release with the calendar date of cultivar maturity, as measured by the reproductive stage R8 (95% pod maturity; Fehr et al., 1971). Breeders compare the R8 date of experimental lines to standard maturity checks to ensure that lines are placed into proper MG sets before widespread agronomic testing. This testing of experimental lines within a relatively narrow maturity is a breeder response to the tendency of late maturing lines to outyield early lines because the late lines have a longer growing season. Selection for later maturity to achieve higher yield is technically not a route one uses to attain improvement in genetic yield potential. Maturity groups are standardized in the public sector by including maturity checks to the Uniform Soybean Tests organized by the USDA. Each MG has a maturity range of ~8 d (Crochet, 2012) based on the relative maturity of the checks.

The data collected from the evaluation of MG II, III, and IV cultivars from different eras indicated that cultivar maturity date had "advanced" by about 1 d every decade, such that cultivars released after 1999 were, as a group, maturing 5 to 6 d later than all cultivars (as a group) released before 1970. As a means to evaluate the extent to which later maturity was influencing the estimation of what might be considered basal genetic yield improvement, a covariate analysis of variance was conducted to adjust cultivar yields for maturity date differences. This adjustment did result in somewhat lower annual linear rates of genetic yield improvement of 19.6, 19.8, and 18.2 kg ha^{-1} yr^{-1} for MG II, III, and IV, respectively. The later maturity of recently released cultivars likely stems from the tendency for breeders to select experimental lines that have high yield but mature later than average within a given MG set test. These late-maturing selections, when combined with a turnover of maturity checks over time, likely resulted because the replacement maturity checks had drifted to slightly later maturity dates. Two more contributing factors are that soybean producers now plant earlier (discussed below), which intrinsically provides more seasonal crop development time, and that warmer seasonal temperatures arising from climate change now extend both the start and end of the soybean growing season (Sacks and Kucharik, 2011). In effect, selection of cultivars with later R8 dates per se are likely needed within each MG zone (given that those zones have remained fixed; see Fig. 12–7B) to take advantage of a now longer growing season. Indeed, one could theoretically take the inverse perspective that the greater magnitude of the genetic yield gain rates not adjusted for maturity (vis-à-vis adjusted ones) reflects, in reality, what breeders had to select for with respect to breeding cultivars with adaptation to the earlier planting dates and slightly warmer fall temperature—both of which became possible because of climate change.

Physiological changes have occurred with time as breeders selected for and released ever higher yielding soybean cultivars to producers. Rincker et al. (2014) documented a significant annual reduction in seed protein of 0.22 and 0.16 g ha^{-1} yr^{-1} in MG II and III, and IV, respectively, but an annual enhancement in seed oil of a significant 0.14, 0.10, and a nonsignificant 0.05 g ha^{-1} yr^{-1}, respectively, in MGs II, III, and IV. Chung et al. (2003), in a quantitative trait loci (QTL) study, also observed this negative relationship of protein with yield (and oil) and documented some plausible genetic and physiological reasons for this three-way pleiotropic association. The reduction in seed protein content as yields move higher is of concern in view of its impact on the constituent value of the soybean seed.

Rowntree et al. (2014) evaluated total dry matter production, harvest index, and growth stage duration in experiments of the MG II and III cultivars from the previously described genetic gain study that included an early and late planting treatment with a 30-d target interval between the planting dates. They found that total dry matter production at reproductive stage R1 (beginning bloom) (Fehr et al., 1971) did not increase over generations of breeding; however, recently released cultivars did have significantly greater dry matter production at R4 (full pod) and at R7 (physiological maturity) than did earlier cultivar releases. Also detected was an increase over time in apparent harvest index at R8 (full maturity), as measured by the weight of grain divided by total biomass (seeds, pods, and stems). Rowntree et al. (2014) observed that recently released cultivars, when compared to older cultivar releases, spent fewer days in the vegetative period of V1 to R1, but spent more days in the reproductive development period of R1 through R7, and in the seed fill period of R5 through R7. In contrast, Sacks and Kucharik (2011) documented that when 1981 to 2005 USDA-NASS state reported data on the seasonal progress of soybean blooming percentage was used as a proxy for the date of R1 each year, the R1 date advanced at a rate of 0.27 d yr^{-1}, but that rate was only 56% of the 0.49 d yr^{-1} rate of advance in soybean planting date. Because of that discordance, these authors concluded that the soybean vegetative period (from planting to R1) had lengthened in this 25-yr period. They also reported little change in the length of the reproductive period from R1 to R7, when USDA-NASS state statistics reported progress in soybean leaf drop was used as an R7 proxy.

Rate of Genetic Improvement in Brazil

Increases in soybean area and productivity have been important growth factors for Brazilian agriculture as a whole (Fig. 12–10). Changes have occurred over time in the geographical distribution of the Brazilian areas of soybean production

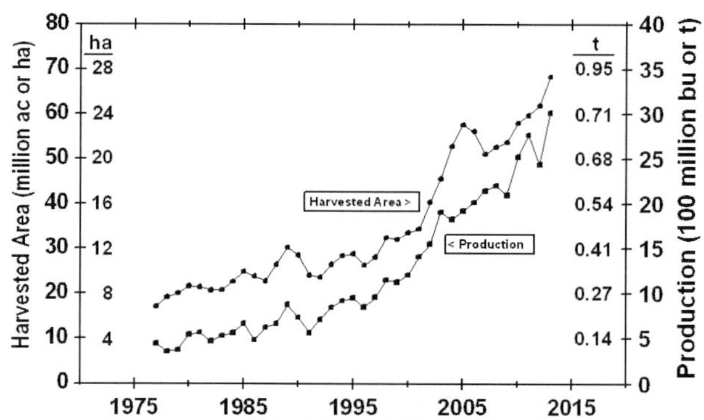

Fig. 12–10. Brazil soybean statistics from 1977 to 2013 for harvested area (left *y* axis; acres or hectares) and production (right *y* axis; tons or bushels). The soybean cropping season in South America is November to May, so the May harvest year was used as the *x* axis in this graph.

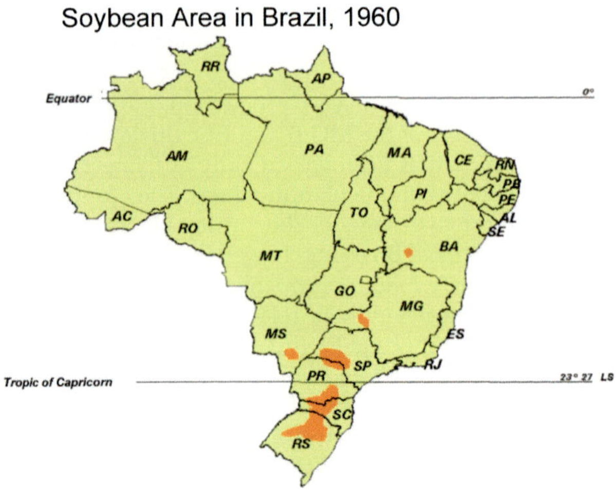

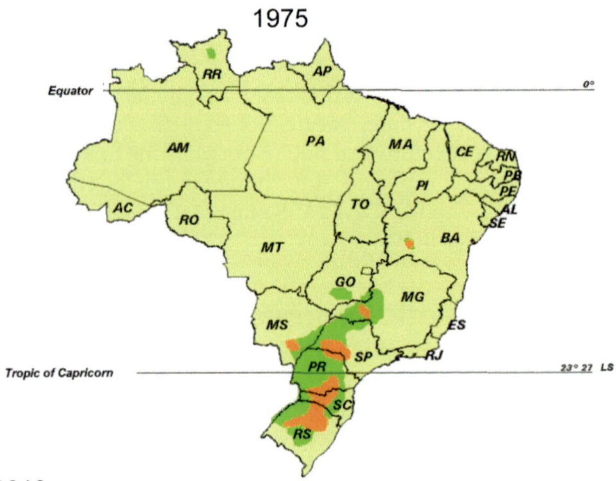

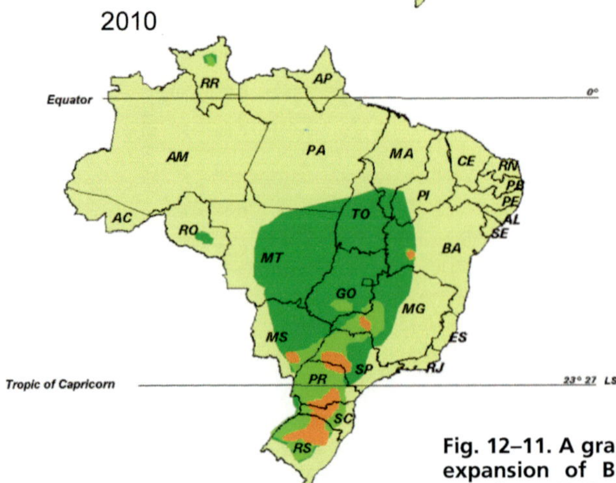

Fig. 12–11. A graphic depiction of the 50-yr expansion of Brazil soybean production area from 1960 to 1975 and then to 2010.

(Fig. 12–11). Annual improvement in estimated on-farm Brazilian yield can be modeled with a simple linear model (Fig. 12–5), and the rate of that on-farm improvement from 1977 to 2013 is estimated to be 43.5 kg ha^{-1} yr^{-1}. Attempts to fit a logistic model, or a segmented linear model, to that data produced AIC values that indicated that these two models were much less probable than the linear regression model (i.e., 28 vs. 72% and 16 vs. 84%, respectively).

Until the early 1980s, soybean cropping in Brazil was primarily located in the South (Fig. 12–11), using American cultivars that were productive in areas between 30° to 22° S. In the 1980s and 1990s, the development of new cultivars adapted to lower latitudes resulted in the movement of production into the tropics, which introduced modern agriculture into the flatland Cerrado area, the high altitude savannahs of Brazil. The soybean crop made agriculture in those areas economically viable, since they were usually cleared of native vegetation, cropped with rice (*Oryza sativa* L.) for a few years, and then used to graze cattle. The Cerrados are formed by old lixiviated, low phosphorous, high aluminum acid soils, which required much investment before they could be cultivated. The initial acquisition price was very low, so these investments were financially viable.

Lime, phosphate, and potassium fertilizer applications were coupled with the development of the no-till system, which had started in the late 1970s in the South but was steadily adopted by soybean growers throughout Brazil, reaching more than 80% of the soybean production area today. Due to the increase in organic matter and fertility of these soils, soybean yields in the Cerrados now have the potential to be as high as yields in the more fertile soils of Brazil where soybean production first began.

The genetic gain estimates obtained in Brazil by comparing old and new soybean cultivars in common environments was assessed in at least three studies. In Parana State (southern Brazil) genetic gains of 45 and 37 kg ha^{-1} yr^{-1} (de Toledo et al., 1990) and 24 and 11 kg ha^{-1} yr^{-1} (Alliprandini et al., 1993) were calculated for the early and semi-early MGs, respectively, between 1981 and 86 and 1986 and 1990. These two MGs corresponded to more than 70% of the cropped area at the time. In Mato Grosso (central Brazil) the calculated genetic gains were of larger magnitude, reaching 60 to 63 kg ha^{-1} yr^{-1} between 1986 and 1990 for the early and semi-early types, which were also sown in the majority of the cropped area.

Agronomic Yield Gains in the Northern United States

Sustained improvement in on-farm crop yield performance (as shown in Fig. 12–3) arises from the combined impact of genetic and agronomic technologies that producers adopt for input into their production systems. Over time, genetic improvement and agronomic improvement have co-evolved in an iterative manner (Messina et al., 2009), which makes it difficult to completely disentangle their individual impacts on on-farm productivity. Plant breeders continually develop and release new cultivars for use in specified targeted production environments, with the goal of enhancing genetic yield potential, while agronomists continually refine existing management practices or devise novel ones, with the goal of optimizing environmental yield potential. The former has the advantage of continuity (new cultivar releases are made available nearly every year), whereas agronomic yield improvement occurs as a result of episodic (one-time) changes in crop management (e.g., converting from conventional tillage to no-till). Producer adoption of any new management practice is typically measured and cautious; thus, the resultant yield improvement is concordantly gradual rather than immediate. Moreover, there is a limit on the degree to which a management practice can improve system productivity; for example, producers advancing the calendar date of their soybean planting will ultimately be constrained by the ever greater probability of a seedling-killing spring freeze, making very early planting too risky to implement. Yet, continual agronomic improvement in the yield potential of modern era production environments is critical. Indeed, the crop management skills of soybean yield contest winners, coupled with favorable weather and fields that have highly fertile soils, have provided the kind of agronomic yield potential that allows them to capture much of the currently available genetic yield potential that is resident in modern era cultivars.

Many agronomic technologies have been adopted by U.S. soybean producers. These include earlier planting dates (Bastidas et al., 2008; Specht et al., 2013), which in turn has led to more use of fungicide-treated seed to deal with colder soil temperatures of late April and early May plantings that slow germination and emergence and thus result in longer exposures of seedlings to fungal pathogens before emergence (Grau and Gaska, 2002). Row spacing has narrowed from 76 to 38 cm, but producer use of 19-cm row spacing has actually declined from 30 to 10% (for Iowa data, see Pedersen, 2008), probably because drill-planters do not work well in heavy residue, no-till production systems (Thorson, 2007; Jeschke, 2012, and references therein). Transgenic-mediated weed control with herbicides that do not need to be till-incorporated into the soil has provided producers with an incentive to convert from conventional to minimum or zero tillage to protect against erosion (Baker, 2011; Johnson, 2013; see also http://www.ctic.purdue.edu/CRM/, accessed 5 Feb. 2014). New planters have more sophisticated row planting units that ensure a more uniform seed placement depth and within-row seed spacing. Platform style combines typically have a 50% greater pod-shattering loss than do soybean-styled row crop header combines (Quick and Buchele, 1974; Elmore et al., 1990), and the latter are the combine of choice in irrigated soybean production systems where ridge-till is a common practice. However, a shift to narrow rows elsewhere has lessened their use in rainfed systems. A general widening of all farm equipment to accommodate more rows per equipment pass, has provided producers with a means for more quickly completing planting and harvesting operations during ever narrower calendar date windows, or to apply pesticides and fungicides over larger areas during critical phenology stage when their application is the most effective.

A key yield-impacting historic agronomic factor of significance was a producer-mediated shift (over many years) from planting soybean into marginal farm fields and corn into more fertile ones, to a corn–soybean crop rotation, which allowed the producer to capture the rotation-engendered synergistic yield increase of each crop that occurs when a legume is cycled with a cereal (Crookston et al., 1991; Porter et al., 2001; Kelley et al., 2003: Pedersen and Lauer, 2004; Temperly and Borges, 2006; Erickson, 2008). For example, in Minnesota and Nebraska, soybean yielded 13 to 18% more in a soybean–corn rotation compared with continuous soybean (Porter et al., 2001). The 1979 to 2011 shift in soybean production to the U.S. corn belt (Fig. 12–2) and the concomitant soybean yield enhancement in those corn belt states (Fig. 12–7) reflect the yield benefit to both crops in the now predominant corn–soybean rotation on the highly fertile loam and silty loam soils in what should really now be called the north-central U.S. corn–*soybean* belt.

A shift from rainfed to irrigated soybean production has occurred over time in those production areas where surface- or groundwater resources are available. U.S. irrigation hectarage is surveyed every 5 yr, and in the 2007 survey (Vilsack and Clark, 2008), approximately 2.12 million hectares of soybean were irrigated (Fig. 12–12), which was about 8% of the total U.S. soybean area harvested that year (Fig. 12–1). These are likely conservative numbers, given the dip in soybean area that occurred in 2007; however, the USDA 2012 survey data will not be released until late 2013. Most of the north-central U.S. irrigated soybean hectarage in 2007 (Fig. 12–12) was located in Nebraska (~0.627 million ha). In other north-central states, the irrigated area was less (in million ha), i.e., Missouri (0.148), Kansas (0.106), Minnesota (0.375), Indiana (0.038), Illinois (0.028), Michigan (0.027), South Dakota (0.21), Iowa (0.020), Wisconsin (0.010), Delaware (0.010), Maryland (0.008), and North Dakota (0.005). In the southern United States, the 2007 soybean area was concentrated in the mid-south states bordering the Mississippi River, notably Arkansas (0.746 million ha in 2007) and its neighboring states of Missouri (0.148), Mississippi (0.195), and Louisiana (0.038), although there were scattered irrigation areas in Georgia (0.014), Texas (0.007), North Carolina (0.006), and Oklahoma (0.005). Although not available for other states, year 2012 data are available from

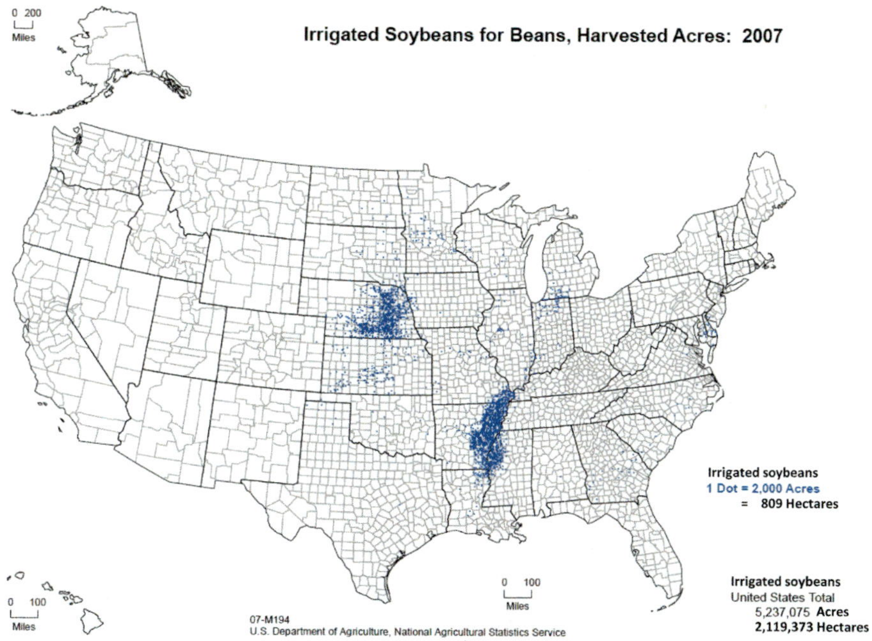

Fig. 12–12. A density map of U.S. irrigated soybean area, on a county basis, for the production year 2007. Of the total U.S. soybean production area of 64.14 million ac or 25.96 million ha grown in 2007 (see Fig. 12–1), 5.24 million ac and 2.12 million ha (i.e., ~8%) were irrigated that year. A soybean–corn rotation is commonly practiced in most north-central U.S. irrigated soybean areas (i.e., compare this 07-M194 irrigated soybean map with the 07-M164 irrigated corn map in Vilsack and Clark, 2008). Source: Vilsack and Clark (2008).

USDA-NASS for Nebraska (0.936 million ha) and Arkansas (1.279 million ha), and that data show a substantial increase in irrigated soybean area that occurred in those two states since 2007. This growth is attributable to the fact that if this agronomic practice is available, it can substantively enhance soybean yields. Using Nebraska irrigated and rainfed soybean yield data from 1960 to 2012 as an example, the difference in yield between the two production systems is clearly evident in the vertical positions occupied by their charted yield trend lines (Fig. 12–13). Scheduling irrigation events to mitigate annual rainfall deficits, or the intra-seasonal timing of these deficits, substantively reduces the interannual yield variation, which is markedly apparent in the higher R^2 and tighter 95% confidence interval surrounding the irrigated trend line. Because of greater certainty in the yield expectation each year in an irrigated soybean production system, the producer is more likely to make long-term investments in novel agronomic practices (especially those aimed at improving soil fertility) that might amplify even more the production environment yield potential. Doing so allows even more of the genetic yield potential (that breeders have continually increased over time) to be expressed in irrigated fields, which likely explains why the Nebraska irrigated–rainfed soybean yield difference documented in Fig. 12–13 is widening at the statistically significant rate of 14.7 ± 2.9 kg ha^{-1} yr^{-1}.

It is difficult to document explicitly, with good experimental data, the specific impact on yield of historic producer adoption of given agronomic technology because of a concurrent commingling with producer adoption of genetic technology, as was noted by Messina et al. (2009). However, we will attempt to document here the yield impact of one agronomic practice—earlier soybean planting, a well-referenced and documented yield enhancement practice (see Bastidas et al., 2008, and references therein).

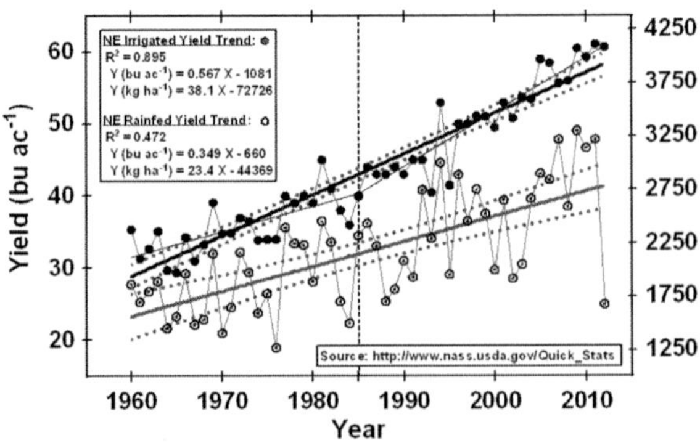

Fig. 12–13. Linear trend lines for Nebraska irrigated (solid circles) and rainfed (open circles) soybean yields from 1960 (start of data record) to 2012 (severe drought year). The dotted lines depict the 95% confidence interval around the estimated best-fit linear trend lines (regression statistics are presented in the text boxes). The irrigated and rainfed linear regression slopes and intercepts are significantly different from each other (P = 0.0004). The irrigated–rainfed yield difference during the 53-yr time frame time has widened at a statistically significant rate of 0.218 ± 0.042 bu ac^{-1} yr^{-1} or 14.7 ± 2.85 kg ha^{-1} yr^{-1}. For the irrigated (but not rainfed) yield data, a two-segment linear regression model (shown as a two-segment trend line, with a breakpoint year of 1985 (vertical dashed line) and a greater slope thereafter of 0.760 bu ac^{-1} yr^{-1}, or 56.1 kg ha^{-1} yr^{-1}), was >99% more probable (based on the Akaike information content test) than the linear model.

Sacks and Kucharik (2011) reported that producers in 13 states had advanced their soybean planting date from 1981 to 2005, by an average of 0.49 d yr^{-1} (the range among states was 0.25 to 1.01 d yr^{-1}). This near 0.5-d advance per year translates into 12-d total advance over 25 yr. For MG II and III cultivars grown in Nebraska, delaying soybean planting beyond May 1 can be expected to reduce the resultant soybean yield by 16.8 to 42.4 kg ha^{-1} d^{-1} of delay (Bastidas et al., 2008, and references therein). These two numbers bracket the estimated yield reduction per day of delayed planting after May 1 documented in other north-central states, such as Ohio, for which the reported yield decline is 35.0 kg ha^{-1} d^{-1} of delay (Lindsey and Sundermeier, 2013). Multiplying the above two Bastidas et al. (2008) numbers by the 0.5-d advance per year in planting date from 1981 to 2005 would suggest that producer-mediated earlier planting could, in theory, have contributed a 8.4 to 21.2 kg ha^{-1} yr^{-1} agronomic yield gain in the MG II and III states, whose long-term on-farm annual yield gain is shown in Fig. 12–8.

To further evaluate this theoretical calculation, we returned to the NASS data set used by Sacks and Kucharik (2011) and expanded their 25-yr analytic time frame to a 33-yr period from 1980 to 2012, but we limited our examination to just the north-central states of Nebraska, Iowa, Illinois, Indiana, and Ohio. A logistic model was fit to the percentage of soybean planting progress collected weekly by USDA-NASS enumerators during the spring of each year in each state. An example of the logistic model fits obtained in each of the 33 yr is presented by Specht et al. (2013). Using an iterative nonlinear estimation method, the day of the year (DOY) corresponding to sigmoid logistic inflection point was determined and was used as an estimate of the DOY when 50% planting progress had occurred in each year and state. Each state's 50% planting progress DOY values were then plotted by production year, and a linear regression trend line was computed for each state (Fig. 12–14). The Nebraska and Iowa, but not the Illinois, Indiana, and Ohio regression coefficients computed for the 1980 to 2012 data were significantly different from zero, but were respectively lower than the −0.54, −0.55,

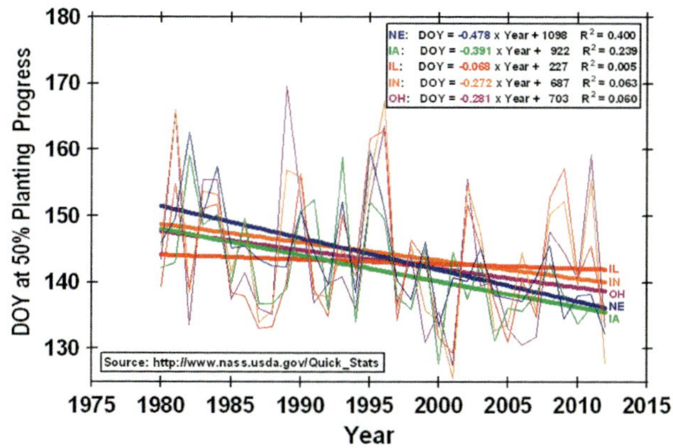

Fig. 12–14. Linear regression of the 50% planting progress day of year (DOY) value on year of production for each of five states in the north-central United States in each year of the past 33 yr from 1980 to 2012. The annual DOY values for each state were estimated in the same manner described by Specht et al. (2013) using USDA-NASS Nebraska weekly planting progress data. Linear regression statistics are provided in the text box.

−0.25, −0.58, and −0.55 values reported by Sacks and Kucharik (2011) for the 1981 to 2005 years of state data. Prolonged rainy periods from late April to late May in three of the seven past years (2006–2012) precluded early soybean planting in those years (Fig. 12–14), despite a likely producer desire to do so. Still, to capture the greater yield potential available when soybean seed is sown earlier in the spring, Fig. 12–14 makes clear that producers now willingly advance their soybean planting date whenever the weather does not preclude them from doing so (Specht et al., 2013). An early planting date obviously cannot be achieved every year due to weather factors, which means that, unlike genetic yield improvement, which is intrinsically cumulative, agronomic yield improvement arising from advancing the planting date is sporadically reversible in those years when the producer-mediated trend of 0.5 d yr^{-1} of earlier planting is thwarted by unfavorable weather.

To more definitively evaluate the potential impact of a hastening of planting date on yield from 1983 to present, the recently developed soybean simulation model known as SoySim (Setiyono et al., 2010) was utilized to estimate the yield potential of three hypothetical (modern) cultivars of MG II, III, or IV, when grown in a hypothetical irrigated production system located near two east-central Nebraska automated weather data station sites located on University of Nebraska–Lincoln research centers near the cities of Mead and Havelock. The sites were chosen because historical growing season weather data—needed by SoySim for multi-year modeling of soybean yield potential—were available for each of the 30 yr spanning 1983 to 2012. The annual SoySim-estimated potential yield values were then regressed on production year in two ways, first by using the actual Nebraska DOY values for 50% planting progress in each of the 30 yr (as shown Fig. 12–14), and second by using the 1983 Nebraska DOY value as a constant late planting date baseline for every year of the 30 yr used in the simulation.

The regression equation data are presented in Table 12–3, and indicate that SoySim-estimated potential yield was substantively reduced in the MG II, III, and IV cultivars (and at both Nebraska sites) when a constant, but now passé, late planting date was input into the model for each year of the 30 yr of simulation. This late planting date was the 1983 DOY of 150 (May 30) that fell on the early part of the regression line in Fig. 4 of the Specht et al. (2013) publication. When averaged over the B factor (CO_2) and across sites (see Table 12–3), and when the input was the actual planting date observed in each year (Fig. 12–14), as opposed to the input of holding planting date constant every year (i.e., 150 DOY), the yield vs. year regression coefficient was reduced from 26 to 13 (MG II), from 19 to 13 (MG III), and from 33 to 11 (MG IV) kg ha^{-1} yr^{-1}. The within-pair difference between regression coefficients reflects the *opportunity cost* in annualized agronomic yield gain that would have accrued if Nebraska soybean producers had not advanced their planting dates by ~0.5 d yr^{-1} during this 30-yr period. The SoySim model is an agronomic management model and does not take into account long-term changes in arising from genetic improvement in yield potential. So, by taking the difference between the foregoing SoySim-based regression coefficient pairs for MG II and III (i.e., 26 − 13 = 13 and 19 − 13 = 6 kg ha^{-1} yr^{-1}), and dividing these by the 1983–2012 on-farm Nebraska irrigated yield regression coefficient of 53 kg ha^{-1} yr^{-1} (last entry in Table 12–3), one could conceivably argue that earlier planting in Nebraska has accounted for as much as 25% (MG II), or as little as 11% (MG III), of the observed on-farm Nebraska irrigated yield gain between 1983 and 2012.

Table 12–3. Potential soybean yields were simulated for maturity group (MG) II, III, and IV cultivars in a Nebraska irrigated production system, using the SoySim model (Setiyono et al., 2010) and annual growing season weather data collected from 1983 to 2012 at the Nebraska weather stations near Mead (41.15° latitude, −96.48° longitude, 366 m asl) and Havelock (40.85, −96.60 and 347 m asl). The 30 annual simulations for each MG at each site were repeated four times to assess the impact on SoySim-estimated yield potential of two levels of two factors, namely: [A] an annual planting date that was (1) set to match the Nebraska day of year (DOY) value of each year shown in Fig. 12–14, or (2) set to a constant baseline of the 1983 Nebraska DOY value of 150 (May 30) for each of the 30 yr, and [B] an annual atmospheric CO_2 concentration, which was (1) set to match the annual NOAA-estimated mean CO_2 concentration of each year, or (2) set to a constant baseline using the 1983 NOAA-estimated CO_2 level for each of the 30 yr. The regression coefficients in the below table have units of kg ha^{-1} yr^{-1} (i.e., 10 kg ha^{-1} = 0.1487 bu ac^{-1}). For comparative purposes, the equation derived from the regression of on-farm Nebraska irrigated yields on the 30 production years for same 1983–2012 period is shown at the bottom of the table.

Soybean maturity group	Automated weather data station location in Nebraska	Regression equation (for *X* range of 1983–2012)	R^2
A1 & B1: Used actual DOY of 50% planting progress + actual Year CO_2 ppmv value			
II	Mead	$Y = 26X - 47210$	0.18
	Havelock	$Y = 26X - 47277$	0.17
III	Mead	$Y = 20X - 34971$	0.18
	Havelock	$Y = 24X - 44117$	0.25
IV	Mead	$Y = 32X - 59517$	0.25
	Havelock	$Y = 34X - 63485$	0.35
A1 & B2: Actual DOY for 50% planting progress + constant CO_2 (1983 baseline ppmv)			
II	Mead	$Y = 26X - 46807$	0.17
	Havelock	$Y = 26X - 46134$	0.16
III	Mead	$Y = 10X - 16073$	0.04
	Havelock	$Y = 20X - 35845$	0.12
IV	Mead	$Y = 30X - 56356$	0.25
	Havelock	$Y = 33X - 62005$	0.36
A2 & B1: Constant DOY (1983 DOY for 50% planting progress) + actual year CO_2 ppmv value			
II	Mead	$Y = 13X - 23202$	0.07
	Havelock	$Y = 18X - 31608$	0.09
III	Mead	$Y = 9X - 14795$	0.03
	Havelock	$Y = 19X - 33087$	0.12
IV	Mead	$Y = 15X - 26766$	0.18
	Havelock	$Y = 8X - 12912$	0.04
A2 & B2: Constant DOY (1983 DOY for 50% planting progress) + constant CO_2 (1983 baseline ppmv)			
II	Mead	$Y = 10X - 16476$	0.05
	Havelock	$Y = 13X - 22462$	0.06
III	Mead	$Y = 13X - 21789$	0.08
	Havelock	$Y = 13X - 20579$	0.05
IV	Mead	$Y = 11X - 19974$	0.16
	Havelock	$Y = 11X - 17687$	0.11
Nebraska irrigated soybean yield trend for the past 30 yr (1983–2012)			
II & III	Source: http://www.nass.usda.gov/Quick_Stats	$Y = 53X - 102489$	0.89

The former is close to, but the latter is less than one-half of, the 26% reported by Sacks and Kucharik (2011). Again, it must be kept in mind that there is a limit in the degree to which planting date can be further advanced, although additional agronomic yield improvement could conceivably be attained by a greater fraction of the population of producers advancing their planting date towards a DOY of 120 (late April) as a goal for 50% planting progress (Fig. 12–14), particularly in years when weather conditions permit that to occur. Although the 1983 to 2012 regression coefficients in Table 12–3 are lower than the on-farm irrigated yield gain regression coefficient (53 g kg^{-1} yr^{-1}) for the same time frame, the simulated ones have a higher elevation context (as can be seen in their intercept values). The regression coefficient differential is a reflection of the RUE of 1.2 Mg g^{-1}, postulated as a limiting maximum for soybean in the review of this subject by Sinclair and Rufty (2012), which is intrinsically incorporated into the SoySim model. Even in the absence of water stress, SoySim-simulated values of soybean yield cannot exceed the dry matter output associated with this RUE maximum.

Interaction of Genetic and Agronomic Improvement in the Northern United States

Most researchers are familiar with the term genotype × environment (G × E) as a statistical interaction component of phenotype. However, the interaction of the *rate* of genetic yield improvement (GYI) with rate of agronomic yield improvement (AYI) is also of great interest as well (as noted by Messina et al., 2009). Has GYI × AYI interaction been documented in soybean, and if so, what is the nature of that interaction? The answer to these two questions is yes. Rincker et al. (2014) noted that the estimated rate of genetic improvement was significantly greater when the historic cultivar sets were evaluated in highly productive environments vs. substantively less productive environments. These authors reported that for MGs II, III, and IV, there was a significant positive slope when genetic gains at individual environments were regressed against average yield of these environments. A zero slope would be expected if a GYI × site productivity interaction did not exist.

In a recently published paper, Rowntree et al. (2013) focused on the interaction of genetic yield gain with a yield gain associated with one major agronomic factor—soybean planting date. Rowntree et al. (2013) evaluated the MG II and MG III historic cultivar sets in additional experiments to assess the impact of a 30-d difference in planting date on the degree of genetic yield gain over time (and vice versa, given that interactions are symmetrical). During 2010 and 2011, the MG II cultivars described above were tested in Wisconsin, and the MG III cultivars were tested in Illinois and Indiana. The targeted planting date treatments at each environment were May 1 (early) and June 1 (late). A significant interaction between the planting date effect for genetic yield gain was not detectable in the Wisconsin-planted MG II tests; that is, the rate of genetic yield gain was identical in each planting date, 18.5 kg ha^{-1} yr^{-1}. However, in the Illinois- and Indiana-planted MG III tests, a statistically significant GYI × planting date interaction was detected—a 22.7 kg ha^{-1} yr^{-1} rate of genetic yield improvement in the May planting that was greater than the 19.6 kg ha^{-1} yr^{-1} genetic yield gain rate for the June planting. This statistically significant GYI × AYI interaction of old (obsolete) vs. new (modern) genotypes interacting with older used (late) vs. newer used (early) planting dates is the first documentation in soybean of the synergism that can arise when modern soybean cultivars of ever higher genetic yield potential are coupled with modern agronomic practices that also offer higher environmental yield potential to allow more expression of the former.

Because a soybean–corn rotation offers a yield benefit to the following crop (at least with respect to the yield obtainable in a monoculture of each crop), and because a long cycle of corn years before a soybean year offers an even greater yield benefit to the soybean, it was of interest to evaluate the soybean genetic yield gain in those two different crop rotation contexts. In fact, Fox et al. (2013) recently examined the interaction of soybean genetic yield gain with respect to the soybean yields attained in the soybean year of a corn–soybean rotation vs. the soybean yields attained in the soybean year of an 11-yr corn–1-yr soybean rotation (the latter was a serendipitous opportunity exploited by the authors, as opposed to being a real crop rotation scenario). In that study, the historic set of MG II, III, and IV cultivars was tested in six environments in Illinois during 2010 to estimate the effects of previous cropping history on the rate of genetic gain. In split-plot experiments, the pre-2010 crop rotation history (as noted above) served as the two main plots, with soybean cultivars as sub-plots. The soybean yields measured after 11 prior years of corn were 360 kg ha^{-1} greater than the soybean yields measured after a single year of corn in a corn–soybean rotation. Although the magnitude of this main effect differential varied among the six locations, it was statistically significant in five. With but two exceptions, the measured rate of genetic yield gain in the two cropping history treatments did not differ, and the in-common genetic yield gain rate estimate was 22.8 kg ha^{-1} yr^{-1}. The two exceptions were two MG II environments that, after adjusting the genetic yield gains for cultivar maturity, exhibited a greater rate of gain in the soybean year after continuous corn than in a soybean year of the soybean–corn rotation. Because soybean is typically bred in fields that are in a soybean–corn rotation, soybean breeders have not made any progress during the past 80 yr in closing the soybean yield gap between the two rotation histories. As noted previously, soybean yield is enhanced when it is grown after many years of continuous corn, but there does not seem to be a detectable GYI $\times$ crop rotation history interaction.

In another study conducted with the historic set of MG II and III cultivars, Suhre et al. (2013, unpublished data) examined the interaction between year of release and plant density relative to cultivar yield. They planted the historic sets at a high (445,000 seeds ha^{-1}) and a low (148,000 seeds ha^{-1}) densities and observed a greater average yield at the high density than the low density. More importantly, they observed a significant interaction between the year of release and plant density, with the newer cultivars taking greater advantage of the high density than older cultivars. At the high density, the yield increase over generations of breeding was largely the result of increased yield on the main stems of plants, whereas at the low density, the yield increases over generations were the result of increased yield on both the main stems and branches.

The historic MG II and III sets also have been evaluated with and without the application of N fertilizer by Wilson et al. (2014). They observed that the rate of yield improvement (and that for seed mass and for seed oil) in the MG II cultivars was not impacted by an N application of 560 kg N ha^{-1}, with 40% applied at planting and 60% applied at V5, although the rate of decline in seed protein content over MG II cultivar release year was less with N than without N. In contrast, the MG III rate of yield gain was 20% greater in the N fertilized plots (27.4 ± 1.6 kg ha^{-1} yr^{-1}) than in the nonfertilized plots (22.8 ± 1.1 kg ha^{-1} yr^{-1}), indicating a greater yield responsiveness of current (in contrast to older) MG III cultivars to supplemental N. This was also true for seed mass, measured as grams per 100 seeds, which the authors speculated was due to a delayed senescence (and thus a longer retention) of green leaves in the fertilized plots, which likely extended the seed-filling period to enable continued photosynthetic carbon acquisition and remobilization of N from vegetative sources to developing seeds. The N supplied from the soil at the MG III test sites in Illinois and Indiana, when coupled with biological N_2 fixation, provided almost the entire amount of N supply needed to maximize the yield potential of the older MG III cultivars, but these two N sources were measurably insufficient for the newer MG III cultivars. Interestingly, the rate of decline in seed protein and rise in seed oil over MG III cultivar release was not impacted by N fertilization.

While genetic improvement alone apparently accounts for about two-thirds of the on-farm yield improvement (as documented above), and agronomic improvement alone also contributes perhaps another 15 to 20%, we can conclude that the *synergistic interaction* of relentless recurrent genetic improvement with intermittently phased periods of agronomic improvement also makes a significant contribution. All three of these factors have ensured a continuance of the persistent rise in on-farm United States and state yields.

Future Changes

Impact of Potential Advances in Biotechnology and Genomics

Soybean is a crop grown around the globe, and currently, transgenic cultivars dominate the planted areas (James, 2012). In the United States, 93% of the soybean area was planted in 2012 to transgenic cultivars, and the vast majority of those cultivars possess a transgene that provides tolerance to the herbicide glyphosate [*N*-(phosphonomethyl)glycine], the active ingredient in Roundup (Padgette et al., 1995). Transgenic cultivars with tolerance to glufosinate (DL-phosphinothricin), the active ingredient in the herbicide Liberty, and with increased seed oleic acid concentration (Waltz, 2010) also have recently gone into commercial production. The soybean biotech and seed industry has many additional transgenic events in the regulatory pipeline, and these transgenes were constructed to provide tolerance to additional herbicides, increase oleic acid concentration, and foster pest resistance (Stein and Rodríguez-Cerezo, 2009).

Tests have been done to evaluate the effect of the Roundup Ready gene on yield. Elmore et al. (2001) tested sister lines that either carried the CP4 EPSPS transgene that conditioned glyphosate tolerance from the breeding line 40-3-2 or did not carry this gene. This breeding line was the source of the Roundup Ready gene for the cultivars until the transgenic event MON89788, known as Roundup Ready 2 Yield, became commercially available. In field tests in Nebraska during 1998 and 1999, they found that the lines with the EPSPS gene from 40-3-2 yielded 5% less than their nontransgenic sister lines. Elmore et al. (2001) concluded that the yield suppression was related to the gene or the insertion process. Alternatively, this yield suppression could have been the result of genes linked to the transgene insertion site that were selected with it during the breeding process. De Vries and Fehr (2011) tested for the effect of EPSPS gene from MON89788 in three backcross populations segregating for the gene. They found no negative or positive association between the gene and yield and lodging score in any population, but did observe an association between the gene and maturity date, plant height, and seed weight.

The Roundup Ready genes are by far the most widely grown transgene in soybean, and the neutral to negative impact of these genes on soybean yield provides evidence that transgenes have not contributed either to genetic based yield gain in soybean, or to on-farm yield improvement per se (Xu et al., 2013). This trend will likely continue as the 22 soybean transgenic events that have received regulatory approval in at least one country (James, 2013) are largely designed to provide herbicide resistance or improved seed composition but not directly increase yield. The only exception is that there are some genes in the development pipeline that target insect and nematode resistance that could capture more of the agronomic yield potential in field environments where these pests are present.

Marker-assisted selection (MAS) is an application of molecular biology that is being employed on a large scale in soybean breeding programs. With MAS, genes are selected in breeding programs indirectly using linked genetic markers. Breeders initially used MAS largely to select for disease and pest resistance such as resistance to soybean cyst nematode (Cahill and Schmidt, 2004). These selections resulted in an increased percentage of experimental lines in breeding programs with disease resistance which has made MAS an important tool in breeding for defensive traits. The large breeding companies have built up their capacity to conduct MAS to the extent that Eathington et al. (2007) flatly stated "Today, molecular marker assisted breeding is becoming our conventional breeding process."

Compared to breeding for disease and pest resistance, the implementation of MAS in breeding for yield has been more difficult. Although a number of QTL that control yield have been identified (Soybase; http://soybean.org, accessed 6 Feb. 2014), it has been a challenge to confirm the QTL in new populations, different genetic backgrounds, and various production environments. Sebastian et al. (2010) introduced the concept of MAS for yield in soybean within the context of specific experimental populations and environments. With this method, markers are used to map and select QTL within each target population. These authors reported success in implementing MAS for yield by selecting for residual heterogeneity within elite soybean cultivars. They also indicated that context-specific MAS would be applicable to biparental populations that are more typical of breeding programs. Research is continuing to develop better methods to implement MAS in breeding populations for complex traits such as seed yield. One approach that holds promise for selection for complex traits is genomic selection. With genomic selection, phenotypic and genotype data from a training population are used to predict the genetic value of lines using genome-wide marker data (Heffner et al., 2011).

Impact of Climate Change on Soybean Production and Productivity

The primary driver of climate change is the unabated annual rise in atmospheric CO_2 concentration (Tans and Keeling, 2013). Greater CO_2 concentration can potentially improve crop dry matter accumulation in a C3 photosynthetic species like soybean, because it mitigates the amount of oxygen-mediated photorespiration.

At the turn of the last century, Specht et al. (1999) used the experimental data published by crop modelers to estimate that the rising levels of atmospheric CO_2 could (conceivably) contribute 1.7 to 3.4 kg ha^{-1} yr^{-1}, or even as much as 5 kg ha^{-1} yr^{-1}, to the rate of on-farm improvement in U.S. soybean yield. Although these annual numbers are small, they would be expected to result in cumulative total soybean yield improvement over time (i.e., just as genetic improvement results in cumulative change in genetic yield potential). Moreover, rising CO_2 concomitantly improves water-use efficiency as well because WUE is simply a ratio of carbon acquisition/fixation vs. water transpired, whether WUE is measured instantaneously or seasonally (Sinclair et al., 1984). If C3 photosynthesis is indeed enhanced with higher CO_2 levels, as most researchers believe, then the numerical value of the numerator in the WUE ratio will become greater as well.

We used the SoySim crop model to determine the degree to which soybean yield potential (in the absence of water stress) was impacted by the change in atmospheric CO_2 concentration from 1983 to 2012 (Table 12–3). Annualized atmospheric CO_2 data for the 30-yr period from 1983 (343.03 ppmv) to 2012 (393.82 ppmv) was obtained from the NOAA web site (Tans and Keeling, 2013). When averaged over the A factor (planting date) and sites, the reduction in the regression coefficient with CO_2 held to a constant 1983 CO_2 ppmv baseline of 343.03 ppmv was 21 to 19 (MG II), 19 to 14 (MG III), and 22 to 21 (MG IV) kg ha^{-1} yr^{-1}. The difference between these three MG regression coefficient pairs of 3, 5, and 1 kg ha^{-1} yr^{-1} is a reflection of the SoySim-based simulated annual rate of on-farm soybean yield improvement between 1983 and 2012 that could be ascribable to the annual rise in atmospheric CO_2 in that same time frame. These numbers are numerically close to rates reported by Specht et al. (1999).

Although rising levels of CO_2 are effectively a form of carbon fertilization for C3 crop species, the relentless increase in CO_2 will certainly result in future negative impacts, such as higher global temperatures and greater weather variability, both of which are critical parameters with respect to the future of crop productivity. Twine and Kucharik (2009) retrospectively examined the impact of climate change on trends in net primary productivity (i.e., in units of g C m^{-2} yr^{-1}) during a 53-yr time frame (1950–2002), and also examined a more recent portion of that time frame (i.e., 1982–2002). The authors, when modeling the impacts, purposefully minimized the effects of human-mediated improvements in crop productivity in their climate-change-only control model runs, but also evaluated the impact of climate change along with producer-mediated management (e.g., cultivar or hybrid selection, planting date) in subsequent model runs. Their analysis, using the Agro-IBIS model, produced results indicating that climate change during the retrospective time frame had increased the net primary productivity of corn/soybean systems at a rate of 6.43/4.06 g C m^{-2} yr^{-1} from 1982 to 2002, which they attributed to a 20-yr period of climate change that provided longer growing seasons, lower than normal summer temperatures, and increased precipitation. This finding of a higher rate of net primary productivity is consistent with the faster rate of post 1983 on-farm yield improvement shown in Fig. 12–3C.

Future climate change is, of course, the focus and concern of breeders and agronomists, so a retrospective examination of recent climate change (such as that just presented in the prior paragraph) is still useful as the basis for modeling future climate change. A recent report (Hassol, 2009) entitled *Global Climate Change Impacts in the United States* provides a lucid and well-illustrated summary of the impact of future climate change on agriculture (an update of this report is expected in 2013). The key points listed were higher temperatures, extreme weather events (prolonged lack of rainfall leading to drought, and/or downpours leading to flooding), potentially greater levels of ozone, and increases in pest, disease, and weed pressure. With regard to soybean, the higher temperatures are of critical concern because future rises in mid- to late summer temperatures will significantly compromise soybean reproductive development (i.e., via high temperature–induced greater floral, pod, and seed abortion). Consequently, soybean breeders may need to search their germplasm stocks and banks for genotypes and genes that provide high temperature tolerance. If this problem is not soluble, the movement of soybean production to the northern United States, as is evident in Fig. 12–2A,B for the 33-yr period of 1979 to 2011, will likely continue, and also will result in soybean production areas moving further north into Canada. In fact, the harvested soybean area in Manitoba rose from a near-zero level in 2000 to 285,316 ha in 2011 (Statistics Canada and Agriculture and Agri-Food Canada, 2012) primarily because of a significant warming trend during the past decade that was coupled with the development of high yielding early-maturing cultivars that offered more seasonal adaptation to this region. Indeed, Manitoba contributed to the 1,601,653 ha of soybean grown in Canada in 2011, and the latter was up 36% from the 1,202,100 ha grown 5 yr earlier in 2006. The *mitigation* of climate change is certainly desirable, but has not yet been politically achievable, yet it is clear that *adaptation* to climate change has already been occurring, and a continuation of that adaptation in the future will have far-reaching consequences on where soybean can and will be grown in the rest of this century.

Summary and Conclusions

- The linear rate of on-farm soybean yield improvement in the United States (1924–2012) is 23.3 kg ha^{-1} yr^{-1}, but in Brazil (1977–2013) it is 43.5 kg ha^{-1} yr^{-1}. A logistic model provides a better fit to the U.S. data than does a straight line and attains an estimated sigmoidal inflection point in year 2008, which suggests that the still-rising U.S. soybean yields may now be in a deceleration phase that will plateau in some distant year at an estimated yield maximum of 5543 kg ha^{-1}. In the absence of any other evidence of a forthcoming yield plateau, a two-segment linear regression model provides the best fit to the U.S. yield data and suggests that the U.S. on-farm yield rate, since the breakpoint year of 1983, has been 29.4 kg ha^{-1} yr^{-1}.
- A 2-yr multisite agronomic trial of a historic set of 60 MG II, 59 MG III, and 49 MG IV cultivar releases (1920–2010) conducted in the north-central states of the United States revealed linear rates of genetic improvement of a respective 23.1, 22.8, and 19.5 kg ha^{-1} yr^{-1}. A two-segment linear model provides the best fit with the data from the historic set for the three MGs. In the period after the breakpoint year, which depending on the given MG, varied from1964 to 1971, the genetic yield gain for MG II, III, and IV increased to rates 31.1, 29.4, and 26.5 kg ha^{-1} yr^{-1}, respectively.
- The contribution of genetic yield improvement to on-farm yield improvement is best assessed in irrigated production systems in which genetic yield potential can be fully expressed in non-water-stressed environments

of high yield potential. The rate of genetic yield improvement in the MG II and III historic cultivar set when grown in Nebraska high yield irrigated production systems is 26.6 kg ha^{-1} yr^{-1} vs. the estimated 42.1 kg ha^{-1} yr^{-1} rate of irrigated on-farm yield improvement, suggesting that breeder-mediated increases in genetic yield potential may be accounting for about two-thirds of increased yield improvement realized on the farm by Nebraska irrigated soybean producers.

- The remaining one-third of on-farm yield improvement likely arises from producer adoption of better agronomic (crop and soil) management, including earlier soybean planting, which may account for as much as 13 to 20% of the irrigated on-farm yield improvement.
- Recent publications have shown that the yield difference between modern and obsolete cultivars widens with incremental improvements in agronomic practices. Thus, this synergistic interaction between genetic and agronomic improvement also is likely a contributor to on-farm yield improvement.
- Climate change has already impacted soybean productivity. Of great concern in the United States is the steadily rising summer temperature, which will likely result in a continued shift of soybean production into northern U.S. states and Canada.

References

Alliprandini, L.F., J.F.F. de Toledo, R.A.S. Kiihl, J. de S. Fonseca Junior, and L.A. de Almeida. 1993. Ganho genético em soja no Estado do Paraná. Pesq. Agropec. Brasil. 28:489–497.

Arantes, N.E., and P.I. de M. Souza. 1993. Cultura da Soja nos Cerrados. Associação Brasileira para Pesquisa da Potassa e do Fosfato – Potafos, Piracicaba, S.P., Brazil.

Baker, N.T. 2011. Tillage practices in the conterminous United States, 1989–2004 Datasets Aggregated by Watershed. USGS Data Series 573. http://pubs.usgs.gov/ds/ds573/pdf/dataseries573final.pdf (accessed 12 Nov. 2013).

Bastidas, A.M., T.D. Setiyono, A. Dobermann, K.G. Cassman, R.W. Elmore, G.L. Graef, and J.E. Specht. 2008. Soybean sowing date: The vegetative, reproductive, and agronomic impacts. Crop Sci. 48:727–740. doi:10.2135/cropsci2006.05.0292

Bernard, R.L., G.A. Juvik, E.E. Hartwig, and C.J. Edwards, Jr. 1988. Origins and pedigrees of public soybean varieties in the United States and Canada. USDA Tech. Bull. 1746. U.S. Gov. Print. Office, Washington, DC.

Bernard, R.L., G.A. Juvik, and R.L. Nelson. 1987. USDA soybean germplasm collection inventory. Vol. 1. International Agricultural Publications. INTSOY Series Number 30. USDA and Univ. or Illinois, Urbana, IL.

Bernard, R.L., G.A. Juvik, and R.L. Nelson. 1989. USDA soybean germplasm collection inventory. Vol. 2. International Agricultural Publications. INTSOY Series Number 31. USDA and Univ. or Illinois, Urbana, IL.

Boerma, H.R. 1979. Comparison of past and recently developed soybean cultivars in maturity groups VI, VII, and VIII. Crop Sci. 19:611–613. doi:10.2135/cropsci1979.0011183X001900050015x

Brown-Guedira, G.L., J.A. Thompson, R.L. Nelson, and M.L. Warburton. 2000. Evaluation of genetic diversity of soybean introductions and North American ancestors using RAPD and SSR markers. Crop Sci. 40:815–823. doi:10.2135/cropsci2000.403815x

Cahill, D.J., and D.H. Schmidt. 2004. Use of marker assisted selection in a product development breeding program. In: T. Fischer et al. New directions for a diverse planet: Proceedings for the 4th International Crop Science Congress, Brisbane, Australia, 26 Sept.–1 Oct. 2004. http://www.cropscience.org.au/icsc2004/symposia/3/4/133_schmidtdh.htm (accessed 12 Nov. 2013).

Carter, T.E., Jr., Z. Gizlice, and J.W. Burton. 1993. Coefficient of parentage and genetic similarity estimates for 258 North American soybean cultivars released by public agencies during 1954–88. USDA Tech. Bull.1814. U.S. Gov. Print. Office, Washington, DC.

Carter, T.E., R.L. Nelson, C.H. Sneller, and Z. Cui. 2004. Genetic diversity in soybean. In: HR. Boerma and J.E. Specht, editors, Soybeans: Improvement, production, and uses. 3rd ed. Agron. Monogr. 16. ASA, CSSA, SSSA, Madison, WI. p. 303–416. doi:10.2134/agronmonogr16.3ed.c8

Chung, J.H., L. Babka, G.L. Graef, P.E. Staswick, D.J. Lee, P.B. Cregan, R.C. Shoemaker, and J.E. Specht. 2003. The seed protein, oil, and yield QTL on soybean linkage group I. Crop Sci. 43:1053–1067. doi:10.2135/cropsci2003.1053

Crochet, W.D., editor. 2012. Uniform soybean tests, northern states. USDA-ARS, West Lafayette, IN.

Crookston, R.K., J.E. Kurle, P.J. Copeland, J.H. Ford, and W.E. Lueschen. 1991. Rotational cropping sequence affects yield of corn and soybean. Agron. J. 83:108–113. doi:10.2134/agronj1991.00021962008300010026x

De Bruin, J.L., and P. Pedersen. 2008. Yield improvement and stability for soybean cultivars with resistance to *Heterodera glycines* Ichinohe. Agron. J. 100:1354–1359. doi:10.2134/agronj2007.0412

de Toledo, J.F.F., L.A. de Almeida, R.A.S. Kiihl, and O.G. Menosso. 1990. Ganho genético em soja no Estado do Parana—Via Melhoramento, No Periodo de 1985/86 A 1989/90. Pesqi. Agropecu. Bras. 25:89–94.

de Toledo, J.F.F., C.G.P. de Carvalho, C.A.A. Arias, L.A. de Almeida, and R.L. Brogin. M.F de Oliveira, J.U.V. Moreira, A.S. Ribeiro, and D.M. Hiromoto. 2006. Genotype and environment interaction on soybean yield in Mato Grosso State, Brazil. Pesq. Agropec. Brasilia. 41:785–791.

de Toledo, J.F.F., D.M. Hiromoto, and C.A.A. Arias. 2000. A união faz a diferença. Agroanalysis, FGV. RJ, Brazil. Janeiro, Rio de Janeiro.

De Vries, B.D., and W.R. Fehr. 2011. Impact of the MON89788 event for glyphosate resistance on agronomic and seed traits of soybean. Crop Sci. 51:1023–1027. doi:10.2135/cropsci2010.11.0642

Dias, W.P., V.P. Campos, R.A.S. Kiihl, C.A.A. Arias, and J.F.F. Toledo. 2005. Genetic control in soybean of resistance to soybean cyst nematode race 4+. Euphytica 145:321–329. doi:10.1007/s10681-005-1944-1

Diwan, N., and P.B. Cregan. 1997. Automated sizing of fluorescent-labeled simple sequence repeat (SSR) markers to assay genetic variation in soybean. Theor. Appl. Genet. 95:723–733. doi:10.1007/s001220050618

Eathington, S.R., T.M. Crosbie, M.D. Edwards, R.S. Reiter, and J.K. Bull. 2007. Molecular markers in a commercial breeding program. Crop Sci. 47:S154–S163. doi:10.2135/cropsci2007.04.0015IPBS

Egli, D.B. 2008a. Comparison of corn and soybean yields in the United States: Historical trends and future prospects. Agron. J. 100:S79–S88. doi:10.2134/agronj2006.0286c

Egli, D.B. 2008b. Soybean yield trends from 1972 to 2003 in mid-western USA. Field Crops Res. 106:53–59. doi:10.1016/j.fcr.2007.10.014

Elmore, R.A., R.S. Moomaw, and R. Selley. 1990. Narrow row soybeans. UNL Extension NEBGuide G90-964. http://digitalcommons.unl.edu/cgi/viewcontent.cgi?article=1741&context=extensionhist (accessed 12 Nov. 2013).

Elmore, R.W., F.W. Roeth, L.A. Nelson, C.A. Shapiro, R.N. Klein, S.Z. Knezevic, and A. Martin. 2001. Glyphosate-resistant soybean cultivar yields compared with sister lines. Agron. J. 93:408–412. doi:10.2134/agronj2001.932408x

Erickson, B. 2008. Corn/soybean rotation literature summary. Purdue Univ., West Lafayette, IN. http://www.agecon.purdue.edu/pdf/Crop_Rotation_Lit_Review.pdf (accessed 12 Nov. 2013).

Fehr, W.R. 1987. Breeding methods for cultivar development. In: J.R. Wilcox, editor, Soybeans: Improvement, production, and uses. 2nd ed. Agron. Monogr. 16. ASA, CSSA and SSSA, Madison, WI. p. 249–293.

Fehr, W.R., C.E. Caviness, D.T. Burmood, and J.S. Pennington. 1971. Stage of development descriptions for soybeans, *Glycine max* (L.) Merrill. Crop Sci. 11:929–931. doi:10.2135/cropsci1971.0011183X001100060051x

Fox, C.M., T.R. Cary, A.L. Colgrove, E.D. Nafziger, J.S. Haudenshield, G.L. Hartman, J.E. Specht, and B.W. Diers. 2013. Estimating soybean genetic gain for yield in the northern USA– Influence of cropping history. Crop Sci. 53:2473–2482. doi:10.2135/cropsci2012.12.0687

Gizlice, Z., T.E. Carter, Jr., and J.W. Burton. 1993. Genetic diversity in North American soybean: I. Multivariate analysis of founding stock and relation to coefficient of parentage. Crop Sci. 33:614–620. doi:10.2135/cropsci1993.0011183X003300030038x

Gizlice, Z., T.E. Carter, Jr., and J.W. Burton. 1994. Genetic base for North American public soybean cultivars released between 1947 and 1988. Crop Sci. 34:1143–1151. doi:10.2135/cropsci1994.0011183X003400050001x

Godoi, C.R.C., J.B. Duarte, and J.B. Pinheiro. 2013. Genetic dissimilarity in tropical soybean genotypes. Paper presented at World Soybean Research Conference, IX. Durban, South Africa. 17–22 February.

Grassini, P., J. Thorburn, C. Burr, and K.G. Cassman. 2011. High-yield irrigated maize in the western U.S. Corn-Belt: I. On-farm yield, yield-potential, and impact of management practices. Field Crops Res. 120:142–150. doi:10.1016/j.fcr.2010.09.012

Grassini, P., H.S. Yang, and K.G. Cassman. 2009. Limits to maize productivity in Western Corn-Belt: A simulation analysis for fully irrigated and rainfed conditions. Agric. For. Meteorol. 149:1254–1265. doi:10.1016/j.agrformet.2009.02.012

Grau, C.R., and J.M. Gaska. 2002. Importance of fungicide seed treatments for soybeans. Univ. of Wisconsin, Madison. http://www.soils.wisc.edu/extension/wcmc/2002proceedings/Grau-Conf-2002.pdf (accessed 12 Nov 2013).

Hartwig, E.E. 1973. Varietal development. In: B.E. Caldwell, editor, Soybeans: Improvement, production, and uses. Agron. Monogr. 16. ASA, Madison, WI. p. 187–210.

Hassol, S.J. 2009. Global climate change impacts in the United States– 2009 Report. U.S. Global Change Research Program. http://nca2009.globalchange.gov/agriculture (accessed 12 Nov. 2013).

Heffner, E.L., J.L. Jannink, H. Iwata, E. Souza, and M.E. Sorrells. 2011. Genomic selection accuracy for grain quality traits in biparental wheat populations. Crop Sci. 51:2597–2606. doi:10.2135/cropsci2011.05.0253

Hiromoto, D.M., and N.A. Vello. 1986. The genetic base of Brazilian soybean (*Glycine max*) cultivars. Braz. J. Genet. 9:295–306.

Hymowitz, T., and J.R. Harlan. 1983. Introduction of soybean to North America by Samuel Bowen in 1765. Econ. Bot. 37:371–379. doi:10.1007/BF02904196

James, C. 2012. Global status of commercialized biotech/GM crops: 2012. ISAAA Brief 44-2012: Highlights. http://www.isaaa.org/resources/publications/briefs/44/highlights/default.asp (accessed 12 Nov 2013).

James, C. 2013. Soybean (*Glycine max* L.) GM events. Crops list—GM approval database. ISAAA. http://www.isaaa.org/gmapprovaldatabase/crop/default.asp?CropID=19 (accessed 12 Nov. 2013).

Jeschke, M. 2012. Crop insights. Row width in soybean. https://www.pioneer.com/home/site/us/agronomy/library/template.CONTENT/guid.654CA746-2C95-A483-C396-D84F1F1DAE2D (accessed 12 Nov. 2013).

Johnson, D.M. 2013. A 2010 map estimate of annually tilled cropland within the conterminous United States. Agric. Syst. 114:95–105. doi:10.1016/j.agsy.2012.08.004

Keim, P., W. Beavis, J. Schupp, and R. Freestone. 1992. Evaluation of soybean RFLP marker diversity in adapted germplasm. Theor. Appl. Genet. 85:205–212.

Kelley, K.W., J.H. Long, and T.C. Todd. 2003. Long-term crop rotations affect soybean yield, seed weight, and soil chemical properties. Field Crops Res. 83:41–50. doi:10.1016/S0378-4290(03)00055-8

Kiihl, R.A. de S., and A. Garcia. 1989. The use of the long-juvenile trait in breeding soybean cultivars. In: A.J. Pascale, editor, Proceedings: World Soybean Research Conference IV, Buenos Aires, Argentina. p. 994–1000.

Kisha, T.J., B.W. Diers, J.M. Hoyt, and C.H. Sneller. 1998. Genetic diversity among soybean plant introductions and North American germplasm. Crop Sci. 38:1669–1680. doi:10.2135/cropsci1998.0011183X003800060042x

Li, Z., L. Qiu, J.A. Thompson, M. Welsh, and R.L. Nelson. 2001. Molecular genetic analysis of the U.S. and Chinese soybean ancestral lines. Crop Sci. 41:1330–1336. doi:10.2135/cropsci2001.4141330x

Lindsey, L., and A. Sundermeier. 2013. Soybean plant date recommendations. Crop Observation and Recommendation Network. Ohio State Extension no. 2013-13. http://corn.osu.edu/newsletters/2013/2013-13-1/soybean-plant-date-recommendations-1 (accessed 12 Nov. 2013).

Messina, C., G. Hammer, Z. Dong, D. Podlich, and M. Cooper. 2009. Chapter 10. Modeling crop improvement in a GxExM framework via gene-trait-phenotype relationships. In: V.O. Sadras and D.F. Calderni, editors, Crop physiology: Applications for Genetic Improvement and Agronomy. Elsevier, Academic Press, New York. p. 235–265.

Mikel, M., B.W. Diers, R.L. Nelson, and H.H. Smith. 2010. Genetic diversity and agronomic improvement of North American soybean germplasm. Crop Sci. 50:1219–1229. doi:10.2135/cropsci2009.08.0456

Miranda, Z. de F.S., C.A.A. Arias, C.E.C. Prete, R.A. de S. Kiihl, L.A. de Almeida, J.F.F. de Toledo, and D. Destro. 2007. Genetic characterization of ninety elite soybean cultivars using coefficient of parentage. Pesqi. Agropecu. Bras. 42:363–369. doi:10.1590/S0100-204X2007000300009

Morse, W.J., and J.L. Cartter. Rev. 1939. Soybean culture and varieties. USDA Farmers' Bull. 1520. U.S. Gov. Print. Office, Washington, DC.

Padgette, S.R., K.H. Klacz, X. Delannay, D.B. Re, B.J. LaVallee, C.N. Tinius, W.K. Rhodes, Y.I. Otero, G.F. Barry, D.A. Eichholtz, V.M. Peschke, D.L. Nida, N.B. Taylor, and G.M. Kishore. 1995. Development, identification, and characterization of a glyphosate-tolerant soybean line. Crop Sci. 35:1451–1461. doi:10.2135/cropsci1995.0011183X003500050032x

Pedersen, P. 2008. Row spacing in soybean. Iowa State University. http://extension.agron.iastate.edu/soybean/documents/RowSpacing.pdf (accessed 12 Nov. 2013).

Pedersen, P., and J.G. Lauer. 2004. Soybean growth and development response to rotation sequence and tillage system. Agron. J. 96:1005–1012. doi:10.2134/agronj2004.1005

Porter, P., S. Chen, C. Reese, and L. Klossner. 2001. Population response of soybean cyst nematode to long term corn–soybean cropping sequences in Minnesota. Agron. J. 93:619–626. doi:10.2134/agronj2001.933619x

Posada, D., and T.R. Buckley. 2004. Model selection and model averaging in phylogenetics: Advantages of Akaike information criterion and Bayesian approaches over likelihood ratio tests. Syst. Biol. 53:793–808. doi:10.1080/10635150490522304

Probst, A.H., and R.W. Judd. 1973. Origin, U.S. history and development, and world distribution. In: B.E. Caldwell, editor, Soybeans: Improvement, production, and uses. Agron. Monogr. 16. ASA, Madison, WI. p. 1–15.

Quick, G.R., and W.F. Buchele. 1974. Reducing combine gathering losses in soybeans. Trans. ASAE 17:1123–1129. doi:10.13031/2013.37044

Rincker, K., R. Nelson, J. Specht, D. Sleper, T. Cary, S. R. Cianzio, S. Casteel, S. Conley, P. Chen, V. Davis, C. Fox, G. Graef, C. Godsey, D. Holshouser, G-L. Jiang, S. K. Kantartzi, W. Kenworthy, C. Lee, R. Mian, L. McHale, S. Naeve, J. Orf, V. Poysa, W. Schapaugh, G. Shannon, R. Uniatowski, D. Wang, and B. Diers. 2014. Genetic improvement of soybean in North American maturity groups II, III, and IV. Crop Sci. (in press).

Rowntree, S., J. Suhre, E. Wilson, V. Davis, B. Diers, S. Casteel, P. Esker, and S. Conley. 2014. Physiological and phenological responses of historical soybean cultivar releases to earlier planting. Crop Sci. (in press). doi:10.2135/CROPSCI2013.06.0428.

Rowntree, S.C., J.J. Suhre, E.W. Wilson, V.M. Davis, B.W. Diers, S.N. Casteel, P.D. Esker, J.E. Specht, and S.P. Conley. 2013. Genetic gain × management interactions in soybean: I. Planting date. Crop Sci. 53:1128–1138. doi:10.2135/cropsci2012.03.0157

Sacks, W.J., and C.J. Kucharik. 2011. Crop management and phenology trends in the U.S. Corn Belt: Impacts on yields, evapotranspiration and energy balance. Agric. For. Meteorol. 151:882–894. doi:10.1016/j.agrformet.2011.02.010

Sakamoto, Y., M. Ishiguro, and G. Kitagawa. 1986. Akaike information criterion statistics. KTK Scientific Publishers, Tokyo, Japan.

Salado-Navarro, L.R., T.R. Sinclair, and K. Hinson. 1993. Change in yield and seed growth traits in soybean cultivars released in the southern USA from 1945 to 1983. Crop Sci. 33:1204–1209. doi:10.2135/cropsci1993.0011183X003300060019x

Salvagiotti, F., K.G. Cassman, J.E. Specht, D.T. Walters, A. Weiss, and A. Dobermann. 2008. Nitrogen uptake, fixation and response to fertilizer N in soybeans: A review. Field Crops Res. 108:1–13. doi:10.1016/j.fcr.2008.03.001

Scott, W.O., and S.R. Aldrich. 1970. Modern soybean production. 1st ed. S&A Publ., Inc, Champaign, IL.

Sebastian, S.A., L.G. Streit, P.A. Stephens, J.A. Thompson, B.R. Hedges, M.A. Fabrizius, J.F. Soper, D.H. Schmidt, R.L. Kallem, M.A. Hinds, L. Feng, and J.A. Hoeck. 2010. Context-specific marker-assisted selection for improved grain yield in elite soybean populations. Crop Sci. 50:1196–1206. doi:10.2135/cropsci2009.02.0078

Setiyono, T.D., K.G. Cassman, J.E. Specht, A. Dobermann, A. Weiss, H. Yang, S.P. Conley, A.P. Robinson, P. Pedersen, and J.L. De Bruin. 2010. Simulation of soybean growth and yield in near-optimal growth conditions. Field Crops Res. 119:161–174. doi:10.1016/j.fcr.2010.07.007

Sinclair, T.R., and T. Horrie. 1989. Leaf nitrogen, photosynthesis, and crop radiation use efficiency: A review. Crop Sci. 29:90–98. doi:10.2135/cropsci1989.0011183X002900010023x

Sinclair, T.R., and T.W. Rufty. 2012. Nitrogen and water resources commonly limit crop yield increases, not necessarily plant genetics. Global Food Security 1:94–98. doi:10.1016/j.gfs.2012.07.001

Sinclair, T.R., C.P. Tanner, and J.M. Bennet. 1984. Water use efficiency in crop production. Bioscience 34:36–40. doi:10.2307/1309424

Skorupska, H.T., R.C. Shoemaker, A. Warner, E.R. Shipe, and W.C. Bridges. 1993. Restriction fragment length polymorphism in soybean germplasm of the southern USA. Crop Sci. 33:1169–1176. doi:10.2135/cropsci1993.0011183X003300060013x

Smith, S., M. Cooper, J. Gogerty, C. Löeffler, D. Borcherding, and K. Wright. 2014. Maize. In: S. Smith, B. Diers, J. Specht, and B. Carver, editors, Yield gains in major U.S. field crops. CSSA Spec. Publ. 33. ASA, CSSA, and SSSA, Madison, WI. p. 125–172. doi: 0.2135/cssaspecpub33.c6

Sneller, C.H. 1994. Pedigree analysis of elite soybean lines. Crop Sci. 34:1515–1522. doi:10.2135/cropsci1994.0011183X003400060019x

Sneller, C.H. 2003. Impact of transgenic genotypes and subdivision on diversity within elite North American soybean. Crop Sci. 43:409–414. doi:10.2135/cropsci2003.0409

Specht, J.E. 2013. NE irrigated and rainfed yield trends. University of Nebraska Extension. http://cropwatch.unl.edu/web/soybeans/yields (accessed 12 Nov 2013).

Specht, J., P. Grassini, and J. Rees. 2013. Data show Nebraskans planting soybeans earlier each year. University of Nebraska Extension. http://cropwatch.unl.edu/web/cropwatch/archive?articleID=5149479 (accessed 12 Nov. 2013).

Specht, J.E., D.J. Hume, and S.V. Kumudini. 1999. Soybean yield potential– a genetic and physiological perspective. Crop Sci. 39:1560–1570. doi:10.2135/cropsci1999.3961560x

Specht, J.E., and J.H. Williams. 1984. Contribution of genetic technology to soybean productivity—Retrospect and prospect. In: W.R. Fehr, editor, Genetic contribution to yield gains of five major crop plants. CSSA Special Publ. 7. CSSA and ASA, Madison, WI. p. 49–74. doi:10.2135/cssaspecpub7.c3

Statistics Canada and Agriculture and Agri-Food Canada. 2012. Overview of the soybean sector in Manitoba. Statistics Canada, and Agriculture and Agri-Food Canada, June 2012. http://www.gov.mb.ca/agriculture/statistics/pdf/crop_soybean_sector.pdf (accessed 12 Nov 2013).

Stein, A.J., and E. Rodríguez-Cerezo. 2009. The global pipeline of new GM crops: Implications of asynchronous approval for international trade. JRC Technical Report EUR 23486 EN. Luxembourg: Office for Official Publications of the European Communities. http://ipts.jrc.ec.europa.eu/publications/pub.cfm?id=2420 (accessed 12 Nov. 2013).

Tans, P., and R. Keeling. 2013. Trends in atmospheric carbon dioxide. NOAA/ESRL http://www.esrl.noaa.gov/gmd/ccgg/trends/#mlo_full (accessed 12 Nov. 2013).

Temperly, R.J., and R. Borges. 2006. Tillage and crop rotation impact on soybean grain yield and composition. Agron. J. 98:999–1004. doi:10.2134/agronj2005.0215

Thompson, J.A., and R.L. Nelson. 1998. Core set of primers to evaluate genetic diversity in soybean. Crop Sci. 38:1356–1362. doi:10.2135/cropsci1998.0011183X003800050034x

Thorson, T. 2007. Soybean objective yield survey data, 1992–2006. http://usda01.library.cornell.edu/usda/current/SoyObjYield/SoyObjYield-07-27-2007.pdf (accessed 12 Nov 2013).

Toledo, J.F.F., M.F. Oliveira, A.C. Tsutida, and R.A.S. Kiihl. 1993. Genetic analysis of growth of determinate soybean genotypes under three photoperiods. Rev. Bras. Genet. 16:713–748.

Twine, T.E., and C.J. Kucharik. 2009. Climate impacts on net primary productivity trends in natural and managed ecosystems of the central and eastern United States. Agric. For. Meteorol. 149:2143–2161. doi:10.1016/j.agrformet.2009.05.012

Ustun, A., F.L. Allen, and B.C. English. 2001. Genetic progress in soybean of the US midsouth. Crop Sci. 41:993–998. doi:10.2135/cropsci2001.414993x

Van Wart, J., K.C. Kersebaumb, S. Peng, M. Milner, and K.G. Cassman. 2013. Estimating crop yield potential at regional to national scales. Field Crops Res. 143:34–43. doi:10.1016/j.fcr.2012.11.018

Vilsack, T., and C.Z.F. Clark. 2008. The 2007 Census of Agriculture. Farm and Ranch Irrigation Survey. Volume 3. Special Studies. Part 1. AC-07-SS-1. http://www.agcensus.usda.gov/Publications/2007/Online_Highlights/Farm_and_Ranch_Irrigation_Survey/fris08.pdf (accessed 12 Nov. 2013).

Voldeng, H.D., E.R. Cober, D.J. Hume, C. Gillard, and M.J. Morrison. 1997. Fifty-eight years of genetic improvement of short-season soybean cultivars in Canada. Crop Sci. 37:428–431. doi:10.2135/cropsci1997.0011183X003700020020x

Waltz, E. 2010. Food firms test fry Pioneer's trans fat-free soybean oil. Nat. Biotechnol. 28:769–770. doi:10.1038/nbt0810-769a

Wilcox, J.R. 2001. Sixty years of improvement in publicly developed elite soybean lines. Crop Sci. 41:1711–1716. doi:10.2135/cropsci2001.1711

Wilson, E.W., S.C. Rowntree, J.J. Suhre, N.H. Weidenbenner, S.P. Conley, V.M. Davis, B.W. Diers, P.D. Esker, S.L. Nave, J.E. Specht, and S.N. Casteel. 2014. Genetic gain × management interactions in soybean: II. Nitrogen utilization. Crop Sci. 54:340–348. doi:10.2135/cropsci2013.05.0339

Xu, Z., D.A. Hennessy, K. Sardana, and G. Moschini. 2013. The realized yield effect of genetically engineered crops: U.S. maize and soybean. Crop Sci. 53:735–745. doi:10.2135/cropsci2012.06.0399

Zhang, L.X., S. Kyei-Boahen, J. Zhang, M.H. Zhang, T.B. Freeland, C.E. Watson, Jr., and X.M. Liu. 2007. Modifications of optimum adaptation zones for soybean maturity groups in the USA. Crop Manage. doi:10.1094/CM-2007-0927-01-RS

Zhu, X.G., S.P. Long, and D.R. Ort. 2010. Improving photosynthetic efficiency for greater yield. Annu. Rev. Plant Biol. 61:235–261. doi:10.1146/annurev-arplant-042809-112206

13

Sugarbeet

Lee Panella,* Stephen R. Kaffka, Robert T. Lewellen, J. Mitchell McGrath, Mike S. Metzger, and Carl A. Strausbaugh

ABSTRACT

Sugarbeet (*Beta vulgaris* L. subsp. *vulgaris*) is an industrial crop grown commercially as a hybrid, with sucrose refined from the root as the plant constituent of interest. Yield of refined white sugar produced per unit area is a complex trait. Sugar yield is calculated by the root weight times the proportion of root that is sucrose, minus the amount of sucrose lost during processing, often expressed as percent loss to molasses or juice purity. Yields have risen in the last 100 yr from about 10% sucrose and 22.4 Mg ha^{-1} (10 t ac^{-1}) to 18% sucrose and more than 65 Mg ha^{-1} (29.3 t ac^{-1}). There also has been a shift in acreage from the western United States (especially California and Colorado) to the northern Midwest (Red River Valley). Although not all of this improvement is attributable to genetic improvement in the crop, a strong collaborative effort between commercial breeders and public breeders (USDA-ARS) has increased yield potential, while improving resistance to many diseases, leading to increases in yield. This continuing collaboration between private and public breeders will be necessary to meet the challenges of climate change and the utilization of beet as a sustainable biofuel and chemical feedstock.

Abbreviations: BCT, beet curly top; BNYVV, Beet necrotic yellow vein virus; CMS, cytoplasmic male sterility; GHG, greenhouse gases; QTL, quantitative trait loci; RRV, Red River Valley.

L. Panella, USDA-ARS, NPA, Crop Research Lab., 1701 Centre Dr., Fort Collins, CO 80526. *Corresponding author (Lee.Panella@ars.usda.gov).
S.R. Kaffka, Dep. of Plant Sciences, Univ. of California, Davis, CA 95616-8780 (srkaffka@ucdavis.edu).
R.T. Lewellen, USDA-ARS, Crop Improvement and Protection Research Unit, 1636 E. Alisal St., Salinas, CA 93905 (Robert.Lewellen@ars.usda.gov).
J.M. McGrath, USDA-ARS, Sugarbeet and Bean Research Unit, Dep. of Plant, Soil, and Microbial Sciences, 494 Plant & Soil Sci. Bldg., 1066 Bogue St., Michigan State Univ., East Lansing, MI 48824-1325 (mitchmcg@msu.edu).
M. Metzger, Minn-Dak Farmer's Cooperative, 7525 Red River Rd., Wahpeton, ND 58075 (mmetzger@mdf.coop).
C.A. Strausbaugh, USDA-ARS, Northwest Irrigation and Soils Research Lab., 3793 North 3600 East, Kimberly, ID 83341-5076 (Carl.Strausbaugh@ars.usda.gov).

doi:10.2135/cssaspecpub33.c13

Yield Gains in Major U.S. Field Crops. CSSA Special Publication 33.
Stephen Smith, Brian Diers, James Specht, and Brett Carver, editors.

Sugarbeet is an industrial crop. Neither the seed nor the foliage is the plant constituent of commercial interest—not even the root itself, but rather the sucrose refined from the root (an enlarged hypocotyl). Yield of refined white sugar produced per unit area is a complex trait. Sugar yield is calculated by the root weight times the proportion of root that is sucrose (usually expressed as the percentage of fresh weight), minus the amount of sucrose lost during processing (often expressed as the percentage of loss to molasses or juice purity). Loss to molasses arises from cations such as Na^+, K+, and amino nitrogen ($R\text{-}NH_3$+) balancing the charge of the anion sucrose at physiological pH. The cations that interfere with the extraction of sucrose are related in a large extent to the plant nutrients in the soil in which the sugarbeet is grown; however, the small amino-nitrogen compounds (e.g., betaine) can be lowered by proper nitrogen fertilization (Follett et al., 1970). All of these components are variable in a population and can be improved through selection and breeding (Biancardi et al., 2010).

Taxonomically, sugarbeet is ordered as Dicotyledoneae, Caryophyllidae (Centrospermae), Amarantheaceae (formerly Chenopodiaceae), *Beta vulgaris* L. subsp. *vulgaris* (McGrath et al., 2011). Originally, it is likely that beet was domesticated as a pot herb, and leaves first harvested from the wild progenitor, sea beet [*B. vulgaris* L. subsp. *maritima* (L.) *Arcang*] for food (Biancardi et al., 2012; Ford-Lloyd et al., 1975; Lange et al., 1999). By the Middle Ages there is reference to the root being used, both as a vegetable and medicinal herb (Biancardi et al., 2012; Goldman and Navazio, 2008). William Morgan provided a historical account of the development of fodder beet in 1822 (cited in Ford-Lloyd et al., 1975), which provided the gene pool for the original sugarbeet (Fischer, 1989). Beet has been grown exclusively for sucrose only since the late 1700s, when it became possible to measure the sucrose present in the beet juice (Francis, 2006; McGrath and Fugate, 2012). Commercial-scale production of sucrose from beet was hastened in Napoleonic France, when that country was blockaded from its colonies and sucrose from sugarcane (Biancardi et al., 2010). The commercial beet crop harvested is the first year's vegetative growth of a biennial plant. Beet has a complex breeding system. Flowering requires 10 to 16 wk at 4°C (i.e., vernalization), at which time the rosette "bolts" a flower stalk with indeterminate inflorescences. Flowers are perfect and wind pollinated. Cross-fertilization is enforced by a complex self-incompatibility system (Biancardi et al., 2010).

Breeding has improved sugar yield markedly over the past 150 yr, from 4 to 6% at the onset of sucrose yield selection to >18% in modern hybrids and from a national average of just more than 20 Mg ha^{-1} in 1909 to a current average more than 60 Mg ha^{-1}. Progeny testing and recurrent selection for sucrose content were employed, likely the first scientific application of these tools in common use today (Vilmorin, 1859, cited in Gayon and Zallen, 1998; Ware, 1880). Selection diversified the sugarbeet germplasm into types with a greater percentage of extractable sucrose (i.e., Z-type sugarbeet for "Zucker," the German word for sugar) and types with increased tonnage (i.e., E-type sugarbeet for "Ernte," the German word for harvest). Sucrose accumulates in sugarbeet roots to about 75% of dry weight (15–20% fresh weight). Sucrose biosynthesis occurs by mechanisms similar to those in other plants. Sucrose is concentrated within the innermost five of 10 to 12 concentric cortical rings, around the point of maximum root girth. It accumulates in vacuoles of parenchyma cells adjacent to the vascular tissue, and the percent sucrose is positively correlated to the amount of solar radiation captured once the leaf canopy reaches full coverage (Elliott and Weston, 1993).

Doney et al. (1981) demonstrated that roots with smaller cells give higher sucrose percentages than those with larger cells. Savitsky (1940) suggested three or four major genes control sucrose percentage in crosses of sugarbeet with divergent types such as fodder and red beet (3–12% sucrose fresh weight each). Others suggest sucrose percentage is a quantitatively controlled trait with high heritability (Culbertson, 1942; Zhao et al., 1997). Recent mapping of sucrose genes suggests at least five loci on five of the nine beet chromosomes contribute to final sucrose content in a sugarbeet × sugarbeet population (Schneider et al., 2002). Root yield, on the other hand, is generally considered a nonadditive trait, and quantitative trait loci (QTL) analyses to date have shown only two loci with significant influence over yield (Schneider et al., 2002).

The improvement in commercial cultivars in percent sucrose over the past 50 yr, although considerable, has not been as great as the increase in root yield. Over the last 100 yr of sugarbeet production, an increase in both sucrose content, and especially, root yield has been achieved through genetic improvement and better crop management (Francis, 2006; Fig. 13–1). It is difficult to estimate, however, exactly how much of that improvement is due to genetic improvement.

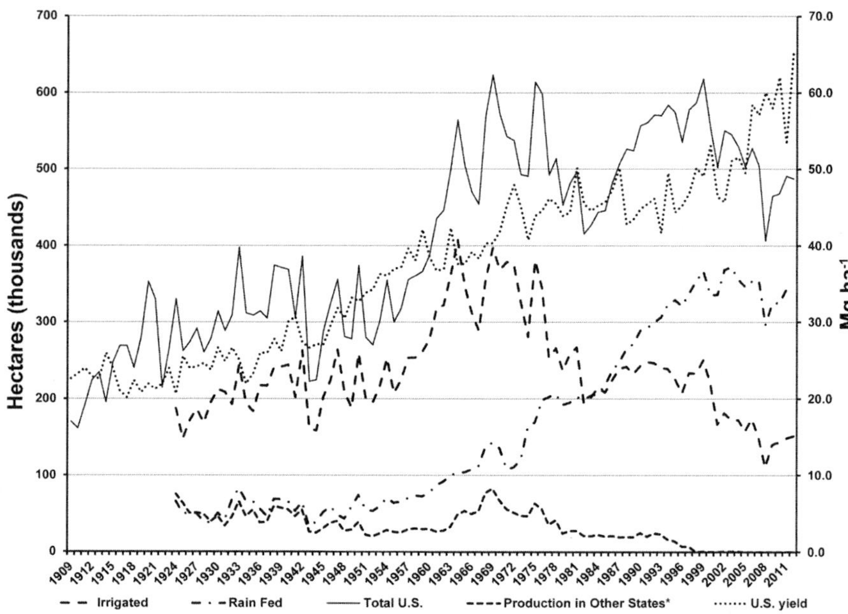

Fig. 13–1. Changes in the area of sugarbeet harvested are shown for the entire United States from 1909 to 2012, as well as changes over time (1924–2012) in the area harvested in the western irrigated regions (California, eastern Oregon, Idaho, Montana, Wyoming, Colorado, and western Nebraska) and the eastern rainfed regions (North Dakota, Minnesota, and Michigan) (left vertical axis). The growth in average root yield per hectare is shown for the entire United States from 1909 to 2012 (right vertical axis). Data are from USDA-NASS. *"Other States" includes Arizona, Illinois, Indiana, Iowa, Kansas, Maine, New Mexico, New York, Ohio, South Dakota, Texas, Utah, Washington, and Wisconsin.

All commercial sugarbeet cultivars currently grown in the United States are hybrid, generated using a single, vulnerable, and complex system of cytoplasmic male sterility (CMS) (Owen, 1945). Modern cultivars also are monogerm (i.e., a single flower borne per axil), which eliminates the need for labor to thin stands, as was required for open-pollinated multigerm sugarbeet cultivars (Savitsky, 1950). New characters can be difficult to introgress into the inbreds that are hybrid parents. The pollinator parent of a hybrid often is selected out of a multigerm population that was selected for disease resistance. It can be easier to fix resistance alleles in pollinators than in the CMS (female) parent lines, because of the time invested in developing a maintainer line (O-type line) and female pair. It is necessary that the O-type maintainer be homozygous recessive at both of the loci that control the restoration of sterility in the female.

Current germplasm developed in U.S. public breeding programs and released to the seed companies is the product of several phases of development: early efforts to improve disease resistance in open-pollinated cultivars, followed by conversion of open-pollinated cultivars into genetic CMS parents for hybrid cultivar production (Biancardi et al., 2005); the mining of existing crop diversity to manage emerging diseases and achieve desired agronomic traits (e.g., reducing soil tare); and more recently the discovery and deployment of novel traits from crop wild relatives and unadapted genetic resources (Panella and Lewellen, 2007). Many sugarbeet breeding activities in the public breeding sector have been focused on improving populations via mass selection of mother roots, harvested and vernalized for subsequent seed production, and releasing the improved populations from these programs in the form of enhanced germplasm (Doney, 1995).

Foundations of U.S. Germplasm

Through World War I, most U.S. seed was imported from Europe or Russia as unadapted, open-pollinated cultivars, and the U.S. beet sugar industry struggled with recurring disease problems using these materials (Lewellen, 1992; Panella and Lewellen, 2007). The USDA's Division of Sugar Plants responded by the development of a sugarbeet research program that began almost 100 yr of collaboration that has been maintained among commercial, federal, and state researchers and plant breeders. Through this collaboration there was the development and release of cultivars resistant to *Beet curly top virus* for western growing regions (Coons et al., 1931) and later, for eastern growers, of cultivars resistant to aphanomyces seedling and root diseases (caused by *Aphanomyces cochlioides* Drechs.) (Coons et al., 1954). This distinction in germplasm persists today in modern hybrids (McGrath et al., 1999). Cercospora leaf spot (caused by *Cercospora beticola* Sacc.) also was a major focus for sugarbeet germplasm enhancement, mainly in areas east of the Rocky Mountains, and the development of this germplasm has been reviewed recently (Panella and McGrath, 2010). Development of the two U.S. germplasm pools has not been summarized recently, and since these are foundational germplasm materials, their development is briefly summarized here.

Beet curly top (BCT) in sugarbeet (caused by a mixture of closely related *Curtovirus* species: *Beet curly top virus*, *Beet mild curly top virus*, and *Beet severe curly top virus*) (Stenger and McMahon, 1997; Strausbaugh et al., 2008), transmitted by the beet leafhopper, *Circulifer tenellus* (Baker), is not found in European production areas (Bennett, 1971). Beet curly top severely impacted yields just as the sugarbeet industry was beginning to expand in the western United States (Bennett, 1971; Carsner, 1924, 1926a; Murphy, 1946). This especially was true in California and Idaho (Fig. 13–2), where the beet leafhopper over-winters in weeds in the surrounding foothills. In California between 1910 and 1930 there was a decrease in root yield, sugar yield per acre, as well as a sharp drop in the area harvested (Duffus, 1978; Wisler and Duffus, 2000) (Fig. 13–3 and 13–4). Survival of the sugarbeet industry in the western United States depended on being able to manage this devastating disease (Bennett, 1971; Panella, 2005). In the 1920s, the USDA-ARS, then called the Bureau of Plant Industry) joined the private sector sugarbeet breeding efforts to develop cultivars resistant to BCT (Bennett, 1971; Bennett and Leach, 1971; Carsner, 1926b; Coons et al., 1931; Coons, 1953; Murphy, 1942, 1946). A series of joint releases between USDA-ARS and the sugarbeet industry began with the release of 'US 1,' a curly top resistant open-pollinated cultivar (Carsner, 1933; Owen et al., 1939). US 1 primarily had been derived from selections within European cultivars, the most important of which was R & G Old Type, also known as 'KleinE' (Lewellen 1992). Further increases in resistance to BCT resulted in 'US 33' and 'US 34' selected from heavily curly top infested fields of 'US 1,' and these eventually were replaced by 'US 12' and 'US 22,' which were further improved as 'US 22/2' and 'US22/3' (Coons et al., 1955).

Aphanomyces became a serious threat to eastern U.S. sugarbeet growing areas by the early 1940s, primarily causing a serious damping-off problem (black-leg), but also a mature root rot. Both remain problems, but their occurrence is limited and sporadic, mainly being a problem in warmer, wetter soils. Bockstahler observed differential reactions to the chronic form of black-leg in 1941 (Henderson and Bockstahler, 1946). In *A. cochlioides* infested plots in Waseca, MN, a polycross cultivar designated as '1942 Minnesota Synthetic No. 1' was created, and additional populations were selected in subsequent years (Bockstahler

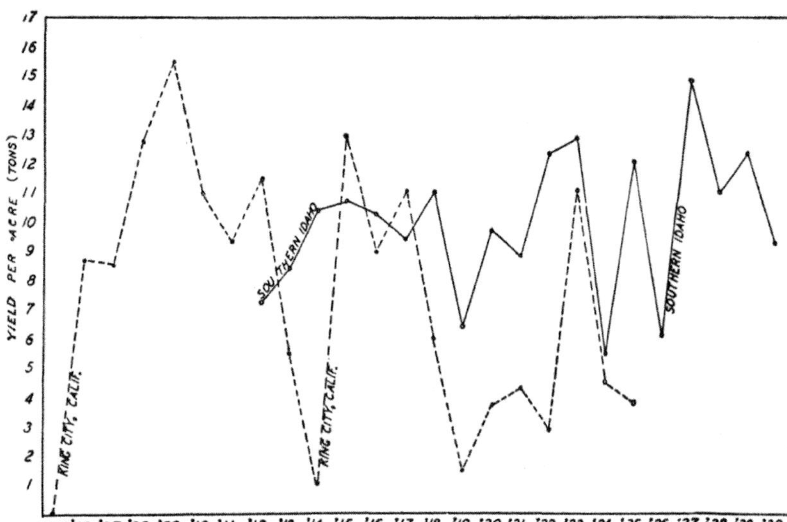

Fig. 13–2. Yield fluctuations from 1905 until 1930 seen in the areas of California and Idaho most affected by beet curly top (from Carsner, 1933, p. 2).

and Reece, 1948). Most aphanomyces-resistant germplasm was derived from selections within populations polycrossed in 1945 and 1946 from 'US 215,' 'US 216,' 'US 200,' 'R-581,' 'NYA,' 'LaRoyale,' 'Cesena,' 'US 22,' 'SL 561,' 'SL 611,' 'SL 618,' 'American 0-701-0,' 'American 1,' 'American 3,' 'Minnesota Synthetic No. 1,' as well as synthetics containing Italian varieties, *Beta maritima* [sic], and sucrose and yield selections. Much of this early germplasm had its roots in European open-pollinated cultivars, such as R&G Pioneer and Old Type, or in Great Western open pollinated cultivars, such as GW304 and GW359, derived from some of these European sources (Lewellen, 1992).

Four hundred four roots from 48 progenies of this family, as well as five related families, were intercrossed in 1947 as '48B3-00.' The direct increase of this seed was released as 'US 1177'. Fifty-four roots of '48B3–00' were reselected in 1949, and named '50B3–0,' and a direct increase was released as 'US 400' in 1954 (Coons et al., 1954), which had better aphanomyces black-leg resistance than 'US 1177.' By 1955, enough commercial seed of 'US 400' was available to plant 12,150 ha (30,000 ac). Selections of 105 roots from '50B3-0' in nurseries at East Lansing, Blissfield, MI and Waseca, MN were interpollinated to produce '53AB3–0,' which was released in 1955 as 'US 401,' with greater aphanomyces and cercospora leaf spot resistance than 'US 400,' and replaced it commercially. A second important lineage of releases from these materials differed in the method of selection. That is, 'US 400' and 'US 401' were derived from mass selection after the 1947 progeny selection, but the alternate lineage relied on selection of the best individual roots in the best families (i.e., mother root selection) from progeny of '53AB1,' which was thrice selected from the 1946 Waseca selections for the highest combined black-leg, leaf spot, sucrose, and yield. These selections included the '02 clone' selected from '53AB1–32' (self-sterile but sib-fertile), which was released subsequently as 'EL 40' (Hogaboam et al., 1982). Collaborative work for cercospora leaf spot selection was centered in Beltsville, MD, where 'SP 5460-0' was released from '53AB1-65' and 'SP5481-0' from combined progenies of '53AB1,' and likely 'SP 5822-0.' SP 5822 likely was the progenitor of 'SP 6322,' which was the original pollinator of the historically important and widely cultivated hybrid 'US H20' (Coe and Hogaboam, 1971a,b).

Rhizomania is another yield limiting virus problem worldwide (Biancardi et al., 2002), caused by the *Beet necrotic yellow vein virus* (BNYVV), which is vectored by the plasmodiophorid, *Polymyxa betae* (Rush et al., 2006; Rush, 2003). Rhizomania infection causes a reduction of up to 50% in root yield and negatively affects sugar yield and processing quality (Büttner et al., 1995). In the United States, the virus first was discovered in 1983 in California (Duffus et al., 1984; Biancardi et al., 2002). The first documented and widely used source of resistance to BNYVV was the gene, *Rz1* (Lewellen et al., 1987). *Rz1* is monogenic and although the dominant allele does not confer immunity, it greatly reduces damage and symptom expression. *Rz1* has been deployed in backcross and germplasm enhancement programs worldwide and has become the major management tool against yield loss by BNYVV (Asher et al., 2002; Francis et al., 1998; Pavli et al., 2011; Rush et al., 2006; Scholten et al., 1997).

Shortly after the discovery of *Rz1*, an additional resistance gene was found in a sea beet (*Beta vulgaris* subspecies *maritima*) accession (WB42–PI 546385), designated *Rz2* (Lewellen, 1997; Scholten et al., 1999; Whitney, 1989). Cultivars with the *Rz1* gene began showing rhizomania symptoms in some fields in the Imperial

Valley of California in the early 2000s. Extensive testing in 2004 and 2005 by USDA-ARS at Salinas, confirmed that resistance conditioned by the *Rz1* gene had been overcome (Liu et al., 2005; Liu and Lewellen, 2007; Rush et al., 2006). The *Rz2* source of resistance, alone or in tandem with *Rz1*, has since become more common in commercial cultivars. Other sources of rhizomania resistance that have been identified are *Rz3* from WB41 (Gidner et al., 2005, Lewellen et al., 2007), *Rz4* derived from a composite of about 60 sea beet accessions from Europe and the Middle East (Grimmer et al., 2007), and *Rz5* from WB258 (PI 546426) (Grimmer et al., 2008). There is debate as to whether or not these sources of resistance represent individual genes or simply alleles at the same locus, but all may be needed to maintain genetic resistance to rhizomania (McGrann et al., 2009).

The U.S. beet seed industry initially was a cooperative effort among sugar factory districts, which always had conducted official cultivar trials for their growers; public breeders (USDA-ARS), who develop enhanced germplasm; private breeders working for U.S. seed companies, often owned by the sugar processors; and approved cultivar multiplication, primarily done by the West Coast Beet Seed Co. of Salem, OR (Johnson and Burt, 1990). With the advent of hybrids, the commercial opportunities afforded by hybrid seed sales, and the reentry of large international seed companies into the U.S. market, a process which began again during the 1970s, public sector breeders have focused more on germplasm enhancement, while commercial sugarbeet seed company breeders have come to dominate cultivar development. Along with the recent assimilation of many small seed companies into larger corporations, this has led to a few companies supplying all of the seed to U.S. growers. This has been an important aspect in rising yields due to their affiliation with international seed companies, which have diverse germplasm collections and genetic marker and genetic engineering capabilities.

Some biolistic- and Agrobacterium-mediated transformation of sugarbeet has occurred in the public sector, but this technology is largely subsumed by the commercial sector, and the only genetically engineered trait currently commercialized is glyphosate [*N*-(phosphonomethyl)glycine] resistance for weed control. This trait initially was deregulated in 1998, but seed from that transformation event was never commercialized. The current glyphosate-resistant sugarbeet resulted from another transformation event (H7-1, U.S. Patent 7335816 B2) and was deregulated in March 2005, grown on a limited basis in Wyoming in 2007, and released across the United States in 2008. Commercial adoption rates quickly reached >98% by 1 yr after its official initial deregulation. A series of legal challenges resulted in partial re-regulation in 2010 and finally full deregulation in 2012. Although other bioengineered traits of interest are in development (varied disease resistances, increased water and nitrogen use efficiency, and winterbeet for more temperate climates), none have been commercialized.

Cultivation and Beet Production

Sugarbeet cultivation came to this country with German and Russian immigrants, who were familiar with the crop from their homelands. After the U.S. Civil War and the abolition of slavery, there was concern that the United States might not remain self-sufficient in sugar production. In addition, the potential economic impact of a northern sugar industry was also an attractive proposition in itself (Grant, 1871). Between 1838 and 1879, 14 beet sugar factories were built—all of which failed (Theis, 1971). Finally in 1879, the factory built in Alvarado, CA in 1870 turned a profit. The industry then began to expand and by 1909, there were about 162,000 ha (400,000 ac) of sugarbeet harvested in the United States, with an average yield of about 22.4 Mg ha^{-1} (10 t ac^{-1}), with 10% sucrose, processed in 41 factories (Harris, 1919; USDA-NASS, 2013) (Fig. 13–1). From 1909 to present (Fig. 13–1), maximum acreage (1969) was about 607,500 ha (1.5 million acres) and currently varies between 400,000 and 510,000 ha (1.00 and 1.25 million acres). Both root yield (Fig. 13–1) and the percentage of sucrose in the root have risen (Francis, 2006), as has the capacity of the 21 currently operating sugarbeet processing factories to process sugarbeet. Today, U.S. farmers and processors are producing much more sugar on less land.

By 1924, when the USDA-NASS began keeping records for most crops, sugarbeet was being cultivated in 20 states. The U.S. sugarbeet crop grew as more land was brought into cultivation in more areas; sugarbeet was cultivated at one time or another in 28 states ranging from California to New York and from Maine to the semiarid southwest. Today, sugarbeet is cultivated in 10 states that often are grouped into five production regions: the West (California), the Intermountain Region (Idaho and eastern Oregon), the Great Plains (Colorado, Wyoming, Nebraska, and Montana), the Upper Midwest or Red River Valley (Minnesota and North Dakota), and the Great Lakes (Michigan). In Fig. 13–1, the area harvested data are shown for "irrigated" (West, Intermountain, and Great Plains regions) and rainfed (Upper Midwest and Great Lakes regions) areas from these 10 states, with the historical production from all the other states combined. It is clear from the changes in the area harvested (Fig. 13–1) that production was initially highest in the western United States, where the sugarbeet crop was irrigated, and that as the area under cultivation in the Red River Valley and Michigan increased, the area cultivated for sugarbeet peaked at the end of the 1960s, when the production in states other than the current 10 states began to fall. Disease pressure in some growing areas, fluctuations in environmental conditions, and economic fluctuations in the price of sugar pushed many areas out of production. A serious lack of investment in factories to keep them efficiently operating with the latest technology was also a major factor. Production also migrated to the rainfed areas, where cost of production is lower by avoiding the need to irrigate. As the area under cultivation has dropped, yields have continued to rise, maintaining production goals (Fig. 13–1).

In the United States root yield was variable and did not increase much during the 20 yr from 1950 to 1970 (Fig. 13–1). However, since then the development and use of superior cultivars (resistance to BCT, bolting, *Rhizoctonia solani* Kühn, *Cercospora*, downy mildew [*Peronospora farinosa* (Fr.:Fr.) Fr. f. sp. *betae* Byford] resistance, and other pests and diseases), the introduction of hybrid cultivars with monogerm seed, greater use of new chemical protectants, better harvesting equipment, and more advanced fertilization management have led to a steady increase of yield in the United States. This increase was mirrored in much of Europe (Zimmermann and Zeddies, 2000). Breeders and producers in each of the U.S. growing regions share best ideas for cultural practices and continued improvements in germplasm for cultivar development; however, each growing region has a unique set of challenges that often differs from other regions in the magnitude of the problem. For instance, curly top and yellowing virus problems are mainly confined to the western U.S. growing regions, fungal leaf spot and root rots are more problematic in the east, and rhizomania is a very serious problem in all growing regions. Thus, solutions achieved in areas of severe impact are transferred throughout the United States (and worldwide) as the situation demands.

California—The Western Region

Although the initial attempts to grow sugarbeet in the United States were made in the eastern part of the country as early as 1838 (Grant, 1871), it was in California that sugar production from beet first became successful. After a number of failures, E.H. Dyer was successful in establishing a factory that made a profit by 1879. This factory in Alvarado, CA is considered the first successful U.S. factory. In 1888, Claus Spreckels built a factory at Watsonville, CA, which produced 907 Mg (1000 t) of sugar in the first year, and by 1889, there were two sugarbeet processing factories operating in the United States, both in central California (Harris, 1919; Theis, 1971). By 1899 California was the leader in beet sugar production, with 37% of the U.S. cultivated area. There was a severe drought from 1897 to 1900, and then tough economic times, which caused dramatic fluctuations in the production and cultivation of sugarbeet in California (Souder, 1970). Nonetheless, by 1910, the area under cultivation had grown to about 36,650 ha (90,500 ac), with a yield of about 24.48 Mg ha^{-1} (11.08 t ac^{-1}) (Fig. 13–3 and 13–4). Over the next 20 yr, the disease pressure from BCT caused a steady drop in root yield, until the first curly top-resistant cultivars became available in the early 1930s (Fig. 13–3 and 13–4). Beet curly top affects the percentage of sucrose in the root as well, causing an even more drastic drop in gross sugar yield over the area harvested (see Fig. 1 of Duffus, 1978; Wisler and Duffus, 2000). With resistant cultivars available to

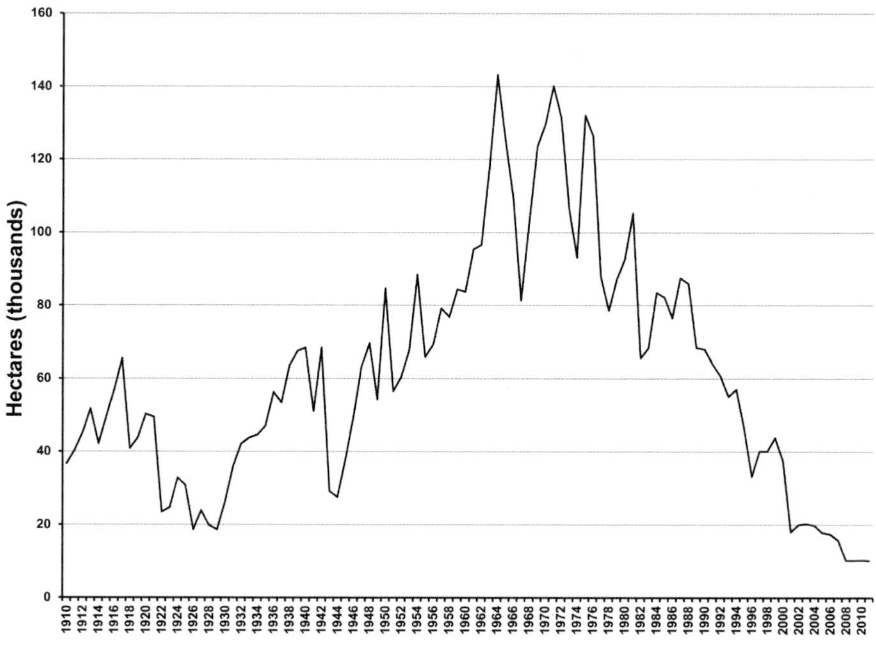

Fig. 13–3. Changes in the area of sugarbeet harvested are shown for the state of California from 1924–2012. Data are from USDA-NASS.

manage BCT, and improved crop cultural practices, the industry rebounded, and the area under cultivation, root yield, and gross sugar yield grew over the next 20 yr (Fig. 13–3 and 13–4) (Duffus, 1978; Wisler and Duffus, 2000) until there were nine sugarbeet processing factories active in California (United States Beet Sugar Association, 1936).

In 1951, the virus yellows complex was reported in California for the first time (Coons and Kotila, 1951). *Virus yellows* is term used to describe a complex of yellows-inducing viruses including, *Beet western yellows virus, Beet chlorosis virus, Beet yellows virus, Beet mosaic virus* and other aphid-transmitted viruses. With heavy infections of more virulent strains, root yield reductions of more than 50% and losses of up to two percentage points of sucrose have been reported (Bennett et al., 1957). Therefore, although U.S. yields trended upward (Fig. 13–1), it was clear that virus yellows limited yield progress in California, much as curly top had depressed yields three decades earlier (Fig. 13–4). The cycle of infection of yellowing viruses was sustained by continual (year-round) beet crops in the areas most affected. With both fall-sown and spring-sown crops, aphid vectors could move the yellowing viruses from one crop to the next. This disease transmission cycle was broken when growers organized regionally to isolate old beet fields from young fields by "beet-free" periods in diverse regions of the state (Wisler and Duffus, 2000). With this cycle broken, the new, higher yielding hybrids; increased access to irrigation water; and economic changes, which favored sugarbeet being grown on more productive land, led to increased yields (Duffus, 1978).

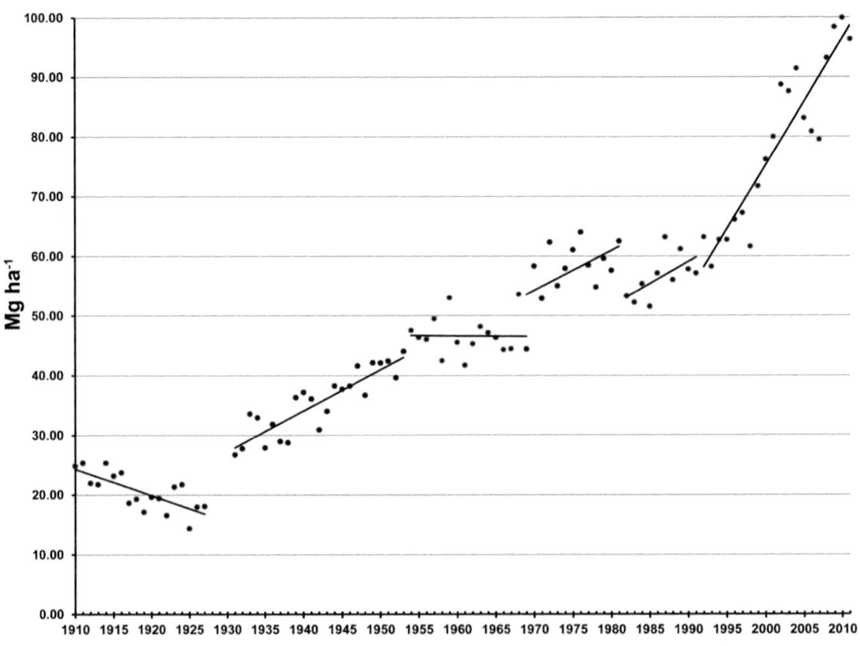

Fig. 13–4. Changes in the average root yield are shown for the state of California from 1910 to 2011. Linear regression lines show yield trends that were present over different time periods. Data are from USDA-NASS, California Field Office.

In addition to the increasing area infested with BNYVV during the 1980s and early 1990s, a number of other problems affected California sugarbeet production. Changing economics of growing sugarbeet, sugarbeet cyst nematode (*Heterodera schachtii* Schmidt) pressure, and a number of whitefly (*Bermisia* spp.) vectored diseases impacted both yields and acreage (Wisler and Duffus, 2000). The less productive areas in northern California were the first to lose sugarbeet in their rotations. California had about 25% of the U.S. sugarbeet area under cultivation in the 1960s and 1970s, but by the end of the 1990s it represented only 10% of the U.S. acreage, and today there only are about 10,100 ha (25,000 ac) in the Imperial Valley, where sugarbeet is grown as a fall-planted, summer-harvested crop (Fig. 13–3) (Wisler and Duffus, 2000).

Imperial Valley of California

The Imperial Valley of California deserves a brief discussion because it is a unique growing area for sugarbeet in the United States and has the highest gross sugar yields per area in the world. Sugarbeet has been grown in the Imperial Valley since 1932, shipped by rail to factories elsewhere in California or Chandler, AZ. The currently operating processing factory in Brawley was built by Holly Sugar Corporation at the end of the World War II and opened in 1947. Shipping of a portion of the crop to other factories continued from then until the early 1990s, when rail shipment became uneconomical or factories closed. The factory in Brawley is the last one operating in California, where 24 (11 successful) factories have been constructed, operated, and eventually closed since the first successful commercial sugarbeet factory was opened in 1870 near San Jose (Harris, 1919; Souder, 1970).

The Imperial Valley is a low desert ecosystem with mild, sunny winters and hot summers. It receives the most sunlight of any location in the United States. Most soils are high in pH and calcareous, with large amounts of naturally occurring gypsum. Most Imperial Valley fields are tile drained at 1.8 m (6 ft) in depth. Subsurface drainage maintains positive salt balances in the Imperial Valley soils, which otherwise would salinize. Water delivery to the Imperial Valley and power production is provided by the Imperial Irrigation District, first organized in 1911. The All American Canal conveys approximately 3.2074 billion m^3 (2.6 million ac ft^{-1}) of water per year to 172,125 ha (425,000 ac) of farmland and seven cities and towns. The canal runs along the Mexican border to the south. Water is then diverted through a series of canals northwards to towns and fields. Because the land slopes northwards toward the Salton Sea, little power is needed to move water. Imperial Irrigation District holds some of the most senior water rights to the Colorado River. The result for growers has been a relatively constant supply of water at low prices (currently ~$20 per 1233.6 m^3 [ac ft^{-1}]) in comparison to many other irrigation districts in California.

Unlike all other U.S. growing regions where sugarbeet is a spring-planted crop, the Imperial Valley sugarbeet crop is planted from the middle of September to early October, and harvest begins ~180 d later starting in early April. Harvest usually is concluded in July, but there has been a tendency in recent years to extend the harvest later in the month, often running into August. This has occurred due to increasingly larger sugarbeet yields, and at times due to operating difficulties in the factory. Harvest campaign length has tended to increase over the last 25 yr, but is not solely responsible for increased yields over that period. Examining the period from 2007, when all sugarbeet production in California was in the

Imperial Valley (Fig. 13–5) compared to the California production since 1989 (Fig. 13–6), it is clear how much higher yields in the Imperial Valley have been. In 2012 in the Imperial Valley, another world-record sugarbeet crop was harvested on a 28.35-ha (70-ac) field. That field averaged 176.31 Mg ha^{-1} (78.71 t ac^{-1}) with 15.89% sugar (fresh weight) for a gross sugar yield of 11,356.4 kg ha^{-1} (25,014 lb ac^{-1}) (Lilleboe, 2012).

Yields in the Imperial Valley of California continue to increase. Sugarbeet crops increase tonnage during the spring period (April–June) at a greater rate than from establishment in fall to start of harvest in April. Dry matter more than doubles by the end of the last 4 mo of growth in late July and early August, compared to the period from establishment in September until the start of harvest in April. A number of factors have combined to contribute to larger yields: improved performance of new hybrids, including improved rhizomania resistance; reduction in losses to *Lettuce infectious yellows virus* starting in the early 1990s; better and wider selection of herbicides for the control of weeds (glyphosate-resistant cultivars had not been grown in the Imperial Valley, but they have been approved beginning in the 2013 growing season); the adoption of newer style harvesters allowing for harvest from wetter soil, permitting irrigation closer to harvest, and better management and timing of late season irrigations during the hottest weather; improved seed quality and stand establishment practices; and a partial shift to production on better quality soils in recent years due to recent increases in the profitability of sugarbeet. These factors have interacted to increase beet dry matter accumulation in the April to June period at a greater rate than observed during previous decades. During the same period, sugar concentration in roots

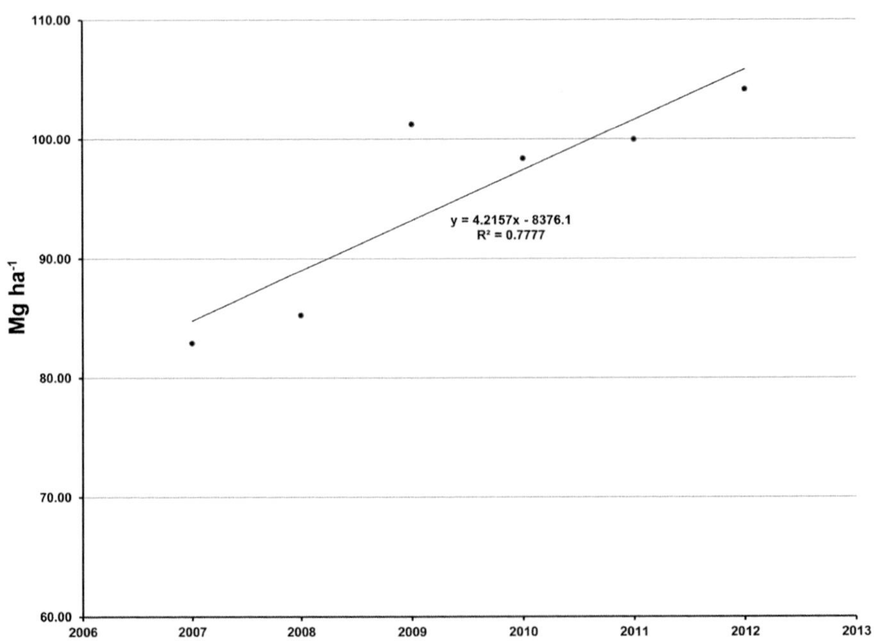

Fig. 13–5. Increased average root yield per hectare in the Imperial Valley of California is shown from 2007 to 2012. Data are from the California Beet Growers Association.

has remained relatively stable, so increased root yields have been primarily responsible for increased sugar yields.

The rate of increase in root yield and decrease in sugar percentage both accelerate over time during the growing season in the IV. This is due to increasing temperatures resulting in several disadvantages as the season progresses. Year-round mild to hot temperatures makes it impossible to store sugarbeet in piles, as is done in other beet-producing regions with spring sown, fall harvested crops. In the IV, logistics are managed to process sugarbeet within 24 h of harvest, with the factory running 24 h, 7 d a week from April to July. Temperatures, especially night time temperatures, and correlated respiratory sugar losses increase as the season progresses, the costs of insect pest management and correlated losses increase, larger amounts of irrigation water are needed, and damage from nematodes and root rots becomes more common, especially in July and August. To operate the factory for a long enough season, late season inefficiencies and costs (especially water use) must be tolerated. Besides future genetic improvements (including the use of transgenic traits), the most likely technology that could help reduce losses late in the season and improve overall efficiency may be the adoption of drip irrigation technology.

Intermountain Region (Idaho, Oregon)

Sugarbeet processing factories operated in the Intermountain Region (Idaho, Utah, eastern Oregon, and eastern Washington) beginning in the late 1800s, and by 1908 the Utah-Idaho Sugar Co. and the Amalgamated Sugar Company had

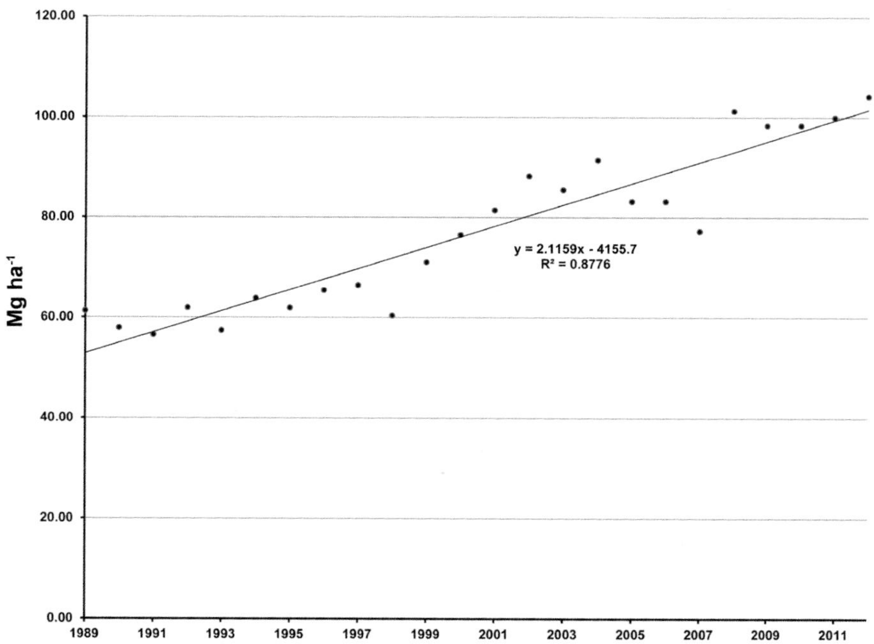

Fig. 13–6. Changes in the average root yield per hectare are shown for the state of California from 1989 to 2012. The line represents the linear regression through these averages. Data are from the California Beet Growers Association.

established themselves as the largest beet sugar processing companies in the region. In the early 1920s, the Utah-Idaho Sugar Co. had 20 factories in four states (ID, UT, WA, OR), and Amalgamated Sugar Co. had eight factories in Idaho and Utah. During the 1920s and 1930s, a number of factories closed or ceased operation because of drought and curly top, which caused large fluctuations in yield (Fig. 13–2) and area cultivated for sugarbeet (Fig. 13–7). Primarily due to BCT, by 1934, Amalgamated had only two operating factories. As BCT-resistant cultivars became available in the early 1930s, yields improved, and Amalgamated resumed operation of five of eight factories. In 1966, the Utah-Idaho Sugar Co. had only five sugar factories left and by the late 1970s had closed all operations. In 1994, the Amalgamated Sugar Co. became a farmer owned cooperative (renamed Snake River Sugar Co.). In 2005, the Nyssa, OR factory was closed leaving only the Nampa, Paul, and Twin Falls factories still operating in the Intermountain region (in Idaho).

Commercial sugarbeet production areas in Idaho, Oregon, and Washington are all coordinated through the Snake River Sugar Co. During 2012, there were 74,058 ha (183,000 ac) (Fig. 13–7) planted in Idaho, all of which were in the southern portion of the state, where over the last 30 yr, there has been a shift in acreage from Treasure Valley (southeast Oregon and southwestern Idaho) to Magic Valley (south-central Idaho) and eastern Idaho. In Oregon, commercial production is concentrated in the eastern portion of the state, where 4411 ha (10,900 acres) were harvested in 2012. (Most seed production for sugarbeet in the United States also occurs in Oregon, but is concentrated in the Willamette Valley, west of the Cascade mountain range.) Sugarbeet production was widespread in central

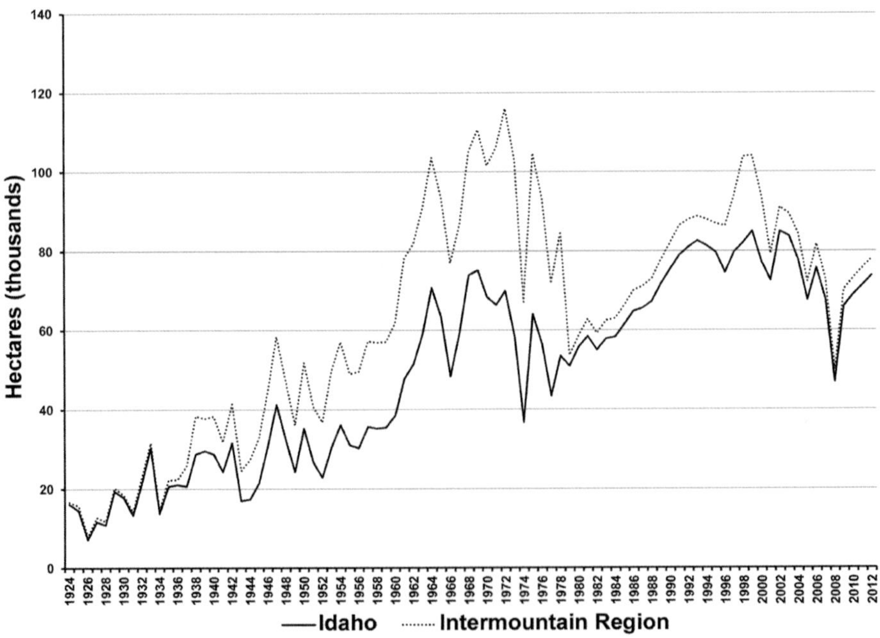

Fig. 13–7. Area planted to sugarbeet in Idaho and the Intermountain region (Idaho, Oregon, and Washington) from 1924 to 2012 based on statistics from the USDA-NASS.

Washington until a major reduction in acreage occurred in the mid-1970s with the closure of the factory at Moses Lake, WA (Fig. 13–7). The USDA-NASS survey last reported data for Washington in 2008 with only 647 ha (1600 ac). Although there still was a small acreage in Washington in 2012, it was too small to be included in the USDA-NASS survey.

Most production fields in this area are prepared by plowing with a moldboard plow, roller harrowed (perhaps more than once), and then marked and/or bedded. Some growers will "drag off" the beds at planting to expose moist soil. Planting in lower elevation areas (610 m [2000 ft]) occurs as early as the first week of March, while higher elevation (1676 m [5500 ft]) areas typically are planted in mid to late April. The crop was thinned by hand or mechanically until the early 1990s, but now is planted to stand (14 cm [5.5 in] between plants). Before 2008, weed control was accomplished through tillage and about a dozen herbicides; however, since the release of glyphosate-resistant sugarbeet, weed control has shifted to the use of glyphosate. The use of glyphosate has eliminated problems associated with herbicide toxicity to sugarbeet seedlings and weather interfering with timely herbicide application. About 4000 ha (10,000 ac) are currently planted using strip tillage (20–30 cm wide), an additional management tool afforded by glyphosate-resistant sugarbeet, which is particularly useful in windy areas with sandy ground prone to soil erosion and plant damage in the spring at planting time and crop emergence. The entire crop is irrigated, but the method of irrigation has changed considerably over the last 30 yr, with a 48% reduction in gravity irrigation and a 49% increase in sprinkler irrigation (Fig. 13–8) due to reduced

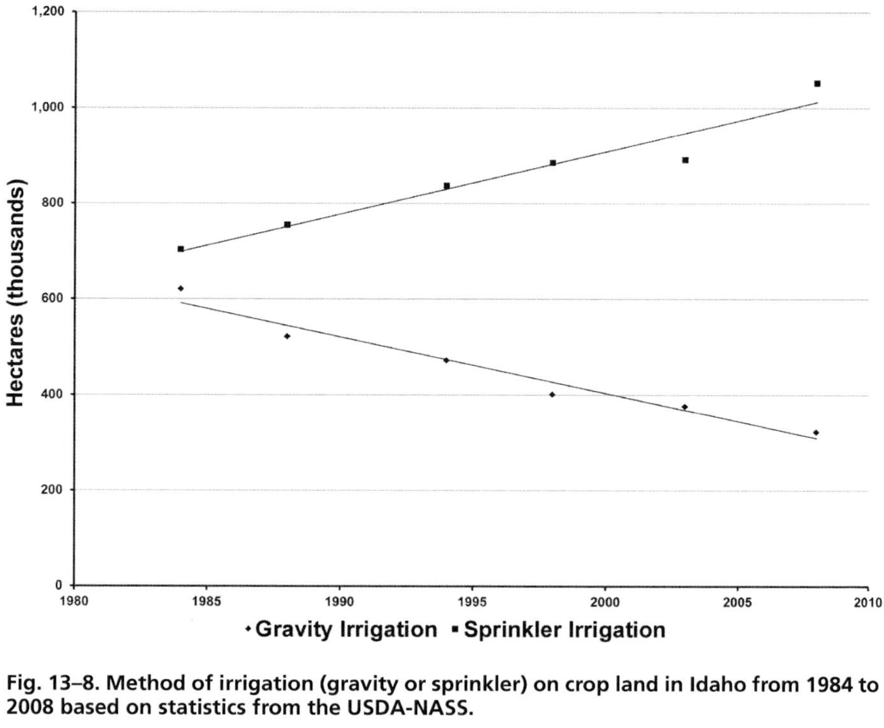

Fig. 13–8. Method of irrigation (gravity or sprinkler) on crop land in Idaho from 1984 to 2008 based on statistics from the USDA-NASS.

labor costs, higher irrigation efficiency, less soil erosion and nutrient loss, and government programs that helped offset the cost of the equipment change.

The primary disease problems in this production area include two virus diseases (BCT and rhizomania), various root rots (*R. solani* Kühn and some bacterial rots), and powdery mildew (*Erysiphe polyfoni* DC). As in the rest of the United States, rhizomania is managed through the use of resistant cultivars. The use of chemical protectants often provides the yield stability needed to introduce new genetic improvements throughout the entire commercial genepool (e.g., rhizomania resistance or glyphosate resistance), while the previous array of disease resistance is being reconstituted.

A survey in the western United States in 2006 and 2007 indicated the primary species of BCT were *Beet severe curly top virus* and *Beet mild curly top virus* (Strausbaugh et al., 2008). Since the mid 1930s host resistance has been the primary control measure for this problem, but in 2006 clothianidin ([C(*E*)]-*N*-[(2-chloro-5-thiazolyl) methyl]-*N*′-methyl-*N*″-nitroguanidine; Poncho, NipsIt), a neonicotinoid insecticide seed treatment, became available for sugarbeet (Strausbaugh et al., 2006). The clothianidin seed treatment along with thiamethoxam (3-[(2-chloro-5-thiazolyl) methyl]tetrahydro-5-methyl-*N*-nitro-4*H*-1,3,5-oxadiazin-4-imine, marketed as Cruiser) has become a valuable supplement to host resistance through control of the beet leafhopper vector (Strausbaugh et al., 2006, 2012). In Idaho, yields with neonicotinoid seed treatments averaged 72.90 ± 3.79 Mg ha^{-1} (32.52 ± 1.69 t ac^{-1}) from

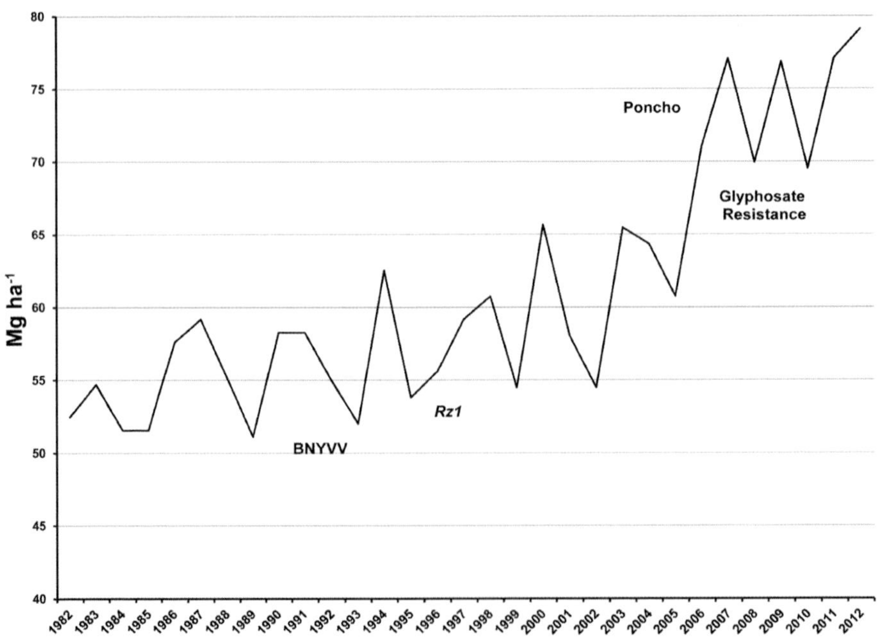

Fig. 13–9. Average root yield trends over the last 30 yr for Idaho sugarbeet production based on statistics from the USDA-NASS. Some of the significant events during this timeframe that likely influenced yields were the introduction of rhizomania, which is a viral disease caused by *Beet necrotic yellow vein virus* (BNYVV), release of rhizomania resistant varieties based on the *Rz1* gene, neonicotinoid seed treatments for curly top vector and pest control (Poncho), and glyphosate resistant varieties.

2006 to 2010, a 17% increase when compared to the preceding 5 yr 60.57 ± 4.51 Mg ha^{-1} (27.04 ± 2.01 t ac^{-1}) (Strausbaugh et al., 2012). In 2008, glyphosate-resistant cultivars also were introduced but did not seem to lead to immediate yield increases, although weather may have been a factor, as well (Fig. 13–9). Idaho yields for 2006–2007 averaged 74.09 ± 4.28 Mg ha^{-1} (33.05 ± 1.91 t ac^{-1}) and the first 3 yr (2008–2010) with glyphosate yields averaged 72.12 ± 4.15 Mg ha^{-1} (32.17 ± 1.85 t ac^{-1}) (Fig. 13–9). In the first commercial glyphosate-resistant cultivars, the level of general disease resistance was not as high as in the previous cultivars. For example, the curly top resistance in some of the initial glyphosate cultivars tested before 2008 was worse than the susceptible check at the later evaluation dates (C.A. Strausbaugh, 2005, data not shown). Thus, the neonicotinoid seed treatments have helped make the transition to glyphosate-resistant cultivars possible (Strausbaugh et al., 2012). Since the initial introduction of glyphosate-resistant cultivars, the level of resistance to curly top and resistance to other diseases has improved steadily. With this improved disease resistance and favorable weather, yields again have increased in Idaho during 2011 [77.11 Mg ha^{-1} (34.4 t ac^{-1})] and 2012 [79.13 Mg ha^{-1} (35.3 t ac^{-1})] (Fig. 13–9).

Another major concern in the Amalgamated production area has been various root rots, such as rhizoctonia crown and root rot (Strausbaugh et al., 2011; Strausbaugh and Gillen, 2009), bacterial root rot (*Leuconostoc mesenteroides* and others) (Strausbaugh and Gillen, 2008), and aphanomyces root rot (caused by *Aphanomyces cochlioides* Drechsler). Rhizoctonia root rot of mature roots is the primary problem, and is caused by *R. solani* AG-2-2 IIIB strains (Bolton et al., 2010; Strausbaugh et al., 2011). However, the fungus initially seems only to penetrate the outer portion of the root (~5% of the root mass) and subsequent bacterial invaders led by *Leuconostoc mesenteroides* can act synergistically to rot the majority of the root mass (Strausbaugh and Eujayl, 2012; Strausbaugh et al., 2011). The Rhizoctonia–bacterial root rot complex is worse in the warmer longer growing season portion of the production area (Strausbaugh et al., 2011). The cooler, higher elevation area in eastern Idaho has very little problem with root rot. Cultivars with resistance to *R. solani* are available in limited numbers, but banding on fungicides, crop rotation, and careful irrigation will continue to be necessary management practices (Bolton et al., 2010; Buhre et al., 2007; Kiewnick et al., 2001; Kirk et al., 2008; Rush and Winter, 1990; Strausbaugh et al., 2013a,b; Stump et al., 2004; Windels and Brantner, 2005). Aphanomyces root rot only recently has become a concern in eastern Oregon and now has been identified in a number of locations in southern Idaho. The management of these root rotting diseases will be critical for continued yield increases in this production region.

The primary pests limiting sugarbeet yield in the Amalgamated production region include: beet leafhopper (vector for curly top), root maggot [*Tetanops myopaeformis* (Roder)], wireworms [*Limonius californicus* (Mann.) and others], cutworms (*Agrotis* spp. and others), root aphid (*Pemphigus betae* Doane), black bean aphid [*Aphis fabae* (Scopoli)], beet leafminer [*Pegomya betae* (Curt.)], and spider mites (*Tetranychus urticae* Koch) (Strausbaugh et al., 2010). The neonicotinoid (clothianidin and thiamethoxam [3-[(2-chloro-5-thiazolyl)methyl]tetrahydro-5-methyl-*N*-nitro-4*H*-1,3,5-oxadiazin-4-imine]) seed treatments provide good early season control of nonchewing pests, but also will provide considerable protection from chewing pests (Strausbaugh et al., 2010). Systemic and foliar insecticides can be used to manage pest problems where the seed treatments are weak. A limited number of commercial cultivars with root aphid resistance also are available. The loss of aldicarb (2-methyl-2-(methylthio)propanal *O*-[(methylamino) carbonyl]oxime) reduced the choice of systemic insecticides and also raised concerns over sugarbeet cyst nematode control. Thus, there is a push to increase the number of commercial cultivars with nematode tolerance available. Currently each seed company has a few cultivars available.

Early season harvest (10% of crop) usually begins in mid September in all areas. Harvest in the longer season, lower elevation areas frequently stretches into November, while other areas try to complete harvest in October. In 2012, Idaho ranked second for total yield (5.829 million tonnes [6.425 million tons]) by state, while Oregon was ranked last (379,203 Mg [418,000 tons]) (Fig. 13–7). In this production area, approximately one-third of the crop is processed directly through the factory, while another third remains in short-term storage (<90 d) under ambient conditions outdoors and the remaining one-third is placed in long-term storage (>90 d) (Peterson et al., 1984). Piles for long-term storage typically are covered and ventilated, and there also is some indoor storage. Piles are cooled with ambient air and prolonged rainy, unusually warm weather in the winter months can have serious, deleterious effects on the quality of the sugarbeet in storage.

Great Plains Region (Colorado, Montana, Nebraska, Wyoming)

The first mention of sugarbeet as a potential crop in Colorado was in the mid 1800s. Because the crop could not be brought from food to table, it had not come with the first settlers, and a certain level of economic development had to be present to provide the capital for the technology and facilities to refine sucrose from the sugarbeet. Nonetheless by the late 1800s, Colorado State Agricultural College (now Colorado State University) had shown in test plots the suitability of the crop to the soils and climate of the Great Plains Region (Hamilton, 2009). The first sugar extracting factories were built right at the beginning of the 20th century, and the Great Western Sugar Company was founded in 1905. In 1907 there were 16 factories operating in Colorado, and by 1926, 22 factories were active in Colorado (Hecker, 2004), and there were more than 81,000 ha (200,000 ac) of sugarbeet under cultivation (Fig. 13–10). Great Western Sugar Company became a major force in the sugar industry and owned 26 factories in six states (Kansas and Ohio in addition to the Great Plains states). It also had a strong research and seed cultivar development program that was very active in the Great Plains.

All sugarbeet grown in the Great Plains region is irrigated, and the climate is excellent for sugarbeet production, with long days, and cool nights due to the altitude. Nonetheless, there are a number of diseases that have had a severe effect on production in the Great Plains. Curly top certainly was a problem in some years and could depress yields across the Great Plains region, especially in the hot dry years of the Dust Bowl; even today, it still can be severe, especially during periods of early season drought. The years until the BCT-resistant cultivars were released showed lack of yield gain (Fig. 13–11) and yield fluctuations in the Great Plains but not as severe as the up and down cycles seen in California and Idaho (Fig. 13–2 and 13–4).

In the early 1920s, breeding efforts were initiated to find sources of resistance to cercospora leaf spot, which was a problem in the Great Plains, especially in the Arkansas Valley and all of eastern Colorado and Nebraska (Skuderna, 1925), where night temperatures were higher, as was the relative humidity. It also was

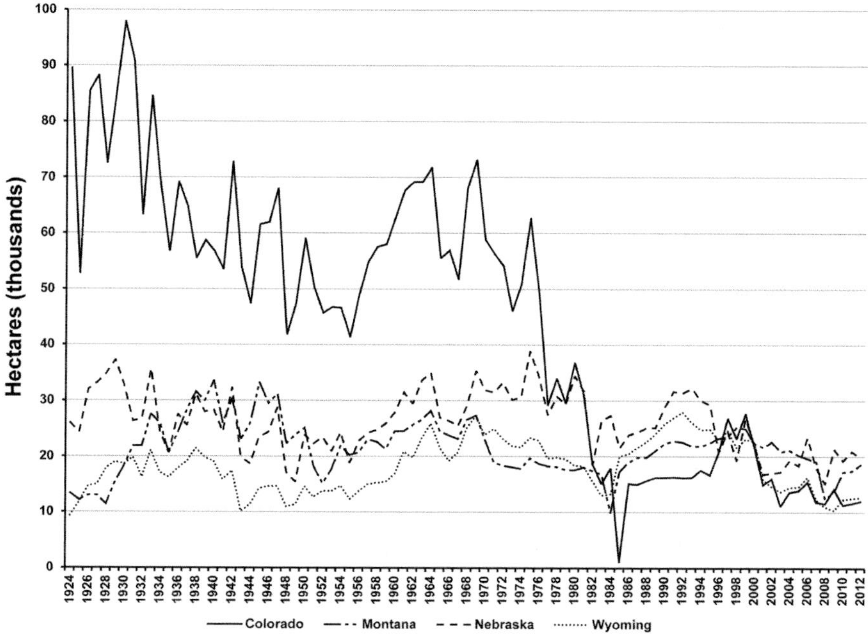

Fig. 13–10. Area of sugarbeet harvested in Colorado, Montana, Nebraska, and Wyoming from 1924 to 2012 based on statistics from the USDA-NASS.

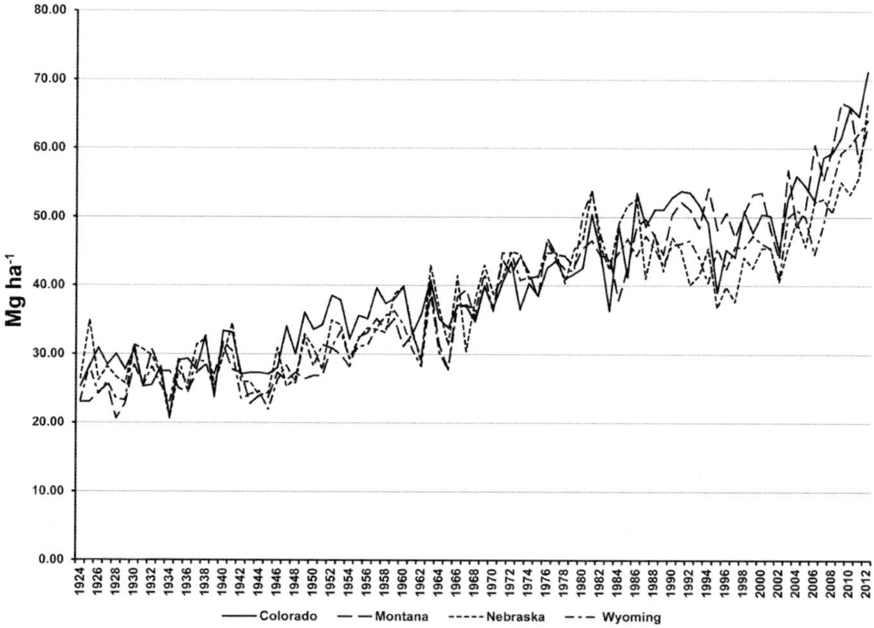

Fig. 13–11. Average root yield for Colorado, Montana, Nebraska, and Wyoming from 1924 to 2012 based on statistics from the USDA-NASS.

a serious problem in Michigan and Ohio (Dahlberg et al., 1940). Resistant lines appeared beginning in 1937 with the release of 'US 217' and continued with a series of synthetic cultivars: 'US 200 × 215,' 'US 215 × 216,' 'US 216 × 225,' which were widely grown in these areas and alleviated to some extent the serious damage done by this disease (reviewed by Panella and McGrath, 2010). From the end of World War II until the early 1980s, there was a general increase in yield as hybrid sugarbeet became available and cultivars with monogerm seed were introduced (Fig. 13–11). However, during this period there was a tremendous decrease in the area cultivated in Colorado (Fig. 13–10), and although area cultivated in the other Great Plains states remained more constant, there has been a general decrease to reach today's levels in this region. In Colorado the decline in sugarbeet cultivation is directly tied to the slow demise of the Great Western Sugar Company, which was a result of the retirement of seasoned management after being sold in 1967 to a less experienced chief executive (Hamilton, 2009). The purchase by the Hunt brothers in 1974 and descent into bankruptcy in 1984 ended the Great Western Sugar Company. It was purchased by Tate & Lyle in 1985, but because growers were not paid timely in 1984 for their crop, in 1985 only about 1000 ha (2500 ac) of sugarbeet was planted. Finally, in 2002, the processing company was sold to the growers in the four state region, and it has been stabilized as the Western Sugar Cooperative (Hamilton, 2009).

The decrease in area cultivated has had severe agronomic consequences, which include the shortening of rotations, leading to increased disease pressure. This was exacerbated by drought in the late 1990s and early 2000s and the appearance of rhizomania in the early 1990s. Rhizomania was confirmed in Colorado and the Great Plains region by 1995, and it is managed almost exclusively with resistant germplasm. Along with resistance to rhizomania, came cultivars with improved resistance to sugarbeet cyst nematode, rhizoctonia root rot, fusarium wilt (caused by *Fusarium oxysporum* f. sp. *betae*), aphanomyces black root, cercospora leaf spot, BCT, and sugarbeet root aphid. During the last 10 yr, there also has been an increase in efficacy and diversity of protective chemical available for disease management. (For many soil borne pests and diseases, fumigation, especially with 1,3-Dichloropropene [Telone II], had been used in the past to prepare for planting sugarbeet. Although effective, there is a high cost to the grower, economically and environmentally.) These improvements and the introduction of glyphosate-resistant cultivars have led to the steady rise of yields over the past 10 yr in the Great Plains of about 1.90 Mg ha^{-1} yr^{-1} (0.85 t ac^{-1} yr^{-1}) (Fig. 13–12). Yields have risen from a four state average of 42.6 Mg ha^{-1} (19 t ac^{-1}) to an average of almost 67.2 Mg ha^{-1} (30 t ac^{-1}) (Fig. 13–12), and although the sucrose percentage has not risen appreciably, it has remained constant.

Upper Midwest Region (Eastern North Dakota, Western Minnesota—Red River Valley)

Sugarbeet has been raised in the Red River Valley (RRV) Region of eastern North Dakota and western Minnesota since 1918 when the first sugarbeet crop was planted on a farm near Crookston, MN. The sugarbeet raised on this small acreage (under contract from the Minnesota Sugar Company) was sent by railcar to Chaska, MN for processing. Over the next few years, the area under cultivation grew steadily (Fig. 13–13), until there was enough sugarbeet production in the RRV to justify building a processing facility near East Grand Forks, MN

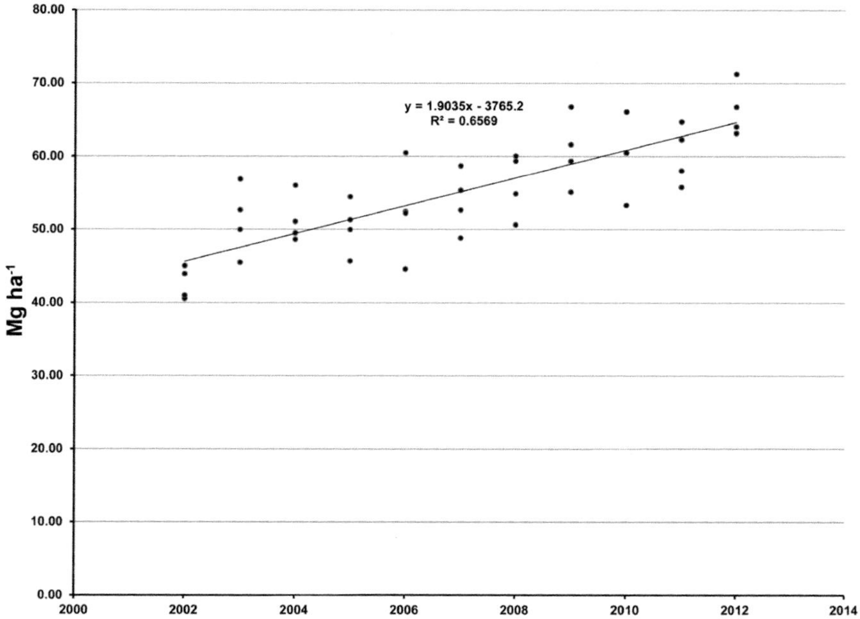

Fig. 13–12. In the past 10 yr average yield in the Great Plains region (Colorado, Nebraska, Montana, and Wyoming) has risen from an average of 42.6 Mg ha^{-1} (19 t ac^{-1}) to an average of almost 67.2 Mg ha^{-1} (30 t ac^{-1}) at a rate of about 1.90 Mg ha^{-1} yr^{-1} (0.85 t ac^{-1} yr^{-1}). Data are from USDA-NASS.

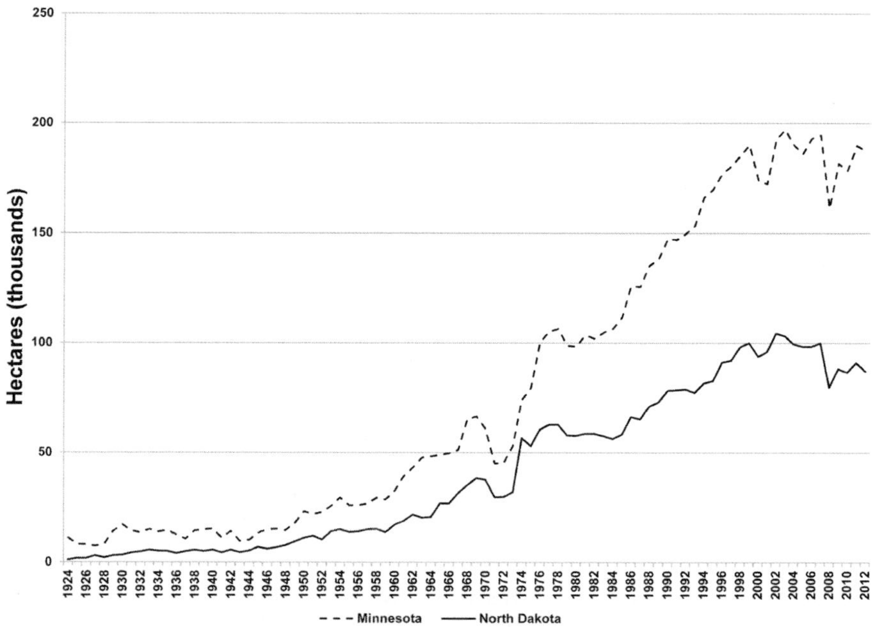

Fig. 13–13. Area planted to sugarbeet in the Upper Midwest region (Red River Valley, North Dakota and Minnesota) from 1924 to 2012 based on statistics from the USDA-NASS.

(completed in 1926). Processing plants in Crookston and Moorhead, MN soon followed in 1948 and 1954, respectively, as sugarbeet acreage continued to expand rapidly, and another factory was built in 1965 near Drayton, ND.

Since the mid 1970s, when three independent grower-owned sugar cooperatives (American Crystal Sugar Company, which purchased the cooperative at Hillsboro after 1 yr of operation; Minn-Dak Farmers Cooperative; and Southern Minnesota Beet Sugar Cooperative) were formed, the RRV has become the largest production area of sugarbeet within the United States. This particular region represented 32% of the U.S. area cultivated to sugarbeet in 1978 (170,455 ha [421,200 ac]) (Fig. 13–13) and has increased steadily over the past three decades to 276,400 ha (683,000 ac), which was 56% of U.S. area sown during the 2012 growing season (USDA-NASS, 2013). One of the main reasons for the increased acreage and continued popularity of this specialty crop is the vital role that sugarbeet production plays in the regional economy. Expansion of acreage also is due to development of long-term storage technology that allows these factories to have the longest processing campaigns in the world—200 d and sometimes more than 250 d in length. The total economic activity (direct and secondary impacts) generated by the sugar industry of the RRV was nearly \$4.9 billion in 2011 alone; or expressed alternatively, each hectare of sugarbeet planted generated about \$18,542 (\$7,507 ac^{-1}) in total local business activity (Bangsund et al., 2012).

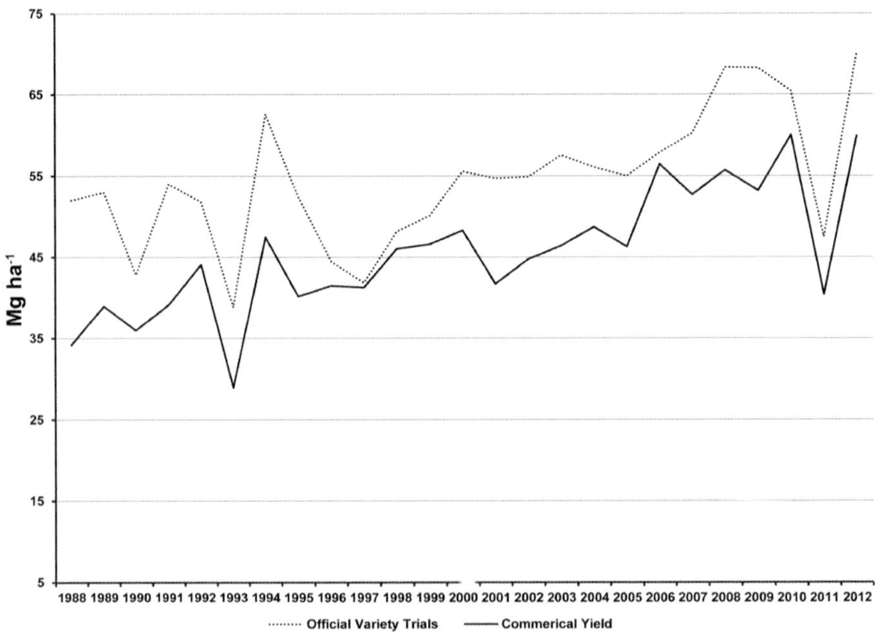

Fig. 13–14. Average yield per harvested area in the Upper Midwest region has increased 1.08 Mg ha^{-1} yr^{-1} (0.48 t ac^{-1} yr^{-1}) in commercial fields and 0.74 Mg ha^{-1} yr^{-1} (0.33 t ac^{-1} yr^{-1}) in the official variety trials over the last 24 yr. Data are from American Crystal Sugar Company, Minn-Dak Farmers Cooperative, and Southern Minnesota Beet Sugar Cooperative.

During the last 25 yr, the sugar industry in the RRV has seen a steady upward trend in overall crop yield and quality. When looking specifically at crop yield, total yield per harvested area has increased 1.08 Mg ha^{-1} yr^{-1} (0.48 t ac^{-1} yr^{-1}) (Fig. 13–14) in commercial fields and 0.74 Mg ha^{-1} yr^{-1} (0.33 t ac^{-1} yr^{-1}) in the Official Variety Trials (replicated field trials established for the evaluation of varietal performance and approval of sale within each cooperative). Combined with an increased level of quality, these increases equate to a 183 and 148 kg ha^{-1} yr^{-1} increase in recoverable white sugar per hectare for commercially grown fields and the official variety trials, respectively (Fig. 13–15). These average annual gains are a reflection of the increased genetic potential of the commercial cultivars, and also of agronomic advancements in technology, efficiency, and production practices seen by the beet sugar industry over the last three decades.

Dynamic changes in the seed cultivars available to sugarbeet growers farming in the RRV have been one of the most important aspects of this upward trend in yield potential. Diseases such as rhizomania, aphanomyces, fusarium wilt, rhizoctonia root rot, and cercospora leaf spot are endemic to the RRV sugarbeet growing region. Although there are many fungicides currently labeled for management of these diseases, they are viewed as supplemental tools to aid the genetic defense package of the individual cultivars. Continued advancements in the field of plant breeding and molecular genomics have made it possible to use "customized" cultivars specifically adapted to growing areas within this region. Without this

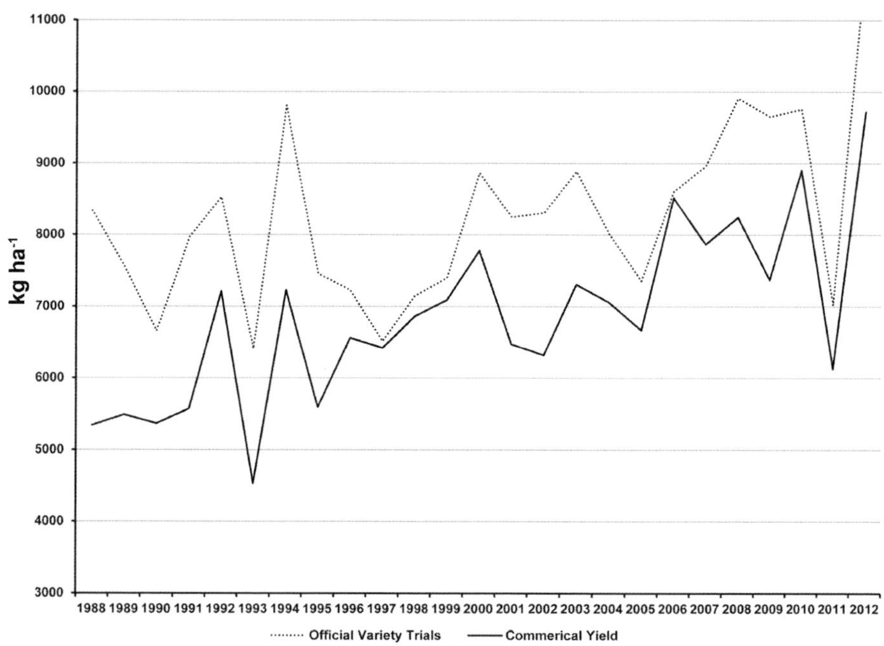

Fig. 13–15. Average root yield combined with an increased level of quality equate to an increase in recoverable white sugar per hectare of 183 and 148 kg ha^{-1} yr^{-1}, for commercially grown fields and the official variety trials over the last 24 yr. Data are from American Crystal Sugar Company, Minn-Dak Farmers Cooperative, and Southern Minnesota Beet Sugar Cooperative.

combination of genetic resistance in the crop, supplemented by chemical protections, diseases like cercospora leaf spot can devastate the region (Khan and Smith, 2005). Continued research and development of genetic resistance to pathogens, insects, and varied environmental conditions has contributed to—and continues to contribute to—the sugarbeet crop's success and progress in yield and quality.

In addition to improved genetics, one of the most notable developments contributing to increased sugar production is that of improved machinery. Over the past 25 yr, farm machines have gotten larger, more powerful, and very much more efficient. In the late 1980s, many growers utilized 8- to 12-row (4.5–6.7 m wide) equipment behind a 74,570 W (100-hp) row crop tractor (Shoptaugh, 1997). Today, sugarbeet growers in the RRV commonly use 24- to 36-row (13.4–20.1 m wide) equipment (a 300% increase in implement size) pulled with row crop tractors that often exceed 246,081 W (330-hp) and are able to guide the tractor from one end of the field to the other with sub-inch (<2.54 cm) accuracy via GPS guidance systems. As a result, the sugarbeet crop is planted earlier, quicker, and is raised far more efficiently than ever before. Since about 1980 target plant population has nearly doubled, which has played an important role in improved yield and quality.

Advancements in crop nutrition and fertilization (variable rate application is growing) also have contributed to improved crop yields. Scientific research has provided the individual grower and their respective cooperative with a much better understanding of the positive effects of fertilizers on sugarbeet quality and the detrimental effects of overuse. This newfound respect for optimized fertility has led to the widespread implementation and refinement of soil sampling. This annual monitoring has transitioned from crude random soil samples to an organized and methodical grid, zone, and topographical soil analysis taking advantage of the integration of GPS technology. The end result is a sugarbeet crop that advances more quickly physiologically and is healthier and more uniform.

Annual sugarbeet production surveys conducted by North Dakota State University from 1983 to 2007 consistently listed weed control as the top response by growers as their "worst production issue" (Carlson et al., 2007). Multiple species of pigweeds (*Amaranthus* spp.), lambsquarters (*Chenopodium album* L.), and Kochia [*Bassia scoparia* (L.) A.J. Scott] were among the weed species causing the largest problem, and they cost the cooperatives millions of dollars in lost revenue each year. Because of this, row crop cultivation and hand labor for weed control was a common practice throughout the growing season. Row crop cultivation was used by 99% of the respondents for each year from 1996 to 2007, while hand labor has steadily declined since the mid 1990s (62% of respondents used hand labor to thin the crop in 1996, 32% in 2002, and 20% in 2007) (Carlson et al., 2007). The addition of several new herbicide chemistries, and new application methodology (i.e., the Micro-Rate Program) (Rothe et al., 2004) were the main causes for the decline in hand weeding.

In 2008, the sugarbeet industry in the RRV transitioned into the use of glyphosate-resistant sugarbeet. Each cooperative held its grower membership to a maximum of 50% of their total allocated acreage during the first season and unlimited use for the second season (2009) and thereafter. The adoption of this new technology by the growers was a nearly 98% use rate by the second season. When comparing commercial yields, the average yield per hectare of sugarbeet grown in this region from 1988 to 2007 was 43.48 Mg ha^{-1}. This average increased almost 9 Mg ha^{-1} after the introduction of glyphosate-resistant sugarbeet (52.46 Mg ha^{-1} average 2008–2012). The impact on quality is just as remarkable with the average kilograms of recoverable white sugar per hectare increasing 1308 kg (1988–2007: 6562 kg ha^{-1}; 2008–2012: 7870 kg ha^{-1}) per harvested hectare.

The Great Lakes Region (Michigan, Ohio, and Ontario, Canada)

Michigan has a long history of sugarbeet promotion and production. In 1881, the Michigan legislature offered a temporary bounty of two cents per pound for sweeteners derived from corn (*Zea mays* L.), sorghum [*Sorghum bicolor* (L.) Moench], and beet. However, it was not until 1890, when the Michigan Agricultural College (now Michigan State University) in East Lansing, MI imported 800 pounds of seed of four sugarbeet cultivars (Kleinwanzlebener), Vilmorin Imperial, Austrian Wohanka, and White Silesian) for distribution to 400 Michigan farmers, that interest was spurred in beet sugar production in Michigan. The Michigan Sugar Bounty Law of 1897 (Act no. 48) codified support for the nascent industry, and with help from the USDA in securing adequate quantities of seed from Europe, sugarbeet was produced in 64 counties with an average of 16% sugar at 83% purity, and interest in this new crop resulted in the distribution of 150 thousand copies of Bulletin 150 (*Sugar Beets in Michigan in 1897*) (Kedzie, 1901; Smith and Kedzie, 1987). Beet for sugar has been grown continuously in Michigan since that time.

Up until World War I, seed was imported from Germany, the majority of seed from the cultivar Kleinwanzlebener (Cox and Hill, 1924). During and after the war, faculty and students at Michigan Agricultural College (e.g., F.A. Spragg, H.B. Smith, and E.E. Down) began selecting within commercial cultivars and European germplasm. By 1923, no fewer than 29 cultivars were being imported from varied suppliers, and eight of these were produced in Michigan or Ontario for the 1921 season (Down, 1925). In 1923, the USDA Bureau of Plant Investigations (currently USDA-ARS) transferred work on sugarbeet from Blissfield, MI to East Lansing, and charged E.E. Down with providing improved lines of sugarbeet for the Michigan industry through a cooperative agreement with Michigan State College. From 1929, Kohls was the sugarbeet breeder for Michigan Agricultural College and was responsible for 'Michigan Hybrid 18,' the original source of genetic monogermy discovered by V.F. Savitsky (Savitsky, 1952). Seed was sent to V.F. Savitsky in about 1945 and two plants he discovered were monogerm (none were ever found in Michigan) (H.L. Kohls, personal notes, circa 1950).

By 1956, most multigerm breeding had ceased at the USDA-ARS station in East Lansing, MI, and efforts were directed towards converting the black-leg- and leaf spot-resistant multigerm populations to monogerm and O-type maintainers for the production of hybrid cultivars. The original source of monogerm and O-type germplasm for East Lansing is unclear; however, it likely involved materials received by H.L. Kohls from F.V. Owen (USDA, Salt Lake City, UT) in 1945. By 1948, these materials were incorporated into the first test hybrids, showing a sugar yield boost of >10% over commercial materials (Kohls, 1951), but they lacked disease resistance. Most or all of the Great Lakes monogerm sources were bred with assistance from USDA-ARS, Beltsville, MD materials (under the direction of G.E. Coe). Releases, 'EL 31' through 'EL 38,' were derived from introgressing monogerm and O-type into the '53AB1' ancestors with continual evaluation and reselection. A combining ability test in 1962 compared 10 CMS parents. Ultimately, the combination of 'SP 6822' as pollinator (Coe and Hogaboam, 1971a) with selections for increased seed set from USDA Salt Lake City females 'SLC 129' and 'SLC 133,' which were released as 'EL 44' and 'EL 45,' respectively, were proved to be the most successful USDA hybrid for the Eastern growing regions, embodied in 'US H20' (Coe and Hogaboam, 1971b). The last USDA hybrid released for the Great Lakes growing region was 'US H23,' which used similar CMS material as 'US H20,' but used 'EL 40' as the pollinator for greater cercospora leaf spot resistance (Doney, 1995). Breeding for resistance to cercospora leaf spot, a perennial problem in the humid Great Lakes Region, has been recently summarized by Panella and McGrath (2010).

Since about 1985, private seed company breeders have contributed all commercial cultivars, while public breeders have focused on release of parental materials and germplasm enhancement. During the early years of commercial hybrids, performance was somewhat variable. Processors demanded higher sugar content, at the expense of good emergence and stand establishment, which is particularly important in the Great Lakes Region, and yields plummeted. Seed priming and greater attention to emergence genetics (McGrath et al., 2000) resulted in improving yields in the 21st century (Fig. 13–16). Plant populations, and hence yield, improved substantially. Since 2008, with the advent of glyphosate resistance and its attendant lack of injury to young beets from multi-application herbicide phytotoxicity and soil compaction, commercial yields have improved

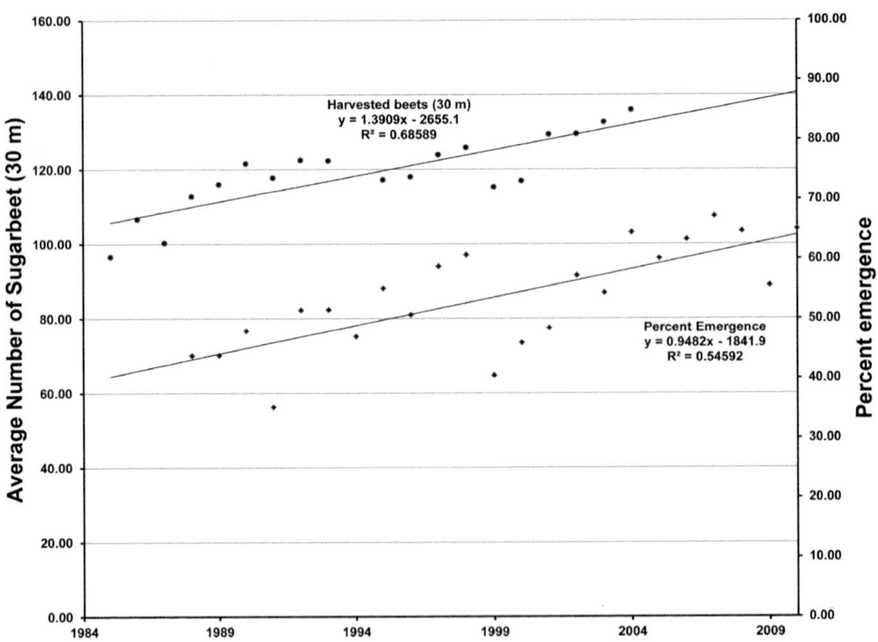

Fig. 13–16. Research in seed priming and attention to the genetics of emergence have resulted in an increased percent emergence over the past 20 yr in Michigan. Increased emergence has resulted in better stands and more beets harvested, which have helped maintain a steady improvement in yield. Data are from the Michigan Sugar Company.

dramatically. Although there has been a steady increase in yield during the last 25 yr, in the years since 2007 when glyphosate-resistant cultivars began to appear in official variety trials, the rate of yield increase has accelerated dramatically (Fig. 13–17). Although acreage has declined by about 10% sugar production actually has increased in recent years, due to improved yields. Sucrose content of beets has remained relatively stable (Fig. 13–17).

The Great Lakes Region sugarbeet processors have consolidated over time. By the 1940s, there were more than 50 factories and 25 companies throughout Michigan, Ohio, and Ontario, Canada that were associated with the now defunct Farmers and Manufacturers Association. By 1980, two companies with five sugar factories remained, and sugarbeet was no longer grown in Ontario. Today, Michigan Sugar Company is a grower-owned cooperative with four factories and more than 1000 growers in Michigan and Ontario producing beet on more than 56,700 ha (140,000 ac) (Fig. 13–18).

What Does the Future Hold?

How will Climate Change Impact Sugarbeet Production?

Climate change will impact every aspect of agriculture, including sugarbeet cultivation and yield (Bindi and Olesen, 2011; Hatfield et al., 2011) by increasing levels of CO_2 and other greenhouse gases (GHG), increasing temperatures, and changing the patterns of precipitation (IPPC, 2008). Considerable uncertainty exists among

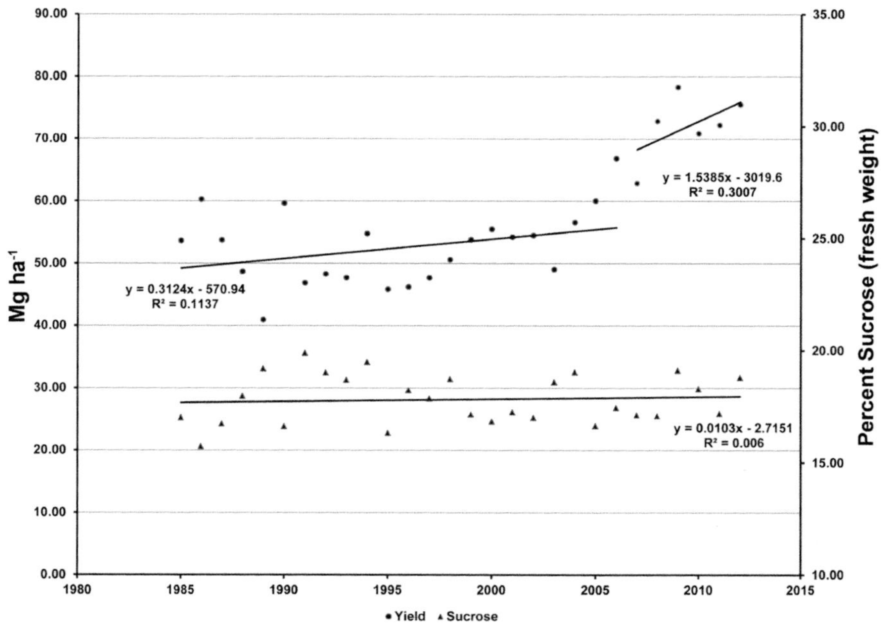

Fig. 13–17. Average yield over the past 25 yr has shown a steady increase in Michigan. Over that time the percent sucrose has remained stable. Data are from the Michigan Sugar Company.

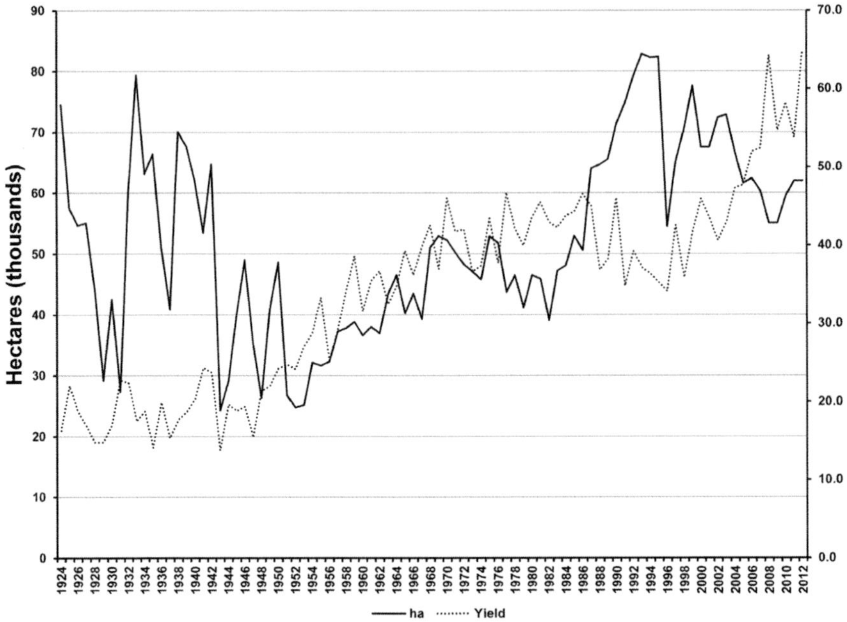

Fig. 13–18. Area of sugarbeet harvested (left vertical axis) and average yield per harvested area (right vertical axis) are shown for the Great Lakes Region (Michigan and Ohio) from 1924 to 2012 based on statistics from the USDA-NASS.

climate models (based on different GHG emission scenarios), although the direct effects of these three main impacts are well supported by modeling (IPCC, 2008). Also predicted are increased variability in rainfall intensity and an increase in the number of sequential record high temperature days (Walthall et al., 2012; IPCC, 2008). Increased temperatures will mean faster snow melt and water that is available earlier in the season, at the expense of water currently stored throughout the season as snow or glacial ice (Bindi and Olesen, 2011; Hatfield et al., 2011; IPCC, 2008; Jaggard et al., 2010; Peltonen-Sainio et al., 2010; Walthall et al., 2012). More intense precipitation events interacting with different soil types may lead to increased erosion and more rapid runoff, affecting water storage in the profile and increasing nutrient contamination in streams and watersheds (Walthall et al., 2012). Flooding may affect both planting and harvest times (Hatfield et al., 2011). Heat stress will affect different crops differently and will be especially important at seed set in grain crops (IPCC, 2008). Increased temperatures may require relocating sugarbeet seed production areas, currently located in Western Oregon and similar environments where sugarbeet roots (stecklings) can be vernalized in situ with little chance of freezing damage (Kockelmann and Meyer, 2006).

Impacts of modifying complex agrobiological interactions are very difficult to predict (Newton et al., 2011). Very little is known about the impact of increased CO_2, temperature, and altered rainfall patterns on the growth and development of crop pests and diseases. Disease will still be a problem (Hatfield et al., 2011; Newton et al., 2011; Savary et al., 2011; Walthall et al., 2012), but the delicately balanced management schemes that have been achieved in the past may change. For example, nematode pressure might increase with the increase in their number of reproductive cycles as a result of a longer growing season (Sticht et al., 2009). The changes also will affect the interactions of crop plants with weed species, native and invasive, and little research has been done to predict these impacts (Bradley et al., 2010; Walthall et al., 2012).

Experiments examining the effects of increased level of CO_2 on growth, photosynthesis, and dry matter accumulation in sugarbeet suggest increased productivity, but only with attention to different levels of inputs (e.g., Bunce, 1992; Demmers-Derks et al., 1998; Ford and Thorne, 1967; Ignatova et al., 2005; Romanova et al., 2002; Wolf, 1998; Ziska et al., 1995). Most results examined productivity in above-ambient CO_2 concentrations in controlled environments, where light intensity, in particular, may not correlate as well with field environments, or experiments in which plants have not been continuously exposed to high CO_2 throughout their life cycle. Also there is the general conclusion that the yield benefit gained by CO_2 enrichment is rapidly lost with nutrient stress, particularly with suboptimal nitrogen levels (e.g., Bunce, 1992; Demmers-Derks et al., 1998; Ford and Thorne, 1967; Ignatova et al., 2005; Romanova et al., 2002; Wolf, 1998; Ziska et al., 1995).

Sugarbeet was grown season long, without enclosure, under an enriched CO_2 concentration in a free air CO_2 enrichment (FACE) study in Braunschweig, Germany (Burkart et al., 2009; Manderscheid et al., 2010). White sugar yield increased about 13% under the higher concentration of CO_2, which was doubled (28%) when the sugarbeets were shaded, preventing a decrease in leaf area index, which often is seen under the elevated CO_2 concentrations. They concluded that the sugarbeet root was sink-limited under higher (than ambient) CO_2 concentrations, and therefore, the potential effect of elevated CO_2 on sugarbeet yield was minimized (Burkart et al., 2009; Manderscheid et al., 2010). Based on the research to date, elevated CO_2 concentrations will be beneficial for sugarbeet production and yields, but there still is considerable uncertainty as to the magnitude of the benefits.

An additional predicted consequence of the rise in the concentration of GHGs is the increase of ozone in the atmosphere (Walthall et al., 2012; IPCC, 208). In a 2-yr study in an open topped chamber, sugarbeet exhibited a significant reduction of root yield of 6% and a slight reduction of sucrose content, resulting in a 9% reduction of sugar yield per hectare (De Temmerman et al., 2007). Another study reported that capacity of sugarbeet to recover from ozone exposure was dependent on the duration of exposure and the time in the plant's life cycle when it was exposed (Köllner and Krause, 2003). Although chlorophyll and sucrose content in the leaves showed some recovery, longer lasting effects were found in taproot glucose levels (Köllner and Krause, 2003). Increased levels of ozone may reduce yield potential of sugarbeet cultivars, although this biennial crop appeared less sensitive than annual crops such as potato (*Solanum tuberosum* L.) or wheat (*Triticum aestivum* L.) (De Temmerman et al., 2007).

Water and heat stress often are considered together when looking at abiotic challenges to sugarbeet production, and reviews are available (Morillo-Velarde and Ober, 2006; Ober and Rajabi, 2010; Stevanato, 2005). Heat and drought are of greatest concern in rainfed areas in the United States (i.e., the Red River Valley and Michigan), where access to irrigation to help alleviate water or heat stress is limited. Both the low and high GHG emission models examined by Walthall et al. (2012) predicted that the rainfed areas will receive more precipitation, while western growing areas are predicted to have reduced precipitation, although both regions will have increased temperatures. Jaggard and coworkers (2007) attributed 0.025 t ha^{-1} yr^{-1} of a total 0.204 t ha^{-1} yr^{-1} increase in yield (1976–2004) to an earlier sowing date. Yield increases were attributed mostly to variation in the weather, and only about 30% of the yield increase over this time period was attributed to technological advances, including plant breeding (Jaggard et al., 2007). Similarly, Richter and coworkers (2006) concluded that growers could compensate for drought-related losses on sandy soils from climate change by planting earlier and harvesting later. The predicted warmer temperatures across the United States should allow for earlier planting dates, which could increase yields in rainfed and irrigated systems if moisture levels are moderate. In a simulation of sugarbeet production in northern and central Italy, the authors concluded that "sugarbeet production in northern and central Italy may not be greatly affected by future climate change, if global warming will be characterized by increased temperature and increased precipitation regimes" (Donatelli et al., 2002).

Each report and review lists a number of research areas that need to be pursued and cropping systems that need to be adjusted and integrated into current production systems to meet climate change challenges. Sugarbeet research always has been an area where private, state, and federal researchers have collaborated well. Current active research projects are focused on drought and heat tolerance in sugarbeet (Bloch et al., 2006; Ober and Rajabi, 2010; Ober et al., 2004; Pidgeon et al., 2006; Romano et al., 2013; Werker and Jaggard, 1998), but there is much more to be done. Walthall et al. (2012) stated, "Increases in temperature coupled with more variable precipitation will reduce productivity of crops, and these effects will outweigh the benefits of increasing carbon dioxide." Areas of research often mentioned (Hatfield et al., 2011; IPPC, 2008, Walthall et al., 2012) and of particular interest to sugarbeet production include the following: We need to better understand the sensitivity of sugarbeet to key direct and indirect climate change effects and their interactions. Adaptive management strategies and climate risk tools need to be developed. More genetic improvement programs targeted to specific climate change factors such as tolerance to drought, heat, and ozone must be initiated and supported to increase yield potential. Accomplishing these goals will require multidisciplinary teams of scientists from industry, university, and federal research programs to work together. Sugarbeet has proven to be a very adaptable crop, and there likely is sufficient genetic variation within the *Beta* gene pool to deploy for future adaptions of the crop to changes in climate (Streibig et al., 2013).

Future Production and Uses of Sugarbeet

The 2008 Farm Bill (and previous farm bills) had previsions to stabilize the cost and production of sugar in the United States (information available online through USDA-ERS, 2013). The program is designed to operate to the greatest extent possible at no cost to the U.S. taxpayer. The approach is to stabilize prices using domestic marketing allotments and tariff-rate quotas to supply the U.S. need for sucrose. Sugar in excess of the domestic food consumption requirement is diverted into ethanol production. A current major challenge of this program is that unlimited, duty-free imports of sugar can now enter the United States from Mexico under the North American Free Trade Act.

Marketing allotments are important because they determine how much sugar is produced domestically. Each year these allotments are set based on USDA forecasts for consumer sugar requirements. There are two conditions that are met in the allotments. First, the overall allotment quantity must be at least 85% of estimated human consumption. Second, the domestic sugar prices must remain above the level that would cause forfeiture of government loans to the sugar processors (which are paid to the growers). The allotments are divided between sugarbeet and sugarcane producers, with sugarbeet producers receiving 54.35% of the allotment. This is divided among beet processors based on their production in crop years 1998 to 2000. Average per capita sugar consumption in the United States from 1988 to 2012 is constant, averaging 46.76 g d^{-1} (SD = 1.169) (1.65 oz.), and the population has grown an average of 0.91% per year since 2000 (The World Bank, 2013). As yield gains outpace population gains, the acreage needed to produce sugar will decline.

If sugarbeet growers are to maintain or expand acreage in the United States, it will be through new uses of beet. "Energy beet" as a feedstock for biofuel production already is being used in many European countries, and there are groups in North Dakota and California exploring its use in the United States (Panella, 2010). By dry weight, roughly 75% of the sugarbeet is sucrose (Harland et al., 2006), which easily is fermented into ethanol, but also could be fermented to a more energy dense methane, biobutanol, ETBE (ethyl tert-butyl ether), or biohydrogen (Antoni et al., 2007; Sanderson, 2006; von Felde, 2008; Zhang et al., 2007). The pulp or marc of sugarbeet, what is left after the sugar has been extracted, is comprised of more than 70% enzyme digestible complex saccharides (C-5/C-6) and only a small percentage of lignin (Eggleston et al., 2010; Harland et al., 2006; Kozak and Laufer, 2009). Many of the current cultivars used as energy beet are sugarbeet cultivars, but as the use of energy beets increases, breeding objectives will diverge from sugarbeet. One of the first objectives will be a suite of disease resistances suited to the climate in which the energy beet will be grown. If the entire beet and not just the pulp and molasses are being used as a feedstock, biomass might become a breeding objective (Panella and Kaffka, 2010). Breeding energy beets is just getting underway.

Very little of the sucrose produced or sugarbeet marc produced is used as value-added feedstock (Eggleston, 2008). Both hemicellulose and pectin are abundant, and there is value for all of these compounds as feedstock chemicals (Harland et al., 2006; Turley, 2008; van Beilen, 2008). Recently, the use of sugarbeet marc as an industrial source of carboxyl methyl cellulose and polysaccharides was reported (Fishman et al., 2010; Hotchkiss et al., 2010). There are a large number of uses for beet and its coproducts as the United States moves toward a more sustainable economy (McGrath and Townsend, 2014; Panella and Kaffka, 2010; Panella 2010), and there is the genetic variation within the beet genepool to be adapted through breeding to both changing climate and changing end uses (Streibig et al., 2013).

Acknowledgments

The authors would like to thank the U.S. sugarbeet grower-owned cooperatives, which provided much of the data used in this study. Mention of trade names or commercial products in this publication is solely for the purpose of providing specific information and does not imply recommendation or endorsement by the United States Department of Agriculture. USDA is an equal opportunity provider and employer.

References

Antoni, D., V. Zverlov, and W. Schwarz. 2007. Biofuels from microbes. Appl. Microbiol. Biotechnol. 77:23–35. doi:10.1007/s00253-007-1163-x

Asher, M.J.C., D.M. Chwarszczynska, and M. Leaman. 2002. The evaluation of rhizomania resistant sugar beet for the UK. Ann. Appl. Biol. 141:101–109. doi:10.1111/j.1744-7348.2002.tb00201.x

Bangsund, D.A., N.M. Hodur, and F.L. Leistritz. 2012. Economic contribution of the sugarbeet industry in Minnesota and North Dakota. Rep. Agribusiness & Applied Economics Rep. 688. North Dakota State University, Department of Agribusiness and Applied Economics. http://ideas.repec.org/p/ags/nddaae/121494.html (accessed 18 Feb. 2014).

Bennett, C.W. 1971. The curly top disease of sugarbeet and other plants. The American Phytopathological Society, St. Paul, MN.

Bennett, C.W., and L.D. Leach. 1971. Diseases and their control. In: R.T. Johnson, J.T. Alexander, G.E. Rush, and G.R. Hawkes, editors, Advances in sugarbeet production: Principles and practices. The Iowa State Univ. Press, Ames, IA. p. 223–285.

Bennett, C.W., C. Price, and J.S. McFarlane. 1957. Effects of Virus yellows on sugar beet with a consideration of some of the factors involved in changes produced by the disease. Proc. Am. Soc. Sugar Beet Technol. 9:479–494. doi:10.5274/jsbr.9.6.479

Biancardi, E., L.G. Campbell, G.N. Skaracis, and M. De Biaggi. 2005. Genetics and breeding of sugar beet. Science Publ., Enfield, NH.

Biancardi, E., R.T. Lewellen, M. de Biaggi, A.W. Erichsen, and P. Stevanato. 2002. The origin of rhizomania resistance in sugar beet. Euphytica 127:383–397. doi:10.1023/A:1020310718166

Biancardi, E., J.M. McGrath, L.W. Panella, R.T. Lewellen, and P. Stevanato. 2010. Sugar beet. In: J.E. Bradshaw, editor, Root and tuber crops. Handbook of Plant Breeding Vol. 7. Springer Science+Business Media, New York. p. 173–219.

Biancardi, E., L.W. Panella, and R.T. Lewellen. 2012. *Beta maritima*, the origin of beets. Springer Science+Business Media, New York.

Bindi, M., and J.E. Olesen. 2011. The responses of agriculture in Europe to climate change. Reg. Environ. Change 11:151–158. doi:10.1007/s10113-010-0173-x

Bloch, D., C.M. Hoffmann, and B. Marlander. 2006. Impact of water supply on photosynthesis, water use and carbon isotope discrimination of sugar beet genotypes. Eur. J. Agron. 24:218–225. doi:10.1016/j.eja.2005.08.004

Bockstahler, H.W., and O.E. Reece. 1948. Progress report on breeding of sugar beet in Minnesota for resistance to black root. Proc. Am. Soc. Sugar Beet Technol. 5:137–141.

Bolton, M.D., L. Panella, L. Campbell, and M.F.R. Khan. 2010. Temperature, moisture, and fungicide effects in managing *Rhizoctonia* root and crown rot of sugar beet. Phytopathology 100:689–697. doi:10.1094/PHYTO-100-7-0689

Bradley, B.A., D.M. Blumenthal, D.S. Wilcove, and L.H. Ziska. 2010. Predicting plant invasions in an era of global change. Trends Ecol. Evol. 25:310–318. doi:10.1016/j.tree.2009.12.003

Bunce, J.A. 1992. Light, temperature and nutrients as factors in photosynthetic adjustment to an elevated concentration of carbon dioxide. Physiol. Plant. 86:173–179. doi:10.1111/j.1399-3054.1992.tb01327.x

Buhre, C., G. Wagner, S. Kluth, C. Kluth, R. Apfelbeck, and M. Varrelmann. 2007. Resistance of sugarbeet varieties as basis for the integrated control of root and crown rot (*Rhizoctonia solani*). Zuckerindustrie 132:50–55.

Burkart, S., R. Manderscheid, and H.J. Weigel. 2009. Canopy CO_2 exchange of sugar beet under different CO_2 concentrations and nitrogen supply: Results from a free-air CO_2 enrichment study. Plant Biol. 11:109–123. doi:10.1111/j.1438-8677.2009.00240.x

Büttner, G., B. Märländer, and R. Manthey. 1995. Breeding for resistance to rhizomania in sugarbeet (*Beta vulgaris* L.). Plant Breed. 114:160–164. doi:10.1111/j.1439-0523.1995.tb00782.x

Carlson, A.L., J.L. Luecke, M.F.R. Khan, and A.G. Dexter. 2007. Survey of weed control and production practices on sugarbeet in Minnesota and eastern North Dakota. Sugarbeet Research and Education Board Extension Rep. 38. p.40–64.

Carsner, E. 1924. Studies on curly-top disease of the sugar beet. J. Agric. Res. 28:297–320.

Carsner, E. 1926a. Resistance in sugar beets to curly top. USDA Circular 388:7.

Carsner, E. 1926b. Seasonal and regional variations in curly-top of sugar beets. Science 63:213–214. doi:10.1126/science.63.1625.213

Carsner, E. 1933. Curly-top resistance in sugar beets and test of the resistant variety U.S. No. 1. USDA Technical Bull. 360. USDA, U.S. Gov. Print. Office, Washington, DC.

Coe, G.E., and G.J. Hogaboam. 1971a. Registration of sugarbeet parental line SP 6322-0. Crop Sci. 11:947. doi:10.2135/cropsci1971.0011183X001100060080x

Coe, G.E., and G.J. Hogaboam. 1971b. Registration of US H20 sugarbeet. Crop Sci. 11:942. doi:10.2135/cropsci1971.0011183X001100060066x

Coons, G.H. 1953. Breeding for resistance to disease. Yearbook of Agriculture 1953. USDA, U.S. Gov. Print. Office, Washington, DC. p.174–192.

Coons, G.H., and J.E. Kotila. 1951. Virus yellows of sugar beets and tests for its occurrence in the United States. Phytopathology 41:559.

Coons, G.H., F.V. Owen, and D. Stewart. 1955. Improvement of the sugar beet in the United States. Adv. Agron. 7:89–139. doi:10.1016/S0065-2113(08)60336-3

Coons, G.H., D. Stewart, H.W. Bockstahler, G.W. Deming, J.O. Gaskill, G.J. Hogaboam, and C.L. Schneider. 1954. Evaluation tests in 1953 of U.S. 400 and related black root- and leaf spot-resistant varieties of the U.S. Department of Agriculture. Proc. Am. Soc. Sugar Beet Technol. 8:112–117.

Coons, G.H., D. Stewart, and H.A. Elcock. 1931. Sugar-beet strains resistant to leaf spot and curly top. Yearbook of Agriculture 1931. USDA, U.S. Gov. Print. Office, Washington, DC. p. 493–496.

Cox, J.F., and B.B. Hill. 1924. Sugar beet growing in Michigan. Spec. Bull. 106. Michigan Agric. Exp. Stn., East Lansing.

Culbertson, J.O. 1942. Inheritance of factors influencing sucrose percentage in *Beta vulgaris*. J. Agric. Res. 64:153–172.

Dahlberg, H.W., A.C. Maxon, and H.E. Brewbaker. 1940. Breeding for resistance to leaf spot and other characters. Proc. Am. Soc. Sugar Beet Technol. 2:169–180.

De Temmerman, L., G. Legrand, and K. Vandermeiren. 2007. Effects of ozone on sugar beet grown in open-top chambers. Eur. J. Agron. 26:1–9. doi:10.1016/j.eja.2006.08.001

Demmers-Derks, H., R.A.C. Mitchell, V.J. Mitchell, and D.W. Lawlor. 1998. Response of sugar beet (*Beta vulgaris* L.) yield and biochemical composition to elevated CO_2 and temperature at two nitrogen applications. Plant Cell Environ. 21:829–836. doi:10.1046/j.1365-3040.1998.00327.x

Donatelli, M., F.N. Tubiello, U. Peruch, and C. Rosenzweig. 2002. Impacts of climate change and elevated CO_2 on sugar beet production in northern and central Italy. Ital. J. Agron. 6:133–142.

Doney, D.L. 1995. USDA-ARS sugarbeet releases. J. Sugar Beet Res. 32:229–257. doi:10.5274/jsbr.32.4.227

Doney, D.L., R.E. Wyse, and J.C. Theurer. 1981. The relationship between cell size, yield, and sucrose concentration of the sugarbeet root. Can. J. Plant Sci. 61:447–453. doi:10.4141/cjps81-060

Down, E.E. 1925. Tests with sugar beet. Circular Bull. 66. Michigan Agric. Exp. Stn., East Lansing.

Duffus, J.E. 1978. The impact of yellows control on California sugarbeets. J. Am. Soc. Sugar Beet Technol. 20:1–5. doi:10.5274/jsbr.20.1.1

Duffus, J.E., E.D. Whitney, R.C. Larsen, H.Y. Liu, and R.T. Lewellen. 1984. First report in western hemisphere of rhizomania of sugar beet caused by *Beet necrotic yellow vein virus*. Plant Dis. 68:251.

Eggleston, G. 2008. Sugar and related oligasaccarides. In: B. Fraser-Reid and T.K. Thiem, editors, Glycoscience. Springer Verlag, Heidelberg, Germany. p. 1164–1183.

Eggleston, G., T. Tew, L. Panella, and T. Klasson. 2010. Ethanol from sugar crops. In: B.P. Singh, editor, Industrial crops and uses. CAB International, Chippenham, UK. p. 60–83.

Elliott, M.C., and G.D. Weston. 1993. Biology and physioloigy of the sugar-beet plant. In: D.A. Cooke and R.K. Scott, editors, The sugar beet crop: Science into practice. Chapman and Hall, London. p. 37–65.

Fischer, H.E. 1989. Origin of the 'Weisse Schlesische Rübe' (white Silesian beet) and resynthesis of sugar beet. Euphytica 41:75–80. doi:10.1007/BF00022414

Fishman, M.L., H.K. Chau, P.H. Cooke, D.R. Coffin, and H. Hotchkiss, Jr. 2010. Sugar beet pulp: A sustainable source of carboxyl methyl cellulose (CMC) and other polysaccharides. American Chemical Society 239th ACS meeting, San Francisco, CA. 22 March 2010.

Follett, R.F., W.R. Schmehl, and F.G.J. Viets. 1970. Seasonal leaf area, dry weight, and sucrose accumulation by sugarbeets. J. Am. Soc. Sugar Beet Technol. 16:235–252. doi:10.5274/jsbr.16.3.235

Ford, M.A., and G.N. Thorne. 1967. Effect of CO_2 concentration on growth of sugar-beet, barley, kale, and maize. Ann. Bot. (Lond.) 31:629–644.

Ford-Lloyd, B.V., A.L.S. Williams, and J.T. Williams. 1975. A revision of *Beta* section *Vulgares* (*Chenopodiaceae*), with new light on the origin of cultivated beets. Bot. J. Linn. Soc. 71:89–102. doi:10.1111/j.1095-8339.1975.tb02448.x

Francis, S.A. 2006. Development of Sugar Beet. In: A.P. Draycott, editor, Sugar beet. Blackwell Publishing Ltd, Oxford, UK. p. 9–29.

Francis, S.A., M. Redfearn, D.M. Chwarszczynska, M.J.C. Asher, and R.T. Lewellen. 1998. Use of molecular markers in breeding for disease resistance in sugar beet (*Beta vulgaris* L.). Aspects App. Biol. 52:279–285.

Gayon, J., and D.T. Zallen. 1998. The role of the Vilmorin Company in the promotion and diffusion of the experimental science of heredity in France, 1840–1920. J. Hist. Biol. 31:241–262. doi:10.1023/A:1004335619901

Gidner, S., B.-L. Lennefors, N.-O. Nilsson, J. Bensefelt, E. Johansson, U. Gyllenspetz, and T. Kraft. 2005. QTL mapping of BNYVV resistance from the WB41 source in sugar beet. Genome 48:279–285. doi:10.1139/g04-108

Goldman, I.L., and J.P. Navazio. 2008. Table beet. In: J. Prohens and F. Nuez, editors, Vegetables I. Asteraceae, Brassicaceae, Chenopodicaceae, and Cucurbitaceae. Handbook of Plant Breeding. Springer, New York. p. 219–238.

Grant, E.B. 1871. Beet-root sugar and cultivation of the beet. Lee, Shepard and Dillingham, New York.

Grimmer, M., S. Trybush, S. Hanley, S. Francis, A. Karp, and M. Asher. 2007. An anchored linkage map for sugar beet based on AFLP, SNP and RAPD markers and QTL mapping of a new source of resistance to Beet necrotic yellow vein virus. Theor. Appl. Genet. 114:1151–1160. doi:10.1007/s00122-007-0507-3

Grimmer, M.K., T. Kraft, S.A. Francis, and M.J.C. Asher. 2008. QTL mapping of BNYVV resistance from the WB258 source in sugar beet. Plant Breed. 127:650–652. doi:10.1111/j.1439-0523.2008.01539.x

Hamilton, C. 2009. Footprints in the sugar: A history of the Great Western Sugar Company. Hamilton Bates, Caldwell, ID.

Harland, J.I., C.K. Jones, and C. Hufford. 2006. Co-products. In: A.P. Draycott, editor, Sugar beet. Blackwell Publ., Oxford, UK. p. 443–463.

Harris, F.S. 1919. The sugar-beet in America. Macmillian, New York.

Hatfield, J.L., K.J. Boote, B.A. Kimball, L.H. Ziska, R.C. Izaurralde, D. Ort, A.M. Thomson, and D. Wolfe. 2011. Climate impacts on agriculture: Implications for crop production. Agron. J. 103:351–370.

Hecker, R.J. 2004. Sugar Beet. In: W.F. Keim, editor, A history of the Department of Agronomy/Soil & Crop Sciences. Citizen Printing Company, Fort Collins, CO. p. 48–54.

Henderson, R.W., and H.W. Bockstahler. 1946. Reaction of sugar beet strains to *Aphanomyces cochlioides*. Proc. Am. Soc. Sugar Beet Technol. 4:237–245.

Hogaboam, G.J., R.C. Zielke, and C.L. Schneider. 1982. Registration of EL40 sugarbeet parental line. Crop Sci. 22:700.

Hotchkiss, A., M. Fishman, and L. Liu. 2010. The role of sugar beet pulp polysaccharides in the sustainability of the sugar beet industry. In: Sustainability of the sugar and sugar ethanol industries. ACS Symposium Series. American Chemical Society, Washington, DC. p. 283–290.

Ignatova, L.K., N.S. Novichkova, V.A. Mudrik, V.Y. Lyubimov, B.N. Ivanov, and A.K. Romanova. 2005. Growth, photosynthesis, and metabolism of sugar beet at an early stage of exposure to elevated CO_2. Russ. J. Plant Physiol. 52:158–164. doi:10.1007/s11183-005-0023-8

IPCC. 2008. Climate change 2007: Synthesis report. Contribution of Working Groups I, II and III to the fourth assessment report of the Intergovernmental Panel on Climate Change. R.K. Pachauri and A. Reisinger, editors. IPCC, Geneva, Switzerland. http://www.ipcc.ch/publications_and_data/ar4/syr/en/contents.html (accessed 13 Feb. 2014).

Jaggard, K.W., A. Qi, and M.A. Semenov. 2007. The impact of climate change on sugarbeet yield in the UK: 1976–2004. J. Agric. Sci. 145:367–375. doi:10.1017/S0021859607006922

Jaggard, K.W., A. Qi, and E.S. Ober. 2010. Possible changes to arable crop yields by 2050. Philos. Trans. R. Soc. Lond. B Biol. Sci. 365:2835–2851. doi:10.1098/rstb.2010.0153

Johnson, R.T., and G.E. Burt. 1990. West Coast Beet Seed Company: 50th anniversary, 1940–1990. West Coast Beet Seed Company, Salem, OR.

Kedzie, R.C. 1901. The sugar beet industry in Michigan. In: Proceedings of the Twenty-second Annual Meeting of the Society for the Promotion of Agricultural Science, Denver, CO. p. 99–106.

Khan, M.F.R., and L.J. Smith. 2005. Evaluating fungicides for controlling Cercospora leaf spot on sugar beet. Crop Prot. 24:79–86. doi:10.1016/j.cropro.2004.06.010

Kiewnick, S., B.J. Jacobsen, A. Braun-Kiewnick, J.L.A. Eckhoff, and J.W. Bergman. 2001. Integrated control of *Rhizoctonia* crown and root rot of sugar beet with fungicides and antagonistic bacteria. Plant Dis. 85:718–722. doi:10.1094/PDIS.2001.85.7.718

Kirk, W.W., P.S. Wharton, R.L. Schafer, P. Tumbalam, S. Poindexter, C. Guza, R. Fogg, T. Schlatter, J. Stewart, L. Hubbell, and D. Ruppal. 2008. Optimizing fungicide timing for the control of Rhizoctonia crown and root rot of sugar beet using soil temperature and plant growth stages. Plant Dis. 92:1091–1098. doi:10.1094/PDIS-92-7-1091

Kockelmann, A., and U. Meyer. 2006. Seed Production and Quality. In: A.P. Draycott, editor, Sugar beet. Blackwell Publ., Oxford. p. 89–113.

Kohls, H. 1951. New objectives in sugar beet breeding. Proc. 6th Regional Meeting of the ASSBT, Eastern US and Canada. American Society of Sugar Beet Technologists, Fort Collins, CO. p. 59–62.

Köllner, B., and G.H.M. Krause. 2003. Effects of two different ozone exposure regimes on chlorophyll and sucrose content of leaves and yield parameters of sugar beet (*Beta vulgaris* L.) and rape (*Brassica napus* L.). Water Air Soil Pollut. 144:317–332. doi:10.1023/A:1022913116566

Kozak, R., and C.S. Laufer. 2009. Addition of a thermostabilized pectin methylesterase significantly enhances the rate of saccharification of sugar beet pulp by the commercial pectinase preparation Pectinex® Ultra SPL. J. Sugar Beet Res. 46:71–72.

Lange, W., W.A. Brandenburg, and T.S.M. de Bock. 1999. Taxonomy and cultonomy of beet (*Beta vulgaris* L.). Bot. J. Linn. Soc. 130:81–96. doi:10.1111/j.1095-8339.1999.tb00785.x

Lewellen, R.T. 1992. Use of plant introductions to improve populations and hybrids of sugarbeet In: Use of plant introductions in cultivar development, Part 2. CSSA, Madison, WI. p. 117–135.

Lewellen, R.T. 1997. Registration of 11 sugarbeet germplasm C79 lines with resistance to rhizomania. Crop Sci. 37:1026. doi:10.2135/cropsci1997.0011183X003700030087x

Lewellen, R.T., H.Y. Liu, A. Gillen, and C. Strausbaugh. 2007. Performance of rhizomania resistant sugarbeet under normal and resistance-breaking strains of beet necrotic yellow vein virus. In: Proceedings from the 34th Biennial Meeting of American Society of Sugarbeet Technologists. 28 Feb.–3 Mar. 2007, Salt Lake City, UT. http://assbt-proceedings.org/ASSBT2007Proceedings/PassbtVol34Agp171PerformanceofRhizomaniaResistantSugarbeetUnderNormalandResistanceBreakingStrainsofBeetNecroticYellowVeinVirus.pdf (accessed 18 Feb. 2014).

Lewellen, R.T., I.O. Skoyen, and A.W. Erichsen. 1987. Breeding sugarbeet for resistance to rhizomania: Evaluation of host-plant reactions and selections for and inheritance of resistance. In: Proc. Proceedings of the 50th Winter Congress of the IIRB. International Institute for Beet Research, Brussels, Belgium. p. 136–156.

Lilleboe, D. 2012. Another world record sugarbeet crop in California. The Sugarbeet Grower 51(3):22.

Liu, H.-Y., and R.T. Lewellen. 2007. Distribution and molecular characterization of resistance-breaking isolates of *Beet necrotic yellow vein virus* in the United States. Plant Dis. 91:847–851. doi:10.1094/PDIS-91-7-0847

Liu, H.-Y., J.L. Sears, and R.T. Lewellen. 2005. Occurrence of resistance-breaking *Beet necrotic yellow vein virus* of sugar beet. Plant Dis. 89:464–468. doi:10.1094/PD-89-0464

Manderscheid, R., A. Pacholski, and H.J. Weigel. 2010. Effect of free air carbon dioxide enrichment combined with two nitrogen levels on growth, yield and yield quality of sugar beet: Evidence for a sink limitation of beet growth under elevated CO_2. Eur. J. Agron. 32:228–239. doi:10.1016/j.eja.2009.12.002

McGrann, G.R.D., M.K. Grimmer, E. Mutasa-Göttgens, and M. Stevens. 2009. Progress towards the understanding and control of sugar beet rhizomania disease. Mol. Plant Pathol. 10:129–141. doi:10.1111/j.1364-3703.2008.00514.x

McGrath, J.M., C.A. Derrico, M. Morales, L.O. Copeland, and D.R. Christenson. 2000. Germination of sugar beet (*Beta vulgaris* L.) seed submerged in hydrogen peroxide and water as a means to discriminate cultivar and seedlot vigor. Seed Sci. Technol. 28:607–620.

McGrath, J.M., C.A. Derrico, and Y. Yu. 1999. Genetic diversity in selected, historical USDA sugarbeet germplasm releases and *Beta vulgaris* ssp. *maritima*. Theor. Appl. Genet. 98:968–976. doi:10.1007/s001220051157

McGrath, J.M., and K.K. Fugate. 2012. Analysis of sucrose from sugar beet. In: V.R. Preedy, editor, Dietary sugars: Chemistry, analysis, function and effects. Food and Nutritional Components in Focus No. 3. Royal Society of Chemistry Publishing, Cambridge, UK. p. 526–545.

McGrath, J.M., L. Panella, and L. Frese. 2011. Beta. In: C. Kole, editor, Wild crop relatives: Genomic and breeding resources, industrial crops. Springer-Verlag, Berlin, Heidelberg. p. 1–28.

McGrath, J.M., and B.J. Townsend. 2014. Sugar beet, energy beet, and industrial beet. In: V.M.V. Cruz and D.A. Dierig, editors, Industrial crops breeding for bioenergy and bioproducts. Handbook of Plant Breeding. Springer, New York. (in press.)

Morillo-Velarde, R., and E.S. Ober. 2006. Water use and irrigation. In: A.P. Draycott, editor, Sugar beet. Blackwell Publishing Ltd, Oxford. p. 221–255.

Murphy, A.M. 1942. Production of heavy curly-top exposures in sugarbeet breeding fields. Proc. Am. Soc. Sugar Beet Technol. 3:459–462.

Murphy, A.M. 1946. Sugar beet and curly top history in Southern Idaho 1912–1945. Proc. Am. Soc. Sugar Beet Technol. 4:408–412.

Newton, A., S. Johnson, and P. Gregory. 2011. Implications of climate change for diseases, crop yields and food security. Euphytica 179:3–18. doi:10.1007/s10681-011-0359-4

Ober, E., and A. Rajabi. 2010. Abiotic stress in sugar beet. Sugar Tech 12:294–298. doi:10.1007/s12355-010-0035-3

Ober, E.S., C.J.A. Clark, M. Le Bloa, A. Royal, K.W. Jaggard, and J.D. Pidgeon. 2004. Assessing the genetic resources to improve drought tolerance in sugar beet: Agronomic traits of diverse genotypes under droughted and irrigated conditions. Field Crops Res. 90:213–234. doi:10.1016/j.fcr.2004.03.004

Owen, F.V. 1945. Cytoplasmically inherited male-sterility in sugar beets. J. Agric. Res. 71:423–440.

Owen, F.V., F.A. Abegg, A.M. Murphy, B. Tolman, C. Price, F.G. Larmer, and E. Carsner. 1939. Curly-top-resistant sugar-beet varieties in 1938. Circular 513. USDA, Washington, DC.

Panella, L. 2005. Curly top. In: E. Biancardi, L.G. Campbell, G.N. Skaracis, and M. de Biaggi, editors, Genetics and breeding of sugar beet. Science Publishers, Enfield, NH. p. 74–76.

Panella, L. 2010. Sugar beet as an energy crop. Sugar Tech. 12:288–293. doi:10.1007/s12355-010-0041-5

Panella, L., and S.R. Kaffka. 2010. Sugar beet (*Beta vulgaris* L) as a biofuel feedstock in the United States. In: G.Eggleston, editor, Sustainability of the sugar and sugar ethanol industries. ACS Symposium Series. American Chemical Society, Washington, DC. p. 163–175.

Panella, L., and R.T. Lewellen. 2007. Broadening the genetic base of sugar beet: Introgression from wild relatives. Euphytica 154:383–400. doi:10.1007/s10681-006-9209-1

Panella, L., and J.M. McGrath. 2010. The history of public breeding for resistance to Cercospora leaf spot in North America. In: R.T. Lartey et al., editors, Cercospora leaf spot of sugar beet and related species. APS Press, St. Paul, MN. p. 141–156.

Pavli, O.I., P. Stevanato, E. Biancardi, and G.N. Skaracis. 2011. Achievements and prospects in breeding for rhizomania resistance in sugar beet. Field Crops Res. 122:165–172. doi:10.1016/j.fcr.2011.03.019

Peltonen-Sainio, P., L. Jauhiainen, M. Trnka, J.E. Olesen, P. Calanca, H. Eckersten, J. Eitzinger, A. Gobin, K.C. Kersebaum, J. Kozyra, S. Kumar, A.D. Marta, F. Micale, B. Schaap, B. Seguin, A.O. Skjelvåg, and S. Orlandini. 2010. Coincidence of variation in yield and climate in Europe. Agric. Ecosyst. Environ. 139:483–489. doi:10.1016/j.agee.2010.09.006

Peterson, C.L., M.C. Hall, and D.J. Traveller. 1984. Sugarbeet storage experiments. ASAE Paper 84-4040. ASAE, Knoxville, TN.

Pidgeon, J.D., E.S. Ober, A. Qi, C.J.A. Clark, A. Royal, and K.W. Jaggard. 2006. Using multi-environment sugar beet variety trials to screen for drought tolerance. Field Crops Res. 95:268–279. doi:10.1016/j.fcr.2005.04.010

Richter, G.M., A. Qi, M.A. Semenov, and K.W. Jaggard. 2006. Modelling the variability of UK sugar beet yields under climate change and husbandry adaptations. Soil Use Manage. 22:39–47. doi:10.1111/j.1475-2743.2006.00018.x

Romano, A., A. Sorgonà, A. Lupini, F. Araniti, P. Stevanato, G. Cacco, and M. Abenavoli. 2013. Morpho-physiological responses of sugar beet (*Beta vulgaris* L.) genotypes to drought stress. Acta Physiol. Plant. 35:853–865. doi:10.1007/s11738-012-1129-1

Romanova, A.K., V.A. Mudrik, N.S. Novichkova, R.N. Demidova, and V.A. Polyakova. 2002. Physiological and biochemical characteristics of sugar beet plants grown at an increased carbon dioxide concentration and at various nitrate doses. Russ. J. Plant Physiol. 49:204–210. doi:10.1023/A:1014897422414

Rothe, I., A.G. Dexter, and J.L. Luecke. 2004. Impact of injury from late season sequential glufosinate applications on sugarbeet yield. J. Sugar Beet Res. 41:119–135. doi:10.5274/jsbr.41.4.119

Rush, C.M., H.-Y. Liu, R.T. Lewellen, and R. Acosta-Leal. 2006. The continuing saga of rhizomania of sugar beets in the United States. Plant Dis. 90:4–15. doi:10.1094/PD-90-0004

Rush, C.M., and S.R. Winter. 1990. Influence of previous crops on Rhizoctonia root and crown rot of sugar beet. Plant Dis. 74:421–425. doi:10.1094/PD-74-0421

Rush, C.M. 2003. Ecology and epidemiology of *Benyviruses* and plasmodiophorid vectors. Annu. Rev. Phytopathol. 41:567–592. doi:10.1146/annurev.phyto.41.052002.095705

Sanderson, K. 2006. US biofuels: A field in ferment. Nature 444:673–676. doi:10.1038/444673a

Savary, S., A. Nelson, A.H. Sparks, L. Willocquet, E. Duveiller, G. Mahuku, G. Forbes, K.A. Garrett, D. Hodson, J. Padgham, S. Pande, M. Sharma, J. Yuen, and A. Djurle. 2011. International agricultural research tackling the effects of global and climate changes on plant diseases in the developing world. Plant Dis. 95:1204–1216. doi:10.1094/PDIS-04-11-0316

Savitsky, V.F. 1940. Genetics of sugar beets. USDA-National Agricultural Library Translation 5825. Translated from Russian, original source unavailable.

Savitsky, V.F. 1950. Monogerm sugar beets in the United States. Proc. Am. Soc. Sugar Beet Technol. 6:156–159.

Savitsky, V.F. 1952. Monogerm Sugar Beets in the United States. Proc. Am. Soc. Sugar Beet Technol. 7:156–159.

Schneider, K., R. Schäfer-Pregl, D.C. Borchardt, and F. Salamini. 2002. Mapping QTLs for sucrose content, yield and quality in a sugar beet population fingerprinted by EST-related markers. Theor. Appl. Genet. 104:1107–1113. doi:10.1007/s00122-002-0890-8

Scholten, O.E., T.S.M. de Bock, R. Klein-Lankhorst, and W. Lange. 1999. Inheritance of resistance to *Beet necrotic yellow vein virus* in *Beta vulgaris* conferred by a second gene for resistance. Theor. Appl. Genet. 99:740–746. doi:10.1007/s001220051292

Scholten, O.E., R.M. Klein-Lankhorst, D.G. Esselink, T.S.M. DeBock, and W. Lange. 1997. Identification and mapping of random amplified polymorphic DNA (RAPD) markers linked to resistance against *Beet necrotic yellow vein virus* (BNYVV) in *Beta* accessions. Theor. Appl. Genet. 94:123–130. doi:10.1007/s001220050390

Shoptaugh, T.L. 1997. Roots of success: History of the Red River Valley sugarbeet growers. Institute for Regional Studies, North Dakota State Univ., Fargo, ND.

Skuderna, A.W. 1925. Sugar beet breeding in the Arkansas Valley of Colorado. J. Am. Soc. Agron. 17:631–634. doi:10.2134/agronj1925.00021962001700100009x

Smith, C.D., and R.C. Kedzie. 1987. Sugar beets in Michigan in 1897. Bulletin 150 Agricultural College, Mich.: Michigan State Agricultural College Experiment Station. S75.E232 no.131–169 1896–1899.

Smith, S., M. Cooper, J. Gogerty, C. Löeffler, D. Borcherding, and K. Wright. 2014. Maize. In: S. Smith, B. Diers, J. Specht, and B. Carver, editors, Yield gains in major U.S. field crops. CSSA Spec. Publ. 33. ASA, CSSA, and SSSA, Madison, WI. p. xxx–xxx. doi: 0.2135/cssaspecpub33.c6

Souder, E., Jr. 1970. The century of the sugar beet. The California sugar beet 1970:34–36.

Strausbaugh, C.A., and I.A. Eujayl. 2012. Influence of sugarbeet tillage systems on the Rhizoctonia-bacterial root rot complex. J. Sugar Beet Res. 49:57–76. doi:10.5274/jsbr.49.3.57

Strausbaugh, C.A., I.A. Eujayl, and P. Foote. 2010. Seed treatments for the control of insects and diseases in sugarbeet. J. Sugar Beet Res. 47:105–125. doi:10.5274/jsbr.47.3.105

Strausbaugh, C.A., I.A. Eujayl, and P. Foote. 2013a. Selection for resistance to the *Rhizoctonia*-bacterial root rot complex in sugar beet. Plant Dis. 97:93–100. doi:10.1094/PDIS-05-12-0511-RE

Strausbaugh, C.A., I.A. Eujayl, L.W. Panella, and L.E. Hanson. 2011. Virulence, distribution and diversity of *Rhizoctonia solani* from sugar beet in Idaho and Oregon. Can. J. Plant Pathol. 33:210–226. doi:10.1080/07060661.2011.558523

Strausbaugh, C.A., I.A. Eujayl, and L. Panella. 2013b. Interaction of sugar beet host resistance and *Rhizoctonia solani* AG-2-2 IIIB Strains. Plant Dis. doi:10.1094/PDIS-11-12-1078-RE.

Strausbaugh, C.A., and A.M. Gillen. 2008. Bacteria and yeast associated with sugar beet root rot at harvest in the Intermountain West. Plant Dis. 92:357–363. doi:10.1094/PDIS-92-3-0357

Strausbaugh, C.A., and A.M. Gillen. 2009. Sugar beet root rot at harvest in the US Intermountain West. Can. J. Plant Pathol. 31:232–240. doi:10.1080/07060660909507596

Strausbaugh, C.A., A.M. Gillen, J.J. Gallian, S. Camp, and J.R. Stander. 2006. Influence of host resistance and insecticide seed treatments on curly top in sugar beets. Plant Dis. 90:1539–1544. doi:10.1094/PD-90-1539

Strausbaugh, C.A., E.J. Wenninger, and I.A. Eujayl. 2012. Management of severe curly top in sugar beet with insecticides. Plant Dis. 96:1159–1164. doi:10.1094/PDIS-01-12-0106-RE

Strausbaugh, C.A., W.M. Wintermantel, A.M. Gillen, and I.A. Eujayl. 2008. Curly top survey in the western United States. Phytopathology 98:1212–1217. doi:10.1094/PHYTO-98-11-1212

Stenger, D.C., and C.L. McMahon. 1997. Genotypic diversity of Beet curly top virus populations in the western United States. Phytopathology 87:737–744. doi:10.1094/PHYTO.1997.87.7.737

Stevanato, P. 2005. Resistance to abiotic stresses. In: E. Biancardi, L.G. Campbell, G.N. Skaracis, and M. de Biaggi, editors, Genetics and breeding of sugar beet. Science Publishers, Enfield, NH. p. 116–119.

Sticht, C., S. Schrader, A. Giesemann, and H.J. Weigel. 2009. Sensitivity of nematode feeding types in arable soil to free air CO_2 enrichment (FACE) is crop specific. Pedobiologia 52:337–349. doi:10.1016/j.pedobi.2008.12.001

Stump, W.L., G.D. Franc, R.M. Harveson, and R.G. Wilson. 2004. Strobilurin fungicide timing for Rhizoctonia root and crown rot suppression in sugarbeet. J. Sugar Beet Res. 41:17–38. doi:10.5274/jsbr.41.1.17

Streibig, J.C., C. Ritz, C.B. Pipper, F. Yndgaard, and J.N. Tomsen. 2013. Sugar beet, bioethanol, and climate change. IOP Conf. Series: Earth and Environmental Science. Climate Change: Global Risks, Challenges and Decisions 6. doi:10.1088/1755–1307/6/24/242040.

The World Bank. 2013. The World Bank: Data: Population growth (annual %). http://data.worldbank.org/indicator/SP.POP.GROW/countries/1W?display=default 8-21-2013 (accessed 18 Feb. 2014).

Theis, T. 1971. A food resource. In: R.T. Johnson et al., editors, Advances in sugarbeet production: Principles and practices. The Iowa State Univ. Press, Ames, IA. p. 3–18.

Turley, D.B. 2008. The chemical value of biomass. In: J.H. Clark and F.E.I. Deswarte, editors, Introduction to chemicals from biomass. Wiley, Sussex, UK. p. 21–46.

United States Beet Sugar Association. 1936. The silver wedge: The sugar beet in the United States. United States Beet Sugar Association, Washington, DC.

USDA-ERS. 2013. Sugar & sweeteners: Policy. http://www.ers.usda.gov/topics/crops/sugar-sweeteners/policy.aspx (accessed 23 Jan. 2013).

USDA-NASS. National Agricultural Statistics Service/USDA. http://quickstats.nass.usda.gov/results/2F65366F-16C2-384B-9B53-ABFE2FE826F1?pivot=short_desc (accessed 9 Apr. 2013).

van Beilen, J.B. 2008. Transgenic plant factories for the production of biopolymers and platform chemicals. Biofuels Bioprod. Bioref. 2:215–228. doi:10.1002/bbb.64

Vilmorin, P.-P.-A.L. 1859. Notice sur l'amélioration de la Carotte sauvage. In: L. Vilmorin, editor, Notices sur l'amélioration des plantes par le semis et considérations sur l'hérédité dans les végétaux. 1st ed. Librairie Agricole, Paris.

von Felde, A. 2008. Trends and developments in energy plant breeding—Special features of sugarbeet. Zuckerindustrie 133:342–345.

Walthall, C.L., J. Hatfield, P. Backlund, L. Lengnick, E. Marshall, M. Walsh, S. Adkins, M. Aillery, E.A. Ainsworth, C. Ammann, C.J. Anderson, I. Bartomeus, L.H. Baumgard, F. Booker, B. Bradley, D.M. Blumenthal, J. Bunce, K. Burkey, S.M. Dabney, J.A. Delgado, J. Dukes, A. Funk, K. Garrett, M. Glenn, D.A. Grantz, D. Goodrich, S. Hu, R.C. Izaurralde, R.A.C. Jones, S.-H. Kim, A.D.B. Leaky, K. Lewers, T.L. Mader, A. McClung, J. Morgan, D.J. Muth, M. Nearing, D.M. Oosterhuis, D. Ort, C. Parmesan, W.T. Pettigrew, W. Polley, R. Rader, C. Rice, M. Rivington, E. Rosskopf, W.A. Salas, L.E. Sollenberger, R. Srygley, C. Stöckle, E.S. Takle, D. Timlin, J.W. White, R. Winfree, L. Wright-Morton, and L.H. Ziska. 2012. Climate change and agriculture in the United States: Effects and adaptation. USDA Tech. Bull. 1935. Washington, DC. http://www.usda.gov/oce/climate_change/effects_2012/CC%20and%20Agriculture%20Report%20(02-04-2013)b.pdf.

Ware, L.S. 1880. The sugar beet: Including a history of the beet sugar industry in Europe. Henry, Carey, Baird, and Co., Philadelphia, PA.

Werker, A.R., and K.W. Jaggard. 1998. Dependence of sugar beet yield on light interception and evapotranspiration. Agric. For. Meteorol. 89:229–240. doi:10.1016/S0168-1923(97)00081-6

Whitney, E.D. 1989. Identification, distribution, and testing for resistance to rhizomania in *Beta maritima*. Plant Dis. 73:287–290. doi:10.1094/PD-73-0287

Windels, C.E., and J.R. Brantner. 2005. Early-season application of azoxystrobin to sugarbeet for control of *Rhizoctonia solani* AG 4 and AG 2-2. J. Sugar Beet Res. 42:1–16. doi:10.5274/jsbr.42.1.1

Wisler, G.C., and J.E. Duffus. 2000. A century of plant virus management in the Salinas Valley of California, 'East of Eden'. Virus Res. 71:161–169. doi:10.1016/S0168-1702(00)00196-9

Wolf, J. 1998. Effects of nutrient (NPK) supply on sugar beet response to elevated atmospheric CO_2. NJAS-Wageningen J. Life Sci. 48:157–170.

Zhang, Y.-H.P., B.R. Evans, J.R. Mielenz, R.C. Hopkins, and M.W.W. Adams. 2007. High-yield hydrogen production from starch and water by a synthetic enzymatic pathway. PLoS ONE 2:e456. doi:10.1371/journal.pone.0000456

Zhao, B., I.J. Mackay, P.D.S. Caligari, and R. Mead. 1997. The efficiency of between and within full-sib family selection in a recurrent selection programme in sugar beet (*Beta vulgaris* L). Euphytica 95:355–359. doi:10.1023/A:1003026800450

Zimmermann, B., and J. Zeddies. 2000. Evolution of sugar beet breeding and its contribution to value added. Univ. of Hohenheim, Dep. Agricultural Economics, Hohenheim, Germany.

Ziska, L.H., R.C. Sicher, and D.F. Kremer. 1995. Reversibility of photosynthetic acclimation of swiss chard and sugarbeet grown at elevated concentrations of CO_2. Physiol. Plant. 95:355–364. doi:10.1111/j.1399-3054.1995.tb00849.x

14

Sugarcane

Serge J. Edmé,* Andru Suman, and Collins Kimbeng

ABSTRACT

To ascertain genetic gains in yield traits obtained through breeding sugarcane (*Saccharum* spp.) for Florida, USA and to assess the correlated impact on genetic diversity, a comprehensive analysis was initiated with three field and a molecular experiments using historical cultivars released from 1930 to 2004. Cultivars were tested in south Florida on muck and sand soils from 2008 to 2012 and in north Florida in 2011 through 2013 to estimate gains in freeze tolerance. Microsatellite and target region amplified polymorphism markers were used to score progenitors and sugarcane cultivars. Genetic gains of 3.2 and 2.8% decade^{-1}, respectively, were realized on muck and sand for recoverable sucrose and gains of 3.5 and 2.6% decade^{-1} for cane yield (CY). No change was observed in height (HT), but diameter (DIAM) was reduced overall at 0.05 mm yr^{-1} on both soils. Modern sugarcane cultivars are more adapted to the soil spectrum in the Florida industry with 70% increase (53 yr^{-1}) in performance index and to the subtropical environment with 18% increase in freeze tolerance. Changes in the phenotypic traits were accompanied with modifications at the gene level, measured as decadal reductions in gene diversity (0.002), in percent polymorphism (6%), in number of alleles (21) and of rare alleles (11), and as an increase in the coefficient of coancestry (0.07). Linkage disequilibrium passed through two intermittent phases of increase (selection) and decay (recombination). Both markers suggested balancing selection as the main driver in maintaining genetic variation in the Florida breeding population. Combining the molecular-phenotypic results indicated that, constrained by unfavorable correlations in the "cane yield–height–diameter–sucrose" genetic architecture, the stronger selection pressure directed at increasing sucrose than at cane yield in Florida might have promoted a higher frequency of loci for this trait over positive pleiotropic loci or of alleles with antagonistic pleiotropy, slowing the upward trend of cane yield. The use of a selection index based on relevant genetic parameters may help bring the composite trait (sucrose–cane yield) closer to the optimum space where a more efficient joint improvement is possible.

Abbreviations: AMMI, additive main effects and multiplicative interaction; CEI, cultivar × environment interaction; CP, Florida Canal Point; CRS, commercially recoverable sucrose; CY, cane yield; DIAM, diameter; HT, height; LD, linkage disequilibrium; Pi, performance index; SSR, simple sequence repeat; SY, sucrose yield; TRAP, target region amplified polymorphism; USSC, United States Sugar Corporation.

S.J. Edmé, USDA-ARS Sugarcane Field Station, 12990 U.S. Hwy. 441 N., Canal Point, FL 33438. *Corresponding author (serge.edme@ars.usda.gov).
A. Suman, DuPont Pioneer, 8305 NW 62nd Ave., Johnston, IA 50131 (sumangene@yahoo.com).
C. Kimbeng, Louisiana State Univ. Agricultural Center, 5755 LSU Ag Rd., St. Gabriel, LA 70776 (CKimbeng@agcenter.lsu.edu).

doi:10.2135/cssaspecpub33.c14

Yield Gains in Major U.S. Field Crops. CSSA Special Publication 33.
Stephen Smith, Brian Diers, James Specht, and Brett Carver, editors.

Sugarcane is one of the high-value industrial crops that has shaped world history and enriched many a country. Its importance has been recognized since time immemorial (~325 BCE; Deer, 1949), and its status is now being elevated to another dimension with regard to generation of renewable energy (Tew and Cobill, 2008). Sugarcane is an important crop in both biomass and sugar production, two of the most prized commodities in the world today (Moore, 1989). Sugarcane is one of the oldest cultivated crops. It is vital to the economies of more than 100 countries spanning both the tropics and subtropics and was planted on 23.7 million ha across the world in 2010 (FAOSTAT, http://faostat.fao.org/home/index.html, accessed 4 Mar. 2013). World production of cane and raw sugar in 2010 was estimated at around 1.70 billion and 156 million tonnes, respectively, corresponding to a 30.8% increase in total production and to a 1.3% increase in yield over the last 40 years (1970–2010) (FAOSTAT).

Since the early 1980s, Brazil has been the world largest producer of sugar, followed by India, China, Thailand, Pakistan, Colombia, Mexico, the United States, and Australia. The USA is a major player (ranked fifth), one of the very few countries to extract sugar from both sugarcane (~45%) and sugarbeet (*Beta vulgaris* L.; ~55%). Seventy-five percent of the sugar delivered on the world market is derived from sugarcane. In the United States, sugarcane is planted commercially in four states (Florida, Louisiana, Texas, and Hawaii), which produced 8.15 million tonnes of raw sugar in 2010, which represents a 56.6% increase in total production and a 4.2% yield increase over the last 30 yr (FAOSTAT). With regard to productivity, the United States is at par with the rest of the world, yielding 75 t ha^{-1} of cane on average as compared with Brazil (79), India (69), China (69), and Australia (85) to cite a few. The increase in world sugar production is associated with an increase in planted area and the adoption of improved varieties and of improved management practices. Tropical countries registered higher yields of cane and sugar because biomass and sugar accumulate continuously in the stalks under these conditions. Florida and Louisiana are the two major sugar-producing U.S. states, respectively, producing around 1.8 and 1.4 million tonnes of raw sugar annually, followed by Hawaii (215,000 t) and Texas (163,000 t) (Haley and Jerardo, 2012).

Genetic improvement of sugarcane started in the USA in the late 1920s, immediately after the releases of the "POJ" (Proefstation Oost Java) varieties from Java, Indonesia and of the "Co" varieties from Coimbatore, India, which were distributed worldwide to replace disease-susceptible commercial varieties (*S. officinarum*, *S. sinense*, and *S. barberi*) and save a failing industry (Stokes and Tisdale, 1962). Those early hybrids were developed by interspecific hybridization among four sugarcane species, namely *S. spontaneum*, *S. officinarum*, *S. sinense*, and *S. barberi*, subsequently extended by several backcrosses with *S. officinarum* clones as female parents to recover (i.e., to "nobilize" as the *S. officinarum* varieties were called "noble" by breeders for their high sucrose content) the high-sucrose genes while maintaining the stress and disease tolerance of the wild species, *S. spontaneum* in particular (Roach, 1972; Tai and Miller, 1978; Sreenivasan et al., 1987). The success of the nobilization technique is recognized as one of the best early examples of the benefits that can be obtained from plant breeding, as it occurred shortly after the rediscovery of Mendel's laws of segregation and before the Green Revolution (James, 2004). All sugarcane breeding programs began with these early hybrids as the foundation for designing new genetic combinations from which modern sugarcane cultivars descended. Deren (1995) identified 17 of these early landrace and hybrid cultivars as founders of the U.S. sugarcane breeding populations.

Breeding programs in the United States capitalized on the philosophy behind the creation of these early hybrids to develop new cultivars with high yield and disease resistance, and the same major players are still active today: the Hawaii Agriculture Research Center, formerly known as Hawaiian Sugar Planters association, in Kunia, HI; the USDA-ARS, with two stations established in Canal Point, FL and in Houma, LA; three universities (Florida, Louisiana, and Texas); the United States Sugar Corporation (USSC) in Clewiston, FL; and the respective Florida, Louisiana, and Texas Sugarcane Leagues. These players have by now unified their efforts into three-way agreements within their respective states, with the Canal Point Station servicing the other two programs with true seed of sugarcane because its proximity to Lake Okeechobee ensures a warming effect that is favorable to natural flowering during winter.

The sugarcane breeding programs in the United States have been successful at developing cultivars for their respective industries, at adapting sugarcane to the subtropical climate and to the different environmental conditions, and at making steady progress towards higher yields (Breaux, 1984; Edmé et al., 2005b; Lingle et al., 2010). The breeding programs in the United States face the same environmental challenges (drought, flood, freeze, and hurricanes), diseases, and insects pressures, with some minor differences in the types of strains involved, and the industries practically utilize the same technologies in the fields and in the mills. While the sugarcane fields in Florida and in Texas are irrigated with water coming from Lake Okeechobee or the Rio Grande, respectively, the fields in Louisiana are mostly rainfed. Moreover, the risks of frequent and severe freezes are higher in Louisiana, imposing a 7 to 9 mo growing season and a shorter harvest period (a 50-d window) on the Louisiana industry, compared to the Florida industry, with 12 to 15 mo of growth and a harvest about 150 d long. Consequently, cultivars tend to be shorter at harvest in Louisiana, thinner in size, but with a larger stalk population. Cane and sugar yields are higher in Florida and Texas because breeders in Louisiana are constrained to select more early-maturing varieties (i.e., those that accumulate the largest amount of sucrose in immature stalks) as an escape mechanism with some relative level of freeze tolerance. In the Louisiana sugarcane breeding program, Breaux (1984) reported that the population improved in sucrose content from 10.8% in 1952 to 14.0% in 1974 after three cycles of selection. Lingle et al. (2010) later quantified absolute gains of 2% in sucrose content, 5% in purity, and of 24 kg Mg^{-1} in sucrose recovery after seven generations of breeding sugarcane for Louisiana, with a plateau in cane yield and juice quality observed by the fifth cycle. Using historical data from commercial fields and from multi-location trials, Edmé et al. (2005b) attributed 69% of the gains in cane and sucrose yields to genetics and the remaining 31% to improved cultural management in the Florida industry. However, the gains observed in Florida needed to be ascertained with new data that assess the performance of the cultivars in joint tests and across the environmental spectrum that characterizes the Florida industry. This chapter will evaluate progress in the Florida Canal Point (CP) breeding program only because the data from the other states were not made available and because the integration of the molecular genetics into the phenotypic evolution was unique to the CP program.

The CP program uses a recurrent breeding and selection approach to develop cultivars for the Florida industry, which allows the entry of varieties in the last stage of selection into the parental pool (Edmé et al., 2013). While broad adaptation to the two major soil types (organic or muck and mineral or sand) is targeted, the CP program is renewing its strategies towards identifying genotypes with specific performance, particularly on sand soils (Edmé et al., 2005b; del Blanco et al., 2010). Drought, flood, and periodic freezes represent the array of environmental risks to sugarcane production in all three states (Legendre et al., 2010; Edmé and Glaz, 2013). Disease pressures occur in the forms of rust (brown rust caused by *Puccinia melanocephala* H.&P. Sydow and orange rust caused by *Puccinia kuehnii* E.J. Butler), smut (caused by *Ustilago scitaminea* H. & P. Sydow), mosaic (caused by the *Sugarcane mosaic virus* strain E), leaf scald (caused by *Xanthomonas albilineans* Ashby, Dowson), ratoon stunt (caused by *Leifsonia xyli* subsp. *xyli* Evtsuhenko et al.), yellow leaf syndrome (caused by an unknown virus), and eyespot [caused by the fungus *Bipolaris sacchari* (E.J. Butler) Shoemaker]. Insects, such as mites, grubs, aphids, and the sugarcane rice borers, together with diseases and abiotic stresses, reduce the gains attained by breeding. In this context, it is becoming increasingly difficult to clearly identify the best selections for release because genetic progress may not be as rapid as it was in the earlier years of breeding, with some of the traits probably approaching their theoretical limits and, as another layer of complexity, the modifications in the genetic makeup of pathogens (e.g., rusts) to overcome cultivar resistance. Since their inception, the sugarcane breeding programs were structured to adopt a certain form of "grower-participatory approach" that strengthens the mutual roles that both genetics and management play to increase yields in the face of these various challenges. This chapter summarizes the contribution of genetic improvement to yield gains obtained by the Florida sugarcane breeding programs from yield data collected across the three main environmental factors (two soil types—organic vs. mineral—and freeze) that affect yield performance in Florida.

Breeding and selection can generate unexpected results, particularly in genetic diversity and in trait correlations (Falconer and Mackay, 1996; Ceccarelli, 2009) if these genetic parameters are not properly accounted for. The sugarcane breeder operates on a complex and large alloploid genome (up to 10 Gb in size, d'Hont and Glaszmann, 2001; $\approx$7 pg/2C, Edmé et al., 2005a), augmented by an aneuploid series (with chromosome numbers ranging from 100 to 130), using, as parents, cultivars that had been derived initially from the merging of four *Saccharum* species. The application of Mendelian genetics to support sugarcane improvement, in designing genotypes for prediction of genetic gains, is hampered by these genomics barriers, and empirical breeding has been the dominant feature in most breeding programs (Jackson, 2005). Moreover, the two most economically important traits in sugarcane, cane yield and sucrose content, tend to be negatively correlated in most studies (Brown et al., 1969; Hogarth, 1971), somewhat dependent on the populations (advanced selection lines vs. progeny populations) and environments under study. This study reexamined the impact of breeding on the evolution of these two and other yield-component traits (diameter and height) in the Florida programs. Genetic correlations among traits were evaluated in an attempt to appraise the direction and strength of the constraints or trade-offs and also to propose strategies for realizing greater and faster gains.

Genetic diversity, as affected by breeding, was quantified by using two molecular marker systems that explore different regions of the sugarcane genome: microsatellites (simple sequence repeat, SSR) to amplify the neutral diversity and target region amplified polymorphism (TRAP) (Li and Quiros, 2001) to target the functional diversity. This chapter evaluates the role of breeding in shaping the patterns of genetic diversity and structure in the CP breeding program and, at the same time, attempts to track the footprints of selection. If genetic gains are being achieved, as indicated by Edmé et al. (2005b), the relevant genomic regions are being effectively targeted. This premise was used to identify putative loci under selection retrospectively and, at the same time, detect the type of selection (directional, stabilizing, or balancing) that potentially affects polymorphisms, genetic variation, and gains from selection (Beaumont and Balding, 2004). Such a population genetics technique is being used comprehensively today to evaluate whether selection, and which type of selection, is active in a population and to detect indirectly and confirm regions targeted by breeding (Beaumont and Balding, 2004).

The objectives of this research were to: (i) ascertain genetic gains in performance and adaptability from the Florida sugarcane breeding programs using multi-location trials for joint testing of historical cultivars released from 1930 to 2004, (ii) follow the co-evolution of traits under selection through their genetic correlations, (iii) assess the impact of breeding on the neutral and functional molecular genetic diversity, (iv) characterize the type of selection as an underlying force acting on the genomic regions of interest in the sugarcane genome, and (v) track the footprints of selection by identifying putative loci as being the targets.

Materials and Methods

Materials and Field Management

Three experiments were conducted using historical cultivars released from 1930 to 2004, including some of the founders like POJ 2725, POJ 3016, and NCo 310, some CL cultivars developed by USSC, three of the F-cultivars (F31-407, F31-436, and F31-819) developed by the University of Florida before the formation of the USDA-ARS station in Canal Point (Table 14–1). Fifty cultivars were planted in the first experiment (Exp. I) on 3 Mar. 2006 as a 5 × 11 triple α-lattice design on a Histosol or Torry muck soil (euic hyperthermic Typic Haplosaprist) on station at Canal Point, FL (26°51′50″ N, 80°37′32″ W), and followed for 3 yr (plant cane, first and second ratoon crops). The second experiment (Exp. II) included 66 cultivars and was planted at Townsite, a USSC farm near Clewiston, FL (26°45′25″ N, 80°55′06″ W) on 6 Jan. 2010 on a Margate sand soil (siliceous, hyperthermic Mollic Psammaquents) as a 6 × 11 triple α-lattice design and sampled in the plant cane and in the first ratoon crops. The third experiment (Exp. III) included the same 66 cultivars as in Exp. II and was planted as a 5 × 15 α-lattice augmented design at the Hague farm (Agronomy Dep. Forage Research Unit of the Univ. of Florida, located in Hague, FL, 29°45′0″ N, 82°25′48″ W), on a Spodosol or Pomona sand soil (sandy siliceous, hyperthermic Ultic Alaquod). All three locations are routine testing sites commonly used in the CP program for selection in the first three stages (Seedlings, Stages 1 and 2, Canal Point), in the last two stages (Stages 3 and 4, Townsite), and for freeze tolerance (Hague).

Experiments I and II used three-row plots, with respective lengths of 2.5 and 4.6 m, with 1.5-m spacing between rows, and with respective 2.5- and 4.6-m dis-

Table 14–1. Historical sugarcane cultivars along with their wild progenitor species grouped into seven decades released from 1930 to 2004 in the Florida industry. Four types of experiments were used to assess genetic gains realized over the period and to evaluate the impact of breeding on genetic diversity.

Genotypes	Species	Parentage	Experiments†	Breeding institution	Country of origin
		Progenitor population			
Badila	*S. officinarum*	–	M	–	–
Cayana10	*S. sinense*	–	M	–	–
Djantoer	*S. spontaneum*	–	M	–	–
La Purple	*S. officinarum*	–	M	–	–
NG 57-059	*S. officinarum*	–	M	–	–
NG 77-081	*S. officinarum*	–	M	–	–
		Founder population (Decade 1: 1921–1930)			
POJ 2725	Hybrid	POJ 2364/EK 28	I, II, III, M	Proefstation Oost Java	Java, Indonesia
POJ 3016	Hybrid	POJ 2878/POJ 2940	I, II, III, M	Proefstation Oost Java	Java, Indonesia
C278	Hybrid	Badila/28 NG 251	M	?	?
Co 1148	Hybrid	POJ 2878/Co 299//Co 301	M	Sugarcane Breeding Institute	India
DB4/20	Hybrid	?	M	Guyana Sugar Corp.	Guyana
DB5/96	Hybrid	?	M	Guyana Sugar Corp.	Guyana
Eros	Hybrid	POJ 2878/31 MQ 228	M		Australia?
Nco 310	Hybrid	Co 421/Co 312	I, II, III, M	Sugarcane Breeding Institute	India
		Decade 2 (1930–1960)			
F159	Hybrid	?	M	Formosa	Taiwan
F31-401	Hybrid	POJ 2725/CP 27-0035?	M	University of Florida	Florida, USA
F31-407	Hybrid	POJ 2725/CP 27-0035	I, II, III, M	University of Florida	Florida, USA
F31-436	Hybrid	POJ 2725/CP 27-0035	I, II, III, M	University of Florida	Florida, USA
F31-46	Hybrid	POJ 2725/CP 27-0035?	M	University of Florida	Florida, USA
F31-762	Hybrid	Co 281/CP 27-0108	M	University of Florida	Florida, USA
F31-962	Hybrid	Co 281/CP 27-0108	M	University of Florida	Florida, USA
F36-819	Hybrid	F31-962/POJ 2878	I, II, III, M	University of Florida	Florida, USA
CL 41-0223	Hybrid	F31-436/F31-452	I, II, III	US Sugar Corp.	Clewiston, FL, USA

Genotypes	Species	Parentage	Experiments†	Breeding institution	Country of origin
CP 50-0028	Hybrid	CP 43-0064/CP 33-0372	I, II, III, M	USDA-ARS	Canal Point, FL, USA
CP 56-0059	Hybrid	CL 47-0083/CP 34-0079	I, II, III	USDA-ARS	Canal Point, FL, USA
CP 57-0603	Hybrid	CL 47-0143/?	II, III	USDA-ARS	Canal Point, FL, USA
Decade 3 (1960–1970)					
CL 61-0620	Hybrid	CP 48-1204/CL 54-1910	I, II, III, M	US Sugar Corp.	Clewiston, FL, USA
CP 62-0374	Hybrid	CP 53-0018/CP 33-0224	I, II, III	USDA-ARS	Canal Point, FL, USA
CP 63-0306	Hybrid	?	I, II, III	USDA-ARS	Canal Point, FL, USA
CP 65-0357	Hybrid	CP 52-0068/CP 53-0017	I, II, III, M	USDA-ARS	Canal Point, FL, USA
CP 68-1067	Hybrid	CP 52-0068/CP 57-0603	I, II, III	USDA-ARS	Canal Point, FL, USA
CL 69-886	Hybrid	–	I, II	US Sugar Corp.	Clewiston, FL, USA
Decade 4 (1970–1980)					
CP 70-1133	Hybrid	CP 56-0063/67P06	I, II, III, M	USDA-ARS	Canal Point, FL, USA
CP 70-1527	Hybrid	CP 62-0374/CP 57-0526	I, II, III	USDA-ARS	Canal Point, FL, USA
CL 72-0231	Hybrid	–	I, II, III	US Sugar Corp.	Clewiston, FL, USA
CP 72-1210	Hybrid	CP 65-0357/CP 56-0063	I, II	USDA-ARS	Canal Point, FL, USA
CP 72-2086	Hybrid	CP 62-0374/CP 63-0588	I, II, III, M	USDA-ARS	Canal Point, FL, USA
CL 73-239	Hybrid	–	I, II, III	US Sugar Corp.	Clewiston, FL, USA
CP 73-1547	Hybrid	CP 66-1043/CP 56-0063	I, II, III, M	USDA-ARS	Canal Point, FL, USA
CL 75-854	Hybrid	–	I, II, III	US Sugar Corp.	Clewiston, FL, USA
CP 75-1082	Hybrid	CP 68-1067/CP 70-1133	I, II, III	USDA-ARS	Canal Point, FL, USA
CP 75-1553	Hybrid	CP 68-1062/73P02	I, II, III	USDA-ARS	Canal Point, FL, USA
CP 77-1776	Hybrid	CP 68-1067/CP 68-1022	I, II	USDA-ARS	Canal Point, FL, USA
CL 78-1120	Hybrid	–	I, II, III	US Sugar Corp.	Clewiston, FL, USA
CP 78-1628	Hybrid	CP 65-0357/CP 68-1026	I, II, III, M	USDA-ARS	Canal Point, FL, USA
CP 78-2114	Hybrid	?	I, II, III	USDA-ARS	Canal Point, FL, USA
Decade 5 (1980–1990)					
CP 80-1743	Hybrid	CP 72-2083/CP 68-1067	I, II, III, M	USDA-ARS	Canal Point, FL, USA
CP 80-1827	Hybrid	CP 70-1133/CP 73-1311	I, II, III, M	USDA-ARS	Canal Point, FL, USA
CL 81-1453	Hybrid	–	I, II, III	US Sugar Corp.	Clewiston, FL, USA

Table continued.

Table 14–1. Continued.

Genotypes	Species	Parentage	Experiments†	Breeding institution	Country of origin
CP 82-1592	Hybrid	CP 72-1210/CP 70-1133	I, II	USDA-ARS	Canal Point, FL, USA
CP 81-1384	Hybrid	–	I, II, III	USDA-ARS	Canal Point, FL, USA
CL 83-3431	Hybrid	–	I, II, III	US Sugar Corp.	Clewiston, FL, USA
CP 84-1198	Hybrid	CP 70-1133/CP 72-2086	I, II, III, M	USDA-ARS	Canal Point, FL, USA
CP 84-1591	Hybrid	CP 72-1370/CP 68-1022	II, III, M	USDA-ARS	Canal Point, FL, USA
CP 85-1432	Hybrid	CP 70-1527/82P26	I, II, III	USDA-ARS	Canal Point, FL, USA
CL 86-4340	Hybrid	–	I, II, III	US Sugar Corp.	Clewiston, FL, USA
CL 86-616	Hybrid	–	I, II, III	US Sugar Corp.	Clewiston, FL, USA
CL 86-4519	Hybrid	–	I, II, III	US Sugar Corp.	Clewiston, FL, USA
CL 88-4730	Hybrid	–	II, III	US Sugar Corp.	Clewiston, FL, USA
CP 88-1762	Hybrid	CP 80-1743/85P06	I, II, III, M	USDA-ARS	Canal Point, FL, USA
CP 89-1509	Hybrid	CP 80-1827/86P19	M	USDA-ARS	Canal Point, FL, USA
CP 89-2143	Hybrid	CP 81-1254/CP 72-2086	I, II, III, M	USDA-ARS	Canal Point, FL, USA
CP 89-2376	Hybrid	?	II, III	USDA-ARS	Canal Point, FL, USA
CP 89-2377	Hybrid	?	I, II	USDA-ARS	Canal Point, FL, USA
		Decade 6 (1990–2000)			
CL 90-4500	Hybrid	–	I, II, III	US Sugar Corp.	Clewiston, FL, USA
CL 90-4725	Hybrid	–	II, III	US Sugar Corp.	Clewiston, FL, USA
CP 92-1213	Hybrid	CL 73-239/CP 85-1498	I, II, III	USDA-ARS	Canal Point, FL, USA
CP 92-1167	Hybrid	CP 84-1591/SP 70-1143	M	USDA-ARS	Canal Point, FL, USA
CP 92-1641	Hybrid	CP 80-1827/CP 84-1322	I, II, III, M	USDA-ARS	Canal Point, FL, USA
CP 92-1666	Hybrid	CP 82-1592/CP 84-1322	II, III	USDA-ARS	Canal Point, FL, USA
CP 94-1100	Hybrid	CP 81-1238/CP 88-2045	I, II, III	USDA-ARS	Canal Point, FL, USA
CP 94-2059	Hybrid	CP 87-1475/CP 85-1308	M	USDA-ARS	Canal Point, FL, USA
CP 94-2203	Hybrid	US90-1072/CP 80-1827	M	USDA-ARS	Canal Point, FL, USA
CP 95-1039	Hybrid	CP 81-1238/CP 84-1198	M	USDA-ARS	Canal Point, FL, USA
CP 96-1252	Hybrid	CP 90-1533/CP 84-1198	I, II, III, M	USDA-ARS	Canal Point, FL, USA
CP 96-1602	Hybrid	CP 81-1425/94P03	II, III, M	USDA-ARS	Canal Point, FL, USA
CP 96-1865	Hybrid	Green German/CP 70-1133	M	USDA-ARS	Canal Point, FL, USA

Genotypes	Species	Parentage	Experiments†	Breeding institution	Country of origin
CP 97-1777	Hybrid	CP 90-1233/CP 57-0603	M	USDA-ARS	Canal Point, FL, USA
CP 97-1944	Hybrid	CP 80-1743/94P15	I, II, III, M	USDA-ARS	Canal Point, FL, USA
CP 97-1979	Hybrid	CP 75-1091/CL 61-0620	I, II, III	USDA-ARS	Canal Point, FL, USA
CP 97-1989	Hybrid	CP 75-1091/CL 61-0620	I, II, III, M	USDA-ARS	Canal Point, FL, USA
CL 97-2730	Hybrid	–	II, III	US Sugar Corp.	Clewiston, FL, USA
CP 98-1029	Hybrid	CP 91-1980/CP 94-1952	II, III	USDA-ARS	Canal Point, FL, USA
CP 98-1118	Hybrid	CL 61-0620/US 87-1006	M	USDA-ARS	Canal Point, FL, USA
CP 98-1325	Hybrid	CP 90-1030/95P08	M	USDA-ARS	Canal Point, FL, USA
CP 98-1335	Hybrid	TCP 87-3388/CP 70-1133	M	USDA-ARS	Canal Point, FL, USA
CP 98-1417	Hybrid	HoCP 85-845/CP 80-1827	M	USDA-ARS	Canal Point, FL, USA
CP 99-1534	Hybrid	CP 89-2377/CP 89-1756	M	USDA-ARS	Canal Point, FL, USA
CP 99-1542	Hybrid	CP 90-1535/95P16	M	USDA-ARS	Canal Point, FL, USA
CP 99-1686	Hybrid	CP 85-1382/CP 70-1133	M	USDA-ARS	Canal Point, FL, USA
CP 99-1889	Hybrid	CP 87-1475/CP 72-1210	M	USDA-ARS	Canal Point, FL, USA
CP 99-1893	Hybrid	CP 87-1475/CP 72-1210	M	USDA-ARS	Canal Point, FL, USA
CP 99-1894	Hybrid	CP 87-1475/CP 72-1210	M	USDA-ARS	Canal Point, FL, USA
CP 99-1896	Hybrid	CP 90-1204/CP 90-1436	M	USDA-ARS	Canal Point, FL, USA
CP 99-2084	Hybrid	CP 93-1634/CP 84-1198	M	USDA-ARS	Canal Point, FL, USA
CP 99-2099	Hybrid	CP 89-2377/CP 84-1198	M	USDA-ARS	Canal Point, FL, USA
CL 99-4455	Hybrid	–	II	US Sugar Corp.	Canal Point, FL, USA
Decade 7 (2000–)					
CP 00-1101	Hybrid	CP 89-2143/CP 89-2143	II, III	USDA-ARS	Canal Point, FL, USA
CP 00-1446	Hybrid	CP 93-1634/CP 92-1167	II, III	USDA-ARS	Canal Point, FL, USA
CP 00-2180	Hybrid	HoCP 91-552/HoCP 91-552	II, III	USDA-ARS	Canal Point, FL, USA
CP 01-1372	Hybrid	CP 94-1200/CP 89-2143	II, III	USDA-ARS	Canal Point, FL, USA

† Experiments I, II, and III were used to assess genetic progress with these particular cultivars under muck, sand soils, and freeze conditions, respectively. M indicates that molecular genotyping was performed on these cultivars.

tance between blocks. Cultivars in Exp. III were tested in nonreplicated single row plots, 2.7 m long, spaced 3 m apart, and with six replications of CP 89-2143, which is the standard for freeze tolerance in the Florida sugarcane industry. The three testing sites represent the three most important environmental conditions (from a soil or temperature point of view) known to impact the yield potential of cultivars in the Florida sugarcane industry. Disease ratings were not monitored in this research.

Experiment I was sampled on 12 Feb 2007, 19 Feb 2008, and 17 Feb. 2009 in the plant cane, first and second ratoon crops, respectively. Experiment II was sampled on 28 Feb. 2011 and on 2 Feb. 2012 in the plant cane and first ratoon crops, respectively. For both tests, 10-stalk samples were cut from each plot each year, weighed and milled to measure Brix, pol, and sucrose content by extrapolation from calibration curves developed using near-infrared reflectance and to calculate, as per Legendre (1992), commercially recoverable sucrose (CRS), which was expressed in grams per kilogram. Cane yield was expressed in kilograms per sample because stalk counts were not taken and sucrose yield (SY), as the product of CY and CRS, was expressed in kilograms per sample. Stalk height (m) and DIAM (mm) were measured as the mean of each sample, considering the millable portion of the stalks. Diameter was measured at mid-length using a digital caliper as the mean of three stalks. Experiment III was sampled four times in each of two crop years: the first year on 9 and 30 Nov. 2011 and on 6 and 25 Jan. 2012 and the second year on 15 Nov. and 14 Dec. 2012 and on 16 Jan. and on 8 Feb. 2013 to follow the time-course deterioration of sucrose (CRS) in the juice as affected by freeze.

At CP, Exp. I received no fertilizer applications and was mostly rainfed-irrigated each crop year as the station has no control over water and depends on the level in the ditch that is regulated by surrounding commercial farms. Cultural practices (herbicides and cultivation) were as needed and according to commercial recommendations. At Townsite, Exp. II was managed in accordance with industry standard practices for irrigation, weed control, and two applications of fertilizer for sand soils. At Hague, Exp. III was planted also on a sand soil and was managed accordingly with split applications of 135 kg ha^{-1} of a 10–10–10 formulation. Sprinkler irrigation and weed control (herbicide application and cultivation) were provided as needed. To establish the ratoon crops, the first two experiments were machine-harvested, and Exp. III was cut down with a heavy-duty field mower.

Statistical Analyses

For Exp. I and II, individual analyses were performed on each year for CY, CRS, SY, DIAM, and HT according to the following statistical model:

$$Y_{(\text{trait})ijk} = \mu_{ijk} + B(R)_{jk} + R_k + C_i + \varepsilon_{ijk}$$

where Y is the phenotypic trait measured on a cultivar in blocks nested within replicates, μ is the overall mean, $B(R)$ is the random effect of blocks within replication, R is the replication effect, C is the random cultivar effect, and ε is the residual variation (error term). A combined analysis of variance was performed subsequently, by taking each year's dataset as an environment to assess the cultivar × environment interaction (CEI), with environments taken as the soil × crop–year combinations. A random-effect model was then applied as follows:

$$Y_{(trait)} = \mu + B(R)_j + R_k + C_i + C \times B(R)_{ijk} + \varepsilon_{ijk}$$

where $C \times B(R)$ is the cultivar–block interaction effect. The cultivar and CEI effects were also tested by the likelihood ratio tests via restricted maximum likelihood.

Experiment III was not included in the combined analysis with the other two, as the design was completely different and only CRS was measured. The CEI estimation was performed in two steps (Federer et al., 2001): in the first step, the adjusted or predicted means for CRS were extracted from individual analyses of each of the eight samplings (4 samples × 2 crop–years) according to an augmented design and using the following random model for both cultivar and block effects:

$$Y_{(CRS)} = \mu + B_j + C_i + e_{ij}$$

A combined analysis of variance was subsequently performed on the adjusted means that were properly weighed by the reciprocal of the error variance of the mean, considering each sampling × crop–year combination as an environment (Federer et al., 2001).

The additive main effects and multiplicative interaction (AMMI) analysis was applied on both combined datasets (all crop–years combinations for both sand and muck tests and all crop–years × sampling combinations for the freeze tolerance test at Hague). Biplots were constructed from the AMMI models to investigate the stability of these cultivars across the range of environmental conditions considered in this study. The superiority measure or performance index (Pi) of Lin and Binns (1988) was estimated for individual cultivars as the squared differences between each cultivar yield (CRS, CY, and SY) and the maximum yield in each environment and averaged across environments. Low Pi values characterize cultivars with broader adaptation. The Pi values were interpreted as indicative of their relative sensitivity to environmental changes or as relative measures of adaptability on a spatial and temporal scale (muck vs. sand and crop–years conditions).

Estimates of Genetic Trends

Adjusted means of the cultivars for the measured phenotypic traits were regressed on years of releases, first across environments and then within individual environments or crop–years. The slopes of the regression lines were interpreted as chronological genetic trends (temporal changes in the traits) if significant at $P < 0.05$. The cultivar Pi values were also regressed on years of release to determine the relative impact of breeding on stability or adaptation of the cultivars across years.

Genetic Correlations

Genetic correlations (r_G) among the traits were measured within environments to understand their co-evolution as constrained or promoted by the interactions among genes. Genetic correlations (r_B) between the same trait measured in different environments were also estimated to ascertain if the trait is under the same or a different genetic control when expressed in different environments (Falconer and Mckay, 1996; Lynch and Walsh, 1998). In the CP breeding program, CY, CRS, and SY are routinely measured and used for advancement and release, complemented with disease surveys. Stalk height (HT) and DIAM, being components of CY, were considered in this research even though there are not among the breeding goals. Genetic correlations were determined as:

$$r_{Gxy} = \mathrm{cov}_{Gxy}/[\sigma^2_{Gx}\ \sigma^2_{Gy}\ ]^{1/2} \text{ and } r_B = \mathrm{cov}_{Gxixj}/[\sigma^2_{Gxi}\ \sigma^2_{Gxj}\ ]^{1/2}$$

where r_{Gxy} is the genetic correlation between traits x and y, r_B is the genetic correlation for the same trait measured in environments i and j, cov_{Gxy} is the genetic covariance between traits x and y, and σ^2_{Gx} and σ^2_{Gy} are the respective genetic variances for each trait. The standard errors associated with the correlations were approximated by the Delta method (Lynch and Walsh, 1998).

Molecular Genetic Diversity

Two molecular-marker systems were used to genotype the cultivars included in this study. Fifty-five sugarcane genotypes, including wild progenitors and cultivars released during the period from 1930 to 2000 period, were described using 12 SSR primer pairs developed by the International Consortium of Sugarcane Biotechnologists (Cordeiro et al., 2000). Another type of markers, known as target region amplified polymorphisms (TRAP) (Li and Quiros, 2001), were designed around six genes putatively associated with sucrose synthase (SuSy), sucrose phosphate synthase (SuPS), soluble acid invertase (SAI), pyruvate orthophosphate dikinase (PODK), starch synthase (StSy), and calcium-dependent protein kinase (CDPK) and assembled as 24 primer combinations (6 fixed × 4 arbitrary; Table 14–2). TRAP is a polymerase chain reaction–based marker system consisting of an 18-bp fixed primer (designed from expressed sequence tags) and of an 18-bp arbitrary primer, with the capability to amplify both intronic and exonic regions of the genome (Hu and Vick, 2003). The SSR primers were used to amplify the neutral regions of the sugarcane genome.

Both markers were treated as dominant (0/1) to evaluate several genetic diversity parameters, such as percent polymorphism (%P), expected heterozygosity (H_E), number of bands and of private alleles, and coefficient of relatedness (r_{ab}). The software Hickory v1.1 (Holsinger and Lewis, 2007) and Popgene v32 (Yeh et al., 2000) were used to calculate these values. A measure of linkage disequilibrium among the markers was estimated based on the standardized index of association (I_A^S) developed by Haubold and Hudson (2000) for dominant markers using the software Lian v.3.5. For the molecular analyses, the genotypes were divided into seven decades (Table 14–1): the first decade or progenitor group was made up of clones from the different species of sugarcane used in wide-cross breeding; the "founder" group comprised some very early hybrids (e.g., POJ, NCo, Eros) used to launch breeding programs worldwide; the third decade was composed of cultivars released by the University of Florida; the fourth, fifth, and sixth decades included cultivars released during three successive decades starting in the 1950s; the seventh decade was made up of some released cultivars and breeding lines from the last stage of selection developed during the 1990s by the CP program. Values for the different parameters of diversity (H_E, %P, private alleles, I_A^S) were regressed on decades and interpreted as influenced by the combined effects of drift and breeding and selection.

Inference of Selection

Coalescent simulations were implemented on the two datasets of markers to infer selection (from the detection of outlier loci) based on the F_{ST} distribution conditional on H_E across marker loci (Beaumont and Balding, 2004). The Bayesian approach, developed by Zhivotovsky (1999) and incorporated into the Dfdist software (provided by Dr. M. Beaumont) for dominant markers, was followed to

Table 14–2. Arbitrary and fixed target region amplified polymorphism (TRAP) primer sequences (≈18-mer) used to assess genetic diversity in a set of 61 sugarcane genotypes in the USDA-ARS Canal Point breeding program.

Primers	Gene name	Sequence (5′ → 3′)	Reference or NCBI accession number
		Arbitrary	
Arbit1		GACTGCGTACGAATTAAT	Li and Quiros, 2001
Arbit2		GACTGCGTACGAATTTGC	Li and Quiros, 2001
Arbit3		GACTGCGTACGAATTGAC	Li and Quiros, 2001
Arbit4		GACTGCGTACGAATTTGA	Li and Quiros, 2001
		Fixed	
SuSy	Sucrose synthase	GGAGGAGCTGAGTGTTTC	AF263384
SuPS	Sucrose phosphate synthase	CGACAACTGGATCAACAG	AB001338
SAI	Soluble acid invertase	AGGACGAGACCACACTCT	AF062735
PODK	Pyruvate orthophosphate dikinase	CGTAAAGATTGCTGTGGA	AF194026
StSy	Starch synthase	GGCAAGAAGAAGTTCGAG	AF446084
CDPK	Calcium-dependent protein kinase	ACAGAACCACCAAAGGAG	CF572977

generate the null distribution for 50,000 simulated loci and to interpret the location of the SSR and TRAP loci relative to a 95% confidence envelope as being: (i) under directional selection for F_{ST} values above the upper 95% line, (ii) under balancing selection for F_{ST} values or loci below the lower limit, or (iii) neutral for loci falling inside the envelope.

Results

Realized Genetic Gains

The organic (muck) and mineral (sand) soils in the Florida sugarcane industry represent the two major environmental conditions that significantly drive the differential performance of cultivars, requiring that specific genotypes be identified for each soil type (Edmé et al., 2005b; del Blanco et al., 2010). While the program targets broad adaptation, the adopted sequential testing and selection strategy (muck–muck+sand in this order) of the new genetic materials (starting with ≈70,000 in the Seedling stage) is somewhat biased against identifying the best genotypes with high-yield performance on sand. It uses an intense selection pressure (0.2–0.3% in three steps of ± 10% each) to advance 135 genotypes across the two soil types and farms in Stage 3. The dataset used by Edmé et al. (2005b) attributed all the genetic gains (69%) to production on muck and none to production on sand. These results needed to be updated with a side-by-side testing of the historical CP cultivars across the environmental spectrum.

Cane Yield: Performance and Gains on Sand and Muck Soils

In the individual and combined analyses, cane yield was significantly ($P < 0.01$) affected by the genetic, the environmental, and their interaction effects, with the environments having a larger contribution (56%) to the variation for this trait and with the interaction being about one-half the magnitude (46%) of the cultivar effect (Table 14–3). On muck, mean CY values were 14.4, 10.45, and 10.05 kg in the plant cane, first ratoon, and second ratoon crops, respectively, and the best cultivars ranking consistently in the top five across the three crops were CP 73-1547, CL 81-1453, F31-407, and F36-819. Under the sand soil conditions, the cultivars produced mean CY of 10.15 and 9.60 kg in the plant cane and first ratoon crops, respectively, with CP 75-1547 being the only cultivar with a consistent rank among the top five cultivars.

Besides a slight and significant decrease (0.014 kg yr^{-1}) observed in weights in the first ratoon crop on muck, CY stayed relatively constant on both soils during the 1930 to 2004 period (Fig. 14–1). Apparently, a more pronounced decrease was observed during the first 20 yr (1930–1950), which was followed by an improvement to the initial level thereafter. The period from 1930 to 1950 coincided with the growing of the early hybrids developed from Java (Indonesia), from Coimbatore (India), and of the F-cultivars developed by the University of Florida, with the later ones being 13% heavier on average than the POJ and Co hybrids. The two breeding programs (CL and CP) started with cultivars somewhat lower in CY than the imported hybrids, ultimately increased CY by 10% by the end of the 1990s and maintained it at the same level since (Fig. 14–1).

The decadal trends in CY (Fig. 14–2) indicated that there was a linear decrease of 7.4 kg during the first two decades on muck, followed by a linear improvement of 2.2 kg through the third decade and by a small decrease of 0.2 kg thereafter; the current level reflects a 19% deficit over the POJ, Co, and F-hybrids. On sand, CY dropped at a linear rate of 1.8 kg by the third decade to continue on a linear increase of 0.23 kg thereafter, with CY in modern cultivars being comparable to that of the POJ and Co hybrids on sand.

Sucrose Content: Performance and Gains on Sand and Muck Soils

The combined analysis indicated there was significant difference in main effects and in their interaction, with 42% of the total phenotypic variation in CRS attributed to the effect of the environments, 15% to cultivar, and 20% to CEI effect (Table 14–3). The variation attributable to cultivar was 74% of that of CEI. On both soils, the ratoon crops produced more sucrose yield than the plant cane crop, with respective mean CRS of 207.2, 250.1, and 241.1 g kg^{-1} measured from the plant cane to the second ratoon on muck and with mean CRS of 255.0 and 275.9 g kg^{-1} measured in the plant cane and first ratoon on sand. CP 89-2143 was the only cultivar with consistent performance among the top five CRS producers across soil types and ratoon crops. Across crop–years, CP 89-2143 and CP 77-1776 were among the best five cultivars on muck, whereas no genotype gave stable CRS on sand in the two crop–years. The mode rank for the early hybrids that originated from Indonesia and India was 45 (range 8–65).

Overall, the CP breeding program increased CRS at a linear rate of 0. 63 g kg^{-1} yr^{-1} ($P < 0.001$), considering production on both soil types (Fig. 14–3). On muck, CRS improved significantly at a rate of 0.58 g kg^{-1} yr^{-1} ($P < 0.001$) across the 75-yr period, supported by improvements in the plant cane (1.0 g kg^{-1} yr^{-1} ($P < 0.001$), in the first ratoon (0.31 g kg^{-1} yr^{-1} ($P < 0.001$), and in the second ratoon (0.42 g kg^{-1} yr^{-1} ($P < 0.001$). On sand, an overall gain of 0.52 g kg^{-1} yr^{-1} ($P < 0.001$) was obtained across crop–years, sustained by improvements of 0.64 g kg^{-1} yr^{-1} ($P < 0.001$) in

Table 14–3. Variance components for three agronomic and two morphological traits measured on historical sugarcane cultivars released and grown by the Florida industry during a 75-yr period (1930–2004). Cultivars were tested in three field tests and for three crops (2007–2010) on a muck soil at Canal Point, FL and for two crops on two sand soils at Townsite (Clewiston, FL; 2011–2012) and at Hague (Alachua, FL; 2012–2013). Environments were analyzed as soil × crop–years combinations for the Canal Point and Townsite tests and as a samplings × crop–years matrix for the Hague test.

Sources of variation	Cane yield	CRS1†	Sugar yield	Diameter	Height	CRS2‡
Environment (E)	6.326	577.482	0.085	0.522	0.042	444.258
Cultivar (C)	2.625	212.455	0.185	3.507	0.023	202.116
C×E	1.202	279.830	0.127	0.849	0.010	48.452
Residual	3.120	260.384	0.235	7.149	0.043	198.290
CV	9.636	5.806	11.882	5.286	5.600	–
LSD	2.210	28.336	0.662	2.662	0.232	–
Grand mean	11.634	247.614	2.827	25.551	2.103	224.178
Heritability	0.854	0.743	0.819	0.844	0.822	0.90
Number of reps	3	3	3	3	3	1‡
Number of environments	5	5	5	5	5	8

† CRS, commercially recoverable sugar, with CRS1 measured in the combined CP-Townsite analysis, and with CRS2 as the only trait measured at Hague to reflect freeze tolerance based on four samples taken in each of two crop–years.

‡ The Hague test was planted as a 5 × 11 α-lattice augmented design.

the plant cane and of 0.40 g kg^{-1} yr^{-1} ($P < 0.001$) in the first ratoon. The decadal profiles suggested faster rates of CRS improvements on muck than on sand and according to a quadratic trend: an initial linear increase of 24.95 g kg^{-1} ($P < 0.001$) through the fourth decade was followed by a plateau around 242 g kg^{-1} through the fifth decade and by a final decrease of 2.65 g kg^{-1} between the fifth and the sixth decades (Fig. 14–2). On sand, the respective trend in CRS was linear (5.95 g kg^{-1} $decade^{-1}$) throughout the entire period.

Sucrose Yield: Performance and Gains on Sand and Muck Soils

Only the cultivar main effect was highly significant ($P < 0.01$) for SY, contributing 39% to the variation in this trait (Table 14–3). The SY means were comparable across soil types and crop–years, ranging from 2.4 kg (second ratoon on muck) to 2.98 kg (plant cane on muck). CP 75-1547 was the most stable and high-yielding cultivar in the tests, being among the five top-ranked cultivars across crop–years and soil types. In addition to CP 75-1547, cultivars CP 89-2143 and CL 86-4519 appeared twice across crop–years on muck among the top five. CP 75-1547 was the only cultivar that was consistent across the two crop-years on sand.

There was no noticeable gain in SY, neither overall nor on a soil type basis, except for a small gain of 0.005 kg yr^{-1} in the plant cane on sand (Fig. 14–4). However, from a decade point of view, a decrease of 1.25 kg was observed on muck soils in the first two decades, which was followed by an increase of 0.45 kg $decade^{-1}$ thereafter. On the sand soils, SY continued to increase significantly on sand soils at a linear rate of 0.05 kg that started after an initial decrease of 0.31 kg from the first to the third decade (Fig. 14–2f).

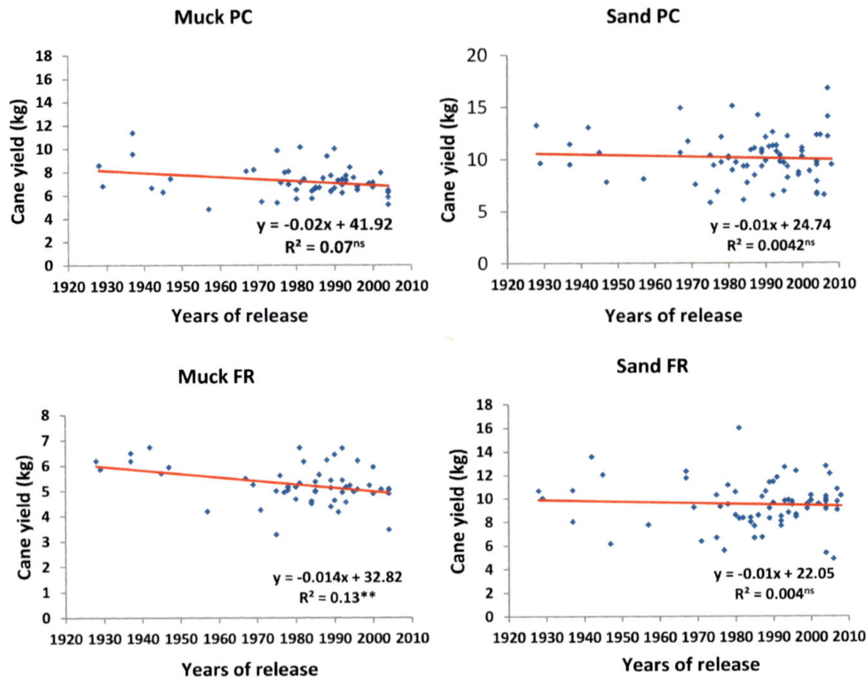

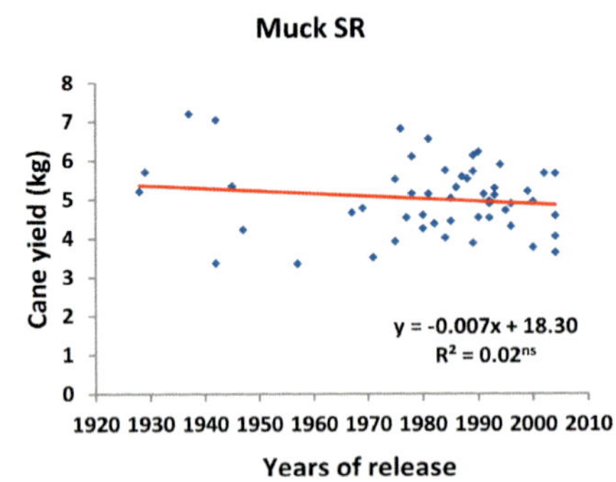

Fig. 14–1. Linear trends in cane yield for historical sugarcane cultivars released during the period 1930 to 2004 in the Florida industry and tested for three crop–years (2007–2009) on a muck soil and for two crop–years (2011–2012) on a sand soil. Cane yield was expressed as stalk weight per sample.

Height and Diameter: Performance and Gains on Sand and Muck Soils

Height and diameter are not normally taken into consideration in the selection process in the CP sugarcane breeding program. They were measured in this research to understand how breeding impacted the yield-component traits in sugarcane. Significant differences were detected among cultivars, environments, and their interactions for HT, which contributed 33, 15, and 6%, respectively, to the variation in HT (Table 14–3). The cultivars were 19% taller on muck (2.32 m on average in the plant cane and first ratoon) than on sand, with similar height (1.95 m on average) observed for the other crop–years. Cultivar had a larger contribu-

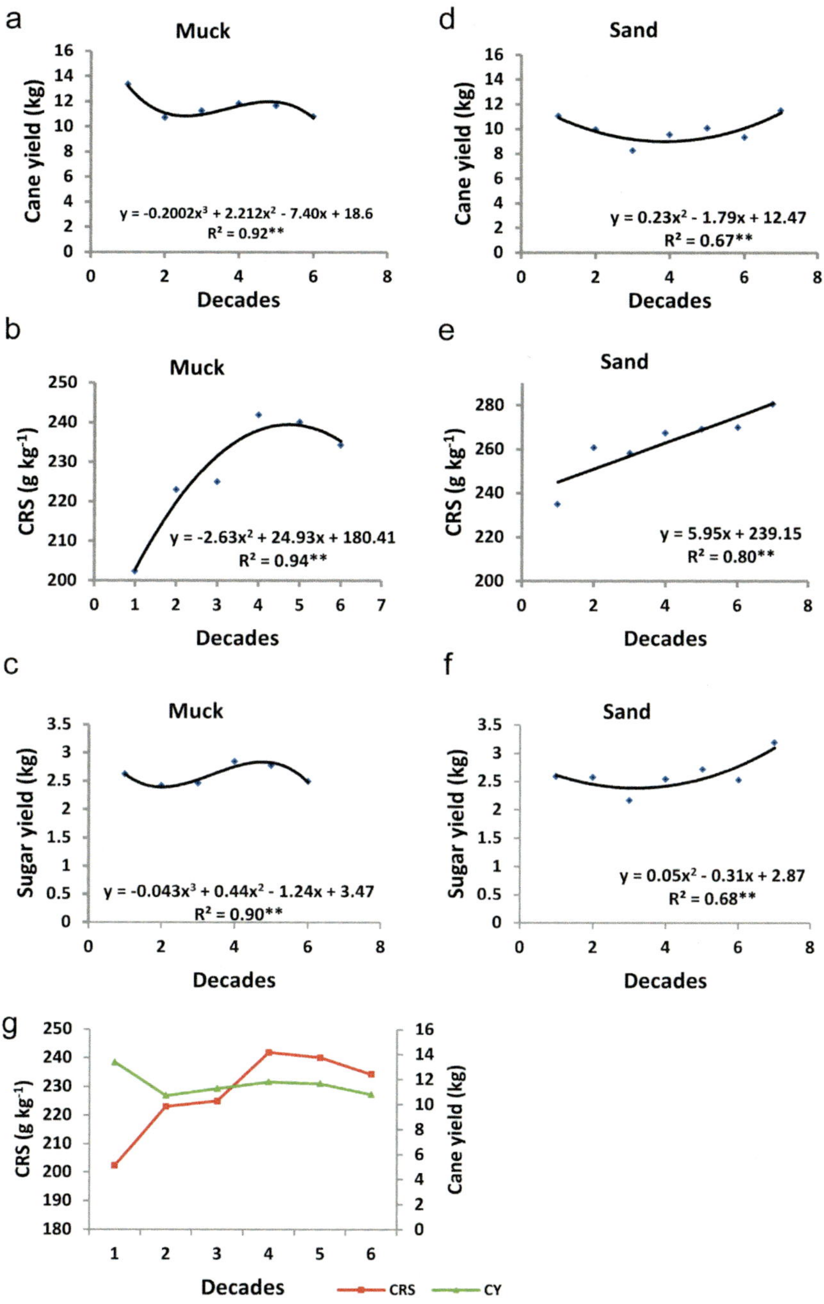

Fig. 14–2. Decadal trends in cane yield, commercially recoverable sugar (CRS), and sugar yield of historical sugarcane cultivars released during the period 1930 to 2004 in the Florida sugarcane industry and tested for (a, b, c) three crop–years (2007–2009) on a muck soil and for (d, e, f) two crop–years (2011–2012) on a sand soil and for CRS and (g) cane yield across soils.

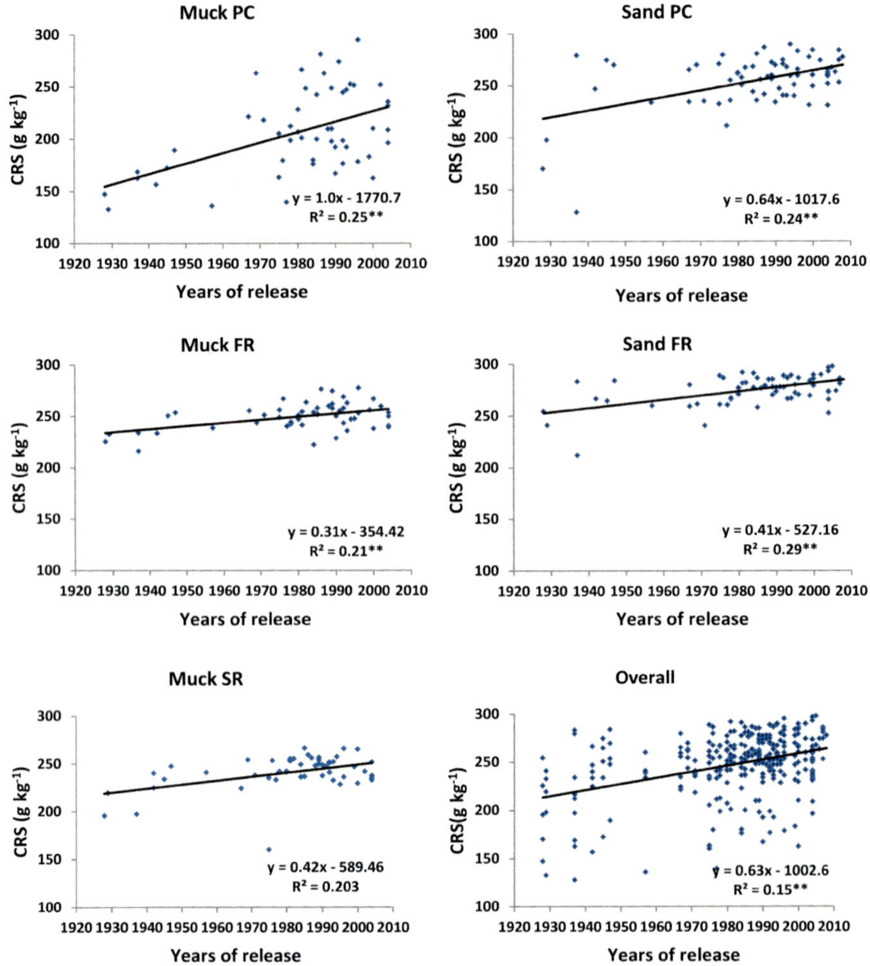

Fig. 14–3. Linear trends in commercially recoverable sugar (CRS) for historical sugarcane cultivars released during the period 1930 to 2004 in the Florida sugarcane industry and tested for three crop-years (2007–2009) on a muck soil and for two crop-years (2011–2012) on a sand soil. Overall was based on means across environments (soil–crop year combinations). PC, plant cane; FR, first ratoon; SR, second ratoon.

tion (25%) to the variation in DIAM than CEI (19%), and no significant differences were observed among environments. Mean DIAM values across environments ranged from 24.6 mm (first ratoon on sand) to 26.7 mm (plant cane on muck).

There was no particular change in HT in the sugarcane populations that could be attributed to the effect of breeding and selection. Diameter, on the other hand, showed significant ($P < 0.01$) linear decreases across soil types and crop-years, with a similar decline of 0.045 mm yr^{-1} (on average) across environments (Fig. 14–5).

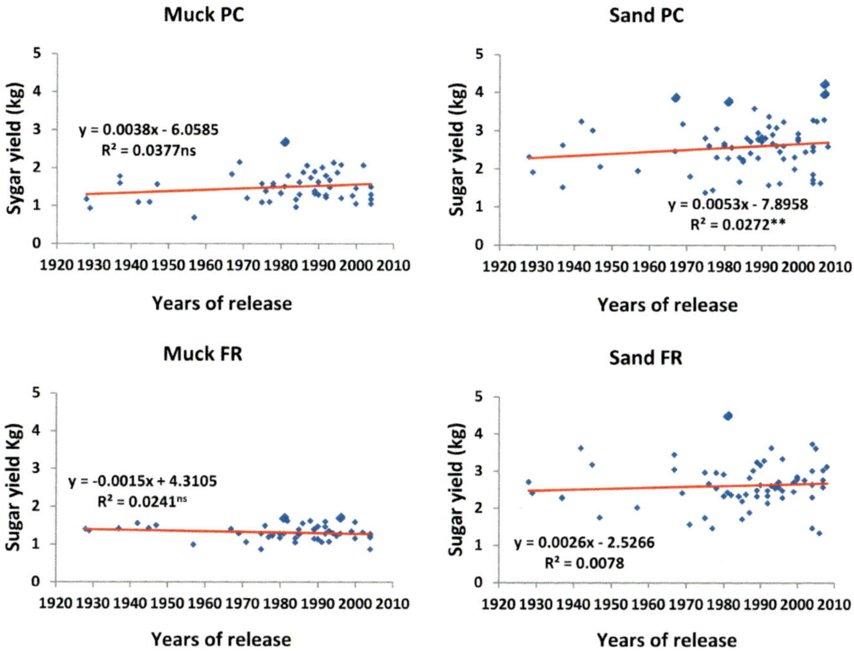

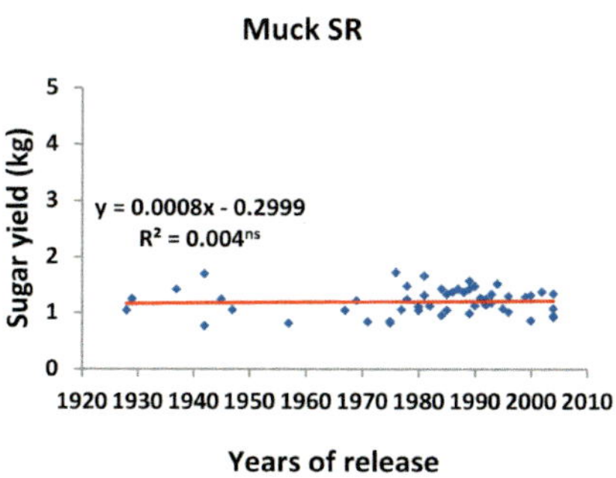

Fig. 14–4. Linear trends in sugar yield (measured on a sample basis) for historical sugarcane cultivars released during the period 1930 to 2004 in the Florida industry and tested for three crop–years (2007–2009) on a muck soil and for two crop–years (2011–2012) on a sand soil. PC, plant cane; FR, first ratoon; SR, second ratoon.

Stability or Adaptation

Regressing the Pi on years of release indicated that more modern sugarcane cultivars were better adapted to the two soil types and to the subtropical environment in the Everglades Agricultural Area, where sugarcane is grown, than the earliest hybrids (POJ) and those of the 1940–1960 era released through this industry. There was no change in adaptability associated with CY or SY across the 75 yr of sugarcane production. The gains in Pi were exclusively associated with CRS, corresponding to an improvement of 52.65 yr^{-1} ($P < 0.001$), which is translated to gains of 70% over the era of the POJ, 43% over the cultivars released in the 1960s, 29% over those released in the 1970s, and to a drop of 28% over those released in the 1980s (Fig. 14–6). CP 89-2143, the most popular and productive cultivar in the Florida industry, had the lowest Pi value (59) and ranked first, with the sec-

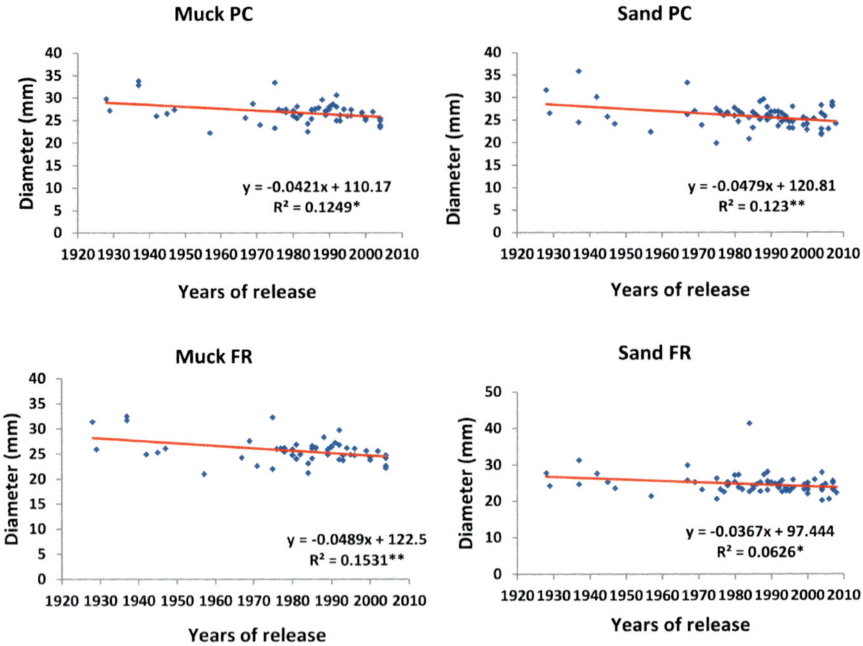

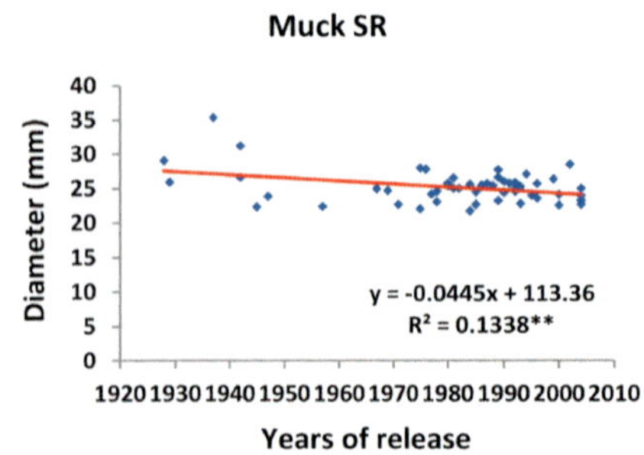

Fig. 14–5. Linear trends in stalk diameter for historical sugarcane cultivars released during the period 1930 to 2004 in the Florida industry and tested for three crop–years (2007–2009) on a muck soil and for two crop–years (2011–2012) on a sand soil. PC, plant cane; FR, first ratoon; SR, second ratoon.

ond and third best ranked and stable cultivars being CP 77-1776 (Pi = 344) and CP 78-1628 (Pi = 384). The 1980s era was dominated by cultivars such as CP 70-1133, CP 72-2086, CP 73-1547, CP 77-1776, CP 78-1628, all five with relatively low Pi values, ranging from 344 (CP 77-1776) to 546 (CP 73-1547).

Freeze Tolerance

Modern cultivars are more freeze tolerant (based on the CRS deterioration profiles) than the earliest hybrids derived from the Indonesian (POJ) and Indian (Co) sugarcane breeding programs and more than the earliest hybrids released by the two breeding programs (CP and CL combined). Starting with the second sampling, the cultivars were exposed to –2°C/28°F (6 h), –3°C (7 h), and –7.7°C/18°F (4 h) in the plant cane, and to –1.5°C/29°F (7 h), –3°C (8 h), and to –1.5°C (8 h) in the first ratoon

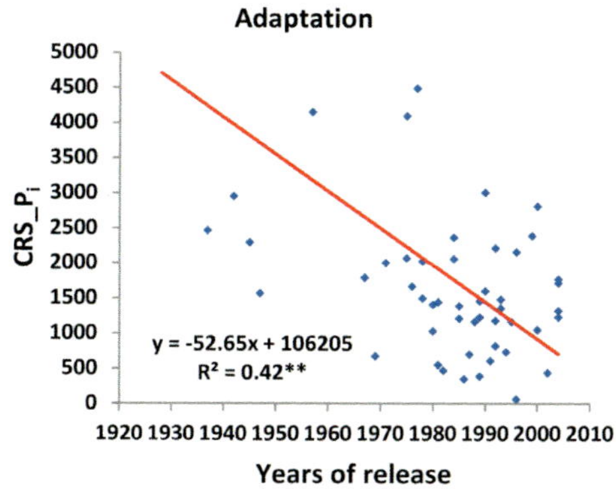

Fig. 14–6. Linear trend in superiority measure or Pi (adaptability) as per Lin and Binns (1988) calculated across environments for historical sugarcane cultivars released during the period 1930 to 2004 in the Florida industry and tested for three crop–years (2007–2009) on a muck soil and for two crop–years (2011–2012) on a sand soil. Lower Pi-values indicate better adaptability. CRS, commercially recoverable sugar.

crop. The CEI effect was highly significant but not substantial, with the environment and cultivar main effects contributing 45 and 17% more (than CEI) to the variation in CRS, respectively (Table 14–3). The plant cane data were chosen for estimating genetic gains. Overall gains in CRS improved at a rate of 0.28 g kg^{-1} yr^{-1} ($P < 0.001$) after exposure to temperatures ≥ -3°C and at a rate of 0.40 g kg^{-1} yr^{-1} ($P < 0.001$) after exposure to temperatures < −3°C (Fig. 14–7). Considering the USDA-ARS CP breeding program separately from the USSC CL program, the realized gains were even greater and were equivalent to an overall gain of 0.44 g kg^{-1} yr^{-1} ($P < 0.001$), based on gains of 0.41 g kg^{-1} yr^{-1} ($P < 0.001$) for temperatures ≥ -3°C and on gains of 0.46 g kg^{-1} yr^{-1} ($P < 0.001$) for temperatures < −3°C.

Genetic Correlations: Among Traits, Among Locations, and Heritability

Across soil types, the r_G values between CRS and CY were low, generally negative and nonsignificant, ranging from −0.032 to 0.097 (Table 14–4). Commercially recoverable sucrose was positively correlated with SY, with the highest value (0.77, $P < 0.01$) obtained under the plant cane on muck, and with values ranging from 0.36 to 0.45 ($P < 0.01$) in the other environments. The CRS–HT relationship was unstable in the muck environment, with a high negative value (−0.45, $P < 0.01$) measured in the first ratoon and with small positive values (0.08–0.14) in the plant cane and second ratoon. Under sand conditions, the CRS–HT relationship was stable (0.20, $P < 0.05$) across crop–years. Negative correlations were found between CRS and DIAM across soil types, with values increasing in strength from the plant cane (−0.08) to the second ratoon (−0.36, $P < 0.01$) on muck and values stabilizing at a mean of 0.38 ($P < 0.01$) in the two crop–years on sand. Commercially recoverable sucrose contributed more strongly to the values of SY (0.77) under the plant cane muck than under the other environments where values ranged from 0.36 to 0.47 ($P < 0.01$). Across soil types, CY contributed more to the magnitude of SY under ratoon crops (r_G ranging from 0.82 to 0.97, $P < 0.01$) than to that of SY under plant cane (0.67, $P < 0.01$), with the values being even more important on sand (0.87–0.97, $P < 0.01$) than on muck (0.66–0.90, $P < 0.01$) (Table 14–5). Of the two yield components, DIAM had a stronger influence (r_G range = 0.75–0.97, $P < 0.01$) on CY across soil types than HT (−0.02 to 0.68). Whereas both

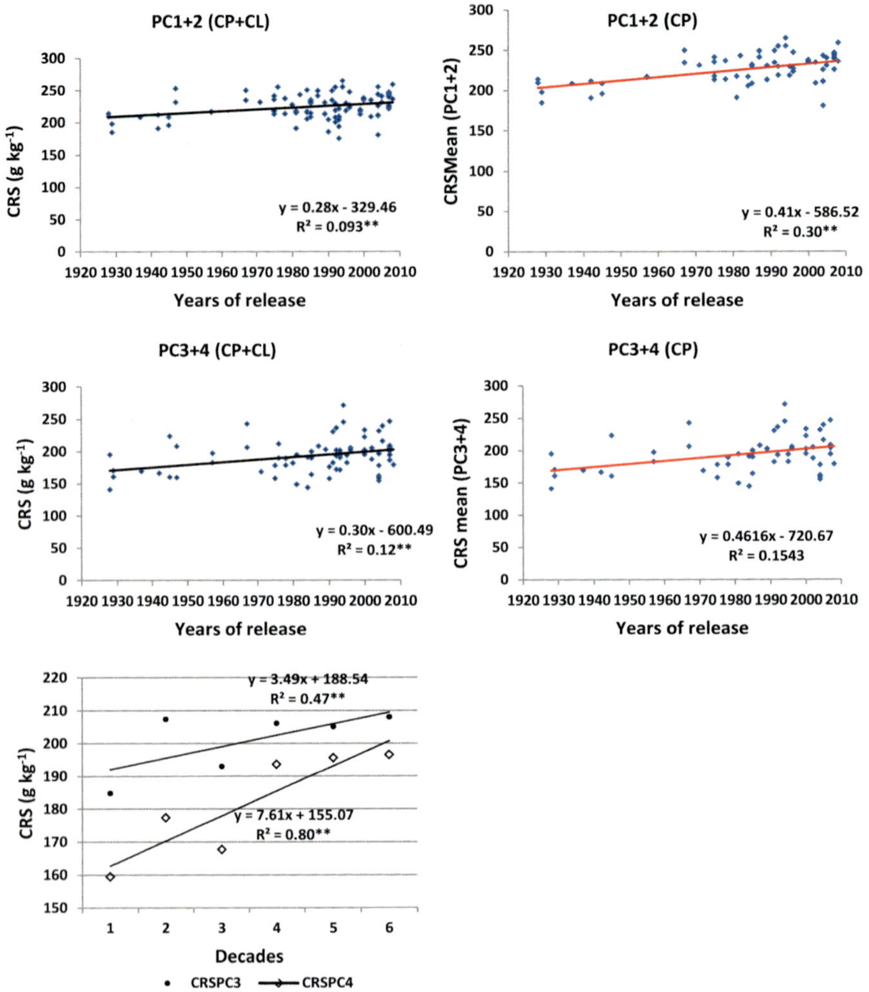

Fig. 14–7. Linear trends in freeze tolerance based on the profiles of commercially recoverable sucrose (CRS) measured before (PC1+2) and after (PC3+4) freeze conditions on historical (1930–2004) sugarcane cultivars released in the Florida industry from a test planted at the University of Florida Agronomy Farm located at Hague, FL. Distinction was made here of the two Florida breeding programs releasing cultivars with the prefix CL (USSC) and CP (USDA-ARS). Four (1–4) samples were taken in the plant cane (PC) crop before or after freeze events. The very bottom graph pictures gains in CRS obtained under more severe subzero temperatures (CRSPC3 refers to the third sample taken when night temperatures fluctuated between –2°C and –3°C and CRSPC4 corresponds to the fourth sample taken when temperatures dropped below –3°C).

HT and DIAM were important for higher CY on sand, HT had a lesser and at times negative influence (r_G = −0.02 to 0.12) on CY than DIAM, particularly in the first two crop–years. Consequently, HT and DIAM were negatively correlated under muck conditions (r_G values ranging from −0.10 to −0.17, $P < 0.05$) because

Table 14–4. Genetic correlations† (r_G) among five traits measured within an environment (taken as soil–crop year combinations) on historical sugarcane cultivars released to the Florida industry and tested jointly on muck (2008–2010) and sand (2001–2012) soils to assess the impact of breeding.

	CRS‡	Cane yield	Sugar yield	Height	Diameter
		Plant cane			
CRS	—	0.048	0.768	0.141	−0.076
Cane yield	−0.049	—	0.660	0.119	0.872
Sugar yield	0.438	0.874	—	0.192	0.470
Height	0.201	0.682	0.735	—	−0.169
Diameter	−0.428	0.767	0.476	0.118	—
		First ratoon			
CRS	—	−0.032	0.401	−0.446	−0.164
Cane yield	0.097	—	0.898	−0.020	0.748
Sugar yield	0.355	0.966	—	−0.101	0.590
Height	0.191	0.725	0.732	—	−0.099
Diameter	−0.336	0.927	0.824	0.313	—
		Second ratoon			
CRS	—	−0.058	0.471	0.075	−0.363
Cane yield		—	0.823	0.439	0.786
Sugar yield			—	0.405	0.495
Height				—	−0.174
Diameter					—

† Values above the diagonals represent correlations within the muck environments and below diagonals represent correlation within the sand soil environments for the respective crop–year.

‡ CRS, commercially recoverable sucrose.

they had different effects on CY, but positively correlated (r_G = 0.12 and 0.31) in the sand environment. Height and diameter were more important drivers of higher SY under sand than under muck conditions.

The same trait measured in two different environments can be regarded as two traits governed by different sets of genes (Falconer and Mackay, 1996). Genetic correlations (r_B) among environments based on the same trait provide a direct measure of that genetic control and an indirect measure of CEI, with values below 0.7 indicating moderate to strong interactions between the two main effects as different sets of genes are activated to enable adaptation of genotypes to a specific environment. For CRS and CY, the highest correlations (r_B values = 0.91 and 0.88) were between the two crop–years grown on sand soils (Table 14–5). The muck soil had low predictability potential (r_B ranging from −0.12 to 0.26) to select genotypes for the sand soil with regards to CRS and better potential with regards to CY (r_B ranging from 0.63 to 0.82). Moderate interactions existed among the crops grown on muck, with r_B values for CRS ranging from 0.58 to 0.67. Stronger and similar correlations (r_B ranging from 0.8 to 0.9) were found among the crops grown on muck and on sand for CY. The r_B values for SY were more reflective of the greater influence of CY on SY than of CRS, with the first ratoon test on muck being more predictive (r_B = 0.81) of SY measured in the second ratoon test on muck than in the plant cane (r_B = 0.59), and more predictive of SY measured on the two sand tests. However, the highest r_B value (0.83) for SY was between the two sand

Table 14–5. Genetic correlations (rB) among environments based on five traits measured on historical sugarcane cultivars released to the Florida industry during the period from 1930 to 2004 period and tested jointly on muck (2007–2010) and sand (2011–2012) soils to assess the impact of breeding.

	MuckFR09	MuckPC07	MuckSR10	SandFR12	SandPC11
			Commercially recoverable sucrose		
MuckFR09	1	0.674	0.575	0.263	0.251
MuckPC07	0.674	1	0.594	−0.121	0.053
MuckSR10	0.575	0.594	1	0.080	0.019
SandFR12	0.263	−0.121	0.080	1	0.908
SandPC11	0.251	0.053	0.019	0.908	1
Heritability	0.863	0.723	0.912	0.815	0.923
			Cane yield		
MuckFR09	1	0.797	0.899	0.629	0.832
MuckPC07	0.797	1	0.816	0.487	0.565
MuckSR10	0.899	0.816	1	0.746	0.657
SandFR12	0.629	0.487	0.746	1	0.874
SandPC11	0.832	0.565	0.657	0.874	1
Heritability	0.672	0.826	0.668	0.811	0.809
			Sugar yield		
MuckFR09	1	0.661	0.807	0.617	0.756
MuckPC07	0.661	1	0.589	0.498	0.538
MuckSR10	0.807	0.589	1	0.604	0.498
SandFR12	0.617	0.498	0.604	1	0.825
SandPC11	0.756	0.538	0.498	0.825	1
Heritability	0.667	0.821	0.726	0.819	0.815
			Height		
MuckFR09	1	1.000	0.653	0.584	0.466
MuckPC07	1.000	1	0.658	0.595	0.498
MuckSR10	0.653	0.658	1	0.550	0.222
SandFR12	0.584	0.595	0.550	1	0.929
SandPC11	0.466	0.498	0.222	0.929	1
Heritability	0.799	0.744	0.766	0.274	0.855
			Diameter		
MuckFR09	1	1.000	0.818	0.406	0.810
MuckPC07	1.000	1	0.813	0.381	0.807
MuckSR10	0.818	0.813	1	0.662	0.935
SandFR12	0.406	0.381	0.662	1	0.786
SandPC11	0.810	0.807	0.935	0.786	1
Heritability	0.799	0.744	0.766	0.274	0.855

crop–years. For HT, the first two crop–years had perfect (r_B = 1.0 muck) or close to perfect (r_B = 0.93 sand) correlations measured on muck or on sand. Low to moderate r_B values (0.22–0.65) were observed for HT between crop–years on muck and its measurements on muck and on sand. For DIAM, the muck tests were good predictors of each other, with r_B ranging from 0.82 to 1.0, and even better predictors (r_B ranging from 0.8 to 0.94) of DIAM measured on sand than the two sand tests were to each other (r_B = 0.79).

Highly significant and very high heritability (or repeatability) estimates (H_B values ranging from 0.27 to 0.92) were estimated in this research (Table 14–5).

Except for HT and DIAM measured on sand in the first ratoon (the two lowest values of 0.27), higher H_B values were obtained on sand than on muck for corresponding crop–years. Commercially recoverable sucrose was, in general, more repeatable within an environment than the CY and component traits. The muck and sand soil environments promoted different strengths in the heritability of the traits with respect to the crops: for example, HB values for CRS, HT, and DIAM were higher in the ratoon crops on muck, but higher in the plant cane on sand.

Molecular Evolution

Genetic Diversity

There was significant variation among decades as indicated by AMOVA, with 4% of the variation partitioned among decades and the remainder 96% being within decades. There were differences in allelic frequencies among decades (Fig. 14–8), with, however, little change in the within-decade level of differentiation. Figure 14–8 reveals differences in SSR or TRAP-derived allele frequencies, with some of the TRAP bands/alleles going to fixation, but none of the SSR alleles. This indicated that the two marker systems amplified different regions of the *Saccharum* genome, the neutral portion with SSR and TRAP and the functional portion with TRAP.

The first, fourth, and seventh decades had the largest mean genetic diversity (H_E = 0.22–0.23). The sixth decade was the most divergent from the progenitor and from all the other groups, followed by Decades 5, 4, and 6 (data not shown). The CP breeding population experienced a decrease across time in gene diversity and percent polymorphism, at respective rates of 0.02 and 6% decade^{-1} (Fig. 14–9a,b). Erosion in genetic diversity, as related to the number of polymorphic bands and of private alleles, was quantified to be evolving at respective rates of 21 and 11 bands decade^{-1} (Fig. 14–9c,d). There was an increase in the coefficient of relatedness values with elapsed generations, going from r_{ab} = 0.126 (progenitor) to r_{ab} = 0.583 (1981–1990) at a rate of 0.07 decade^{-1} (Fig. 14–9f). As indicated by the other diversity parameters (H_E, or number of private alleles), there was also a single large split in r_{ab} between the progenitor population and the cultivars in the second decade.

Highly significant ($P < 0.0002$) linkage disequilibrium between loci was detected within each decade and also at the population level (data not shown). The variance of departures from random associations (V_D) was always 2 to 10 times higher than that expected for linkage equilibrium (V_E), indicating that each decade had some degree of linkage disequilibrium (LD). The highest number of associations was recorded in the progenitor population and during the 1960s, with I_A^S on the order of 0.0121 and 0.0139, respectively (Fig. 14–9e). The lowest number of associated loci was observed in the cultivars of the 1980s at 0.0009. Values of I_A^S or of LD fluctuated across the breeding generations in the Florida program, with a large drop (from 0.0121 to 0.0026) from the wild progenitor to the founder group, a sharp increase thereafter up to 0.0139 in the 1960s, and finally another drop to 0.0009 in the 1980s or to 0.004 in the 1990s (Fig. 14–9e).

Inference of Loci Under Selection

Plotting the F_{ST} against heterozygosity (H_E) levels indicated that the majority of the TRAP and SSR markers were neutral as they fell within the 95% envelope that delineated the region for loci with neutral behavior (Fig. 14–10). Twenty-five loci (15 TRAP and 10 SSR) were detected as outliers, falling below the lower limit

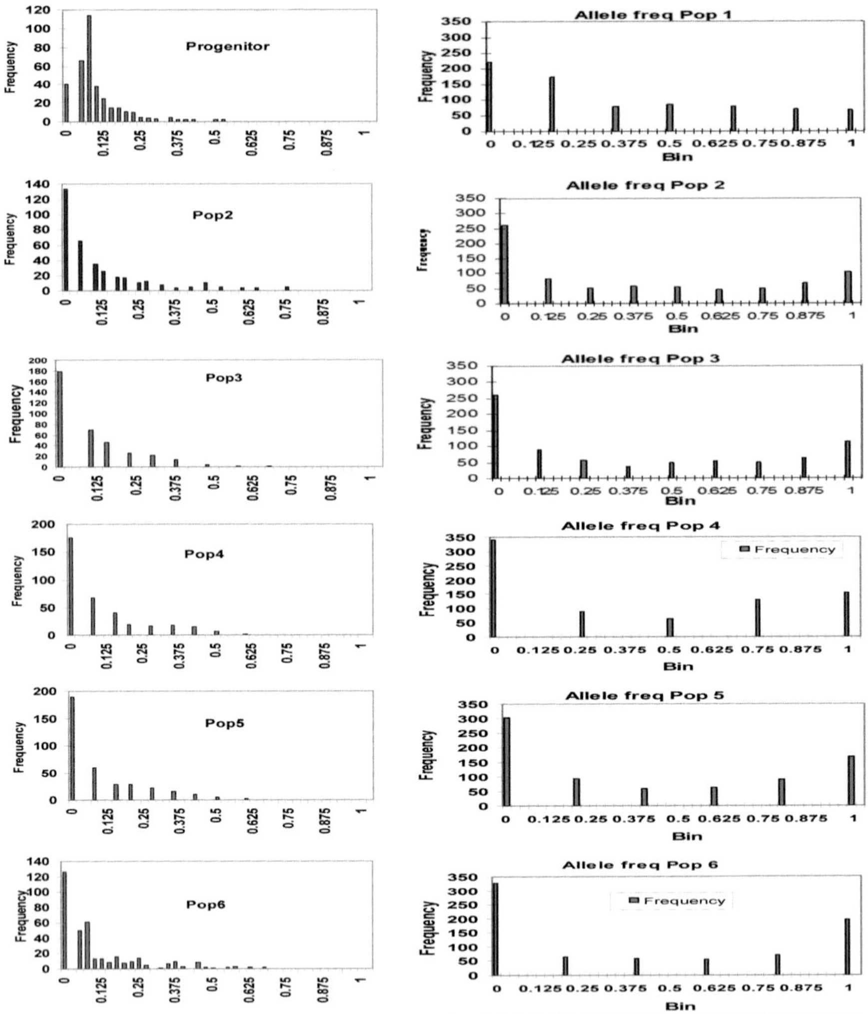

Fig. 14–8. Allele frequency distributions of microsatellites (left) and of target region amplified polymorphisms (right) used to survey historical sugarcane cultivars released in the Florida industry from 1930 to 2004. Cultivars were divided by decades of release, symbolized as Pops for decades) on the graph.

of the 95% quantile, that is, outside the expected distribution region for neutral loci. Some of the loci had a consistently low level of differentiation across decades even though they exhibited high heterozygosity values. The position of the 25 loci indicated that balancing selection was the underlying force governing polymorphisms and the evolution of traits in the CP breeding program (Beaumont and Balding, 2004).

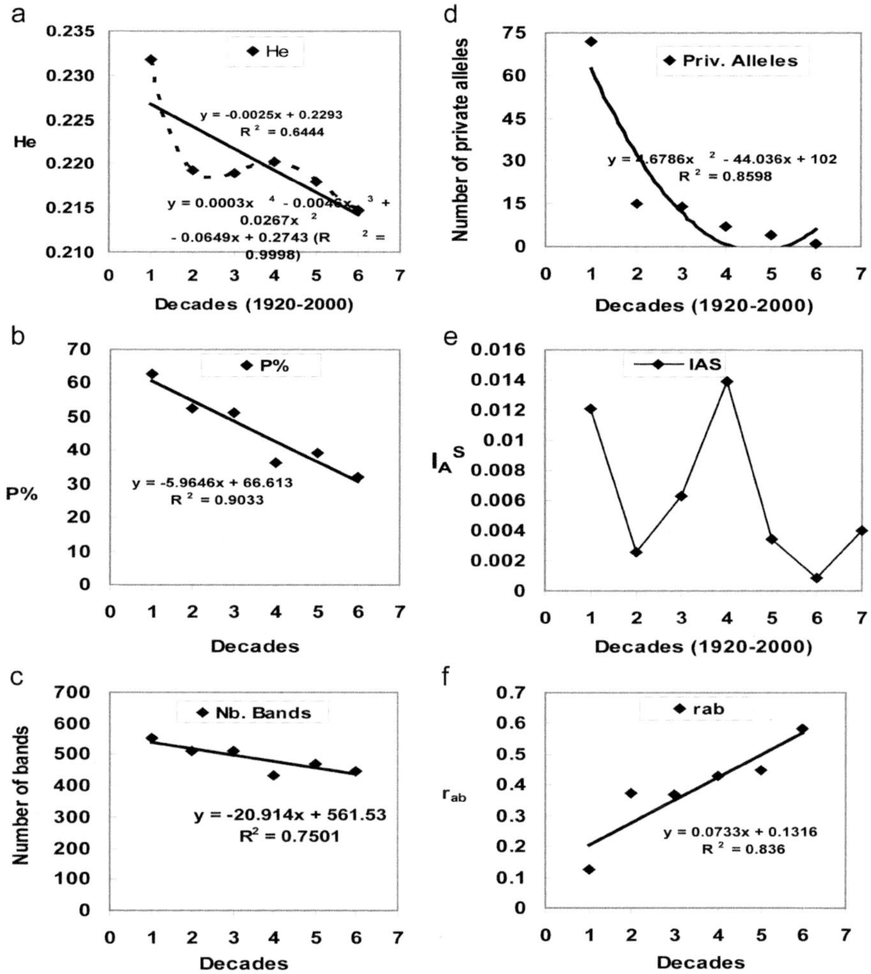

Fig. 14–9. Changes in genetic diversity parameters across decades measured on historical sugarcane cultivars released from 1930 to 2004 in the Florida industry. (a) H_E = expected heterozygosity or gene diversity, (b) percentage polymorphism, (c) allelelic richness, (d) number of private alleles, (e) standardized index of association as a measure of linkage disequilibrium, and (f) r_{ab} = coefficient of relatedness among genotypes. Decades 1 = pre-1920 progenitor group; 2 = 1920–1930; 3 = 1930–1940; 4 = 1950–1970; 5 = 1970–1980; 6 = 1980–1990; 7 = 1990–2000.

Discussion

Sugarcane breeding has a relatively recent history, particularly in the USDA Canal Point program where it was initiated in the 1930s after the first hybrids were released from Java (Indonesia) and from Coimbatore (India) to develop hybrids for Louisiana and in the 1950s to support the Florida industry. Modern sugarcane cultivars are direct descendants of these few early founders, with most

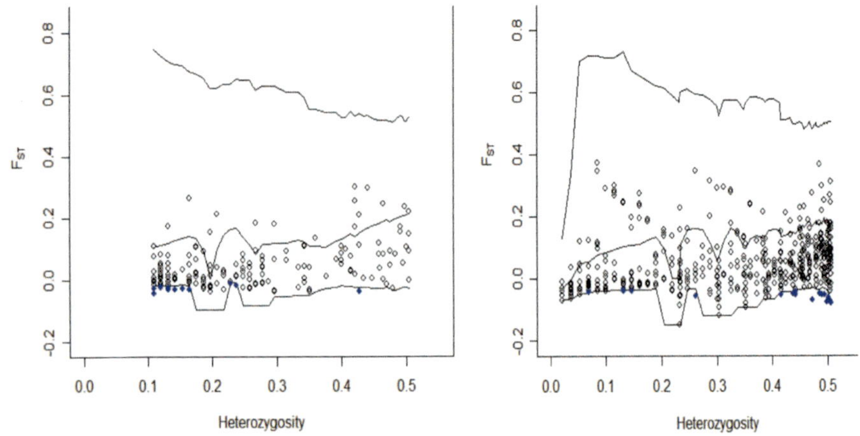

Fig. 14–10. Simulation results plotting F_{ST} values, estimated from microsatellites (left) and from target region amplified polymorphisms (TRAP) loci (bands) under a coalescent model, as a function of heterozygosity (H_E). The solid lines represent the upper (95%), median (50%), and lower (5%) limits of the 95% distribution envelope for the neutral loci. Blue dots indicate outlier loci said to be under balancing selection for being below the lower confidence limit.

programs relying exclusively on intercrossing among intergenerational hybrids to develop new genetic combinations for their target environments. It is crucial to re-evaluate, at this time, the gains that had been achieved with this strategy, to assess the effects of this philosophy on the genetic diversity of the breeding population, and to determine its impacts on the evolution of some yield component traits (sucrose, cane yield, height, and diameter).

Phenotypic Evolution

The present study established that the Florida sugarcane breeding programs wer e more successful at improving CRS across years than at increasing CY. Genetic gains in CRS were substantial (0.64 g kg^{-1} yr^{-1}) across soil types and crops and even sharper (24.9 g kg^{-1}) on a decade basis. Cane yield decreased somewhat in the early years of breeding with the first released cultivars (CL 41 or CP 50), but it increased thereafter at a slow pace to be now close to the CY level of the POJ hybrids, leaving an overall stagnation in CY improvement. Looking at progress from a decadal point of view revealed that there was no yield gain in CY for either crop year (plant cane or ratoon).

It is important to recognize that, in the CP breeding program, CY is usually extrapolated from plot stalk weight and count and expressed in Mg ha^{-1}, and there is a possibility that a somewhat different CY profile could have been found, had stalk counts been available in this study. However, inspecting the yield statistics data indicated that CY decreased from the 82-99 Mg ha^{-1} range in the 1960s to around 70-85 Mg ha^{-1} since 1980 in the USA (Haley and Ali, 2007). It must be acknowledged that the success of breeding sugarcane has taken the modern cultivars to more marginal land since the 1970s and that a greater plethora of diseases present more serious challenges to yield potential than in the early years of breeding (Edmé et al., 2005b). Moreover, considering Fig. 1b in Edmé et al. (2005b), it is clear that CY had been increasing at a very slow pace or not at all in commercial fields. That same figure (1a and b) indicated also that the industry has been instantly capturing and cashing in the gains made in CRS, while there is a big gap between the gain in CY shown in test plots and that shown in commercial fields. The routine method of basing CY on stalk counts in test plots, which is prone to errors even with trained personnel, may not reflect the ultimate performance of selected cultivars in commercial fields (thereby obscuring improvement), and may need to be revisited.

Sugarcane genotypes grown on sand soils have higher sucrose content than on muck, because the former are considered to provide more stressful environments (drought and heat) than the latter soils. Faster gains (25%) were recorded on muck (the more optimum environment) overall than on sand when considering both the plant cane and first ratoon. It is generally accepted that optimum environments more often enable greater and faster progress than suboptimal environments. While Edmé et al. (2005b) did not find any gains associated with breeding for the sand soils environment in Florida, this study revealed a consistently upward trend for all three economically important traits: ≈6 g kg^{-1} $decade^{-1}$ in CRS, and, starting with the third decade, 0.23 kg in CY and 0.05 kg in SY. It is difficult to speculate at this time if the restrictions of fertilizer application in the Everglades Agricultural Area, made in the context of its protection and restoration (McCray et al., 2012), could inhibit the newer cultivars from attaining their full cane yield potential. It is clear, however, that management will have to be reexamined to capture the potential gains in both CRS and CY with regards to the sand soils.

Even though the regression analyses did not reveal a particular trend for CY and SY, the two breeding programs in Florida have, throughout the years, released high-tonnage cultivars, such as CL 61-620, CP 70-1133, CP 75-1547, CP 78-1628, and CP 80-1827, which had, in general, found a production niche on the sand soils, but usually with low sucrose content. Most of these cultivars were finally removed from commercial production because of diseases (rust and/or smut, particularly) or were replaced by newer and more productive cultivars. Increases in sucrose content are preferred in the Florida industry over increases in cane tonnage to reduce transportation costs and damage (with high fiber cane) to the mills. That may have, unconsciously, driven the program to that stagnation in CY. A clearer indication of how much of an increase in CY versus CRS can be tolerated is necessary as the trend in the last three decades seems to suggest the program may be approaching that elusive "theoretical limit" (Moore, 1989) in CRS where the joint improvement can coexist. Lingle et al. (2010) also observed a plateau in both CY and CRS in the last three cycles of selection in the Louisiana breeding program. Jackson (2005) reported a plateau in sucrose content since inception of the Australian sugarcane breeding program. The plateau, as shown in Fig. 14–2, is expressed in the joint CRS–CY improvement, not in the trajectory of the individual yield components. This feature is similar to the grain size–grain number relationship that exists in cereal breeding (Bradshaw, 1965), whereby a trade-off or a cost in fitness is imposed in moving the two traits in the same direction by breeding and selection because of their unfavorable genetic correlation.

Based on growers' accounts, record yields (cane and sugar) were obtained in Florida in the last two seasons (2011–2012 and 2012–2013) as the weather (no hurricanes, no freezes, and good rainy season) provided ideal growing conditions for sugarcane. Several important changes in the sugar industries have accompanied the breeding efforts and have reinforced the genetic gains obtained through breeding. The U.S. sugar (from sugarcane) industry now operates as a fully vertically integrated business model (Moss and Schmidtz, 2002; Shine, 2002) with the capacity to grow the plants, process and refine the sugar, particularly in Florida, and to increase the profit margin of the growers. The shift to mechanical harvesting in the 1970s has shortened the time from the field to the processing plant, thereby allowing a faster and higher recovery of the sucrose in the cut stalks. Through investment in machinery at the mill, recovery of sucrose has also increased from 9% in the 1950s to around 12% since 2000 (Haley and Jerardo, 2012). Application of best management practices in the form of crop rotation, fertilizer application, pest (diseases, insects, and weeds) control, and irrigation have also contributed to sustaining the yield potential of the released varieties. In an earlier study, Edmé et al. (2005b) attributed 69% of this success to genetics and 31% to management practices.

Based on the graphs shown, the breeding programs have been sampling, within the infinitesimal number of genotypes possible when making a cross in this highly polyploid background, cultivars with different combinations of tonnage and sucrose content, mostly in the mid-range values of the two traits. The joint improvement of these two traits is, however, highly constrained by negative correlations, ranging from −0.032 to 0.097 across soil types in this study, but at times as high as −0.50 in other studies (Brown et al., 1969; Ahmed and Obeid, 2012). This negative CRS–CY association appeared to be driven by DIAM, itself the largest contributor (as compared with HT) to CY and always negatively linked with CRS. Diameter was found to be always negatively related to HT (r_G ranging from −0.10 to −0.17, $P < 0.05$) on the muck soils, but positively associated on sand. Whereas HT was kept unchanged, DIAM was reduced consistently and significantly across soil types and crop years at an overall annual rate of 0.045 mm during the period considered. These intricate relationships among these yield traits (CRS–HT–DIAM–CY) seem to be the key determining factors in constraining the joint improvement of CRS and CY. Comparing the experience in Australia (Jackson, 2005) with that in the United States in breeding sugarcane is revealing. These two breeding programs are relatively of the same age, began with similar materials, have the same objectives to jointly improve CRS and CY, but still ended up with two different results. The Australian breeders increased CY while keeping CRS constant (unintentionally), whereas the U.S. breeders were successful at increasing CRS while keeping CY relatively constant (unintentionally) after an initial decrease. These results suggest that the negative associations between CRS and CY may be stronger than the breeding designs allow to measure or that breeding might have changed the correlation structure of these two traits. It is fair to say that the Australians breed for a larger and more diverse environmental continuum than the Americans do, and there are also some differences in the disease spectrum between the two countries.

The Florida breeding programs have certainly changed the phenotypes of the cultivars, particularly diameter and stalk numbers (here implied based on the negative genetic correlation that exists between stalk numbers and diameter) and by so doing have modified, that is, nullified, the r_G between the two main traits under positive directional selection. With greater selection pressure put on

CRS and based on the fact that response to selection is multivariate even with a single trait under selection, those shifts in the complex trait relationships might have modified the genetic variance–covariance matrix to a position away from the optimum space where the best or extreme genotypes (high CY and high CRS) are relatively abundant, even in the presence of standing genetic variation for the individual component traits. This phenomenon is the basis of renewed debates around evolutionary genetics of traits under selection. Superimposing the trajectories of CRS and CY (Fig. 14–2g) highlights the constraint imposed by selection on the two traits, and reorienting the G-matrix will require acting on the genetic variance of the composite trait (CRS–CY) as an index of selection, and re-examining its size and its usefulness.

Molecular Evolution

Changes at the DNA level have accompanied the phenotypic modifications brought about by breeding and selecting sugarcane to fit the production systems in Florida. Genetic diversity (H_E) has changed little (at a rate of 0.002 decade^{-1}). The fact that the number of bands (alleles) decreased dramatically (21 bands decade^{-1}) throughout generations (a reduction of 6% in %P) would suggest that the breeding populations went through a bottleneck phase, particularly from the first to the second decade. However, the increase in genome size (Edmé et al., 2005a) in the cultivars (by inheriting the full and half somatic numbers from the *S. officinarum* and *S. spontaneum* parents, respectively) combined with the polyploid background might have shielded the breeding populations from experiencing a serious impact from a drop in genetic diversity. The study provided evidence also that allele frequencies have diverged since the progenitor population (Fig. 14–8), with a predominance of low-frequency alleles, an increase in more common alleles (>85%), and a marked decrease in rare alleles (44 decade^{-1}). While the distribution of SSR alleles stayed as L-shaped (Luikart et al., 1998) throughout generations, the outline of the TRAP alleles changed from an L- to a U-shaped one as some of them went to fixation. The reduction in the number of alleles might have resulted also partly from elimination of chromosomal segments (frequent in sugarcane meioses) when the divergent and competing constituent genomes interact (Burner, 1997; Edmé et al., 2006), thereby creating aneuploids.

Modern sugarcane cultivars represent genomic mosaics as they are composed of different levels of the constituent genomes involved in their makeup. Sugarcane breeders have put strong selection pressure on the *officinarum* cytoplasm to the detriment of the *spontaneum* and, by so doing, might have generated different fixation probabilities of rare alleles between genetic backgrounds (Barton, 1995). Considering Fig. 14–11, weak selection was imposed on the regions of the *Saccharum* colored green and red as these sections predominated in the progenitor and founder populations, whereas selection was stronger for the other regions as they tended to persist across generations. Extensive use of proven parents or crosses with superior performance had the added effect to overlap generations and more specifically to homogenize (r_{ab} increased by 0.07 decade^{-1} by sib mating) the allele frequencies of the populations because of re-introduction and reshuffling of lost alleles, but still generating some differentiation and genetic variance as a result of recombinations and gains of new alleles.

One of the most important features of the effects of breeding and selection in the Florida breeding populations is shown in the trajectory of LD. Transient in its

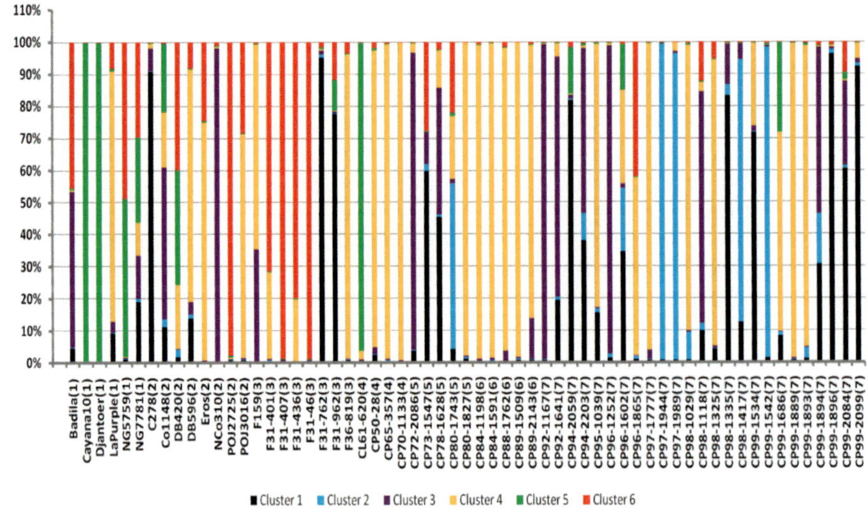

Fig. 14–11. Proportion of the *Saccharum* genome as inferred by the program Structure v2.3.1 (Pritchard et al., 2000) shared by historical sugarcane genotypes (wild clones and cultivars released from the 1930 to 2000) based on a TRAP molecular survey of temporal changes in genetic diversity in the Florida programs. Length and color of bars within a genotype determine relative genomic composition to respective clusters. Numbers in parentheses represent successive groups or decades of use or development of the genotypes, with 1 = pre-1920 progenitor group; 2 = 1920–1930; 3 = 1930–1940; 4 = 1950–1970; 5 = 1970–1980; 6 = 1980–1990; 7 = 1990–2000.

nature, it closely reflects the philosophy adopted by sugarcane breeders in Florida and some other programs (Jackson, 2005): the first to second decade was a period of active interspecific crossings that emulated the success following the creation of the POJ and Co sugarcane hybrids and that decreased LD from the progenitor species. The second phase (Period 2 to 4) capitalized on the available variation in the new hybrids to intercross among them, molding the genome by concentrating the frequency of alleles associated with sucrose content and cane yield primarily, thereby creating a new upsurge of LD. Period 4 to 6 is characterized by the introgression of new alleles (either by intercrossing new cultivars with an *S. officinarum* clone or with an immediate F_1 from an interspecific cross) that disrupted and reshuffled the linkage blocks created in the previous period. Using this new variation to impose selection pressure on the regions of the genome linked with sucrose content and cane yield, the breeding population is poised to a new round of increasing LD (Period 6–7). This new round has led to the creation of some of the most popular cultivars in the Florida industry—CP 80-1743, CP 88-1762, and CP 89-2143, to name a few. These LD results are consistent with both lineage sorting and transspecies polymorphisms, that is, alleles or haplotypes in a species being genealogically closer to those of a divergent than to those of a conspecific species (Charlesworth et al., 2006). These causes are actually invoked when balancing selection (Fig. 14–10) is driving the inheritance of alleles, in this case from the ancestral clones to the descendant sugarcane cultivars.

Linkage disequilibrium can be caused by several factors, including pleiotropy (same alleles or genes affecting two or more traits) and selection (Bulmer, 1971). Stabilizing or balancing selection that underlines the breeding philosophy in Florida can deplete genetic variation (known as the Bulmer effect) in the traits under selection and can change the variance–covariance matrix and allele frequencies (Bulmer, 1971). As pointed out by Jackson (2005) and indicated here, the genetic parameters (variance–covariance matrix, genetic correlations, heritability) can change with time, environments, breeding populations, and selection strategy and are themselves dependent on allele frequencies. The transient nature of LD in this study is fundamentally at the basis of these changes, and when the r_G is negative (e.g., the complex CRS–HT–DIAM–CY genetic architecture revealed in this study), the selection pressure applied on the individual or pleiotropic loci may have different results if not supported by accurate genetic parameters. The comparison of the Australian and U.S. experiences indicates that the genes/loci affecting one of the two traits (CRS in the U.S. case and CY in the Australian case) were under more selection pressure than the pleiotropic loci, and those favored individual loci may have had opposite effect on the other trait, thereby slowing progress in the latter and in the composite trait.

References

Ahmed, A.O., and A. Obeid. 2012. Correlation pattern among morphological and biochemical traits in relation to tillering capacity in sugarcane (*Saccharum* spp). Acad. J. Plant Sci. 5:119–122.

Barton, N.H. 1995. Linkage and the limits to natural selection. Genetics 140:821–841.

Beaumont, M.A., and D.J. Balding. 2004. Identifying adaptive genetic divergence among populations from genome scans. Mol. Ecol. 13:969–980. doi:10.1111/j.1365-294X.2004.02125.x

Bradshaw, A.D. 1965. Evolutionary significance of phenotypic plasticity in plants. Adv. Genet. 13:115–155. doi:10.1016/S0065-2660(08)60048-6

Breaux, R.D. 1984. Breeding to enhance sucrose content of sugarcane varieties in Louisiana. Field Crops Res. 9:59–67. doi:10.1016/0378-4290(84)90006-6

Brown, A.H.D., J. Daniels, and B.D. Latter. 1969. Quantitative genetics of sugarcane. II. Correlation analysis of continuous characters in relation to hybrid sugarcane breeding. Theor. Appl. Genet. 39:1–10. doi:10.1007/BF00283078

Bulmer, M.G. 1971. The effect of selection on genetic variability. Am. Nat. 105:201–211. doi:10.1086/282718

Burner, D. 1997. Chromosome transmission and meiotic behavior in various sugarcane crosses. J. Am. Soc. Sugar Cane Technol. 17:38–50.

Ceccarelli, S. 2009. Evolution, plant breeding and biodiversity. J. Agric. Environ. Int. Dev. 103:131–145.

Charlesworth, D., E. Kamau, J. Hagenblad, and C. Tang. 2006. Trans-specificity at loci near the self-incompatibility loci in *Arabidopsis*. Genetics 172:2699–2704. doi:10.1534/genetics.105.051938

Cordeiro, G.M, G.O. Taylor, and R.J. henry. 2000. Characterization of microsatellite markers from sugarcane (*Saccharum* sp.), a highly polymorphic species. Plant Science 155:161–168.

del Blanco, I.A., B. Glaz, and S.J. Edmé. 2010. Improving efficiency of sugarcane genotype selection for Florida. Crop Sci. 50:1744–1753. doi:10.2135/cropsci2009.09.0539

Deren, C.W. 1995. Genetic base of U.S. mainland sugarcane. Crop Sci. 35:1195–1199. doi:10.2135/cropsci1995.0011183X003500040047x

Deer, N. 1949. The history of sugar. Vol. I. Chapman and Hall, London.

d'Hont, A., and J.C. Glaszmann. 2001. Sugarcane genome analysis with molecular markers, a first decade of research. Proc. Int. Soc. Sugar Cane Technol. 24:556–559.

Edmé, S.J., J.C. Comstock, and J.D. Miller. 2005a. Determination of DNA content and genome size in sugarcane. J. Am. Soc. Sugar Cane Technol. 25:1–16.

Edmé, S.J., R.W. Davidson, R.A. Gilbert, N.C. Glynn, D. Zhao, J.C. Comstock, B. Glaz, S. Sood, J.D.M., and P.Y.P. Tai. 2013. Registration of CP 04–1935′ Sugarcane. J. Plant Reg. 7:288–295. doi:10.3198/jpr2012.12.0057crc

Edmé, S.J., and B. Glaz. 2013. Field response of sugarcane genotypes to freeze stress with genotype x environment effects of quality traits. J. Crop Improv. 27:1–30. doi:10.1080/15427528.2012.720653

Edmé, S.J., J.D. Miller, B. Glaz, P.Y.P. Tai, and J. Comstock. 2005b. Genetic contribution to yield gains in the Florida sugarcane industry across 33 years. Crop Sci. 45:92–97. doi:10.2135/cropsci2005.0092

Edmé, S.J., N. Glynn, and J. Comstock. 2006. Genetic segregation of microsatellite markers in *Saccharum officinarum* and *S. spontaneum*. Heredity 97:366–375. doi:10.1038/sj.hdy.6800888

Falconer, D.S., and T.F.C. Mackay. 1996. Introduction to quantitative genetics. 4th ed. Longman Group, Ltd., London.

Federer, W.T., M. Reynolds, and J. Crossa. 2001. Combining results from augmented designs over sites. Agron. J. 93:389–395. doi:10.2134/agronj2001.932389x

Haley, S., and M. Ali. 2007. Sugar backgrounder. USDA-ERS Report. USDA. SS-249-01.

Haley, S., and A. Jerardo. 2012. Sugar and sweetener outlook. USDA-ERS Rep. USDA SSS-M-287.

Haubold, H., and R. Hudson. 2000. Lian 3.0: Detecting linkage disequilibrium in multilocus data. Bioinformatics 16:847–848. doi:10.1093/bioinformatics/16.9.847

Hogarth, D.M. 1971. Quantitative inheritance studies in sugarcane. Correlations and predicted responses to selection. Aust. J. Agric. Res. 22:103–109. doi:10.1071/AR9710103

Holsinger, K.E., and P.O. Lewis. 2007. Hickory: Package for the analysis of population genetic data. Version 1.1. Dep. Ecology and Evolutionary Biology, Univ. of Connecticut, Storrs.

Hu, J.G., and B.A. Vick. 2003. Target region amplification polymorphism: A novel marker technique for plant genotyping. Plant Mol. Biol. Rep. 21:289–294. doi:10.1007/BF02772804

Jackson, P.A. 2005. Breeding for improved sugar content in sugarcane. Field Crops Res. 92:277–290. doi:10.1016/j.fcr.2005.01.024

Legendre, B.L. 1992. The core/press method for predicting the sugar yield from cane for use in cane payment. Sugar J. 54(9):2–7.

Legendre, B., H. Birkett, J. Stein, M. Duet, T. Tew, and G. Eggleston. 2010. Stalk cold tolerance of commercial and candidate varieties. Sugarcane Research Annual Rep. 2009. Louisiana State Univ. Agric. Ctr., Baton Rouge, LA.

Li, G., and C.F. Quiros. 2001. Sequence-related amplified polymorphism (SRAP), a new marker system based on a simple PCR reaction: Its application to mapping and gene tagging in Brassica. Theor. Appl. Genet. 103:455–461. doi:10.1007/s001220100570

Lin, C. S. and M.R. Binns. 1988. A superiority measure of cultivar performance for cultivar × location data. Can. J. Plant Sci. 68:193–198.

Lingle, S.E., R. Johnston, T. Tew, and R. Viator. 2010. Changes in juice and sugarcane yield with recurrent selection for sucrose. Field Crops Res. 118:152–157. doi:10.1016/j.fcr.2010.05.002

Luikart, G., F.W. Allendorf, J.M. Cornuet, and W.B. Sherwin. 1998. Distortion of allele frequency distributions provides a test for recent population bottlenecks. J. Hered. 89:238–247. doi:10.1093/jhered/89.3.238

Lynch, M., and B. Walsh. 1998. Genetics and analysis of quantitative traits. Sinauer Associates, Sunderland, MA.

McCray, J.M., R.W. Rice, and A.L. Wright. 2012. Phosphorus fertilizer recommendations for sugarcane production on Florida organic soils. SS-AGR-348. IFAS Extension, University of Florida, Gainesville.

Moore, P.H. 1989. Physiological basis for varietal improvement in sugarcane. In: K.M. Naidu, T.V. Sreenivasan, and M.N. Premachandran, editors, Sugarcane varietal improvement. Sugarcane Breeding Inst., Coimbatore, India. p. 19–55.

Moss, C.B., and A. Schmidtz. 2002 Vertical integration and trade policy: The case of sugar. Agribusiness 18:49–60.

Pritchard, J.K., M. Stephans, and P. Donnelly. 2000. Inference of population structure using multilocus genotype data. Genetics 155:945–959.

Roach, B.T. 1972. Nobilization of sugarcane. Proc. Int. Soc. Sugar Cane Technol. 14:206–216.

Shine, J.M. 2002. Vertical integration in the Florida sugar industry. Sugar J. July 6:23.

Stokes, I.E., and H.M. Tisdale. 1962. Significant in genealogies of Canal Point varieties of sugar cane. Proc. Int. Soc. Sugar Cane Technol. 11:456–464.

Sreenivasan, T.V., B.S. Ahloowalia, and D.J. Heinz. 1987. Cytogenetics. In: D.J. Heinz, editor, Sugarcane improvement through breeding. Elsevier, Amsterdam. p. 211–253.

Tai, P.Y.P., and J.D. Miller. 1978. The pedigree of selected Canal Point (CP) varieties of sugarcane. Proc. Am. Soc. Sugar Cane Technol. 8:34–39.

Tew, T.L., and R.M. Cobill. 2008. Genetic Improvement of Sugarcane (*Saccharum* spp.) as an Energy Crop. In: W. Vermerris, editor, Genetic improvement of bioenergy crops. Springer Science+Business Media, New York. p. 249–272.

Yeh, F.C., R. Yang, T.J. Boyle, and J.M. Xiyan. 2000. POPGENE32. Microsoft Window-based freeware for population genetic analysis, Version 1.32. Molecular Biology and Biotechnology Centre, Univ. of Alberta, Edmonton.

Zhivotovsky, L.A. 1999. Estimating population structure in diploids with multilocus dominant DNA markers. Mol. Ecol. 8:907–913. doi:10.1046/j.1365-294x.1999.00620.x

15

Sunflower

Brent S. Hulke* and Larry W. Kleingartner

ABSTRACT

Sunflower (*Helianthus annuus* L.) is a relatively new crop to North America, with production beginning in earnest in the 1970s in the United States. There have been many challenges to production, including agronomic issues such as seed placement and herbicide systems adequate for weed control, disease issues caused by many different pathogens, insect infestation caused by several, mainly seed-based insect pests, and even predation by blackbirds. Genetics-based solutions have been provided in the forms of herbicide resistance systems, vertical resistance to downy mildew and sunflower rust, and to some extent, quantitatively inherited Sclerotinia resistance. The possibility of a genetic solution to sunflower insect issues has been considered and is the topic of current research. Genetic gain in sunflower has been judged more by the ability of the plant to overcome these problems via genetic means (by defending existing yield gains) than by improving yield itself. Employing modern genomic tools now and in the future should allow breeders to spend less time breeding defensive traits and more time concentrating on making yield gains in target environments. However, this will probably need to be accomplished without the use of transgenic modification, as genetically modified (GM) lines will likely face difficult hurdles in regulatory and market acceptance.

Abbreviations: BLUPs, best linear unbiased predictions; C18:1, oleic acid; C18:2, linoleic acid; CMS, cytoplasmic male sterility; GM, genetically modified; NSA, National Sunflower Association; SNP, single nucleotide polymorphism; VNIIMK, All-Russian Research Institute of Oil Crops.

B.S. Hulke, USDA-ARS, Northern Crop Science Laboratory, 1307 18th St. N., Fargo, ND 58102-2765. *Corresponding author (brent.hulke@ars.usda.gov).
L.W. Kleingartner, National Sunflower Association, 2401 46th Ave. SE., Suite 206, Mandan, ND 58554-4829 (larry.kleingartner@icloud.com).

doi:10.2135/cssaspecpub33.c15

Yield Gains in Major U.S. Field Crops. CSSA Special Publication 33.
Stephen Smith, Brian Diers, James Specht, and Brett Carver, editors.

Sunflower is a species native to North America. Wild or native sunflower is characterized as branched with multiple small heads and small seeds. Native Americans in the southwestern part of the United States may have been the first to cultivate the plant about 3000 BCE (Semelczi-Kovacs, 1975). There is archaeological evidence in numerous regions of North America that sunflower was a part of Native American crops and may have predated maize (*Zea mays* L.) and common bean (*Phaseolus vulgaris* L.) as a cultivated crop. Native Americans developed sunflower with single heads. The domesticated sunflower was an important food source. The seed was often ground into flour and rolled into balls. This was easily carried by hunting parties and was a good source of protein and fat, and, thus, energy (Heiser, 1951). Native American sunflower varieties from the southwestern United States have been evaluated for resistance to major diseases of cultivated sunflower. These varieties were grown over a long period and in diverse environments "... and in the process have acquired unique traits through natural and human-aided selection" (Gulya and Mancl, 1993).

Sunflower made its way to Europe via the Spanish galleons in the 16th century. There were a number of European botanists who published descriptions of the plant. Sunflower spread over Europe and was both a garden ornamental and a source of food (Heiser, 1951). By 1740, sunflower seeds were noted as a snack product. This provided incentive to develop one large head with large seeds (Semelczi-Kovacs, 1975). Sunflower arrived in Russia at the time of Peter the Great in the 18th century. The popularity of sunflower was enhanced among the Russian people when the seeds and oil of sunflower were omitted from a list of foods prohibited during Lent by the Russian Orthodox Church (Heiser, 1955). Sunflower production and breeding made great progress in Russia in the late 19th and early 20th centuries. Local farmers or peasants are credited with selections in both oil and confection types for early maturity (Semelczi-Kovacs, 1975).

The improved Russian sunflower seeds found their way to North America in the late 19th and early 20th century. It is likely that German-Russian immigrants brought seed with them on their long journey to North America. It is also noted by Heiser (1976) that seed catalogues in the United States were offering 'Russian Mammoth' seeds as early as 1880. It is interesting to note that Russian Mammoth can still be found in seed catalogues today and is a popular garden variety. The Government of Canada was instrumental in developing a sunflower breeding program. Canadian officials wanted to increase the production of vegetable oils, and sunflower was an excellent oilseed candidate. The first breeding program was established in 1936 with introductions from Russia and local varieties developed in gardens of Mennonite farmers who had migrated from Russia. Of most interest was developing early maturing varieties with high oil content (Putt, 1997).

Breeders in the former Soviet Union made great strides in breeding for higher yield and oil content in the mid 20th century. Vasilii Stephanovich Pustovoit (1886–1972) is best known for increasing oil content to more than 40% in a number of cultivars and making other genetic improvements to oil sunflower (Dvoryadkin, 1976). Pustovoit is still revered among international sunflower scientists. The highest international sunflower breeding award, presented by the International Sunflower Association, is the Pustovoit Award. Another Russian breeder, K.I. Soldatov, developed a high oleic variety of sunflower seed which became the basis of NuSun (registered trademark of National Sunflower Association [NSA]), a mid-oleic sunflower oil that is a marketed in the United States, as well as high oleic hybrids around the world (Soldatov, 1976; Watkins, 2006).[1]

Sunflower is a relatively new crop in the United States, with commercialization starting in the 1950s. High oil content sunflower seed changed the economics of producing and crushing sunflower seed for oil. Early adopters grew the improved Canadian open pollinated varieties, which migrated across the international border to North Dakota and Minnesota (Putt, 1997). During this time of the Cold War between the former Soviet Union and the United States, obtaining improved Russian sunflower seed directly was difficult for American scientists and breeders. The landmark Soviet open pollinated 'Peredovik' was introduced to U.S. farmers via a Canadian–Soviet seed exchange (Dvoryadkin, 1976; Putt, 1997). As time passed and political tension decreased, seed and information exchanges directly with the United States were initiated, which continue to the present day between the All-Russian Research Institute of Oil Crops (VNIIMK; Krasnodar, Russia) and USDA-ARS (Fargo, ND).

The early 1970s were the beginning of hybrid variety breeding in sunflower, and American and European scientists collaborated in the development of this technology. Before this time, most successful varieties were open-pollinated varieties. Self-compatibility of sunflower varieties was generally poor, due to the natural sporophytic self-incompatibility system in sunflower (Hamilton, 1926; Habura, 1957). Three key developments were necessary to make hybrid sunflower a reality. First, because the species has perfect flowers, a cytoplasmic male sterility (CMS) system was needed in the designated "female" heterotic group. The first CMS system was developed by Leclercq in France (Leclercq, 1969), and was based on an introgression of a cytoplasm from *H. petiolaris* L. Since then, multiple other CMS systems have been developed (e.g., Miller and Wolf, 1991). Second, nuclear genes restoring male fertility in the hybrid variety were required in the designated "male" heterotic group. The Rf_1 locus was the first discovered, which restored fertility to Leclercq's CMS when used as a male in a hybrid cross. It was derived from a wild sunflower found in Texas (Kinman, 1970). A third requirement is self-compatibility of the sunflower lines because several rounds of inbreeding are necessary to produce male and female inbred lines that are highly homogeneous (Hamilton, 1926). This continues to be a problem in some germplasm pools, but the most highly developed sunflower lines are highly self-compatible. A combination of these factors resulted in the first set of hybrids in the 1970s, of which 'Hybrid 894' from USDA-ARS is particularly noteworthy because it was widely marketed across the United States by several seed companies in the 1970s (Berglund, 1978). One major limitation to this hybrid system is the fact that male sterile female inbreds cannot be self-pollinated. To overcome this, an A–B line system is required in which a B line is first developed that is male fertile and lacks the restorer gene. A desirable B line is backcrossed into CMS lineage, and after several backcrosses, results in an A line that has the same nuclear genome as the B line of interest, together with the male sterile cytoplasm. Increases of the A line are accomplished in cages or field-sized isolation plots together with the accompanying B line and an appropriate pollinator, such as the honey bee (*Apis mellifera* L.) (Fehr, 1987).

[1] Mention of trade names or commercial products in this article is solely for the purpose of providing specific information and does not imply recommendation or endorsement by the U.S. Department of Agriculture.

Land area of the crop increased dramatically in the 1970s due to strong European demand and the need for a rotation crop with wheat (*Triticum aestivum* L.) and barley (*Hordeum vulgare* L.) in North Dakota, South Dakota, and Minnesota. This expansion led to the formation of two distinct production regions in the United States. The northern region encompassing North Dakota, South Dakota, and Minnesota produces about 85% of the annual production today. A second region that includes Kansas, Nebraska, Colorado, and Texas produces the remaining portion (USDA-NASS, http://www.nass.usda.gov/Data_and_Statistics/index.asp, accessed 15 Mar. 2013). The U.S. land area planted to sunflower is much smaller compared to that occupied by other major crops. In total, the land area planted to sunflower in 2012 in the United States was 776,609 ha (1,919,000 ac), compared to 37,132,740 ha (91,755,000 ac) of maize and 31,241,603 ha (77,198,000 ac) of soybean [*Glycine max* (L.) Merr.] (USDA-NASS). Sunflower is, however, a major world crop, with planted area centered in southern Europe and the Black Sea region, and Argentina. In 2012, a total of 10,860,000 ha (26,835,060 ac) were planted in the European Union and Russia, and 1730,000 ha (4274,830 ac) in Argentina (Oil World, http://www.oilworld.biz/statistics?direct_call=ista&ista=3126e3a41d469ed26f1fd81d07c5d4ba, accessed 15 Mar. 2013).

Today, the land area planted to sunflower is split between the two market types, oilseed and non-oil (confectionery). Oilseeds were responsible for the major expansion of the crop in the United States, and this type includes the high-oil forms, typically with black hulls, that were introduced in the early days for vegetable oil extraction. Currently, the oilseed type makes up 86% of the total land area of sunflower in the United States (USDA-NASS). The remainder is planted to non-oil sunflower, which is characterized by black and white striped hulls, longer and/or plumper seed, and lower oil content. The non-oil type is directly consumed as a snack food domestically and around the world, with no universal ideal for seed shape and size. Alas, the breeder must be prepared to deal with very specific market preferences in the various non-oil seed consuming nations of the world, which includes most of the developed nations (Bill Sullivan, personal communication, 2013).

Introducing a new crop has infrastructure challenges. This can include a lack of research institutions, trained personnel, grower knowledge, field equipment, crop protection products, and adapted, improved genetics. These factors inhibit yield on a regional level, although it is not possible to quantify accurately how much yield potential is actually lost due to inferior infrastructure. Since 2002, the NSA has sponsored an annual Crop Survey to determine which factors most negatively impact sunflower yield and quality (NSA, 2013a). The survey also allows the sunflower research community to identify areas where progress has been made and topics that need additional research. The survey is conducted by trained personnel in all of the production states in the United States and the province of Manitoba, Canada. The surveyors are required to find one field per 10,000 acres (4047 ha) of planted production in each county or municipality. Surveyors are asked to designate and document two yield-impacting factors for each field surveyed, as well as to take plant samples, estimate yield, and comment on any production issues that are uncommon and may affect yield and quality. In addition, several states also sponsor commercial sunflower variety trials. North Dakota and South Dakota are the most active in this regard, with the North Dakota trials published as far back as 1978 (Berglund, 1978).

Cultural Factors Influencing Production

There are certain aspects of the production system that can be influenced directly by the actions of the producer and/or improved technologies. According to the NSA Crop Survey, the most significant cultural factor limiting yields is the lack of uniform plant stands (NSA, 2013a). This refers to plant "skips" within the row, or double or triple plants competing in the same place. There has been a consistent effort among researchers, farmers, and crop advisers to reduce the problem (Smith and Kocher, 2008). Farmers often encounter planting deadlines or impending weather events, resulting in tractor speed greater than recommended when planting. Seed is sometimes planted into dry or extremely wet soil, further inhibiting uniform germination. Another factor is seed placement and planter technology. Improvements have been made by equipment manufacturers, but it is important to note that planters are designed for maize, soybean, and small grain production (Smith and Kocher, 2008). Sunflower seed is not always uniform in size and weight. "Coated" planting seed has been proposed as a solution to this issue. The uniformly sized seed aids in more consistent seed placement, resulting in improved plant stands (Lilleboe, 2012). Seed treatments have aided in the reduction of seedling diseases and soil insects, contributing to more uniform stands. Downy mildew, caused by *Plasmopora halstedii* (Farl.) Berl. et de Toni, a seedling disease, has been reduced with seed-applied fungicides and genetic resistance. However, this disease continues to be a concern (Viranyi and Spring, 2011). Insects, such as cutworms and wireworms, also produce large stand gaps (Knodel and Charlet, 2007). Considering the low planting density typical of sunflower (37,000–60,000 plants ha^{-1} [15,000–24,000 plants ac^{-1}]) compared to other major field crops, stand losses due to disease and insects easily can result in an irrecoverable loss of yield early in the growing season. Seed-applied or broadcast-sprayed insecticides can prevent or reduce damage from these insects.

Another major cultural constraint in achieving increased sunflower yields from 1970 to 2000 was the absence of herbicide choices. During this 30-yr period, dinitroaniline (DNAn) herbicides were the only weed control products available. Sunflower did not command the acreage base to spur crop protection companies to dedicate research for specific sunflower herbicides. Yield-robbing weeds of most concern were small seeded broadleaf weeds like Kochia (*Kochia* spp.). Most broadleaf weeds were only suppressed (not controlled) by DNAn herbicides, the most popular preplant broadleaf herbicide class on sunflower and other broadleaf crops at the time (Blamey et al., 1997; Lilleboe, 1997). A further limitation to this DNAn group was the tillage and timely rainfall required for activation. The DNAn herbicides worked best when shallow tillage was performed twice after application, the so-called double soil incorporation (Blamey et al., 1997). This greatly reduced soil moisture. This was of particular concern as the sunflower crop continued to move westward in North Dakota and South Dakota, where rainfall is typically limited. During this era, between-row cultivation was a common weed control system (Lamey et al., 1992). This also reduced soil moisture and did not eliminate established weeds within the row.

Additional herbicides became available for sunflower beginning in 2000. These include preplant herbicides (sulfentrazone, *N*-{2,4-dichloro-5-[4-(difluoromethyl)-3-methyl-5-oxo-4,5-dihydro-1H-1,2,4-triazol-1-yl]phenyl}methanesulfonamide) and post-emergent herbicides such as imidazolinone (1,5-dihydro-4H-imidazol-4-one) and sulfonylurea herbicides. The latter two herbicides required development and incorporation of herbicide resistance genes into sunflower (Miller and Al-Khatib, 2002, 2004). These products provided increased management of formerly difficult-to-control small seeded broadleaves and grassy weeds as well. The herbicides also allowed farmers to utilize reduced tillage systems (Ashley and Tanaka, 2007). This provided additional incentive for the crop to be planted in more arid regions of western North and South Dakota, western Kansas, eastern Colorado, and the Texas panhandle, where no-till practices are seen as an essential part of the production system to limit soil moisture and erosion losses. With the availability of additional herbicides, weeds have not been considered a significant yield-limiting factor over the last 12 production seasons (NSA, 2013a).

Environmental Factors Influencing Production

As with any other crop, there are many aspects of the growing environment that are out of the control of producers or researchers. For sunflower, the single greatest environmental factor influencing producer yields has been weather, in particular, the highly variable precipitation in the growing regions. The average annual North Dakota precipitation since 1975 ranged from a low of 27.7 cm (10.9 in) in 1988 to a high of 61.4 cm (24.16 in) in 2010 (Fig. 15–1). Drought has been a major contributing factor to yield decline in at least one of the production regions each year since 2002, and excess rainfall in crop years 2009 through 2011 contributed to decreased yields due to diseases in the northern region (NSA, 2013a; Fig. 15–1).

Sunflower acreage has declined in general in the United States and has increasingly moved westward in North and South Dakota. For example, Barnes County, ND (east-central) had 54,788 ha (125,500 ac) in 1993 compared to 2671 ha (6600 ac) in 2011. This compares to Emmons County, ND (west-central) with 1255 ha (3100 ac) in 1993 and 29,946 ha (74,000 ac) in 2011 (e.g., Fig. 15–2; USDA-NASS). The western Dakotas receive less rainfall, and yields of all crops are lower in this more arid region compared to the eastern Dakotas. There are two major reasons for this westward shift in production. One reason is the series of abnormally wet years observed in eastern and central North and South Dakota starting in 1999 and extending to 2011 (Fig. 15–1). The above normal rainfall that contributed to sunflower diseases in the eastern regions also increased yields of maize and soybean. The second relates to the addition of glyphosate [*N*-(phosphonomethyl) glycine] and glufosinate [(RS)-2-Amino-4-(hydroxy(methyl)phosphonoyl)butanoic acid] resistance technology to soybean and maize. The ease of farming with these herbicide technologies and strong demand for maize and soybean made sunflower and other crops less competitive for acres. Meanwhile, the addition of herbicide systems conducive to reduced tillage provided growers the opportunity to expand their sunflower production in the west.

The diseases of greatest economic impact to sunflower generally occur under wet field conditions. The Sclerotinia diseases (caused by *Sclerotinia sclerotiorum* [Lib.] deBary) are the most important diseases of sunflower and include a head rot disease, mid-stalk rot disease, and a basal stalk rot disease. A common and devastating form of Sclerotinia is head rot. Yields can be reduced from 1 to 100% (Masirevic and Gulya, 1992). Head infection is the result of ascospore infection and requires moist soil and high humidity for up to 40 h during the flowering stage in late July and August (Payen, 1983; Ratkos and Nagy, 1992).

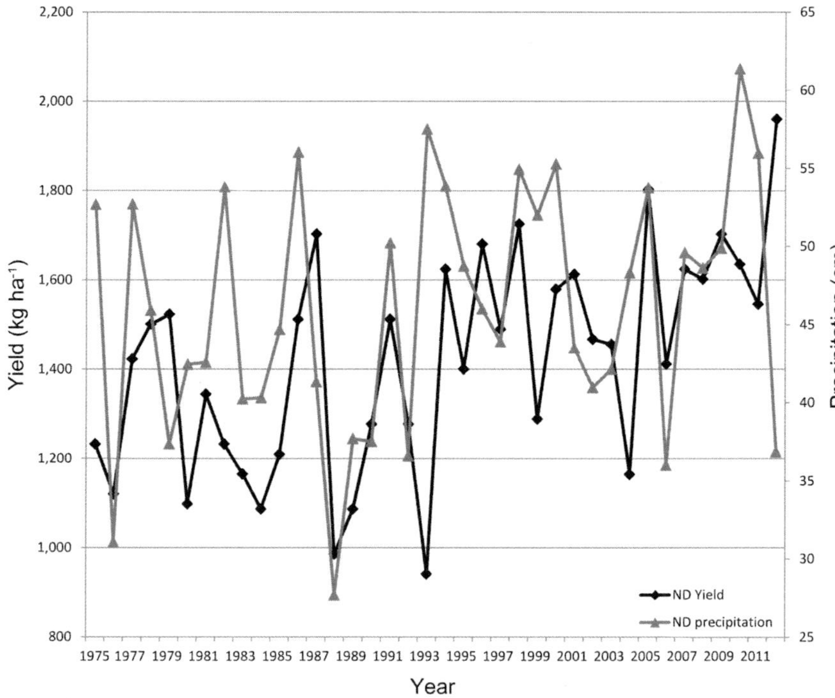

Fig. 15–1. Yield and yearly precipitation for the state of North Dakota, 1975–2012 (NOAA National Climatic Data Center, 2013; USDA-NASS).

According to NSA Crop Surveys, Sclerotinia head rot was a significant factor in yield reduction in 7 out of 11 yr in North Dakota and Minnesota between 2002 and 2012. In 2009, on average, 20% of sunflower plants in Minnesota were infected, 22% of the plants were infected in Manitoba, and 5% of the plants were infected in North Dakota (NSA, 2013a). Infected plants produce no yield in most cases because the floral parts and seeds are converted by the fungus into hardened, black mycelial masses called sclerotia (Gulya et al., 1997). Sclerotized non-oil seeds are a concern because they may actually look like a normal seed on the surface, but may contain a sclerotia, which is hard enough to cause injury to a consumer when eaten. The decrease in seed yield is the primary concern with oilseed sunflower. In the case of non-oil sunflower, the presence of head rot in fields could cause rejection of the harvested seed by grain marketing firms, or at the very least, the need for the grain marketer to use specialized sorting equipment to remove sclerotized seed.

The Sclerotinia mid-stalk rot, also caused by ascospore infection, and basal stalk rot, caused by direct mycelial infection of roots in the soil, can also have an effect on yield and quality by causing a plant wilt symptom, thus reducing the flow of nutrients to developing seed and reducing total yield and quality of the seed in both oilseed and non-oil market classes (Dorrell and Huang, 1978). An additional issue with these two diseases is stalk lodging before harvest. Strong winds and/

a

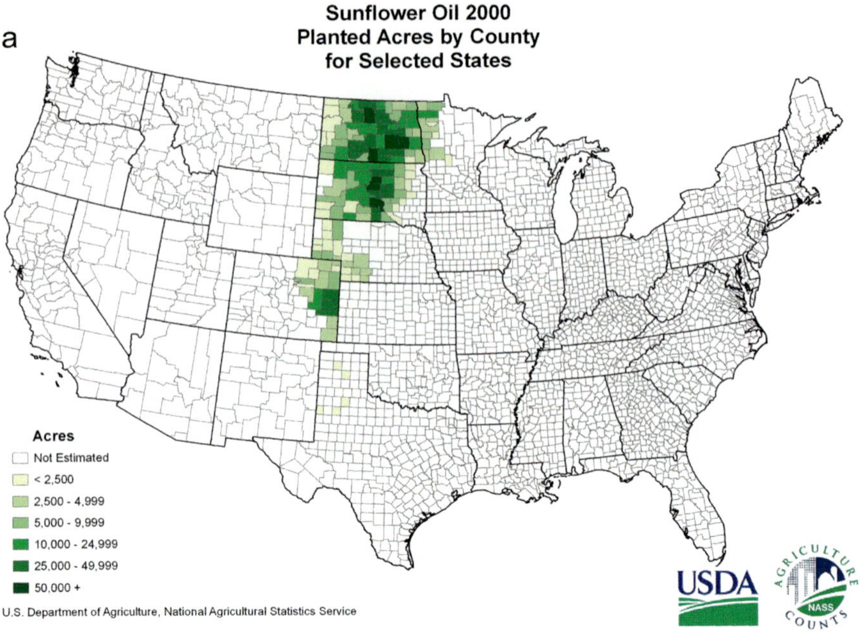

b

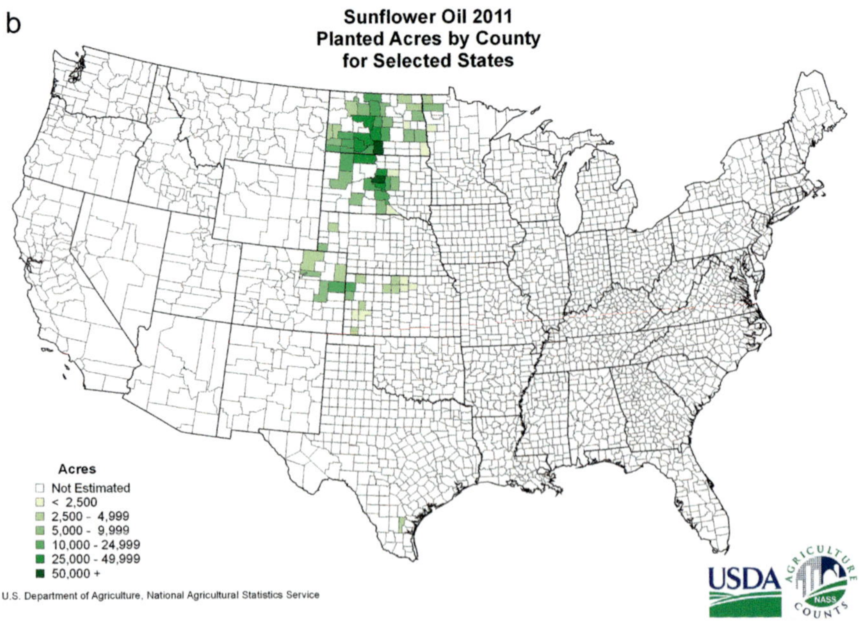

Fig. 15–2. Land area devoted to sunflower cultivation by county in (a) 2000 and (b) 2011.

or early season snow on severely infected plants can result in basal or mid-stalk breakage of the plant, making the plant unharvestable (Gulya et al., 1997).

Sclerotinia inoculum has increased due to narrow sunflower rotations, other susceptible crops in rotations, and reduced tillage. There are no fungicides available that have proven effective. Genetic resistance, deep tillage (if the production system permits), and broad rotations with non-broadleaf crop species are the best control methods for producers (Bradley et al., 2007).

Another historically significant disease is downy mildew [caused by Plasmopara halstedii (Farl.) Berl. et de Toni]. For 10 yr, the fungicide metalaxyl [methyl *N*-(methoxyacetyl)-*N*-(2,6-xylyl)-DL-alaninate] provided nearly 100% disease control when applied as a seed treatment. However, resistance to the fungicide occurred in 1995, and no single fungicide seed treatment has been found to replace metalaxyl (Albourie et al., 1998). The combination of genetic resistance and multiple fungicide seed treatments is providing acceptable protection at the present time. Infected seedlings either die or develop into stunted plants with abnormal flowers that take soil resources but do not produce seed (Bradley et al., 2007). Thus, the disease has contributed to issues of inadequate plant population and stand uniformity. These issues have been identified as significant yield limiting factors in the NSA Crop Survey (NSA, 2013a). As Table 15–1 demonstrates, downy mildew increased in severity after fungicide resistance developed in the pathogen.

Phomopsis stem canker, caused by *Diaporthe* (*Phomopsis*) *helianthi* Munt.-Cvetk., Mihaljč. & M. Petrov, *D. gulyae, D. kongii*, or *D. kochmanii*, causes a mid-stalk rot similar to Sclerotinia mid-stalk rot. This soilborne disease is quite common in France and in the Black Sea region, where it was originally discovered, and the latter three species have been identified in sunflower in Australia (Maric and Masirevic, 1980; Bert et al., 2002; Thompson et al., 2011). In recent years, an increase in the disease has occurred in the northern production area of the United States, with up to 90% of the fields and 45% of the plants within a field infected in Minnesota in 2011 (NSA, 2013a). Damage is similar to Sclerotinia mid-stalk rot, including reduced seed yield and quality, as well as increased lodging before harvest (Gulya et al., 1997). Control is difficult because efficacy of fungicides requires forecasting systems, and best management practices are similar to that of Sclerotinia: broad crop rotation and deep tillage of plant debris (if the production system allows) (Delos et al., 1995; Bradley et al., 2007). Currently, no fungicides are labeled for use in Phomopsis control in the United States, but quantitative genetic (horizontal) resistance is available to breeders (Skoric, 1985; Bert et al., 2002).

Other sunflower diseases have been severe regionally or have not contributed to year-on-year yield declines. Sunflower rust, caused by *Puccinia helianthi* Schw., is a common pathogen in the United States and Canada because it is native and found

Table 15–1. Incidence and severity of sunflower downy mildew recorded during field surveys in North Dakota (adapted from Gulya, 2005).

Indicator	Year of survey				
	1984	1991	1995	2001	2005
Incidence, % of fields	59	16	18	43	52
Severity, % of plants	1	0.05	0.3	1.8	4.9
Percent of fields with >15% downy mildew	5	0	0	3	14

on native sunflower species. It is also found in Europe and Argentina and is a widespread problem in Australia (Gulya et al., 1997; Sendall et al., 2006). As a macrocyclic rust, this pathogen only completes its sexual stages when conditions early in the growing season are ideal for the fungus, resulting in pycnial and aecial pustules on seedling plants or adjacent wild sunflower species (Laundon and Waterston, 1965; Markell et al., 2009). It is relatively rare for these fungal growth stages to appear in a widespread area, but this scale of event does occasionally occur, as it did in 2010, when pathologists were concerned that rapid race evolution could occur as a result (Markell et al., 2009). Appropriately timed fungicide applications can control this disease. Vertical genetic resistance is also available (Hulke et al., 2010a). Verticillium wilt, caused by *Verticillium dahliae* Kleb., is a common problem in Argentina, where rapid evolution of the pathogen has occurred (Romano et al., 1994). In the United States, however, Verticillium wilt has been controlled until very recently by one or two vertical resistance genes (Putt, 1964; Fick and Zimmer, 1974). A new pathotype virulent to the resistance gene(s), while present in the United States, is not currently causing major disease outbreaks.

There are many insects that attack maturing sunflower and are capable of causing significant yield loss. These insects are native to North America and have co-evolved with native sunflower species. In general, damage from a particular insect species is limited to a specific part of the growing region. Insect damage has been minimized over the years due to increased knowledge of the pests, improved crop protection products, and adherence to agronomic guidance by growers.

In the northern states, two insect species are common pests. The banded sunflower moth (*Cochylis hospes* Walsingham) is a Lepidopteran insect whose larvae feed on developing seeds and is a major pest in the North Dakota and Manitoba growing environments. Eggs are typically laid before bloom (Stage R4; Schneiter and Miller, 1981) on the bracts of the developing flower (Beregovoy and Riemann, 1987; Beregovoy et al., 1989). During feeding, the larva will cover the front of the sunflower head with a web-like frass, which is easily visible. The amount of damage can be significant enough to warrant chemical control measures, especially in non-oil sunflower where product quality is of the utmost importance and the insect damage is a visible defect in the product. In both oilseed and non-oil sunflower, common practice is to spray the field with an insecticide, such as a pyrethroid, at the onset of bloom (R5.1 growth stage; Schneiter and Miller, 1981; Knodel and Charlet, 2007). Because of the stature of the crop at this growth stage, insecticide is usually applied aerially. Similar damage can be caused by the red sunflower seed weevil (*Smicronyx fulvus* LeConte), which is most commonly found in central South Dakota. A specific concern with this pest is that the red sunflower seed weevil will deposit eggs directly in the seed with the result that larvae will sometimes remain in the seed beyond harvest (Oseto and Braness, 1979; Knodel and Charlet, 2007). This behavior causes a large concern for non-oil growers because a damaged seed may also have an insect in it, which may be consumed if left in the product. Control methods and timing for this pest are similar to banded sunflower moth (Knodel and Charlet, 2007).

In the southern plains, the Lepidopteran pest sunflower head moth (*Homoeosoma electellum* Hulst), is particularly harmful to yield and quality of the seed product (Randolph et al., 1972). In a manner similar to banded sunflower moth in the north, the moth will lay eggs on the head, and the resultant larvae will eat the maturing seeds. There are some key differences between the sunflower head moth and banded sunflower moth, however. The most important is that sunflower head moth cannot overwinter in cold, temperate climates, which relegates this insect mainly to the southern growing region, which is closer to the overwintering site (Arthur, 1978). Second, the sunflower head moth will lay eggs a little later than banded sunflower moth, usually waiting until after the R5.1 growth stage to begin (Satterthwait and Swain, 1946). Third, the primary targets of chemical control are the adults rather than larvae, so an insecticide treatment should be based on adult population estimates (versus egg counts for banded sunflower moth), and timed to the R5 growth stage (Knodel and Charlet, 2007; J. Prasifka, personal communication, 2013). Fourth, the sunflower head moth usually strikes in larger numbers, overwhelming plants with larvae and causing large losses (Randolph et al., 1972). This fact often causes producers to apply multiple insecticide treatments to the same field. Similar insecticide chemistries can be used on the sunflower head moth as the northern insects.

Two stem boring insects are also pests of sunflower, particularly in the southern plains. Dectes stem borer (*Dectes texanus* LeConte) infests both sunflower and soybean and causes preharvest lodging in sunflower by girdling the base of the stalk after drydown commences (Michaud et al., 2009). This species is somewhat peculiar in that, in general, only one larva is present in each stalk, and if more than one is present, then the extra larva will be eaten in an act of cannibalism (Charlet et al., 1987). Sunflower stem weevil (*Cylindrocopturus adspersus* LeConte) larvae are, by contrast, much smaller in size, but overwhelm plants in number. If multiple tunneling larvae are present in any plant, then there is a possibility of preharvest lodging due to wind or even the action of a combine as it approaches a plant (Rogers and Jones, 1979). Both of these insect pests were adequately controlled by the use of the insecticide Furadan (2,3-dihydro-2,2-dimethyl-7-benzofuranyl methylcarbamate), which was recently eliminated from the market by the USEPA. Current agronomic guidance calls for the use of "stay green" hybrids, which have slower stalk dry down in the autumn, and the use of lower plant densities to increase the stalk diameter (Knodel and Charlet, 2007; Michaud et al., 2009). More research is underway to find better recommendations, as these are incomplete measures at best.

The red-winged blackbird (*Agelouis phoeniceus*), among other species, has been a consistent pest on ripening sunflower. Damage estimates vary between regions. The highest damage occurs within the Prairie Pothole region that lies within parts of North Dakota and South Dakota, as well as portions of the Canadian provinces of Saskatchewan and Manitoba (Hagy et al., 2008; Linz et al., 2011). Damage assessment research conducted over a 3-yr period in North Dakota showed an average damage to ripening sunflower of 0.95% in 2008, 2.17% in 2009, and 3.29% in 2010 (Klosterman et al., 2011). The annual NSA Crop Survey assesses bird depredation. The following is a 5-yr average of yield reduction in crop years 2002 to 2007 (excluding 2004): 3.85% in North Dakota, 2.71% in South Dakota, and 3.66% in Minnesota (NSA, 2013a). In areas of consistently high damage, which can sometimes amount to nearly a complete loss of harvestable seed, farmers have eliminated sunflower in their rotation. Because birds are a higher organism and very mobile, they are virtually impossible to control by conventional agronomic practices, and, thus, innovative techniques will be required to prevent damage. The use of repellant chemicals sprayed before harvest, together with untreated Wildlife Conservation Sunflower Plots near roosting habitat (prairie potholes with emergent vegetation such as cattails), appears to be a logical loss prevention method, although it requires further scientific study, as well as improved cost structure, before such recommendations can be made on a large scale basis (Hagy et al., 2008; Linz et al., 2011).

Clearly, much progress has been made on cultural or environment-influenced problems that sunflower shares with other crops, particularly the control of insects and diseases. Insecticides designed for general control of insects with some residual activity have been useful in controlling the most pervasive head-infesting pests. Herbicides with developed systems in other crops such as corn, wheat, and soybean have been adapted for use in sunflower. For those diseases that are controllable by fungicides (e.g., downy mildew and rust), fungicides from small grain and vegetable crop systems have been adapted with some success. The problems that are too complex for simple agronomic changes to mitigate, as well as those that are specific to sunflower, have largely been dealt to the realm of the plant breeder and geneticist to solve.

Genetic Gain for Defensive Traits

Breeding for defensive traits has been the single most important advance in the genetic improvement of sunflower, because many biotic stresses plague this crop and ease of production through mitigation of risk is essential for keeping producers interested in the crop. The most significant genetic advance in sunflower for the eastern growing region of North Dakota, South Dakota, and Minnesota is resistance to downy mildew conditioned by single, dominant genes, also called vertical resistance. After the defeat of metalaxyl fungicides by the causal organism circa 1995, there was a critical need for genetic resistance to fill in the void (Albourie et al., 1998). Vertical resistance was desirable for several reasons. First, it required no risk or additional cost on the part of the farmer. Use of fungicides does carry some exposure risks to the user, and most of the metalaxyl alternatives are not completely effective against downy mildew. Second, it was relatively easy to include in new hybrids because it is dominant (i.e., only needed in either the female or male line to be expressed in the hybrid) and 100% heritable with a fairly simple greenhouse phenotyping procedure to find resistant progeny (Hulke et al., 2010b). Third, multiple unique genes are freely available from public sources to include in inbred lines with no restriction (Miller and Gulya, 1988; Hulke et al., 2010a,b). This latter point is important because the most critical weakness of vertical resistance, downy mildew resistance included, is the high risk of altered virulence in the causal organism, which would overcome a resistance gene and render it ineffective. Breakdown of genetic resistance was witnessed in the sunflower production region of France, where the entire region rapidly adopted the Pl_6 and Pl_7 resistance loci from public inbred lines 'HA 335' and 'HA 336' (Miller and Gulya, 1988; Tourvieille de Labrouhe et al., 2000). Initially, these genes provided universal resistance to all known races, but then races virulent to the resistance appeared, which is a result of race evolution and/or proliferation of rare races. More recently, races of downy mildew virulent to these same resistance genes were discovered in North Dakota. It quickly became apparent to the world community that several different resistance genes were needed at the ready to minimize the downy mildew threat (Tourvieille de Labrouhe et al., 2010).

Over the last 25 yr, multiple different downy mildew resistance genes were identified and released by the USDA-ARS breeding program in Fargo, ND. To this day, widely deployed downy mildew resistance genes are limited to only a few, including Pl_8, derived from an interspecific cross with *H. argophyllus* L. (Miller and Gulya, 1988) and Pl_{15} from inbred line 'RNID' (Bertero de Romano et al., 2010). However, many other genes are available and effective against all currently known races, even those virulent to the Pl_6 and Pl_7 resistance genes. Broader deployment of these genes in the future, perhaps in a pyramiding arrangement as suggested by Joshi and Nayak (2010) and Tourvieille de Labrouhe et al. (2010), may slow evolution of the pathogen and reduce the threat of downy mildew to sunflower producers. At the very least, broader deployment will provide multiple sources of resistance that could be distributed as soon as a new threat is known.

Another key advance in sunflower genetics was the development of herbicide resistance genes that provided the sunflower plant with resistance to broad spectrum herbicides. Although glyphosate and gluphosinate resistance have not been developed in sunflower, breeders have successfully incorporated resistance to imidazolinone and sulfonylurea herbicides into inbred lines, and then into commercial hybrids (Miller and Al-Khatib, 2002, 2004). While not as broad spectrum as glyphosate, these herbicides control many of the key broadleaf species that are known to be problems in sunflower, as well as many grasses. The resistance was originally derived from wild conspecifics that were known to have field resistance to a herbicide. The resistance was then isolated by traditional breeding, and a standard field herbicide challenge protocol to develop and release resistant lines was implemented. The IMISUN releases (Miller and Al-Khatib, 2002) were the first public imidazolinone-resistant sunflower lines, and the SURES releases (Miller and Al-Khatib, 2004) were the first public sulfonylurea-resistant lines. In both cases, a major gene conferred resistance to the herbicide, but in the case of the IMISUN releases, there were modifier loci that could reduce the resistance and cause "flashing" of the leaf tissue at labeled rate of the herbicide (Bruniard and Miller, 2001). This, combined with a tendency of imidazolinone-resistant lines to be later maturing and lower yielding than nonresistant relatives, decreased desirability of the resistance initially. However, additional breeding with the IMISUN resistance to reduce linkage drag with other traits, as well as release of alternative, proprietary resistance genes to this class of herbicide, have increased desirability and effectiveness of this particular herbicide system (Sala et al., 2012). Despite these improvements, any commercial hybrid seeking to be labeled as "herbicide resistant" for these two classes must be homozygous for the dominant resistance to reduce the risk of injury in unpredictable field environments. Currently, no hybrids are being marketed that are resistant to both imidazolinone and sulfonylurea herbicides.

While the previous two examples of genetic gain for defensive traits have centered on single dominant genes with nearly 100% heritability, Sclerotinia resistance is quantitative (horizontal) with less than 100% heritability. There are no known vertical resistances to any of the three Sclerotinia diseases of sunflower. Further, the quantitative resistance is generally made up of loci of small or moderate effect, and different loci are responsible for resistance to each of the three diseases (Gulya et al., 1997; Bert et al., 2002; Yue et al., 2008; B.S. Hulke, unpublished data, 2013). The implications of this from the practical business perspective of a seed company is that no marketing claims of reliable Sclerotinia resistance can be made on any hybrid, while downy mildew resistance or herbicide resistance claims can easily be made because of the simple traceability and high heritability of those traits. Nonetheless, the danger of Sclerotinia outbreaks on the sustainability of the crop has always been a concern, and thus, a great deal of research has been in place in the public sector to identify Sclerotinia resistance. Consequently, Sclerotinia resistance has been introgressed into modern breeding lines through traditional breeding, molecular breeding, and even cytogenetic techniques if a wide cross is used (e.g., Miller and Gulya, 2006; Yue et al., 2008; Liu et al., 2011). The public lines and associated molecular information are then used by the private sector to improve resistance in modern hybrids. This work is coordinated and funded by the National Sclerotinia Initiative, which was created in 2000 by USDA-ARS, in cooperation with the NSA and other commodity groups, to provide competitive grants to researchers on multiple crops affected by Sclerotinia. This Initiative has resulted in the release of 28 oilseed sunflower germplasm lines in both the female and male heterotic groups, mostly improved by traditional breeding with cultivated sunflower resistance donors that are outside the immediate commercial breeding pool. These releases possess improved head rot and/or basal stalk rot resistance in an improved agronomic and yield background (Miller and Gulya, 1999; Miller and Gulya, 2006; Miller et al., 2006b,c; B.S. Hulke, unpublished data, 2013). These lines have been the first attempt at optimizing Sclerotinia resistance with yield and agronomic traits. Additional efforts on molecular breeding methods and tools, such as single nucleotide polymorphism (SNP) markers, will be required to adequately address the negative correlation between Sclerotinia resistance and yield observed by commercial breeders (Lilleboe, 2011).

Four other defensive traits have been the focus of improvement around the world, but have historically had a lesser impact on sunflower breeding in the United States. Among these is Verticillium resistance, which has largely been controlled by the V_1 locus, a dominant, single gene (vertical) resistance locus that was quickly integrated into breeding programs and has provided nearly 100% control in the United States until very recently (Putt, 1964; Fick and Zimmer, 1974; Gulya, 2007). Even though virulent forms of the pathogen are present, the disease is not an imminent threat to production in the United States (NSA, 2013a). This scenario contrasts with the situation in Argentina, where the V_1 locus was defeated very quickly after deployment and additional breeding with new vertical resistances was then necessary to control Verticillium, which is one of the primary threats to production in that country (de la Vega et al., 2007; Gonzalez et al., 2007). Phomopsis disease is a stalk rot similar to Sclerotinia mid-stalk rot that is controlled by quantitative (horizontal) resistance loci. This disease is particularly damaging in the Black Sea region of eastern Europe, where much emphasis has been placed on breeding for resistance in public and commercial germplasm (Vranceanu et al., 1983; Vranceanu and Soare, 1988). The increase of this disease recently in the United States has prompted the USDA-ARS laboratory to reach out to Russian scientists to exchange Phomopsis-resistant germplasm and expertise as part of a former Soviet Union exchange, sponsored by the USDA-Foreign Agricultural Service. As a part of this work, resistance is being mapped using association mapping technology in a diverse germplasm panel (B.S. Hulke and T.J. Gulya, unpublished data, 2013). Sunflower rust is another disease that, while present in the United States, is not a current focus of improvement. USDA-ARS scientists have released and later molecularly mapped vertical resistance genes for this disease (e.g., Hulke et al., 2010a; Gong et al., 2013), but most commercial hybrids do not contain resistance genes. This could be because rust is only occasionally a major threat to the sunflower growing region, and when a rust outbreak is imminent, it is easily controlled by aerially applied fungicides (Bradley et al., 2007; NSA, 2013a). In Australia, by contrast, rust is an ever-present threat as both sunflower and rust are introduced species, and breeding for resistance has been a goal for many years (Sendall et al., 2006). The recent genetic resistance mapping efforts may make breeding and selection efforts easier for rust resistance, which may increase resistance gene usage in American hybrids and cause a concomitant decrease in the amount of fungicide used to control rust in sunflower. Broomrape infestation, caused by the parasitic plant *Orobanche cumana* Wallr., is a major threat to production in southern Europe and the Black Sea region of eastern Europe. What is peculiar about this pest–crop interaction is that the native range of the pest and the crop do not overlap. Thus, broomrape infestation has been opportunistic. It is the main threat to sunflower production in this European region, and resistance is governed by vertical resistance genes augmented with application of imidazolinone herbicides (Velasco et al., 2007). Presence of this pest has never been documented in the United States.

Genetic Gain for Seed Quality Traits

Historically, the vast majority of American sunflower production, both non-oil and oilseed, has been exported. Sunflower oil was not able to find a consistent market in the United States. Traditional sunflower oil is high in linoleic acid (C18:2; NSA, 2013b). This essential fatty acid is also found in other vegetable oils, most notably soybean oil. Soybean oil has dominated the United States, with market share reaching as high as 85% (Oil World). Soybean oil has been inexpensive, functional for most uses when hydrogenated, and always available because of the large land area devoted to its cultivation. However, these conditions began to change as the issue of trans fatty acids emerged in the early 1990s. A series of human nutrition studies led by Judd et al. (1994) found that trans fatty acids, created in the hydrogenation process, were detrimental to heart health. Members of the NSA recognized this development as an opportunity to gain domestic market share if naturally stable sunflower oil could be developed. High oleic (C18:1) sunflower oil, which is considerably more stable than traditional sunflower oil and nonhydrogenated soybean oil, was already in the market, but that oil profile and seed were under patent. Food companies were not keen to pay a high premium for this oil with only one seed supplier. But the market was interested in any oleic-based sunflower oil that could be produced in place of the traditional linoleic sunflower oil. A mid-oleic sunflower oil appeared to be the best option to obviate the patent and potentially satisfy a part of the domestic market as a replacement for partially hydrogenated soybean oil. Thus, NuSun mid-oleic sunflower was born and by 2009, 90% of U.S. produced sunflower oil was consumed in North America (NSA, 2013b).

Redirecting an entire industry to NuSun production was a monumental undertaking that required a great deal of planning and coordination. The entire market chain from the producer, seed handlers, and the crushing plants had to work together to develop an infrastructure to segregate NuSun production. From the breeder's perspective, the major challenge was creating NuSun hybrids with the critical agronomic characteristics of yield, oil content, disease resistance, and oleic acid content stability. Breeders from USDA-ARS provided important breeding lines to the private sector seed companies. These lines possessed the high oleic mutation derived from the Russian 'Pervenets' (Soldatov, 1976). There were two ways that a stable NuSun hybrid could be produced. One was by the crossing of two or more parental inbred lines that possessed oleic acid levels around 60%. Producing genetically stable compositions of this nature required the mutation from Pervenets together with the same allelic state of modifier loci that reduced the oleic level, all as homozygotes (Miller and Vick, 2002). This was difficult to do in practice and required incorporation of the required genes into all breeding germplasm. The second method involved incorporating the Pervenets mutation together with positive modifier alleles for oleic acid, producing lines with oleic acid around 85%. These lines were then crossed to existing traditional composition sunflower lines, and the resulting hybrid variety had a stable oleic composition midway between the parents, at about 60% (Miller et al., 2006a). The latter technique was adopted by the sunflower industry because they could maintain the broadness of the genetic base of their breeding programs, including all of the other necessary traits for a hybrid, by including genes for those traits primarily in the "traditional" parent, at least initially. With time, increased oleic acid content was combined in the same lines with Sclerotinia resistance, downy mildew resistance, herbicide resistance, and other traits (Miller et al., 2006b). The evolutionary process of completely switching the industry to NuSun took 10 yr to accomplish, and today traditional sunflower is rarely grown.

The genetics of NuSun and high oleic acid are more complex than the single gene from Pervenets. Inheritance studies have indicated that one or more modifier loci can affect oleic composition (Fernandez-Martinez et al., 1989; Lacombe et al., 2001). Because a reduction in one fatty acid must result in a relative (percentage basis) increase of one or more others, a combination of loci for low saturated fat together with the Pervenets mutation and other modifiers can result in oleic acid concentrations of 90 to 95% (Gerdes et al., 2011). Thus, depending on genetic background, the number of segregating modifiers in concert with the Pervenets mutation can vary, which supports the assertion by breeders that certain genetic backgrounds are better suited to making NuSun and high oleic lines and hybrids. Another possible source of complication for breeders is the fact that the Pervenets mutation itself is a tandem duplication of the *Fad2-1* gene, encoding an important desaturase, and the tandem duplication is thought to cause RNA silencing (Schuppert et al., 2006). It is possible but not yet proven that the effects of RNA silencing may be more variable than the effects of a recessive-acting point mutation.

During the NuSun conversion era, the confection industry was going through its own process of changing hybrids. The world in-shell market for direct human consumption demanded larger seed size. Premium seed is determined by sieve size. Average seed size in the 1999 to 2002 crop years was 60% of seed >7.9-mm (20/64-in) sieve size. The average seed size in the 2008 to 2012 crop years was 83% of seed over that same sieve (NSA, 2013b). This genetic improvement provided enhanced efficiency in the production and processing of confection sunflower with more useable seed for the in-shell market. Because the end market is influential on the commercial success of a hybrid, the non-oil breeding programs needed to de-emphasize other traits and concentrate on improving size and shape of the seeds to meet the general world market desire of larger seed, as well as the demands of specific markets for specific size, shape, and even coloration of the seeds. To accomplish this, the seed companies sourced seed from open-pollinated varieties and landraces in places where large seeded sunflower existed due to the selective work of farmers over hundreds of years, such as Turkey. While these varieties were large seeded, they lacked a number of important traits that are required in modern hybrids, including adaptability to modern production practices, disease resistance, and self-compatibility. These were crossed to existing non-oil lines that were enhanced for yield, self-compatibility, and agronomic performance, in particular. While this work resulted in larger seed and more seed type diversity in the commercial germplasm, the tradeoff is the general lack of development in defensive traits and yield of the caliber of elite oilseed hybrids.

Genetic Gain for Yield

Yield for field crops is often expressed in the terms for which the seed product is marketed. Oilseed sunflower is complicated by the fact that both seed mass and oil percentage factor into the pricing scheme for the commodity. Thus, yield can be expressed in two different ways, the components of seed mass per hectare and percentage oil, as well as the composite total, oil mass per hectare, depending on the customary method of determining crop value in a particular region or country. In the United States, the components of yield are usually reported because payment is not directly tied to oil mass harvested, but rather to seed mass primarily with a premium or discount for each percentage of oil above or below, respectively, 400 g kg^{-1}. Thus, producers are looking first for hybrids that yield large amounts of seed, but also for those with a high percentage oil. In the Black Sea region, producers are paid by seed mass only, while the Argentine system is similar to the U.S. system, but with a slightly different premium–discount structure (J. Swanson and M. Canepa, personal communication, 2013).

The NSA Crop Survey does provide insight into yield potential when pests, adverse weather, and cultural practices do not impact on-farm fields. Every year, approximately 10% of the surveyed fields are designated as having "no problem." These fields tend to have a higher potential yield as determined by an objective yield assessment by each surveying team. It is worthy to note that yields in the "no problem" designated fields averaged 2540 kg ha^{-1} (2267 lb ac^{-1}) compared to the average USDA-NASS yields of 1620 kg ha^{-1} (1457 lb ac^{-1}) over the entire production region (data from a 3-yr average of the 2010–2012 seasons; NSA, 2013a). Considerable yield potential is clearly not being realized due to a combination of pests, poor cultural practices, and weather. These data are typical and have put sunflower breeders on the defensive for decades, as we have already demonstrated. The strategies employed thus far by sunflower breeders have resulted in yield gains, in terms of the U.S. average seed mass harvested, of 7.72 kg ha^{-1} yr^{-1} (6.89 lb ac^{-1} yr^{-1}) and 9.96 kg ha^{-1} yr^{-1} (8.90 lb ac^{-1} yr^{-1}) for both non-oil and oilseed sunflower, respectively (Fig. 15–3; USDA-NASS). However, rates of increase are not as great for oilseed and non-oil sunflower as they are for maize and soybean. Figure 15–3 allows a comparison of yield increases for the four crops by standardizing yield into a percentage of the 38 yr average. On average, maize yield increases in the USA are double those of both oilseed sunflower and non-oil sunflower.

One weakness of using actual production data is that it does not allow for the relative contributions of genetic gain and gain from improved cultural practices to be calculated. At this time, there is no such analysis completed for U.S. data, but this work is currently underway (B.S. Hulke, unpublished data, 2013). However, such analyses are available from other nations. In Argentina, de la Vega and others (2007) demonstrated that best linear unbiased predictions (BLUPs) of commercial and experimental hybrids over the course of 20 yr were increasing at an annual rate of 11.9 kg ha^{-1} yr^{-1} for oil yield, but seed yield BLUPs did not increase from the mid 1990s to the mid 2000s. These BLUPs are interpreted as the genetic component of total observed yield gain. Improvement of oil yield over seed yield was believed to be due to the philosophy of breeders in the region that oil yield was the main target of selection. The authors also differentiated the hybrid types into conventional (yield-targeted), backcross conversions (herbicide resistant, high oleic), and Verticillium resistant, the latter being the main disease threat in the region. The backcross conversion hybrids diverged greatly from the average of the conventional hybrids (11.9 kg oil ha^{-1} yr^{-1}), to the extent of performing similar to hybrids developed 20 yr prior. Such results could be a warning that breeding for defensive traits primarily with yield as a secondary goal may slow overall genetic gain for yield. However, Verticillium-resistant hybrids had a higher genetic gain of 14.4 kg ha^{-1} yr^{-1}, suggesting that major threats to production need to be mitigated to fully enhance genetic gain for yield, similar to the observations of the NSA Crop Survey.

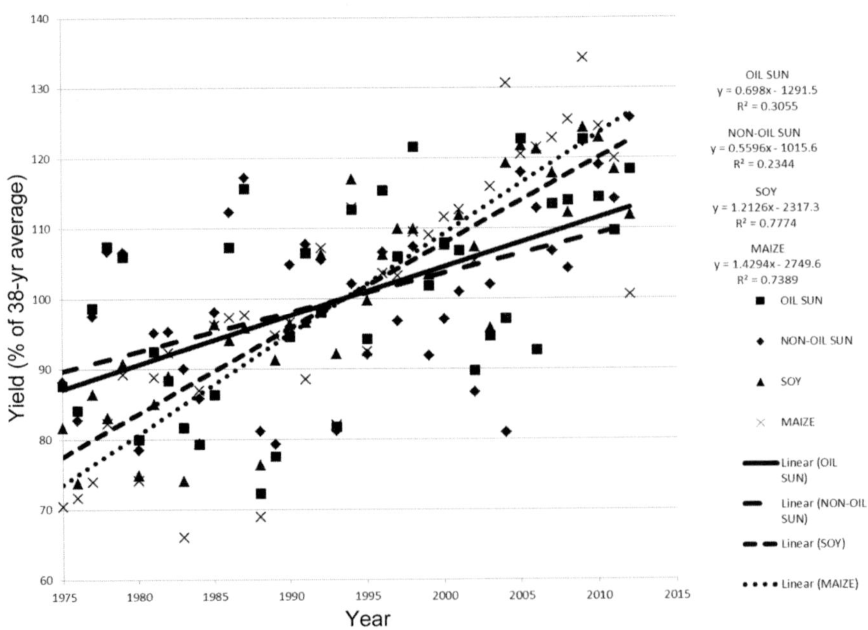

Fig. 15–3. Yields of maize, soybean, oilseed sunflower, and non-oil sunflower in terms of percentage of the 38-yr average, 1975–2013.

Similar work was conducted recently in South Africa, a country with a similar land area devoted to sunflower cultivation. Chigeza et al. (2012) grew reference hybrids from the last 40 yr of production in replicated trials over the course of 2 yr and used the data to determine the relative contribution of genetic gain to total observed gain in national statistics data. Linear increases were observed for both oil yield and seed yield, but a curvilinear trend with a plateau was observed for percentage oil increase. Oil yield genetic gain was calculated at 12 kg ha^{-1} yr^{-1}, a rate very similar to that of de la Vega and others (2007). Unlike the Argentinean data, seed yield appears to be driving the increase in oil yield in South Africa, but as in Argentina and the United States, South African production has been shifting away from prime farmland to semiarid regions, which may partially explain why the estimated contribution of genetic gain to total production increases has dropped from 56.3% for 1970 to 1989 to 23.3% from 1990 to 2009. Selection and adaptation of hybrids to a particular production region directly can increase genetic gain more rapidly for that region than indirect selection in a poorly correlated environment, which makes these shifts in production a roadblock to genetic gain (de la Vega and Chapman, 2006).

Future Prospects

The U.S. sunflower crop has had a very rocky history. There were predictions in the 1970s that acreage would reach 4 million hectares (10 million acres). This is in contrast to later assessments that sunflower would not continue to be viable into the 21st century. In this short 40-yr history in the United States, nearly every possible obstacle has impacted the sunflower crop. This has ranged from federal farm programs that "…favor planting wheat, feed grains, and cotton and discourage domestic production of oilseeds" to major disease outbreaks to competition with GM crops (Ash et al., 1995). Although sunflower acreage has declined considerably since the 1970s, the crop has maintained a consistent grower base in central and western North Dakota and South Dakota.

The future of sunflower in the United States is dependent on increasing yields and consistent annual levels of production (yield stability). Yield potential will continue to increase annually if previous breeding and production trends continue (USDA-NASS; de la Vega et al., 2007). The critical question is whether yield enhancement can be maintained at the same or higher pace than competing crops such as soybean and maize. Sunflower is at a disadvantage with these crops for three primary reasons. First, sunflower is not the crop of choice on the best agricultural land and is increasingly being cultivated in semiarid regions (Fig. 15–2). Second, it is not the recipient of the same level of research investment, public or private, as maize or soybean. This delays development of key technologies that enhance breeding, such as validated, functional genetic markers for important traits, transformation systems, and well characterized genomes and metabolomes. It also influences the quality of these studies. Scale and capabilities of breeding programs are also at a disadvantage to these crops with higher investments; for example, in sunflower, much of the work is still being accomplished by traditional breeding approaches, and while genomic approaches are beginning to take hold, the adaptation to the new approaches is not as mature as in other crops. Third, sunflower will have some difficulty getting regulatory acceptance of a GM event in the USA because of the close proximity of the production land to weedy conspecifics and closely related and partially interfertile congeneric species. The gene flow potential has been thoroughly studied and confirms this fear (Mercer et al., 2006a,b; Ureta et al., 2008).

Despite the issues inherent with being a crop with smaller investment, several advancements have been made to modernize sunflower improvement. High throughput genotyping technologies, such as SNPs, are currently being used to develop trait-marker associations and improve the efficiency of breeding (Bachlava et al., 2012; Pegadaraju et al., 2013). Whole genome sequence, while not a reality today for sunflower, is in development, and a completed genome is expected soon, which would greatly simplify targeted genomics studies (Kane et al., 2011). Doubled haploid technology is currently being researched in a few laboratories worldwide, and this technology, if adaptable to sunflower, would provide efficiencies in line development and perhaps even CMS (A line) conversion if systems analogous to those available in maize are realized (Carlson, 1988).

Some of sunflower's weaknesses can also be considered strengths. The large population of weedy conspecifics and congenerics that are interfertile with cultivated sunflower, while a potential barrier to GM commercialization, are a valuable resource for gene mining and introgression (Miller and Gulya, 1988; Miller and Wolf, 1991; Velasco et al., 2007; Liu et al., 2011). Multiple public laboratories have historically had an interest in practical application of cytogenetics to sunflower improvement. One of the current goals of this work is development of sunflower immune to Sclerotinia via interspecific gene transfer from *Helianthus* species already possessing this immunity naturally (Liu et al., 2011). For a crop with a quality product but small landscape footprint, even roadblocks to the development and acceptance of GM can be positive developments. Fatty acid improvements, such as NuSun and high oleic, have already added value to sunflower oil and spurred demand. Additional fatty acid profiles with greatly reduced saturated fat levels, in combination with high oleic acid, will soon to be introduced in the United States, further enhancing value and demand. While other oilseeds are attempting to make the same advancements, all have the potential market negative of being GM. Some states and even some retailers are considering mandatory labeling of all GM ingredients, potentially making this something greater than a niche market issue in the near future (Choi and Perrone, 2013). It could also be advantageous to have crops like sunflower with alternative herbicide systems available for the producer to use in a rotation, with the glyphosate-resistant weed threat expanding across the United States.

While market demand could potentially increase in the near future, the challenge for the American sunflower industry is consistency of supply. Based on past experience, the food industry views sunflower production as unreliable. This can be changed with improved hybrids possessing disease and insect resistance and by expanding the production base from North Dakota to Texas. Sunflower production in the High Plains region, in particular, could be enhanced as irrigation water is further rationed, and economically sound production of other common, less drought tolerant crops becomes difficult. Sunflower has proven to be a viable crop under limited irrigation in these water scarce regions (Stone et al., 1996). Sunflower may have a niche in prime farmland to the east of the current production region should droughts, such as those observed in 2012 in the central United States, become more commonplace as a result of climate change. Production data for 2012 suggest that this poor year for corn and soybean was a better than average year for sunflower (Fig. 15–3). Sunflower researchers have made good progress in providing genetic resistance to a number of diseases that have had real impact on yield, but insect resistance has historically been ignored despite its impact on yield and quality, particularly in the central and southern plains. Recent data suggest that resistance to the red sunflower seed weevil and sunflower head moth may be highly heritable with simple genetic basis, potentially making insect resistance easy to commercialize in the future, stabilizing yields (J. Prasifka, unpublished data, 2013).

Consistently improved yields via enhanced genetics hold the key to the future of U.S. sunflower production. This will involve adaptation of new breeding technology to improve defensive, quality, and yield traits simultaneously, in the context of the production environment.

References

Albourie, J.-M., J. Tourvieille, and D. Tourvieille de Labrouhe. 1998. Resistance to metalaxyl in isolates of the sunflower pathogen *Plasmopara halstedii*. Eur. J. Plant Pathol. 104:235–242. doi:10.1023/A:1008691123239

Arthur, A.P. 1978. The occurrence, life history, courtship, and mating behavior of the sunflower moth, *Homoeosoma electellum* (Lepidoptera: Phycitidae), in the Canadian prairie provinces. Can. Entomol. 110:913–916. doi:10.4039/Ent110913-9

Ash, M., G. Douvelis, J. Castaneda, and N. Morgan. 1995. Oilseeds: Background for 1995 Farm Legislation. Agric. Econ. Rep. 715. USDA-ERS, Beltsville, MD.

Ashley, R., and D. Tanaka. 2007. Tillage, seedbed preparation, and planting In: Sunflower production. Ext. Bull. A-1331, rev. North Dakota State Univ. Ext. Serv. and North Dakota Agric. Exp. Stn., Fargo, ND. p. 18–22.

Bachlava, E., C.A. Taylor, S. Tang, J.E. Bowers, J.R. Mandel, J.M. Burke, and S.J. Knapp. 2012. SNP discovery and development of a high-density genotyping array for sunflower. PLoS ONE 7:e29814. doi:10.1371/journal.pone.0029814

Beregovoy, V.H., G.L. Hein, and R.B. Hammond. 1989. Variations in flight phenology and new data on the distribution of the banded sunflower moth (Lepidoptera:Cochylidae). Environ. Entomol. 18:273–277.

Beregovoy, V.H., and J.G. Riemann. 1987. Infestation phenology of sunflowers by the banded sunflower moth, *Cochylis hospes* (Cochylidae:Lepidoptera) in the Northern Plains. J. Kans. Entomol. Soc. 60:517–527.

Berglund, D.R. 1978. North Dakota 1978 hybrid sunflower national performance trials. Circular A-652. North Dakota Coop. Ext. Serv., Fargo, ND.

Bert, P.-F., I. Jouan, D. Tourvieille de Labrouhe, F. Serre, P. Nicolas, and F. Vear. 2002. Comparative genetic analysis of quantitative traits in sunflower (*Helianthus annuus* L.) 1. QTL involved in resistance to *Sclerotinia sclerotiorum* and *Diaporthe helianthi*. Theor. Appl. Genet. 105:985–993. doi:10.1007/s00122-002-1004-3

Bertero de Romano, A., C. Romano, M. Bulos, E. Altieri, and C. Sala. 2010. A new gene for resistance to downy mildew in sunflower. Proceedings of the International Symposium "Sunflower Breeding on Resistance to Diseases," Krasnodar, Russia, 23–24 June 2010.

Blamey, F.P.C., R.K. Zollinger, and A.A. Schneiter. 1997. Sunflower production and culture. In: A.A. Schneiter, editor, Sunflower technology and production. Agron. Monogr. 35. ASA, CSSA, and SSSA, Madison, WI. p. 595–670. doi:10.2134/agronmonogr35.c12

Bradley, C., S. Markell, T. Gulya. 2007. Diseases in sunflower. Sunflower production. Ext. Bull. A-1331, rev. North Dakota State Univ. Ext. Serv. and North Dakota Agric. Exp. Stn., Fargo, ND.

Bruniard, J.M., and J.F. Miller. 2001. Inheritance of imidazolinone herbicide resistance in sunflower. Helia 24:11–16.

Carlson, W.R. 1988. The cytogenetics of corn. In: G.F. Sprague and J.W. Dudley, editors, Corn and corn improvement. Agron. Monogr. 18. ASA, CSSA, and SSSA, Madison, WI. p. 259–343. doi:10.2134/agronmonogr18.3ed.c4

Charlet, L.D., D.D. Kopp, and C.Y. Oseto. 1987. Sunflowers: Their history and associated insect community in the Northern Great Plains. Bull. Entomol. Soc. Am. 33:69–75.

Chigeza, G., K. Mashingaidze, and P. Shanahan. 2012. Seed yield and associated trait improvements in sunflower cultivars over four decades of breeding in South Africa. Field Crops Res. 130:46–56. doi:10.1016/j.fcr.2012.01.015

Choi, C., and M. Perrone. 2013. Whole Foods: Products will carry GMO labeling. Record Number 2750387c219a71bf2cf2d8685743e7d3. The Associated Press News Service, New York.

de la Vega, A.J., and S.C. Chapman. 2006. Defining sunflower selection strategies for a highly heterogeneous target population of environments. Crop Sci. 46:136–144. doi:10.2135/cropsci2005.0170

de la Vega, A.J., I.H. DeLacy, and S.C. Chapman. 2007. Progress over 20 years of sunflower breeding in central Argentina. Field Crops Res. 100:61–72. doi:10.1016/j.fcr.2006.05.012

Delos, M., J. Moinard, and D. Jacquin. 1995. Study and monitoring of sunflower *Phomopsis*. Phytoma 473:25–27.

Dorrell, D.G., and H.C. Huang. 1978. Influence of *Sclerotinia* wilt on seed yield and quality of wilted sunflower at different stages of development. Crop Sci. 18:974–977. doi:10.2135/cropsci1978.0011183X001800060017x

Dvoryadkin, N.I. 1976. Results of research work in sunflower in the USSR. p. 45–62. In: Proceedings of the Seventh International Sunflower Conference, Krasnodar, USSR. 27 June–3 July 1976. Vol. 1. International Sunflower Association, Paris.

Fehr, W.R. 1987. Development of hybrid cultivars In: Principles of cultivar development. Vol. 1. Macmillan, New York. p. 428–438.

Fernandez-Martinez, J., A. Jimenez, J. Dominquez, J.M. Garcia, R. Garces, and M. Mancha. 1989. Genetic analysis of the high oleic acid content in cultivated sunflower (*Helianthus annuus* L.). Euphytica 41:39–51. doi:10.1007/BF00022409

Fick, G.N., and D.E. Zimmer. 1974. Monogenic resistance to Verticillium wilt in sunflowers. Crop Sci. 14:895–896. doi:10.2135/cropsci1974.0011183X001400060037x

Gerdes, J.T., C.J. Kahl, A.L. Erickson, and R.M. Benson. 2011. Low saturated fat sunflower and associated methods. U.S. Patent Application 13/024,002.

Gong, L., B.S. Hulke, T.J. Gulya, S.G. Markell, and L.L. Qi. 2013. Molecular tagging of a novel rust resistance gene R_{12} in sunflower (*Helianthus annuus* L.). Theor. Appl. Genet. 126:93–99. doi:10.1007/s00122-012-1962-z

Gonzalez, J., N. Mancuso, P. Ludueña, and A. Ivancovich. 2007. Verticilosis en germoplasma de girasol (*Verticillium* wilt of sunflower germplasm). Helia 30(47):121–126. doi:10.2298/HEL0747121G

Gulya, T.J. 2005. The downy mildew situation on the 2005 North Dakota sunflower crop. National Sunflower Association, Mandan, ND. http://www.sunflowernsa.com/uploads/research/250/Gulya_Mildew_06.pdf (accessed 15 Mar. 2013).

Gulya, T.J. 2007. New strain of *Verticillium dahliae* in North America. Helia 30(47):115–120. doi:10.2298/HEL0747115G

Gulya, T.J., and M. Mancl. 1993. Disease resistance of Native American traditional sunflower varieties. National Sunflower Association, Mandan, ND. http://www.sunflowernsa.com/uploads/research/730/1993_gulya_diseaseresistancenativeamerican.pdf (accessed 15 Mar. 2013).

Gulya, T.J., K.Y. Rashid, and S.M. Masirevic. 1997. Sunflower diseases. In: A.A. Schneiter, editor, Sunflower technology and production. Agron. Monogr. 35. ASA, CSSA, and SSSA, Madison, WI. p. 263–380. doi:10.2134/agronmonogr35.c6

Habura, E.C.H. 1957. Parasterilitat bei sonnenblumen. Z. Pflanzenzuecht. 37:280–298.

Hagy, H.M., G.M. Linz, and W.J. Bleier. 2008. Optimizing the use of decoy plots for blackbird control in commercial sunflower. Crop Prot. 27:1442–1447. doi:10.1016/j.cropro.2008.07.006

Hamilton, R.I. 1926. Improving sunflowers by inbreeding. Sci. Agric. 6:190–192.

Heiser, C.B., Jr. 1951. The sunflower among the North American Indians. Proc. Am. Philos. Soc. 95:432–448.

Heiser, C.B., Jr. 1955. Origin and development of the cultivated sunflower. Am. Biol. Teach. 17:161–167. doi:10.2307/4438706

Heiser, C.B., Jr. 1976. The sunflower. Univ. Oklahoma Press, Norman, OK.

Hulke, B.S., J.F. Miller, and T.J. Gulya. 2010a. Registration of the restorer oilseed sunflower germplasm RHA 464 possessing genes for resistance to downy mildew and sunflower rust. J. Plant Reg. 4:249–254. doi:10.3198/jpr2010.02.0114crg

Hulke, B.S., J.F. Miller, T.J. Gulya, and B.A. Vick. 2010b. Registration of the oilseed sunflower genetic stocks HA 458, HA 459, and HA 460 possessing genes for resistance to downy mildew. J. Plant Reg. 4:93–97. doi:10.3198/jpr2009.08.0426crgs

Joshi, R.K., and S. Nayak. 2010. Gene pyramiding—A broad spectrum technique for developing durable stress resistance in crops. Biotech. Mol. Biol. Rev. 5:51–60.

Judd, J.T., B.A. Clevidence, R.A. Muesing, J. Wittes, M.E. Sunkin, and J.J. Podczasy. 1994. Dietary trans fatty acids: Effects on plasma lipids and lipoproteins of healthy men and women. Am. J. Clin. Nutr. 59:861–868.

Kane, N.C., N. Gill, M.G. King, J.E. Bowers, J. Berges, J. Gouzy, E. Bachlava, N.B. Langlade, Z. Lai, M. Stewart, J.M. Burke, P. Vincourt, S.J. Knapp, and L.H. Rieseberg. 2011. Progress towards a reference genome for sunflower. Botany 89:429–437. doi:10.1139/b11-032

Kinman, M.L. 1970. New developments in the USDA and state experiment station sunflower breeding programs. p. 181–183. In: Proc. 4th Int. Sunflower Conf., Memphis, TN. 23–25 June 1970. Int. Sunflower Assoc., Paris.

Klosterman, M.E., G.M. Linz, A.A. Slowik, and W.J. Bleier. 2011. Assessment of bird damage to sunflower and corn in North Dakota. National Sunflower Association, Mandan, ND. http://www.sunflowernsa.com/uploads/research/575/klosterman_11.pdf (accessed 15 Mar. 2013).

Knodel, J., and L. Charlet. 2007. Pest management. in sunflower production. Ext. Bull. A-1331, rev. North Dakota State Univ. Ext. Serv. and North Dakota Agric. Exp. Stn., Fargo, ND. p. 26–90.

Lacombe, S., F. Kaan, S. Leger, and A. Berville. 2001. An oleate desaturase and a suppressor loci direct high oleic acid content of sunflower (*Helianthus annuus* L.) oil in the Pervenets mutant. Life Sci. 324:839–845.

Lamey, H.A., D.K. McBride, R.K. Zollinger, J.L. Luecke, M.P. McMullen, and D.R. Berglund. 1992. 1990 Sunflower grower survey of pest problems and pesticide use in North Dakota. NDSU Ext. Rep. 9. North Dakota State Univ. Coop. Ext. Serv., Fargo, ND.

Laundon, G.F., and J.M. Waterston. 1965. *Puccinia helianthi*. CMI Descriptions of Pathogenic Fungi and Bacteria 55.

Leclercq, P. 1969. Une sterilite cytoplasmique chez le tournesol. Ann. Amelior. Plant. 19:99–106.

Lilleboe, D. 1997. Weapons search. The Sunflower 23(2):8–9.

Lilleboe, D. 2011. 40 year career in sunflower. The Sunflower 37(5):8–10.

Lilleboe, D. 2012. Seeing double—and liking it. The Sunflower 38(6):8–9.

Linz, G.M., H.J. Homan, S.J. Werner, H.M. Hagy, and W.J. Bleier. 2011. Assessment of bird-management strategies to protect sunflowers. Bioscience 61:960–970. doi:10.1525/bio.2011.61.12.6

Liu, Z., X. Cai, G.J. Seiler, T.J. Gulya, K.Y. Rashid, and C.-C. Jan. 2011. Transferring Sclerotinia stalk rot resistance genes from wild *Helianthus* species into cultivated sunflower. National Sunflower Association, Mandan, ND. http://www.sunflowernsa.com/uploads/research/581/liu_sclerotinia_11.pdf (accessed 15 Mar. 2013).

Maric, A., and S. Masirevic. 1980. Pojava sive pegavosti stabla (*Phomopsis* sp.) do sada neopznale bolesti suncokreta. Zaštita Bilja 12:421–423.

Markell, S., T. Gulya, K. McKay, M. Hutter, C. Hollingsworth, V. Ulstad, R. Koch, and A. Knudsvig. 2009. Widespread occurrence of the aecial stage of sunflower rust caused by *Puccinia helianthis* in North Dakota and Minnesota in 2008. Plant Dis. 93:668. doi:10.1094/PDIS-93-6-0668C

Masirevic, S., and T.J. Gulya. 1992. *Sclerotinia* and *Phomopsis*—Two devastating sunflower pathogens. Field Crops Res. 30:271–300. doi:10.1016/0378-4290(92)90004-S

Mercer, K.L., R.G. Shaw, and D.L. Wyse. 2006a. Increased germination of diverse crop–wild hybrid sunflower seeds. Ecol. Appl. 16:845–854. doi:10.1890/1051-0761(2006)016[0845:IGODCH]2.0.CO;2

Mercer, K.L., D.L. Wyse, and R.G. Shaw. 2006b. Effects of competition on the fitness of wild and crop–wild hybrid sunflower from a diversity of wild populations and crop lines. Evolution 60:2044–2055.

Michaud, J.P., P.W. Stahlman, J.L. Jyoti, and A.K. Grant. 2009. Plant spacing and weed control affect sunflower stalk insects and the girdling behavior of *Dectes texanus* (Coleoptera: Cerambycidae). J. Econ. Entomol. 102:1044–1053. doi:10.1603/029.102.0324

Miller, J.F., and K. Al-Khatib. 2002. Registration of Imidazolinone herbicide-resistant sunflower maintainer (HA 425) and fertility restorer (RHA 426 and RHA 427) germplasms. Crop Sci. 42:988–989. doi:10.2135/cropsci2002.988a

Miller, J.F., and K. Al-Khatib. 2004. Registration of two oilseed sunflower genetic stocks, SURES-1 and SURES-2 resistant to Tribenuron herbicide. Crop Sci. 44:1037–1038. doi:10.2135/cropsci2004.1037

Miller, J.F., and T.J. Gulya. 1988. Registration of six downy mildew resistant sunflower germplasm lines. Crop Sci. 28:1040–1041. doi:10.2135/cropsci1988.0011183X002800060073x

Miller, J.F., and T.J. Gulya. 1999. Registration of eight Sclerotinia-tolerant sunflower germplasm lines. Crop Sci. 39:301–302. doi:10.2135/cropsci1999.0011183X003900010075x

Miller, J.F., and T.J. Gulya. 2006. Registration of two restorer (RHA 439 and RHA 440) and one maintainer (HA 441) Sclerotinia tolerant oilseed sunflower germplasms. Crop Sci. 46:482–483. doi:10.2135/cropsci2005.04-0009

Miller, J.F., T.J. Gulya, and B.A. Vick. 2006a. Registration of three maintainer (HA 444 to HA 446) and one restorer (RHA 447) high oleic oilseed sunflower germplasms. Crop Sci. 46:484–485. doi:10.2135/cropsci2005.04-0011

Miller, J.F., T.J. Gulya, and B.A. Vick. 2006b. Registration of two maintainer (HA 451 and HA 452) and three restorer (RHA 453-RHA 455) Sclerotinia tolerant oilseed sunflower germplasms. Crop Sci. 46:2727–2728. doi:10.2135/cropsci2006.06.0436

Miller, J.F., T.J. Gulya, and B.A. Vick. 2006c. Registration of three maintainer (HA 456, HA 457, and HA 412 HO) high-oleic oilseed sunflower germplasms. Crop Sci. 46:2728. doi:10.2135/cropsci2006.06.0437

Miller, J.F., and B.A. Vick. 2002. Registration of four mid-range oleic acid sunflower genetic stocks. Crop Sci. 42:994. doi:10.2135/cropsci2002.0994

Miller, J.F., and S.L. Wolf. 1991. Registration of three cytoplasmic male-sterile and three restorer sunflower germplasm lines. Crop Sci. 31:500. doi:10.2135/cropsci1991.0011183X003100020087x

National Sunflower Association (NSA). 2013a. USA sunflower crop survey. National Sunflower Association, Mandan, ND. http://www.sunflowernsa.com/growers/yield-and-survey/sunflower-crop-survey/ (accessed 6 Mar. 2013).

National Sunflower Association (NSA). 2013b. U.S. sunflower crop quality report. National Sunflower Association, Mandan, ND. http://www.sunflowernsa.com/stats/us-crop-quality/ (accessed 15 Mar. 2013).

NOAA National Climatic Data Center. 2013. North Dakota precipitation, December to November 1896–2012. NOAA. http://www.ncdc.noaa.gov/temp-and-precip/time-series/index.php?parameter=pcp&month=11&year=2012&filter=12&state=32 (accessed 15 Mar. 2013).

Oseto, C.Y., and G.A. Braness. 1979. Bionomics of insects on cultivated sunflower, *Helianthus annuus*. Ann. Entomol. Soc. Am. 72:524–528.

Payen, D. 1983. Modelisationde l'epidemiologie de *Sclerotinia sclerotiorum* sur tournesol. Bull. Org. Europ. Plant Prot. 13:277–281.

Pegadaraju, V., R. Nipper, B. Hulke, L. Qi, and Q. Schultz. 2013. De novo sequencing of sunflower genome for SNP discovery using RAD (restriction site associated DNA) approach. BMC Genomics 14(1):556.

Putt, E.D. 1964. Breeding behavior of resistance to leaf mottle or Verticillium in sunflowers. Crop Sci. 4:177–179. doi:10.2135/cropsci1964.0011183X000400020016x

Putt, E.D. 1997. Early history of sunflower. In: A.A. Schneiter, editor, Sunflower technology and production. Agron. Monogr. 35. ASA, CSSA, and SSSA, Madison, WI. p. 1–20. doi:10.2134/agronmonogr35.c1

Randolph, N.M., G.L. Teetes, and M.C. Baxler. 1972. Life cycle of the sunflower moth under laboratory and field conditions. Ann. Entomol. Soc. Am. 65:1161–1164.

Ratkos, J., and J. Nagy. 1992. A study on the sporogenesis and pathogenesis of *Sclerotinia sclerotiorum*. Helia 15–16:59–72.

Rogers, C.E., and O.R. Jones. 1979. Effects of planting date and soil water on infestation of sunflower by larvae of *Cylindrocopturus adspersus*. J. Econ. Entomol. 72:529–531.

Romano, A.B., A. Vazquez, S. Piubello, and C. Sala. 1994. Quantifying the relationship between *Verticillium* wilt intensity and yield losses of sunflower and feasibility of utilizing the hypodermic inoculation technique as a selection method. Helia 17(20):49–54.

Sala, C.A., M. Bulos, E. Altieri, and B. Weston. 2012. Response to imazapyr and dominance relationships of two imidazolinone-tolerant alleles at the *Ahasl1* locus of sunflower. Theor. Appl. Genet. 124:385–396. doi:10.1007/s00122-011-1713-6

Satterthwait, A.F., and R.B. Swain. 1946. The sunflower moth and some of its natural enemies. J. Econ. Entomol. 30:575–580.

Schneiter, A.A., and J.F. Miller. 1981. Description of sunflower growth stages. Crop Sci. 21:901–903. doi:10.2135/cropsci1981.0011183X002100060024x

Schuppert, G.F., S. Tang, M.B. Slabaugh, and S.J. Knapp. 2006. The sunflower high-oleic mutant *Ol* carries variable tandem repeats of *FAD2-1*, a seed-specific oleoyl-phophatidyl choline desaturase. Mol. Breed. 17:241–256. doi:10.1007/s11032-005-5680-y

Semelczi-Kovacs, A. 1975. Acclimatization and dissemination of the sunflower in Europe. (In German.) Acta Ethnogr. Acad. Sci. Hung. 24:47–88.

Sendall, B.C., G.A. Kong, K.C. Goulter, E.A.B. Aitken, S.M. Thompson, J.H.M. Mitchell, J.K. Kochman, W. Lawson, T. Shatte, and T.J. Gulya. 2006. Diversity in the sunflower: *Puccinia helianthi* pathosystem in Australia. Australas. Plant Pathol. 35:657–670. doi:10.1071/AP06071

Skoric, D. 1985. Sunflower breeding for resistance to *Diaporthe/Phomopsis helianthi*. Helia 8:21–24.

Smith, J.A., and M.F. Kocher. 2008. Evaluate planter meter and seed tube systems for seed spacing performance of confection sunflower seed to improve plant spacing in the field. National Sunflower Association, Mandan, ND. http://www.sunflowernsa.com/uploads/research/330/Smith_PlanterMeter_08.pdf (accessed 15 Mar. 2013).

Soldatov, K.I. 1976. Chemical mutagenesis in sunflower breeding. p. 352–357. In: Proc. of the 7th Int. Sunflower Conf., 27 June–3 July 1976, Krasnodar, USSR. Vol. 1. Int. Sunflower Assoc., Paris.

Stone, L.R., A.J. Schlegel, R.E. Gwin, Jr., and A.H. Khan. 1996. Response of corn, grain sorghum, and sunflower to irrigation in the High Plains of Kansas. Agric. Water Manage. 30:251–259. doi:10.1016/0378-3774(95)01226-5

Thompson, S.M., Y.P. Tan, A.J. Young, S.M. Neate, E.A.B. Aitken, and R.G. Shivas. 2011. Stem Cankers on sunflower (*Helianthus annuus*) in Australia reveal a complex of pathogenic *Diaporthe (Phomopsis)* species. Persoonia 27:80–89. doi:10.3767/003158511X617110

Tourvieille de Labrouhe, D., A. Bordat, J. Tourvieille, E. Mestries, P. Walser, N. Sakr, M. Ducher, and F. Vear. 2010. Impact of major gene resistance management for sunflower on fitness of Plasmopara halstedii (downy mildew) populations. OCL-Oleagineux Corps gras Lipides 17:56–64.

Tourvieille de Labrouhe, D., S. Lafon, P. Walser, and I. Raulic. 2000. Une nouvelle race de plasmopara halstedii, agent du mildiou du tournesol. OCL-Oleagineux Corps gras Lipides 7:404–405.

Ureta, M.S., A.D. Carrera, M.A. Cantamutto, and M.M. Poverene. 2008. Gene flower among wild and cultivated sunflower, *Helianthus annuus* in Argentina. Agric. Ecosyst. Environ. 123:343–349. doi:10.1016/j.agee.2007.07.006

Velasco, L., B. Perez-Vich, C.C. Jan, and J.M. Fernandez-Martinez. 2007. Inheritance of resistance to broomrape (*Orobanche Cumana* Wallr.) race F in a sunflower line derived from wild sunflower species. Plant Breed. 126:67–71. doi:10.1111/j.1439-0523.2006.01278.x

Viranyi, F., and O. Spring. 2011. Advances in sunflower downy mildew research. Eur. J. Plant Pathol. 129:207–220. doi:10.1007/s10658-010-9683-0

Vranceanu, A.V., N. Csep, N. Pirvu, and F.M. Stoenescu. 1983. Genetic variability of sunflower reaction to the attack of *Phomopsis helianthi*. Helia 6:23–25.

Vranceanu, A.V., and G. Soare. 1988. International trials with sunflower hybrids (the sixth cycle 1986–1987). Helia 11:5–9.

Watkins, C. 2006. Operation Oilseed. The Sunflower. 32(2):24–25.

Yue, B., S.A. Radi, B.A. Vick, X. Cai, S. Tang, S.J. Knapp, T.J. Gulya, J.F. Miller, and J. Hu. 2008. Identifying quantitative trait loci for resistance to Sclerotinia head rot in two USDA sunflower germplasms. Phytopathology 98:926–931. doi:10.1094/PHYTO-98-8-0926

16

Wheat

Robert Graybosch,* Harold E. Bockelman,
Kimberly A. Garland-Campbell, David F. Garvin, and Teshome Regassa

ABSTRACT

Grain yield of wheat (*Triticum* spp.), a vital component of world food supplies, rose steadily in the 20th century, due both to improved agronomic practices and plant breeding. Increases in wheat grain yield (kilograms harvested per hectare) over time derive from improved agronomic practices and from enhanced genetic potential of wheat cultivars. Estimation of the rate of genetic improvement through time is important because it allows prediction of whether future grain supplies will meet future demands of world population. In this study, wheat grain yields, relative to check cultivars in USDA-ARS coordinated regional performance trials, were calculated for the time period 1980 through 2010, and regression analysis was used to evaluate changes in relative grain yield potential versus year of the trial. The results were compared to those of previous investigations in which an identical approach was used for the time period 1958 through 1980. Averaged across all trials, the genetic potential for grain yield increased from 1980 to 2010 at a rate of approximately 0.8% per year, lower than the value of 1.2% reported for the time period 1958 through 1980. The observed rates of increase were statistically different ($P = 0.05$) from zero for all USDA-ARS performance nurseries, with the exception of the Southern Regional Performance Nursery and the Pacific Northwest Soft Winter Wheat Nursery. These two trials represent the lowest and highest yielding environments, suggesting that genetic progress has been difficult when moisture is often limiting or when wheat yields are approaching biological limits. Future prospects for enhanced genetic potential for grain yield in wheat might very well depend on the application of the tools of genetic engineering, if the exploitable genetic variation for grain yield already has been utilized.

Abbreviations: DH, doubled haploid; HRS, Great Plains Hard Red Spring Wheat Nursery; MAS, marker assisted selection; MP, most productive entry; NRPN, Great Plains Northern Regional Performance Nursery; PNW HW, Pacific Northwest Hard Winter Regional Trial; PNW SW, Pacific Northwest Soft Winter Regional Trial; QTL, quantitative trait loci; SRPN, Great Plains Southern Regional Performance Nursery; UESRWWN, Eastern Uniform Soft Winter Wheat Nursery; UNL, University of Nebraska; USSRWWN, Southern Uniform Soft Winter Wheat Nursery.

R. Graybosch, USDA-ARS, 137 Keim Hall, East Campus, UNL, Lincoln, NE 68583-0937. *Corresponding author (Bob.Graybosch@ars.usda.gov).
H.E. Bockelman, Small Grains and Potato Germplasm Research, 1691 S 2700 W, Aberdeen, ID, 83210 (Harold.Bockelman@ars.usda.gov).
K.A. Garland-Campbell, Wheat Genetics, Quality Physiology and Disease Research, Johnson Hall, 291D, Washington State Univ., Pullman, WA 99164-6420 (kim.garland-campbell@ars.usda.gov).
D.F. Garvin, USDA-ARS, 411 Borlaug Hall, 1991 Upper Buford Circle, Saint Paul, MN 55108-6026 (David.Garvin@ars.usda.gov).
T. Regassa, Dep. of Agronomy & Horticulture, 175 Keim Hall, Univ. of Nebraska, East Campus, Lincoln, NE 68583 (Tregassa2@unl.edu).

doi:10.2135/cssaspecpub33.c16

Yield Gains in Major U.S. Field Crops. CSSA Special Publication 33.
Stephen Smith, Brian Diers, James Specht, and Brett Carver, editors.

Common wheat (*Triticum aestivum* L.) and durum wheat (*T. turgidum* L. var. *durum*) remain among the world's staple food products. Global wheat production in 2010 was approximately 670 million tonnes (24.6 billion bushels) (USDA-FAS, 2010), and U.S. wheat production is responsible for approximately 10% of the annual global wheat crop. In 1980, only six countries—the United States, Canada, Australia, New Zealand, Argentina, and South Africa—produced more wheat than they consumed, and U.S. exports constituted 50% of the world wheat trade (Johnson, 1986a). The United States remains one of the world's largest wheat exporters, still responsible for more than 20% of exported wheat (USDA-ERS, 2013). At present, the United States, Canada, Australia, the European Union, the former Soviet Union (primarily Russia, Ukraine, and Kazakhstan), and Argentina produce approximately 90% of world wheat exports. Thus, current production by U.S. wheat farmers remains a significant source of the world's food supply, although the contribution is less than one-half that observed in 1980.

Total U.S. harvested wheat area has steadily declined since 1980 (Fig. 16–1; USDA-NASS, http://www.nass.usda.gov/Data_and_Statistics/index.asp , accessed 12 Mar. 2013). Harvested area declined from approximately 30 million hectares (8.15×10^7 acres) in 1980 to 19 million hectares (4.69×10^7 acres) in 2010. Total wheat production during the same period (Fig. 16–2) also declined, but not as rapidly as harvested area. Average U.S. wheat production during the time period 1980 to 1985 was approximately 70 million tonnes; during the time period 2005 to 2010 average production was approximately 60 million tonnes. The loss of harvested wheat hectares from 1980 to 2010 occurred in all regions of the United States (Table 16–1). Only five states—Delaware, Louisiana, Maryland, Montana, and Wisconsin—harvested more hectares in 2010 than in 1980. The most dramatic

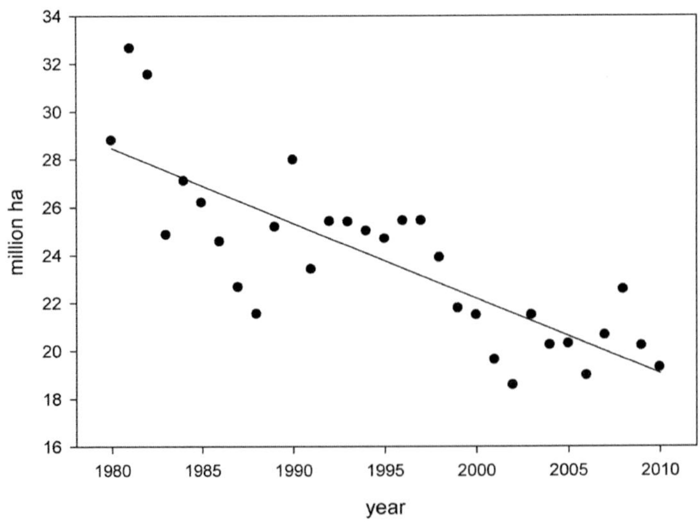

$$y = -0.31(\text{year}) + 650.1,\ r^2 = 0.64,\ p < 0.0001$$

Fig. 16–1. U.S. wheat production, harvested hectares, 1980–2010. Source of data: USDA-NASS.

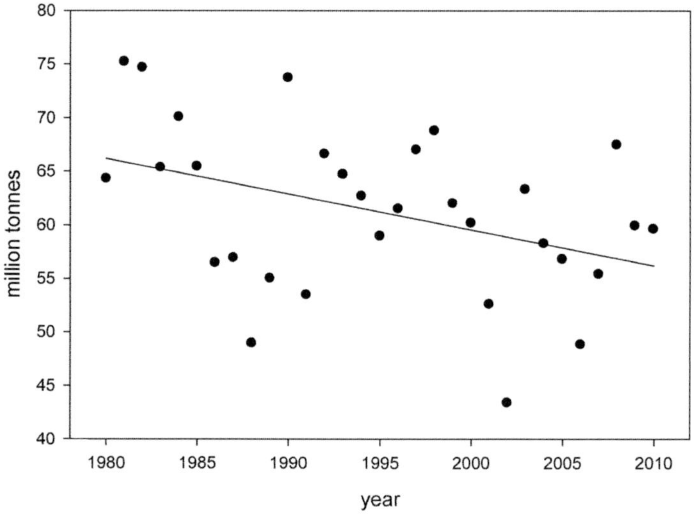

$$y = -0.33(\text{year}) + 727.0,\ r^2 = 0.16,\ p = 0.03$$

Fig. 16–2. U.S. wheat production, million tonnes, 1980–2010. Source of data: USDA-NASS.

loss of harvested area occurred in the northeastern (exclusive of Delaware), southeastern, and Corn Belt states (exclusive of Wisconsin, where acreage more than doubled), with some states, notably Arkansas, Georgia, Iowa, Illinois, Indiana, and Missouri, losing approximately 80% of their harvested area. The loss of wheat production area primarily has been due to large increases in areas seeded to maize (*Zea mays* L.) and soybean [*Glycine max* (L.) Merr.] in the Ohio Valley, southeastern United States, and Great Plains states. Similar trends were observed in production (tonnes), but due to both improved agronomic practices and genetic improvement in grain yield potential, the declines were not as pronounced.

Wheat producers in the United States have consistently realized increased productivity (mean kg ha^{-1}) during the past 30 yr (Fig. 16–3). Average on-farm wheat grain yields, tabulated across all locations and market classes, have increased at a statistically significant rate of approximately 35 kg ha^{-1} (0.52 bu ac^{-1}) since 1980. According to Johnson (1986a), average U.S. wheat yields increased from approximately 1000 kg ha^{-1} (15 bu ac^{-1}) in 1945 to 2000 kg ha^{-1} in 1979. From 1980 to 2010, average U.S. wheat yields again nearly doubled to approximately 3800 kg ha^{-1} (57 bu ac^{-1}) (Fig. 16–3). Linear regression (Table 16–1) revealed that most states demonstrated statistically significant increases in productivity (kg ha^{-1}), although it should be noted that statistically significant ($P = 0.05$) increases were not observed in the key wheat producing states Colorado, Oklahoma, Kansas, Montana, Texas, and Oregon. Wheat production in these states is mostly concentrated in arid regions where plant-available water is limiting. Wheat yields in the Great Plains states also are limited by disease outbreaks and late winter freezes, both of which can cancel or obscure any recent gains from breeding efforts. Lin and Huybers (2012) evaluated worldwide wheat production zones and found yields have stagnated in 27 of 50 zones evaluated, including both the central and western United

Table 16–1. Total U.S. wheat production and productivity, 1980–2010 (source USDA-NASS), and linear changes estimated from regression analysis.

Region/ state	Harvested area			Production		Production change (1980–2010)			Productivity		Productivity change (1980–2010)		
	1980	2010	Change	1980	2010	Slope	R^2	P	1980	2010	Slope	R^2	p
	ha	ha	%	t	t	t yr^{-1}			kg ha^{-1}	kg ha^{-1}	kg ha^{-1} yr^{-1}		
Northeast													
DE	10,927	18,211	66.67	29,393	71,034	1,912	0.33	<0.01	2690	3901	64	0.60	<0.01
NJ	17,402	9,308	−46.51	50,322	30,672	−237	0.05	0.25	2892	3295	30	0.37	<0.01
NY	60,703	40,469	−33.33	163,296	182,347	−159	0.01	0.81	2690	4506	39	0.54	<0.01
PA	101,172	60,703	−40.00	251,748	240,862	482	0.02	0.46	2488	3968	46	0.67	<0.01
Southeast													
AL	105,218	46,539	−55.77	180,442	172,141	−9,060	0.26	<0.01	1749	3699	58	0.56	<0.01
AR	350,053	60,703	−82.66	894,590	220,450	−24,125	0.22	<0.01	2556	3632	37	0.38	<0.01
GA	242,812	49,372	−79.67	538,877	132,814	−22,836	0.56	<0.01	2219	2690	37	0.42	<0.01
KY	141,640	101,172	−28.57	376,261	449,064	4,267	0.07	0.16	2690	4439	68	0.59	<0.01
LA	27,114	44,515	64.18	51,057	149,688	−191	0.01	0.94	1883	3363	52	0.55	<0.01
MD	39,255	54,633	39.18	100,318	220,450	4,547	0.28	<0.01	2556	4035	56	0.57	<0.01
MS	121,406	40,469	−66.67	253,109	127,915	−9,249	0.15	0.03	2085	3161	55	0.56	<0.01
NC	161,874	153,781	−5.00	381,024	382,657	5,163	0.07	0.14	2354	2488	32?	0.29	<0.01
SC	77,700	52,609	−32.29	188,117	127,371	−6,798	0.33	<0.01	2421?	2421?	27?	0.22	<0.01
TN	198,296	72,843	−63.27	506,762	259,641	−5,990	0.09	0.11	2556	3564	48	0.49	<0.01
VA	115,740	62,726	−45.80	288,000	215,142	−711	0.01	0.68	2488	3430	54	0.57	<0.01
WV	4,047	2,023	−50.00	10,342	7,348	−66	0.05	0.23	2556	3632	45	0.61	<0.01
Corn Belt													
IA	37,231	4,047	−89.13	95,147	12,519	−2,520	0.47	<0.01	2556	3094	42	0.40	<0.01
IL	635,357	119,382	−81.21	2,093,727	449,608	−27,224	0.23	<0.01	3295	3766	33	0.31	<0.01
IN	445,155	93,078	−79.09	1,466,942	375,581	−23,355	0.51	<0.01	3295	4035	49	0.49	<0.01
MI	323,749	206,390	−36.25	958,003	971,611	6,611	0.08	0.12	2959	4708	61	0.58	<0.01

Region/ state	Harvested area			Production		Production change (1980–2010)			Productivity		Productivity change (1980–2010)		
	1980	2010	Change	1980	2010	Slope	R^2	P	1980	2010	Slope	R^2	p
	ha	ha	%	t	t	t yr^{-1}			kg ha^{-1}	kg ha^{-1}	kg ha^{-1} yr^{-1}		
MN	1,282,450	651,544	−49.20	2,791,164	2,396,913	−31,462	0.18	0.02	2152	3699	32	0.23	<0.01
MO	837,700	113,312	−86.47	2,422,496	342,922	−42,201	0.42	<0.01	2892	3026	30	0.33	<0.01
OH	554,420	303,515	−45.26	1,827,010	1,245,132	4,082	0.02	0.49	3295	4102	54	0.57	<0.01
WI	44,920	93,078	107.21	118,798	400,620	11,509	0.58	<0.01	2623	4304	54	0.55	<0.01
Great Plains													
CO	1,375,932	961,939	−30.09	3,001,925	2,945,697	−31,985	0.21	0.01	2152	3094	1?	0.01	0.9
KS	4,856,232	3,237,488	−33.33	11,430,720	9,797,760	−46,200	0.04	0.27	2354	3026	16	0.12	0.05
MT	2,063,899	2,108,414	2.16	3,260,477	5,861,238	39,754	0.12	0.05	1614	2757	14?	0.11	0.08
ND	3,893,079	3,399,362	−12.68	4,889,354	9,839,945	47,823	0.05	0.22	1278	2892	25?	0.27	<0.01
NE	1,153,355	6,02,982	−47.72	2,947,493	1,743,729	−25,177	0.32	<0.01	2556	2892	14	0.13	0.04
NM	202,343	117,359	−42.00	285,768	220,994	−6,305	0.31	<0.01	1412	1883	14	0.12	0.05
OK	2,630,459	1,578,275	−40.00	5,307,120	3,290,414	−63,110	0.33	<0.01	2018	2085	<1	<0.01	0.99
SD	1,313,206	1,102,769	−16.02	1,698,959	3,360,496	34,844	0.17	0.02	1278	3026	47	0.58	<0.01
TX	2,104,367	1,517,573	−27.88	3,538,080	3,470,040	−53,090	0.24	<0.01	1681	2287	8?	0.06	0.16
WY	127,476	58,679	−53.97	234,602	126,282	−4,187	0.56	<0.01	1816	2152	3	<0.01	0.66
Mountain West													
AZ	87,007	34,398	−60.47	468,115	259,505	−2,543	0.06	0.18	5380	7532	37	0.57	<0.01
CA	465,389	184,132	−60.43	2,326,968	1,068,228	−37,872	0.52	<0.01	4977	5784	14	0.09	0.09
ID	627,263	544,303	−13.23	2,613,552	2,923,271	10,112	0.08	0.12	4170	5380	42	0.53	<0.01
NV	11,736	4,856	−58.62	48,989	34,564	−685	0.24	0.01	4170	7129	87	0.77	<0.01
OR	546,326	383,238	−29.85	2,106,518	1,730,557	−17,536	0.29	<0.01	3833	4506	−11	0.04	0.31
UT	110,075	53,014	−51.84	243,365	173,611	−2,631	0.36	<0.01	2219	3295	27	0.4	<0.01
WA	1,278,808	924,708	−27.69	4,360,548	4,024,974	−9,494	0.02	0.43	3430	4371	17	0.13	0.04
Total	28,783,291	19,270,742	−33.1	64,280,076	59,581,965	−333,754	0.16	0.02	2641	3660	35	0.82	<0.01

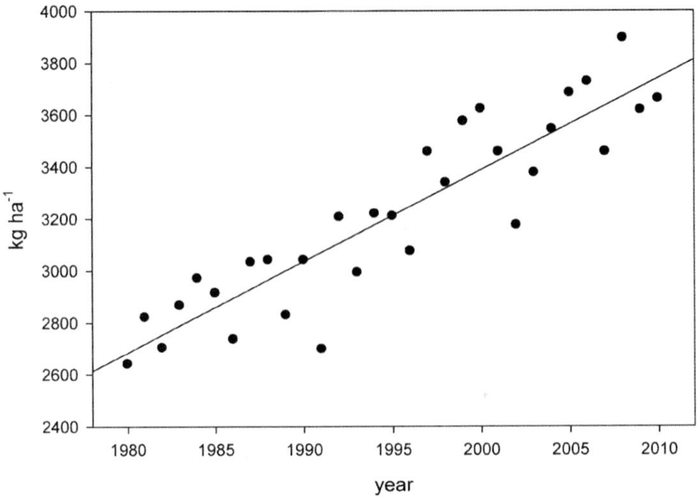

$$y = 35.3(\text{year}) - 67014,\ r^2 = 0.82,\ p < 0.0001$$

Fig. 16–3. Mean grain yield of U.S. wheat, all classes, 1980–2010. Source of data: USDA-NASS.

States. The states noted above as lacking significant productivity increases since 1980 fall within these two listed U.S. zones. Lin and Huybers (2012) further noted diverse causes might explain the leveling of wheat grain yields in different regions. Sociological practices, environmental or economic considerations limiting nitrogen use, or biological constraints on current levels of wheat production in specific zones contribute to the observed lack of progress in on-farm productivity.

Genetic Contributions to Productivity Increases

Changes in grain yield arise from both improved production practices, including enhanced fertility (primarily due to supplementation with N, but also P, K, and micronutrients), weed, pathogen, and pest control, crop rotation, tillage and seeding techniques, and improved genetic potential. In fact, these two main factors, production practices and genetics, are intimately linked to create yield increases. When improved production practices are used, they allow increased expression of genetic potential, and likewise, improved genetics allows the use of advanced production practices. Three approaches have been used to estimate the contribution of improved genetics to increased wheat grain yield (de la Vega et al., 2007; Graybosch and Peterson, 2010). de la Vega et al. (2007) used best linear unbiased predictors (BLUPS) to remove environmental effects from multilocation and multiyield trials and estimate "relative peak performance" of sunflower (*Helianthus annuus* L.) hybrids. Plotted against year of release, the relative peak performances were used to estimate genetic gain in sunflower grain and oil yields over time. A second procedure involves yield trials of entries released at various eras compared to a historic check variety, as in Cox et al. (1988). Finally, one can use yearly data from long-term nurseries or yield trials in which check cultivars remain constant to calculate relative performance values (Schmidt, 1984; Graybosch and Peterson, 2010). Yields relative to the check cultivars can be calculated, and regression analysis used to evaluate changes in relative grain yield potential versus year of the trial. This approach was used by Schmidt (1984) to evaluate the contribution of U.S. wheat breeding programs to increased genetic potential for grain yield for the time period 1958 to 1980. Schmidt evaluated data from USDA-ARS coordinated regional performance nurseries. The present chapter continues the analysis of Schmidt (1984) using data from some of the same long-term nurseries as a means to estimate genetic improvement in wheat yield potential for the time period 1980 through 2010. This approach may have an inherent limitation—long-term trials may not fully distinguish yield improvement due to genetic gain from that due to production practices, especially if check cultivars and improved cultivars differ in their responses to the latter. A good example would be the use of a tall variety for a long-term check that lodges under improved fertility, while the semidwarf comparison cultivars can fully express yield without lodging. Thus, the approach could underestimate genetic gains. However, coupled with the earlier analysis of USDA regional trials by Schmidt (1984), the method offers an assessment, albeit perhaps an imperfect one, of the results of 50 yr of wheat breeding.

The USDA-ARS coordinated regional performance nurseries are conducted in four major U.S. wheat production zones, namely Eastern/Southern Soft Wheat, Great Plains Spring Wheat, Great Plains Winter Wheat, and Pacific Northwest Winter Wheat. Public and private breeding programs enter their highest performing experimental breeding lines. Long-term check cultivars are included in all the nurseries, although the check cultivars have changed with time in some of the trials. Test locations within each nursery may change through time, but all remain within the original broad agroecological zones. Thus, absolute grain yields, while of interest, are not as informative as relative (to long-term check cultivars) grain yields. Absolute grain yields are herein presented primarily to inform the reader of the typical yield levels encountered in the respective nurseries, with relative grain yields being the focus of this chapter. The use of relative grain yields reduces the contribution of fluctuating environmental effects and provides a more accurate assessment of the contribution of improved genetics.

Another possible limitation of this approach is that it requires the use of average performance of entries across rather broad geographic areas. This might dilute results should they be calculated only from data within each cultivar's specific area of origin. To partially alleviate this possible problem, relative grain yields were calculated both for all entries, and for the most productive entry within each trial. Also, trends in wheat improvement were evaluated within a narrower geographic region. Data from the University of Nebraska (UNL) Wheat Variety trials (http://cropwatch.unl.edu/web/varietytest/, accessed 12 Mar. 2014) were analyzed along with results from the USDA nurseries. The UNL trials were conducted across the state of Nebraska and include test sites exposed to the typical continental climate of the Great Plains of North America, but ranging in average annual precipitation from 660 mm in the easternmost sites, to less than 300 mm in the western portion of the state. In addition, the trials have maintained the cultivar Scout 66 (CI 13996) as a long-term check for more than 45 yr. The UNL trials also have a higher percentage of entries of lines that eventually are released as cultivars than the regional nurseries. Results from similar studies available in the literature are summarized in a later section.

Procedures used were those of Graybosch and Peterson (2010). For each nursery, both average grain yield and average grain yield relative to a check were calculated. Relative grain yields were calculated using the long-term check cultivars specific to the trial in question. For the Great Plains Northern and Southern Regional Performance Nurseries (NRPN and SRPN) the long-term check used was the Crimean land-race 'Kharkof' (CItr 1442). For the Great Plains Spring Wheat Regional Performance Nursery Great Plains Hard Red Spring Wheat Nursery (HRS) 'Chris' (CItr 13751) was used as the long-term control. Kharkof and 'Stephens' (CItr 17596) served as respective checks for the Pacific Northwest Hard Winter and Soft Winter Regional Trials (PNW HW and PNW SW). Stephens was released in 1977, and, at the time of its release, its typical grain yield in the PNW trials was approximately 165% of Kharkof. Kharkof was dropped from the PNW SW in the early 1980s. In the Eastern and Southern Uniform Soft Winter Wheat Nurseries (UESRWWN and USSRWWN), long-term checks changed over time. Checks, however, overlapped by two or more years. Sliding scales were developed in which relative grain yields of breeding lines in given years were adjusted using relative grain yields of overlapping check cultivars. Check cultivars used for the UESRWWN were 'Knox' (CItr 12798), 'Caldwell' (CItr 17897), 'Roane' (PI 612958), and 'INW0411.' For the USSRWWN, checks used were 'FL301' (CItr 17769), 'FL302' (PI 601163), and 'AGS2000'. The UNL trials were analyzed using Scout 66 as a long-term check. Data from the UNL trials were evaluated both statewide and independently for four intrastate ecogeographic zones, namely, southeastern, south-central, southwest, and western rainfed. Within each zone, relative yields were calculated independently using Scout 66 as the long-term control.

Data collection for all nurseries terminated after 2010 when the HRS Nursery was temporarily suspended. Data were converted to relative yield values by expressing each entry mean as a percentage of the yield of the check cultivar(s) from the same nursery and year. For each nursery and each year, mean relative yields were determined from the mean yields of all experimental entries. Regression analysis was used to evaluate relationships between both absolute and relative yield values and calendar year of nursery. Again, the focus of this chapter is on the relative grain yields, with trends in absolute yields presented for information only. Separate regression analyses were performed, for each nursery, using: (i) all experimental lines and (ii) the experimental line with the highest yield (i.e., most productive entry, MP). SAS Institute (2003) Version 9.1 programs and procedures were used for all statistical analyses.

Three-year mean grain yields and relative grain yields for the earliest (1980–1982) and most recent (2008–2010) years evaluated are presented in Table 16–2. Relative grain yields for all evaluated trials in the more recent 3-yr periods exceeded those of the earliest years. Absolute and relative grain yields over time are presented for each trial in Fig. 16–4 through 16–10, and regression equations are presented in Table 16–3. Relative grain yields, as a percentage of the check, increased from 1980 to 2010 (as indicated by positive slopes in regression equations) in all USDA-ARS performance trials. Relative rates of increase per year ranged from 0.21% (PNW SW) to 1.46% (USSRWWN) when all nursery entries were considered. When only the MP entry in each trial was evaluated, rates of increase ranged from 0.05% yr^{-1} in the PNW SW to 1.57% yr^{-1} in the SSRWWN.

Table 16–2. Mean grain yields, and relative (to respective check cultivars) grain yields of entries in USDA-ARS coordinated regional wheat performance nurseries and University of Nebraska wheat variety test, first and last 3 yr of analyses.

Nursery/trial†	Entries‡	Mean yield§		Relative yield§	
		1980–1982	2008–2010	1980–1982	2008–2010
		kg ha^{-1}			
UESRWWN	All	3740	4940	97	123
	MP	4400	5520	114	138
USSRWWN	All	3420	4240	85	121
	MP	3950	4900	98	140
SRPN	All	3440	3820	133	162
	MP	3820	4210	148	179
NRPN	All	2890	4120	111	149
	MP	3280	4530	126	164
HRS	All	3210	4290	128	138
	MP	3440	4780	137	155
PNW SW	All	4820	6970	96	101
	MP	5390	7750	107	113
PNW HW	All	3630	5960	122	144
	MP	3940	6680	139	161
Nebraska					
Statewide	All	3230	3720	109	127
	MP	3730	4350	130	154
Southeast District	All	3040	3620	109	130
	MP	3720	4390	138	158.0
South-central District	All	3560	3370	111	145
	MP	4260	4200	139	188
West-central District	All	3260	4250	113	126
	MP	3720	4780	129	141
Western District	All	3000	3540	106	114
	MP	3230	4010	1147	130

† HRS, Great Plains Hard Red Spring Wheat Nursery; Nebraska, University of Nebraska Winter Wheat Variety Trial; NRPN, Northern Regional Performance Nursery; PNW HW, Pacific Northwest Hard Winter Wheat Nursery; PNW SW, Pacific Northwest Soft Winter Wheat Nursery; SRPN, Southern Regional Performance Nursery; UESRWWN, Uniform Eastern Soft Red Winter Wheat Nursery; USSRWWN, Uniform Southern Soft Red Winter Wheat Nursery. Locations have changed over time within some nurseries. PNW SW and HW: earliest years available = 1984, 1986, 1987.

‡ Entries: All = all trial entries, exclusive of control cultivars; MP = most productive entry, or entry with the highest average trial grain yield.

§ Mean yield should be viewed as an approximate estimate of changes in production within each trial over time. Relative yield is the rate of change in grain yield, as percentage per year of long-term control cultivars. See text for additional explanation.

The observed rates of increase were statistically different (P = 0.05) from zero for all USDA-ARS performance nurseries, with the exception of the SRPN (when either all entries or MP only were considered) and the PNW SW (MP entries only). The fluctuating nature of the climate in the SRPN region may have influenced results obtained from regression analysis. The relative grain yield of SRPN entries, especially in the time period 1995 through 2010, vacillated widely from

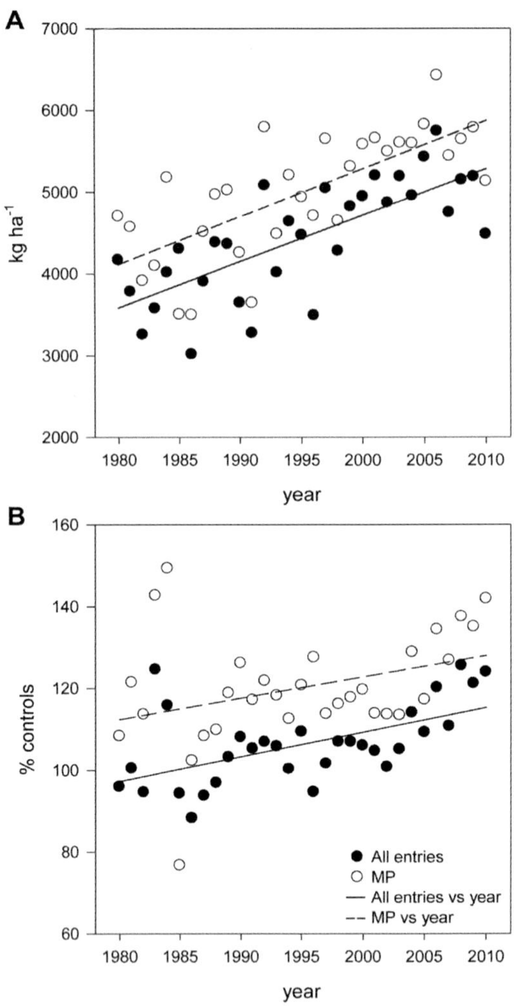

Fig. 16–4. (A) Average and (b) relative (to check cultivars) grain yields of all entries and the most productive entry (MP), Uniform Eastern Soft Red Winter Wheat Nursery, 1980–2010.

year to year. Thus, the lack of statistical significance of the slope from regression could be attributed to either no gain from genetics, or an inability to detect such gain, due to the limiting or confounding effects of environmental conditions in some years.

Results from the UNL trials were similar to those observed for the USDA-ARS nurseries (Fig. 16–11 and 16–12; Table 16–3). The statewide average rate of increase relative to Scout 66 was 0.51% yr^{-1} when all entries were considered and 0.73% yr^{-1} when only the MP entry was included; in both cases the slopes significantly differed from zero. Within the state, however, results differed in the four agroecological zones. In the more humid southeast and south-central districts, statistically significant increases in relative grain yields were documented. In the drier west-central and western districts, no significant increases in relative grain yields were observed. Fufa et al. (2005) used yield trials containing historic and modern wheat cultivars to evaluate the rate of genetic improvement in grain yield

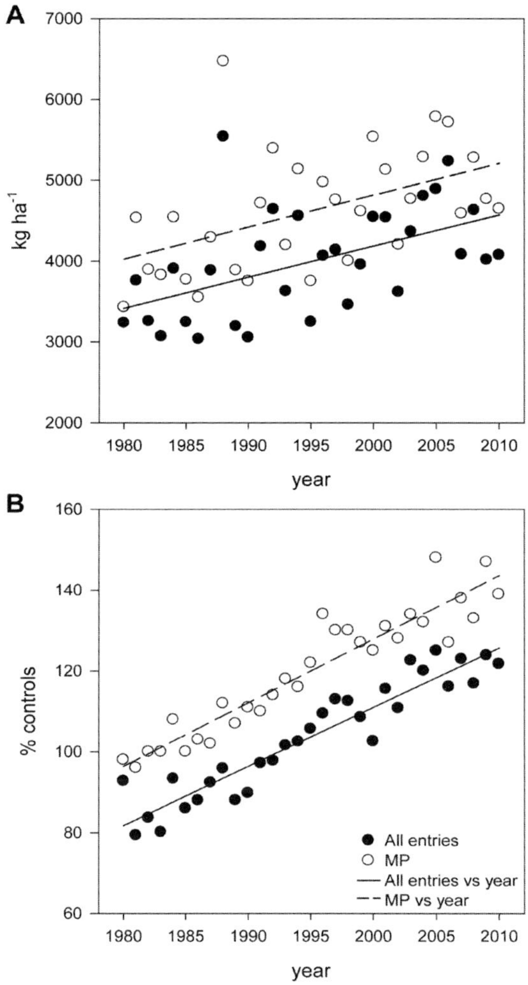

Fig. 16–5. (A) Average and (b) relative grain yields of all entries and the most productive entry (MP), Uniform Southern Soft Red Winter Wheat Nursery, 1980–2010.

in Nebraska and estimated the mean rate of increase, for the time period 1874 through 2005 to be 0.60% per year.

Graybosch and Peterson (2010) previously had suggested that the rate of increase of relative grain yield had slowed, and perhaps peaked, in the SRPN in the mid 1990s. As noted above, however, this observation could be influenced by the widely fluctuating climate within the geographic region in which the SRPN trials are staged. To determine if a diminishing rate of increase occurred in the additional USDA nurseries or in the UNL trials, regression equations were calculated for the most recent 20-yr period (1990–2010) (Table 16–4). Again, the SRPN, PNW SW and west-central and western Nebraska districts did not demonstrate significant increases in relative grain yields. In addition, the rate of increase of relative grain yield in the PNW HW nursery did not differ from zero for the time period 1990 through 2010 (Table 16–4). In the HRS nursery, the rate of increase for

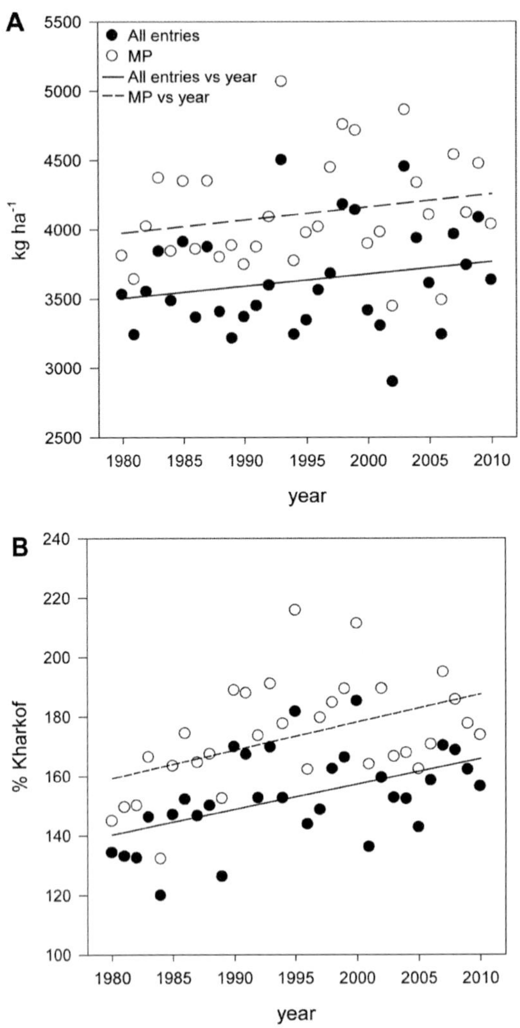

Fig. 16–6. (A) Average and (b) relative grain yields of all entries and the most productive entry (MP), Southern Regional Performance Nursery, 1980–2010.

the MP entry was not significant; when all entries were included, a significant slope was observed (Table 16–4).

Results from the USDA-ARS Regional Performance Nurseries from the time period 1980 through 2010 were compared to those observed by Schmidt (1984) for the time period 1958 through 1980 (Fig. 16–13). Averaged across the tested USDA-ARS regional performance nurseries, Schmidt (1984) observed a 1.10% yr^{-1} increase, greater than the average 0.80% increase observed herein for the time period 1980 through 2010. Only the USSRWWN and the NRPN demonstrated higher rates of genetic improvement for grain yield over the latter time period, although no statistical test can be applied to these results. Taken together, results of the two investigations demonstrate U.S. wheat breeders, in most regions, have steadily increased the genetic potential for grain yield since the late 1950s.

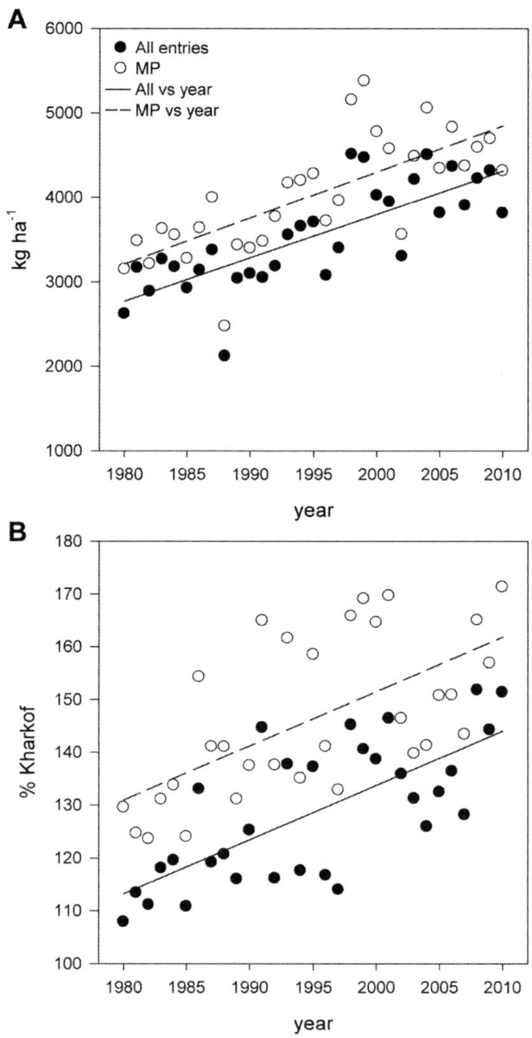

Fig. 16–7. (A) Average and (b) relative grain yields of all entries and the most productive entry (MP), Northern Regional Performance Nursery, 1980–2010.

However, Fig. 16–13 does suggest that, for some gene pools and regions of the United States, the rate might be diminishing.

Genetic Gains: Global Perspective

Estimates of rates of genetic gain for wheat yield from breeding programs around the world are fairly consistent (Table 16–5). The observed average rate of improvement in the worldwide survey was 0.8%, identical to the rate observed for the USDA-ARS Regional Performance Nurseries for the time period 1980 through 2010 (Fig. 16–13). Over various time periods, and in various breeding programs, 70% of the observed rates of genetic gain fell within the range of 0.5 to 1.2% yr^{-1} (Table 16–5). Rates ranged from a low of 0.2% in India to a high of >4.0% within a more narrowly defined selection response region in Oklahoma. The observed low

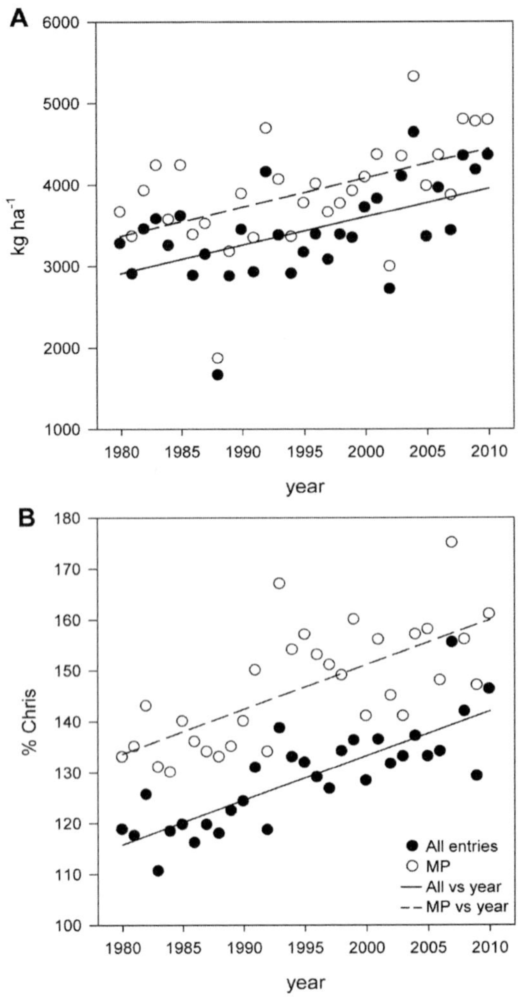

Fig. 16–8. (A) Average and (b) relative grain yields of all entries and the most productive entry (MP), Hard Red Spring Regional Performance Nursery, 1980–2010.

rate of improvement in the study conducted in India is concerning, as India long has been recognized as one of the success stories of the Green Revolution (Parayil, 1992). Lin and Huybers (2012) included India among the regions experiencing a leveling of wheat grain yields, attributing their observations to the approach of a genetic yield potential ceiling in this nation.

Factors Influencing Observed Genetic Gains

Wheat breeding and production in the United States essentially operates in six or more distinct agroecological zones (roughly conforming to the zones covered by the USDA-ARS regional nurseries), each utilizing what might be characterized as separate gene pools, facing different environmental limiting factors and using somewhat different production practices. Generally, alternative wheat gene pools are accessed for specific genes for disease and insect resistance, or, to a more limited extent, quality factors. Thus, differential rates of observed genetic

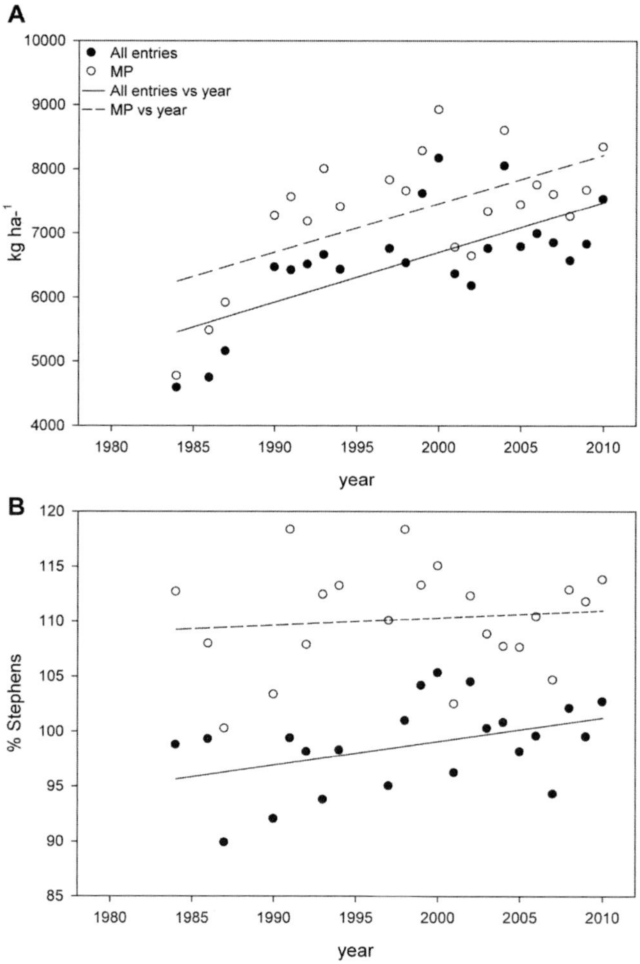

Fig. 16–9. (A) Average and (b) relative grain yields of all entries and the most productive entry (MP), PNW Soft Winter Wheat Regional Performance Nursery, 1980–2010.

progress among the regions perhaps should be expected. Rates of genetic gain have been reported to be larger in more favorable environments (Austin et al., 1980; Slafer et al., 1994). In some circumstances, the observed genetic progress in U.S. wheat yield potential supports this observation. In the UNL trial, significant genetic progress was observed in the eastern, higher rainfall districts (Fig. 16–11 and 16–12; Table 16–3). In the more arid and typically lower yielding west-central and western districts, no statistically significant increase in relative grain yield over Scout 66 was observed. In the USDA-ARS Regional Performance Nurseries, the lowest rates of genetic progress were observed in the nursery with the lowest average grain yield (SRPN) but also in the nursery (PNW SW) with the highest average yields, contradicting the proposed association of yield level and yield gain. The zone covered by the SRPN faces both periodic drought, and a continuous evolution of new pathogen races, especially leaf rust (caused by *Puccinia triticina* Eriks.), and more recently stripe rust (caused by *P. striiformis* f. sp. *tritici*)

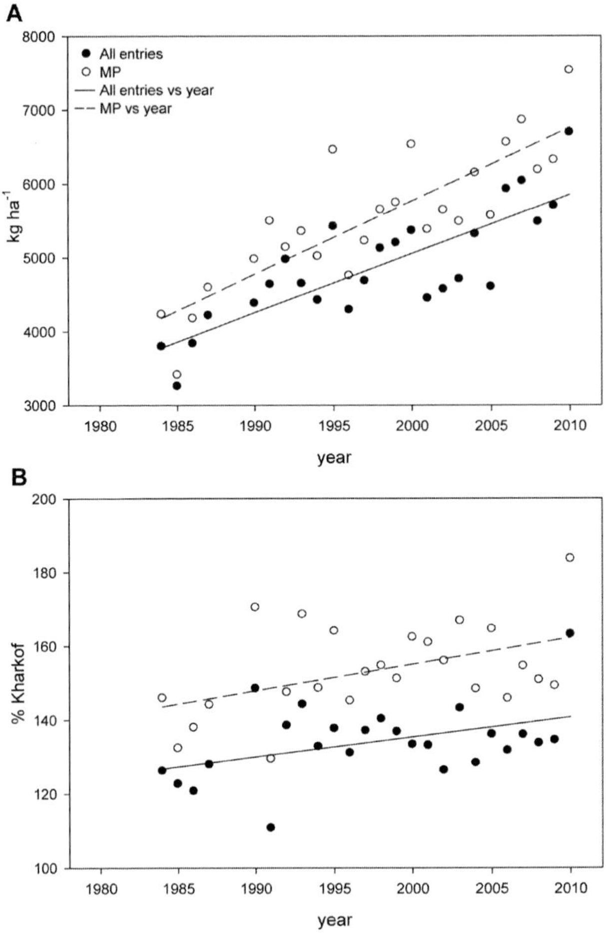

Fig. 16–10. (A) Average and (b) relative grain yields of all entries and the most productive entry (MP), PNW Hard Winter Wheat Regional Performance Nursery, 1980–2010.

(Eversmeyer and Kramer, 2000; Chen, 2005). Thus, breeders in this region contend with both abiotic and biotic limiting factors.

In the PNW SW nursery, grain yields of the long-term check used in the present investigation, Stephens, were approximately 165% of Kharkof 30 yr ago. Breeders in this region are operating at the highest level of wheat production in the United States, and perhaps, have neared the current biological threshold for wheat yields, at least in this environment (Reynolds et al., 2012). This area does face continuous evolution of new pathogen races, especially stripe rust (Chen, 2005); hence, maintenance of the current high yield levels represents significant continuous genetic improvement in terms of pathogen resistance. However, pathogen resistance, like prevention of all disease, pest, weed, and other problems, does not increase yield, but only preserves yield potential, thus preventing yield loss. Pathogen resistance is of economic benefit because it replaces other input costs, such as for fungicides, but does not build yield potential.

Table 16–3. Regression analysis of change in production and productivity, USDA-ARS coordinated regional wheat performance nurseries, and University of Nebraska Wheat Variety trials, 1980–2010.

Nursery/trial†	Trait‡	Entries§	Regression equation	SE (slope)	R^2	P
UESRWWN	Yield, kg ha^{-1}	All	y = 56.72(year) – 108714	9.8	0.54	<0.01
		MP	y = 58.69(year) – 112093	10.7	0.51	<0.01
	Rel. yield, kg ha^{-1}	All	y = 0.60(year) – 1088	0.2	0.31	<0.01
		MP	y = 0.52(year) – 917	0.3	0.12	0.06
USSRWWN	Yield, kg ha^{-1}	All	y = 38.40(year) – 72616	11.7	0.27	<0.01
		MP	y = 39.52(year) – 74228	13.2	0.24	<0.01
	Rel. yield, kg ha^{-1}	All	y = 1.46(year) – 2811	0.1	0.89	<0.01
		MP	y = 1.57(year) – 3020	0.1	0.89	<0.01
SRPN	Yield, kg ha^{-1}	All	y = 8.76(year) – 13856	7.5	0.05	0.25
		MP	y = 9.46(year) – 14768	7.9	0.05	0.24
	Rel. yield, kg ha^{-1}	All	y = 0.84(year) – 1540	0.3	0.25	<0.01
		MP	y = 0.94(year) – 1703	0.3	0.22	<0.01
NRPN	Yield, kg ha^{-1}	All	y = 51.27(year) – 98761	7.9	0.59	<0.01
		MP	y = 54.39(year) – 104494	5.8	0.54	<0.01
	Rel. yield, kg ha^{-1}	All	y = 1.03(year) – 1922	0.2	0.51	<0.01
		MP	y = 1.03(year) – 1903	0.2	0.39	<0.01
HRS	Yield, kg ha^{-1}	All	y = 34.80(year) – 65986	10.3	0.28	<0.01
		MP	y = 35.87(year) – 67669	11.7	0.25	<0.01
	Rel. yield, kg ha^{-1}	All	y = 0.87(year) – 1613	0.1	0.66	<0.01
		MP	y = 0.88(year) – 1602	0.2	0.48	<0.01
PNW SW	Yield, kg ha^{-1}	All	y = 77.67(year) – 148617	18.5	0.47	<0.01
		MP	y = 75.69(year) – 143857	22.1	0.37	0.01
	Rel. yield, kg ha^{-1}	All	y = 0.21(year) – 330	0.10	0.18	0.05
		MP	y = 0.06(year) + 10.2	0.13	0.01	0.63
PNW HW	Yield, kg ha^{-1}	All	y = 79.69(year) – 154324	12.0	0.66	<0.01
		MP	y = 99.15(year) – 192533	13.3	0.71	<0.01
	Rel. yield, kg ha^{-1}	All	y = 0.54(year) – 938	0.2	0.18	0.04
		MP	y = 0.72(year) – 1269	0.3	0.21	0.02

Table continued.

Table 16–3. Continued.

Nursery/trial†	Trait‡	Entries§	Regression equation	SE (slope)	R^2	*P*
Nebraska						
Statewide	Yield, kg ha^{-1}	All	y = 19.78(year) – 36048	8.4	0.16	0.03
		MP	y = 27.74(year) – 51274	8.8	0.26	<0.01
	Rel. yield, kg ha^{-1}	All	y = 0.51(year) – 900	0.2	0.18	0.02
		MP	y = 0.73(year) – 1315	0.3	0.18	0.02
Southeast District	Yield, kg ha^{-1}	All	y = 29.70(year) – 55843	13.3	0.15	0.03
		MP	y = 29.57(year) – 54870	13.2	0.15	0.03
	Rel. yield, kg ha^{-1}	All	y = 0.86(year) – 1589	0.3	0.22	<0.01
		MP	y = 0.76(year) – 1378	0.4	0.10	0.09
South-central District	Yield, kg ha^{-1}	All	y = 34.59(year) – 65481	23.5	0.09	0.16
		MP	y = 51.51(year) – 98320	25.7	0.15	<0.1
	Rel. yield, kg ha^{-1}	All	y = 0.79(year) – 1456	0.3	0.29	<0.1
		MP	y = 1.19(year) – 2233	0.5	0.22	0.02
West-central District	Yield, kg ha^{-1}	All	y = 19.08(year) – 34432	14.5	0.06	0.2
		MP	y = 18.56(year) – 32744	16.5	0.04	0.27
	Rel. yield, kg ha^{-1}	All	y = 0.19(year) – 258	0.3	0.02	0.56
		MP	y = 0.12(year) – 94	0.5	0.01	0.81
Western District	Yield, kg ha^{-1}	All	y = 12.68(year) – 22224	9.9	0.05	0.21
		MP	y = 18.11(year) – 32655	10.8	0.09	0.10
	Rel. yield, kg ha^{-1}	All	y = 0.23(year)- 360	0.2	0.04	0.31
		MP	y = 0.40(year)- 675	0.3	0.07	0.16

† HRS, Great Plains Hard Red Spring Wheat Nursery; Nebraska, University of Nebraska Winter Wheat Variety Trial; NRPN, Northern Regional Performance Nursery; PNW HW, Pacific Northwest Hard Winter Wheat Nursery; PNW SW, Pacific Northwest Soft Winter Wheat Nursery; SRPN, Southern Regional Performance Nursery; UESRWWN, Uniform Eastern Soft Red Winter Wheat Nursery; USSRWWN, Uniform Southern Soft Red Winter Wheat Nursery. Locations have changed over time within some nurseries.

‡ Yield should be viewed as an approximate estimate of changes in production within each trial over time. Relative yield is the rate of change in grain yield, as percentage per year of long-term control cultivars. See text for additional explanation.

§ Entries: All, all trial entries, exclusive of control cultivars; MP, most productive entry, or entry with the highest average trial grain yield.

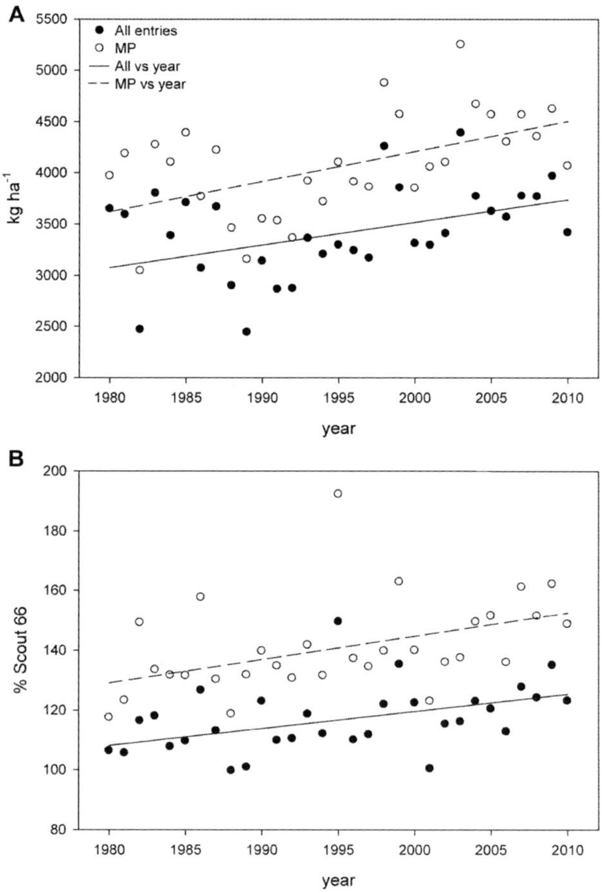

Fig. 16–11. (A) Average and (b) relative grain yields of all entries and the most productive entry (MP), Nebraska Winter Wheat Variety Trials, 1980–2010.

Graybosch and Peterson (2012), after evaluating several production zones within the SRPN and NRPN over the time period 1987 through 2010, noted that the trend of greater genetic progress in the most favorable locales was not universal. They found the highest yielding production zone in the SRPN and NRPN did yield significant rates of genetic improvement, and these rates were, in fact, higher than those observed in most other zones. However, in the SRPN, the highest rate of genetic improvement actually was observed in the lowest yielding Southern High Plains production zone. This perhaps supports the observation of Gourdji et al. (2012), who suggested that programs designed specifically to elevate grain yield in hotter environments can result in genetic progress. Nonetheless, temperature and precipitation patterns might limit the ability of wheat to express innate genetic potential, especially in the more arid zones evaluated such as those covered by the SRPN and the western Nebraska trials. In zones such as the southeastern United States, in which pathogens are the primary yield limiting factor, genetic progress for grain yield has been steady (see results for SSRWWN, Table 16–3, Fig. 16–4), but has relied, no doubt, on continuous discovery and deployment of resistance genes. The observation that none of the slopes of the regression

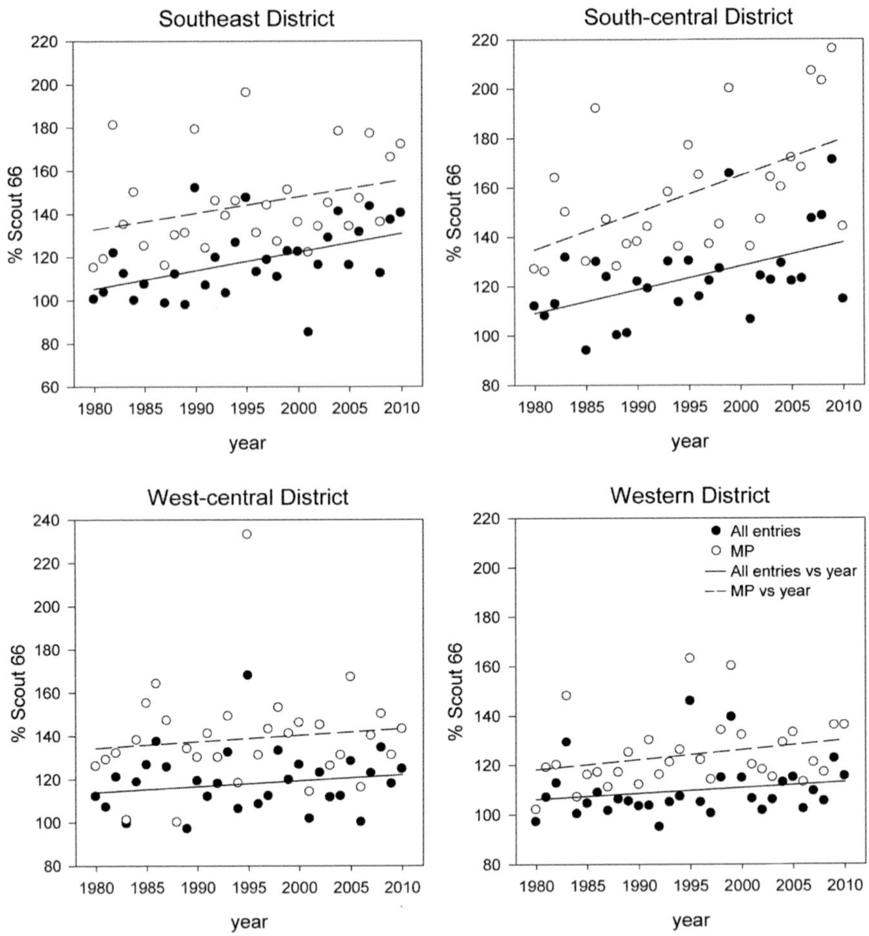

Fig.16–12. (A) Relative grain yields for all entries and the most productive entry (MP) only, University of Nebraska Winter Wheat Variety Trials, 1980–2010, for four crop reporting districts.

lines from any nursery were significantly negative suggests breeders have made continuous progress in deployment of pathogen resistance genes.

Factors Contributing to Genetic Gains for Grain Yield

Introgressions from Related Species

Due to both its polyploid nature, and the existence of an array of cross-fertile wild and domestic relatives, wheat improvement programs have successfully introduced numerous genes from related members of the Triticieae via chromosomal engineering (Sears, 1972). No statement more succinctly summarizes the contribution of wild and domesticated relatives of wheat to genetic gains than the following: "The distant relatives of wheat provide a vast reservoir of genetic variation for agronomically important traits and—given that wide crossing is

Table 16–4. Parameters from regression equations of change in production and productivity, USDA-ARS coordinated regional wheat performance nurseries, and University of Nebraska Wheat Variety trials, 1990–2010.

Nursery/trial†	Trait‡	Entries§	Slope	SE (slope)	R^2	*P*
UESRWWN	Yield, kg ha^{-1}	All	65.55	18.5	0.40	<0.01
		MP	65.82	18.1	0.41	<0.01
	Rel. yield, kg ha^{-1}	All	0.88	0.2	0.46	<0.01
		MP	0.80	0.3	0.31	<0.01
USSRWWN	Yield, kg ha^{-1}	All	37.18	18.8	0.17	0.06
		MP	41.13	3.7	0.14	<0.01
	Rel. yield, kg ha^{-1}	All	1.43	0.2	0.81	<0.01
		MP	1.43	0.1	0.42	<0.01
SRPN	Yield, kg ha^{-1}	All	9.67	15.4	0.02	0.54
		MP	3.75	16.2	0.01	0.82
	Rel. yield, kg ha^{-1}	All	−0.26	0.5	0.02	0.58
		MP	−0.60	0.5	0.06	0.22
NRPN	Yield, kg ha^{-1}	All	51.14	14.1	0.41	<0.01
		MP	45.66	17.5	0.26	<0.02
	Rel. yield, kg ha^{-1}	All	0.83	0.4	0.20	<0.05
		MP	0.52	0.5	0.06	0.27
HRS	Yield, kg ha^{-1}	All	47.88	16.6	0.30	<0.01
		MP	41.33	19.3	0.18	<0.05
	Rel. yield, kg ha^{-1}	All	0.74	0.2	0.35	<0.01
		MP	0.43	0.3	0.08	0.20
PNW SW	Yield, kg ha^{-1}	All	30.07	20.4	0.11	0.16
		MP	12.87	22.0	0.02	0.56
	Rel. yield, kg ha^{-1}	All	0.23	0.13	0.16	0.10
		MP	−0.06	0.17	0.01	0.75
PNW HW	Yield, kg ha^{-1}	All	69.80	17.4	0.46	<0.01
		MP	81.47	18.3	0.51	<0.01
	Rel. yield, kg ha^{-1}	All	0.29	0.4	0.04	0.42
		MP	0.38	0.4	0.04	0.33
Nebraska						
Statewide	Yield, kg ha^{-1}	All	19.78	8.4	0.16	0.03
		MP	27.75	8.9	0.26	<0.01
	Rel. yield, kg ha^{-1}	All	0.51	0.2	0.18	0.02
		MP	0.73	0.3	0.19	0.02
Southeast District	Yield, kg ha^{-1}	All	69.99	22.6	0.33	<0.01
		MP	29.57	13.2	0.15	0.03
	Rel. yield, kg ha^{-1}	All	0.47	0.6	0.03	0.43
		MP	0.76	0.4	0.10	0.09
South-central District	Yield, kg ha^{-1}	All	49.02	37.9	0.11	0.22
		MP	51.51	25.7	0.15	0.06
	Rel. yield, kg ha^{-1}	All	0.58	0.4	0.11	0.21
		MP	1.19	0.5	0.22	0.02

Table continued.

Table 16–4. Continued.

Nursery/trial†	Trait‡	Entries§	Slope	SE (slope)	R^2	*P*
West-central District	Yield, kg ha^{-1}	All	24.25	21.1	0.07	0.26
		MP	18.56	16.5	0.04	0.26
	Rel. yield, kg ha^{-1}	All	−0.11	0.5	0.01	0.83
		MP	0.12	0.5	0.01	0.81
Western District	Yield, kg ha^{-1}	All	24.83	17.9	0.09	0.18
		MP	18.11	10.8	0.09	0.11
	Rel. yield, kg ha^{-1}	All	0.28	0.5	0.02	0.55
		MP	0.40	0.3	0.07	0.15

† HRS, Great Plains Hard Red Spring Wheat Nursery; Nebraska, University of Nebraska Winter Wheat Variety Trial; NRPN, Northern Regional Performance Nursery; PNW HW, Pacific Northwest Hard Winter Wheat Nursery; PNW SW, Pacific Northwest Soft Winter Wheat Nursery; SRPN, Southern Regional Performance Nursery; UESRWWN, Uniform Eastern Soft Red Winter Wheat Nursery; USSRWWN, Uniform Southern Soft Red Winter Wheat Nursery. Locations have changed over time within some nurseries.

‡ Yield should be viewed as an approximate estimate of changes in production within each trial over time. Relative yield is the rate of change in grain yield, as percentage per year of long-term control cultivars. See text for additional explanation.

§ Entries: All, all trial entries, exclusive of control cultivars; MP, most productive entry, or entry with the highest average trial grain yield.

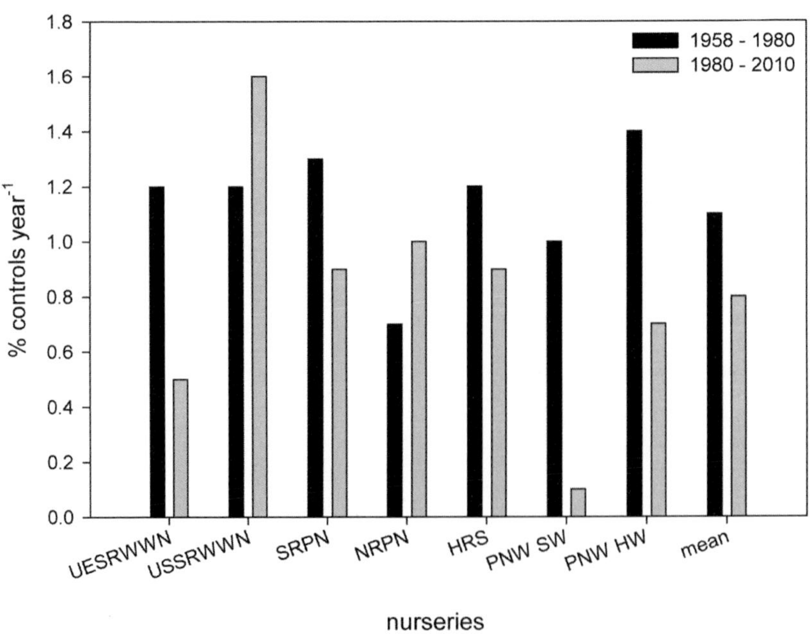

Fig. 16–13. Rate of genetic improvement for grain yield of U.S. wheat estimated from USDA-ARS coordinated regional performance nurseries.

Table 16–5. Recent estimates of genetic gain for wheat grain yield potential.

Citation	Breeding program(s) or nation	Time frame	Estimated annual rate of improvement
Acreche et al., 2008	Spain	1940–2005	0.6%e†
Battenfield, 2011	Oklahoma, USA	1971–2008	0.5%
Brancourt-Hulmel et al., 2003	France	1949–1993	0.8%e
Cargnin et al., 2008	Brazil (irrigated)	1976–2005	1.8%
Clarke et al., 2010	Canada (durum)	1946–2009	0.70%
DeVita et al., 2010	Italy (durum)	1900–2003	0.9%e
Fufa et al., 2005	Nebraska, USA	1874–2005	0.60%
Green et al., 2012	Eastern USA	1919–2009	0.5–1.4%
Khalil et al., 1995	Oklahoma, USA	1969–1993	4.2%
Khalil et al., 2002	Oklahoma, USA	1919–1997	0.9–1.3%
Khodarahmi et al., 2010	Iran	1942–2007	1.0%
Lopes et al., 2012	CIMMYT "Historic" Set	1977–2008	0.7%
Manés et al., 2012	CIMMYT Semi-Arid Wheat Trial	1994–2010	1.0%
Miri, 2009	Iran	1940–2000	0.9%
Morgounov et al., 2010	Russian Federation	1900–2008	0.7%
O. Rodrigues et al., 2007	Brazil (spring wheat)	1940–1992	1.7%
Peltonen-Sainio et al., 1994	Finland	1970–2005	0.7%e
Royo et al., 2007	Italy (durum)	1945–2000	0.4%
Royo et al., 2007	Spain (durum)	1945–2000	0.4%
Sadras and Lawson, 2011	South Australia	1958–2007	0.5%
Sener et al., 2009	Turkey	1975–1999	0.5%
Sharma et al., 2012	CIMMYT Elite Spring Wheat Trial	1995–2009	0.7%
Tian et al., 2011	Yangtze Basin, China	1950–2000	0.9–1.2%e
Underdahl et al., 2008	North Dakota, USA	1986–2006	1.3%
Yasin, 2011	India	1969–2005	0.2%e
Zhang et al., 2011	Sichuan, China	2000–2007	0.7%
Zhang et al., 2011	Yunnan, China	1945–2008	0.3%
Zhang et al., 2011	Gansu, China	1964–2004	0.6%
Zhang et al., 2011	Xinjiang, China	1975–2007	1.4%
Zheng et al., 2011	Henan, China	1981–2008	0.6%
Zhou et al., 2007	China	1960–2000	0.5–1.2%
		Avg.	0.8%

† e, estimated from data provided by authors.

already a mature discipline in wheat—represent a key resource in the pipeline for increasing yield potential" (Reynolds et al., 2012). Near and distant relatives have contributed numerous useful genes especially for those conditioning pest and pathogen resistance, but also for drought and stress tolerance for environments that are water limited (Jones et al., 1995; Friebe et al., 1996; Gill et al., 2011). The vast majority of U.S. wheat cultivars likely carry genes derived from related members of the Triticeae.

DNA Marker Assisted Selection

Although studies have investigated the use of DNA marker assisted selection (MAS) in the improvement of wheat grain yield (Shah et al., 1999; Campbell et al., 2003), replacement of direct wheat grain yield evaluation and selection by MAS have been rare (Mergoum et al., 2009). Resource limitations, lack of consistent response of quantitative trait loci (QTLs) across populations, and high genotype by environment effects on grain yield and associated QTLs have restricted the application of MAS to grain yield. Marker assisted selection has seen primary application in wheat breeding programs in the selection of chromosomal fragments from related species carrying desirable genes (Gupta et al., 2010) and in the selection of specific genes conditioning phenotypes [e.g., resistance to Fusarium head blight, caused by *Fusarium graminearum* (Schwabe) and *F. culmorum* (W. G. Smith) Sacc.) that are difficult to identify with direct phenotypic assays (Buerstmayr et al., 2009]. The use of MAS to identify and select introgressions is a powerful complement to a proven and useful technology. The high level of polymorphism between wheat and distant relatives, and the low levels of recombination often associated with introduced chromatin present in wheat backgrounds, nearly guarantees successful MAS for alien genes. The majority of examples of successful application of MAS in wheat improvement cited by Gupta et al. (2010) involve alien introgressions. Future developments in the area of genome-based selection (Heffner et al., 2009) might lead to more direct application of DNA-based selection in grain yield improvement.

Doubled Haploids

Doubled haploid (DH) technology allows rapid attainment of homozygous, homogeneous lines (Chu and Chumley, 2010; Basu et al., 2011). Production of DH lines can be especially useful in winter wheat breeding programs, where, due to the requirements for up to 8 wk of vernalization, and the necessity for cool temperatures for optimal fertility and seed set, cultivar development routinely requires 10 to 12 yr from initial crossing to farm production. Doubled haploid technology currently is being applied in a number of breeding programs, and has already resulted in cultivar registrations (Elhaddoury et al., 2012; Fowler, 2012). While DH production does not aid directly in genetic improvement for wheat yield, it can speed the rate at which new lines can be generated and evaluated. As noted by Baenziger and DePauw (2009), however, doubled haploid technology has some disadvantages, namely a reduction in both initial and cumulative recombination frequencies, and financial constraints that limit the number of crosses and derived lines evaluated.

Mutagenesis

Mutagenesis has indirectly contributed to genetic enhancement of wheat grain yield via isolation of mutants conferring herbicide resistance (Newhouse et al., 1992). Genes conditioning tolerance to imidazolinone herbicides have been deployed in several Clearfield (registered trademark of BASF) wheat cultivars (Baenziger et al., 2006; Haley et al., 2006). The trait reduces competition from weedy relatives of wheat, especially jointed goat-grass (*Aegilops cylindrica* Host), and can assist in stabilizing yields. Mutagenesis long has been used to generate novel genetic characters in wheat (Konzak, 1956). The FAO/IAEA database (http://mvgs.iaea.org/AboutMutantVarities.aspx , accessed 13 Mar. 2014) lists more than

250 wheat varieties that have been released after using mutagenesis in their development, though the majority of such lines derive from China. Mutagenesis is experiencing a bit of a renaissance, with recent advances such as TILLING (Slade et al., 2005) rendering identification of new mutants in targeted genes of interest, even in polyploid species such as wheat. Recent studies also have demonstrated the use of mutagenesis to generate novel disease resistance genes (Boyd and Minchin, 2001; Campbell et al., 2012).

Transgenics

Johnson (1986b) made the following observation: "The likely role of genetic engineering techniques…to wheat improvement is unclear." Johnson might be surprised to learn today the role remains unclear. Development of transformation techniques in wheat lagged behind those of other row crops, with the first successful applications occurring in the early 1990s (Weeks et al., 1993). This technology lag, coupled with a view by the major wheat exporting nations that the first to deploy transgenic wheat might subsequently be the first to lose their export markets, has prevented any commercial application of transgenic wheat. The science has progressed, and studies on wheat transformation are widespread in the literature (see review by Li et al. (2012). Genetic transformation of wheat has yet to contribute to genetic improvement for grain yield in commercial settings.

Future Prospects

While wheat awaits potential gains from genetic engineering, wheat breeders can continue to generate, evaluate, and deploy useful genetic variation from wild and domesticated relatives and mutagenesis to aid in grain yield improvement and prevention of yield losses due to pests and weeds. This genetic improvement must be deployed in concert with improved production practices to achieve yield increases in all production environments. Whether the observed rates of genetic improvement since 1980 will continue at the same pace remains uncertain. Schmidt (1984), evaluating yield trends from 1958 to 1980, observed, "There is some suggestion... that increases in yield potential since 1975 have been minimal," and he raised the question of whether wheat yields had peaked. It is safe to say such concerns were unwarranted. An objective observer today, however, raising this same question, might answer "not yet, but maybe soon," and information presented above suggests a diminished rate of genetic improvement for grain yield in some environments and gene pools. Whether or not one believes yields will peak shortly in all wheat production zones, the current rate of gain does seem to fall short of that needed to meet world demands (Reynolds et al., 2012). Reynolds et al. (2012) proposed one way to elevate the rate of genetic gain would be through increased solar radiation use efficiency. While traditional approaches used in wheat breeding, including the use of genes from wheat relatives, could aid in this goal, many of the approaches outlined by Reynolds et al. (2012) require the use of genetic engineering techniques. Hopefully, our successors 30 yr in the future will find application of genetic engineering techniques far more common than at present.

Acknowledgements

The authors wish to thank the hundreds of public and private researchers who have participated in the USDA-ARS Regional Nursery Programs and the University of Nebraska Wheat Variety Testing Program since their inception.

References

Acreche, M., G. Briceño-Félix, M. Sánchez, and G.A. Slafer. 2008. Physiological bases of genetic gains in Mediterranean bread wheat yield in Spain. Eur. J. Agron. 28:162–170. doi:10.1016/j.eja.2007.07.001

Austin, R.B., J. Bingham, R.D. Blackwell, L.T. Evans, M.A. Ford, C.L. Morgan, and M. Taylor. 1980. Genetic improvements in winter wheat yields since 1900 and associated changes. J. Agric. Sci. 94:675–689. doi:10.1017/S0021859600028665

Baenziger, P.S., B. Beecher, R.A. Graybosch, D.D. Baltsenperger, L.A. Nelson, J.M. Krall, Y. Jin, J.E. Watkins, D.J. Lyon, A.R. Martin, M. Chen, and G. Bai. 2006. Registration of 'Infinity CL' wheat. Crop Sci. 46:975–979. doi:10.2135/cropsci2005.05-0044

Baenziger, P.S., and R.M. DePauw. 2009. Wheat breeding: Procedures and strategies. In B.F. Carver, editor, Wheat: Science and trade. Wiley-Blackwell, Ames, IA. p. 275–308.

Basu, S.K., M. Datta, M. Sharma, and A. Kumar. 2011. Haploid production technology in wheat and some selected higher plants. Aust. J. Crop Sci. 5:1087–1093.

Battenfield, S.D. 2011. Improvement in genetic yield potential of semi-dwarf wheat in the Great Plains of the USA. M.S. Thesis, Oklahoma State Univ., Stillwater. http://dc.library.okstate.edu/cdm/singleitem/collection/theses/id/4052/rec/9 (accessed 13 Mar. 2014).

Boyd, L.A., and P.N. Minchin. 2001. Wheat mutants showing altered adult plant disease resistance. Euphytica 122:361–368. doi:10.1023/A:1012948702416

Brancourt-Hulmel, M., G. Doussinault, C. Lecomte, P. Bérard, B. Le Buanec, and M. Trottet. 2003. Genetic improvement of agronomic traits of winter wheat cultivars released in France from 1946 to 1992. Crop Sci. 43:37–45. doi:10.2135/cropsci2003.0037

Buerstmayr, H., T. Ban, and J.A. Anderson. 2009. QTL mapping and marker-assisted selection for Fusarium head blight resistance in wheat: A review. Plant Breed. 128:1–26. doi:10.1111/j.1439-0523.2008.01550.x

Campbell, B.T., P.S. Baenziger, K.S. Gill, K.M. Eskridge, H. Budak, M. Erayman, I. Dweikat, and Y. Yen. 2003. Identification of QTLs and environmental interactions associated with agronomic traits on chromosome 3A of wheat. Crop Sci. 43:1493–1505. doi:10.2135/cropsci2003.1493

Campbell J., H. Zhang, M.J. Giroux, L. Feiz, Y. Jin, M. Wang, X. Chen, and L. Huang. 2012. A mutagenesis-derived broad-spectrum disease resistance locus in wheat. Theor. Appl. Genet. 125:391–404. doi:10.1007/s00122-012-1841-7

Cargnin, A., M.A. De Souza, and V. Fronza. 2008. Progress in breeding of irrigated wheat for the Cerrado region of Brazil. Crop Breed. Appl. Biotech. 8:39–46. doi:10.12702/1984-7033.v08n01a06

Chen, X. 2005. Epidemiology and control of stripe rust on wheat. Can. J. Plant Pathol. 27:314–337. doi:10.1080/07060660509507230

Chu, C., and F. Chumley. 2010. Using doubled haploids to speed up genetic analysis for resistance to FHB and other complex traits in wheat. In: Proc. National Fusarium Head Blight Forum, Milwaukee, WI. 7–9 Dec. 2010. Univ. of Kentucky, Lexington. p. 15.

Clarke, J.M., F.R. Clarke, and C.J. Pozniak. 2010. Forty-six years of genetic improvement in Canadian durum wheat cultivars. Can. J. Plant Sci. 90:791–801. doi:10.4141/cjps10091

Cox, T.S., R. Shroyer, L. Ben-Hui, R.G. Sears, and T.J. Martin. 1988. Genetic improvement in agronomic traits of hard red winter wheat cultivars from 1919 to 1987. Crop Sci. 28:756–760. doi:10.2135/cropsci1988.0011183X002800050006x

de la Vega, A., I.H. DeLacy, and S.C. Chapman. 2007. Progress over 20 years of sunflower breeding in central Argentina. Field Crops Res. 100:61–72. doi:10.1016/j.fcr.2006.05.012

De Vita, P., A.M. Mastrangelo, L. Matteu, E. Mazzucotelli, N. Virzì, M. Palumbo, M. Lo Storto, F. Rizza, and L. Cattivelli. 2010. Genetic improvement effects on yield stability in durum wheat genotypes grown in Italy. Field Crops Res. 119:68–77. doi:10.1016/j.fcr.2010.06.016

Elhaddoury, J., S. Lhaloui, S.M. Udupa, B. Moatassim, R. Taiq, M. Rabeh, M. Kamlaoui, and M. Hammadi. 2012. Registration of 'Kharoba': A bread wheat cultivar developed through doubled haploid breeding. J. Plant Reg. 6:169–173. doi:10.3198/jpr2011.07.0385crc

Eversmeyer, M.G., and C.L. Kramer. 2000. Epidemiology of wheat leaf and stem rust in the Central Great Plains of the USA. Annu. Rev. Phytopathol. 38:491–513. doi:10.1146/annurev.phyto.38.1.491

Fowler, D.B. 2012. Sunrise soft red winter wheat. Can. J. Plant Sci. 92:195–198. doi:10.4141/cjps2011-107

Friebe, B., J. Jiang, W.J. Raupp, R.A. McIntosh, and B.S. Gill. 1996. Characterization of wheat alien translocations conferring resistance to diseases and pests: Current status. Euphytica 91:59–87. doi:10.1007/BF00035277

Fufa, H., P.S. Baenziger, B.S. Beecher, R.A. Graybosch, K.M. Eskridge, and L.A. Nelson. 2005. Genetic improvement trends in agronomic performances and end-use quality characteristics among hard red winter wheat cultivars in Nebraska. Euphytica 144:187–198. doi:10.1007/s10681-005-5811-x

Gill, B.S., B.R. Friebe, and F.F. White. 2011. Alien introgressions represent a rich source of genes for crop improvement. Proc. Natl. Acad. Sci. USA 108:7657–7658. doi:10.1073/pnas.1104845108

Gourdji, S.M., K.L. Mathews, M. Reynolds, J. Crossa, and D.B. Lobell. 2012. An assessment of wheat yield sensitivity and breeding gains in hot environments. Proc. Biol. Sci. 280. doi:10.1098/rspb.2012.2190

Graybosch, R.A., and C.J. Peterson. 2010. Genetic improvement in winter wheat yields in the Great Plains of North America, 1959–2008. Crop Sci. 50:1882–1890. doi:10.2135/cropsci2009.11.0685

Graybosch, R.A., and C.J. Peterson. 2012. Specific adaptation and genetic progress for grain yield in Great Plains hard winter wheats from 1987 to 2010. Crop Sci. 52:631–643. doi:10.2135/cropsci2011.08.0412

Green, A.J., G. Berger, C.A. Griffey, R. Pitman, W. Thomason, M. Balota, and A. Ahmed. 2012. Genetic yield improvement in soft red winter wheat in the Eastern United States from 1919 to 2009. Crop Sci. 52:2097–2108. doi:10.2135/cropsci2012.01.0026

Gupta, P., P. Langridge, and R. Mir. 2010. Marker-assisted wheat breeding: Present status and future possibilities. Mol. Breed. 26:145–161. doi:10.1007/s11032-009-9359-7

Haley, S.D., J.J. Johnson, F.B. Peairs, J.S. Quick, P.H. Westra, J.A. Stromberger, S.R. Clayshulte, B.L. Clifford, J.B. Rudolph, and A. Giura. 2006. Registration of 'Bond CL' wheat. Crop Sci. 46:993–994. doi:10.2135/cropsci2005.0031

Heffner, E.L., M.E. Sorrells, and J.L. Jannink. 2009. Genomic selection for crop improvement. Crop Sci. 49:1–12. doi:10.2135/cropsci2008.08.0512

Johnson, V.A. 1986a. World wheat production. In: E.L. Smith, editor, Genetic improvement in yield of wheat. CSSA Spec. Publ. 13. CSSA, Madison, WI. p. 1–5. doi:10.2135/cssaspecpub13.c1

Johnson, V.A. 1986b. Future prospects in genetic improvement in wheat. In: E.L. Smith, editor, Genetic improvement in yield of wheat. CSSA Spec. Publ. 13. CSSA, Madison, WI. p. 109–114. doi:10.2135/cssaspecpub13.c8

Jones, S.S., T.D. Murray, and R.E. Allan. 1995. The development of disease resistance in wheat. Annu. Rev. Phytopathol. 33:429–443. doi:10.1146/annurev.py.33.090195.002241

Khalil, I., B.F. Carver, E.G. Krenzer, C.T. MacKown, and G.W. Horn. 2002. Genetic trends in winter wheat yield and test weight under dual-purpose and grain-only management systems. Crop Sci. 42:710–715. doi:10.2135/cropsci2002.0710

Khalil, I.H., B.F. Carver, and E.L. Smith. 1995. Long-term genetic gains in two selection phases of a wheat breeding program. Plant Breed. 114:117–120. doi:10.1111/j.1439-0523.1995.tb00774.x

Khodarahmi, M., A. Nabipour, and K. Zargari, K. 2010. Genetic improvement of agronomic and quality traits of wheat cultivars introduced to temperate regions of Iran during 1942–2007. African J. Agric. Res. 5:947–954.

Konzak, C.F. 1956. Induction of mutations for disease resistance in cereals. Brookhaven Symp. Biol. 9:157–176.

Li, J., X. Ye, B. An, L. Du, and H. Xu. 2012. Genetic transformation of wheat: Current status and future prospects. Plant Biotech. Rep. 6:183–193.

Lin, M., and P. Huybers. 2012. Reckoning wheat yield trends. Environ. Res. Lett. 112:1–7. doi:10.1016/j.envres.2011.11.006

Lopes, M.S., M.P. Reynolds, Y. Manes, R.P. Singh, J. Crossa, and H.J. Braun. 2012. Genetic yield gains and changes in associated traits of CIMMYT spring bread wheat in a "historic" set representing 30 years of breeding. Crop Sci. 52:1123–1131. doi:10.2135/cropsci2011.09.0467

Manés, Y., H.F.J. Gomez, M.P. Puhl, H.J. Reynolds, H.J. Braun, and R.M. Trethowan. 2012. Genetic yield gains of the CIMMYT international semi-arid wheat yield trials from 1994 to 2010. Crop Sci. 52:1543–1552. doi:10.2135/cropsci2011.10.0574

Mergoum, M., P.K. Singh, J.A. Anderson, R.J. Pena, R.P. Singh, S.S. Xu, and J.K. Ransom. 2009. Spring wheat breeding. In: M.J. Carena, editor, Handbook of plant breeding. Vol. 3. Cereals. Springer, New York. p. 127–156.

Miri, H.R. 2009. Grain yield and morpho-physiological changes from 60 years of genetic improvement of wheat in Iran. Exp. Agric. 45:149–163. doi:10.1017/S001447970800745X

Morgounov, A., V. Zykin, I. Belan, L. Roseeva, Y. Zelenskiy, H.F. Gomez, H. Budak, and F. Bekes. 2010. Genetic gains for grain yield in high latitude spring wheats grown in Western Siberia in 1900–2008. Field Crops Res. 117:101–112. doi:10.1016/j.fcr.2010.02.001

Newhouse, K.E., W.A. Smith, M.A. Starrett, T.J. Schaefer, and B.K. Singh. 1992. Tolerance to imidazolinone herbicides in wheat. Plant Physiol. 100:882–886. doi:10.1104/pp.100.2.882

Parayil, G. 1992. The Green Revolution in India: A case study of technological change. Technol. Cult. 33:737–756. doi:10.2307/3106588

Peltonen-Sainio, P., and J. Peltonen. 1994. Progress since the 1930s in breeding for yield, its components, and quality traits of spring wheat in Finland. Plant Breed. 113:177–186. doi:10.1111/j.1439-0523.1994.tb00721.x

Reynolds, M., J. Foulkes, R. Furbank, S. Griffiths, J. King, E. Murchie, M. Parry, and G. Slafer. 2012. Achieving yield gains in wheat. Plant Cell Environ. 35:1799–1823. doi:10.1111/j.1365-3040.2012.02588.x

Rodrigues, O., J.C.B. Lhamby, A.D. Didonet, and J. Marchese. 2007. Fifty years of wheat breeding in southern Brazil: Yield improvement and associated changes. Pesquisa Agropecu. Bras. 42:817–825. doi:10.1590/S0100-204X2007000600008

Royo, C., F. Álvaro, V. Martos, A. Ramdani, J. Isidro, D. Villegas, and L.F. García Del Moral. 2007. Genetic changes in durum wheat yield components and associated traits in Italian and Spanish varieties during the 20th century. Euphytica 155:259–270. doi:10.1007/s10681-006-9327-9

Sadras, V.O., and C. Lawson. 2011. Genetic gain in yield and associated changes in phenotype, trait plasticity and competitive ability of South Australian wheat varieties released between 1958 and 2007. Crop Pasture Sci. 62:533–549. doi:10.1071/CP11060

SAS Institute. 2003. SAS 9.1 Foundation for 64-bit Microsoft® Windows®, Copyright. SAS Institute Inc., Cary NC.

Schmidt, J.W. 1984. Genetic contributions to yield gains in wheat. p. 89–101. In: W.R. Fehr, editor, Genetic contributions to yield gains of five major crop plants. CSSA and ASA, Madison, WI. doi:10.2135/cssaspecpub7.c5

Sears, E.R. 1972. Chromosome engineering in wheat. In: G. Kimber and G.R. Redei, editor, Stadler Symposium. University of Missouri, Columbia, MO. p. 23–28.

Sener, O., M. Arslan, Y. Soysal, and M. Erayman. 2009. Estimates of relative yield potential and genetic improvement of wheat cultivars in the Mediterranean region. J. Agric. Sci. 147:323–332. doi:10.1017/S0021859609008454

Shah, M.M., K.S. Gill, P.S. Baenziger, Y. Yen, S.M. Kaeppler, and H.M. Ariyarathne. 1999. Molecular mapping of loci for agronomic traits on chromosome 3A of bread wheat. Crop Sci. 39:1728–1732. doi:10.2135/cropsci1999.3961728x

Sharma, R.C., J. Crossa, G. Velu, J. Huerta-Espinoso, M. Vargas, T.S. Payne, and R.P. Singh. 2012. Genetic gains for grain yield in CIMMYT spring bread wheat across international environments. Crop Sci. 52:1522–1533. doi:10.2135/cropsci2011.12.0634

Slade, A.J., S.I. Fuerstenberg, D. Loeffler, M.N. Steine, and D. Facciotti. 2005. A reverse genetic, nontransgenic approach to wheat crop improvement by TILLING. Nat. Biotechnol. 23:75–81. doi:10.1038/nbt1043

Slafer, G.A., E.H. Satorre, and F.H. Andrad. 1994. Increases in grain yield in bread wheat from breeding and associated physiological changes. In: G.A. Slafer, editor, Genetic improvement of field crops. Marcel Dekker, New York. p. 1–68.

Tian, Z., Q. Jing, T. Dai, D. Jiang, and W. Cao. 2011. Effects of genetic improvements on grain yield and agronomic traits of winter wheat in the Yangtze River Basin of China. Field Crops Res. 124:417–425. doi:10.1016/j.fcr.2011.07.012

Underdahl, J.L., M. Mergoum, J.K. Ransom, and B.G. Schatz. 2008. Agronomic traits improvement and associations in hard red spring wheat cultivars released in North Dakota from 1968 to 2006. Crop Sci. 48:158–166. doi:10.2135/cropsci2007.01.0018

USDA-ERS. 2013. U.S. wheat trade. http://www.ers.usda.gov/topics/crops/wheat/trade.aspx (accessed 15 Mar. 2014).

USDA-FAS. 2010. Grain: World markets and trade. Circ. Ser. FG 05-10. http://www.fas.usda.gov/grain/circular/2010/05-10/grainfull05-10.pdf (accessed 12 Mar. 2014).

Weeks, J.T., O.D. Anderson, and A.E. Blechl. 1993. Rapid production of multiple independent lines of fertile transgenic wheat (*Triticum aestivum*). Plant Physiol. 102:1077–1084.

Yasin, B.A. 2011. Genetic improvement in yield, yield attributes and leaf rust resistance in semi-dwarf wheat varieties developed in India from the last 40 years. Int. J. Agric. Res. 6:747–753. doi:10.3923/ijar.2011.747.753

Zhang, Y., S.Z. Li, Z.L. Wu, W.X. Yang, Y.X. Yu, D.H. Zhao, X.C. Xia, and Z.H. He. 2011. Contribution of CIMMYT wheat germplasm to genetic improvement in grain yield in spring wheat of Sichuan, Yunnan, Gansu and Xinjiang Provinces. Acta Agron. Sin. 37:1752–1762.

Zheng, T.C., X.K. Zhang, G.H. Yin, L.N. Wang, Y.L. Han, L. Chen, G. Huang, J.W. Tang, X.C. Xia, and Z.H. He. 2011. Genetic gains in grain yield, net photosynthesis and stomatal conductance achieved in Henan Province of China between 1981 and 2008. Field Crops Res. 122:225–233. doi:10.1016/j.fcr.2011.03.015

Zhou, Y., H.Z. Zhu, S.B. Cai, Z.H. He, X.K. Zhang, X.C. Xia, and G.S. Zhang. 2007. Genetic improvement of grain yield and associated traits in the Southern China Winter Wheat Region: 1949 to 2000. Euphytica 157:465–473. doi:10.1007/s10681-007-9376-8